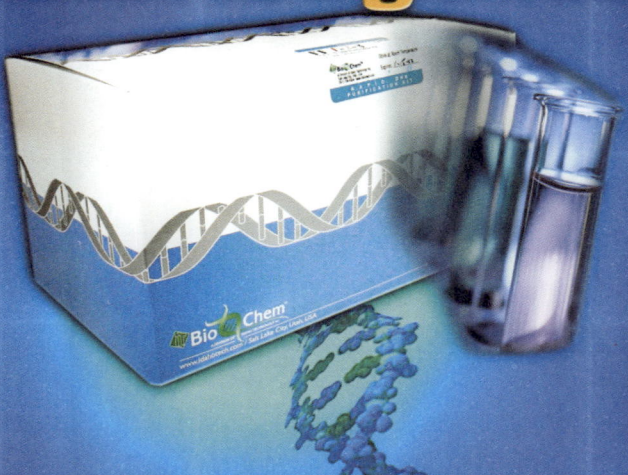

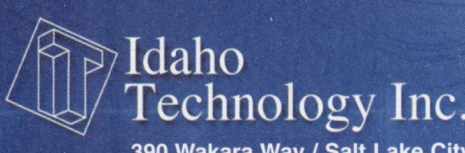

Jane's

NUCLEAR, BIOLOGICAL AND CHEMICAL DEFENCE

Edited by John Eldridge

Fifteenth Edition
2002-2003

Total number of entries 1,555   New and updated entries 637
Total number of images 1,087   New images 116

Bookmark Jane's homepage on
http://www.janes.com

Jane's award-winning web site provides you with continuously updated news and information.
As well as extracts from our world renowned magazines, you can browse the online catalogue,
visit the Press Centre, discover the origins of Jane's, use the extensive glossary,
download our screen saver and much more.

Jane's now offers powerful electronic solutions to meet the rapid changes in your
information requirements. All our data, analysis and imagery is available on CD-ROM
or via a new secure web service – Jane's Online at http://www.janes.com.

Tailored electronic delivery can be provided through Jane's Data Services.
Contact an information consultant at any of our international offices to
find out how Jane's can change the way you work or e-mail us at

info@janes.co.uk *or* **info@janes.com**

ISBN 0 7106 2442 5
"Jane's" is a registered trade mark

Joint Service General Purpose Mask (JSGPM) / XM50

![Soldier wearing JSGPM/XM50 mask]

Avon Rubber & Plastics Inc., 805 West Thirteenth Street, Cadillac, Michigan 49601-9282 USA
Tel.: +1 231 775 6571   Fax.: +1 231 775 7304

Avon Technical Products, Hampton Park West, Melksham, Wilts. SN12 6NB England
Tel.: +44 (0) 1225 896375   Fax +44 (0) 1225 896301 E-mail: protection@avonrubber.co.uk
Website: www.avon-rubber.com

Contents

How to use *Jane's Nuclear, Biological and Chemical Defence*

The background analysis and main section entries

Jane's Nuclear, Biological and Chemical Defence is designed to offer the decision maker up-to-date background on the challenges for NBC defence, a country by country analysis of the world NBC scene and a comprehensive catalogue of NBC defence equipment. After the analysis sections, the book presents a series of main sections, each of which relates to a specific topic in the nuclear, biological and chemical defence arena. The sections form a logical sequence, starting with Detection. Individual section entries are ordered alphabetically by country of origin and then alphabetically by equipment name. Each equipment entry contains a description of the product, followed by its availability status and the names of manufacturers or agencies. Pictures or diagrams of the equipment are included wherever possible and the photographs are dated and acknowledged where appropriate.

Details on the layout of *Jane's Nuclear, Biological and Chemical Defence*

The book is laid out in a way which allows the reader to move logically between sections. The early part of the book deals with how the main Weapons of Mass Destruction (WMDs) work, who has them, and the measures that are being taken to prevent their spread. The sections on weapons and their effects deal with the topics from a defender's point of view, aiming to point the user towards better methods of dealing with each environment when designing defensive measures.

The Technical developments section analyses all the key areas: detection, protection, contamination control and training to identify the areas where science has succeeded, or failed, to identify solutions. This section is designed to assist decision-makers to tailor solutions according to their needs.

To help users of this title evaluate the published data, Jane's Information Group has divided entries into three categories. A full list of all entries indicating their current status is provided in the index.

● **VERIFIED** The editor has made a detailed examination of the entry's content and checked it's relevancy and accuracy for publication in the new edition to the best of his ability. This designation also includes those equipments and services which the editor believes to remain available or considers of significant importance but for which hard information has been difficult to obtain.

● **UPDATED** During the verification process, significant changes to content have been made to reflect the latest position known to Jane's at the time of publication.

● **NEW ENTRY** Information on new equipment and or equipment which is appearing for the first time in the title.

All new pictures are dated with the year of publication. New pictures this year are dated 2002. Some are followed by a seven digit number for ease of identification by our image library.

| Total number of entries | 1,555 | New and updated entries | 637 |
| Total number of images | 1,087 | New images | 116 |

Copyright enquiries
Contact: Keith Faulkner, Tel/Fax: +44 (0) 1342 305032, e-mail: keith.faulkner@janes.co.uk

British Library Cataloguing-in-Publication Data.
A catalogue record for this book is available from the British Library.

Printed and bound in Great Britain by Biddles Ltd, Guildford and King's Lynn

The main part of the book lists equipment and services logically, grouped by function and then by country, to reflect the way NBC is generally viewed by defence planners and procurement authorities. The entries are shorter, referring the user to the Glossary for acronyms (where they appear in more than one entry) and to the Contractors index for contact details.

Improvements to all the graphics are planned and manufacturers are invited to supply new visual material as soon as it becomes available. Each entry is maintained on a database to speed the updating process and allow a choice of media – printed or CD.

Users can also contact Jane's Information Group for advice on how to access more precisely targeted electronic data packages. While this book is updated annually, the same data is refreshed monthly at Jane's Online. Future editions of this yearbook will also include a wider range of NBC-related equipment and services. The equipment categories are:

- Detection (sensor systems) – Nuclear
- Detection (sensor systems) – Biological
- Detection (sensor systems) – Chemical
- Detection (C³I systems)
- Detection (reconnaissance systems)
- Protection (individual) – Masks (general issue)
- Protection (individual) – Masks (aircrew)
- Protection (individual) – Filters
- Protection (individual) – Body protection
- Protection (individual) – Medical countermeasures
- Protection (collective)
- Decontamination
- Demilitarisation
- Training and simulation

The Detection (reconnaissance systems) section includes products for marking contaminated areas.

The Protection (individual) – Masks categories include all forms of individual respiratory protection, mask testing equipment and communications equipment. The masks and filters sections include facelets as well as closed-circuit breathing apparatus and fan-augmented filtration systems. The aircrew category includes special forces systems where information is available.

The Body protection category includes suits, footwear and gloves.

Products which act on the body's internal systems, such as injectors, or provide relief, such as dressings and casualty protection, are now included under Protection (individual) – Medical countermeasures.

Products that act on the agent itself, such as lotions and personal decontamination systems, are now listed under Decontamination.

Protection (collective) will also include water-purification plant and testing and miscellaneous equipment with a purely NBC function.

Demilitarisation deals with the array of different technologies being explored.

The Contractors list follows the same sequence as the main entries but is further sub-divided where necessary. For example, the section dealing with individual body protection is sub-divided into NBC clothing, gloves and footwear.

List of Contractors

After the main sections, the list of Contractors groups manufacturers and distributors for each equipment or service shown in the main sections, following the same order. The list is further subdivided, in order, where appropriate. For example, manufacturers of COLPRO shelters, filter systems, and test equipment are identified separately within the COLPRO group. Similarly, NBC gloves and footwear manufacturers are separately identified from those who make NBC clothing. In principle, the actual manufacturer of the equipment is listed, rather than any owning group or holding company. Reference is made to the latter, where appropriate, as a note, below the manufacturer's address. Look here, therefore, if you seek manufacturers of a particular type of NBC equipment or service.

Organisations, with their contact details, are listed alphabetically within each group. E-mail and website details are included. URLs are generally written in lower case using the European alphabet. E-mail addresses are sometimes case-sensitive.

Manufacturers Index

The Manufacturers Index follows the list of Contractors. For each alphabetically listed manufacturer, equipment is shown in alphabetical title order together with the relevant page number. So, knowing the name of an organisation, you can view its list of products and go straight to the correct page for more details.

Alphabetical Index

Finally, there is a simple list of equipment titles in the Alphabetical Index.

Glossary

AAV	Armoured Amphibious Vehicle
ABC	Atomic, Biological and Chemical (archaic. Normally 'NBC' today)
ABM	Anti-Ballistic Missile (as in the ABM Treaty)
AC	Hydrogen Cyanide
ACAS	Airfield Chemical Alarm System
ACF	Activated Carbon Fabric
ACU	Air Conditioning Unit
ACW	Abandoned Chemical Weapons
AFB	Air Force Base
AFU	Air Filtration Unit
AFV	Armoured Fighting Vehicle
AHCP	Ad Hoc Collective Protection
ALRV	Armoured Light (NBC) Reconnaissance Vehicle
ANP	*Appareil Normal de Protection*
APC	Armoured Personnel Carrier
APD	Advanced Portable Detector
AQAP	Allied Quality Assurance Programme (a NATO standard designation)
ARRV	Armoured Repair and Recovery Vehicle
ATP	Adenosine TriPhosphate
BADS	Biological Agent Detection System
BBC	BromoBenzylCyanide
BCD	Bio-Chemical Detector
BD	Biological Detector
BDS	Biological Detection System
BIDS	Biological Integrated Detection System
BITE	Built-In Test Equipment
BTWC	Biological and Toxins Weapons Convention
BW	Biological Warfare
BWR	Basic Wind Report(s)
BZ	BenactyZine
BZS	*Bundesamtes fur ZivilSchutz*
CAD	Chemical Agent Decontaminant
CAM	Chemical Agent Monitor
CARM	Chemical Agent Resistant Material
CAT	Chemical Agents Tracer
CATM	Chemical Agent Training Mixture
CBD	Chemical and Biological Defence
CBDA	Chemical and Biological Defense Agency
CBDE	Chemical and Biological Defence Establishment
CBR	Chemical. Biological and Radiological (US). (Also 'CBRN' which includes 'Nuclear')
CBW	Chemical and Biological Warfare
CCA	Contamination Control Area
C³I	Command, Control, Communications and Information (or Intelligence)
CCIS	Command, Control and Information System
CDR	Chemical Downwind Report(s)
CEOD	Chemical Explosive Ordnance Disposal
CEP	Civil Emergency Planning
CG	Phosgene
Cgy	Centigray (Unit of measurement of the absorbed dose of ionising radiation)
CK	Cyanogen chloride
CN	Chloroacetophone, an RCA
CNB	Codename for tear agent
CNC	Codename for tear agent
CNCL	CyaNogen ChLoride
CNS	Codename for tear agent
COLPRO	COLlective PROtection
COTS	Commercial-Off-The-Shelf
CP	Command Post
CPE	Collective Protection Equipment
CPS	Collective Protection System
cpm	counts per minute
CPR	Cardio-Pulmonary Resuscitation
Cr	ChRomium
Cryofracture	DEMIL process whereby munitions are frozen solid with liquid nitrogen, broken apart and the solidified CW agent disposed of
CS	O-chlorobenzamalonitrile, an RCA

CTBT	Comprehensive Test Ban Treaty. Applies to testing nuclear weapons
Cu	Copper
CVR(T)	Combat Vehicle Reconnaissance (Tracked)
CW	Continuous Wave
CWC	Chemical Weapons Convention
CX	Phosgene oxime
DA	Diphenylchloroarsine
DAP	Decontamination Apparatus, Portable
DC	Diphenylcyanoarsine
DEFSTAN	DEFence STANdard *(US and UK)*
DEMIL	DEMILitarisation. (The process of compliance with international treaties on the disposal of weapons of mass destruction)
DM	Adamsite
DMMP	DiMethyl MethylPhosphonate
DP	DiPhosgene
Dunnage	Material involved in handling or transport of CW weapons, including miscellaneous solid materials such as packing or protective gear
ECS	Environmental Control System
ECU	Environmental Control Unit
ED	EthylDichloroarsine
EDR	Effective downwind report(s)
ELISA	Enzyme-Linked ImmunoSorbent Assay
ELSS	Environmental Life Support System
EMA	Emergency Management Agency (or Authority) (US)
EMP	ElectroMagnetic Pulse
EOD	Explosive Ordnance Disposal
EPD	Electronic Personal Dosemeter
FBU	Filter Blower Unit
FCM	Flow CytoMeter
FINABEL	Group within NATO which allows full French participation with others in joint project decision-making (France, Italy, Netherlands And BELgium).
FMGC	*Filtre Mixte de Grande Capacité*
G	Nerve agent
GA	Tabun
GB	Sarin
GC	Gas Chromatography
GCC	Gulf Co-operation Council. A loose economic and security alliance between the more pro-western Gulf states: Bahrain, Kuwait, Oman, Qatar, Saudi Arabia, United Arab Emirates
GD	Soman
GM	Geiger-Müller
GPS	Global Position System
GRD	*Gruppe fur Rustungdienste*
GZ	Ground Zero
H	Mustard; Levenstein mustard
HCN	Hydrogen CyaNide
HD	Distilled mustard
HE	High Effect or High Explosive
HEPA	High Efficiency Particulate Air (filters)
HL	Mustard-Lewisite mixture
HMMWV	High-Mobility Multipurpose Wheeled Vehicle
HN	Nitrogen mustard
HVAC	High-Volume Air Conditioning
HVM	High-Velocity Missile
IA	Irrespirable Atmosphere
IAEA	International Atomic Energy Agency
IBAD	Interim Biological Agent Detector
IBDS	Integrated Biological Detection System
ICBM	Inter Continental Ballistic Missile
IDF	Israel Defence Forces
IDU	Ionisation (Ionization) Detector Unit
IFV	Infantry Fighting Vehicle
IMS	Ion Mobility Spectroscopy (or Spectometry)

INR	INitial Radiation (nuclear)
INS	INertial Navigation System
IOC	Initial Operational Capability
IPE	Individual Protection Ensemble (or Equipment)
IRBM	Intermediate Range Ballistic Missile
IRDA	Infra-Red Data Access (a wireless method of exchanging data between a computer and a peripheral device)
JGSDF	Japanese Ground Self-Defence Force
L	Lewisite
LAN	Local Area Network
Lance	Long hand-held or fixed rod with nozzle at the end designed for spraying decontaminant without hazard to the user.
LAPS	Light Addressable Potentiometric Sensor
LCD	Liquid Crystal Display
LED	Light-Emitting Diode
LIDAR	LIght Detection And Ranging
LMS	Lightweight Multipurpose Shelter
LR-BSDS	Long-Range Biological Standoff Detection System
LS	Liquid sampler
LSID	Large Scale Issue Detector
MAV	Micro Air Vehicle
MBT	Main Battle Tank
MCCU	MicroClimate Cooling Unit
MCMV	Mine CounterMeasures Vessel
MD	MethylDichloroarsine
MICAD	Multipurpose Integrated Chemical Agent Alarm
MILES	Multiple Integrated Laser Engagement System
MIRV	Multiple Independently targeted Re-entry Vehicle
MLF	MultiLaminar Film
MOPP	Mission Oriented Protective Posture
MPDS	MultiPurpose Decontamination System
MPS	Motor Pump Set
MR	Simulated mustard
MS	Methyl Salicylate
NAEDS	Non-Aqueous Equipment Decontamination System
NAIAD	Nerve Agent Immobilised enzyme Alarm and Detection
NAPS	Nerve Agent Pretreatment Tablets
NATO	North Atlantic Treaty Organisation
NBC	Nuclear, Biological and Chemical
NBCF	Nuclear, Biological, Chemical and Fire
NBCPC	Nuclear, Biological and Chemical Protective Cover
NBCRS	Nuclear, Biological and Chemical Reconnaissance System
NBCSS	Nuclear, Biological and Chemical Shelter System
NCO	Non-Commissioned Officer
NCU	Network Control Unit
Ni/Cd	Nickel Cadmium
NiMH (or NiH)	Nickel Metal Hydride (battery technology)
NNPT	Nuclear Non-Proliferation Treaty
OP	Observation Post
OPCW	Organisation for the Prohibition of Chemical Weapons (Den Haag-based organisation created by the CWC)
OPV	Offshore Patrol Vessel
PAO	PolyAlphaOlefine
PAS	PhotoAcoustic Spectroscopy
PCAS	Persistent Chemical Agent Simulant
PCR	Polymerase Chain Reaction
PD	PhenylDichloroarsine
PDRM	Portable Dose Rate Meter
PE	PolyEthylene (also Plastic Explosive in other context)

PEG	Polyethylene glycol
PES	PolyESter
PFIB	Perfluoroisobutene
POL	Petrol, Oil, Lubricants
ppb	Parts per billion
PPE	Personal Protective Equipment
PS	Chloropicrin
PTFE	PolyTetraFluoroEthylene
PTU	Protection Training Unit
PVA	Poly Vinyl Acetate
RAM	Random Access Memory; Riot Agent Monitor
RAS	Replenishment At Sea
RCA	Riot Control Agent (for example CS or CN)
RCU	Remote-Control Unit
RF	Radio Frequency
RFP	Request For Proposals
ROM	Read-Only Memory
RPV	Remotely Piloted Vehicle
SA	Arsine
SAW	Surface Acoustic Wave (a technology used in CW detection)
SAWE	Simulated Area Weapons Effects
SCBA	Self-Contained Breathing Apparatus
SCW	Stockpiled Chemical Weapons
SDK	Skin Decontamination Kit
SEB	Staphylococcal Enterotoxin B
SFOR	UN Stabilisation FORce (former Yugoslavia)
SIBCA	Sampling and Identification of Biological and Chemical Agents
SICAS	Ship Installed Chemical Alarm System
SICS	Ship Installed Chemical System
SIRS	Ship Installed Radiac System
SITREP	Situation report
STANAG	STAnding NATO AGreement
START	Strategic Arms Reduction Talks (the US-Russian forum which seeks mutual balanced reduction of the nuclear warhead stockpile)
STB	Super Tropical Bleach
T	Codename for sulphur and chlorine compound
TAP	Toxic Agent Protection
TDCC	*Trousse de Detection Chimique de Controle*
TEDA	Triethylenediamine
TEG	ThermoElectric Generator
TFA	Toxic-Free Area
TIC	Toxic Industrial Compounds
TIM	Toxic Industrial Materials
TREE	Transient Radiation's Effects on Equipment
UAE	United Arab Emirates
UAV	Unmanned Aerial Vehicle
UCPS	Unhardened Collective Protection System
UNSCOM	United Nations Special COMission (related specifically to compliance with UN resolutions by Iraq on weapons of mass destruction)
URL	Uniform Resource Locator – the unique name by which a website is identified
UV	Ultra-Violet
Vector	(Bio) Means of transmitting a biological agent for example, by aerosol, birds or insects
VOC	Volatile Organic Compound (or, sometimes, 'Content')
VPU	Voice Projection Unit
VX	Codename for nerve agent
WHO	World Health Organisation
WMD	Weapons of Mass Destruction

EDITORIAL AND ADMINISTRATION

Publishing Director: Alan Condron, e-mail: Alan.Condron@janes.co.uk

Managing Editor: Mike Bryant, e-mail: Mike.Bryant @janes.co.uk

Group Content Manager: Anita Slade, e-mail: Anita.Slade@janes.co.uk

Content Editing Manager: Jo Agius, e-mail: Jo.Agius@janes.co.uk

Pre-Press Manager: Christopher Morris, e-mail: Christopher.Morris@janes.co.uk

Team Leaders: Sharon Marshall, e-mail: Sharon.Marshall@janes.co.uk
Neil Grace, e-mail: Neil.Grace@janes.co.uk

Production Editor: Laura Kew, e-mail: Laura.Kew@janes.co.uk

Production Controller: Victoria Powell, e-mail: Victoria.Powell@janes.co.uk

Content Update: Jacqui Beard, Information Collection Co-Ordinator
Tel: (+44 20) 87 00 38 08 Fax: (+44 20) 87 00 39 59
e-mail: yearbook@janes.co.uk

Jane's Information Group Limited, Sentinel House, 163 Brighton Road, Coulsdon,
Surrey CR5 2YH, UK
Tel: (+44 20) 87 00 37 00 Fax: (+44 20) 87 00 37 88
e-mail: jnbc@janes.co.uk

SALES OFFICE

Send Europe, Middle East and Africa enquiries to: *Mike Gwynn – Head of Information Sales*
Jane's Information Group Limited, Sentinel House, 163 Brighton Road, Coulsdon, Surrey CR5 2YH, UK
Tel: (+44 20) 87 00 37 00 Fax: (+44 20) 87 63 10 06
e-mail: info@janes.co.uk

Send USA enquiries to: *Robert Loughman – Sales Director*
Jane's Information Group Inc, 1340 Braddock Place, Suite 300, Alexandria, Virginia 22314-1651, USA
Tel: (+1 703) 683 37 00 Fax: (+1 703) 836 02 97 Telex: 6819193
Tel: (+1 800) 824 07 68 Fax: (+1 800) 836 02 971
e-mail: info@janes.com

Send Asia enquiries to: *David Fisher – Group Business Manager*
Jane's Information Group Asia, 5 Shenton Way , #01-01 UIC Building, Singapore 068808
Tel: (+65) 6410 1240 Fax: (+65) 6226 1185
e-mail: info@janes.com.sg

Send Australia/New Zealand enquiries to: *Pauline Roberts – Business Manager*
Jane's Information Group, PO Box 3502, Rozelle Delivery Centre, New South Wales 2039, Australia
Tel: (+61 2) 85 87 79 00 Fax: (+61 2) 85 87 79 01
e-mail: info@janes.thomson.com.au

ADVERTISEMENT SALES OFFICES

(Head Office)
Jane's Information Group
Sentinel House, 163 Brighton Road,
Coulsdon, Surrey CR5 2YH, UK
Tel: (+44 20) 87 00 37 00
Fax: (+44 20) 87 00 38 59/37 44
e-mail: defadsales@janes.co.uk

Richard West, Senior Key Accounts Manager
Tel: (+44 1892) 72 55 80 Fax: (+44 1892) 72 55 81
e-mail: richard.west@janes.co.uk

Kate Hamlin, Advertising Sales Manager
Tel: (+44 20) 87 00 38 53 Fax: (+44 20) 87 00 38 59/37 44
e-mail: kate.hamlin@janes.co.uk

Joni Beeden, Advertising Sales Executive
Tel: (+44 20) 87 00 39 63 Fax: (+44 20) 87 00 38 59/37 44
e-mail: joni.beeden@janes.co.uk

Steve Soffe, Advertising Sales Executive
Tel: (+44 20) 87 00 39 43 Fax: (+44 20) 87 00 38 59/37 44
e-mail: steven.soffe@janes.co.uk

(USA/Canada office)
Jane's Information Group
1340 Braddock Place, Suite 300,
Alexandria, Virginia 22314-1651, USA
Tel: (+1 703) 683 37 00
Fax: (+1 703) 836 55 37
e-mail: defadsales@janes.com

USA and Canada
Katie Taplett, US Advertising Sales Director
Tel: (+1 703) 683 37 00 Fax: (+1 703) 836 55 37
e-mail: katie.taplett@janes.com

Northern USA and Eastern Canada
Harry Carter, Northeast Region Advertising Sales Manager
Tel: (+1 703) 683 37 00 Fax: (+1 703) 836 55 37
e-mail: harry.carter@janes.com

Southeastern USA
Kristin D Schulze, Advertising Sales Manager
PO Box 270190, Tampa, Florida 33688-0190
Tel: (+1 813) 961 81 32 Fax: (+1 813) 961 96 42
e-mail: kristin@intnet.net

Western USA and Western Canada
Richard L Ayer
127 Avenida Del Mar, Suite 2A, San Clemente, California 92672
Tel: (+1 949) 366 84 55 Fax: (+1 949) 366 92 89
e-mail: ayercomm@earthlink.com

Australia: *Richard West* (see UK Head Office)

Benelux: *Steve Soffe* (see UK Head Office)

Brazil: *Katie Taplett* (see USA address)

Corporate Accounts: Simon Kay
33 St John's Street, Crowthorne, Berkshire RG45 7NQ, UK
Tel: (+44 1344) 77 71 23 Mobile: (+44 7702) 54 96 84
Fax: (+44 1344) 77 58 85
e-mail: simon.kay@btclick.com

Eastern Europe: MCW Media & Consulting Wehrstedt
Dr Uwe H Wehrstedt
Hagenbreite 9, D-06463 Ermsleben, Germany
Tel: (+49) 0700/WEHRSTEDT / (+49) 03 47 43/620 90
Fax: (+49) 03 47 43/620 91
e-mail: info@Wehrstedt.org

France: Patrice Février
BP 418, 35 avenue MacMahon,
F-75824 Paris Cedex 17, France
Tel: (+33 1) 45 72 33 11 Fax: (+33 1) 45 72 17 95
e-mail: patrice.fevrier@wanadoo.fr

Germany and Austria: *MCW Media & Consulting Wehrstedt* (see Eastern Europe)

Greece: *Steve Soffe* (see UK Head Office)

Hong Kong: *Joni Beeden* (see UK Head Office)

India: *Joni Beeden* (see UK Head Office)

Israel: Oreet – International Media
15 Kinneret Street, IL-51201 Bene Berak, Israel
Tel: (+972 3) 570 65 27 Fax: (+972 3) 570 65 27
e-mail: admin@oreet-marcom.com
Defence: Liat Shaham
e-mail: liat_s@oreet-marcom.com

Italy and Switzerland: Ediconsult Internazionale Srl
Piazza Fontant Marose 3, I-16123 Genoa, Italy
Tel: (+39 010) 58 36 84 Fax: (+39 010) 56 65 78
e-mail: genova@ediconsult.com

Japan: Skynet Media, Inc
748, 1-7 Akasaka 9-chome, Minato-ku, Tokyo 107-0052, Japan
Contact: Mr Osamu Yoneda
Tel: (+81 3) 54 74 78 35
Fax: (+81 3) 54 74 78 37
e-mail: skynetme@wonder.ocn.ne.jp

Middle East: *Steve Soffe* (see UK Head Office)

Pakistan: *Joni Beeden* (see UK Head Office)

Russia: Vladimir N Usov, PO Box 98, Nizhniy Tagil,
Sverdlovsk Region, 622018, Russia
Tel/Fax: (+7 3435) 23 02 68
e-mail: uvn125@uraltelecom.ru

Scandinavia: The Falsten Partnership
PO Box 21175, London N16 6ZG, UK
Tel: (+44 20) 88 06 23 01 Fax: (+ 44 20) 88 06 81 37
e-mail: sales@falsten.com

Singapore: *Richard West/Joni Beeden* (see UK Head Office)

South Africa: *Richard West* (see UK Head Office)

South Korea: JES Media Inc
2nd Floor, ANA Building, 257-1 Myungil-Dong, Kandong-Gu, Seoul 134-070, Korea
Contact: Mr Young-Seoh Chinn, President
Tel: (+82 2) 481 34 11/34 13
Fax: (+82 2) 481 34 14
e-mail: jesmedia@unitel.co.kr

Spain: Via Exclusivas SL
Contact: Julio de Andres
Viriato 69SC, E-28010 Madrid, Spain
Tel: (+34 91) 448 76 22 Fax: (+34 91) 446 02 14
e-mail: j.a.deandres@viaexclusivas.com

Turkey: *Richard West* (see UK Head Office)

ADVERTISING COPY
Linda Letori (Jane's UK Head Office)
Tel: (+44 20) 87 00 37 42 Fax: (+44 20) 87 00 38 59/37 44
e-mail: linda.letori@janes.co.uk

For North America, South America and Caribbean only:
Shanee Johnson (Jane's USA/Canada Office)
Tel: (+1 703) 683 37 00 Fax: (+1 703) 836 55 37
e-mail: shanee.johnson@janes.com

Alphabetical list of advertisers

DISCLAIMER

Jane's Information Group gives no warranties, conditions, guarantees or representations, express or implied, as to the content of any advertisements, including but not limited to compliance with description and quality or fitness for purpose of the product or service. Jane's Information Group will not be liable for any damages, including without limitation, direct, indirect or consequential damages arising from any use of products or services or any actions or omissions taken in direct reliance on information contained in advertisements.

Information Services & Solutions

Jane's is the leading unclassified information provider for military, government and commercial organisations worldwide, in the fields of defence, geopolitics, transportation and law enforcement.

We are dedicated to providing the information our customers need, in the formats and frequency they require. Read on to find out how Jane's information in electronic format can provide you with the best way to access the information you require.

Jane's Online

Why not choose the online format for your Jane's information?

Choosing to subscribe to Jane's information online will allow you to get the maximum use of the detailed information, as you will have:
- Instant access 24-hours a day.
- Advanced power search tools to take you directly to the information you are seeking.
- The opportunity to browse the information section-by-section and cut and paste the information for use in your own internal presentations.
- High quality colour JPEG images to support recognition and for use in your internal presentations.

- Active interlinking, allows you to navigate via hyperlinks in records, within the viewed documents, to other related information, thus reducing your search time down to minutes.
- Regular monthly updates to ensure you always have the most current information available.

Jane's information is accessible by IP address for networking within your organisation or by unique username and password, allowing you access from anywhere in the world.

Check out this site today: **http://www.janes.com**

Jane's CD-ROM Libraries

Quickly pinpoint the information you require from Jane's

Choose from nine powerful CD-ROM libraries for quick and easy access to the defence, geopolitical, space, transportation and law enforcement information you need. Take full advantage of the information groupings and purchase the entire library.

Libraries available:
Jane's Air Systems Library
Jane's Defence Equipment Library
Jane's Defence Magazines Library
Jane's Geopolitical Library
Jane's Land and Systems Library
Jane's Market Intelligence Library
Jane's Police and Security Library
Jane's Sea and Systems Library
Jane's Transport Library

Key benefits of Jane's CD-ROM include:
- Quick and easy access to Jane's information and graphics
- Easy-to-use Windows interface with powerful search capabilities
- Online glossary and synonym searching
- Search across all the titles on each disc, even if you do not subscribe to them, to determine whether you would like to add them to your library
- Export and print out text or graphics
- Quarterly updates
- Full networking capability
- Supported by an experienced technical team

Jane's Data Service – Intranet Solution

Get Jane's Data behind your intranet

Access over 200 sources of defence, security, law enforcement and transport data, integrated behind your intranet or closed network

When you need mission-critical information, searching across multiple sources retrieves your answers quickly and easily. Integrate Jane's data with your own intelligence sources and your users can rely on the impartiality, accuracy and authority of Jane's information as a benchmark. With more users able to access Jane's directly from their desktop, you can centrally streamline

your information requirements to better monitor and respond to needs as they arise.
- Flexibility of choice with your selection of Jane's data
- Full integration of data into a secure environment
- Frequent updates via e-mail or ftp
- All data can be exported into other desktop applications
- High quality JPEG images for recognition training, internal briefing or analysis

For further information contact your local Jane's office or e-mail: jds@janes.co.uk

Jane's Consultancy

Jane's Consultancy draws on a unique international network of experts to undertake special research on your behalf, to your specifications. Simply contact us, in confidence, with your requirements and we will provide the expert and authoritative research you need.
- Unrivalled access to hard-to-find information
- Impartial expert analysis

- Unique global reach providing a balanced view
- Cost and time effective solutions
- Complete confidentiality

Visit consultancy.janes.com today and put our experts to work for you. Alternatively contact your nearest Jane's office or e-mail: consultancy@janes.com

The information you require, delivered in a format to suit your needs.

Jane's Chemical-Biological Defense Guidebook

Desktop reference to support your chem-bio defence strategy

This vital reference is intended for anyone responsible for meeting emerging chem-bio threats directly. You will find detailed information on chemical and biological agents, their characteristics and potential delivery mechanisms along with authoritative surveys of the available detection and protective equipment. In addition, expert workings of potential threat scenarios together with security checklists make this a crucial resource for your chem-bio defence strategy.

Contents include • Chemical and biological agents • CB agent weaponisation • Munitions and delivery systems • CB agent detection and warning, contamination control and protection • Threats to buildings • Domestic preparedness • World review of threat capabilities • Detailed appendix • Action plan and security checklist

For further information, please visit our online catalogue at janes.com or telephone your local sales office using one of the numbers below

UK/Europe/Middle East/Africa
Tel: +44 (0) 20 8700 3700
Fax: +44 (0) 20 8763 1006
e-mail: info@janes.co.uk

North/Central/South America
Tel: 1 800 824 0768
1 703 683 3700
Fax: 1 800 836 0297
1 703 836 0297
e-mail: info@janes.com

US West Coast
Tel: 1 714 850 0585
Fax: 1 714 850 0606
e-mail: janeswest@janes.com

Canada
Tel: 1 613 288 0189
Fax: 1 613 288 0190
e-mail: geoff.mizen@janes.com

Asia (Singapore)
Tel: +65 6410 1240
Fax: +65 6226 1185
e-mail: info@janes.com.sg

India
Tel: +91 (0)11 651 6105
Fax: +91 (0)11 651 6105
e-mail: janesindia@sify.com

Japan
Tel: +81 3 5218 7682
Fax: +81 3 5222 1280
e-mail: norihisa.fukuyama@janes.jp

Australia
Tel: +61 (02) 8587 7900
Fax: +61 (02) 8587 7901
e-mail: info@janes.thomson.com.au

Jane's
www.janes.com

Users' Charter

This publication is brought to you by Jane's Information Group, a global company with more than 100 years of innovation and an unrivalled reputation for impartiality, accuracy and authority.

Our collection and output of information and images is not dictated by any political or commercial affiliation. Our reportage is undertaken without fear of, or favour from, any government, alliance, state or corporation.

We publish information that is collected overtly from unclassified sources, although much could be regarded as extremely sensitive or not publicly accessible.

Our validation and analysis aims to eradicate misinformation or disinformation as well as factual errors; our objective is always to produce the most accurate and authoritative data.

In the event of any significant inaccuracies, we undertake to draw these to the readers' attention to preserve the highly valued relationship of trust and credibility with our customers worldwide.

If you believe that these policies have been breached by this title, you are invited to contact the editor.

A copy of Jane's Information Group's Code of Conduct for its editorial teams is available from the publisher.

INVESTOR IN PEOPLE

Quality Policy

Jane's Information Group is the world's leading unclassified information integrator for military, government and commercial organisations worldwide. To maintain this position, the Company will strive to meet and exceed customers' expectations in the design, production and fulfilment of goods and services.

Information published by Jane's is renowned for its accuracy, authority and impartiality, and the Company is committed to seeking ongoing improvement in both products and processes.

Jane's will at all times endeavour to respond directly to market demands and will also ensure that customer satisfaction is measured and employees are encouraged to question and suggest improvements to working practices.

Jane's will continue to invest in its people through training and development, to meet the Investor in People standards and changing customer requirements.

Jane's

Detect

UV-APS

The Ultraviolet Aerodynamic Particle Sizer® Spectrometer (UV-APS) continuously monitors air for possible BW agents. When it detects an aerosol that fits the profile of possible agents, the UV-APS provides a trigger so your analytical instrument can identify and confirm its presence.

The UV-APS sensor measures particle size and provides size distribution of the aerosol. Each particle is measured with ultraviolet fluorescence technology to measure its biological characteristics. Measurements are made on each particle (not a batch of particles) so you have size and fluorescence for each particle.

Deployed worldwide, both the Canadian CIBADS and U.S. Army BIDS systems use the TSI UV-APS because of its proven sensitivity and discrimination.

Protect

The type M41 Protection Assessment Test System (PATS) uses a unique aerosol technology to test the fit and integrity of NBC protective masks. It measures the leakage around the face piece seal while the mask is being worn.

Include the M41 as part of your NBC training program. Verify that your personnel have the optimal mask size and that they know how to get the best possible protection from their masks. Most modern military masks give the wearer protection factors up to 10,000. The TSI M41 PATS measures protection factors to 50,000. Other fit test methods, such as CS chambers, only verify that the protection is greater than 100.

A proven instrument, the M41 has been fully fielded by the U.S. Army, U.S. Marine Corps, U.S. Air Force and the German Bundeswehr.

M41 PATS

With TSI UV-APS and M41 PATS, you can feel confident in the protection you are providing your troops. For more information, contact Mr. Gerald Gerard.

TSI Incorporated
500 Cardigan Road , Shoreview, MN 55126 USA
Tel: *+1 651 490 3871* **Fax:** *+1 651 490 3824* **E-mail:** *nbc@tsi.com* **Web:** *www.nbc.tsi.com*

Foreword

NUCLEAR, BIOLOGICAL AND CHEMICAL DEFENCE – A REVIEW

The past year in perspective

I introduced the previous edition of *Jane's Nuclear, Biological and Chemical Defence* by describing how terrorist groups, especially fundamentalist organisations at the extreme ends of the belief structures of the common faiths, may well consider Weapons of Mass Destruction (WMD) as valid tools to achieve their aims. In the year since then, cataclysmic events have brought home to us, firstly, how much terrorism has migrated towards the fundamentalist model and, secondly, how bio-terrorism in particular was able to cause enormous panic and fear across the United States of America. The anthrax used in the October 2001 attacks could hardly qualify, by the results anyway, as a weapon of mass destruction. The events may well underscore an established historical trend in bio-terrorism: that the majority of bio-terrorist incidents remain the preserve of the disgruntled loner rather than the terrorist cell. Nevertheless doomsday scenarios were painted in the US (Exercise *Dark Winter*) and in the UK on television (The BBC's *Smallpox 2002: Silent Weapon* and Channel 4 television's *Gas Attack* are examples).

'. . . . the majority of bio-terrorist incidents remain the preserve of the disgruntled loner rather than the terrorist cell.'

Clearly, the jury is still out on the US anthrax events, but there is more than a finger of suspicion pointing at a highly qualified insider within the US Federal research community, rather than a recognised terrorist group. Much of the media hype focussed on the phrase 'weapons-grade anthrax', implying a link to a state sponsor (nearly always Iraq). It emerged that the anthrax spores had not only been dried and milled to the precise particle size required to ensure the dust reached into the victims lungs but that it had been mixed with an anti-clumping material (Bentonite) to make it disperse more effectively. This was traced to a US laboratory. What surprised the public most of all was how long it took to establish exactly what the material was and we were all reminded that identifying the precise strain of a disease is vital in driving the correct treatment. Failure to do so may hasten the patient's demise. It is, therefore, in the area of biodetection where the most frenetic activity in research and development is taking place and some of the challenges are explored here (see Technical Developments).

It is hard to identify any benefit from the awful events and discoveries of the last 12 months. One possible benefit has been a vastly increased understanding by the general public of the science and mechanics behind the creation, weaponisation and delivery of WMD. Here, the more responsible press cover has helped greatly. We now know how difficult it is to make pure sarin and how close *Aum Shinrikyo* came to discovery by the police. We know that the quantity of agent required to qualify as a weapon of mass destruction is much higher than we used to think. We know that several of the diseases commonly identified as biological weapons are highly infectious. They are therefore extremely difficult to control when released into the wild for example, by a self-infected terrorist.

'The terrorist organisation is as likely to infect all of its friends as much as its enemies'

The terrorist organisation is as likely to infect all of its friends as much as its enemies. We know that the science required to produce a nuclear fission weapon may be simple in theory but that it is extremely risky and technically complex to make one in secret. Fear of the unknown is giving way, not to complacency, but to the realisation that the terrorist, always seeking simplicity in the creation of an event, faces some pretty tough challenges if he is to use WMD effectively.

The media and Weapons of Mass Destruction

Over the past 20 years a constant barrage of medical, chemical and lifestyle scare stories has imperceptibly but steadily raised the baseline fear level in society and made people increasingly risk averse. This baseline fear was raised by several orders of magnitude after both 11 September 2001 and the anthrax attacks. It continues to grow, especially in the open society of the US, fuelled by reports of 'tanker lorry weapons', 'dirty bombs' or revelations from the Guantanamo Bay prisoners on the

Accurate identification of C or BW agents involves the analysis of samples from the surrounding atmosphere or the surface. However, optical or radar systems can scrutinise the atmosphere for anomalies by comparing them against the background. This system (RAPID) uses fast Fourrier transform analysis at the infra-red wavelength via a flex-pivot interferometer (Bruker RockSolid™) to watch for possible agent release (Bruker Daltonics©) 2002/0137487

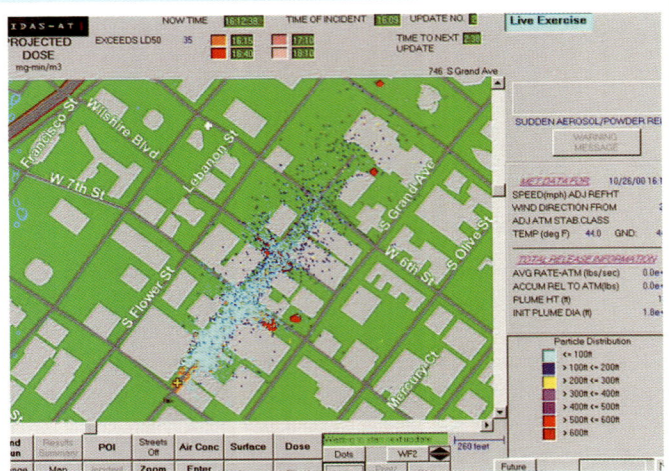

This screen shot shows how command support software has developed over recent years. Real-time detection and meteorological data from a network of sensors is overlaid on topographical data and calculations are made to predict future movement of the plume of agent (MIDAS-AT – urban incident example from PLG Inc) **2002/0137748**

interest displayed by Al-Qaeda in the acquisition of weapons of mass destruction. Some of the announcements have been vague or unhelpful. There seems little point in stating in public that we 'can expect another attack probably using chemical weapons' without either suggesting the likely timing or location, or recommending specific precautions we should take. The public is constantly being told to be vigilant. Vigilant for what? How much more vigilant can they be than they already are? Public concerns are real and they need continued help, with both facts and sound guidance, to set WMD issues into context and allay their alarm.

'Public concerns are real and they need continued help, with both facts and sound guidance, to set WMD issues into context . . . .'

The general public is now healthily sceptical of the irresponsible, scaremongering headlines which appeared, especially in the UK tabloid media, on the topic of WMD. Large format pictures of the mushroom cloud from a 1950s megaton Bikini Atoll test accompanied articles on 'dirty bombs' in several papers. The slightest hint of further revelations involving Al-Qaeda and WMD inevitably had picture editors searching for yet more shots of people rushing around in respirators and NBC suits. The demand for 24-hour news coverage is rapidly becoming the enemy of truth and the public is depressed by a constant barrage of mainly bad news. Politicians and pundits alike, harried by the media, are now made to feel they have to say something about nothing in order to fill time.

In fairness to the tabloid media, apart from the widespread use of WMD in wars of the last century (1914 to 1918 and the 1980s Iran-Iraq war) there is little hard information with which to set WMD use in context. Experts are drawn in. They disagree on air and leave the public confused. There was one exasperated exchange on the BBC 'Today' radio programme over the US anthrax attack. One pundit argued that making anthrax was pretty easy – school grade science – another, that the technical problems were so tough to overcome that it must have come from a sophisticated laboratory.

The enemy within

Western governments, thrown completely onto the back foot by the events of 11 September 2001 and afterwards,

have reappraised their entire national effort towards countering fundamentalist terrorism and its concomitant risk of WMD use. Society is based on permission and trust. Governments realise that trust has been sorely tested and that they have to look deep within the structure of society itself to identify ways in which that trust might be abused. They need to seek countermeasures which, whilst effective, do not change society in fundamental ways. In the WMD context it has led to a reappraisal of the storage locations and volumes of those toxic materials which are essential to the very existence and conduct of modern society. How could the terrorist use these materials against society itself? Should we store and transport so much at once?

The fear is that the terrorist has asked himself whether it's worth the risk to try and produce large quantities of nerve agent, when there are tanker loads of poorly protected toxic chemicals on our roads and in our cities every day. This fear is justified. In February this year, four Moroccans were arrested in Rome in possession of 4 kg of potassium ferrocyanide crystals. A search of the flat they rented revealed detailed maps of the Rome water supply and the US Embassy. Lethal diseases, haemorrhagic fevers for example, are endemic in many parts of the world and their malicious introduction into urban society may not even need a lab. It makes our external and internal defence policies from NMD downwards appear to have been looking in the wrong direction altogether.

'Lethal diseases, haemorrhagic fevers for example, are endemic in many parts of the world . . . . malicious introduction . . . . may not even need a lab.'

Building regulations in Israel require the creation of a 'sanctuary' room in every building which can be supplied with filtered air to protect citizens against a WMD attack. Domestic COLPRO systems are commercially available comprising blast-hardened intake and exhaust valves, filters and power packs. Following a total electrical failure (the Rainbow 36 System shown has a 10-hour duration battery), air can be drawn through the filter unit using a hand-powered bellows (Beth-El Zikhron Yaaqov Industries Ltd) **2002/0137272**

Protection for operational personnel uses permeable systems which are designed to trap any CW agent laden vapour in a carbon layer. A variety of coloured or camouflaged outer layers are available. The suit shown here is specified for OPCW verification inspectors (Blücher)   **2002**/0102792

Nevertheless, there is cause for some optimism, quite apart from the fact that authorities have begun to think more radically in tackling the problem. One of the spectacular benefits of the Internet is its ability to alert epidemiologists to unusual outbreaks of disease instantaneously. Surveillance at every level, from the satellite to the hidden camera, has begun to tighten the screws considerably on organised crime's ability to traffic in everything from pot to plutonium. The most glaring gap, however, remains human intelligence ('HUMINT') – the need for dogged, determined and effective penetration of the terrorist cells.

New bombs for old

Something is stirring in Los Alamos. Many of the scientists who designed the current range of US nuclear warheads are reaching the end of their working lives. In fact all the nuclear weapons laboratories face the same dilemma. How do you attract new talent to keep the deterrent effective against a shrinking warhead requirement? Constant testing is vital to guarantee the integrity of the stockpile and ensure effective detonation if ever required. This is, of course, banned. The industry is therefore forced to try and assure the operability of entire systems by a mixture of individual component testing and computer simulation of weapon effects. This kind of work requires the highest levels of personal integrity. National security is at stake. It is highly specialised, with little application outside the industry, making it difficult to attract a new generation of recruits. Both defence analysts and the laboratories themselves wonder how to keep the expertise alive and the work exciting. There is therefore more than a little interest in the concept of 'small nukes': follow-on systems comprising

highly specialised low-yield nuclear weapons designed for specific tactical tasks. This trend upsets the old-timers. Brought up in the culture which argued that nuclear weapons should be so big and awful that, ideally, nobody would ever dream of using them, they are alarmed at the trend which suggests that they might become part of the conventional war-fighting inventory. How useful it would be if you could bury a whole mountainside of Al-Qaeda cave networks with a single weapon, goes the argument. It is no surprise, therefore that the US administration's 2002

Water and food are vulnerable to WMD attack. This high-capacity water purification system (WPU(NBC)S) in use with the British Army filters out NBC agents (Stella-Meta Filtration Systems)   **2002**/0102995

The safe demilitarisation of chemical weapons is a major challenge to the largest stockpile holders (the USA and Russia). The SILVER II™ system uses low-temperature chemical oxidation of the organic content (most CW agents are volatile organic solvents) whereby aqueous silver ions progressively and irreversibly convert the material to CO_2, water and residual salts in a series of steps (Accentus plc) **2002**/0102235

Nuclear Posture Review hints at a new approach for the nuclear 'triad' (manned bombers, Submarine-Launched Ballistic Missiles and Inter-Continental Ballistic Missiles). See *NBC Capabilities – USA.*

'nuclear weapons should be so big and awful that, ideally, nobody would ever dream of using them.'

In addition to this new thinking on offensive nuclear weapons, the recent bilateral agreement between Russia and the US has secured Russia's reluctant agreement to discard the old antiballistic missile treaty (ABM). This clears the way for the US to pursue its National Missile Defence Plan (NMD). Whilst the accord identifies considerable further reductions in both stockpiles, the US will not reduce its nuclear arsenal to below a level which it perceives necessary to counter the growing nuclear weapons menace from the so-called rogue states (North Korea, Iraqi, Iran), as well as from Russia and China. The reduction in warhead stocks will ease the load on the poorly-funded Russian maintenance programme and, hopefully, lead to better safety in Russia's stockpile.

'Dirty bombs'

Despite the improved surveillance mentioned above, there is still a considerable traffic in nuclear materials. They continue to find their way out into the wider world via the southern and western states of the former Soviet Union. In July last year, 2 kg of highly enriched uranium, packed in glass jars, was confiscated from a hotel in Batumi in Georgia. Uranium (U^{235}) has a half-life of thousands of years (half-life is the time it takes half of the isotope to

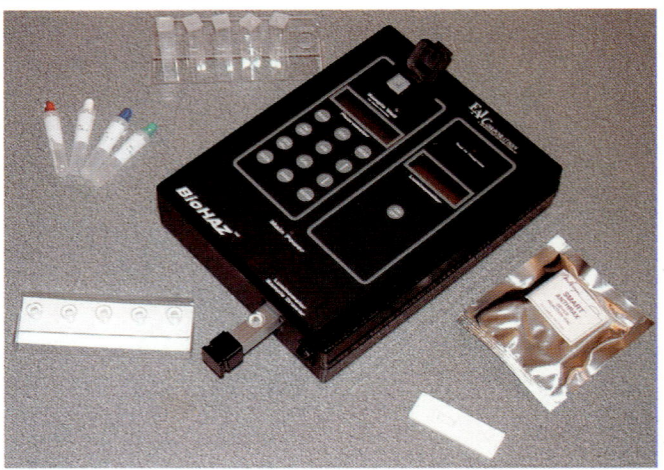

The BioHaz Luminometer from EAI Corporation is an example of a number of bio detection devices being developed to meet the future 'first responder' market (John Eldridge) **2002**/0137924

radioactively decay – an exponential process). Caesium (Cs^{137}) has a half-life of 37 years. At Goiânia, Brazil, in 1987 scrap metal thieves retrieved a 20 g capsule of Cs^{137} from an abandoned radiotherapy clinic. The scraps were dissolved in water to make paint for their children's faces and make them glow in the dark. Four died and 10 others suffered radiation burns. Another 249 were contaminated. More than 110,000 people had to be monitored for exposure for months afterwards. Cs^{137} is widely used in radiotherapy in chloride form as a γ source.

Radioactive isotopes are commonly held in small quantities in every large hospital and many engineering works. They are essential for x-ray generation, cancer radiotherapy and for measuring the thickness of materials. Even your domestic smoke alarm has a minute quantity of Americium (Am^{241}), essential to create the ionisation process which detects the smoke.

The so-called 'dirty bomb' is simply a quantity of explosive with a quantity of radioactive material wrapped around it. The radioactive isotope's explosive dispersion may not cause instant death and injury but will serve to deny a wide area to human occupation for an extremely long time.

'The radioactive isotope's explosive dispersion will deny a wide area to human occupation for an extremely long time.'

Security at storage sites has traditionally been poorly paid and with a high turnover of staff. This situation calls into question complacent claims that targeted theft is unlikely.

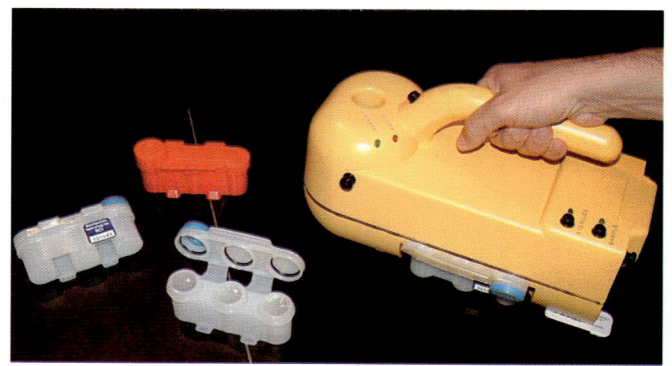

This hand-held sample collector from Mesosystems Incorporated can concentrate air samples (intake at top left of the unit), read from container samples (bottom centre) or from the BTA™ Test Strips from Tetracore Incorporated (John Eldridge) **2002**/0137923

The US Armed Forces HMMWV-mounted BIDS (Biological Integrated Detection System) programme is designed to give commanders a comprehensive field BW collection and analysis capability (TSI Incorporated) 2002/0137927

This will change, as the consequences of acquisition by terrorist organisations dawns on the authorities and the industry.

Russia, DEMIL and G8

Concerns over Russians chemical demilitarisation obligations have not gone away. Lack of money, bureaucratic tangles, and cultural differences have all contributed to slow progress in the various programmes designed to support Russia in complying with her requirements under the CWC. The seriousness of the situation even brought it into the agenda of the June 2002 meeting of the Group of Eight Industrialised Nations at Calgary, Canada. After a considerable battle, Western leaders have been persuaded to provide increased funding support and leadership, US$19 billion has been pledged. This vital programme requires the safe disposal of some 40,000 tons of toxic munitions from seven sites across Russia at a cost estimated at nearly US$4,000 billion over a 15-year timescale (putting this in perspective, Russia's 2002 GDP is estimated at US$350 billion).

From G8 to GM

Arguments continue to rage over whether we should sanction genetic modification to our food. This may appear to have little relevance to WMD. However, our food chain is extremely vulnerable to malicious spread of disease. In many countries, agricultural industry is poorly regulated and the regulations poorly policed. Natural outbreaks such as the UK's foot and mouth disease epidemic will continue to place great strain on an under-resourced veterinary community. Whilst strong moral arguments are raised against GM, of crops in particular, it is a process which has been going on naturally for years. Man has selectively bred the wolf over three millennia such that most breeds scarcely resemble their common ancestor.

'Our food chain is extremely vulnerable to malicious spread of disease.'

The same basic techniques have been used in the production of everything from hybrid roses to albino guinea-pigs. The two key issues here are intervention and secrecy. Science now allows us to intervene to the extent that we can *instantaneously* introduce modified species into the wild – without the time to assess the long-term effects. Biosystems do not exist in isolation. Each has an impact, however small, on the others and we would be wise to check this out first. Setting aside, therefore, the nonsensical claim

that GM pollen will remain within the test site area, the public sees an inconsistency in the policy. Why the hurry? Secondly, it is highly suspicious that something is being covered up – by the biotech industry. Nevertheless, GM seems to offer some spectacular results in the prevention of the type of disease, which could maliciously be introduced to destroy wheat crops or kill cattle. Agroterrorism has largely been ignored, in terms of countermeasures funding. That needs to change as well.

Industry consolidation

There has been considerable consolidation in the NBC sector of the worldwide defence industry. In detection, UK-based Smiths Industries has acquired a full range of chemical and biological know-how through its acquisition of UK's Graseby Dynamics, as well as both Barringer and ETG Incorporated in the USA. In radiological detection the French conglomerates COGEMA and Saint-Gobain Plastics and Ceramics seem to have the lion's share in radiological detection and survey. A large proportion of the market for field-use personal dosimetry is serviced by UK's Siemens Environmental Systems Limited, Rados of Finland and France's MGP Instruments. Acquisition trends seem set to continue with considerable activity in the biodetection area as young technologies mature under the impetus of events such as the anthrax attacks in the US.

'Biodetection will mature under the impetus of events such as the anthrax attacks . . . .'

Summary

The world is more jittery than this time last year, a trend which will only reverse if terrorism is firmly dealt with and society more confidently defended against the spectre of WMD. People understand the risks better and, although there may be further events, they are mentally better prepared to cope. Government has been caught out by the audacity of attacks, the vulnerability of urban society and the ferocity with which terrorism is pursued. Arms control and demilitarisation appears in dangerous recession, there being a trend to tear up old treaties and renege on hard-won deals. Alongside all this, biotechnology, for both good and ill, will grow exponentially as we seek better detection and protection from diseases and chemical attack.

Acknowledgements

It is always risky to single people out in a team from which I have derived such unstinting, cheerful support. However, I must especially thank Mike Bryant, Managing Editor Land and Systems, and the production team who have been extremely tolerant of my shortcomings – Jo Agius, Jacqui Beard and Laura Kew – as they strove to lead this edition to successful publication.

John H Eldridge
July 2002

User comments

Readers' constructive views are greatly appreciated and assist in the aim of continuing the improvement process of this product. You can telephone, fax or e-mail your comments, additions and alterations or new material for the next edition to: The Editor, Jane's Nuclear, Biological and Chemical Defence, Sentinel House, 163 Brighton Road, Coulsdon, Surrey CR5 2YH, UK. Tel: (+44 208) 700 3700 Fax: (+44 208) 700 3900 e-mail address: jnbc@janes.co.uk

JOHN ELDRIDGE

Commander John Eldridge MBE FNI MRIN MIMgt spent 25 years in the Royal Navy, rising to become the Naval Staff expert on shipping safety, survivability and toxic defence.

During Operation Desert Storm he served in the Ministry of Defence as Principal Adviser in his specialist field to the Assistant Chief of Naval Staff, with a central role in the preparation and management of defensive NBC measures for Coalition land, sea and air forces.

Earlier, after a flying career, service as second-in-command of *HMS Penelope* during the Falklands operation, and command of his own ship, he ran the ship survivability and safety training department at the Fleet training centre, working closely with NATO, other Western and Gulf Co-operation Council forces.

He left the Royal Navy in 1992 to set up his own consultancy company offering risk management, research and training in crisis management for marine and land organisations.

He has since launched a successful company offering IT-based training and point-of-information solutions, systems management and web-based products.

He was appointed editor of *Jane's NBC Defence Systems* in November 1997 and elected a Fellow of the Nautical Institute in December 2001.

FREE ENTRY/CONTENT IN THIS PUBLICATION

Having your products and services represented in our titles means that they are being seen by the professionals who matter – both by those involved in procurement and by those working for the companies that are likely to affect your business. We therefore feel that it is very much in the interests of your organisation, as well as Jane's, to ensure your data is current and accurate.

■ **Don't forget** – You may be missing out on business if your entry in a Jane's book, CD-ROM or Online product is incorrect because you have not supplied the latest information to us.

■ **Ask yourself** – Can you afford not to be represented in Jane's printed and electronic products? And if you are listed, can you afford for your information to be out of date?

■ **And most importantly** – The best part of all is that your entries in Jane's products are TOTALLY FREE OF CHARGE.

Please provide (using a photocopy of this form) the information on the following categories where appropriate:

1. Organisation name: _____

2. Division name: _____

3. Location address: _____

   _____

4. Mailing address if different: _____

   _____

5. Telephone (please include switchboard and main department contact numbers, for example Public Relations, Sales, and so on):

   _____

6. Facsimile: _____

7. E-mail: _____

8. Web sites: _____

9. Contact name and job title: _____

   _____

10. A brief description of your organisation's activities, products and services: _____

   _____

   _____

11. Jane's publications in which you would like to be included: _____

   _____

Please send this information to:
Jacqui Beard, Information Collection, Jane's Information Group,
Sentinel House, 163 Brighton Road, Coulsdon, Surrey, CR5 2YH, UK
Tel: (+44 20) 87 00 38 08
Fax: (+44 20) 87 00 39 59
E-mail: yearbook@janes.co.uk

Copyright enquiries:
Contact: Keith Faulkner
Tel/Fax: (+44 1342) 30 50 32
E-mail: keith.faulkner@janes.co.uk

Please tick this box if you do not wish your organisation's staff to be included in Jane's mailing lists ☐

JNBC

NBC Capabilities

This section examines the NBC policy and offensive capability of nearly 100 key nations. It is based on information gathered over the year from sources in the public domain. Where precise information has been unavailable, expert views have been sought to allow as clear a picture as possible. Assessments, validated on 20 June 2002, are offered as a guide to future capabilities and intentions.

AFGHANISTAN

NBC policy
Taleban forces are not thought to have a coherent strategy for the use of toxic weapons, nor do they appear to have any materials. Other forces may have retained some former Soviet weapons although there is no evidence for this. The nature of the regime, the terrain and the climate make co-ordinated use by the unsophisticated protagonists unlikely in the foreseeable future.

Nuclear weapons
None.

Biological weapons
None reported.

Chemical weapons
Probably none remaining.

Assessment
Unconfirmed reports of the use of chemical weapons against populated areas, thought to be havens for rebel forces, emerged during the period of occupation by the former Soviet Union. If stocks remained after the withdrawal, they would by now be in a state of deterioration and probably ineffective. There is no evidence to date of a programme of biological or chemical development.

The use of Afghanistan by fundamentalist terrorist groups for training and, possibly, for the stockpiling of weapons is likely to continue. Some groups may incorporate CBW weapons into their strategy. Including the main base at Khost, the US bombed six suspected terrorist training camps in Afghanistan, as apparent retaliation for attacks on their embassies in Nairobi and Dar es Salaam. The bases were thought to have been used by the organisation of Osama Bin Laden, a wealthy terrorist of Saudi Arabian origin. Bin Laden's organisation has been linked to CBW development at the Ashifa site in Sudan, also targeted by the US in August 1998. See entry on Sudan in this section.

VERIFIED

ALBANIA

NBC policy
Albania ratified the CWC in May 1994 and signed the BTWC in June 1992. Albania has no identifiable policy to acquire an NBC capability.

Nuclear weapons
None. Albania is a signatory to the NNPT and CTBT.

Biological weapons
None.

Chemical weapons
Only riot control irritants have been identified.

Assessment
As with every other sector of national defence, Albania's fears over its neighbours, especially Serbia, could influence decisions about CW acquisition in the future. Albanian incursions into western Macedonia via Kosovo have become a continuing feature of the regional conflict. There may be new interest in the development of offensive NBC capability, either by the Albanian government or by the rebels. However, financial constraints, the terrain and the prevailing

climate make it unlikely that CW would be considered as having any advantage over conventional weapons.

VERIFIED

ALGERIA

NBC policy
Algeria ratified the CWC and acceded to the NNPT in 1995 but has yet to sign the BTWC. The government has not made any significant statements about WMDs in recent years.

Nuclear weapons
The Algerian government has been provided with a 1.5 MW research reactor by Argentina. The facility is IAEA compliant and Algeria has declared that it is for entirely peaceful purposes. The government denied suspicions raised by sources close to the Spanish security service that the project far exceeded civilian needs.

Biological weapons
None in the inventory; none planned.

Chemical weapons
Algeria maintains a stockpile of non-persistent chemical weapons for battlefield use. In addition, the Gendarmerie and the national police have several types of anti-riot non-persistent irritants available for crowd control. There is no confirmation of chemical weapon development programmes but the country's petrochemical expertise would allow such development were government policy to change.

Assessment
While a covert chemical weapons programme is possible, there is no current evidence for it, nor for any military nuclear development. There is no evidence of a BW programme.

VERIFIED

ANGOLA

NBC policy
Angola has yet to accede to the CWC or the BTWC and appears ambivalent towards the possession of NBC weapons.

Nuclear weapons
None. Angola acceded to the NNPT and the CTBT in 1996.

Biological weapons
None known.

Chemical weapons
Cuban support for Angola during the conflict with South Africa led to frequent reports of the use of CW agents. Several of these reports were confirmed by reconnaissance. Medical examination of affected victims produced evidence of the isolated use of nerve agent. These reports ceased after the withdrawal of Cuban advisers and it seems unlikely that local government forces retained any CW capacity. However, factions involved in the civil war continue to make unconfirmed claims of use by the opposing side.

Assessment
The legacy of sporadic CW use and failure to sign the CWC lead to the possibility that Angola will continue to keep open the option of CW and BW development.

VERIFIED

ARGENTINA

NBC policy
Argentina's signature of the 1967 Treaty of Tlatelolco indicated an intention not to develop nuclear weapons. The treaty established an 18-nation nuclear weapon free zone in South America. Amendments were made in 1992 and agreed by Argentina, Brazil and Chile, further strengthening the agreement. Argentina's policy appears directed towards peaceful use of nuclear technology. However, continued interest in longer-range missile technology may indicate long-term intentions to keep NBC options open.

Nuclear weapons
None. Argentina has two 935 MW reactors with a third nearly complete (2000).

Biological weapons
None known.

Chemical weapons
There is no evidence of either focussed CW research or the development of production facilities. Riot-control agents are available to internal security forces and the armed forces are equipped and trained for NBC defence.

Assessment
Argentina remains one of the most technically advanced South American nations, providing a fertile seedbed for future development of NBC weapons, either for export or regional use. However, any development or expansion of this capability would be led by perceptions of regional conflict and involve the breaking of existing treaty obligations.

VERIFIED

AUSTRALIA

NBC policy
Australia's policy is to maintain strong and overt support towards the elimination of WMDs. It remains highly active in the arms control and disarmament arena and its strategy for countering chemical weapons formed the basis of the current CWC. Warships carrying nuclear weapons (and nuclear-powered vessels) are still (2001) not permitted to use Australian harbour facilities.

Nuclear weapons
None.

Biological weapons
None.

Chemical weapons
None.

Assessment
Australia continues to offer a strong general lead to the world in the destruction of all toxic weapons and

has been conspicuously active in the development of the CWC and the future strengthening of the BTWC. The armed forces are investing in the modernisation of their NBC defence capability.

VERIFIED

AUSTRIA

NBC policy
Austria was an early signatory to the CWC and other treaties constraining offensive use of NBC weapons, whilst maintaining a strong industry in the provision of NBC defensive measures.

Nuclear weapons
None.

Biological weapons
None.

Chemical weapons
The only chemical weapons on Austrian soil are a variety of ex-German chemical warheads of various types remaining from the Second World War. These are held in long-term storage pending safe destruction. Apart from riot control agents, Austria has no offensive CW capability.

Assessment
A neutral defence policy and continued strong support for intrusive international inspection of potential offensive NBC activity combine to ensure that Austria is highly unlikely to desire an NBC capability.

VERIFIED

BAHRAIN

NBC policy
Bahrain has no NBC Policy to develop or acquire strategic weapons.

Nuclear weapons
There are no nuclear weapons on Bahrain territory; the country signed the Non-Proliferation Treaty in 1988.

Biological weapons
Bahrain became a signatory to the BWC in 1988 and ratified the CWC in April 1997.

Assessment
No known NBC activity. GCC members such as Bahrain are concerned about the potential BW and CW threat from Iraq and Iran and are investing in NBC defensive measures for armed forces and civilians.

VERIFIED

BANGLADESH

NBC policy
No known NBC activity, nor any declaration to acquire NBC weapons. Bangladesh ratified the CWC in April 1997.

Nuclear weapons
None.

Biological weapons
None.

Chemical weapons
Some police units are equipped with riot control agents such as CS.

Assessment
No known NBC activity.

VERIFIED

BELGIUM

NBC policy
Belgium provided host-nation facilities to NATO (US) nuclear forces in the 1950s and provided national delivery systems for US-controlled warheads. As a key member of NATO, Belgium adheres to NATO NBC deterrence policy.

Nuclear weapons
None, although NATO nuclear weapons may pass through or be stored on Belgian territory.

Biological weapons
None known.

Chemical weapons
None known, although chemical weapons may pass through or be stored on Belgian territory.

Assessment
Belgium will continue to retain NATO-agreed NBC defensive measures. Her armed forces are well equipped for NBC defence and she maintains a small but high-quality industrial capability in the NBC protection sector.

VERIFIED

BOSNIA-HERZEGOVINA

NBC policy
No declared policy.

Nuclear weapons
None, no production capability.

Biological weapons
None admitted.

Chemical weapons
A comprehensive range of incapacitants is available to both Serb and Muslim-Croat Federation (MCF) internal security forces, including CS. Limited stocks of chlorine and precursor agents were identified at a chemical plant in Tuzla. A large facility at Mostar was dismantled in 1991 and dispersed to sites at Lucani, Cecak and to Belgrade in Serbia before the region became part of the MCF area under the terms of the Dayton Peace Agreement. There may be other locations on the territory of Bosnia-Herzegovina. Release of toxic industrial chemicals for military purposes was threatened prior to 1992 in an area denial role but there appears no coherent plan to create CW facilities for military use.

Assessment
WMDs were not a significant military factor in operations to stabilise the country after the break-up of the Yugoslav Federation. Claims of their possession and allegations of their use featured regularly in propaganda generated by the opposing factions.

UPDATED

BOTSWANA

NBC policy
Botswana is not militarily active in the region and her armed forces are currently tasked for internal policing and the support of humanitarian activities. Botswana

has not participated in nuclear non-proliferation activities but has signed the BTWC (February 1992) and acceded to the CWC in August 1998.

Nuclear weapons
None.

Biological weapons
None.

Chemical weapons
Limited to riot control munitions supplied via South Africa.

Assessment
No known NBC activity and no apparent aspiration to acquire a capability.

VERIFIED

BRAZIL

NBC policy
Brazil ratified the CWC in March 1996. In the nuclear arena, she is a co-signatory of the 1967 Treaty of Tlatelolco which established an 18-nation nuclear weapon free zone in South America. Amendments to the treaty in 1992 preclude Argentina, Brazil and Chile from any further moves towards military nuclear capability.

Nuclear weapons
None. Brazil acceded to the NNPT and ratified the CTBT in late 1998.

Biological weapons
None known although reports of a development capability have been made.

Chemical weapons
None known. Riot agents for internal policing are manufactured locally.

Assessment
Offensive CBW is not identified as a useful capability. However, Brazilian armed forces are reasonably well-equipped and trained for NBC defence.

VERIFIED

BRUNEI DARUSSALAM

NBC policy
Brunei signed the NNPT in 1985 and abides by its nuclear safeguard provisions. The country was an early signatory to the CWC and ratified the convention in July 1997.

Nuclear weapons
None.

Biological weapons
None; no capability for manufacture. Brunei signed the BWC in 1991.

Chemical weapons
Brunei is not a signatory of the Geneva Protocol, which includes provisions on chemical weapons.

Assessment
Brunei has a small conventional army and no prospect of developing strategic weapons, although there is no doubt that the country's financial wealth would allow swift acquisition should a need be identified.

VERIFIED

BULGARIA

NBC policy
Bulgaria has signed the BTWC and ratified the CWC in August 1994. Bulgaria was a foundation signatory of the nuclear CTBT and has adhered to the NNPT since 1968. Defence policy appears to be moving away from retention of offensive NBC capability.

Nuclear weapons
Bulgaria has six nuclear reactors, with a combined output of 3.5 GW. There are no reports of nuclear weapons or warheads remaining on Bulgarian soil. However, Bulgaria remains a convenient route for the clandestine transfer of nuclear weapon materials from the former Soviet Union states to end-users with a desire to acquire a capability in secret, such as Iraq and Serbia.

Biological weapons
There are no reports of BW storage, development or manufacture. However, the Bulgarian secret service was implicated in the death of Georgi Markhov, a dissident Bulgarian journalist, in London in 1967. The BW agent used was *Ricin* (see under Biological Warfare).

Chemical weapons
Stockpiles of chemical weapons are limited to riot control and non-persistent irritants.

Assessment
Bulgaria continues to manufacture a wide range of NBC protection equipment. The Warsaw Pact legacy may well keep alive the view amongst defence planners that NBC weapons have their part to play in Bulgarian defence. However, treaty obligations and the lean towards NATO and EU acceptance will probably steer this view away from offensive NBC aspirations.

VERIFIED

BURMA (MYANMAR)

NBC policy
Burma (Myanmar) has been governed for years by a succession of military regimes and such a secretive society is difficult to analyse. Her priority for offensive NBC capability probably centres on her internal security requirements, whereby the fiercely independent hill people remain a thorn in the side of the national government. Surrounded by strong neighbours — China, India and Thailand — she may well see a need to develop a more sophisticated offensive NBC capability.

Nuclear weapons
None.

Biological weapons
Photographs have been taken of parachute- and balloon-supported containers which were dropped from low-flying aircraft by night over villages, resulting in local outbreaks of a virulent strain of Cholera. In addition, there have been numerous unconfirmed reports of similar attacks by government forces in remote areas against the dissident Karen population.

Chemical weapons
Numerous documented reports state that chemical agent attacks have been made by government forces against dissident Karen tribesmen and other groups in remote areas. Weapons reported to have been involved include mortar and artillery projectiles.

Assessment
A biological warfare capability appears to exist, a fact supported by various well-documented reports, including photographs of air-dropped weapons. It is likely that a chemical warfare capability also exists.

VERIFIED

CAMBODIA

NBC policy
The country is not a party to the CWC but participates in other arms control treaties.

Nuclear weapons
There is no evidence of nuclear weapons in the country, nor any known programme to acquire them. Cambodia signed the NNPT in 1972.

Biological weapons
Cambodia has no biological weapons, no plans to acquire them and signed the BTWC in 1983.

Chemical weapons
It is possible that both sides in the civil war have access to non-persistent chemical weapons, but there is no proof. Cambodia signed the Geneva Protocol in 1983.

Assessment
It is very unlikely that Cambodia will have the financial resources or development facilities to venture into the WMD field.

VERIFIED

CANADA

NBC policy
As a key member of NATO, Canada adheres to the NBC policy of the alliance. There is a strong and sophisticated defensive NBC research effort, directed towards improving the capability of the armed forces to withstand NBC attack. Canada is highly respected in NATO for her contribution to development and manufacture of high-quality NBC defence equipment.

Nuclear weapons
None.

Biological weapons
None.

Chemical weapons
None. Stocks of various chemical agents held over from the Second World War have been destroyed by incineration and the research at DRDC/RDDC Suffield is entirely defensive. Riot control agents are available to police forces.

Assessment
Canada will maintain its defensive NBC posture, continuing to offer research data and equipment to the NATO alliance.

UPDATED

CHAD

NBC policy
Undetermined. Traditional French defence support would not include release of NBC agents or materials to Chadian forces.

Nuclear weapons
None.

Biological weapons
None.

Chemical weapons
None known. During the various incursions carried out by Libya during the 1980s, there were reports of chemical agents being released in some areas but these reports have not been confirmed.

Assessment
No known activity in the BW or CW arena.

VERIFIED

CHILE

NBC policy
Defence policy before the end of the Pinochet regime may well have included research and development of B and CW weapons both for external and internal security roles. However, Chile ratified the CWC in July 1996 and the new democracy appears committed to conventional defence. Chile was a co-signatory of the 1967 Treaty of Tlatelolco which established an 18-nation nuclear weapon free zone in South America. Amendments made during 1992 strengthened the treaty, precluding Argentina, Brazil and Chile from military nuclear activity.

Nuclear weapons
None.

Biological weapons
None known.

Chemical weapons
Reports during the 1980s indicated the existence of a facility capable of manufacturing militarily significant quantities of VX and other nerve agents, apparently with outside assistance (unspecified). The reports remain unconfirmed and may have been incorrect.

Assessment
As one of the most technically advanced South American nations, Chile may have aspired to own nuclear weapons and to develop BW and CW agents under the 1973 to 1990 dictatorship. The current democratic government deploys strictly conventional defence forces, with a reasonable NBC defence capability.

VERIFIED

CHINA, PEOPLE'S REPUBLIC

NBC policy
An important element in shaping Chinese NBC policy has been the advent of the new administration in the United States. In 1999, the US published evidence (the Cox Report) that Chinese agents had been continuously passing important elements of western nuclear programmes and technology to the Chinese government from the 1970s to the present. Concurrently, the easing of sino-western political relations in the 1990s made the information exchange environment more relaxed and a variety of government-sanctioned technological programmes were developed with China. Chinese policy remains consistent with its aims: to acquire the widest possible technological expertise from western sources, allowing it to accelerate the full range of its WMD programmes. The country is benefiting from almost explosive economic growth, even under a centralised communist regime. It remains to be seen whether this leads to a more confident and aggressive regional policy towards issues such as Taiwan. In addition, the continued warmth towards Russia could lead to closer military ties and co-operation in the NBC area.

Nuclear weapons
The modernisation programme for China's nuclear capability continues apace, helped considerably by the extremely successful intelligence gathering programme described in the US Cox Report. Not only has the data on the entire history of the US test programme from 1950 onwards been passed across but also details on all currently deployed US nuclear warheads. Details of the enhanced radiation weapon (known as the neutron bomb), additional data on IT-based test simulation techniques (which obviate the

need to carry out live testing) and data on new systems for SSBN detection were also gained. Signs of the success of this operation will begin to emerge over the next few years as the latest information technology and significant funding combine to see warhead design moving to smaller size, more accurate delivery, enhanced radiation effects and the development of MIRV. US intelligence sources assess (June 2002) that China's strategic missile force will be able to deploy 75-100 nuclear warheads (see *Jane's Strategic Weapon Systems* for information on launch capability). China acceded to the NNPT in 1992 and signed the CTBT in 1996. However, it retains a significant, sophisticated and comprehensive nuclear capability.

Biological weapons
Prior to (and probably after) its signature of the BTWC in 1984, evidence emerged of China's significant offensive BW research programme, including the weaponisation and stockpiling of agent delivery systems. The programme remains in place.

Chemical weapons
Despite signature of the CWC, China is known to maintain a significant CW research, production and delivery programme. The Chinese government protested at the imposition of trade sanctions by the US in 1997 against two companies suspected of giving assistance to Iran in the development of a CW capability. No hard evidence has emerged except the arrest of an Israeli businessman implicated in the deal. US intelligence believes that China continues to maintain a CBW programme, based on evidence that two facilities, declared as civilian, are in fact administered by the Chinese Ministry of Defence. Both are known to have been involved in BW research in the 1980s and one establishment was expanded in 1991. China ratified the CWC in 1997.

Assessment
China's renewed confidence and regional aspirations have led to a massive upgrade programme for nuclear warheads, strategic delivery systems and CBW agents. The nuclear programme has been significantly accelerated by the intelligence gathered from the US over the last 20 years. There is also a thriving NBC defence industry, supplying the full range of equipment capability.

UPDATED

CROATIA

Nuclear weapons
None.

Biological weapons
None admitted, although, as part of the former Yugoslavia, some unconfirmed BW research capability was suggested. By now, this would have been absorbed by Serbia.

Chemical weapons
No sophisticated capability. Stocks of riot control agent are deployed with internal security forces.

Assessment
A CW research and production facility at Mostar in Bosnia was dismantled and removed by the Serbian authorities (1991). Mostar is now part of the Muslim-Croat Federation territory in Bosnia under the terms of the Dayton Peace Agreement. Croatia does not have an indigenous NBC capability. However, perceptions of regional instability and the availability of technology from the countries of the former Soviet Union could, in the future, place acquisition of an offensive capability on the defence agenda in the future.

VERIFIED

CUBA

NBC policy
It is believed that Cuba may maintain stocks of CW agent. There is no evidence of serious interest in military nuclear acquisitions. The Cuban government ratified the CWC on 29 April 1997.

Nuclear weapons
None.

Biological weapons
None admitted.

Chemical weapons
Cuba has a limited chemical weapons capability, using conventional delivery platforms such as medium-range tactical systems, aerial bombs or artillery shells, although interest in this form of offensive warfare appears to be in decline. The Revolutionary Armed Forces are highly trained in the defensive aspects of NBC warfare. Evidence from Angola, where Cuba was the sole defence supplier in the conflict with South Africa, showed that militarily useful CW systems were probably deployed.

Assessment
Cuba's offensive NBC capability remains limited and little new research is likely to be funded by the Castro regime.

VERIFIED

CZECH REPUBLIC

NBC policy
The Czech Republic does not admit to possession of an offensive NBC capability and, as one of the three new NATO members, will subscribe to the alliance NBC defence policy. The Republic ratified the BTWC and CWC ratification was achieved on 6 March 1996.

Nuclear weapons
None.

Biological weapons
None.

Chemical weapons
None admitted. Some ex-Soviet stocks may remain but their state of preservation and military utility will be doubtful.

Assessment
The Republic maintains a thriving and capable NBC protection industry and the armed forces are well-equipped and well-trained in NBC defence. There is no evidence of military aspirations to acquire an offensive capability.

VERIFIED

DENMARK

NBC policy
Denmark has a strong policy against WMDs in general and, as a key member of NATO and a signatory to all the conventions against proliferation, continues to work towards a worldwide ban on NBC weapons.

Nuclear weapons
None.

Biological weapons
None.

Chemical weapons
None. However, Denmark is still concerned over the decaying condition of old and abandoned CW agent munitions dumped in the Baltic following the First and Second World Wars. Some were dumped as late as 1965 close to her shores and are still being disturbed by Danish and German fishermen from time to time.

There is a small but technically sophisticated NBC defence industry, especially in the areas of command and control, NBC protection and CW detection.

Assessment
Denmark remains a strong voice against WMDs.

VERIFIED

ECUADOR

NBC policy
With no policy for offensive NBC use, Ecuador is a signatory of the 1967 Treaty of Tlatelolco which established an 18-nation nuclear weapon free zone in South America.

Nuclear weapons
None.

Biological weapons
None admitted.

Chemical weapons
During the 1995 border clashes with Peru, there were unconfirmed reports of CW use, including various toxic smokes and flame weapons. The reports have been denied by both sides.

Assessment
Ecuador is not considered an NBC risk.

VERIFIED

EGYPT

NBC policy
Egypt is a declared non-nuclear weapon state but has yet to sign the CWC. The move away from the confrontational politics of the 1970s and early 1980s leaves Egypt's national policy centred on self-defence. Her strategically sensitive geographical position drives a need for long-range regional deterrent capability.

Nuclear weapons
Although there is no evidence to support the claims, there were rumours in the 1960s and early 1970s that Egypt had attempted to develop Nuclear Weapons with which to attack Israel. Egypt's policy change came with the Camp David Accord and the signing of the NNPT in 1981. A decade later, Egypt and Israel developed a joint programme to provide water to Sinai by using a nuclear desalination programme; the project is backed by the UN International Atomic Energy Agency.

Biological weapons
There is no evidence that the current Egyptian administration is developing or has stockpiled biological weapons. Uncorroborated accusations were made in President Nasser's time of an Egyptian plan to unleash BW agent over Tel Aviv and other cities in Israel.

Chemical weapons
In the 1960s and early 1970s Egypt possessed considerable quantities of chemical weapons. During the brief war with North Yemen, Egyptian aircraft dropped non-persistent nerve agents on royalist troops in that country. There have been claims that Israeli troops found chemical stockpiles in the Sinai Desert after the June 1967 war and again in 1973.

In 1989, Israel is reported to have persuaded the US to advise Egypt against continued development of CW capability.

Anti-riot agents are widely available to security forces but there is no evidence, despite accusations from extremist groups, that anything more powerful has been used against the civilian population in the south.

Assessment
It is difficult to assess the extent of Egypt's current offensive NBC capability. The motives for retention certainly exist. Regional perceptions on peace, security and leadership suggest a need for deterrence but there is a shortage of resources. These are factors which may make a capability appear attractive.

VERIFIED

EL SALVADOR

Nuclear weapons
None. El Salvador signed the Nuclear Non-Proliferation Treaty in 1972 and the Partial Test Ban Treaty in 1964.

Biological weapons
None. El Salvador is a signatory to the BTWC (1972), the Geneva Protocol and the CWC (October 1995).

Chemical weapons
None other than riot control agents.

Assessment
There is no indication of covert activity in the development of CBW systems.

VERIFIED

ETHIOPIA

NBC policy
There is no coherent NBC policy by the government or by armed opposition forces and Ethiopia is a signatory to the key WMD arms control agreements. Since the First World War, Ethiopia is one of the few nations to have suffered comprehensive and sustained CW attack on its people, notably by the Italians in the late 1930s. Since then, the country has been in a state of virtually continuous internal strife.

Nuclear weapons
None.

Biological weapons
None.

Chemical weapons
From time to time reports have been made regarding the use of chemical weapons by government forces against dissidents operating in remote areas. These reports remain unconfirmed.

Assessment
As far as can be determined, Ethiopia does not have an NBC capability and does not appear likely to acquire one for the foreseeable future.

VERIFIED

FINLAND

NBC policy
Finland is bound by treaty obligations to adopt a neutral stance and forswear any form of offensive NBC capability. Finland is a BTWC signatory and ratified the CWC in February 1995.

Nuclear weapons
None. NNTP signed in 1968.

Biological weapons
None.

Chemical weapons
None known. An NBC defence research establishment is located at Lakiala and military training schools are maintained.

Assessment
Finland has no NBC capabilities other than defensive measures. She has a well respected NBC defence industry.

VERIFIED

FRANCE

NBC policy
France continues to pursue an independent line on nuclear deterrence and is investing strongly in a wide-ranging modernisation programme for her *Force de dissuasion*, focussing exclusively on submarine-launched and air-launched cruise missiles. She also appears ambivalent to disarmament issues in this sensitive area. Months before her declared intention to sign the Comprehensive Test Ban Treaty in 1996, she drew unfavourable international attention by conducting a series of fresh tests at Mururoa Atoll. These were apparently designed to augment existing computer models of warhead function, aimed at warhead enhancements for the new M4 and M5 ballistic systems. France continues to see a need for a national nuclear deterrent strategy and her defence forces will continue to benefit from investment into the foreseeable future.

Nuclear weapons
The core deterrent systems of the *Force Nucléaire Stratégique* consist of the S3 land-based ICBM and the submarine-launched series which will include the M51. The latter is a 6-10 MIRV system, with each warhead reported as 100 kT. Air-launched weapons include the ASMP system. See *Jane's Strategic Weapon Systems* for further details of delivery systems.

Biological weapons
None known. Having ratified the BTWC, France does not admit to a BW programme but has the resources and capacity to produce such weapons if required.

Chemical weapons
A French facility manufacturing nerve agent was in production until the 1960s. NBC countermeasures form a relatively high priority for defence and, because of its huge land area, for France's regional authorities alike. Riot control munitions and a full range of NBC defensive equipments are manufactured by French industry and are strongly advertised on the export market. France ratified the CWC in March 1995.

Assessment
France's historically independent line over nuclear deterrence is unlikely to change and the state will continue to give the highest priority to the maintenance of a viable nuclear capability. France remains one of the largest volume suppliers of defence equipment, especially to markets in Africa and the Middle East. France is a member of NATO and pursues alliance policy on NBC matters, whilst remaining outside the integrated military structure.

UPDATED

GERMANY

NBC policy
Germany, a key NATO member, adheres to alliance policy on offensive NBC matters. In the past, amid some protest, US NATO-declared nuclear systems were based on German soil. Following German unification, stocks of various ex-East German chemical agents and weapons may still remain. These are gradually being destroyed under existing treaty obligations. Germany is a signatory to the BWC and the CWC.

Nuclear weapons
None.

Biological weapons
None admitted.

Chemical weapons
None. Together with other western nations, German industry was suspected of the commercial provision of equipment and precursor chemicals to regimes with active NBC acquisition programmes, including Libya, Iraq and Iran. Germany's high-technology industry is an attractive potential source for clandestine NBC programmes but the provisions of the CWC allow states to intervene if there is suspicion of such activity. The German industrial sector dealing with NBC defensive measures includes leading manufacturers of body protection, decontamination and COLPRO. Demilitarisation is a key national concern and German companies are centrally involved in clearance operations amongst the chemical weapons storage and testing facilities in former East Germany. Some former East German NBC defence equipment has been adopted by the Federal armed forces. Riot control agents are readily available to the security forces.

Assessment
Germany does not aspire to an offensive NBC capability. It is one of the best-equipped and trained NATO members in NBC defence. Germany also has a sophisticated CW DEMIL facility at Munster. See under Demilitarisation.

VERIFIED

GREECE

NBC policy
Greece has no policy to acquire NBC weapons.

Nuclear weapons
There are no nuclear weapons stored on Greek soil.

Biological weapons
There is no evidence of a BW programme.

Chemical weapons
Other than standard anti-riot irritants, there are no known sources of chemical weapons in Greece.

Assessment
As an important NATO member, Greece adheres to current NBC policy, equipping her armed forces with adequate defence measures. She is unlikely to develop or procure an offensive NBC capability. Greece-based terrorist organisations, notably the so-called 'November 17' group have carried out a number of unchallenged terrorist attacks in recent years but are not thought to have considered acquisition of N, B or CW devices.

VERIFIED

GUATEMALA

NBC policy
Guatemala is a co-signatory of the 1967 Treaty of Tlatelolco, which established an 18-nation nuclear weapon free zone in South America. It has also signed the usual round of international non-proliferation and other treaties.

Nuclear weapons
None.

Biological weapons
None.

Chemical weapons
None known other than riot control agents.

Assessment
As far as can be determined, Guatemala has no NBC capability, nor does she appear likely to acquire one.

VERIFIED

HAITI

NBC policy
Haiti is a co-signatory of the 1967 Treaty of Tlatelolco which established an 18-nation nuclear weapon free zone in South America.

Nuclear weapons
None. Haiti ratified the NNPT in 1970.

Biological weapons
None. BTWC signatory.

Chemical weapons
None. Haiti has yet to sign the CWC.

Assessment
Haiti shows no evidence of an NBC capability, other than riot control munitions for internal security.

VERIFIED

HUNGARY

NBC Policy
Hungary is a signatory to all the major WMD arms control treaties and, now a NATO member, it adheres to alliance policy on NBC defence.

Nuclear weapons
Hungary has four nuclear reactors, but there is no evidence of any weapons-related development.

Biological weapons
None.

Chemical weapons
None. All systems deployed by former Soviet forces have been removed.

Assessment
Hungary is unlikely to have the political will or the necessary funds to develop WMDs. Its defensive industry is well developed and covers most areas of defensive equipment.

VERIFIED

INDIA

NBC policy
Regional security perceptions see both India and Pakistan actively and vigorously seeking long-range delivery systems and the development of WMDs. This is considerably heightened by the activity in Afghanistan in the so-called 'war against terrorism' (November 2001). India's NBC policy appears to support international disarmament initiatives, exemplified by its ratification of the non-nuclear treaties. However, its nuclear aspirations are now clear. Delivery system development is advanced. Based on experience with the in-service Prithvi

vehicle, a much longer-range (2,500 km) system, Agni, brings Pakistan and China well within range and is credited with the option of an N, B or C warhead. See *Jane's Strategic Weapon Systems* for more detail.

Nuclear weapons
In 1997, India announced the development of an advanced data-processing facility, credited with the capacity to run warhead performance simulations and validate designs for both delivery systems and warheads. The same year, it completed its fuel-processing facility at Kalpakkam. On 11 May 1998, three low-yield underground nuclear tests were conducted, followed, on 13 May, by a further two detonations in the kiloton range. India is not a signatory to the NNPT or the CTBT – signals of its perception of a dual standard by others in non-proliferation as well as reflecting its security needs and ambitions. Its delivery systems include the *Prithvi* series of ballistic missiles with ranges up to 350 km and the *Agni* system with a range of 1,500 km. In 1996, development of a satellite launch vehicle, designed to maintain a geosynchronous orbit, was reported and a new 300 km submarine-launched missile, the *Sagarika*, is nuclear-capable. India had previously detonated a 12 kT nuclear test device in 1974, an event which achieved the opposite of the intended effect amongst its non-aligned colleagues, eroding its position as their champion.

Biological weapons
None admitted, but the potential to produce such weapons is present. India has ratified the BWC.

Chemical weapons
None admitted, but the potential to produce such weapons is present. Riot control munitions are manufactured locally for police and paramilitary forces. In 1992, an Indian company attempted to deliver to Syria raw materials capable of being used to manufacture chemical weapons, although it seems unlikely that this was carried out with Indian government approval. India ratified the CWC in 1996. Suspicions remain that it may be engaged in CW agent development.

Assessment
Despite political changes in Pakistan, the nuclear rivalry between the two countries remains set on a dangerous path of escalation, risking involvement from more volatile regional powers such as Iran and China. This is aggravated by Indian fears of the leakage of nuclear materials and capability towards the Taliban regime in Afghanistan and thence to Al-Qaeda. Whilst there appears some reduction in the temperature of the Kashmir dispute, the danger of a nuclear miscalculation remains. In WMD development, India has a strong technical base and the indigenous production of offensive NBC systems is likely to continue. In NBC protection, India appears to rely heavily on import. In May 2001, India conducted its largest military exercise in 13 years along the border with Pakistan. The exercise scenario included an enemy Nuclear, Biological and Chemical (NBC) strike. Considerable emphasis was placed on the management and conduct of NBC defensive measures.

UPDATED

INDONESIA

NBC policy
The armed forces remain a strong political force as the fragile democracy struggles to establish itself. Whilst there is no history of aspiration to an offensive NBC capability, the need could be identified for internal security reasons in future.

Nuclear weapons
Indonesia has no nuclear weapons and is a signatory of the Partial Test Ban Treaty (1964). The country has also undertaken not to receive nuclear weapons under the terms of the Non-Proliferation Treaty. Indonesia has no long-range delivery vehicles.

Biological weapons
Indonesia signed the BTWC in 1992. There is no evidence of plans to acquire an offensive BW capability.

Chemical weapons
Indonesia is a signatory to the CWC. Reports emerged in 1991 of the use of non-persistent agents (unspecified) against the East Timorese. Additionally, during the unrest of recent years there has been no evidence of the use of CW agent.

Assessment
The country's attitude to weapons of mass destruction is ambivalent and causes local concern. Although regional loyalties remain high, especially within the ASEAN forum, politicians and military officials say privately that they are concerned about ballistic missile acquisition as well as other signs of WMD interest. There is international concern, principally from the USA, that fundamentalist terrorist activity by the Al-Qaeda network could establish itself in the country, bringing the risk of WMD development.

UPDATED

IRAN

NBC policy
The Iranian government publicly declares all weapons of mass destruction as fundamentally anti-Islamic and has been careful to ratify most existing treaties on control of WMDs. However, it also clearly perceives a potential threat from neighbouring states, especially Iraq, and exhibits aspirations to regional leadership. Iran's experiences of the war with Iraq have heightened her awareness of the utility of WMDs. These factors explain a highly ambitious acquisition programme in all areas of WMD development: N, B and C.

Nuclear weapons
Mohammad Mohaddessin of the National Council of Resistance of Iran (NCR) foreign affairs committee believes that the Iranian nuclear programme began in 1985, when an embryo development by the Shah, abandoned six years previously, was reactivated. The search for expertise is widespread and comprehensive. Large numbers of Iranian students currently study nuclear physics in Western European universities at post-graduate level. In addition, valuable nuclear intelligence obtained continuously over a 20 year period by China may be selectively shared with Iran in a bid to expand anti-Western influence in the region.

The Iranian Revolutionary Guard Corps (IRGC) is responsible for overseeing military nuclear developments and Mohammad Mohaddessin claims that transfer of dual-use technologies from Argentina, China, France and Pakistan has assisted the development programme. In 1992 Tehran signed a civilian nuclear power agreement with the Russian Federation to assist in the development of two 440 MW reactors. Apparently, a Calutron-type uranium enrichment plant has been obtained from China. Other projects and facilities include:

- Bandar Abbas project: on the integration of nuclear systems with ballistic missiles.
- Bushehr project: a facility damaged in the Iran-Iraq war and currently under reconstruction.
- Darkhovin site: the country's dual-purpose nuclear site in southern Iran currently under construction with Chinese assistance. It is 55 km northeast of Abadan.
- Gorgan project: located on the shores of the Caspian Sea, this site will be used for nuclear reactor development.
- Isfahan project: the centre of the nuclear industry with its own reactor and a Chinese neutron-sparker. The site is 40 per cent complete and includes underground facilities.
- GAMA Energy Centre: located in northeast Iran, at Banab. No further details are available.

- Moalem Kelayeh project: located at Qazvin, 120 km northwest from Tehran. In order to keep its purpose secret, this facility uses no foreign expertise.
- Yazd project: built underground close to a uranium extraction site in 1989-90; its purpose is unknown.

The progress of President Khatami's relaxations may, on public mandate, encourage him to gain ground towards a more open approach but the president faces serious opposition from other centres of power within the country. Relaxations may well enocurage Western nations to exchange more sophisticated technology with Tehran.

Western analysts believe that if Iran acquired weapons of mass destruction and the means to deliver them, it would pose a significant threat to regional stability. A nuclear-equipped Iran may seek to reverse the Saudi and OPEC policy of cheap oil and perhaps seek revenge on the Saudis and the Gulf Arabs for their support of Iraq in the 1980-88 Gulf War.

Israel is most certainly very concerned. In May 1993 it extracted assurances from Beijing that ballistic missile technology was no longer being transferred to Iran.

CIA Director Robert Gates, head of the US intelligence agency under President Bush, drew attention to evidence of Iran's involvement in acquisition of both nuclear and chemical weapons. He claimed that if Iran's attempts to buy components from Kazakhstan, Hungary, the UK and Germany were successful, viable nuclear systems could well be deployed by 2002. There are persistent reports (1999) of practical co-operation between Russia and Iran on nuclear development.

The International Atomic Energy Agency (IAEA) is unable to confirm nuclear weapons development and evidence from resistance groups could, of course, be self-serving and unreliable.

Iran, while accepting that it is involved in the development of nuclear power facilities, dismisses claims of weapon development as American paranoia. She views the claims as attempts to justify Western arms sales to the Gulf states, where expenditure on arms is far greater than in Iran.

Iran's economic position remains weak and government officials say that their priority, far from clandestine weapon acquisition, is the repair of the country's ravaged infrastructure and the development of the domestic economy.

Biological weapons
The UN believes Iran has yet to develop a serious BW production capability, but Mohammad Mohaddessin indicates that biological weapons research is being carried out at one site.
- Razi: this is the government's serum and vaccine production centre to the northwest of Karaj, on the Qazvin-Hessarak road.

Chemical weapons
US authorities claim that Iran already has stocks of up to 2,000 tonnes of weaponised CW agent, including choking, blister and blood agents. German intelligence reports show that Iran was in possession of the blueprint used to build the Rabta chemical arms plant in Libya.

The NCR claims that, using technology gleaned from various sources in China, Germany and North Korea, the IRGC, which appears to have control of strategic weapons, can adapt chemical warheads for the country's ballistic missiles of the Scud family. The government department responsible for all chemical development facilities is the Engineering Research Centre of the Construction Crusade (Jahad-e-Sazandegi). Facilities include:
- Bandar-Khomeini: this chemical production complex in the southwest of Iran was set up during the Iran-Iraq War to provide chemical agents for the battlefield. It is managed by the Razi Chemical Corporation, which, although co-located with the Petrochemical Industries Establishment of the Oil Ministry, is nevertheless independent.
- Isfahan: about 45 km from the city, the Poly-Acryl Corporation's commercial plant has been developed into a major chemical weapons production facility.
- Karaj programme: a chemical weapons site about 14 km from Tehran, in the direction of Karaj. Mohammad Mohaddessin claims that Chinese engineers and technicians have been involved in the site's development.

- Marvdasht Centre: the mustard gas production facility for the IRGC during the Iran-Iraq War, situated in Fars Province.

Assessment
It seems certain that, with good relations between Iran and potential suppliers such as China, North Korea, and Russian Federation states, the independent Iranian nuclear deterrent is either here now or will be soon. The USA has expressed its continued concern to Russia about the Bushehr project and has identified Iran as part of an 'axis of evil', believing it to be active in the offensive WMD area. It is hard to assess progress on the BW and CW front but the effectiveness of CW during the war with Iraq will not be lost on Iranian defence planners. Iranian students continue to benefit from specialist courses around the world with the probable aim of returning to supply the core of indigenous development programmes.

UPDATED

IRAQ

NBC policy
The attention drawn to the advantages and disadvantages of BW and CW before and since the Gulf War of 1991 has effectively strengthened Iraq's resolve to continue development by clandestine means. On delivery systems, there is evidence of expertise entering the country to assist with ballistic system development. The current SCUD-based vehicle has been deployed in extended-range versions and work on alternative systems continues. See *Jane's Strategic Weapons Systems* for further detail. Prior to their withdrawal in December 1998, the efforts of UNSCOM to achieve full destruction of Iraqi programmes were regularly hampered by internal disagreement between UN Security Council members over the sanctions issue. At the diplomatic level, Iraq played hard on these differences while continuing to develop an NBC capability, according to UNSCOM. Allegedly uncovering collusion between UNSCOM, the CIA and Mossad, the regime skilfully exploited international reaction, leading to inexorable pressure on the UN to withdraw the teams. Iraq has ratified the BWC but has yet to accede to other international WMD control treaties. The nation is a rival to Iran for regional domination and pursues a policy which cleverly combines the establishment of friendly relationships with neighbouring Arab states with the hidden threat of an effective NBC capability. The leadership is autocratic and unpredictable, posing the greatest threat to regional stability in Western eyes. However, it has been remarkably astute at manipulating the lack of resolve by the international community to take firm action against NBC acquisition.

Nuclear weapons
Iraq's nuclear weapons development programme was substantially damaged and probably temporarily halted, by the coalition air raids of 1991. The Al Rabiya facility at Zaafarniyah was also attacked and damaged by US Navy Tomahawk cruise missiles in January 1993.

There is increasing evidence, however, that UN inspection teams did not achieve total destruction of the nuclear facilities and it is believed that at least one facility remains operational, east of Baghdad. International Atomic Energy Agency teams have sealed, but not necessarily destroyed, tools and fixtures.

In June 1993, there were reports of samples of Highly Enriched Uranium (HEU) being offered for sale by Kurds in the border region with Iran. The UN inspection teams believe that Iraq had two HEU plants in 1991, one operating and the other coming on stream. Both were subsequently reported as damaged beyond repair.

Iraq is known to remain committed to acquiring a nuclear capability and has about 7,000 nuclear scientists and technicians who retain the knowledge required to rebuild the programme.

Prior to the invasion of Kuwait, Iraq is believed to have employed over 100,000 people in the nuclear weapons programme, including those in Western universities and on exchange with North Korea.

Following the various visits of UNSCOM inspectors, the CIA estimates that Iraq has made considerable clandestine, to rebuild the nuclear weapons industry. The CIA does acknowledge that two facilities have been rebuilt: Saad 16 and Al Rabiya. IAEA claims that the nuclear weapons programme has been dismantled may be over-optimistic. According to Scott Ritter, former inspector with UNSCOM, there is evidence that Iraq had pursued production of explosive lenses, neutron generators and electronic initiators between 1992 and 1996. Most of the work on lenses had taken place at AlQaQa prior to 1992 whence it shifted to Al Kewthar, probably without the knowledge of the IAEA.

Iraq, under an earlier regime, ratified the NNPT in 1969 and a routine inspection by IAEA under the terms of the treaty, 22-25 January 2000, drew media attention. The inspectors visited the Tuwaitha site to verify declared material and were apparently satisfied that the rules had not been breached. The IAEA pointed out at the time that the inspection was no substitute for the resumption of its inspection activities under the terms of relevant Security Council resolutions.

In 2001 a defector claimed, to the UK Sunday Times, that the Iraqi nuclear programme was more sophisticated than the West had hitherto believed. He produced diagrams illustrating the weapons and stated that tests had been conducted in a sophisticated underground plant at Lake Rezazza, southwest of Baghdad, which was designed to prevent seismic detection of the tests. These claims have since been discredited and the true extent of the current Iraqi programme is difficult to judge. However, US intelligence sources continually suggest that Iraq will have a renewed ability to deliver a nuclear warhead within the region and the wider world in the near future.

Biological weapons
Evidence of the large-scale production of Anthrax and Botulinum toxin continues to be found and UNSCOM believes that work on other BW agents may well be advanced. The regime has also shown interest in aflatoxin, a material that can cause long-term carcinogenic diseases, especially to the liver.

Chemical weapons
In March 1994, the UNSCOM said there was no evidence to support allegations that Iraq used chemical weapons against opposition groups in the southern marshes in September 1993.

In early July 1994, UNSCOM announced that its two-year operation to destroy all of Iraq's declared chemical agent stockpile was complete. The work was carried out by 100 specialists of the Chemical Destruction Group who were seconded from 23 member nations of the UN. However, since then, and following assiduous work, there is no doubt about the existence of further extensive development programmes, with concealment a high priority.

The Commission supervised the destruction of more than 480,000 litres of chemical weapon agents (including mustard gas and the nerve agents, sarin and tabun), over 28,000 specially designed munitions (including short-range battlefield rockets and artillery shells), plus nearly 1.8 million litres of 45 different chemical precursors for weapons manufacture.

The US CIA reported in early 1993 that it would take less than 12 months for Baghdad to restore its former chemical weapons capability.

There are numerous reports that Iraq now uses phosphoric bleach as a chemical agent against opposition groups; the substance is not banned by international convention, yet is highly effective.

Assessment
By the end of August 1998 and the untimely departure of UNSCOM, little doubt remained that, despite the best efforts of these highly experienced and capable inspectors, several covert programmes remain unaccounted for. Iraq has several secret storage locations in the eastern provinces and a number of buildings belonging to otherwise innocent government departments have been used as

laboratories and storage places. There is considerable evidence that much of the programme has moved underground, beyond the gaze of surveillance satellites and a fresh UN arms inspection regime seems beyond achievement at the moment. However, it is interesting to note that, in mid-July 1999 under an existing agreement accepted by the regime, a team of experts from OPCW destroyed a small stockpile of VX left behind after the hurried departure of UNSCOM.

In February 2001, the German intelligence agency expressed concern over the continuing rebuilding of Iraq's NBC capability. On delivery systems, it cited its manufacture of solid rocket-fuel at the Al Mamoun plant and the enhancements of the Al Samoud missile (based on the existing Russian-designed SCUD system) with precision guidance technology.

Russia remains under suspicion of supplying help to the regime and, in February 1999, vehemently denied claims that it had signed a £100 million arms deal with Iraq.

Evidence emerged in the late 1990s of close collusion between the Milosevic regime in Belgrade and Baghdad. The new government in Belgrade has assured the West that the secret involvement of Serbian technical experts in Iraqi weapons development programmes has ceased.

In 2002, western intelligence, especially in the US, believe strongly that Iraq's offensive nuclear, BW and CW programmes have been able to develop and expand and although there is little hard evidence in the public domain, it has created a strong momentum towards elimination of this capability by all means.

Iraq has begun to develop its own NBC defence industry, making protective suits and masks.

UPDATED

ISRAEL

NBC policy
Although Israel has not declared itself as a nuclear power, the evidence that she maintains a sizeable and sophisticated capability is overwhelming. Her IRBM systems (Jericho II) are capable of reaching most centres of Arab population. Although Israel has signed the NNPT she does not accede to other WMD arms control treaties, notably the CWC, and it is believed that she holds stocks of C and BW. However, the latter remains uncorroborated.

Nuclear weapons
Israel does not admit to a nuclear capability. However, there is clear evidence of a comprehensive and sophisticated programme, probably with US support. Israel is understood to hold up to 300 nuclear warheads of various types, the bulk of which are available for about 100 Jericho II and III missiles. More warheads are understood to be available for free-fall bombs and 155 or 203 mm artillery projectiles.

Biological weapons
Although the evidence is circumstantial, Israel certainly has the capacity and resources to develop and produce biological weapons and has long been suspected of maintaining an advanced research programme.

Chemical weapons
The capacity and resources are available to develop and produce chemical weapons. The Jericho I system is credited with a possible CW payload. Riot control agents are readily available and frequently used against opposing forces in the occupied territories. The anticipation of chemical attacks by Iraq during the 1991 Gulf War prompted a full review of civilian and military NBC defence and protective measures. As a result, these were updated to include distribution of protective equipment, medical resources and the preparation of shelters. The Israeli population receives almost as much attention in this regard as the Israeli Defence Forces.

Assessment
Israel, the most technically advanced regional state, is capable of the manufacture and delivery of most types of NBC weapon. Israel, like Denmark, is one of the few nations with a well supported civil NBC defence plan, devolved to city level. A sophisticated NBC defence industry, particularly strong in the provision of body protection for civilians of all ages, delivers good quality protection to the population.

VERIFIED

ITALY

NBC policy
Italy adheres to NATO alliance policy on NBC defence.

Nuclear weapons
None.

Biological weapons
None known.

Chemical weapons
None known other than riot agents for police and paramilitary use.

Assessment
Italy has no offensive NBC capability. Her industry provides a range of good-quality defensive equipment, especially in the DECON field. See under Decontamination for details.

VERIFIED

JAPAN

NBC policy
To date, Japan is the only nation to have suffered both a nuclear attack and recent CW terrorist attacks. She maintains a principled position against nuclear development for military use. By statute, nuclear weapons cannot be based on Japanese soil. This policy embraces other WMDs and Japan has been an active participant in the arms control process. Since 1995, civil emergency organisations have been retraining and re-equipping to cope with the toxic threat. Under the terms of the CWC, Japan also faces responsibility for the disposal of ACWs on Chinese soil. There was widespread use of CW in Manchuria in the 1930s by the Japanese army. Dumps of decaying ordnance have been unearthed. For example, nearly 700,000 munitions in various states of decay are buried at Haerbaling, close to the border with North Korea.

Nuclear weapons
None. Japan is the third largest provider of nuclear-generated electricity and has a sophisticated peacetime nuclear industry.

Biological weapons
None. Japan was particularly active in the development of both BW and CW in the 1930s. An establishment at Harbin (Unit 731) is reported to have been used for live human trials involving prisoners from China, USSR, Korea, UK and elsewhere. Today, an offensive BW capability neither exists nor would be permitted.

Chemical weapons
The Chinese and Japanese governments are working closely to tackle the demilitarisation of ACWs. Japan has a sound CW defence capability and riot control agents are retained for police and paramilitary use. Tokyo's experience at the hands of the Aum Shinrikyo sect has focussed national attention on the vulnerability of an unprotected civilian population to CW attack. A thorough review of the entire range of protective measures for defence and internal security

personnel has been undertaken and a comprehensive equipment upgrade programme continues.

Assessment
Japan does not have an offensive NBC capability.

VERIFIED

JORDAN

NBC policy
Jordan is a BWC signatory and ratified the CWC in October 1997. Neither Jordan's strategic location nor the accession of King Abdullah appear to have materially changed Jordanian strategy on WMDs or long-range delivery systems. However, maintenance of the nation's position as the stabilising influence between opposing forces will require sensitive handling by the new regime. There are no early indications of a change in strategy over NBC.

Nuclear weapons
None. Jordan is a signatory to the CTBT and the NNPT.

Biological weapons
None.

Chemical weapons
None known, although past understandings with neighbouring Iraq may have resulted in the transfer of chemical weapons or expertise.

Assessment
If Jordan does have any form of NBC potential it is likely to be limited to chemical weapons.

VERIFIED

KENYA

NBC policy
Kenya does not have an NBC capability and does not appear to have made any moves to acquire one.

Nuclear weapons
None.

Biological weapons
None.

Chemical weapons
None known other than riot control agents.

Assessment
Kenya is unlikely to seek NBC weapons in the medium term.

VERIFIED

KOREA, NORTH

NBC policy
The government of North Korea presides over a state whose infrastructure and ability to feed and support its people is increasingly poor. One of the few remaining hardline communist states, it has continued to develop all forms of WMD system and has put successful effort into the design and export to internationally irresponsible regimes of strategic delivery systems. North Korea and the US remain in protracted negotiation to trade intrusive inspection of the regime's nuclear programme for help in developing its agriculture. The so-called Agreed Framework, signed in 1994 between both parties, aimed to freeze North Korea's nuclear programme in

exchange for supplies of fuel oil and the promise of two light-water nuclear power plants. North Korea remains an active producer of CBW systems. The June 2000 visit by the South Korean President Kim Dae Jung to North Korea has encouraged speculation about a softening of the North's hard attitude to the outside world. Whether mutually agreed confidence-building measures will lead to a reduction in President Kim Jong II's WMD programme remains to be seen.

Nuclear weapons
The US is particularly suspicious of North Korea's reluctance to allow inspection of a suspected underground nuclear test site at Kumchang-ri. This is a complex about 25 miles northeast of Yongbyon, monitored by US satellite surveillance and revealing a daily traffic of personnel in and out of the mountainside. The US views this development as evidence for North Korea's abrogation of the Agreed Framework. Washington continues to believe that North Korea already possesses nuclear devices suitable for installation on its indigenous Scud derivatives (see *Jane's Strategic Weapon Systems* for further details). Further, it fears proliferation of ballistic capability to other regular clients of the Pyongyang government such as Iran and Syria. However, more recent research shows that reports on Kumchang-ri have been overstated. The research reactor site, located at Yongbyon since 1986, appears to form the basis for offensive nuclear development, together with a plutonium enrichment facility.

Biological weapons
Although the government has ratified the BTWC, North Korea is widely believed to be pursuing an active clandestine BW research and development programme. The need for a mass area denial capability is identified as the aim of the programme. North Korea may see uses for its alleged plague, cholera, botulinum toxin and anthrax programmes in preventing an invasion by US-backed South Korean forces.

Chemical weapons
North Korea has yet to sign the CWC (as at June 2002). The regime was forced to continue indigenous CW development following the withdrawal of Soviet assistance in 1962. It looked to China and Japan for the help it eventually received to cover the gap until better relations with the Soviet Union resumed in 1966. CW production is the province of the so-called 32nd Department, part of the Nuclear-Chemical Defence Bureau headquartered in Pyongyang. This department runs chemical agent and weapon factories at Kanggye and Sakju, both of which produce nerve, blood and choking agent. A third factory (Factory 279) at Sokam-Ri produces decontamination agents and NBC defensive equipment. Current CW munition stockpiles are estimated at between 1,000 and 5,000 tons, comprising mortar, artillery shells and FROG-7 rockets.

NBC defence is a high priority and the North Korean forces are among the best trained in the region in this respect. Each North Korean Army corps has its own NBC defence battalion and frequent exercises are conducted with dilute live agent.

Assessment
North Korea's secretive and apparently paranoid regime is moving ahead on all fronts in pursuing a comprehensive range of WMD capability. The USA has identified North Korea as part of an 'axis of evil', believing it to be active in the offensive WMD area.

UPDATED

KOREA, SOUTH

NBC policy
Ever since the Korean War of the 1950s, South Korea has remained on a high state of alert, maintaining a powerful military infrastructure supported by the US. It does not have any indigenous NBC capability.

Korea has ratified the BTWC. CWC was ratified on 28 April 1997.

Nuclear weapons
None.

Biological weapons
None known.

Chemical weapons
None known. The People's Democratic Republic of Korea (generally known as North Korea or PDRK) has repeatedly accused South Korea of maintaining chemical weapon stocks, an accusation strongly denied. Locally developed riot control munitions are available to security forces. The armed forces are well equipped and drilled in NBC defence. The civilian population in major urban centres also carries out NBC exercises.

Assessment
South Korea does not appear to have an indigenous NBC capability. However, suspicions remain over the development of a clandestine offensive CBW capability. The country has a small but well-developed defensive NBC equipment industry.

VERIFIED

KUWAIT

NBC policy
Kuwait is a party to almost all international treaties and conventions on strategic and non-conventional weapons.

Nuclear weapons
None.

Biological weapons
None.

Chemical weapons
Although the National Guard and National Police have anti-riot irritants, there are no chemical weapons for battlefield use in the Kuwaiti inventory.

Assessment
Kuwait does not have an NBC capability. Following heightened awareness of the CBW threat from Iraq in 1991, the GCC, which includes Kuwait, encourages an active acquisition programme for NBC defence measures for its armed forces and populations.

VERIFIED

LAOS

NBC policy
Laos has been a signatory to the Non-Proliferation Treaty since 1970.

Nuclear weapons
None.

Biological weapons
Laos does not appear to possess a biological warfare capability and is a party to the BWC.

Chemical weapons
No evidence of an offensive capability. Laos ratified the CWC on 25 February 1997.

Assessment
The country has suffered from use of CW agents in the past and the armed forces are equipped with rudimentary NBC protective measures.

VERIFIED

LIBYA

NBC policy
Evidence is strong that Libya's policy is to continue development of an offensive NBC capability by clandestine means. Outwardly, Colonel Gadaffi's regime appears to have mellowed with time, welcoming arms control initiatives. Libya has acceded to the NNPT and BTWC. However, the regime has yet to accede to the CWC, possibly believing that the intrusive inspection regime that it brings would interfere with clandestine development.

Nuclear weapons
CIA reports indicate that Libya is considered to be a second rate threshold nuclear power. If funding and technical expertise continues to be invested at the current rate, it will take Libya only until 2005 to have an operational nuclear weapon. This will probably take the form of a warhead for the SS-1C or its derivatives.

Biological weapons
Research and production of BW agents is thought to be small scale. Permanent or temporary interruptions to CW development (see below) and the ease with which Western-supplied dual-use BW development facilities can be acquired may prompt greater effort in this area.

Chemical weapons
Gadaffi's forces are believed to have used H agent against Chadian guerrilla forces in 1987 and against Sudanese guerrillas at Nasir and Mayom in 1988. UN sanctions against Libya for terrorist activities served to strangle progress on a major CW production facility at Rabta. According to US intelligence, a fire there in 1990 was probably started deliberately to act as cover for the transfer of key operations to a much larger complex being constructed at Tarhunah, near Tripoli, away from the threat of further US air strikes. In January 1992, the former CIA Director Robert Gates offered evidence on the scale of the Libyan programme to a US Senate hearing. The Tarhunah facility, if completed, would have an output greater than that planned at Rabta. H agent production is estimated at 1,000 tons and G agent 1,400 tons per year. Libya had reportedly stockpiled over 100 tonnes of H agent as well as several varieties of nerve agent.

It is understood that Libya has acquired non-persistent nerve agent warheads for its SS-1c inventory and that it has made efforts to acquire 155 mm long-range howitzer shells with chemical warheads.

Assessment of covert programmes
Libya is thought to have developed a series of covert strategic weapons programmes. The projects range from ballistic missiles to long-range artillery and most involve BW, CW and nuclear warhead technology. Media reports in 1998 suggested a strong link with the Iraqi regime whereby Iraqi scientists were assisting the Libyan regime with BW development.

UPDATED

MACEDONIA

NBC policy
Macedonia has no policy on NBC acquisition.

Biological weapons
There is no evidence of biological weapons stockpile or development.

Chemical weapons
A part of Yugoslavia, in the 1960s, Mount Krivolak in central Macedonia was used as a testing site for CW-filled munitions. Today, apart from riot control agents, no significant CW materials are held in Macedonia.

Assessment
Even with the return of stasis to Kosovo, Macedonia remains highly vulnerable to the political fallout from the current situation in the region and the recently

signed peace agreement between the Albanian rebel and indigenous Serb factions, appears fragile. The agreement, however, does allow NATO (mainly UK) forces to monitor the situation and supervise the retrieval of rebel weapons. No CBW weapons are likely to emerge from this, although there may be signs of interest in such a capability by the rebels. The Macedonian government appears unlikely to have the political will, cohesion or financial resources to develop NBC weapons in the medium or long term.

UPDATED

MALAYSIA

Nuclear weapons
Malaysia has no nuclear capability and there is no evidence that it is preparing one. It is a signatory to the Non-Proliferation Treaty (signed in 1970).

Biological weapons
In 1991 Malaysia signed the BWC. It almost certainly does not possess biological weapons.

Chemical weapons
There is also no evidence of Malaysia having used or intending to develop a chemical warfare capability. Malaysia became a state party to the CWC in 2000.

Assessment
Malaysia appears unlikely to acquire NBC weapons. As part of a group of southeast Asian nations, including Burma (Myanmar), Cambodia and Thailand (qv) whose regional security perceptions appear to remain an obstacle to CWC accession, she has set an example for the others to follow.

VERIFIED

MALI

NBC policy
The country has no coherent offensive NBC capability.

Nuclear weapons
None.

Biological weapons
None.

Chemical weapons
Considerable stockpiles of anti-riot agents are available to the police, gendarmerie and army.

Assessment
Mali, one of the poorest of the sub-Saharan African nations, appears unlikely to aspire to an offensive NBC capability in the medium term.

VERIFIED

MAURITANIA

NBC policy
Undetermined. Unlikely to have a need for an offensive NBC capability in the near term.

Nuclear weapons
None.

Biological weapons
None.

Chemical weapons
The army and gendarmerie have small stockpiles of riot control agent. There is no hard evidence of a CW stockpile but support by Libya and Iraq was offered in 1988 during the conflict with Senegal and may have included CW weapons.

Assessment
Mauritania does not appear to have an NBC capability.

VERIFIED

MOROCCO

NBC policy
Morocco is a signatory to the major WMD arms control agreements.

Nuclear weapons
None.

Biological weapons
None.

Chemical weapons
The Polisario guerrillas allege that the Moroccan government used CW weapons against the Sahrawi People's Liberation Army fighters, but the reports remain unconfirmed. The national police, gendarmerie and army hold stocks of non-persistent riot control agents.

Assessment
It appears unlikely that Morocco retains a viable offensive NBC capability.

VERIFIED

MOZAMBIQUE

NBC policy
Internal conflict continues to dog the region, possibly creating the right conditions for CW use, but the Mozambique government has shown no inclination to acquire WMDs. Although a signatory to nuclear arms control agreements, the regime has yet to sign either the CWC or the BTWC.

Nuclear weapons
None.

Biological weapons
None known.

Chemical weapons
The Renamo guerillas have repeatedly accused government forces of using chemical weapons, including nerve agents, attracting UN investigation in 1992. The subsequent account was inconclusive but indicated that at least some reports were sourced reliably and may have been correct.

Assessment
Hard evidence of CW use during the internal conflict remains uncorroborated but non-adherence to the CWC and BTWC would support international suspicion over Mozambique's NBC acquisition plans.

VERIFIED

NAMIBIA

NBC policy
With no identifiable NBC Policy, Namibia does not appear to aspire to an NBC capability.

Nuclear weapons
None.

Biological weapons
None. Namibia is not a signatory to the BTWC.

Chemical weapons
Namibia ratified the CWC on 27 November 1997.

Assessment
Namibia is not known to have any offensive NBC capability.

VERIFIED

NETHERLANDS

NBC policy
The Netherlands is a key member of the NATO alliance and adheres fully to the organisation's NBC policy.

Nuclear weapons
None.

Biological weapons
None known.

Chemical weapons
None known to be held, other than riot control munitions and agents.

Assessment
There is considerable research effort into NBC defensive measures, notably by the TNO laboratory and the armed forces are well equipped and trained. The headquarters of the OPCW, the CWC's inspection and monitoring organisation is located in Den Haag.

VERIFIED

NEW ZEALAND

NBC policy
New Zealand remains one of the most ardent supporters of non-proliferation and arms control development.

Nuclear weapons
None.

Biological weapons
None.

Chemical weapons
None.

Assessment
New Zealand forces are well equipped with effective defensive NBC measures, purchased mainly from UK and USA.

VERIFIED

NICARAGUA

NBC policy
There is no identifiable offensive policy or plan.

Nuclear weapons
None. Nicaragua is a co-signatory of the 1967 Treaty of Tlatelolco, which established an 18-nation nuclear weapon free zone in South America, and the two main nuclear arms control treaties.

Biological weapons
None. Nicaragua has also signed the BTWC.

Chemical weapons
None known other than riot control agents for police and paramilitary use. Nicaragua is part of a group of neighbouring nations which appears reluctant to accede to the CWC (Belize, Guatemala and Honduras).

Assessment
Nicaragua is not known to have any NBC capability.

VERIFIED

NIGERIA

NBC policy
Nigeria has no declared NBC Policy. Although she has yet to sign the CTBT she is a signatory to the other agreements on WMDs.

Nuclear weapons
None.

Biological weapons
None.

Chemical weapons
None known other than riot control agents for police and paramilitary use.

Assessment
Nigeria is not known to have any offensive NBC capability. Providing NBC protection for the armed forces remains a government priority.

VERIFIED

NORWAY

NBC policy
Norway is a strong supporter of the abolition of WMDs, through the key arms control treaties and remains a key member of NATO.

Nuclear weapons
None.

Biological weapons
None known.

Chemical weapons
None known to be held.

Assessment
Norway's armed forces remain well equipped and trained for NBC defence. In DEMIL terms, the country remains concerned at the legacy of ACWs dumped in the Skaggerak channel and off her west coastline after the end of the Second World War.

VERIFIED

OMAN

NBC policy
Oman is a non-aligned nation whose proximity to Iran prompts a keen interest in supporting NBC arms control.

Nuclear weapons
None. NNPT signatory.

Biological weapons
None. Accedes to the BTWC.

Chemical weapons
The Royal Oman Police have a small stock of CS and other anti-riot irritants. Oman has no plans for a CW programme and ratified the CWC on 8 August 1995.

Assessment
Oman has no offensive NBC capability. The country is actively seeking better NBC defence measures for its armed forces and population. Concern remains over Iran. Abu Musa Island, some 150 km from her border is, reportedly, a storage site for Iranian CW weapons.

VERIFIED

PAKISTAN

NBC policy
Regional security, especially the relationship with India, has stimulated Pakistan to develop a military nuclear capability. It also shows considerable interest in acquisition of technology in BW and CW. Pakistan is a signatory to most international WMD control treaties. It has yet to agree to the Nuclear Non-Proliferation Treaty but has signed the Chemical Weapons Convention.

Nuclear weapons
Since the May 1998 underground nuclear test series, Pakistan has continued to accelerate its military nuclear programme. Although a signatory to most international WMD control treaties, including the NNPT, Pakistan remains in serious breach of obligations under the CTBT, having conducted tests in Spring 1998. Pakistan has arranged to purchase a number of long-range missiles from the People's Republic of China. Each is reported to be capable of delivering nuclear or chemical warheads. In 1983 and 1987, China supplied full design details for a 25 kT device and, between 1993 and 1996, procured a facility from Germany to purify tritium. China also supplied gas centrifuge equipment for uranium enrichment and on 28 May 1998, clearly in response to the Indian nuclear tests, Pakistan detonated five devices, reported to be in the sub-kiloton range, at the Chagai site near the Afghan border. A further two tests of 25 and 12 kT yield followed. On 30 May 1998, a sixth test was carried out. There are unconfirmed reports of preparations for a fusion weapon test. No further tests had been carried out to date 3 March 2000. Potential key nuclear delivery systems include the *Hatf* and *Gauri* series of ballistic missiles, with ranges up to 1,500 km. On 2 June 1998, Samar Mobarik Mand, head of Pakistani missile development, announced a test programme for an advanced new missile series, *Shaheen*, with a range of 1,200 km. See *Jane's Strategic Weapons Systems* for further details of Pakistani delivery capabilities.

The recent campaign in Afghanistan posed significant risks for nuclear proliferation. The former head of the Pakistan nuclear development programme, Dr Abdul Qadeer Khan, is reported to be a strong supporter of the Taliban. Although the government has made assurances that warheads, fuses and delivery systems are stored in separate locations, with the aim of making theft of a complete system more difficult, the Al-Qaeda group has declared a clear intention to 'acquire weapons of mass destruction'.

Biological weapons
None known although the potential to develop and produce such weapons is present. It shows considerable interest in acquisition of technology in both BW and CW.

Chemical weapons
None known. In recent years Pakistan has taken steps to acquire an NBC protective clothing and equipment manufacturing capacity.

Assessment
Hard evidence of Pakistani BW and CW programmes is scarce but it has the potential and probably the intention to develop and produce offensive NBC systems and has a shown considerable interest in acquisition of NBC defensive technology. Together, with the 1998 Indian tests and the most serious conventional clashes over Kashmir, its nuclear test programme has heightened regional security fears in south Asia. During 1999 and the early part of 2000, some progress was made on the border dispute issue, but the recent potentially explosive stand-off may have eased somewhat (June 2002) although tensions will remain extremely strong. Following the military *coup d'état*, the government of General Pervez Musharraf appears preoccupied with re-establishing stable management of the country and the economy. Pakistan remains a key element in the so-called 'war against terrorism' and internal pressure, together with the Kashmir dispute maintains the volatility of the situation. The smuggling or removal by force of nuclear materials is of special concern to the west.

UPDATED

PARAGUAY

NBC policy
Paraguay is a co-signatory of the 1967 Treaty of Tlatelolco, which established an 18-nation nuclear weapon free zone in South America and both of the main international nuclear arms control treaties.

Nuclear weapons
None.

Biological weapons
None. BTWC signatory.

Chemical weapons
None known other than riot control agents for police and paramilitary use. Paraguay ratified the CWC 1 December 1994.

Assessment
Paraguay is not known to have any NBC capability, nor is it likely to generate one in the near term.

VERIFIED

PHILIPPINES

Nuclear weapons
The Philippines has no nuclear capability and would seem to have little desire to gain one. It has been a signatory to the Non-Proliferation Treaty since 1972.

Biological weapons
There is no evidence of biological weapons having been developed and the Philippines acceded to the BTWC in 1973.

Chemical weapons
There has been no use of chemical weapons throughout the long struggle with the communist insurgents. The CWC was ratified 11 December 1996.

Assessment
A lack of funding and political need will prevent the Philippines from becoming a nation with non-conventional weapons this decade. However, the political turmoil in the region may encourage the government to review its strategy on WMDs.

VERIFIED

POLAND

NBC policy
Poland does not admit to possession of an offensive NBC capability and, as a new NATO member, she will subscribe to alliance NBC defence policy.

Nuclear weapons
None. Signatory to CTBT and NNPT.

Biological weapons
None admitted. Poland has ratified the BWC.

Chemical weapons
None admitted. Any Warsaw Pact legacy stocks of CW weapons that remain on Polish soil are likely to be removed by current collaborative DEMIL efforts to destroy all proscribed stocks of NBC matériel in the former communist territories. Riot-control agents are available to all security forces. CWC ratification was achieved on 23 August 1995.

Assessment
The Polish armed forces were trained and equipped similarly to all former Warsaw Pact nations. The expertise and defensive equipment remains. In fact, along with many of the former Warsaw Pact nations, NBC defence training took a relatively high priority and it is to be expected that NATO will learn some valuable lessons from the experience of the Polish armed forces. Changes in approach will be required, as NATO membership demands, in NBC readiness and host-nation facilities. Polish industry is active in the provision of NBC defensive equipment such as masks and decontamination facilities. A residual, unassessed, offensive CW capability may remain but its effectiveness is likely to be low.

VERIFIED

PORTUGAL

NBC policy
Portugal, as a member of NATO, adheres to alliance policy on NBC matters.

Nuclear weapons
None. Portugal signed the CTBT in September 1996 (yet to ratify) and is a signatory to the NNPT.

Biological weapons
None known. Portugal has ratified the BWC.

Chemical weapons
None admitted. Portugal ratified the CWC in September 1996.

Assessment
Portugal is highly unlikely to aspire to an offensive NBC capability. The armed forces are trained and equipped for NBC defence to NATO standards.

VERIFIED

QATAR

NBC policy
Qatar has similar concerns to other Gulf states, especially over Iraq and Iran's potential offensive NBC capability.

Nuclear weapons
None.

Biological weapons
None. BTWC signatory.

Chemical weapons
With the exception of anti-riot irritants, there are no chemical weapons in the Qatari inventory. CWC ratified September 1997.

Assessment
Qatar does not have an offensive NBC capability. The country is active in seeking better NBC defence measures for its armed forces and population.

VERIFIED

ROMANIA

NBC policy
Romania does not admit to possession of an offensive NBC capability.

Nuclear weapons
None present and no plans to develop or procure them. NNPT and CTBT signatory.

Biological weapons
None admitted. Romania has ratified the BWC.

Chemical weapons
None admitted, except riot control agents, such as CS. Romania appears to have no plans to develop or procure persistent weapons and ratified the CWC in February 1995.

Assessment
Romania lies at a potential flashpoint between Serbia and the Black Sea. Defence planners will be conscious of the risks of sudden conflict and the danger of its rapid spread. However, plans do not appear to involve offensive NBC weapons but the armed forces are equipped with effective NBC defensive measures. Romanian industry makes a range of NBC equipments from detectors to decontamination vehicles. The Romanian government will be well aware of the well-founded suspicion that Serbia maintains at least a research capability in offensive NBC.

VERIFIED

RUSSIAN FEDERATION AND ASSOCIATED STATES (CIS)

NBC policy
Russian post-Soviet WMD limitation programmes are suffering from a critical lack of cash. Moscow apparently plans to cut a further 450,000 jobs in its nuclear industry, re-igniting Western fears over the exodus of former soviet nuclear expertise. Under the Co-operative Threat Reduction programme (CTR), the US Congress authorised US$20,000,000 to assist in the role conversion of Russian nuclear scientists and engineers to more peaceful employment. The US plans to spend a further US$30,000,000 over the next five years to assist with the conversion programme. Despite this, there have been several protests in two of the 10 'nuclear' closed cities (Arzamas-16 and Chelyabinsk-70) by scientists who have not been paid by the government.

CTR also addresses the CWC and other arms control requirements. It has been moderately successful at government level but frustrations at local level, caused by cultural differences as well as the opaque Russian bureaucracy, have slowed actual progress. It has also been complicated by the apparent shift to a more confrontational stance on foreign relations with the West following the rise to power of Vladimir Putin. However, the key delaying factor has been lack of cash and, in July 1999, Moscow met with EU and NATO nations with a view to encouraging not only more cash input, but also changes in local liaison (from outside the US) to rejuvenate the process.

On CW, Russia's DEMIL task is formidable, with over 40,000 tons (declared) of stockpiled munitions. Russia estimates the cost of the programme at between US$5-6 billion.

The BW programme remains an area of suspicion and there is clear defector evidence of a continued programme.

Nuclear weapons
The new arms agreement signed between Presidents Putin and Bush may have little effect on the reality, serving to enshrine an already agreed mutual stockpile reduction. The US aim to persuade Russia to accept the US missile defence plan and declare void the old Anti-Ballistic Missile treaty appears to have been satisfied. Russia, on the other hand,

appears to have gained little from the new arrangement. Overall stockpile reduction will lower the maintenance costs for Russia's deterrent force. There are also gains in other, non-NBC related areas.

Russia has ratified the NNPT and signed the CTBT. Under the mid-1991 START I agreement, the strategic Rocket Forces were limited to 6,449 nuclear warheads against an original 10,237. The list comprises declared totals of 308 RS-20 (SS-18) heavy ICBM launchers (START limit, 154) 60 RS-22 (SS-24), 225 RS-12M (SS-25) mobile launchers, 805 other ICBM (350 SS-11, 60 RS-12 (SS-13), 75 RS-16 (SS-17) and 320 RS-18 (SS-19)). The weapons were located at about 20 sites in seven operational regions in European Russia, the Urals and along the route of the trans-Siberia railway. Soviet nuclear strategic deployments also included 100 air-launched cruise missile-armed manned bombers, comprising 85 Tu-95MS and 15 Tu-160, plus a further 95 Tu-95 carrying conventional High Explosive (HE) weapons.

START II, signed by Presidents Yeltsin and Clinton in early 1993, aims to commit both superpowers to reduce their nuclear warheads still further (to a maximum of 3,500 each in two stages by 2003) with the elimination of land-based MIRVed missiles. Early 1993 CIA reports claimed that Russia planned to reduce its strategic nuclear warhead total to between 2,000 and 2,500 by 2003 and that deployment of road-mobile and silo-based single-warhead versions of the RS-12M (SS-25) has continued. Some 105 RS-18 (SS-19) would be retained, modified to single-warhead configuration. The remaining 154 RS-20 (SS-18) would be eliminated under START II, with 90 of their silos available for RS-12M (SS-25) deployment. Up to 100 heavy bombers are authorised for reassignment to non-nuclear roles and are excluded from the overall fixed levels. However, START negotiations were thrown into a neutral gear by the Kosovo crisis in 1999 and remain frustrated by Russia's suspicion over the new US ABM strategy.

By 2003, Russian strategic nuclear forces are expected to deploy 504 ICBM, 752 bomber and 1,744 submarine-launched warheads. According to Lev Volkov, one of the architects of a 10-year Russian strategic nuclear deterrent programme, total delivery systems by 2005 will comprise solely 900 RS-12M, against START II limits of 1,300. For more details on delivery systems, see *Jane's Strategic Defence Systems*.

On the environmental front, there is concern both inside and outside Russia over the environmental legacy of nuclear programmes. Recent reports point to an impending toxic disaster in the Irkutsk region (especially Lake Baikal) and the pressure from concerned Russian scientists for better government environmental regulation of spent nuclear fuel.

Biological weapons
Russia has, as the legal successor to the Soviet Union, pledged to adhere to the BTWC. However, she maintained a continuous and advanced BW research effort throughout the post-war period. Despite the changed political face of the country, many suspect that this research continues. The focus has largely been on an organisation known as the Biopreparat, located at St Petersburg. Russia inherited the Soviet Ministry of Defense sponsored BW programme which operated five primary and seven secondary laboratories with a combined staff of approximately 6,000 personnel. The Ministry for the Medical and Microbiological Industry (often referred to as Biopreparat or Glavmikrobioprom) administers 20 other laboratories. The Soviet programme is credited with the development of: dried anthrax, tularaemia, plague, Q fever, cholera, botulinum toxin, enterotoxin, and mycotoxins. In 1992 President Boris Yeltsin admitted that the Soviets had violated the BTWC to which the USSR was a party, by carrying out a vast and covert BW programme. He also confirmed the 1992 Sverdlovsk incident, during which a quantity of anthrax was accidentally or deliberately released from a military facility, causing the deaths of at least 66 people. Although Yeltsin issued a decree halting the programme, the Russian BW effort remains a cause for concern. A trilateral (UK/US/Russian) process, designed to alleviate those concerns has been moribund since 1994. At the fourth BTWC Review Conference in late 1996, John Holum, Director of the US Arms Control and Disarmament Agency (ACDA),

argued that Russia remains challenged to demonstrate eradication of the problem. The UK urged Moscow to re-establish confidence that it is in compliance with its BTWC obligations.

Russian scientists have reportedly helped the Iraqi regime to develop an offensive BW capability and one explanation for the reluctance of Russia to support tough UNSCOM verification may have been fear of implication, according to some observers.

Chemical weapons

In 1992, Russia declared that all chemical weapons produced in the former Soviet Union had been placed within its borders, under Russian control. In July of the same year the Moscow parliament adopted a resolution on Russia's international obligations on CBW weapons. This resolution places responsibility for adherence to the CWC, to which the Soviet Union was a signatory, on Russia's shoulders.

However, accusations of continued experimentation with lethal CW persist, with substantial evidence from Russia's own scientists. Moscow has pointed out in its own defence that it has pledged to end CW production, not development. The stockpile disposal problem is severe and it appears unlikely that Russia will be able to achieve full safe destruction of the declared 40,000 tons within the 10-year timescale demanded by the CWC. In addition to this, the long-term environmental impact of the development programmes will emerge over time.

Assessment

It must be assumed that, despite the various treaties, Russia will retain an active NBC capability for the foreseeable future.

UPDATED

SAUDI ARABIA

NBC policy

Saudi Arabia is pledged to prevent the proliferation of WMDs.

Nuclear weapons

Saudi Arabia has no nuclear weapons. However, in July 1994, a defecting diplomat from the Saudi Mission to the UN claimed that Saudi Arabia had provided US$5 billion over 20 years to Iraq to assist with the development of nuclear weapons. The diplomat alleges that, in return, Saudi Arabia would acquire the technology.

Biological weapons

There are no biological weapons on Saudi soil. Officially, the kingdom has no plans to acquire or develop them. Reports in the late 1980s said that plans had been made for the establishment of a biotechnology centre by 1994-95. In the event of tension, this plant would be capable of producing BW agents. These details remain unconfirmed. The Saudi Arabian government has ratified the BWC.

Chemical weapons

There have been reports that chemical warheads for the CSS-2 missiles have been developed and are in the current inventory. The police and other internal security agencies have anti-riot irritants and other non-persistent chemical agents. The CWC was ratified in August 1996.

Assessment

The alleged Saudi support for the Iraqi nuclear weapons programme is said to have ceased in 1990, after the invasion of Kuwait. There are no reports of any current covert programmes. Along with other GCC states (*qv*), the kingdom is active in re-equipping its armed forces with a wide range of defensive NBC matériel and training. Measures are also under consideration for essential civilian personnel.

VERIFIED

SERBIA

NBC policy

Serbian policy under the Milosevic regime and earlier identified CW and possibly BW as part of the war-fighting armoury. Nuclear research with a military flavour continues. Serbia has failed to ratify the CWC and there is a strong and consistent history of CBW research and development dating from the late 1950s. However, the post-Milosevic era may see more enthusiasm for participation in NBC arms control agreements.

Nuclear weapons

None at present. Serbia is a signatory to the major nuclear arms control agreements. However, there are stocks of weapons quality material and at least two attempts have been made in the past to attain a nuclear capability. The latest attempt occurred in 1974, immediately after the first Indian nuclear test, although the programme appears to have been abandoned in the 1980s. Facilities at the Vinca Institute of Nuclear Science include a 6.5 MW research reactor, 48 kg of weapons-grade material (80 per cent enriched U^{235}) and a wealth of expertise in enrichment and processing of uranium and plutonium.

Biological weapons

None admitted. Serbia is a BTWC signatory.

Chemical weapons

CW research and development has been conducted in a wide variety of locations in former Yugoslavia, including sites round Belgrade, the Prva Iskri facility at Baric, Krusevac, Lucani and sites in Macedonia. A site near Mostar, now occupied by the Muslim-Croat Federation under the terms of the Dayton Peace Agreement, was a major research location. This facility, at Potoci, 10 miles north east of Mostar, was dismantled and relocated inside Serbia in 1991.

A 1976 Yugoslav National Army (JNA) programme to produce 200 kg of GB and H agent per day was planned at Baric. There, and at Lucani (the M Biagojevic Powder Mill), production was due to have been completed in 1970. The Mostar plant has been active for the last 30 years. A phosgene production line established there in 1959 had produced 15 tons by 1965. H and GB production facilities, with a target output of 40 tons (GB) and 30 tons (H) came on stream in 1976. In the same year, the variety of agents increased, with the production of test quantities to fill 155mm artillery shells as well as MLRS, bombs and land mines. Vomiting agents included Adamasite and diphenylcyanoarsine. Phosgene and disphosgene choking agents, bromosilcyanide, chloropicrin, cyanogen chloride and benactyzene (BZ) were also researched. The output of chloropicrin was planned at 10 kg per day and BZ, 5 kg per day. A major programme, referred to as the Jastrebac Project began in 1976. Under this scheme, weaponisation trials were conducted with 122mm and 155mm shells containing 3.5 litres of sarin, 128mm mortars (2 litres), BAD-100 aerial bombs (20 litres) and land mines (0.5 litres). Production is thought to have started in 1986 with a trial run of 250 122mm shells (placed in storage). A storage facility was planned at Hazici near Sarajevo and 122mm shells were reported ready for transfer to Mostar from another plant (Vogosce UNIS) as late as 1990. The Jastrebac Project plan was not completed and, in 1991, destruction of the older stockpile commenced. Some 220 rockets, 15 artillery shells and a quantity of unfilled munitions were apparently destroyed.

The Bosnia and Kosovo crises may have put a halt to the programme and, since March 1999, CW research, development and production facilities were high priority targets in the NATO degradation plan. However, accurate predictions of the quantities of CW agent stockpiled and still available to the army are difficult to make.

There was considerable collusion between the Milosevic regime and that of Saddam Hussein. Finding themselves in similar predicaments, with the same opponents, co-operation was an obvious choice. There are striking similarities between the Iraqi CW facility at Muthana and the original Mostar site. Also, visits to Iraq by senior Serbian officials and

CBW experts were frequent. Western media noted in March 1999 a visit to Baghdad by the Serbian Deputy defence Minister Lt Gen Jovan Djukovic and by Ivan Ivanovic, a Serb CBW expert. The supply of oil and cash by Baghdad was to be balanced by Serbian refurbishment of Iraqi air defence facilities and aircraft which continue to be degraded by the coalition forces. The programme is reported to have been stopped under President Vojislav Koštunica's new regime in Belgrade.

Assessment

The NATO degradation programme against Milosovic's armed forces served to pause any coherent CBW development. However, stockpiles probably remain, allowing future deployment should the political situation become unstable once again. Analysts were in fact surprised that Milosovic, unburdened by anxiety over the views of the international community, did not use CW during Operation Horseshoe, his plan to cleanse Kosovo of Albanian personnel.

Whilst nuclear aspirations appear distant, the close links that exist with both Russia and Belarus may involve exchanges at working level on CBW issues. Serbia is still assessed as having a military interest in the development of long-range delivery systems, probably based on a variant of the RFAS-acquired SCUB B. Although degraded, Serbian industry makes a wide range of effective NBC defensive equipment, including nuclear and chemical detection instruments, protective clothing and masks (see main entries in this publication) and is keen to export. EU help is agreed for a re-build of Serbia's industrial base following the NATO air bombing campaign. Although the defensive NBC equipment manufacturing sector was affected, most equipment appears available via the Yugoimport organisation. The regular Serbian armed forces are equipped for NBC defence and conduct NBC training drills.

UPDATED

SINGAPORE

NBC policy

None declared.

Nuclear weapons

Singapore is a signatory to the NNPT and the CTBT.

Biological weapons

Singapore has ratified the BTWC.

Chemical weapons

CWC ratified 21 May 1997.

Assessment

Regional security remains a concern and the NBC defence capability of the Singaporean armed forces is being upgraded. Singapore is unlikely to change her policy of support for non proliferation of WMDs.

VERIFIED

SLOVENIA

NBC policy

No declaration.

Nuclear weapons

There is no evidence of the storage or development of nuclear weapons, although Slovenia does have a civilian nuclear power generation capability (632 MW).

Biological weapons

There are no known plans to develop BW weapons. BWC ratified.

Chemical weapons

Other than riot control agents, there is no CW component in the Slovene inventory. NBC protection equipment is available and new equipment is under development. CWC ratified 11 June 1997.

Assessment

Slovenia is unlikely to seek development of an offensive NBC capability. Like other former communist regimes, the nation's armed forces are traditionally very aware of the need for NBC protection and train accordingly. Although obsolescent, NBC defensive equipment is assessed as effective against 'traditional' agents in the short term.

VERIFIED

SOMALIA

NBC policy

Somalia has no declared offensive NBC policy.

Nuclear weapons

None. Somalia is an NNPT signatory.

Biological weapons

None. Somalia is a signatory to the BWC.

Chemical weapons

None known, although unconfirmed reports have emerged on the possible use of CW against anti-government forces in remote areas. Some riot control agents are held by police and paramilitary forces. Somalia is not a signatory to the CWC.

Assessment

Somalia currently has no NBC capability. The nature of the conflicts and the climate may encourage some experimentation with BW but the government's continued border disputes with Ethiopia and the internal civil war continue to maintain the country in abject poverty.

VERIFIED

SOUTH AFRICA

NBC policy

A long period of sanctions-induced isolation forced the white South African government to attempt development of all forms of military equipment from internal resources. This policy included the development of a nuclear industry, with the suspicion of nuclear weapons development and CBW weapons for internal security and external counter-insurgency. At the Truth and Reconciliation Council, set up under the Mandela government, evidence emerged of an extraordinary programme of secret CBW research and development (see below).

Nuclear weapons

South Africa has two nuclear reactors with a total output of 1,842 MW. During the 1960s and 1970s, the white government is reported to have manufactured up to six nuclear warheads. The US applied considerable pressure on the Vorster regime at the time, with a view to preventing a planned underground testing programme at a site in the Kalahari Desert. The warheads were apparently dismantled. The nuclear plant at Valindaba opened in 1978, producing highly enriched uranium. It was decommissioned in 1990 under IAEA supervision. South Africa is now a signatory to the NNPT (1991) and the CTBT (1996).

Biological and chemical weapons

The Head of the CBW programme, Dr Wouter Basson gave evidence to Bishop Tutu's Council in 1998 that the whole range of CBW agents had been tested for a variety of roles and delivery systems. Aimed primarily at the secret emasculation of the internal opposition, schemes comprised cigarettes laced with anthrax, cyanide treated chocolate and drugs aimed at interfering with the fertility of the population. Common delivery methods were examined, such as air guns, hypodermic needles and umbrellas – the latter reminiscent of the method used by the Bulgarian secret service to assassinate the dissident Georgi Markhov in London. Internal security requirements have driven the production of a large stockpile of riot control agents which are readily used by the police.

There is no confirmation of the reported external use of CW by South African forces in Angola and elsewhere.

South Africa is a signatory to the BTWC and the CWC.

Assessment

The government continues its struggle to maintain law and order in an increasingly tense society. With a considerable research and development effort in place, it would not be a surprise if CBW agents were used again in the IS role. Additionally, an advanced NBC defensive infrastructure has been established, ranging from the local production of protective clothing to advanced CW agent detection equipment. South Africa has embarked on a comprehensive programme of re-equipment for NBC defence

VERIFIED

SPAIN

NBC policy

As a NATO member, Spain adheres to alliance NBC policy.

Nuclear weapons

None.

Biological weapons

None admitted.

Chemical weapons

None admitted, although there is some evidence of defence interest. Riot control agents are available to the security forces.

Assessment

Spain does not aspire to offensive NBC acquisition. The defence forces are committed to NATO policy, being well equipped and drilled in NBC defence. Spanish industry does not offer a large range of NBC defensive equipment but does develop effective NBC protective clothing.

VERIFIED

SUDAN

NBC policy

Sudan has not announced an NBC policy but there is evidence she sees utility in CBW. The fundamentalist ruling regime has strong links with other fundamentalist autocracies such as Afghanistan and appears content to offer facilities and support to anti-western terrorist organisations.

Nuclear weapons

There is no evidence that Sudan has access to nuclear weapons. She is a signatory to the NNPT (1976), but not the CTBT.

Biological weapons

Although Sudan has not signed the BTWC, there is no evidence of offensive BW acquisition.

Chemical weapons

The Sudan has shown distinct interest in CW. It was reported in 1990 that Iraq sent CW expertise and materiel to Sudan to remove them from domestic UNSCOM scrutiny. Munitions were reportedly stored at the Yarmouk military complex at Sheggara, south of Khartoum. Sudan also Sudan became a base for the efforts of Osama bin Laden to train and equip terrorists. The US cruise missile attack on 20 August 1998 was aimed at the El Shifa pharmaceutical facility where VX precursors had apparently been identified, although the US has yet to release hard evidence. Rebel sources report that CW has been used against them, especially in the south and it seems likely that it may be used again. While the ability of Sudan to include CBW in its armed forces doctrine appears low, the capability exists. Sudan has yet to sign the CWC.

Assessment

Sudan's internal politics make it unlikely that a coherent offensive NBC capability will emerge in the near term, but continued use of bases in Sudan for international terrorist activities involving CBW and internal security needs may drive towards a stockpile of offensive CW and, possibly, BW weapons. In the longer term, the region has experienced the use of CW, notably by Italy between the two world wars. It would therefore not be surprising if the use of WMD was considered, especially internally against the rebels in the south and, externally, in the border dispute with Ethiopia.

VERIFIED

SWEDEN

NBC policy

Sweden continues to pursue a successful neutral defence policy which eschews any involvement in offensive use of WMDs. In fact, the government has consistently led in the area of disarmament and the prohibition of NBC weapons.

Nuclear weapons

None held although, at one point during the 1950s, at least one nuclear device was constructed as a purely technical venture. It was not tested but was instead dismantled, leaving only the potential expertise as a future insurance against some possible need. The 12 nuclear power plants within Sweden could provide the basis for the raw materials for nuclear devices.

Biological weapons

None. The technical expertise and facilities would be well within reach.

Chemical weapons

None admitted. There is a continuing programme of NBC defence modernisation. The Swedish government remains concerned about the quantities of old German and former Soviet war stocks of various agents which have been dumped in the Baltic and are still occasionally disturbed by fishing activities. Considerable efforts have been made to provide the Swedish population with shelters and refuges capable of providing protection against chemical and other weapons.

Assessment

Sweden has the technical potential to develop and produce the full array of NBC agents and weapons but has chosen not to do so. The Swedish armed forces are equipped and train regularly in NBC defence procedures. The wide use of reserve forces means that the general level of expertise among the civilian population is also high.

VERIFIED

SWITZERLAND

NBC policy

Switzerland maintains a strictly neutral defensive posture and has shown no signs of acquiring any form of strategic weapon or offensive NBC capability.

CAPABILITIES

Nuclear weapons
None, although the potential to develop and produce such weapons is present. There is a traditionally high awareness of nuclear defence and national building construction regulations still demand attention to nuclear protection.

Biological weapons
None admitted.

Chemical weapons
None admitted. Considerable efforts have been made to provide the Swiss population with shelters and refuges capable of providing protection against chemical and other weapons.

Assessment
Switzerland is not assessed as aspiring to offensive NBC capability. The armed forces are well equipped and trained for NBC defence. Swiss industry has a well developed NBC defence sector, offering a wide range of detection, protection and contamination control equipment.

VERIFIED

SYRIA

NBC policy
Syria's aspirations of regional leadership and its strategic position have led the government to actively seek an offensive NBC capability. It is not a signatory to any of the international WMD control treaties and maintains an improving arsenal of over 60 SCUD variants. See *Jane's Strategic Weapons Systems* for further details.

Nuclear weapons
None known, but there have been persistent rumours of an undercover nuclear weapons programme. Syria has ratified the NNPT.

Biological weapons
None admitted, but Syria's interest and expertise renders the establishment of a BW programme highly likely. Syria is a signatory to the BTWC.

Chemical weapons
Hard evidence of an offensive CBW programme, long suspected by US and Israeli intelligence, is scarce. Certainly, as a part of the former Soviet Union's sphere of influence, Syria will have received considerable help with its delivery systems and may well have embarked on the acquisition of CW agent. The Israelis cite three facilities — one near Damascus and the others near Hamah and at the village of Safira in the Habat region — which are strongly suspected of producing nerve and blister agents. Irritation over Israel's occupation of the Golan Heights could see deployment of Syrian CBW agent, either to depopulate the region prior to reoccupation, or to stem further incursion by Israeli forces. The Syrian regime faces a similar challenge in the north from the Kurds and, like Iraq, may in future consider the use of CBW against them. Recent acquisition of earth-boring equipment may indicate a plan to bury CBW facilities away from surveillance in order to maintain a development programme. Syria is not a signatory to the CWC.

Assessment
Evidence indicates that Syria does have a CBW capability and it is strongly suspected of attempting to develop a nuclear capability. The rapprochement with Israel continues to founder on the rocks of her continued sponsorship of Hezbollah and it would not come as a surprise if CW became a feature in the medium term.

UPDATED

TAIWAN

NBC policy
Taiwan's strategic position and defence perceptions have kindled a strong interest in some aspects of offensive CBW.

Nuclear weapons
None known. Taiwan does not have a well-developed nuclear industry. Taiwan acceded to the NNPT in 1995 but has yet to sign the CTBT.

Biological weapons
None admitted. Taiwan has not ratified the BTWC and there is a suspicion that the potential for a BW programme exists.

Chemical weapons
Taiwan, no longer recognised as representing 'China' by the UN, is only 150 km from the mainland of the People's Republic of China and therefore feels vulnerable to overwhelming opposing forces. Notably not a signatory to the CWC, Taiwan is strongly suspected of maintaining an active CBW research programme. Evidence for this is based on the character of Taiwan's participation in international scientific and technical activities and the size of co-operation programmes with western nations in microbiology and genetic engineering. As a technically advanced nation, the capability to develop and produce such weapons has to be assumed. Riot control munitions are produced for local police and paramilitary use and have been offered for export sales.

Assessment
There is a strong suspicion that Taiwan is developing a clandestine CBW programme. Most defensive NBC equipment is imported or manufactured under licence from the USA. The increasingly aggressive stance by China over 'unification' may prompt the government to consider CW, in an area denial role, as a feature of future defence policy.

VERIFIED

TANZANIA

NBC policy
Tanzania has no coherent NBC offensive NBC policy.

Nuclear weapons
None. Tanzania has acceded to NNPT but is not a signatory to the CTBT.

Biological weapons
None. Tanzania has signed the BTWC.

Chemical weapons
None known other than riot control agents for police and paramilitary use. Tanzania ratified the CWC in 1998.

Assessment
Tanzania is not a nation of concern in offensive NBC terms.

VERIFIED

THAILAND

NBC policy
Thailand appears to have no offensive NBC acquisition policy.

Nuclear weapons
Thailand has no nuclear capability, having signed the Nuclear Non-Proliferation Treaty in 1972 and the CTBT in 1976.

Biological weapons
The country signed the BTWC in 1975. There is no evidence of weapons or weapon-related research and no history of their use.

Chemical weapons
Thailand has signed the Geneva Protocol but has yet to sign the CWC.

Assessment
Thailand appears unlikely to develop WMDs in the near or medium term.

VERIFIED

TUNISIA

NBC policy
Tunisia appears not to aspire to offensive NBC acquisition. However, regional considerations may alter the view.

Nuclear weapons
Tunisia has no nuclear weapons in its inventory and no plans to develop them. The headquarters of the Arab Atomic Energy Agency is located in Tunisia, which is also the location of a feasibility study, currently in progress, into the use of nuclear power for desalination.

Biological weapons
There are no biological weapons in the inventory and no plans to acquire them. BWC ratified.

Chemical weapons
Tunisia's only chemical weapons are anti-riot irritant agents held by the National Guard and the national police. Tunisia ratified the CWC on 15 April 1997.

Assessment
There is no evidence implicating Tunisia in the clandestine development of WMDs.

VERIFIED

TURKEY

NBC policy
Turkey has one of the largest ground forces available to NATO and publicly adheres to alliance policy on offensive NBC use.

Nuclear weapons
None.

Biological weapons
None admitted.

Chemical weapons
None admitted. There have been persistent but unconfirmed reports of Turkish troops and aircraft deploying CW agents against the dissident Kurdish population in remote areas. The injuries reported are apparently consistent with CW use. A limited NBC protection equipment programme has been initiated, but will be difficult to fulfil due to the large size of the Turkish armed forces and limited funding. Riot control munitions are available for police and paramilitary forces.

Assessment
Turkey may have developed some CW capability, although this cannot be confirmed. In defence, the armed forces are adequately equipped and trained to NATO standards. The government has recently embarked on a comprehensive upgrade programme.

VERIFIED

UGANDA

NBC policy
Uganda has no declared NBC policy.

Nuclear weapons
None. Uganda acceded to the NNPT in 1982 and the CTBT in 1976.

Biological weapons
None. Uganda has ratified the BWC.

Chemical weapons
None known other than riot control agents for police and paramilitary use. Uganda has yet to sign the CWC.

Assessment
Uganda is not known to have any NBC capability.

VERIFIED

UNITED ARAB EMIRATES

NBC policy
There is very little attention given to strategic issues in Dubai, other than an apparent desire not to be involved in any procurement plans. The UAE has yet to sign the BWC or the CWC.

Nuclear weapons
None. UAE acceded to the NNPT 1995 and signed the CTBT in 1996.

Biological weapons
The UAE does not plan to develop biological weapons and has signed the BTWC.

Chemical weapons
Except for anti-riot irritants, the UAE possesses no chemical weapons and has no plans to acquire them. However, the union has yet to sign the CWC.

Assessment
UAE has no offensive NBC aspirations.

VERIFIED

UNITED KINGDOM

NBC policy
The UK maintains an independent nuclear deterrent and is a key member of NATO. Nuclear deterrent weapons are deployed in a fleet of dedicated nuclear-powered submarines. The UK has declared that it does not intend to develop or stockpile CBW weapons, although it does maintain an advanced defensive capability against such weapons. The current policy on CBW weapons identifies a four-track approach: arms control, prevention of supply (of precursors and technology to suspect organisations), deterring use of CBW and defending against its use. The UK has ratified international WMD treaties.

Nuclear weapons
The three 'Trident' class nuclear submarines form the backbone of the UK nuclear deterrent. Each submarine can carry up to 16 Trident D-5 missiles with a deployed total warhead capability of 48. Nuclear warhead development is conducted by AWE at the Burghfield site.

Biological weapons
There is no offensive BW development, although the UK's traditional and widely respected expertise in defensive research allows small amounts of agents to be maintained under strict control.

Chemical weapons
The UK does not maintain a stock of chemical weapons, although some riot control agents and their delivery systems are held and have been used actively. In the past, a wide range of chemical warfare agents has been produced but, since the mid-1960s, all stockpiles have been withdrawn and destroyed and manufacturing facilities have been dismantled. However, the capability to produce such agents is well within the nation's industrial capacity, although political and other considerations would make such a course of action most unlikely.

The UK has a well-developed industry for the provision of defensive measures against chemical warfare agents and continues to investigate methods, equipments and techniques for protection against current and future CW agents. The national centre of chemical (and biological) protection and associated expertise is based at the DSTL (Chemical and Biological Defence) site at Porton Down in Wiltshire, the oldest establishment of its type in the world and one widely acknowledged to be the leader in its field.

Assessment
The UK will continue to maintain an independent nuclear deterrent for the foreseeable future. A comprehensive NBC defence posture is actively maintained and will continue, reinforced by the formation of the Army's NBC Reconnaissance Regiment. This battalion-strength formation includes mainly Territorial Army volunteers and will provide detection and decontamination capabilities. Based at Manston in Kent, it is equipped with the Fuchs armoured NBC reconnaissance vehicle and the Interim Biological Defence System (IBDS). Elements have already seen service in Kuwait. UK is a key exporter of the full range of NBC defence products, expertise and training.

UPDATED

UNITED STATES OF AMERICA

NBC policy
The Bush administration ordered a fundamental review of all its WMD-related policies across the board and this process was well in train prior to 11 September 2001. In May 2001, the political complexion of the Senate changed and Democrats found themselves chairing both the influential Senate Foreign Relations and Senate Armed Services Committees, placing them in a position to modify some elements of government programmes, including the NMD plan. There is no doubt that 11 September 2001 has served to harden up the desire for a defensive shield against the nuclear threat and the determination to pursue nations and organisations seen as posing an increasing threat to national security through the development of WMD. In private, the US is also continually disappointed by what it sees as equivocation by its allies in the 'war against terrorism' and it also sees neither concrete development nor consensus for decisive action in the arms control arena.

NBC defence policy is nevertheless founded on a mixture of national, allied and disarmament initiatives. Nuclear deterrence forms the cornerstone of national policy, supported by strong conventional forces. Progress in the disarmament field (SALT and START treaties) has led to significant equivalent reductions in the US and Russian nuclear inventories. This process has been accelerated by the recent bilateral nuclear arms reduction agreement with Russia. Having ratified most WMD-related treaties, it is undergoing a 10-year programme to demilitarise its stockpiled CW agent munitions. In addition, the US is assisting Russia with its DEMIL programme through the Co-operative Threat Reduction (CTR) programme. Alliance responsibilities, especially towards European NATO, ensure that a sound capability in CBW defence is maintained.

Nuclear weapons
US policy on nuclear deterrence includes pursuit of the National Missile Defence plan (NMD).

This proposes a shield of technologically advanced detection and interception systems designed to counter the nuclear-tipped ballistic missile threat. However, sceptics point to its high technological risk and the fact that is unlikely to deal with the 'lower-tech' nuclear threat posed by the so-called 'rogue states' (North Korea, Iraq, Iran and others).

The USA has identified that fact that the original ABM treaty struck with the former USSR has now been largely overtaken by events. The federal government has set out in an attempt to convince the most strident critics, including Russia, China and NATO that NMD will not disturb the delicate nuclear deterrent balance nor provide the trigger for a new arms race. The US 2002 Nuclear Posture Review (NPR) has crystallised the future needs for a nuclear capability as follows:

- The development, beyond the end of the decade, of a 'New Triad' which will continue to maintain the three launch vehicle options of ICBMs, manned bombers and SLBMs but also incorporate some non-nuclear options.
- The eventual reduction of the numbers of nuclear warheads on front-line systems to between 1,700 and 2,200 (the 'active stockpile'). The FY07 target is 3,800. 1,300 will be withdrawn as *Peacekeeper* systems are retired, in addition to some *Trident* SLBM and other systems. This does not mean they will all be destroyed. They would revert to an 'inactive stockpile' which could be re-activated within "weeks, months, even years depending on the system and the threat" according to the US Secretary of Defense for International Security Policy, J D Crouch (June 2002). The removal of some short-lifed materials, such as tritium or neutron generators, would qualify them for the inactive stockpile. Reactivation could not deal with a tactical threat but could be ordered if the administration sensed a major adverse change in the security environment.

However, the NPR does not address (at least publicly), the close attention that the US administration and the key laboratories are paying to the design and development of nuclear warheads. Certainly a keen need was felt by CENTCOM to have a weapon with sufficient penetration to deal with the network of deep mountain tunnels and caves known to be occupied by opposing forces in Afghanistan. There is evidence of a shift in approach towards the development of smaller nuclear warheads. These would have significant earth-moving capability or the ability to destroy enemy personnel over a wide area without wreaking the kind of damage to the infrastructure which results from the blast and heat in current nuclear weapons.

In the arms control environment, START, CTBT and the NNPT combine to maintain a continued downward pressure on the US nuclear arsenal of strategic and tactical weapons. Nevertheless, the stockpile includes ICBMs, cruise missiles, nuclear depth charges and free-fall bombs. About 350 Minuteman missiles remain. Trident C-4 and D-5 ICBMs are deployed in nuclear-powered submarines. Airborne delivery systems include the B-2 bomber. The US Army retains a Lance tactical nuclear missile strike capability and large numbers of 155 mm artillery projectiles remain available for deployment. (See *Jane's Strategic Weapon Systems* for more details).

Biological weapons
US technical capability to develop BW agents is probably unrivalled. Low volumes of all types of agent are created for research and testing of BW defensive measures but work on offensive programmes ceased in the 1960s. The US is investing heavily in improved BW agent detection (see under DETECTION (sensor systems) – Biological), given renewed impetus by the anthrax attacks in October 2001. The administration's review of BW arms control led it to withdraw from the BTWC working Group late last year, leading the collapse of further efforts by the states parties to strength the BTWC through more intrusive no-notice inspection.

Chemical weapons
The 5-year US chemical DEMIL programme started with a budget of US$1.7 billion. Today, 15 years on, the programme budget is nearly US$16 billion.

The CWC imposes a 10-year timescale for destruction of declared Chemical Weapons – a significant technical challenge. The 30,000 ton US declared stockpile includes land mines, artillery projectiles and artillery rocket warheads. It does not include other munitions, such as binary weapons, which are also subject to destruction under a separate programme. A pilot plant on Johnston Island in the Pacific was built to test the incineration process. Technical difficulties and pressure from the environmental lobby has driven a new programme to evaluate other candidate technologies, including chemical neutralisation. However, the site has been successful in completing destruction of H agent munitions.

In the past, chemical agent production was carried out on an industrial scale at the Rocky Mountain Arsenal, Colorado and other sites. Offensive CW policy changed and production ceased soon after an accident in 1968 at Dugway Proving Ground, Utah, in which over 6,000 sheep were killed following a leak of VX agent from an aircraft in flight.

On 1 October 1994, the US Army Chemical and Biological Defence Command (CBDCOM) was established, headquarters at Aberdeen Proving Ground, Maryland with the aim of co-ordinating CBD requirements across all the armed services. CBDCOM, the successor to the former Chemical Warfare Service, is responsible for research, development and acquisition of CBD measures. The organisation was retitled SBCCOM (1997) to reflect a new, holistic responsibility for protecting service personnel.

Following the sarin attack in Japan, the federal government has set in train a scheme to equip and train emergency services in major cities to manage terrorist or disaster relief incidents with a CBW component. The Federal Emergency Management Agency (FEMA) has been set up to oversee the development of successful performance by 'first responders' to an incident

Riot control agents are routinely issued to internal security forces.

Assessment
The US is determined to maintain its nuclear deterrent, while reducing force levels. Further, against the background of its NMD initiative it has agreed with Russia to make significantly greater cuts in offensive nuclear capability than previously envisaged. However, the US Senate's continued reluctance to ratify the CTBT remains a stumbling block. In CW and BW defence, the DoD effort is aimed at capability improvements across the board, especially in bio-detection. The huge impact of events from September 2001 onwards has led to a second fundamental review of homeland defence and much of the efforts of the new initiatives will focus on the risks of nuclear, and CBW use by states and non-state actors alike.

UPDATED

VENEZUELA

NBC policy
Venezuela is a signatory to the 1967 Treaty of Tlatelolco which established an 18-nation nuclear weapon-free zone in South America.

Nuclear weapons
None. Venezuela has ratified the NNPT and signed the CTBT in 1997.

Biological weapons
None admitted. Venezuela has ratified the BWC.

Chemical weapons
None admitted other than riot control munitions. Venezuela ratified the CWC in December 1997.

Assessment
Venezuela does not appear to have any NBC capability.

VERIFIED

YEMEN

NBC policy
Yemen does not aspire to acquisition of an offensive NBC capability.

Nuclear weapons
Yemen does not possess and has no plan to develop nuclear weapons. Yemen ratified the NNPT in 1979 and signed the CTBT in 1996.

Biological weapons
Yemen has ratified the BTWC, does not have any biological agents and has no plans to acquire them.

Chemical weapons
Persistent allegations that a CW agent was used during recent civil wars remain unconfirmed. CW warheads for SS-1c and FROG-7 systems may have been provided by former Soviet states. Internal security forces have access to riot control agents. Yemen has yet to sign the CWC.

Assessment
Yemen may have some offensive CW capability but no nuclear or BW programme.

VERIFIED

ZAMBIA

NBC policy
Zambia has no NBC policy for offensive or defensive NBC capability.

Nuclear weapons
None. Zambia acceded to the NNPT in 1991 and to the CTBT in 1996.

Biological weapons
None.

Chemical weapons
None known other than riot control agents for police and paramilitary use.

Assessment
Zambia is not known to have an offensive NBC capability. However, it has yet to sign either the BTWC or the CWC. This is viewed more as a sign of Zambia's focus on internal rather than international affairs than a sign of a desire to acquire an offensive WMD capability.

VERIFIED

ZIMBABWE

NBC policy
Zimbabwe has no NBC policy for offensive or defensive NBC capability.

Nuclear weapons
None. Zimbabwe acceded to the NNPT in 1971 and has yet to sign the CTBT.

Biological weapons
None. The BWC has been ratified.

Chemical weapons
None known other than riot control agents for police and paramilitary use. Zimbabwe ratified the CWC on 25 April 1997.

Assessment
Zimbabwe is not known to have any NBC capability.

VERIFIED

NUCLEAR WEAPONS AND THEIR EFFECTS

NUCLEAR WEAPONS AND THEIR EFFECTS

Introduction

Is there still a threat?

Actual proliferation of nuclear weapons technology has increased and the trend is running against the ability of arms control activists to exercise control. It is not only nuclear expertise that is leaking its way into the universities and defence ministries of what the USA describes as 'rogue states'. Despite the provisions of the nuclear non-proliferation treaty, uranium, plutonium and nuclear equipment are also passing across. The sceptical reactions of China, Russia, NATO and the European Union to the US-Russia agreement on nuclear stockpile reduction and the US NMD comes as no surprise. Advocates of NMD envisage successful achievement of a missile shield which would instantly render most current ballistic systems obsolete. Others point to the fact that the huge investment will not address the 'low tech' nuclear threat. The technological mountain to climb is enormous. Recent interceptor tests and trials in the USA have been more successful but have also reinforced how difficult it is to achieve. Against this background, it is not only diplomatically tough for the arms control community to rein in the spread of nuclear weapons capability, but it is much more urgent. A continuing area of concern for the IAEA is the weak management of nuclear material. This is an issue of control. The world has simply lost track of large quantities of weapons-grade nuclear material – and it has a high value. For those active nuclear specialists, whose jobs evaporated as a result of perestroika, life is hard and they are free now to seek their fortune elsewhere. There is clandestine material and expertise at large, despite the efforts by the international community to regain control. The break-up of the former Soviet Union into a number of states, many with a long history of mutual hostility, has made the world more likely to have to deal with the spectre of nuclear detonation than was the case before the fall of the Berlin Wall.

Poor housekeeping affects the material generated for peaceful use as well. During the rise of nuclear electrical power after the end of the Second World War, obsessive secrecy became an ingrained feature of the industry, born of the military implications in the early days, preventing public scrutiny. Therefore, evolution of the effective and visible checks and balances required for controlling the material never got off the ground. This is a power source with more manmade potential to destroy human life than anything in history.

Nuclear weapons testing continued apace throughout the 1950s and 1960s. The largest Bikini Atoll test was the BRAVO experiment on 28 February 1954, unleashing the equivalent of 15 million tons of conventional explosive power into the atmosphere. At the time, big was beautiful. Today, the technology has moved towards smaller yield warheads and the development of enhanced radiation weapons: the so-called neutron bombs. With the accuracy of long-range delivery systems now refined to the order of metres, the need for large area warhead coverage is not so great. In addition, the neutron warhead allows the attacker to deliver the wider destructive effect on the people rather than the defenders' buildings and equipment. Today, this high-end technology remains largely in the hands of the five declared nuclear powers but it will not be many years before it is available to those nations actively seeking a military nuclear capability.

The internet reveals a wealth of information on nuclear issues, from strategic analysis to details on how nuclear weapons are constructed. This plethora of information is not lost on terrorist organisations and peaceful nations need to remain alert to this new challenge. Measures need to be put in place to deal not only with the actual threat but also the declared threat to detonate. There is public danger also from the simple release (or threat of release) of fissile material as an attention-seeking act. These issues have generated a welcome trend in defence and interior ministries to address the issue of low-level release of fissile material and to prepare plans and equipment accordingly. Apart from its use in a bomb, Plutonium itself is one of the most potent poisons known.

Delivery systems in the leading nuclear arsenals have evolved over the years to capitalise on the advantages of the radiation effects of the technology, particularly in the development of tactical weapons. The distinction between strategic and tactical may seem academic to the thoughtful defender facing an unexpected sunrise. Nevertheless, strategic weapons generally comprise those large warheads designed to deter aggression by the threat of massive destruction of the aggressor's land and people – making his aggression less than worthwhile. The principles of mutual assured destruction – the keystone in the defence arch attributed with preventing a third world war – became the subject of deeper and more convoluted argument in the 1960s. Ideas such as rational irrationality, selectively declared targeting and the shifting around of empty launchers became an elegant game for politicians and defence planners. The division of the Soviet empire has now crumbled one side of the MAD arch; not perhaps with respect to the potential destructive capability still available in eastern Europe but through less coherent control. Other ambitious countries, not in the nuclear club, may feel the uncomfortable gaze shifting away from Russia and her former empire and onto themselves. This is a pressure which has quickened the desire to acquire their own weapons of mass destruction.

In summary, the hazard has become more difficult than ever to analyse and therefore, more difficult to prepare for. Security services in peaceful nations need

A shallow underwater nuclear water burst at Bikini Atoll. Twenty-three atmospheric tests took place here between 1946 and 1958   0011398

to be aware, in detail, of the type of technology an organisation might seek in order to construct nuclear weapons and the IAEA publish circulars on the things to look for. Additionally, defence planners and purchasing authorities still need to remain aware of the basic physics involved, the orders of magnitude and the materials used in these weapons. This section offers a brief guide.

Warhead types

Strategic deterrent weapons remain generally in the megaton range, delivered by ballistic missile from nuclear-powered submarines, underground silos or mobile launchers. They are likely to remain the cornerstone of defence policy for China, France, Russia, UK and the USA. Tactical concepts for the use of nuclear weapons evolved as technology provided highly accurate delivery systems and allowed warhead design to deliver less indiscriminate damage. Cruise missiles, 'smart' bombs, torpedoes and depth charges can carry nuclear warheads aimed at more limited area denial or gain. To move mountains, nuclear demolition charges have also been developed.

Fundamentally, all nuclear weapons depend for their effect on the near-instantaneous release of enormous energy from the break-up or fusing together of atoms. In fission weapons, the original technology, the nuclei of elements at the heaviest end of the periodic table are broken apart, releasing high-energy neutrons to bombard neighbouring nuclei. This catastrophic chain reaction is initiated by the use of conventional explosives which cause two sub-critical masses of heavy elements (normally the isotopes of uranium: U^{235} or plutonium: Pu^{239}) to merge together. The critical mass is that above which such a chain reaction is likely to occur, leading to the release of massive amounts of energy. Some 1 kg of plutonium, for example, can release 10^{14} joules of energy or 10 equivalent kTs of conventional explosive. The world's first such nuclear device was detonated from a steel gantry in the Alamogordo Desert at 05.30 on 16 July 1945. The first military use occurred over the Japanese city of Hiroshima on 7 August 1945, to be followed by a second nuclear strike on Nagasaki on 9 August. Early weapons were between 10 and 20 kT in yield.

The fusion weapon, developed later, is much more energetic. Often known as the hydrogen bomb, it requires the colossal energy from a fission reaction to generate the initial conditions (50-100,000,000°C) for fusion of nuclei to occur. The common nuclei are the Hydrogen isotopes: Tritium and Deuterium. The megaton-range weapons commonly trialled in the atmosphere and underground by Russia and the USA were fusion weapons. The initial radiation products from the two types of weapon differ in composition but, from the defender's viewpoint, the other effects, such as shock and heat, predominate.

The height (or depth) of the detonation changes the resultant effect of the weapon considerably. Some weapons, such as nuclear depth charges, focus their energy primarily into the shock wave, which collapses ship structures and causes percussion injuries on the human frame. The noise effect can render sonar detection systems useless for days. At the other end of the scale, a high-altitude burst may cause little shock at the surface but, by ionising the upper atmosphere, will have a lengthy and catastrophic effect on radio and radar performance. The electromagnetic pulse (EMP) from such a detonation can effect an area of 10,000 km or more. Any weapon whose fireball comes into contact with the surface will draw up thousands of tons of material, which becomes irradiated. Some of this material loses its radioactivity very quickly but, equally, some may stay dangerous for years. This material can remain in the atmosphere for days, descending to the surface and causing injury to living organisms. This is radioactive fallout. The initial or direct effects of nuclear explosions can be detected and measured to give an indication to the defender where the burst occurred. Together with the time of the burst and the prevailing meteorological profile, the EFW can be calculated and the likely fallout deposition footprint mapped out. The information assists authorities in the preparation of recovery plans and this book lists a number of such detection, measurement and prediction systems.

Effects of nuclear detonations

The chain reaction which develops inside the kilogram or so of active material in a nuclear weapon is, for all practical purposes, instantaneous. The raw energy released covers the entire range of the electromagnetic spectrum. The radioactive particles generated by the reaction and the physical effects are devastating, but all follow the basic laws of physics. In this context, with an event of such scale, alpha and beta particles created at the detonation can be ignored. Initial gamma radiation and fast neutrons (INR) can reach to a range of 5 km or so from the Ground Zero (GZ) position (further, depending on the height of the detonation). However, their effects on electronic equipment can reach further. In effect, the detonation creates a pulse (EMP) of extremely high, short-duration energy – an electric field – with tens of kV/m at considerable ranges. Jumping across conventional surge protectors and other devices, the damage inflicted by the EMP is devastating as it enters sensitive antennas designed to pick up the minutest levels of energy and amplify them to give useful information. This transient radiation (TREE) burns out resistors and permanently changes the characteristics of semiconductors. The INR and EMP together form about 15 per cent of the total energy of the weapon.

The scientists who developed the bomb were really searching for the enhanced physical effects to do the work that an equal sized conventional weapon could not deliver. They succeeded. The physical effects are dramatic. The light energy reaching the retina of a soldier looking at the burst will cause blindness, especially at night when the contrast is greatest. The soldier may recover, but will certainly be temporarily blinded and therefore ineffective. The intense light will adversely affect electro-optical detection and direction systems, in fact any light-sensitive component. The nuclear flash does not last long, becoming obscured by the density of the solid particles and gases emitted. Heat energy, further down the EM spectrum, is significant – about 4.0^5 kW h per every kT of weapon yield. Briefly, the intense heat radiation from the Hiroshima bomb ignited flammable materials instantly, at a distance of 5 km from GZ. Differentially absorbed by darker objects and reflected back and forth from lighter ones, the energy caused first degree burns on exposed people out to 7 km. Finally, the sheer blast effect of the detonation causes an immense shockwave to spread out from GZ to great distances. With a ground, or low air burst, the intensity is reinforced by the reflection of the shockwave from the ground itself. The peak overpressure can be enormous and the wind can reach 300 m/s, flattening everything in its path. After a few seconds, the surrounding air rushes back in towards GZ to fill the vacuum, causing a reversal of the wind direction at a lower velocity but lasting twice as long. Some 5 seconds after the Hiroshima detonation, a person standing 3 km away would have felt an overpressure of 0.4 bar – quite enough to burst an eardrum. Delicate structures are highly vulnerable to blast. A peak overpressure of just 1.5 bar can wreck a railway train or a frigate at sea. Given that the effect reverses after a few seconds, the damage can be terminal. All the potential physical effects (light, heat and blast) can be calculated by knowing the weapon yield, detonation height and distance from GZ. Many of them follow a logarithmic scale. Some 35 per cent of the total energy is transmitted as light and heat. Blast and shock effects form the remaining 50 per cent.

Apart from the physical devastation, the residual effects of a nuclear detonation – nuclear fallout – can cause problems for years. The radiation emitted from this material can be measured, allowing predictions to be made about the safety of personnel. There are 3 forms of radiation. The most powerful and penetrating is γ-radiation. This can travel up to 5 km in air and has a devastating effect on the human body. The other two types (α and β) have a limited range in air and, service clothing provides adequate protection from permanent damage. However, although these types of radiation have short ranges, they can represent an extremely harmful long-term threat. For example, α-particles, if ingested by entering the nose or mouth or passing into a wound or body orifice, can cause severe and long-term radiation sickness. β-emitting material is likely to be the main cause of early fallout casualties because of its ability to penetrate bare skin.

Radioactivity decays exponentially, leading to the concept of half-life as an indication of the likely duration of this activity. Half-life is defined as the time taken for the measurable radioactivity to decay to half its original level. This time varies tremendously between isotopes. For example, the half-life of radioactive Iodine (I^{128}) is about 25 minutes, whereas that of Uranium (U^{238}) is 4.5×10^9 years.

Both the INR and the radiation from fallout can poison humans, depending on its intensity (the Dose – measured in cGy) and the rate at which it is absorbed (the Dose Rate). In Hiroshima, the *Hibakusha* (survivors) are still affected, although monitoring over the years has revealed testimony to nature's powers of recovery. The short-term effects can be dramatic. 600 cGy causes death in 50 per cent of those exposed. At 300 cGy, people will probably recover in 3 to 8 weeks depending how fit they are. All will suffer nausea, diarrhoea and vomiting. Their hair will fall out and they will suffer sores and lesions. The body's immune system is greatly debilitated, leading to some early deaths from secondary infection.

Protection

The destructive forces involved make it almost impossible to defend against a direct hit. Burial of key C^3I facilities underground (collective protection) is a policy undertaken by most sophisticated nations. By use of reinforced structures and the provision of filtration plant and stocks of provisions, organisations can survive to emerge and commence the recovery process. However, ground forces, ships and airfields remain vulnerable. System design that takes early account of TREE, can significantly reduce the damage to electronic equipment on which all defence forces rely today. Individual measures are largely managerial, supplemented by medical prophylaxis to delay the symptoms of radiation poisoning. Medical science offers anti-emetic drugs and displacers such as Potassium Iodide tablets. Measurement of the effects of fallout can be used to calculate how long people can stay and how often they should be exposed in a given period of time. In management terms, defence personnel generally receive training on the effects of nuclear weapons and some of the measures they can adopt to reduce the effects. The latter usually cover the principles of shelter construction, defensive posture and the use of IPE (MOPP in the US), whilst noting that IPE is totally ineffective against penetrating g -radiation. Civil emergency planners are increasingly seized with the need to prepare – recreating a capability lost through cost-cutting and perceptions of a reduced threat in recent years.

Summary

Nuclear weapons are here to stay. The threat from them has become more diffuse and the knowledge and components have become dangerously available for unscrupulous use. The control of strategic weapons following the break-up of the Soviet Union has become less certain and ambitious neighbours may see acquisition of a nuclear capability as a badge of sophistication and the path to regional power. Defending against the effects is tough, but for those outside the primary area of damage, there is much that can be done. Planners need to be aware of the risks and ensure that materiel design for nuclear survivability is built in early, when it is cheapest. In the civil protection arena, authorities should develop contingency plans, both for the possibility of nuclear detonation and for the clandestine release of fissile material into the atmosphere.

UPDATED

BIOLOGICAL WARFARE

BIOLOGICAL WARFARE

Introduction

Background

The military use of living organisms to cause casualties to the enemy is as old as warfare itself. From biblical times, water-holes have been poisoned, infected corpses fired across enemy lines and diseased prisoners returned to infect their colleagues. In the 19th century, unscrupulous traders distributed blankets contaminated with smallpox among the American Indians. This reduced their fighting strength through epidemics against which they had no immunity. Today, science allows the development of new BW agents, synthesised in the laboratory. They are derived from nature but do not occur naturally. Genetic engineering and, especially, the availability of human genetic data in the public domain through the Human Genome Project increases the risk of the clandestine development of harmful organisms for military purposes. Additionally, this new science allows existing organisms to be mutated either to give a more specific, more intense or quicker pathogenic effect. Organisms can be modified to survive longer, take advantage of different vectors or target specific ethnic groups. Recent public unease over genetic engineering has raised the debate and improved general public understanding of the issues. Information specifically on Iraq's BW development programme has reinforced the belief that modern, manipulated, biological weapons are more likely to be used than in the past.

The modern environment

Our lifestyles have become more complex and therefore more vulnerable. The 2001 outbreak of foot and mouth disease in the UK has highlighted the extraordinary interdependence of agriculture, transport and commerce. Recently, again in the UK, there have been several outbreaks of tuberculosis — an 'old' disease. This infection is thought to have been brought back to the UK by pilgrims attending the 2001 Haj in Mecca. Current scientific research reinforces the fact that over-use of antibiotics and the increasingly 'clean' environment of the western citizen, from birth onwards, combine to render people increasingly susceptible to both new and old diseases. Our interdependence brings the risk of the instant pandemic and experts are by no means sure that technology could be brought to bear in time to prevent millions of deaths.

There have also been natural changes in some of the diseases which affect man: caused by mutations in viruses and other organisms which allow them to cross the species barrier from animals to man. In January 1998, a fresh outbreak of Rift Valley Fever occurred in north-eastern Kenya, showing the characteristics of a new and more virulent strain. The cross-species transfer has been helped by some appalling errors made by man himself. The advent of Bovine Spongiform Encephalopathy (BSE) in cattle and the resultant potential for an epidemic of the new-variant, Kreutzfeldt-Jacob Disease (NVCJD), is one example. Other diseases have jumped the gap for other reasons or have come to prominence as a result of lifestyle, such as those caused by HIV and *escherichia coli* (E Coli). The emergence of the Ebola virus and the extremely rapid tissue deterioration caused by Necrotising Fasciitis have also focused attention on the microbiologist for some answers to the way these diseases change and adapt. From time to time cultures of these dangerous organisms escaped from research facilities or, even worse, become the subject of clandestine 'trials'. A blatant example of this has been reported in Iraq where it is believed that prisoners were tied to stakes and deliberately infected with Anthrax for trial purposes.

Finally, the advances in biotechnology which have occurred this decade are momentous. Publishing data on the entire human genome into the public domain allows, at the same time, not only spectacularly better disease management but also genetic manipulation for offensive military use. The genie is well and truly out of the bottle.

The CBW Spectrum

It is convenient to divide non-radiological WMDs into either biological and chemical warfare agents. In fact, there is a range of agents of interest which falls in between these two strict definitions. They include the insect and reptile venom and synthesised materials designed for very specific effects.

The features of offensive BW

There are several important general features of BW which make its offensive use problematical. Firstly, it is difficult to deliver effectively. It can be dispensed by bomb or missile warhead. Once released, its spread is unpredictable. Unlike Chemical Warfare (CW) agents, only a minute number of organisms are needed to cause infection. The particles, like feathers, are more unpredictably affected by currents of air. Some pathogens can be killed by the ultraviolet radiation that is present in strong sunlight. In other words, the BW weapon can be highly successful or completely ineffective. Only regular reconnaissance or a change in the opponent's performance will reveal its eventual effect. Secondly, its unpredictability can lead to an 'own goal'. Unless the user's own population is adequately protected, the military commander cannot be certain that their own troops, or worse, the population they are charged to protect, will not succumb to the very illness pressed upon the enemy. Thirdly, the effect is not instantaneous. The body's immune system needs time to fail against the pathogenic assault. A disease may take several days to make significant inroads against the opponent's capability in the field. It is therefore too slow and unpredictable to be effective in a rapid tactical action or during an offensive amphibious landing. It falls best to the special forces to deliver, perhaps as a precursor to a conventional attack.

Clandestine delivery and the delayed effects offer another benefit — the act can be denied. There may be strong circumstantial evidence for blame, but the UN and

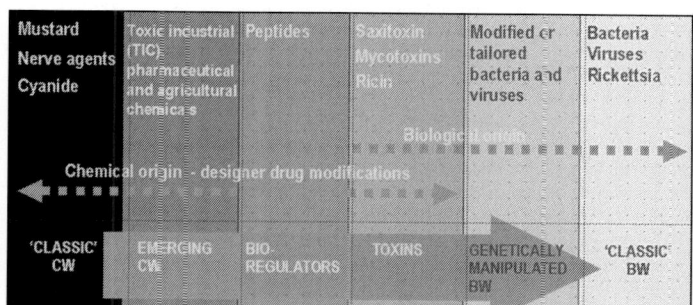

The distinction between 'Biological Warfare agents' and 'Chemical Warfare agents' is increasingly blurred. Advanced bio-engineering allows genetic manipulation of BW agents. Laboratory synthesis of those chemicals which are naturally produced by living organisms such as Saxitoxin, is also common. Toxins are listed under Biological Weapons   2001/0101334

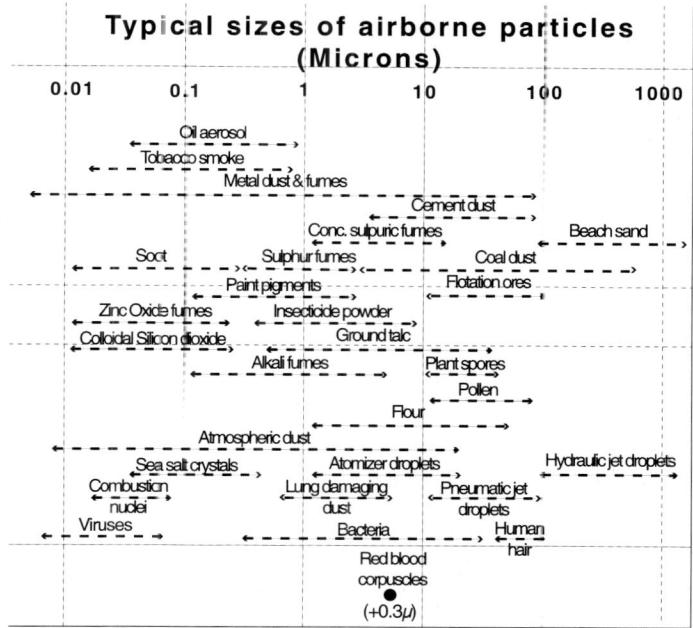

This diagram shows the relative sizes of potentially harmful agents relative to common particulates   2001/0102164

arms-control authorities will need proof before they react. Hence the need for safe collection and transfer of evidence. Fourthly, BW agents, while difficult to store in a viable form, are easy to make. The process is extremely low-tech, with more than a passing similarity to the production of beer, but it does take a long time to produce in militarily useful quantities. The more sophisticated processing equipment is identical to that found in legitimate pharmaceutical laboratories, as UNSCOM inspectors have found in Iraq. Despite the drawbacks, BW can appear a cost-effective option. It is also a potent terrorist risk.

Defence — problems

For the defender, things are no better. Science is struggling to produce solutions which can effectively warn for BW agents. There are so many types. Being biological themselves, they are so close to the biology of the human being that they are often difficult to differentiate from organisms which are harmless or essential to life itself, such as those that live in the digestive tract. Current detection technology uses schemes which mimic the mechanism by which the pathogen acts on the human system. Although the time to identify an agent, from the moment an organism alights on the collector, may be down to just minutes, it is still not a predictive technology. The operators may raise the alarm at the same time as their colleagues are collapsing from disease. Conversely, the detector operator may hear that nearby colleagues are falling ill whilst his detection equipment has not been challenged. The spread pattern is unpredictable. Provision of effective warning, remains a problem.

Concerns are increasing about the engineering of micro-organisms, either to defeat detection and protection or to increase toxicity and speed of effect. Defence interest centres currently on benign agents which have been altered to produce toxins, on bioregulator compounds and on 'ruggedised' agents with increased survivability in storage, on explosive dispersion and in the target environment.

What are the most likely challenges?

Although, it is worth highlighting the ease of development of BW, the military threat can effectively be reduced to those organisms which are the most robust, potent, quick-acting, easy to make, store and deliver. Whilst a distinction can be made in this respect between BW and CW agents, the edges are blurred. It is often more useful to talk about a spectrum of agents. At one extreme lie the organic chemicals such as the organo-fluorines (nerve-type agents) and, at the other: diseases. There is a range of agents in the middle which offer a mixture of drawbacks and benefits from both ends. These mid-spectrum agents comprise materials such as toxins and peptides and it is assessed that the most militarily useful groups of agents at present comprise bacteria, viruses, rickettsiae and toxins.

Bacteria are single-celled micro-organisms. They rapidly reproduce in the right conditions and many bacteria form nature's mainstay in the decomposition of waste material and the formation of soil. Man was quick to use their voracious appetites in fermentation to make alcoholic drinks and cheese. More recently, benign bio-engineering has allowed the development of organic waste disposal, dispersal of oil at sea, the industrial production of enzymes and vitamins and as a source of antibiotics to counter disease. Some species are naturally toxic to man. These cause diseases such as Cholera, Typhoid and Tuberculosis. In military terms, *bacillus anthracis* (Anthrax) is the most stable, long-lifed and toxic. It is therefore of primary concern to defence authorities. Others include Brucellosis, Plague and Tularaemia. Details of the effects and life cycles of these agents follow shortly. Agent-specific vaccines are available for these, normally as inoculations of attenuated, dead or detoxified versions of the target bacterium.

Viruses act differently. They are a group of much smaller micro-organisms which act by entering a host cell and change its instruction set (DNA) so as to reproduce the virus. Often the host cell commits suicide on instruction also. Some viruses cause the cell itself to reproduce uncontrollably. Cancers are an example of this. Viruses exhibit instability, sufficient to allow them to mutate to defeat immunisation programmes or to transfer to other species. Defence is tougher than against bacteria, relying on the stimulation of the body's immune system to react strongly enough for the virus to be defeated. Militarily significant viral agents include Encephalitis, Influenza, Rift Valley Fever and Yellow Fever.

Rickettsiae, although much larger than viruses, operate in a similar way. Not to be confused with the disease Rickets, which is caused by a deficiency of vitamin D, rickettsiae are highly toxic to man. They usually live in the gut of blood-sucking insects such as lice, fleas and ticks and infect man during the parasite's meal. They cause disease such as Typhus, Rocky Mountain Spotted Fever and Q Fever (often known to epidemiologists as Nine Mile Fever or Queensland Fever). The fevers generally respond to antibiotic treatment. Chlamydia are similar to rickettsiae, growing within living cells like viruses.

Toxins are essentially the non-living products of micro-organisms such as bacteria. Botulism, one of the most potent killers, is a disease caused by the toxin from *clostridium botulinum*. The organism produces spores that are not destroyed by cooking. It is often up to 80 per cent fatal. While occurring naturally in poorly prepared or processed food, the botulinum toxin is attractive militarily as a weapon of mass destruction. It can be purified, forming crystals which provide the physical basis for use in a weapon warhead. *Staphylococcal enterotoxin B* is another, less toxic material which nevertheless can cause epidemic illness. Many toxins are significantly more potent in causing illness than bacteria or viruses. They are stable, tough and quick-acting.

There are other types of BW agent which may have military implications, such as fungi. Of particular research interest are peptides and bioregulators — products which can be programmed to target the body's natural control mechanisms.

Defence — prospects
There are some classic general hygiene measures which reduce the effectiveness of basic BW agents. Sampling water and food for evidence of harmful agents is a sensible precaution and should be practised by all operationally deployed forces. Boiling water, careful food distribution and preparation, personal hygiene, avoidance of close contact to avoid infectious or contagious transmission of disease all play their part. However, bio-engineered agents are designed to defeat these measures and prophylaxis is necessary.

The first weapon is intelligence. By maintaining constant vigilance towards those organisations that have shown undue interest in the offensive use of BW or mid-spectrum agents, defenders can identify the potential killers. Good risk analysis will lead to targeted prophylaxis treatment.

Despite the continuing row over 'Gulf War Syndrome', coalition forces developed and distributed a carefully researched and well balanced series of measures against the key hazards thought to have been weaponised by the Iraqis. UNSCOM inspections have reinforced the view held by the coalition forces at the time that Iraq had indeed developed militarily significant quantities of both Anthrax and Botulinum Toxin.

Leading democratic nations are all engaged in programmes to research and develop prophylaxis and treatment for these hazards. For example, the US DoD has recently undertaken a US$800 million programme of mass immunisation for its forces, targeted against most of the significant viruses and toxins mentioned earlier.

Detection and warning is a huge challenge. The aim is to develop a standoff capability that offers true warning, thereby obviating the need to take full protective measures at the slightest hint of a BW incident. Some of the products listed in *Jane's Nuclear, Biological and Chemical Defence* address this issue.

The BW Convention
Finally, in this introduction, there is the need to take worldwide agreed measures for the control and destruction of BW agents, supported by intrusive inspection and tough sanction regimes. The main vehicle for this is the Biological Weapons Convention (1972). The agreement came into full force in 1972, attracting 138 of the nearly 200 nation states in the world. Conventions need teeth and the principles of control established in the agreement continue to come up against the plans of those states determined to leave their own doors open for the development of toxic weapons. The easy development of BW continues to present difficulties to inspection regimes, as it did to UNSCOM. The key to success is to keep the

monitoring process going and maintain the pressure to increase the effectiveness of policing.

Summary
The BW hazard remains a strategic threat. Its potential use during the 1990-91 Gulf War has beneficially given impetus to a number of defensive measures and raised awareness and support for better prophylaxis and treatment measures. Warning remains a problem until technology allows reliable stand-off detection. This is on the horizon and being actively pursued by a number of countries.

UPDATED

BACTERIA

Bacillus Anthracis (Anthrax or Woolsorter's Disease)

Bacillus Anthracis is a rod-shaped, gram-positive, aerobic sporulating micro-organism with the spores constituting the usual infective form.

Anthrax may appear in three forms in man; cutaneous, pulmonary and intestinal. The cutaneous or skin form is also referred to as malignant pustule. Occurring most frequently upon the hands and forearms of persons working with infected livestock, it is characterised by carbuncles and swelling at the site of infection. Sometimes this local infection will develop into systemic infection. The intestinal form, which is rare in man, is contracted by the ingestion of insufficiently cooked meat from infected animals. The pulmonary form is the type of most interest in a BW sense. Known also as Woolsorter's Disease, it is an infection of the lungs contracted by inhalation of the spores and occurs mainly among workers handling infected hides, wool and furs.

Cattle, sheep and horses are the chief animal hosts, but other animals may be infected. The disease may be contracted by the handling of contaminated hair, wool, hides, flesh, blood and excreta of infected animals and from manufactured products such as bone meal. Transmission is made through scratches or abrasions of the skin, wounds, inhalation of spores, eating uncooked infected meat or by flies.

Incubation is from one to seven days. It is usually less than four days and may be less than 24 hours in pulmonary cases.

All human populations are susceptible, although it is rare in man. When delivered as a BW weapon, the inhaled spores appear to cause disease differentially in age. Data from the 1979 Sverdlovsk incident, which was almost certainly due to the release of anthrax in BW agent form, indicates that the young (under 25) were especially resistant to inhaled pathogens. Generally, recovery from an attack of the disease may be followed by immunity.

In man, the mortality of untreated cutaneous anthrax ranges up to 25 per cent; in pulmonary cases, it is almost 100 per cent, while the rare intestinal cases are usually fatal.

Artificial active immunisation measures have been developed and are presently being evaluated for man. Anti-anthrax serum confers some passive immunity.

Cutaneous anthrax can be treated effectively with some antibiotics, including penicillin, aureomycin, terramycin and chloromycetin; sulfadiazine and immune serum. Similar treatment for respiratory and intestinal infections may be useful in the very early stages, particularly if combined with the use of immune serum, but is of uncertain value after the disease is well established.

The disease is not epidemic in man. Control of the disease is accomplished by the disposal of carcasses (by burning or deep burial) and by the decontamination of animal products.

The spores are very stable and may remain alive for many years in soil and water. They will resist sunlight for several days. Steaming under pressure or exposure to dry heat above 159°C for 1 hour are necessary to kill spores. Effective decontamination can also be accomplished by boiling contaminated articles in water for 30 minutes or by using some of the common disinfectants. Iodine and chlorine are most effective in destroying spores and vegetative cells.

As part of a BW agent system, the spores are dried and milled to 1-10 μ particle size for delivery in dried form or in a slurry medium.

Infective dose: 8,000 to 50,000 spores.

VERIFIED

Brucella Group

This group includes three closely related organisms: Brucella melitensis, Brucella abortus and Brucella suis. All are non-motile, non-sporulating, gram-negative, rod-shaped bacilli.

Brucellosis or Undulant Fever in humans, a general infection, is characterised by irregular prolonged fever, profuse sweating, chills, pain in joints and muscles and fatigue. The illness lasts for months, sometimes for years and may be caused by any one of the three related organisms. Brucella (Br) abortus is a parasite of milk cows, producing contagious abortion in cattle; the organism has also been reported in mares, sheep, rabbits and guinea pigs. Brucella melitensis is primarily a strict parasite of goats and sheep; Brucella suis is a parasite of swine. Br melitensis and Br suis are more virulent for man than Br abortus.

Brucella organisms are found in the tissues, milk and dairy products of infected goats, cattle and swine.

These diseases are transmitted to man by the ingestion of contaminated milk and other dairy products, pickled meats, uncooked foods and water contaminated by the excretions of infected animals; and by direct contact with infected animals or animal products. Infection has also occurred by inhalation and by accidental inoculation among laboratory workers.

Incubation is from 6 to 60 days or more, averaging 14 days.

Most individuals have some degree of resistance or acquired partial immunity to the abortus strains of the organism, probably from the ingestion of small doses. The susceptibility of man to Br melitensis infection ranges from 50 to 80 per cent although it may range from 75 to 80 per cent; susceptibility to Br suis appears to be approximately equal to that of Br melitensis; susceptibility to Br abortus appears to be not more than 50 per cent.

Brucellosis is prevalent in most areas where cattle, goats and swine are raised. The infection of humans occurs more often in males than in females, particularly among persons working with cows, hogs, goats and dairy products, or among those using unpasteurised milk from cows or goats.

Mortality of untreated infections is said to average 2 to 3 per cent with Br abortus and 3 to 6 per cent with Br suis and Br melitensis.

Immunisation methods are unsatisfactory for man. The immunisation of calves is effective as a control measure.

The course of the disease may be shortened by appropriate treatment with antibiotics, particularly by a combination of streptomycin and terramycin. However, some cases are resistant to all forms of therapy.

The disease is not communicable from man to man. Epidemics could result only from the wide scale consumption of contaminated, unpasteurised dairy products.

Brucella organisms will remain alive for weeks in water, unpasteurised dairy products and soil and are very resistant to low temperatures. Contaminated materials are easily sterilised or disinfected by common methods. Pasteurisation is effective for contaminated dairy products.

Infective dose: 10-100 organisms.

VERIFIED

Corynebacterium Diphtheriae (Diphtheria)

This bacterium (a slender, often slightly curved rod) is gram-positive, non-motile, non-sporulating and non-acid fast. It varies in size from 2 to 7 μm in length and from 0.5 to 1 μm in diameter. The rod-like forms are usually arranged in palisades and often exhibit club-shaped terminal swellings. They stain irregularly, displaying bars or granules as a result of irregular distribution of protoplasm within the cell. Although the organism is normally aerobic, it is often capable of anaerobic cultivation. It produces a highly potent exotoxin, both in the body and in culture.

Diphtheria, an acute febrile disease, is generally characterised by local infection, usually involving the air passages. The systemic manifestations are due to absorption of the soluble toxin into the bloodstream. The bacteria multiply rapidly in the tonsils, nose and throat, where greyish membranous patches appear on the mucous membranes, causing sore throat, swelling and stoppage of air passages. Skin and wound infections are not uncommon in tropical and subtropical climates. Early diphtheria is usually a surprisingly mild disease, unless symptoms of obstruction develop. During the first few days of infection the throat is not particularly sore, there is only slight fever and there are no severe constitutional symptoms. This lack of obvious symptoms is characteristic of diphtheria in the adult and is especially dangerous when infection occurs in the nasal passages. This is because the infection is not recognised or treatment is not begun until sufficient exotoxin has been absorbed to cause irreparable damage to other parts of the body.

Discharges from the nose and throat of infected persons and healthy carriers or from skin lesions are sources of infection.

The disease is contracted by direct contact with patients or carriers, by droplet infection, or through articles freshly contaminated with nose and throat discharges of infected individuals.

The incubation period is usually from two to five days but may occasionally be longer.

Susceptibility is more widespread in the absence of previous contact with the organism or its toxin. In the past there was a very high percentage of immunity in the adult population because of repeated contact with usually unrecognised sources of infection. Due to the widespread practice of diphtheria vaccination during infancy, there is less frequent opportunity for natural exposure to doses sufficient to produce immunity. Therefore, the adult population is now more susceptible to the disease than in the past. This susceptibility can be accurately measured by means of the Schick test. Recovery from the disease does not necessarily result in immunity.

The disease is endemic and epidemic around the world. It is more common in temperate zones than elsewhere, during the autumn and winter. Age distribution of cases and deaths depends largely upon childhood immunisation practices.

The fatality rate is variable, depending upon the virulence of the infecting strain; among untreated cases it may range from 10 to 50 per cent. In cases receiving anti-toxin treatment, this rate is lowered to 2 to 8 per cent.

Diphtheria toxoid is extremely effective. Permanent immunity may be maintained by means of booster inoculations at regular intervals.

Diphtheria anti-toxin is effective when given promptly and in adequate dosage. Penicillin as a supplementary treatment suppresses secondary invaders, shortens the period of illness and reduces the number of convalescent carriers.

Epidemicity is high, depending on the immunity status of the population and degree of exposure to the disease. A large proportion of the cases occur in children under five years of age.

The diphtheria organism is more resistant to light, drying and freezing than most non-sporulating bacilli, remaining viable for a long time in air and dust. It is capable of surviving many hours on a cotton swab and has been cultured from dried bits of

diphtheritic pseudo-membrane after 14 weeks. It can be destroyed by ordinary antiseptics and by being boiled for 12 minutes or being heated to 75°C for 10 minutes.

VERIFIED

Malleomyces Mallei (Glanders)

This organism is a slender, non-motile, non-sporulating, gram-negative, aerobic, rod-shaped bacterium.

Glanders, an infection occasionally communicated to humans, is characterised by nodular, ulcerative lesions of the skin, mucous membranes and viscera. It is an acute or chronic disease mainly of horses, mules and asses, communicable to dogs, goats and sheep. The acute form is limited to the nasal mucosa and upper respiratory tract; the chronic form, called Farcy, is characterised by Farcy buds, ulcers and pus-forming lesions in the joints and muscles.

Infected horses, mules and asses are sources of infection.

Transmission is usually made by droplet infection (inhalation) or through breaks in the skin; it is sometimes made through the gastro-intestinal tract.

Incubation is from three to five days.

Humans are highly susceptible to Glanders and the disease does not confer immunity against a second attack.

Glanders is prevalent among horses, mules and asses in the Balkans, Russia, Southeast Asia and India. It is uncommon elsewhere.

In untreated cases, the acute form has a mortality of nearly 100 per cent, while mortality of the chronic form ranges from 50 to 70 per cent.

Satisfactory immunisation procedures have not been developed.

Prompt and radical surgery is the most effective treatment for the chronic form. Sulfadiazine and streptomycin are reported to be effective. The value of newer antibiotic drugs is not yet clearly established.

Although the disease is contagious, an epidemic spread in man is improbable.

The organism resists drying for two or three weeks but is killed by direct sunlight in a few hours. It may remain alive in decaying matter for two to three weeks. It is easily killed by the common disinfectants and by being heated at 72°C for 10 minutes.

VERIFIED

The UK Prototype Biological Detection System (PBDS) is now in development (Hunting Engineering)   2001/0050701

Malleomyces Pseudomallei (Melioidosis)

This bacterium is motile, non-sporulating, gram-negative, aerobic, rod-shaped and small (1 to 2 μm long and 0.5 μm wide). It is often marked by bipolar staining and closely resembles Malleomyces Mallei.

Melioidosis, also known as Whitmore's Disease, is a Glanders-like disease primarily of rodents but occasionally found in man. It tends to run a more rapid course than Glanders and, in man, is almost always acute and rapidly fatal, death occurring usually in three to four weeks, often within 10 days. Reports of chronic infections involving lungs and lymph glands, bones, joints and legs have been made. The disease is characterised by the sudden onset of severe chills, high fever, rapid prostration, headache, muscle and joint pains, coughing, laboured breathing, nausea and vomiting. In a short time, numerous small abscesses form in the skin, bones, lymph nodes, lungs and other internal organs.

Probable sources of infection are food or other materials contaminated with rodent excreta and possibly rat fleas. Transmission takes place apparently by the ingestion of food contaminated with excreta from infected rats and by rat flea bites.

Although not accurately known, the incubation period is probably only a few days.

Susceptibility is general but the organism is apparently not extremely infective under natural conditions, considering the low prevalence of human disease in endemic areas. However, man appears to have little or no resistance to it once infection is established.

Cases of the disease have been found chiefly in Malaysia, Southeast Asia and Sri Lanka. They have also been reported in Guam, the Philippine Islands and the USA.

Acute Melioidosis is usually fatal.

No vaccine has been developed. Little is known of any immunity acquired through infection.

Chloromycetin alone or in combination with terramycin or aureomycin has been used with marked success. Sulfadiazine is also effective in the acute stage.

The disease is not normally contagious. Spreading by droplet infection might occur in a cold climate, which is more suitable to this type of transmission.

The organism is extremely resistant to drying and may survive a month or more in dried soil, in excreta and in water. It is easily killed in 10 minutes by 1 per cent phenol or 0.5 per cent formalin and by moist heat at 74°C.

VERIFIED

Mycobacterium Tuberculosis (Tuberculosis)

Tubercle bacilli are slender, straight or slightly curved rods with rounded ends. They vary from 0.2 to 0.5 µm in width and from 1 to 4 µm in length. They are acid-fast, non-motile, gram-positive and are strictly aerobic.

Pulmonary Tuberculosis is characterised by severe lung involvement accompanied by coughing, fever, fatigue and loss of weight. This form of the disease is the chief cause of morbidity and mortality. The primary type is acute, healing or progressing in a relatively short time. It is most commonly seen in infants and children and occasionally in adults who have escaped childhood infection. Adults acquiring their first infection may manifest the post-primary type, passing through the primary phase inconspicuously because of its rapid development. The post-primary (reinfection) type is more stable and more chronic than the primary type and is associated with a significant, but inadequate, degree of resistance. Tuberculosis infection in the bones, joints, skin, or other tissues is usually caused by the bovine variety of M. Tuberculosis, although this type may also invade the lungs.

Infection is acquired from persons with draining lung cavities. Tuberculous cattle and particularly their raw milk, are the source of the bovine variety.

Transmission usually occurs through the discharges of the respiratory tract, by direct or indirect personal contact. Primary infection is almost always a result of inhalation of the bacilli in droplet form. The post-primary type of Tuberculosis may be caused either by organisms which have survived in primary lesions or by newly inhaled bacilli. The bovine type is acquired by contact with tuberculous cattle or ingestion of their raw milk but can be transmitted also by the same route as the human type, from person to person. Natural infection usually requires continued and intimate exposure.

This period is variable, depending on dosage, age and other factors, but probably is not less than one month. The period may be reduced considerably by exposure to heavy concentrations of the organism.

Susceptibility to the disease is dependent upon age, race, family characteristics and previous exposure to the organism. It is lowest in persons from 3 to 12 years of age and is greater in the undernourished, the elderly and fatigued and among peoples who have not previously been exposed to the disease. People suffering from immune deficiency disease such as HIV are particularly susceptible. In the USA the rapidly progressive primary type is seen more often in young adult Negroes than in Caucasian adults. Recovery from the disease leaves no solid immunity, but resistance is altered so that reinfection is not as acute as the previous infection. The chance of contracting progressive clinical Tuberculosis is higher in tuberculin-negative than in tuberculin-positive individuals.

Tuberculosis is one of the most common of the infectious diseases of man. It occurs in all parts of the world, although it has never appeared in some isolated groups of people. Infection is widespread in the urban population, as is shown by the 50 to 95 per cent tuberculin reactor rate, but the progressive disease develops in only a small proportion of those infected.

The untreated fatality rate is about 10 per cent. There are 30 million cases worldwide and the disease is advancing, despite the efforts of the World Health organisation.

Despite the many improvements in case-finding and the newer therapeutic procedures, Tuberculosis is still among the 10 leading causes of death in countries such as the USA. The recent outbreak of the disease in the UK is thought to have been transmitted by returning pilgrims who had become infected in the Middle East.

Vaccination of tuberculin-negative persons with living strains of attenuated tubercle bacilli of Calmette and Geurin (BCG) confers some protection against naturally acquired tuberculous disease (primary and post-primary). However, prevention and control of Tuberculosis depends largely upon detection and isolation of carriers and the general improvement of environmental and economic conditions.

Streptomycin, particularly when combined with para-aminosalicylic acid (PAS) or isonicotinic acid hydrazine (INH), is valuable in arresting the disease, particularly in its acute manifestations, but this treatment in itself is not curative. Success in treatment depends upon supportive therapy, including rest, good food (supplementing the diet with vitamin A, C and D) and fresh air. Local rest of the lungs is promoted by several methods. Surgery is required in some cases.

The spread of Tuberculosis occurs in large part through continued family or household case association, the disease being transmitted slowly from one generation to the next. Under favourable conditions, epidemic outbreaks may take place. Occurrence is influenced by occupation, such as continued exposure to mineral (silica) dust, which predisposes to infection.

When organisms are exposed to direct sunlight, in artificial culture they are killed in 2 hours but in sputum under the same conditions they may survive 20 to 30 hours.

When organisms are protected from the sun, they will live in putrefying sputum for weeks and in dried sputum for as long as six to eight months. They are resistant to the usual chemical disinfectants, 24 hours being required for the decontamination of sputum by 5 per cent phenol, but possess no greater resistance to moist heat than other bacteria, being killed in 15 to 20 minutes at 78°C. Pasteurisation is effective in destroying the organisms in milk.

VERIFIED

Pasteurella Pestis (Plague)

Pasteurella Pestis is a rod-shaped, non-motile, non-sporulating, gram-negative, aerobic bacterium.

Plague, or Black Death, occurs as three clinical types in man: bubonic, pneumonic and septicaemic. Another type of plague, Sylvatic Plague, is an infectious disease of wild rodents; it is transmissible to man by flea bites. In general, Plague is characterised by a rapid clinical course with high fever, extreme weakness, glandular swelling, pneumonia and/or haemorrhages in the skin and mucous membranes.

(1) Bubonic Plague, the most common, is transmitted to man by the bite of an infected flea, the disease being perpetuated by the rat-flea-rat transmission cycle. The flea bites are usually on the lower extremities where the bacilli spread rapidly through the lymphatic system, enlarging the lymph nodes (buboes) in the groin. The bacilli escape from the nodes, invade the bloodstream and produce a generalised infection. Other parts of the body affected are the spleen, lungs and meninges.

(2) Pneumonic Plague, transmitted by inhalation, spreads rapidly until the entire lung is involved in a haemorrhagic, pneumonic process. The disease is usually fatal, the patient dying of suffocation and/or general toxaemia.

(3) Septicaemic Plague occurs as the result of gross invasion of the bloodstream by Plague bacilli, which cause small haemorrhages in the skin and mucous membranes. Death occurs before buboes or pulmonic manifestations appear.

(4) Sylvatic Plague, transmitted by wild rodent fleas, is somewhat different from the Plague transmitted by the rat flea (Bubonic). The flea bites usually occur in the upper extremities and buboes originate in the armpits rather than in the groin. It frequently changes over to the pneumonic type.

Infected rodents and human patients with Pneumonic Plague are sources of infection. The primary source of the disease is Plague endemic in wild rodents, including the ground squirrel, pack rats and harvest mice of the USA and various species of similar wild rodents in other parts of the world. Infection may reach man from these sources or more often through the medium of the domestic rat.

Pneumonic Plague is usually transmitted directly from man to man by droplet infection. Bubonic Plague is generally transmitted by the bites of fleas from infected rats and other rodents.

Incubation is from one to seven days for Pneumonic Plague, four to seven days for Bubonic Plague.

Susceptibility is general, particularly to the Pneumonic form. Recovery is followed by temporary, relative immunity.

The disease is rare in North America and island possessions of the USA. Occasional cases of the bubonic type occur west of the Mississippi River from bites of fleas from infected wild rodents. The disease has foci of infection in various parts of the world, particularly in Asia.

Untreated Bubonic Plague has a mortality of 30 to 60 per cent, while untreated Pneumonic Plague kills from 90 to 100 per cent of its victims.

Anti-Plague serum produces an artificial passive immunity of two weeks' duration. Active immunisation with killed bacterial vaccines is protective for some months when administered in two or three doses at weekly intervals, repeated stimulating doses being necessary. According to some authorities, vaccines prepared from living virulent strains confer a better and longer immunity than vaccines from other sources.

Prompt treatment with sulfonamides and streptomycin (chloromycetin and aureomycin may be used if resistance to streptomycin develops) combined with serum therapy is essential and is effective if used early. Supportive treatment for pneumonic and septicaemic forms is required. These measures shorten the duration of the disease and reduce its mortality.

Bubonic Plague is not directly communicable from man to man, but Pneumonic Plague is intensely communicable during the acute period. Strict area quarantine and sanitation, in addition to other measures such as rat flea extermination (DDT), are essential to control outbreaks.

The organism probably will remain viable in water from 2 to 30 days and in moist meal and grain for about two weeks. At near freezing temperatures, it will remain alive from months to years but is killed by 15 minutes exposure to 72°C. It also remains viable for some time in dry sputum, flea faeces and buried bodies but is killed by 3 to 5 hours exposure to sunlight. Decontamination is effected by boiling, use of dry heat above 72°C or steam and, treatment with lysol or chloride of lime.

Infective dose: 100-500 organisms.

VERIFIED

Pasteurella Tularensis (Tularemia, Rabbit or Deer-Fly Fever)

This is a small, aerobic, gram-negative cocco-bacillus, often varying in size and shape. It is non-motile and non-sporulating.

Tularemia is also known as Rabbit Fever and Deer-Fly Fever. It is a fatal septicaemic (blood-poisoning) disease of wild rodents, accidentally communicable to man in whom it is characterised by a sudden onset of chills, fever and prostration and by a tendency to pneumonic complications. It is an acute, severe, weakening disease, later becoming chronic. It may be accompanied by enlargement of the regional lymph glands with or without a lesion at the site of infection, or by typhoid-like symptoms with no local lesion or enlargement of the local lymph glands.

Wild rabbits or hares, deer flies, ticks and many other animals (including the woodchuck, coyote, opossum, tree squirrel, skunk, cat, deer, fox, hog, sage hen and some snakes) are sources of infection.

Transmission is made by infection through the skin, eyes, or lungs from handling infected animals, as in skinning or dressing the animals or performing autopsies; by bites of infected flies and ticks; by eating insufficiently cooked rabbit meat; or by drinking contaminated water. Laboratory infections are not infrequent.

Incubation is from one to ten days, usually about three days.

All ages are susceptible and recovery from an attack is followed by permanent immunity. The infectivity rate is from 90 to 100 per cent.

The disease is present throughout North America and in many parts of continental Europe and Japan. It occurs in every month of the year in the USA.

Untreated cases have a death rate of 4 to 8 per cent, averaging 5 per cent.

Vaccination greatly reduces the severity of the disease and may prevent infection in some cases.

Antibiotics, particularly streptomycin, aureomycin and chloromycetin, are effective.

The disease is essentially sporadic, but may be epidemic when modes of transmission (see above) are prevalent. It is not transmitted directly from man to man.

The organism remains viable for weeks in water, soil, carcases and hides and for years in frozen rabbit meat. It is resistant for months to temperatures of freezing and below. It is easily killed in 2 to 3 minutes at 38°C or above and by 0.5 per cent phenol in 15 minutes.

Infective dose: 10-50 organisms.

VERIFIED

Salmonella Paratyphi and Salmonella Schottmuelleri (Paratyphoid)

The organisms are short, plump, rod-shaped, motile, non-sporulating, gram-negative bacteria. S. paratyphi is known as type A of the group; S. Schottmuelleri is also known as S. paratyphi B.

Paratyphoid Fever (an acute, febrile, generalised infection) is very similar to Typhoid Fever (indistinguishable clinically), but its symptoms are usually milder. It is characterised by continued fever, severe diarrhoea and abdominal pain, with involvement of the lymphoid tissues of the intestines, enlargement of the spleen and sometimes rose-coloured spots on the trunk. S. Schottmuelleri (type B) is responsible for more cases of the disease than type A and may also produce Gastroenteritis. (Salmonella Hirschfeldii (S. Paratyphi C) may also produce Paratyphoid Fever.)

Ready for anything, an EOD operative fully protected by an impermeable NBC suit and an Interspiro Spiromatic breathing apparatus – although such complete measures are not normally necessary to protect against biological agents, some form of overall protection must be worn during a BW attack

The contaminated faeces and urine of patients and carriers are sources of infection.

Transfer of organisms is the same as for Typhoid.

Incubation is variable, from 1 to 10 days, depending on the strain of organism but averaging less than a week.

Susceptibility is general and recovery is followed by permanent immunity.

The disease is worldwide, but incidence has declined with that of Typhoid. Outbreaks are sporadic or limited and are due to contact with or consumption of contaminated foods such as milk or water. There are probably many unrecognised cases. Paratyphoid Fever caused by S. Hirschfeldii has been found relatively frequently in parts of Asia, Africa and southeast Europe.

Fatalities are low, perhaps between 1 and 2 per cent.

Paratyphoid vaccine for types A and B is usually incorporated with typhoid vaccine.

Chemotherapy with 'sulfa' drugs and use of antibiotics, such as chloromycetin, aureomycin and streptomycin, shorten the period of communicability and hasten cure of the disease.

The epidemicity is similar to that for Typhoid, depending on presence of carriers; inadequate sanitary controls for water, food and milk supplies and absence of immunisation.

Stability is the same as for S. typhosa. Decontamination measures include chlorination, pasteurisation, boiling and cooking.

VERIFIED

Salmonella Typhimurium (Salmonella)

This bacterium is a short plump rod which occurs singly and measures 0.5 µm in width and from 1 to 1.5 µm in length. It is gram-negative, non-sporulating and motile.

Salmonella food-poisoning (gastroenteritis) is most frequently caused by S. typhimurium in man. The onset of the infection is nearly always sudden, characterised by headache, chills and usually by abdominal pains. This is followed by nausea, vomiting, severe diarrhoea with a rise in temperature and prostration. Recovery is usually complete within two to four days.

The sources of infection are usually rodents, especially rats and mice, human carriers who handle eggs and meat from diseased animals.

The disease is usually obtained by the ingestion of contaminated food (particularly meat), water or milk. It may also be obtained by direct contact with infected persons or carriers, by direct contact with articles contaminated by discharges (faeces, urine and vomitus) of infected persons or carriers, or from flies.

Food-poisoning occurs after an incubation period ranging from 6 to 24 hours but seldom after more than 48 hours. The short interval suggests that large numbers of the organisms are usually ingested.

Susceptibility is general. The disease is more severe in infants and young children than in adults. Natural immunity is believed to exist in some persons, while acquired immunity is usually permanent after recovery from the disease. (The bacterium is primarily pathogenic for animals; in mice it produces a Typhoid-like disease with high mortality.)

The disease is widely distributed geographically and occurs in almost all warm-blooded animals.

Fatalities range from 1 to 2 per cent in epidemics.

Vaccination is not practical.

Treatment is mainly physiologic. A saline purge is indicated if the infected food has not been eliminated by nature. The restoration of fluid balance is most important, particularly in the very young and old. Streptomycin, chloromycetin, terramycin and aureomycin reduce the number of organisms in the intestinal tract.

The infection is contagious. Epidemics usually occur when the mass consumption of contaminated food occurs. Explosive epidemics occur in animals, particularly rodents, and many surviving animals become chronic carriers. Spread of the infection can be halted by the elimination of carriers as food handlers, proper sanitation, pasteurisation of milk, elimination of rodents and flies where food is prepared, careful handling and adequate cooking of food.

VERIFIED

Salmonella Typhosa (Typhoid)

This organism is a rod-shaped, motile, non-sporulating, gram-negative bacterium. It is also known as S. typhi.

Typhoid Fever is a systemic infection characterised by continued fever, lymphoid tissue involvement, ulceration of the intestines, enlargement of the spleen, rose-coloured spots on the skin, diarrhoea and constitutional disturbances.

The faeces and urine of infected individuals and carriers are sources of infection.

Transfer of organisms is made through the alimentary tract by direct or indirect contact with a typhoid patient or a chronic carrier, by consumption of contaminated water, food, milk or shellfish and by flies.

Incubation is from 3 to 38 days, usually 7 to 14 days.

Susceptibility is general, except that some adults have an acquired immunity from unrecognised infection. Recovery is usually followed by permanent immunity.

The disease is widespread throughout the world. Once endemic in most large cities of North America, it has been steadily falling in incidence, particularly in areas supplied with safe water and pasteurised milk and where modern sewage disposal facilities are used. It is still endemic in some rural areas of the USA, usually as sporadic cases or in small carrier or contact epidemics.

The mortality in untreated cases ranges from 0 to 10 per cent.

Inoculation with typhoid vaccine produces an artificial active immunity of about two years' duration. High protection lasting for about a year can be maintained by an annual booster injection of vaccine.

Prompt use of appropriate antibiotics (chloromycetin, aureomycin) shortens the period of communicability and rapidly cures the disease.

Epidemicity is high in the presence of carriers; in the absence of sanitary control for water, food and milk supplies; and where individuals are not protected by immunisation.

The organism remains viable for two to three weeks in water, up to three months in ice and snow and for one to two months in faecal material. Pasteurisation, exposure to 73°C for 20 minutes, exposure to 5 per cent phenol or 1:500 bichloride of mercury for 5 minutes, cooking and boiling are effective decontamination measures.

VERIFIED

Shigella Dysenteriae (Dysentry)

Shigella dysenteriae is a rod-shaped, gram-negative, non-motile, non-sporulating bacterium.

Bacillary dysentery, an infectious disease of man, is characterised by mild or severe irritation of the lower gastro-intestinal tract; it is usually accompanied by fever, abdominal pain, diarrhoea, weakness or prostration and ulceration of the mucous membranes of the intestine.

Faeces of infected human patients and carriers are sources of infection.

Transmission is made by the ingestion of contaminated food, water, or milk; by hand-to-mouth transfer of contaminated material soiled with faeces of a patient or carrier; or by flies.

Incubation is from one to seven days, usually less than four days.

Most persons are susceptible, but the disease is more common and severe in children than in adults. Recovery from the disease is followed by a relative, transitory immunity. Washing, ordinary sterilisation methods and use of some of the common disinfectants are effective decontamination measures.

The disease is endemic throughout the world and epidemics or sporadic outbreaks occur where sanitation is lacking or inadequately applied or enforced, particularly in relation to sewage disposal, food handling and preparation and infant hygiene. Outbreaks are most common during the summer months and occur frequently in large institutions.

Mortality is variable, ranging from 2 to 20 per cent in untreated cases, depending on the particular strain of organism.

Beneficial results may be obtained from treatment with sulfadiazine, terramycin, chloromycetin and aureomycin.

The disease is highly contagious, particularly in insanitary conditions. Control measures include rigid sanitation (careful handling of excreta and adequate sewage disposal) and fly control.

The dysentery organisms remain viable for considerable periods in water, ice and mucous discharges but are readily killed by sunlight. Sterilisation by steam and common disinfectants are effective decontaminants.

VERIFIED

Vibrio Cholera

This micro-organism is a short, slightly bent, motile, gram-negative, non-sporulating rod.

Cholera, an acute infectious gastro-intestinal disease of man, is characterised by sudden onset of nausea, vomiting, profuse watery diarrhoea with 'rice-water' appearance, the rapid loss of body fluids, toxemia and frequent collapse.

Faeces and vomitus of patients, faeces of convalescents and temporary carriers are sources of infection.

Transmission is made through direct or indirect faecal contamination of water or foods, by soiled hands or utensils or by flies.

Incubation is from one to five days, usually three days.

All populations are susceptible, while natural resistance to infection is variable. Recovery from an attack is followed by a temporary immunity which may furnish some protection for years.

Endemic centres exist in India and Southeast Asia, from where the disease may spread along human communication lines to more remote countries and cause epidemics. It is normally absent from the Western Hemisphere.

The mortality rate ranges from about 3 to 30 per cent in treated cases to 50 per cent in untreated cases.

Artificial immunisation with vaccines is of variable degree and uncertain duration (6 to 12 months). Acquired immunity lasts for many years.

The first consideration in the treatment of cholera is to replenish the fluid and mineral losses of the body. Drug therapy has little or no effect upon the clinical course of the disease. However, chloromycetin, aureomycin and terramycin, given by mouth, cause rapid disappearance of the vibrio organisms, thus reducing the spread of the disease.

Epidemicity is very high under insanitary conditions, especially those concerned with water supplies, foods and fly control.

The organism is easily killed by drying. It is not viable in pure water, but will survive up to 24 hours in sewage, and as long as six weeks in certain types of relatively impure water containing salts and organic matter. It can withstand freezing for three to four days. It is readily killed by dry heat at 117°C, by steam and boiling, by short exposure to ordinary disinfectants and by chlorination of water.

VERIFIED

FUNGI

Coccidioides Immitis (Valley or San Joaquin Fever)

In man and animals, this fungus occurs as thick-walled endospore-filled spherules, 20 to 80 µm in diameter; in artificial culture, it appears as a fluffy white cotton-like mould.

Coccidioidomycosis is a highly infectious disease. The usual primary form (known as Valley or San Joaquin Fever) is an acute, disabling, self-limiting respiratory infection resembling influenza, usually with a low grade fever of 35 to 36°C and a slight cough. The secondary, progressive form (known as Coccidioidal Granuloma) is a chronic, malignant, disseminated infection which involves any and all organs of the body, including the skin and bones and produces numerous abscesses. The progressive form has a high fatality rate. A primary localised form of infection may occur on the exposed surfaces of the skin; it sometimes develops into the progressive, disseminated form of infection.

Dust, soil and vegetation contaminated with spores of this fungus are sources of infection.

Transmission is made by inhalation of spores in dust from soils and dry vegetation and possibly through skin scratches or wounds.

The incubation period for the primary pulmonary form is 10 to 21 days, the average being about 12 days. The clinically active cases in about 1 per cent of Caucasians and a much higher percentage of Negroes develop into the progressive form of the disease, usually within a period of weeks or months after onset of the primary infection. The progressive form is not necessarily preceded by symptoms of primary infection.

Susceptibility to the primary infection is general. With the progressive form of the disease, Negroes and Orientals appear to be much more susceptible than Caucasians and females are less apt to develop granulomatous lesions than are males. Probably up to 60 per cent of infected persons have no symptoms but become immune.

Recognised endemic areas are the arid southwestern USA, northern Mexico, Central America and the Chaco region of South America, involving Argentina, Bolivia and Paraguay. Incidence is at its highest in hot, dry weather.

Fatalities are about 50 per cent in the secondary progressive form.

Vaccine therapy is only in the experimental stage of development and prospects for a good vaccine seem poor.

Supportive treatment, with complete rest, is the standard procedure. Progressive coccidioidomycosis is very resistant to specific chemotherapy, but some cases have shown improvement under treatment with prodigiosin, stilbamidine or ethyl vanillate. Localised lesions may be treated by X-ray or surgery. Prognosis is excellent in the primary pulmonary infection, good in the primary cutaneous type and very poor in the disseminated, progressive disease, especially in the Negro.

The disease is non-contagious. Small epidemics may occur in hot, dry seasons when large numbers of individuals, such as military units, are stationed in endemic areas or engaged in manoeuvres in these areas. Dust control, such as paving roads and runways, oiling sports fields and planting lawns in endemic areas is effective in reducing the number of infections.

Spores of the fungus are highly resistant to drying and will live for months or years in culture or in the soil (dust). They are destroyed by autoclaving for 15 minutes or by exposure to formaldehyde fumes for 48 hours. Decontamination of large areas, even if possible, would not appear to be practicable.

VERIFIED

Histoplasma Capsulatum (Histoplasmosis)

The fungus appears as small (1 to 5 µm), oval, yeast-like intracellular bodies in the tissues of man and animals and in cultures on sealed, blood agar slants at 53 to 54°C. In culture at room temperature, it forms a white, cotton-like, filamentous mould-like growth.

Histoplasmosis is a chronic, local or systemic, infectious disease of man and animals. It is characterised by low grade granulomatous lesions of the skin or mucous membranes and/or Tuberculosis-like lesions of the lungs and by involvement of internal organs, especially the spleen and liver. Less than 1 per cent of naturally occurring primary infections develop into the progressive and usually fatal form of the disease.

Dust contaminated with spores of this fungus is a source of infection. The agent has been recovered from man, dogs, cats, rodents, skunks, opossums and from the soil and air.

Transmission is usually by inhalation of spores in dust from soils and dried organic matter; it may also be transmitted by ingestion or through skin scratches.

In the few reported epidemics, symptoms appeared within 5 to 18 days following exposure.

Susceptibility is general. The occurrence and severity of clinical symptoms depend upon the dosage of infectious agent. The extreme infectiousness of the disease is shown by a positive histoplasmin skin test in 80 per cent or more of the adult population in local endemic areas. The effect of dosage is illustrated by essentially 100 per cent epidemic involvement among groups exposed to highly contaminated dust in enclosed places.

Histoplasmosis is endemic in the USA in regions including the central Mississippi and Ohio River valleys and in other scattered areas including the St Lawrence River valley, eastern North Carolina and parts of Mexico and Panama. However, cases have been reported from many other parts of the world, including South America, Europe, Africa, Australia and the Pacific islands.

The acute pulmonary form of the disease has a very low fatality rate unless it disseminates. The progressive form is usually fatal.

Immunisation has not been developed.

Supportive treatment is used. Ethyl vanillate has been effective in many cases. Among other promising drugs being developed is hydroxystilbamidine. Surgery is curative in localised lesions.

The disease is non-contagious.

Spores are viable for several months, probably years in dry soil and for a period of months in tap water at temperatures ranging from near freezing to 54°C. They are destroyed by heat at 73°C for 15 minutes and by 1 to 2 per cent formalin in 24 hours.

VERIFIED

Nocardia Asteroides (Nocardiosis)

This aerobic fungus has characteristics of both moulds and bacteria and has been classified in an intermediate position. It usually forms orange colonies consisting of gram-positive, weakly acid-fast, branching filaments, 1 μm or less in diameter. In both culture and tissues it tends to fragment readily into bacillary and coccoid forms. The acid-fast rod-like fragments may be mistaken for tubercle bacilli.

Nocardiosis, a severe pulmonary infection, is similar in many respects to Tuberculosis but tends to form numerous abscesses instead of tubercules and is characterised by chronic pneumonia. It tends to spread to other organs of the body, especially the brain, where abscesses are formed. The disease often takes other forms, such as localised subcutaneous abscesses and tumours. Pulmonary infection is characterised by a general malaise, fever, a productive cough, night sweats, loss of appetite and loss of weight. Brain involvement (tumour or abscess) presents symptoms of headache, nausea and vomiting.

Soil, dust or vegetation contaminated with the organism are the main sources of infection.

The disease is transmitted by contaminated dust and possibly by droplet infection or through pus and other discharges from infected individuals. Skin infections usually result from contamination of wounds or scratches.

The incubation period in man is unknown; experimental infection in guinea pigs is usually fatal within a week.

Susceptibility to pulmonary infection is probably general and is dependent to a great extent upon dosage. Special predisposing factors may be required for production of progressive infection, since the organism is apparently widely distributed, but relatively few cases are reported. It is not known whether recovery provides any immunity.

The disease is worldwide.

The death rate is very high in untreated or advanced cases of generalised infection, perhaps close to 100 per cent. Early diagnosis and vigorous treatment with sulfonamides result in the cure of most cases.

Active immunisation in man has not been tried; in animals, experimental immunisation has produced an active protective immunity.

The sulfonamides, particularly sulfadiazine alone or combined with sulfamerazine, offer effective treatment. Supportive treatment is essential.

The disease is not known to spread from man to man or from animal to man, but infection by droplet spray or through the infection of wounds by contaminated dressings and other material is possible.

As the fungus exists normally in soil, it is presumed to be quite resistant to adverse environmental conditions.

VERIFIED

RICKETTSIAE

Coxiella Burneti, (Rickettsia Burneti) (9-Mile, North Queensland or 'Q' Fever)

This rickettsia is a bacterium-like, gram-negative organism which may vary in size and shape from a lanceolate rod 0.25 by 0.5 μm to a diplobacillus 0.25 by 1.5 μm. As the small forms will pass through bacterial filters, they are considered by some to be viruses.

Q-Fever, also known as Nine Mile Fever and North Queensland Fever, is an influenza-like disease that is moderately incapacitating and is characterised by an acute fever of sudden onset, headache, chills, weakness and severe perspiration. Pulmonary involvement occurs in the majority of cases, accompanied by a mild cough, scanty expectoration, and chest pains. Q-Fever is distinguished from other rickettsial diseases in its failure to cause a skin rash.

Cows, sheep, goats and wild animals appear to be natural reservoirs, with numerous ticks transmitting the disease among them. The organism may be found in the milk and mammary glands of infected cows, in the milk of sheep and goats and in the dust-laden air of dairy cattle barns and goat pens which harbour infected animals.

In man the disease appears to be transmitted by the inhalation of infected dust from one source or another as well as by ingestion. Raw milk from cows and goats, dried milk, raw wool, hides, infected meat, goat hair and tick faeces, as well as cultures of infected tissues, have been involved in infections.

The incubation period is from 14 to 26 days with a mean of 19 days.

Susceptibility is general. An attack confers immunity of long duration.

The disease has been reported in many parts of the USA, particularly in California; it has also appeared in Australia, the Mediterranean area, Europe including the Balkans, Spain and the UK. Also, Panama and scattered areas of Africa.

Fatalities are rare, but may be up to 4 per cent during epidemics.

Vaccines have been effective when used by laboratory personnel, slaughterhouse and stockyard workers.

Appropriate antibiotics (aureomycin, chloromycetin, and terramycin) may be effective, although they have not been completely evaluated. Supportive treatment is indicated.

The disease is relatively non-contagious. Outbreaks have occurred among slaughterhouse and stockyard workers in the USA.

The micro-organism is resistant to temperature changes from +40 to -52°C, 0.5 per cent phenol, and is relatively resistant to desiccation. It is killed by 0.5 per cent formalin. It probably persists on surfaces from 5 to 60 days.

VERIFIED

Rickettsia Mooseri (Rickettsia Typhi) (Murine Typhus)

This rickettsia is similar to Rickettsia Prowazeki, but with less variation in appearance.

Murine (endemic, rat, or flea-borne) typhus is similar to classic epidemic typhus except that the disease is milder and has a slower onset.

Infected rodents, particularly rats, are sources of infection. The disease is transmitted from rodents to man by the bite of the rat flea.

The incubation period ranges from 6 to 14 days, usually 12 days.

Susceptibility is general. One attack confers immunity, which sometimes is not permanent.

The disease is widely distributed in temperate, subtropical and tropical countries. It is transmissible to man throughout the year, with an increase during Summer and Autumn.

Case fatality rate is about 2 per cent, increasing in older people (over 70 years of age).

Vaccine similar to that for classic epidemic typhus is probably equally effective.

Treatment is similar to that for classic epidemic typhus. Delayed recognition of the disease may result in the optimum period of treatment being passed.

The disease is not contagious. Rodent and flea eradication are the most successful control measures.

Stability is the same as for classic epidemic typhus.

VERIFIED

Rickettsia Prowazeki (Typhus)

This micro-organism is a non-motile, minute, coccoid or rod-shaped rickettsia, occurring sometimes in pairs or chains, with a diameter of about 0.3 μm and with varied sizes and shapes.

The disease is known as a classic epidemic (human or louse-borne) typhus. It is an acute infectious disease of man and is characterised by severe headache, sustained high fever, general pains and a skin rash.

Infections are acquired from persons with the disease.

Transmission is mainly by body lice which have fed upon infected persons. The individual becomes infected usually by scratching or rubbing louse faeces into a wound made by the louse bite; by crushing an infected louse on the skin; by rubbing louse faeces into the eyes. It is possible that louse faeces from dirty clothing may be transmitted through the air to the respiratory tract.

Incubation is from 6 to 15 days, averaging 12 days.

All peoples are susceptible. One attack confers immunity, which is not always permanent.

Mortality is from 10 to 80 per cent, varying in different epidemics and with the age of individuals. Some vaccines confer considerable protection of uncertain duration; immunisation should be repeated every four months where the danger of typhus is present. The vaccine reduces the risk of infection, modifies the course of the disease and lowers the mortality.

The course of the disease can be shortened by use of appropriate chemotherapy and antibiotics (chloromycetin, aureomycin, terramycin). Supportive treatment and prevention of secondary infections are essential.

Epidemics usually occur in winter under crowded and insanitary conditions, particularly during famine and war and when the population is heavily infested with body lice. Louse eradication is used to control epidemics.

The organism is destroyed by heat at 52°C in 15 to 30 minutes and inactivated by use of 0.1 per cent formalin and 0.5 per cent phenol.

VERIFIED

Rickettsia Rickettsii (Rocky Mountain Spotted Fever)

The rickettsia is a diplococcus-like micro-organism, with the distal ends tapered so that the pairs resemble minute pneumococci. The average is 1 μm in length and 0.2 to 0.3 μm in width.

Rocky Mountain Spotted Fever (Sao Paulo Fever in South America), an acute infectious disease, is characterised by fever and joint and muscular pains. A skin rash usually appears on the third or fourth day, rapidly spreading from the ankles and wrists to the legs, arms and chest.

Infected ticks such as the common dog tick in eastern and southern USA, the wood tick in the northwestern USA, and occasionally, the lone star tick in the southwest, are sources of infection. In other countries the vectors may be other species of ticks. The infection is passed from generation to generation in ticks and is probably maintained by larvae feeding on susceptible wild rodents.

The disease is transmitted to man by the bite of an infected tick, by contamination of abraded skin with infected tick tissues or faeces, or by contact of tick tissues with unbroken skin.

The incubation period is from 3 to 10 days.

Susceptibility is probably general; recovery from an attack is followed by immunity which may or may not be permanent.

The disease occurs throughout North America and in some parts of South America. Predominant occurrence is in the Spring and early Summer, when adult ticks appear.

The fatality rate may vary from 20 to 60 per cent, depending on the locality.

An effective egg-yolk vaccine is available, which lessens the chance of infection and lowers the mortality rate. Yearly booster doses are needed to attain maximum protection.

Appropriate antibiotics (chloromycetin, aureomycin, and terramycin) are effective in reducing fatality and shortening the course of the disease. Supportive treatment also is indicated.

The disease is not communicable from human to human. Tick eradication and preventive immunisation are the best control measures.

The rickettsia can be killed by exposure to a temperature of 62°C for 10 minutes and by drying for 10 hours. It is inactivated by 0.1 per cent formalin and 0.5 per cent phenol.

VERIFIED

Rickettsia Tsutsugamushi (Scrub Typhus)

This rickettsia appears as small diplococcus-like structures or short rods which are 0.3 to 0.5 µm in length and 0.2 to 0.4 µm in width.

Scrub typhus (tsutsugamushi disease or mite-borne typhus), an acute infection in man, is characterised by sudden onset with chills, fever and headache which may increase in intensity. A dull red eruption appears on the trunk of the body on the fifth to eighth day and may extend to the arms and legs. Coughing and other signs of pneumonia are also frequently present.

The rickettsia is perpetuated in infected mites, where it is passed on from generation to generation and is maintained by the mites feeding on susceptible wild rodents, particularly mice and rats of different species.

Transmission to man is made by the bite of infected larval mites.

The incubation period is from 7 to 10 days, sometimes it may be as long as 14 days.

Susceptibility is general. One attack confers immunity, which is not always permanent.

The disease occurs in Southeast Asia, including India, Burma (Myanmar), Malaysia, Vietnam, islands of the West and South Pacific, Japan, Taiwan, Indonesia, New Guinea and Australia. Transmission is limited to the summer months in Japan but may occur throughout the year near the Equator.

Mortality varies with age and locality and may go as high as 50 per cent.

There is no effective vaccine available for general use. The immunity produced in man (by scrub typhus) is not as permanent as that produced by other rickettsiae.

Prompt use of an appropriate antibiotic (chloromycetin) is highly effective.

The disease is not communicable from man to man. Control is by the elimination of rodents and eradication of mites.

Stability is the same as for typhus rickettsiae.

VERIFIED

TOXINS

Botulinum Toxin (Botulism)

This is the protein-like exotoxin formed by the botulinum bacillus. Through repeated purification procedures, it has been obtained in a crystalline form and is the most powerful poison known. The crude material or 'mud' is a brownish, amorphous mass. There are at least five distinct types, A, B, C, D and E, of which types A, B and E are known to be toxic to man; C and D are toxic to animals and probably to man.

Botulism is a highly fatal, acute poisoning, it is characterised by vomiting, constipation, thirst, general weakness, headache, fever, dizziness, double vision, dilation of the pupils, paralysis of the muscles involved in swallowing and difficulty of speech. Respiratory paralysis is the usual cause of death.

Sources of the toxin are bacteria *Clostridium botulinum* and *Clostridium parabotulinum*, which are rod-shaped, slightly motile, sporulating, gram-positive, anaerobic bacilli. The principal reservoir of the bacteria is the soil. The bacteria grow and form their toxin under anaerobic conditions, usually in improperly canned, non-acid foods such as meats, sausages and some vegetables including corn, string beans, spinach and olives.

Transmission is through eating food contaminated with botulinum toxin. The bacteria do not grow or reproduce in the human body and poisoning is due entirely to the toxin already formed in the ingested material. Fresh foods are not involved; fresh well-cooked foods are not involved, as heating destroys the toxin. The toxin could possibly be introduced through breaks in the skin or by inhalation.

Symptoms of poisoning do not usually appear until between 12 and 72 hours after food containing the unnecessary toxin has been consumed, the length of time depending upon the amount of toxin contained in the consumed food.

All persons are susceptible to poisoning. The few who recover from the disease have an active immunity of uncertain duration and degree.

The disease has worldwide distribution; it is prevalent wherever improperly canned food products are consumed.

Mortality is directly related to the amount of toxin consumed.

Passive immunisation with anti-toxin appears to be encouraging as a protective measure for humans but is of little therapeutic value. Active immunisation with botulinum toxoid is of proved protective value.

Treatment is mainly supportive. Anti-toxin therapy is of doubtful value, particularly where large doses of the poison have been consumed.

The disease is not contagious. Epidemics occur only where widespread distribution and consumption of a contaminated food product have occurred.

The toxin is stable for about one week in non-moving water where it is not aerated. It persists for a long time in food when it is not exposed to air. The toxin is destroyed when boiled for 15 minutes, but botulinum spores resist boiling for 6 hours. Pressure cooking will destroy the spores. Botulinum toxin differs from other bacterial toxins in that it is not destroyed by gastro-intestinal secretions.

Infectious aerosol dose: 0.001 µg/kg (Type A).

VERIFIED

Ricin

Ricin is a constituent toxin (5 per cent by weight) of the mash left over from the processing of castor beans to extract oil. Castor oil production is widespread (1,000,000 tons annually). The toxin is relatively easy to isolate and is very stable at ambient temperature and pressure. Physically, it can be prepared in liquid or crystalline form or as powder. It not as toxic as Botulinum toxin or SEB (*qv*).

Symptomatically, some 3 hours after inhalation, the victim experiences a cough, tightness of the chest, breathing difficulties, nausea and muscle aches. The symptoms progress to severe pulmonary distress: severe inflammation of the lungs and airways. Cyanosis is also characteristic and death is likely within 36 to 48 hours from respiratory and circulatory failure.

The toxin can also be ingested through the digestive tract, where the symptoms include nausea and vomiting, followed by progressive deterioration of the stomach, intestines, liver, spleen and kidneys. Injection of ricin additionally causes tissue necrosis at the site of injection.

There is no vaccine or antitoxin against ricin although symptomatic management, based on the method of ingestion, can be effective.

Infectious aerosol dose: 320 mg.

VERIFIED

Saxitoxin

Saxitoxin and its derivatives are water-soluble compounds that bind to the voltage-sensitive sodium channel, blocking propagation of nerve muscle action potentials. As a family of related neurotoxins, they occur naturally in marine dinoflagellates and have also been isolated in green algae, crabs and blue-ringed octopus.

The human effects are characterised as paralytic shellfish poisoning (PSP), most commonly caused by eating bivalve molluscs which have ingested dinoflagellates during filter feeding. PSP is severe and life-threatening.

Absorption of toxin by the gastrointestinal tract is rapid, leading to the onset of symptoms typically between 10 and 60 minutes after exposure. Initial symptoms include numbness or tingling of the lips, tongue and fingertips, followed by numbness of the extremities and general muscular unco-ordination.

Other symptoms may include dizziness, and other neurological effects such as aphasia, incoherence, visual disturbance and memory loss. Cranial nerves are involved, leading to effects on eye movement, speech and swallowing. Respiratory distress and flaccid muscular paralysis, as terminal phases, can occur as soon as two hours after intoxication. Death results from respiratory paralysis.

Treatments include early induction of vomiting to reduce the ingestion of the toxin. Symptoms can be confused with those of nerve agent poisoning. Use of atropine, following such a misdiagnosis would be fatal.

VERIFIED

Staphylococcal Toxins (Enteritis)

This toxin is produced in food by certain strains of staphylococci. It is an enterotoxin, since it has a specific action on the cells of the intestinal mucosa. Unlike most bacterial exotoxins, it is stable at boiling temperature and antigenically is irregular in eliciting the formation of immune bodies.

Food poisoning (not infection) is produced following the ingestion of food in which various strains of staphylococci have been growing. Staphylococcal Enterotoxin B (SEB) and *Escherichia Coli* (E. Coli) are frequently implicated. Sub-types B through E of staphylococcal enterotoxin sourced from the bacterium *Staphylococcus Aureus* have been identified. Food poisoning is usually characterised by sudden, sometimes violent onset, with severe nausea, vomiting, stomach cramps, severe diarrhoea and prostration. Patients usually feel normal 24 hours after the attack begins.

The source of contamination is not known in most cases but is probably of human origin. Food implicated as sources of food poisoning are chiefly pastries (creamy), milk (raw), milk products and meat. Food handlers who are nasal or skin carriers of pathogenic staphylococci or who have an open staphylococcal lesion on their hands, arms, or face have been traced as sources of poisoning. The implicated foods are usually allowed to remain at a warm temperature before consumption, thus providing an incubation period for formation of the toxin.

The consumption of contaminated custard-filled pastry, processed meats (particularly ham) and perhaps milk from cows with infected udders are the usual modes of transmission. Improper food handling is responsible for many outbreaks.

Incubation is relatively short; 30 minutes to four hours, usually two to four hours, elapse between the taking of food and the appearance of symptoms.

Most persons are susceptible, but individual reactions are variable.

The toxin is worldwide and probably the principal cause of acute food poisoning. Fatalities are rare but can occur.

There is no immunisation.

Treatment is supportive.

Food poisoning is non-contagious. Most outbreaks are small and are confined to persons who have eaten the same contaminated food.

The toxin is resistant to freezing, boiling for 30 minutes and to potable quantities of chlorine. The organism remains viable after 67 days of refrigeration.

Infectious aerosol dose: 30 ng per person (incapacitating); 1.7 µg per person (lethal).

VERIFIED

Tricothecene mycotoxins

Tricothecene mycotoxins are low molecular weight, non-volatile compounds produced by filamentous fungi (moulds) of the genera *Fusarium*, *Myrotecium*, *Trichoderma*, *Stachbotrys* and others. The structures of almost 150 tricothecene derivatives have been identified.

Mycotoxins have allegedly been used in aerosol form ('yellow Rain') to produce lethal and non-lethal casualties in Laos (1975-1981), Kampuchea and Afghanistan (1979-1981). Deaths from mycotoxin poisoning include 6,300 in Laos, 1,000 in Kampuchea and 3,900 in Afghanistan. The allegations have been hard to prove owing to the difficulties of sample retrieval from remote and dense jungle or mountainous areas. Mycotoxins enter through the skin and aerodigestive epithelium. They are potent inhibitors of protein and nucleic acid synthesis. The toxins affect bone marrow, skin, mucosal epithelia and germ cells. In a successful BW attack, the toxins will adhere to and penetrate the skin, or be inhaled or swallowed. These materials appear to be more effective through the dermal than the respiratory route.

Early symptoms begin within minutes of exposure and include skin redness, burning and tenderness, blistering and progression to skin necrosis. Leathery blackening and sloughing of large areas of skin occurs in lethal cases. Nasal contact is manifested by nasal itching and pain, sneezing, epistaxis (nose-bleeds) and rhinorrhea (water discharge from the nose). Pulmonary and tracheobronchial exposure is characterised by dyspnea (breathing difficulty) and cough. Mouth and throat exposure leads to pain and blood-tinged saliva. Anorexia, nausea, vomiting and watery or bloody diarrhoea occurs with gastro-intestinal toxicity. Systemic toxicity is manifested by weakness and overall loss of co-ordination. Death can occur within minutes on heavy ingestion or may take several days.

There are no antidotes or therapeutic regimes which can create other than symptomatic relief.

Effective aerosol dose: 25-50 mg (respiratory route) and 2.4-8 mg (dermal route) per kg body weight.

VERIFIED

VIRUSES

African Swine Fever Virus

The virus is small and very resistant to environmental changes.

African Swine Fever, also known as Wart Hog Disease and East African Swine Fever, is a highly contagious and excessively acute disease of domestic swine. It is characterised by fever, pronounced haemorrhages of the lymphatic glands, the kidneys and the mucosa of the alimentary tract and by marked cyanosis (a patchy or diffuse purplish colour) of areas on the skin. Clinically, African Swine Fever is similar to, but more acute than, American Hog Cholera; its immunology is distinct from that of Hog Cholera.

The bush pig and wart hog harbour the virus but do not usually manifest symptoms. Contact with these wild pigs apparently initiates the disease in domestic swine and once established it spreads rapidly by contact.

The disease is transmitted by contact with urine and faeces from infected pigs, the feeding of raw garbage, or pigs actually eating carcases of virus-carrying wild pigs; it is mechanically spread by caretakers and others who pass from infected premises without taking proper precautionary measures. It is thought that infection may be transmitted from wart hogs to pigs by an insect vector; however, the vector has not been established. In a few cases where hogs have survived the disease, there is evidence that they may act as carriers for as long as 10 months after recovery.

The incubation period is usually four to seven days when exposure is made by contact with infected pigs.

In epidemics the disease may occur in 100 per cent of susceptible domestic swine. It is of such a highly contagious nature that usually, in a few days after the first case is noticed, the majority of the pigs in a herd have become infected. However, it has been noted that the virus coming from the wart hog or wild pig takes some time to attain high virulence; the occurrence of an isolated death being followed a week or so later by one or two more deaths and then, after a further interval, the appearance of many cases, spreading rapidly.

African Swine Fever is not known to exist outside the African continent; most of the cases have been confined to East and South Africa.

Fatalities range from 95 to 98 per cent and may be up to 100 per cent.

Attempts to immunise swine against this virus have been for the most part unsuccessful. Even pigs that have survived infection are not consistently immune to subsequent exposures. Hog Cholera anti-serum affords no protection against this disease, although in a few cases it appeared to delay the reaction. Although not proven, it is felt that hyper-immune serum prepared from the virus at any given time might be found to be of no effect in subsequent outbreaks, owing to frequent spontaneous changes in the antigenic character of the virus.

At present there is no effective treatment for the disease.

The disease is highly contagious. It is prevalent in areas where opportunities exist for cohabitation between wild and domestic hogs. Control of the disease has been successful in East and South Africa by immediate slaughter of all infected and exposed animals disposal of all carcases by burning or deep burial, disinfection of the infected premises and quarantine of infected areas. Measures to prevent contact of domestic swine with wild pigs have resulted in a marked decrease in the incidence of the disease.

The virus is very stable at low temperatures; it has remained viable after storage in a cold dark room for six years. It is killed by sunlight but will remain viable in infected carcases, food and water for some length of time.

VERIFIED

Chikungunya Virus

The virus is transmitted to humans principally by the *Aedes aegypti* species of mosquito and recent outbreaks have occurred in southern and central Africa and in southeast Asia. The incubation period for this virus is 3 to 12 days. Its effects are incapacitating but rarely fatal (less than 1 per cent).

Symptoms last between 3 and 7 days plus an extended convalescent period. They include arthralgic-exanthemic syndrome, chills, fever, headache, nausea and vomiting.

The virus has a sudden onset, sometimes biphasic.

Similarities between Chikungunya and Dengue fevers account for misclassification, misreporting and mistreatment in areas where Dengue fever is endemic. Laboratory confirmation of reported cases is therefore important.

VERIFIED

Dengue Fever Virus

This is a very small virus, about 17 to 25 milli-microns in size.

Dengue Fever, an acute, extremely disabling disease usually of sudden onset, is characterised by fever, chilliness, intense headache, backache, pain behind the eyes, joint and muscle pains, weakness and prostration and an irregular rash. The fever, rarely exceeding 38°C, lasts for 5 or 6 days and usually terminates abruptly after reaching a peak. Loss of appetite and constipation are common during the entire illness and abdominal discomfort with colic type pains and tenderness may be manifested. The acute phase lasts only about a week, but convalescence may take several weeks. Dengue Fever is said to be temporarily the most incapacitating although the least fatal of epidemic diseases.

Sources of infection are the blood of infected persons 1 day before and up to 5 days following onset; infected mosquitos; and, in some regions possibly the blood of infected monkeys.

The disease is transmitted by the bite of the *Aedes aegypti* mosquito and certain other species of *Aedes* which have become infected by biting a patient.

Incubation is from 3 to 15 days, most often 5 to 8 days.

All persons, except natives of endemic areas, appear to be fully susceptible. Immunity to the invading strain is apparently permanent, but reinfection with a different strain is possible. However, if this occurs within a few months of the primary infection, the disease is much milder because of cross-immunity.

The disease is found mainly in the tropics and subtropics, extending to the Gulf Coast of the USA, but it may occur wherever the vector mosquitos exist.

The fatality rate is very low.

A mouse-adapted virus vaccine has been found to be stable, safe and effective in studies on human volunteers and will probably be useful in control of epidemics and for personnel travelling from non-Dengue to Dengue Fever areas.

There is no specific therapy, but supportive treatment is essential.

The virus does not spread directly from person to person. Epidemics occur in areas where vector mosquitos are present in large numbers. The infectivity rate is extraordinarily high, with 75 to 100 per cent of the inhabitants of a locality being attacked. Any spread of the disease can be prevented by diligent mosquito control.

Blood from a patient remains infectious after storage in a refrigerator for several weeks. The virus may be preserved in a frozen and dried state at 23°C for at least five years. Mosquitos do not transmit the infection through their eggs to the young, but the infected insects probably remain so for life. The virus is deactivated by ultra-violet light and by 0.5 per cent formalin.

VERIFIED

Ebola Virus (also Lassa fever)

Ebola virus is one of the most pathogenic viruses known to science. It causes death in 50 to 90 per cent of all clinically ill patients. Other haemorrhagic fevers endemic to Africa are Lassa Fever and Marburg Fever (see separate entry in this section). The Ebola virus is transmitted through direct contact with the blood, secretions, organs or semen of infected people. Transmission of the virus has also occurred by handling ill or infected chimpanzees. Health care workers have been infected while attending patients.

The virus has an incubation period of between 2 and 21 days and is characterised by the sudden onset of fever, weakness, muscle pain, headache and sore throat. This is followed by vomiting, diarrhoea, rash and progressive liver and kidney failure. A feature is the internal and external bleeding (haemorrhage) which generally begin around the fifth day.

There is no vaccine for this disease and only symptomatic relief can be offered. Strict barrier nursing and training of health care staff in the management of an outbreak are vital.

The lethality, contagious nature and health management burden imposed on defenders make Ebola a possible choice for BW use.

The Junin virus causes a disease known as Argentine Haemorrhagic Fever with very similar symptoms to Marburg and Ebola. It has a slightly longer incubation period (7 to 16 days) and is fatal in about 18 per cent of cases.

VERIFIED

Encephalitis and Encephalomyelitis Viruses

These viruses are small neurotropic organisms. They are distinguishable antigenically, but some are so closely related that they appear to be variants of a common ancestor.

There are several types of encephalitic diseases, each produced by a specific virus. Their clinical pictures are similar, varying mainly in severity and the rate of progress of symptoms. They are characterised by inflammation of the meninges of the brain, headache, fever, dizziness, drowsiness or stupor, tremors or convulsions, severe prostration, occasional paralysis and poor muscular co-ordination. The diseases are usually acute, prostrating and of short duration. The following are some of the more common types:
(1) St Louis Encephalitis is endemic and epidemic in the central and western USA and prevails in summer and autumn when the arthropod vectors are most numerous.
(2) Eastern Equine Encephalomyelitis and Western Equine Encephalomyelitis usually occur in summer in Canada, the USA and Central and South America. They are viral diseases of lower animals, especially of horses and mules and are transmissible to man.
(3) Venezuelan Equine Encephalomyelitis occurs in Central and South America. It is a disease primarily of equine animals; when the disease is transmitted to man, it usually induces a mild case with varied symptoms.
(4) Japanese B-type Encephalitis occurs in Japan, Korea, China and some Pacific islands.
(5) Russian Far East Encephalitis occurs in spring and early summer, mainly in the Far East provinces of the former Soviet Union and less frequently in Europe and Siberia.

Wild and domestic birds are the principal sources of mosquito infection for the virus types occurring in the USA. The equines, or horses, are also hosts but are less important sources of mosquito infection than birds.

Mosquitos transmit all types except the Russian spring/summer type, which is tick-borne.

The incubation period is variable, ranging from 2 to 15 days.

Susceptibility ranges from 90 to 100 per cent and is usually high in the very young and the aged. Recovery from an infection results in an excellent short-term immunity to the specific virus, but not to any other type.

The diseases are usually prevalent in summer and early autumn; they are usually limited to areas and years of sustained high temperature and large numbers of mosquitos. The highest rates are in rural and suburban localities.

Mortality is unknown; it is probably 5 to 60 per cent with all types. Japanese B and Eastern equine types have the highest case fatality.

Some effective virus vaccines have been developed on a small scale, but their widespread use is not practicable at present.

Treatment is supportive only; chemotherapy and antibiotic treatments have not been developed.

Transmission between humans is not known to occur but may be possible by means of respiratory droplet infection. Transmission of the Venezuelan Equine Encephalomyelitis by inhalation has occurred accidentally in laboratories.

Stability varies among the different types. All may be preserved frozen at −52°C. St Louis and Japanese B viruses are inactivated at 74°C in 30 minutes; Russian Far East virus is inactivated at 78°C in 10 minutes, but the Western and Eastern equine viruses withstand this treatment and also resist 0.2 per cent chloroform, 1 or 2 per cent phenol and 0.05 per cent mercuric chloride. St Louis and Russian viruses are inactivated by 1 per cent formalin within a day; the Russian virus is inactivated by 1 per cent phenol in 10 days, but the St Louis virus resists this treatment for at least 25 days.

Infectious aerosol dose: 10-100 organisms.

VERIFIED

Foot-and-Mouth Disease Virus

The *picorna* virus which causes Foot-and-Mouth Disease is extremely small, ranging from 8 to 12 nm (1×10^{-9} m) in diameter. At least six strains of the virus have been recognised. Also known as Aphthous Fever, Foot-and-Mouth Disease is an acute, contagious, highly infectious, febrile disease of cloven-footed animals (cattle, sheep and swine). The disease is not generally fatal but causes a marked and rapid weight loss, a rapid decrease in milk flow and a severely lowered reproductive capacity. It is characterised by an acute fever and by vesicle formation on the feet and mucous surfaces of the animal's mouth and cheeks and on the udder. The vesicle sites are intensely painful. In most cases, a long time elapses before the animals return to normal. A major outbreak in the UK in 1967 led to the slaughter of 430,000 animals and the destruction of carcasses by burning or lime-pit disposal. Man is only slightly susceptible and, if infected, shows only mild symptoms. However, one human death was attributed to the disease during the 1967 outbreak. The major 2000/2001 UK outbreak was much more widespread, closing whole areas of the northwest and southwest of England to the general public and imposing Draconian movement control measures as well as the slaughter of over 3 million animals. This outbreak heightened the debate about the measures that the modern complex society can take to control animal and crop diseases. The effective vaccination used early was discounted, largely through commercial considerations.

Humans can be infected with the virus by ingestion through an open wound or, via the digestive tract, by drinking infected milk. Meat consumption is not implicated in transmission. Cooked food is generally safe as high temperature destroys the virus (see below).

Infected animals and contaminated food, water, milk and pastures are the sources of infection.

Transmission between animals is by the ingestion of food or fluids contaminated with urine, saliva, vesicular fluid, faeces and by direct contact.

Incubation is usually from 24 hours to seven days, occasionally two or three weeks. Animals can therefore be infectious for about a week before symptoms are apparent to the owner.

Cloven-footed animals have a susceptibility of from 80 to 100 per cent.

The disease is enzootic in various parts of Europe, Asia, Africa and South America; it has also occurred in Mexico and Canada.

While the mortality in mild cattle epizootics is said to range from 3 to 5 per cent, malignant forms have caused a mortality as high as 50 to 70 per cent. The mortality for sheep is around 5 per cent and for swine is from 50 to 70 per cent.

Treatment is supportive only.

The disease has very high communicability. It tends to spread rapidly over a wide geographic area. Wind-blown transmission of the virus is possible as well as carriage by infected humans, wild birds or animals. Effective vaccines are available but are strain-specific. European Union and other international quarantine regulations for eradication of an outbreak are draconian, focussing on the invariably successful strategy of strict control of all human activity and the slaughter of all animals in the affected area.

The organism is highly resistant under natural conditions. Contaminated hay and bran may remain infective for weeks and the organism may persist in moist soil for months. It is resistant to low temperatures but is destroyed within 30 minutes at 78°C or above. It resists drying for several weeks. For decontamination 2 per cent sodium hydroxide solution is preferred.

Although not generally harmful to humans, artificial introduction of the disease would have an enormous impact on an agriculture-dependent economy. The minute size of the virus and therefore its capability for wide infection over long distances make it particularly pervasive and pernicious. This, and the relatively late onset (and therefore recognition) of symptoms in target species (typically a week), make it one of the most effective agents for the spread of disease.

As a BW weapon, the *picorna* virus might be used by an aggressor to cripple agricultural activity and communication. No evidence exists for the modification of the virus to injure humans but bioengineering techniques may make this possible.

UPDATED

Fowl Plague Virus

The virus is small, measuring 0.06 to 0.09 µm.

Fowl Plague, also known as Fowl Pest, is an acute, contagious, highly fatal disease of fowl (particularly chickens and turkeys), characterised by haemorrhages in various tissues of the body, swelling and blood poisoning.

Infected fowl are sources of infection.

Transmission is by the ingestion of food, water and soil contaminated by the blood, urine, faeces and eye and nasal secretions from infected fowl or birds; by infection through wounds; and possibly by bloodsucking insects.

Incubation is usually from two to seven days although it may be as short as 24 hours.

The disease has an infectivity rate of 50 to 80 per cent.

The disease is worldwide.

The mortality rate is very nearly 100 per cent.

No satisfactory method of immunisation has been established.

No treatment has been developed.

Fowl Plague is highly contagious. Once the disease is introduced into a flock, it tends to spread rapidly and kills all or nearly all of the flock in a short period of time. Control measures include rigid quarantine and slaughter and burning of infected fowl. All houses and equipment should be disinfected and the contaminated soil treated with caustic soda.

The virus is stable in dried blood or tissues for many months but is destroyed by exposure to 83°C for 5 minutes. It is also killed by sunlight in a short time. It survives for 18 days on feathers of birds dying of the disease; it will remain viable for months if kept in a cool, dark place, free from air. It is destroyed by 0.1 per cent bichloride of mercury in 30 minutes or by 5 per cent phenol in 10 minutes.

VERIFIED

Hepatitis Viruses

There are two strains of Hepatitis virus, virus A or Infectious Hepatitis and virus B or Serum Hepatitis (Homologous Serum Jaundice).

Infectious and Serum Hepatitis are characterised first by fever, loss of appetite, nausea, fatigue, headache and abdominal discomfort. After a few days the fever subsides; then, because of liver damage, bile may be present in the urine and jaundice (yellowing of the skin) appears.

Sources of infection of virus A are discharges from the nose, mouth and gastro-intestinal tract of infected persons; sources of infection of virus B are blood, serum, or plasma from infected persons.

The usual mode of transmission for virus A is unknown but may be by contaminated water, food, or milk, or by direct personal contact. Virus B is transmitted by transfusions of infected blood, serum, or plasma, or in immunising serums prepared from infected human serum. The sharing of infected medical instruments, such as by drug addicts, is a common cause of transmission.

Incubation is long and variable – 15 to 40 days for virus A, 40 to 150 days for virus B.

Susceptibility is general. Immunity following an attack is undetermined, but second attacks are infrequent. Only man is susceptible, animals are not affected.

The disease is common and apparently worldwide in distribution. It is most prevalent in the autumn and winter months.

Mortality is less than 0.5 per cent.

There is no vaccine available at the present time. Gamma globulin provides protection against virus A for several weeks.

Treatment is supportive and includes diet therapy involving vitamins K and B.

Epidemics are most common in institutions and rural areas. The general incidence is about the same in rural and urban areas and is most common in children and young adults. Epidemics occur in early winter, sometimes reaching a peak in late winter and extending into spring. Epidemics are controlled by the practice of good sanitation and personal hygiene, particularly in the sanitary disposal of respiratory discharges and faeces; by the elimination of flies; and by the use of disease-free blood or blood products in transfusions and in the preparation of vaccine.

The virus withstands a temperature of 74°C for at least 30 minutes. It is destroyed by boiling or autoclaving for 15 minutes. Virus A is unaffected by the normal concentration of chlorine added to drinking water. Virus B survives in the frozen state for several years and resists desiccation at room temperatures or exposure to 0.25 per cent phenol for at least a year. Virus B has been inactivated in serum by exposure to ultra-violet light for 45 minutes.

VERIFIED

Hog Cholera Virus

The virus is 0.03 to 0.035 μm in size.

Hog Cholera, also known as Swine Fever, is a highly acute, contagious, febrile disease of swine, usually chronic in older swine. It is characterised by high fever, yellowish discharges from the eyes, diarrhoea, loss of appetite, severe haemorrhages, viremia (the existence of viruses or viral particles in the blood stream) and extreme weakness.

Infected swine are the sources of infection.

Transmission is by contaminated garbage (infected pork) and through food, water, hog wallows and pens that have been contaminated from eye and nasal secretions, urine, blood and faeces of infected swine. Aerosol transmission is possible. Hogs that have recovered from the disease are possible carriers.

Incubation is from five to six days.

The percentage of exposed swine that are susceptible is 70 to 100 per cent. Animals recovering from the disease have an immunity of uncertain duration.

Hog Cholera has a worldwide distribution and is the greatest menace with which swine-raisers have had to contend. It is widespread in Africa, North America, France, Germany and (potentially) the UK.

Fatalities are estimated to be 80 to 90 per cent in the USA and Africa. In European countries the death rate is often lower.

Several excellent active and passive immunisation methods have been developed and are available.

Administration of hyper-immune serum is the best treatment and is moderately effective.

This highly contagious disease spreads rapidly and easily among swine. It tends to spread gradually over a wide geographic area. Control of the disease is by rigid measures of isolation and disinfection whenever an outbreak occurs and by crystal violet vaccine inoculation.

The virus is stable for months at low temperatures. It is destroyed by use of 1 to 2 per cent caustic soda, 2 per cent cresol for 1 hour and 5 per cent milk of lime and by exposure to 83°C for 1 hour. It is resistant to phenol and glycerin.

UPDATED

Influenza Virus

The influenza virus appears capable of subtle mutation, rendering vaccines aimed at previous outbreaks ineffective. Epidemics have reappeared and will continue to reappear from time to time. The outbreak of 'Spanish Flu' shortly after the end of the First World War is reliably estimated to have killed 50 million people. In the 1980s a potential pandemic of 'Bird Flu' emanating from Hong Kong, was avoided by the quick action of the Hong Kong and Chinese authorities in slaughtering the majority of infected poultry through which the disease was transmitted to humans. Increased ease of travel and contact may make future epidemics more difficult to manage, despite improvements in epidemiology and vaccination. For example, a laboratory-developed vaccine to counter a specific new strain of the virus requires an enormous financial and logistic effort to produce and distribute in sufficient quantity to stop an epidemic outbreak.

Influenza or *la grippe* is characterised by catarrhal inflammation of the respiratory tract; sudden onset; fever of one to seven days duration; marked prostration; and generalised aches and pains in back, limbs and muscles. Sore throat, bronchitis and pneumonia are complications of secondary bacterial infections.

Soiled articles and discharges from the mouth and nose of infected persons are the main sources of infection.

Transmission is probably by direct contact, by droplet infection (respiratory), or by articles freshly contaminated with nose and throat discharges of infected individuals.

The virus is large, having strains which range between 0.07 and 0.1 μm in size. Types A and B have been identified and there are indications that Type C may come to be recognised.

The incubation period for Type A is from one to two days; for Type B it is from 12 to 18 hours.

The virus is highly infective and contagious; susceptibility is general, there being relatively little natural resistance. The resistance of some individuals exposed to the disease during epidemics appears to be due to previous infection. Acquired immunity following recovery from an attack lasts only for a few months to a year and is effective only against specific strains of the virus.

Prevalence is variable; the disease occurs sporadically as epidemics and pandemics. Epidemics may affect up to 50 per cent of a given population within four to six weeks.

The disease has a mortality range from 0 to 1 per cent, but it is often followed by complicating respiratory infections which can result in high mortality.

Vaccination with specific strains of the virus leads to an effective immunity of several months duration against the same or closely related strains. The extreme antigenic variability of the influenza viruses makes it difficult to protect against all strains.

Treatment is supportive only, with control of secondary infections. There is no specific effective treatment.

Epidemics are widely prevalent under favourable conditions. Pandemics occur irregularly.

The virus is killed by being heated at 73°C for 20 to 30 minutes; it is resistant to freezing for several weeks. It can also be easily destroyed by steam, boiling and ordinary antiseptics.

VERIFIED

Marburg Virus

Marburg virus is another example of a highly contagious and cause of haemorrhagic fever. It is closely related to the Ebola Virus (see separate entry in this section) and was first recognised during an outbreak of severe haemorrhagic fever associated with the importation of African green monkeys from East Africa to Germany.

There is a five to seven day incubation period. The virus produces headache, fever, muscle pain, vomiting, diarrhoea, haemorrhagic diathesis, conjunctivitis, photophobia, skin rash and jaundice. While not as lethal as Ebola, the virus causes fatalities in about 25 per cent of cases. Isolated human cases have been reported but little is known about the epidemiology of the disease.

There is no effective vaccine and only supportive symptomatic treatment can be offered. Strict barrier nursing and the effective training of health care support workers should be practised to avoid the contagious spread of the disease.

The Junin virus causes a disease known as Argentine Haemorrhagic Fever with very similar symptoms to Marburg and Ebola. It has a slightly longer incubation period (7 to 16 days) and is fatal in about 18 per cent of cases.

VERIFIED

Newcastle Disease Virus

The virus is 0.08 to 0.12 μm in size.

Newcastle Disease, Pseudo-pest or fowl, or Pneumoencephalitis, is an acute, highly contagious, febrile disease of fowl (chickens, turkeys and pheasants). The course of the disease is of short duration. It is characterised by severe respiratory and nervous symptoms, including difficult breathing, depression and stupor, twitching of the head and neck, marked weakness, a twisted neck and perhaps paralysis. Egg production is interrupted for weeks.

Infected fowl are sources of infection.

Transmission is by direct contact, by the ingestion of food and water contaminated with faeces and by the inhalation of contaminated dust.

Incubation is usually from four to eight days, although it may be up to 13 days.

The infectivity rate ranges from 50 to 90 per cent. Chickens of all ages and turkeys are highly susceptible; pigeons, geese, ducks and other barnyard fowl are less susceptible.

The disease is worldwide.

Fatalities range from 10 to 100 per cent. The disease found outside the USA is nearly 100 per cent fatal; strains found in the USA and the UK are usually less pathogenic.

Vaccines prepared from live virus of low virulence produce an immunity of a few months duration; revaccination of layers should be at 12 to 16 weeks of age. Recovery from the disease is accompanied by some immunity.

No treatment has been developed.

The disease is highly communicable and tends to spread gradually over wide geographic areas. The greatest incidence of the disease is in the winter months. Control measures consist of vaccination and maintenance of a high standard of sanitation.

The virus remains stable for years when dried in the frozen state (lyophilised); it is readily destroyed by being heated at 40°C (pasteurisation). Highly effective disinfectants are 3 per cent phenol, 3 per cent cresol and 2 per cent sodium hydroxide solutions.

VERIFIED

Psittacosis Virus

This organism, usually classed as a virus, has been classed by some as a rickettsia but more recently has been placed tentatively in a group intermediate between viruses and rickettsiae. It is small and coccoid, is 0.2 to 0.4 µm in length and is gram-negative. Although the name Miyagawanella psittacii has been applied to this virus, it has not been generally accepted.

Psittacosis or Parrot Fever is also called Ornithosis. The disease is often confused with Influenza, Atypical Pneumonia, and Typhoid Fever. It is a severe febrile disease in man; the onset may be sudden or may be gradual and insidious. It is characterised by acute pulmonary infection, chills, fever, anorexia, sore throat, severe headache, backache, constipation, great weakness and prostration and is sometimes accompanied by delirium.

Parrots, parakeets, budgerigars, canaries, pigeons and other birds are sources of infection.

The respiratory tract is the principal entry path. Transmission is made by contact with infected birds or by breathing air contaminated by faeces, urine, nasal discharges or the soiled feathers of sick, dying, or latently infected birds. Contaminated air may travel a considerable distance before it loses its infectivity. Apparently well birds (carriers) can transmit the infection. Transmission from man to man has occurred, and in rare instances man has become a carrier.

Incubation is usually from 6 to 15 days with 10 days as the mean, but it may be as long as 30 days.

All ages and races are susceptible. Children are readily infected but rarely develop severe illnesses. The disease is more severe in adults than in children. Recovery from an attack is generally followed by immunity, although reinfections have been reported.

The disease is widely distributed throughout the world. It is most prevalent in areas where parrots, parakeets and similar birds are found. It usually occurs as a sudden outbreak among persons exposed to infected birds. Mild cases may occur from slight exposure to infected pigeons or to birds showing no symptoms of the disease.

Mortality is variable, from 9 to 20 per cent in untreated cases and usually occurs among persons over 30 years of age.

Sulfadiazine, penicillin, aureomycin, chloromycetin and terramycin shorten the course of the disease and greatly decrease the mortality.

The disease tends to be self-limiting. Transfer from man to man may occur by close contact, such as from patient to attendant.

The virus will remain viable on surfaces for 20 to 30 days and perhaps longer in dry faecal material. It is inactivated in 24 to 36 hours by 0.1 per cent formalin and 0.5 per cent phenol. Decontamination may be accomplished by burning infected material or by the disinfection of infected faeces. An effective disinfectant is 2 per cent cresol.

VERIFIED

Rift Valley Fever Virus

The virus is 23 to 35 milli-microns in size and will pass through most filters.

Rift Valley Fever, or Enzootic Hepatitis, is a highly infective and fatal disease of sheep. Goats and cattle are sometimes affected and the disease is easily transmitted to man, in whom it usually takes a mild form. In sheep, the infection is characterised by a rapid course, high fever, loss of appetite and progressive weakness; it may produce abortion in pregnant females. The disease is more severe in lambs than in older animals.

A new outbreak of the disease, occurred in northeast Kenya in January 1998. The symptoms were more severe than encountered earlier indicating a possibly more virulent strain of the disease. Patients presented with bleeding from the ears, nose and mouth in addition to the other symptoms had become infected through eating carcasses carrying the virus. The disease was fatal in 50 per cent of cases.

Infected animals are the sources of infection.

The disease is transmitted by the bites of mosquitoes and possibly by the ingestion of infected food.

Incubation is from one to four days for adult animals; 12 to 24 hours for lambs.

Sheep are highly susceptible, while goats, cattle and man are less so. Mice, rats, cats and apes are reported to be susceptible, while horses, rabbits, guinea pigs and birds seem to be immune.

The disease is found in southern and central Africa, particularly in Kenya.

Mortality is 90 to 95 per cent in lambs, 35 to 45 per cent in adult sheep and 5 to 20 per cent in cattle.

A living virus vaccine is in use in South Africa.

Treatment is supportive only.

Epidemicity is dependent on the presence of diseased animals and the specific mosquito vector.

The virus is said to be destroyed by exposure to a temperature of 73°C for 40 minutes.

VERIFIED

Rinderpest Virus (Cattle Plague)

The Rinderpest virus is very small. Rinderpest is also known as Cattle Plague. It is an acute, febrile, highly contagious and highly fatal disease of bovine animals (cattle, oxen and water buffalo), sheep and goats. It is characterised by a sudden onset of croupous inflammation of the digestive tract, inflammation and erosion of the mucous membranes of the mouth and bloody diarrhoea. The virus is present in practically all the body tissues and fluids of infected animals.

Infected animals are the sources of infection.

Transmission is by ingestion of food and water contaminated with urine, faeces, saliva, eye and nasal secretions and by direct contact.

Incubation is from three to nine days.

Cattle, oxen, water buffalo, sheep and goats are almost completely susceptible.

The disease is worldwide, except for the Western Hemisphere.

Mortality ranges from 25 to 100 per cent.

Excellent specific vaccines have been produced which induce an active immunity. A good immunising serum can be prepared which confers a passive immunity of short duration. Recovery from the disease is followed by an immunity of indefinite duration.

Massive doses of anti-serum may be used to protect susceptible animals but are of little value after symptoms have appeared.

Epidemicity is very high in non-immunised animals, as the disease is highly infective, is easily transmitted from animal to animal and tends to spread gradually over wide areas.

The virus will survive for nearly three months under controlled optimum conditions. Exposure to 78°C kills it within a few minutes. It will resist drying and sunlight for only a few days but when frozen and then dried will remain viable for months. Effective disinfectants are 2 to 5 per cent phenol and 0.1 per cent bichloride of mercury.

VERIFIED

Variola Virus (Smallpox)

The virus will pass through most filters and ranges from 0.15 to 0.2 µm in size.

Smallpox, or Variola, a highly contagious and often fatal disease, is characterised by severe fever and small blisters of the skin. The blisters later contain pus and form crusts which fall off in 10 to 40 days after the first lesions have appeared, leaving pink scars which gradually fade. Complications of the disease are secondary bacterial infections.

Lesions of the mucous membranes and skin of infected persons are sources of infection.

Transmission is made through contact with patients having the disease or with articles or persons freshly contaminated by discharges from lesions and skin of infected individuals.

Incubation is from 7 to 21 days, commonly 12 days.

The disease is highly contagious and infectious, with universal susceptibility. Recovery is usually followed by permanent acquired immunity.

According to the World Health Organisation, Smallpox has now been eradicated throughout the world but the chance is always present that it might recur. The frequency is greatest in Winter and least in Summer.

Fatalities range from 1 per cent with the mild type of disease to 30 per cent with the more severe types.

Artificial immunity by vaccination may be completely effective for 2 to 20 years, but revaccination every three years is advisable to maintain minimal immunity. A high degree of immunity is required against severe strains of the virus.

No specific treatment is available. Antibiotics and sulphur drugs are used to treat secondary infections.

Epidemicity is high, depending on the immunity status of population and exposure to the disease.

The virus is viable in water for several years at 22 to 27°C and is resistant in dry or wet form to very low temperatures. In the dry state it is more resistant than in the wet state. Decontamination can be effected by exposure to alcohol and acetone for 1 hour at room temperature, but the virus is resistant to some other disinfectants. Moist heat above 78°C and dry heat above 118°C are effective in 10 minutes.

VERIFIED

Vesicular Exanthema Virus

The virus is 0.07 to 0.1 µm in size.

Vesicular Exanthema is a contagious, febrile, weakening disease of swine. It is characterised by the formation of vesicles or lesions in the mouth and nostrils, on the snout, feet and udders and around the coronary band. The vesicles produced by this virus are very similar to those produced by Foot-and-Mouth Disease and Vesicular Stomatitis. Animals become lame and hooves are sometimes shed. Sows may abort in advanced pregnancy. Cattle, sheep and guinea pigs are not susceptible to the disease.

Infected swine and uncooked garbage containing pork trimmings are the sources of infection.

Transmission is by direct contact, by the ingestion of contaminated food and water and by persons going from farm to farm.

The incubation period is from two to seven days.

The infectivity rate ranges from 50 to 100 per cent in swine.

The estimated herd mortality is 0 to 5 per cent for swine.

No immunisation has been developed to date. Immunity following an attack of the disease is comparatively short.

Treatment is supportive only.

The disease is controlled to some extent by quarantine and slaughter but probably will not become completely manageable until the practice of feeding raw garbage is stopped.

See Rift Valley Fever Virus for stability.

VERIFIED

Vesicular Stomatitis Virus

The virus is 0.07 to 0.1 µm in size.

Vesicular Stomatitis, also known as Pseudo Foot-and-Mouth Disease or Mouth Thrush, is a contagious, weakening, febrile disease, primarily of horses and mules and occasionally of cattle and swine. It is characterised by vesicular eruptions of the mucous membranes, particularly of the mouth. Occasionally, lesions may be found on the feet and udders. This disease has great similarity to Foot-and-Mouth Disease; the main difference is that the horse is not susceptible to Foot-and-Mouth Disease.

Horses, mules and cattle infected with the disease are sources of infection.

Transmission is by ingestion of contaminated food and water and by direct contact.

Incubation is from two to nine days.

The percentage of horses and mules estimated to be susceptible is 70 to 100.

The disease is probably worldwide.

Mortality is from 0 to 10 per cent.

Immunisation is in the experimental stage only.

Treatment is unsatisfactory and is only supportive at the present time.

The disease tends to spread gradually over a limited geographical area.

Contaminated food remains infective over long periods. The organism is resistant to low temperatures. The virus remains viable in 0.5 per cent phenol for 23 days. It is destroyed by 2 per cent lye and by 0.05 per cent crystal violet, both within 1 minute. In general, it has much the same resistance as that of Foot-and-Mouth virus.

VERIFIED

Yellow Fever Virus

The virus particles are estimated to range from 0.017 to 0.028 µm in size and pass through most filters.

Yellow Fever (classic urban) is characterised by a sudden onset of chills and fever, prostration, headache, backache, muscular pain, congestion of mucous membranes, severe gastro-intestinal symptoms and jaundice (yellowing of skin) from liver damage. Haemorrhage from the stomach and gums often occurs. The disease is of short duration, with either death or complete recovery occurring within two weeks of onset. A forest or sylvatic form in South America and Africa is known as Jungle Yellow Fever and appears to be clinically identical to the classic type.

The blood from people and monkeys infected with Yellow Fever is the source of infection.

Transmission is usually by the bite of the female *Aedes aegypti* mosquito; in the forests of South America and Africa, it is usually by some other types of mosquito.

Incubation is from three to six days, rarely longer.

Susceptibility is general. Recovery from an attack is followed by lasting immunity. Infants born to immune mothers are immune for three to four months.

The disease is endemic in man, monkeys and some other mammals in Western and Central Africa and among certain apes and monkeys in the tropical forests of Central and South America. Occasional human cases and epidemics occur in these areas and neighbouring urban or rural localities. The disease is unknown in the Pacific basin.

The overall case fatality rate, taking into account undiagnosed mild infections (shown by the later presence in the blood of specific antibodies), is probably about 5 per cent. However, it is much higher in cases in which jaundice appears.

Inoculation with a modified living virus vaccine confers an active immunity which may last for at least four years and probably longer.

There is no specific treatment. Supportive treatment (bed rest and fluids) is essential for even the mildest cases.

Occasional epidemics still occur in Africa and South and Central America. These epidemics can be lessened by mosquito control and active immunisation.

The virus is resistant to freezing and drying but is destroyed by being heated at 78°C or above for 10 minutes. It is easily inactivated by common antiseptics. Mosquito control measures, including use of insecticides, are the preferred methods of control.

VERIFIED

CHEMICAL WARFARE

CHEMICAL WARFARE

Introduction

Background

The previous section raised the idea of a spectrum of toxic agents passing, at one extreme, from the pure Chemical Warfare (CW) agents, through a mid-spectrum, including toxins, to the purely biological pathogens at the other. This section deals with the CW agents – perhaps best defined as those elements or chemically-derived compounds which are capable of military delivery to incapacitate or kill people. CW is younger than BW.

Military commanders throughout history have seen the need for a weapon of last resort. When faced with a frontal attack of unstoppable momentum, it would be helpful to deploy a weapon which would stop the enemy in its tracks – something that affected the entire force at once. Physical barriers, lines of stakes, ditches and moats were surmountable but CW appeared to offer this capability. During the first part of this century, the German high command realised that bulk stocks of chlorine, a choking agent, presented a readily available resource. Chlorine was kept at every swimming bath, for water purification. At Ypres on the morning of 22 April 1915, German troops, desperate to achieve a breakthrough, cracked open the valves of 6,000 cylinders of compressed chlorine gas. The wind was favourable and the brown cloud moved quickly over the battlefield towards the French and British lines. The use of gas was unexpected and the men were totally unprotected. Casualties were severe and by July 1916, the allied forces had retaliated in kind with chlorine and a new choking agent – Phosgene. Earlier attempts had been made to deploy CW, notably by the French, who used tear gas against the Germans with little success (August 1914). In 1917, the Germans, seeing the utility of CW, had accelerated its research programme and deployed a new agent: Sulphur Mustard (now known as agent H). This had an added terror effect through the sight of the huge obscene blisters caused by the action of the agent on the skin. The 125,000 tons of CW agent deployed in the First World War caused 300,000 chemical casualties of whom 91,000 died. Many of the survivors were disabled for life. The real CW victims in this war were the Russians who, generally ill-equipped and with no chemical protection, suffered well over half a million casualties, some 62 per cent of the total.

Between the wars

In 1936, the German scientist Gerhardt Schräder had successfully explored the military usefulness of organo-phosphorous compounds, with their ability to affect the nervous system. One outcome was Tabun, the first nerve agent (agent GA).

Between the wars, the lessons from use of CW on the battlefield began to impact more and more on military planning and research. The Europeans and the Americans embarked on both offensive and defensive CW programmes, leading to the creation of stockpiles of weapons. Few of the brushfire wars in this period did not see the use of CW. It was used against dissident tribesman by the colonial powers in Ethiopia, Algeria and by the Japanese during their excursions into China in the 1930s.

Second World War to the present day

By 1939, both the Axis and the Allied Forces had chemicals ready for use, including the powerful new nerve agents. Why the Germans did not use CW agents during the Second World War remains a mystery. The stockpiles certainly existed. There were a number of occasions when CW use could have been expected – to frustrate the D-day landings or the final assault on Berlin in 1944 for example. A fear of successful CW retaliation may have been a factor. The allied forces certainly had planned for offensive use and protection in defence. Even evacuee children were issued with gas masks. Since 1945, there has been sporadic use. A powerful new nerve agent, VX, was discovered in the 1950s. More recently, defoliants were used in an attempt to unmask Communist guerrillas in Vietnam in the 1960s and the use of 'Yellow Rain' (strictly a toxin – *trichothecene mycotoxin*) was also a feature. CW incidents were reported during the Russian occupation of Afghanistan. One such report described corpses whose flesh had darkened, but appeared otherwise uninjured – the so-called 'black body agent'. None of these reports revealed any consistent use of CW and it was not until near-defeat at the hands of the Iranians that Iraq deployed CW agents on the battlefield. The Iraqis had also previously used nerve agents against unprotected Kurdish civilians at Hallabah. By the time of the occupation of Kuwait in 1991, Iraq had significant stockpiles of both CW and BW agents available. Fortunately, they were never used – probably for fear of nuclear retaliation by the US. Today, following the ignominious departure of the UNSCOM inspection teams, Iraq remains committed to the development of a sophisticated CW capability and it will be some time, if ever, before the kind of detailed surveillance required to monitor such activity can be agreed between the UN and the Iraqi regime.

CW agents, as the Japanese found out in 1995, are of great interest to terrorists. The sarin attack on the Tokyo underground on 20 March that year focused public attention on a fear that had been seizing the minds of defence planners for some years. The five makeshift devices caused the death of 12 people and the final casualty toll rose to 5,500. Of these, four more died within the month. What is little

Under the terms of the CWC, Japan and China are working closely together on a programme to destroy large stockpiles of old chemical weapons abandoned by the Japanese on Chinese soil in the 1930s (Japanese Government)

2001/0089991

reported is the fact that the Aum Shinri Kyo sect had made a previous successful attack in June 1994 on a prosperous suburb in the town of Matsumoto, about 150 km northwest of Tokyo. A van, travelling through the streets, released sarin into the air. There were six hundred casualties in this incident and seven deaths. Incredibly, outside Japan, this incident received little media coverage. In any case, it was perhaps a surprise to defence experts that CW had not been a feature of earlier terrorist activity. A result of these high-profile incidents has been a healthy renewal of interest in the need for effective civil emergency planning, long neglected in some areas. City fathers in all capitals are actively seeking advice and support from specialists in the CW field as they strive to train and equip response teams to operate in a CW hazard area.

The features of offensive CW
There needs to be a reason to use CW. Strategically, as a weapon of mass destruction, it has a significant deterrent effect. The alternative, nuclear weapons, is too expensive and international controls are tougher to circumvent. As Iraq found in 1983, CW helped stop the relentless Iranian advances in the war. CW uniquely creates large numbers of painful casualties who require intensive and long-term care, absorbing enormous medical resources. Secondly, it is a weapon which causes fear. It cannot be seen until too late and, once it has permeated a crowd or a platoon, all are affected at once. The injuries are distressing to watch. This makes just the threat of its use enough to cause alarm, alter plans or slow momentum. Finally, by choosing the appropriate agent, the attacker can quickly overrun an affected position and retrieve the defender's resources without injury to his own forces.

Balanced against these factors there are some serious drawbacks to the ownership and use of an offensive CW capability. Clandestine acquisition will, if international policing is relentless and effective, become tougher and tougher. Greater effort will have to be put into disguise and subterfuge. A CW production facility is not easy to camouflage and the UNSCOM team in Iraq gained great collective experience in identifying the tell-tale signs. The OPCW database will allow better and better training and surveillance by inspectors using all means at their disposal.

Secondly, CW frightens the horses. In other words, neighbouring nations, once neutral, may show sufficient alarm to take the other side were CW to be used. Former allies may fall neutral at the prospect. Thirdly, it is risky to make, store and distribute. Fourthly, as with BW, there is always the risk of an 'own goal'. The wind and the temperature may change. Lastly, contrary to popular belief, tons of agent are required if a general battlefield effect is sought and it needs to be delivered as accurately as HE. It may be appropriate here to say that the raw First World War statistics, mentioned above, imply that it took a third of a ton of agent to cause a single casualty. Clearly today, delivery systems are more sophisticated and accurate but the chemical warhead needs some highly specialised design to be really effective.

Defence – the challenge
CW is a tactical weapon, delivering a relatively localised effect, depending on the meteorological conditions and armed services are better defended than against BW. One of the most potent deterrents to CW use by an enemy is the public demonstration of a sound and comprehensive range of defensive measures, likely to render an attack worthless. Most democratic forces are well equipped. While clandestine research continues on ways to defeat detection and protection, the principal CW agents are known and science continues to refine and improve existing performance. In fact, there is almost a risk of complacency among some defence departments but it is still difficult to work in IPE and there is no doubt that being forced into an NBC defensive stance will slow things down. The key to success is better intelligence, not only through surveillance but by developing better prediction methods and integrating them into existing C³I. Vigilance is essential as break-out from CWC, offensive research and regional aims all strive to keep CW in the frame as a potent and dangerous warfare environment.

Agent types
The principal agents are examined in detail later. They broadly fall into three categories of which the nerve agents are the most potent. The latter acts generally by interfering with the central nervous system. Crudely, a nerve impulse can start an action but not stop it. This leads quickly to paralysis, asphyxia and death. Even at extremely low concentrations, the agent creates miosis (dimness of vision). This effect is clearly significant in flight, monitoring of sensors like radar and other functions where good vision is essential. There is a choice of persistency in this range of agents. In a rapid advance, a low persistency attack would allow follow-on forces to operate unprotected sooner. In area denial, high persistency will be appropriate – from an agent such as VX. The second category of agents act to prevent oxygen being taken up by the bloodstream. These are the blood agents of which a variant, Zyklon B, was used in the extermination camps by the Nazis during the Second World War. Blood agents are highly volatile but quick-acting. The third group comprise those agents which can kill but, more importantly, cause extreme distress and create masses of high-dependency casualties. The vesicants act on the skin (blister agents) while other types choke, cause vomiting and a range of other incapacities. The less toxic variants form the riot control agents such as CS. New types of agent are constantly being identified by research, either offering increased potency or the ability to defeat common-range filtration and skin protection regimes.

Developments in arms control
Before the break-up of the Soviet Union, western intelligence sources predicted some 360,000 tons of stockpiled CW agent. Since the various arms control treaties of recent years and the emergence of the CWC, a comprehensive DEMIL

programme has been put in place for the destruction of all stockpiles of CW agents. This is a mammoth task which will take a number of years to complete.

Throughout the modern era, efforts have been made by peacemakers to outlaw CW altogether. Early attempts were partially successful. The Hague Conventions of 1899 and 1907 banned the use of 'asphyxiating gases' and, with the perspective of the horrors of the First World War, renewed international effort culminated in the 1925 Geneva Protocol. Some 29 nations signed an agreement to prohibit the use of 'poisonous gases' (and 'bacteriological methods of warfare'). It was weak, representing essentially a 'no first use' agreement. It did not prohibit research, development or stockpiling of weapons. By the end of 1988, there were 112 parties to this accord.

Today, a new agreement is in force, the Chemical Weapons Convention. Signed in Paris in January 1993, it represents the desire by signatories to tackle all aspects of CW, from research to use in war. While still not by any means watertight, it has at least created structures to intervene and deliver effective policing. With the creation of the OPCW in The Hague, inspectors now have powers to arrange short-notice inspections of suspect facilities. The effects are far-reaching and will affect any industry that makes or uses listed chemicals.

Summary
CW is a powerful tactical challenge in defence. While there are good counters to it through comprehensive detection and protection, it still has the power to frustrate operations and slow momentum. Some agents kill quickly, others cause horrifying and lingering injuries. It is the business of defence planners to ensure that, through well-funded research and procurement programmes, the defensive guard is not dropped. Complacency is therefore an enemy. The spectre of terrorist use should continue to occupy the minds of defence and civil emergency planners everywhere. The conventions now in place are one way to try and keep this threat under firm control and they remain an important function of the international community, in seeking the emasculation of the power of rogue regimes to wreak CW havoc on the civilised world.

VERIFIED

BLISTER AGENTS (VESICANTS)

Vesicating (blister-causing) agents were developed very early in the modern history of CW agents. The principal strategic advantage they offer to an aggressor is the degree to which they can clog the defender's logistic chain with casualties. Vesicants are generally highly persistent. Disadvantages include their slow speed of effect and the adverse PR their use would attract. The appallingly painful wounds require a level of nursing care which increases with time. Death is almost certain, although distressingly slow and painful, if the material is ingested. This visual impact of vesication increases the level of fear in carers and onlookers and promotes hysteria amongst the defender's personnel. Therefore the principal utility of blister agents is in deterrence and area denial. The threat to use, or the implication of recent use therefore remains an effective deterrent.

During the First World War, sulphur mustard (H agent) was the principal vesicant. Later in the war, its delayed effect was crudely overcome by mixing the vesicant with faster acting choking agents and 'discipline breakers' whose high discomfort caused troops to seek urgent relief by removing their respirators, thereby exposing them to the full effect of the agent.

Protection from blister agents is difficult as the vapour is pervasive and can render poor designs of permeable suit ineffective. However, modern respiratory filtration and impermeable body protection are highly effective.

Disposal of the material is difficult and its impact on the immune system and on the environment is significant. Research on the long-term effects of vesicants on unprotected civilians in Iraq (Hallabjah 1989) reveals that mutagenesis is probably more severe than in nuclear radiation. Profligate disposal of vesicant munitions in the Denmark Strait after the Second World War continues to cause injuries to

US Air Force medical personnel in full NBC protective kit

fishermen. Unlike most other types of CW agent, vesicants, especially H agent, are extremely slow to hydrolyse in salt water and the agent forms a 'cake', denser than seawater, which remain highly toxic. Fishing trawls continue to retrieve the material and cause burn injuries.

UPDATED

Arsenicals

The arsenicals are a group of related compounds in which Arsenic is the central atom. The arsenical chemical agents discussed under blister agents include Lewisite (L), Mustard-Lewisite mixture (HL), phenyldichloroarsine (PD), ethyldichloroarsine (ED) and methyldichloroarsine (MD). Among the disadvantages of these agents are that they hydrolise rapidly and are not as toxic as other blister agents.

In terms of the destruction of old or abandoned chemical weapons, arsenic-based materials are particularly difficult to dispose of.

VERIFIED

Distilled Mustard (HD)

Distilled Mustard (HD) is H which has been purified by washing and vacuum distillation. HD, however, has less odour and a slightly greater blistering power (by a degree negligible in the field) than H, and is more stable in storage.

Chemical name: 2,2'-dichloro-diethyl sulfide.
Rate of hydrolysis: Very slow at ordinary temperatures.
Odour: Garlic-like.
Skin and eye toxicity: Eyes are very susceptible to low concentrations; higher concentrations are required to produce incapacitating effects by skin absorption rather than by eye injury.
Rate of action: Delayed; usually 4 to 6 hours until first symptoms appear. Latent periods have been observed, however, up to 24 hours and, in extreme cases, up to 12 days.

HD acts first as a cell irritant and finally as a cell poison on all tissue surfaces contacted. The first symptoms of HD poisoning usually appear in 4 to 6 hours; the higher the concentration, the shorter the interval of time between the exposure to the agent and the first symptoms. The physiological action of HD may be classified as local and general. The local action results in conjunctivitis or inflammation of the eyes; erythema (redness of the skin) which may be followed by blistering or ulceration; and inflammation of the nose, throat, trachea, bronchi and lung tissue. Susceptibility also varies with individuals. Injuries produced by HD heal much more slowly and are more liable to infection than burns of similar intensity produced by physical means or by other chemicals. This is due to the action of HD making the blood vessels incapable of carrying out their functions of repair and by the fact that necrotic (dead or dying) tissue acts as a good medium for bacterial growth.

Duration of effectiveness: Dependent on the munitions used and the weather. Heavily splashed liquid persists for one to two days under average weather conditions and a week or more under very cold conditions.

VERIFIED

Ethyldichloroarsine (ED)

Ethyldichloroarsine (ED) was introduced by the Germans in March 1918, in an effort to produce a volatile agent with a short duration of effectiveness which would act faster than DP or HD and be more lasting in its effects than PD.

Rate of hydrolysis: Rapid.
Odour: Fruity, but biting and irritating.
Skin and eye toxicity: Vapour is irritating but not harmful to eyes and skin except on prolonged exposure. Liquid ED has approximately one twentieth the blistering action of liquid L.

As with other chemical agents containing arsenic, ED is irritating to the respiratory tract and will produce lung injury upon sufficient exposure. The vapour is irritating to the eyes and the liquid may produce severe eye injury. The absorption of either vapour or liquid through the skin in sufficient amounts may lead to systemic poisoning or death. Blistering of the skin is produced by prolonged contact with either liquid or vapour.

Duration of effectiveness: Evaporates at approximately the same rate as water. Short duration under wet conditions.

VERIFIED

Levinstein Mustard (H)

This is Mustard made by the Levinstein process. It contains about 30 per cent sulphur impurities, which give it a pronounced odour. The properties of H are essentially the same as those listed for HD (see separate entry in this section).

The so-called 'dusty' Mustard, also referred to as micronised Mustard, is a refined Mustard agent absorbed into a diatomaceous earth.

VERIFIED

Lewisite (L)

Chemical name: Dichloro (2-chlorovinyl) arsine.
Rapidly hydrolysed in liquid or vapour state.
Odour: Geranium-like; very little odour when pure.
Skin and eye toxicity: L has approximately the same blistering action to the skin as HD, even though the lethal dosage for L is much higher.
Rate of action: Rapid.

L produces effects similar to HD but, in addition, acts as a systemic poison, causing pulmonary oedema, diarrhoea, restlessness, weakness, subnormal temperature and ow blood pressure. In order of appearance and severity of symptoms it is: a blister agent, a toxic lung irritant and, when absorbed in the tissues, a systemic poison. Liquid L causes an immediate searing sensation in the eye and permanent loss of sight if not decontaminated within 1 minute. It produces an immediate and strong stinging sensation to the skin; reddening of the skin starts within 30 minutes. Blistering does not appear until after about 13 hours. Like HD, it is a cell poison. Skin burns are much deeper than with HD. When inhaled in high concentrations it may be fatal in as short a time as 10 minutes.

Duration of effectiveness is somewhat shorter than HD. Very short duration under humid conditions.

VERIFIED

Methyldichloroarsine (MD)

Methyldichloroarsine (MD) is similar to ethyldichloroarsine.

Rate of hydrolysis: Very rapid.
Odour: None.
Skin and eye toxicity: Blistering action slightly less than that of HD. The effect of MD on eyes is similar to that of L (produces corneal damage) but less severe. Concentration required for blistering effect is too high to attain in the field.
Rate of action: Immediate irritation of eyes and nose. Blistering effect is delayed several hours.

As with L and the other similar arsenicals, MD is irritating to the respiratory tract and produces lung injury upon sufficient exposure. The vapour s irritating to the eyes and the liquid may produce severe eye injury. The absorption of either vapour or liquid through the skin in sufficient amounts may lead to systemic poisoning or death. Blistering of the skin is produced by prolonged contact with either liquid or vapour.

Duration of effectiveness is relatively short. Evaporates much faster than water.

VERIFIED

Mustard-Lewisite Mixture (HL)

Mustard-Lewisite Mixture (HL) is a variable mixture of HD and L which provides a low-freezing mixture for use in cold weather operations or as a high-altitude spray. Properties are listed for the eutectic mixture (the mixture having the lowest possible freezing point) which is 63 per cent L and 37 per cent HD by weight. Other mixtures, such as 50:50, may be prepared to meet predetermined weather conditions and have advantages over the eutectic mixture because of the increased HD content. Mixtures of H and L are not satisfactory due to the poor storage characteristics.

L is rapidly hydrolysed in the liquid or vapour state; HD hydrolyses slowly at ordinary temperatures.

Odour: Garlic-like.
Skin and eye toxicity: Very high.
Rate of action: Produces immediate stinging of skin and redness within 30 minutes; blistering delayed about 13 hours.

The raw material, an American Type D 1 ton chemical agent pressure container; all such containers and their contents are being destroyed under the terms of the Chemical Weapons Convention

HL liquid causes severe damage to the eyes. Contamination of the skin is followed after a short time by reddening, then by blistering which tends to cover the entire area of the reddened skin. The respiratory lesions are similar to those produced by Mustard, except that in the most severe cases pulmonary oedema may be accompanied by pleural effusion. Liquid on the skin, as well as inhaled vapour, is absorbed and may cause systemic poisoning. This change is manifested in capillary permeability which permits the loss of sufficient fluid from the blood stream to cause blood thickening, shock and death.

Duration of effectiveness depends on the munitions used and the weather. Somewhat shorter than HD when heavily splashed, liquid persists one to two days under average weather conditions and a week or more under very cold conditions.

VERIFIED

Mustard-T Mixture (HT)

HT is a mixture of 60 per cent HD and 40 per cent T. T, a sulphur and chlorine compound similar in structure to HD, is a clear yellowish liquid with an odour similar to HD. HT has a strong blistering effect, has a longer duration of effectiveness, is more stable and has a lower freezing point than HD. Its low volatility makes effective vapour concentrations in the field difficult to obtain. Properties are essentially the same as those of HD.

HT causes blisters, irritates the eyes and is toxic when inhaled.

Duration of effectiveness depends on munitions used and the weather. Somewhat longer than HD when heavily splashed, liquid persists one to two days under average weather conditions and a week or more under very cold conditions.

VERIFIED

Nitrogen Mustards

The Nitrogen Mustards are a group of related compounds which may be considered as derivatives of ammonia because the hydrogen atoms are replaced by various organic radicals. In each of these chemical agents, nitrogen is the central atom. Three members of the group are described individually in the following entries.

VERIFIED

Nitrogen Mustard (HN-1)

Chemical name: 2,2'dichloro-triethylamine.
Rate of hydrolysis: Quite slow.
Odour: Faintly fishy or musty.
Skin and eye toxicity: Eyes are very susceptible to low concentration; higher concentrations are required to produce incapacitating effects by skin absorption rather than by eye injury.
Rate of action: Delayed; 12 hours or longer.
Irritates the eyes in dosages which do not significantly damage the skin or respiratory tract, insofar as single exposures are concerned. This irritation appears in a shorter time than that from HD. After mild vapour exposure, there may be no skin lesions. After severe vapour exposure, or after exposure to liquid HN-1, erythema may appear earlier than in HD contamination. There may be irritation and itching as with HD. Later, blisters may appear in the erythematous areas. The skin lesions are similar to those caused by HD. Effects on the respiratory tract include irritation of the nose and throat, hoarseness progressing to loss of voice and a persistent cough. Fever, laboured respiration and moist rales (an abnormal crackling sound from the lungs caused by the accumulation of fluids) may develop. Broncho-pneumonia may appear after the first 24 hours. Following ingestion or systemic absorption, the HN-1s cause inhibition of cell mitosis resulting in depression of the blood-forming mechanism and injury to other tissues. Severe diarrhoea, which may be haemorrhagic, occurs. Lesions are most marked in the small intestine and consist of degenerative changes and necrosis in the mucous membranes. Ingestion of 2 to 6 mg causes nausea and vomiting.

Duration of effectiveness depends on the munitions used and the weather. Somewhat shorter than HD when heavily splashed, liquid persists one to two days under average weather conditions and a week or more under very cold conditions.

VERIFIED

Nitrogen Mustard (HN-2)

HN-2 is highly unstable and is no longer seriously considered as a chemical agent. It is rated as somewhat more toxic than HN-1.
Rate of hydrolysis: In winter, hydrolysis is fairly rapid until 50 per cent complete. Alkalies induce hydrolysis.
Skin and eye toxicity: HN-2 has the greatest blistering power of the nitrogen mustards in vapour form but is intermediate as a liquid blistering agent. Toxic eye effects are produced more rapidly than by HD.
Rate of action: Skin effects delayed 12 hours or longer.
Irritates the eyes in dosages which do not significantly damage the skin or respiratory tract, insofar as single exposures are concerned. This irritation appears in a shorter time than that from HD. After a mild vapour exposure, or after exposure

to liquid HN-2, erythema may appear earlier than in HD contamination. There may be irritation and itching as with HD. Later, blisters may appear in the erythematous areas. The skin lesions are similar to those caused by HD. Effects on the respiratory tract include irritation of the nose and throat, hoarseness progressing to loss of voice and a persistent cough. Fever, laboured respiration and moist rales may develop. Broncho-pneumonia may appear after the first 24 hours. Following ingestion or systemic absorption, the HN-2s cause inhibition of cell mitosis resulting in depression of the blood-forming mechanism and injury to other tissues. Severe diarrhoea, which may be haemorrhagic, occurs. Lesions are most marked in the small intestine and consist of degenerative changes and necrosis in the mucous membranes. Ingestion of 2 to 6 mg causes nausea and vomiting.

Duration of effectiveness depends upon the munitions used and the weather. Somewhat shorter than HD when heavily splashed, liquid persists one to two days under average weather conditions and a week or more under very cold conditions.

VERIFIED

Nitrogen Mustard (HN-3)

HN-3 is the most stable in storage of the three Nitrogen Mustards and would seem to be admirably suited for use in artillery shells.
Chemical name: 2,2',2 – trichloro-triethylamine.
Rate of hydrolysis: Very slow.
Skin and eye toxicity: Eyes are very susceptible to low concentrations; higher concentrations are required to produce incapacitating effects by skin absorption.
Rate of action: Most symptoms delayed 4 to 6 hours as after exposure to HD but in some cases lacrimation, eye irritation and photophobia develop immediately.
HN-3 irritates the eyes in dosages which do not significantly damage the skin or respiratory tract, where single exposures are concerned. This irritation appears in a shorter time than that of mustard. After mild vapour exposure there may be no skin lesions. After severe vapour exposure, or exposure to liquid HN-3, erythema may appear earlier than in HD contamination. There may be irritation and itching as with HD. Later, blisters may appear in the erythematous areas. The skin lesions are similar to those caused by HD. Effects on the respiratory tract include irritation of the nose and throat, hoarseness progressing to loss of voice and a persistent cough. Fever, laboured respiration and moist rales may develop. Broncho-pneumonia may appear after the first 24 hours. Following ingestion or systemic absorption, the HN-3s cause inhibition of cell mitosis resulting in depression of the blood-forming mechanism and injury to other tissues. Severe diarrhoea, which may be haemorrhagic, occurs. Lesions are most marked in the small intestine and consist of degenerative changes and necrosis in the mucous membranes. Ingestion of 2 to 6 mg causes nausea and vomiting.

Duration of effectiveness is considerably longer than HD (see HD).

VERIFIED

Phenyldichloroarsine (PD)

Although Phenyldichloroarsine (PD) is classed here as a blister agent, it also acts as a vomiting agent.
Rate of hydrolysis: Rapid.
Odour: None.
Skin and eye toxicity: Approximately 30 per cent as toxic as HD to the eyes. On bare skin PD is about 90 per cent as blistering as HD but it is decomposed immediately by wet clothing.
Rate of action: Immediate effect on eyes; effects on skin delayed from 30 minutes to 1 hour.
Similar to the vomiting agents but with added blistering action. Its limited use during the First World War did not indicate any marked superiority of PD over the other vomiting agents used.

Duration of effectiveness depends on the munitions used and the weather. Somewhat shorter than HD under dry conditions; short duration when wet.

VERIFIED

Phosgene Oxime (CX). Dichloroformoxime

CX may appear as a colourless, low-melting point (crystalline) solid or as a liquid. It has a high vapour pressure, slowly decomposes at normal temperatures (depending on the temperature and humidity), boils at 53 to 54°C, melts at 39 to 40°C and is readily soluble in water. It has a disagreeable penetrating odour.

CX is a powerful irritant which produces immediate pain varying from a mild prickling sensation to a feeling resembling a bee sting. It causes violent irritation to the mucous membrane of the eyes and nose. When it comes in contact with the skin, the area becomes blanched in 30 seconds and is surrounded by a red ring. A wheal forms in about 30 minutes and the blanched area turns brown in approximately 24 hours, with a scab forming in a week. The scab generally falls off after three weeks. Itching may be present throughout healing, which in some cases may be delayed beyond two months.

VERIFIED

Simulated Mustard (MR)

After the First World War a number of substances having physical properties similar to HD and generally containing a dye and an odoriferous constituent, were tested in an attempt to find a substitute for mustard in the testing of dispersion apparatus and munitions and for training purposes. Almost all of the substances had some disadvantageous feature, such as instability, corrosiveness to metals, staining to fabrics, relatively high cost or, in the case of those containing aniline and nitrobenzene, some toxicity. The use of molasses residuum solution (MR) as a simulant for HD was first suggested in March 1937. Since that time MR has been used successfully in tests of aircraft smoke tanks, thin-case bombs and chemical land mines. MR is well adapted to training needs.

MR is a mixture consisting of a 25 per cent solution by volume of molasses residuum in water. This is obtained in the manufacture of ethyl alcohol from molasses. As obtained from industry, the concentrated molasses residuum is a dark brown viscous liquid with a characteristic molasses odour. It is soluble in water to the extent of 90 per cent. The undiluted material is more viscous than HD, but when one volume is diluted with three volumes of water it forms a dark brown liquid of thin, syrupy consistency which, although it has a lower specific gravity than either pure or crude mustard, has a viscosity and surface tension sufficiently close to those of HD to ensure comparable flow characteristics.

MR has a low freezing point and can be used in moderately cold weather without danger of freezing.

The solution has a distinctive molasses odour, which when sprayed from an aircraft can be detected on the contaminated area from approximately 45 minutes to 1 hour. The solution has sufficient colour to produce easily defined patterns. The patterns obtained by dispersion from aircraft smoke tanks, chemical land mines and thin-case bombs are similar to those produced by HD.

The solution is harmless by contact to man and animals but, because of the cresol used as a stabilising agent, should not be swallowed.

VERIFIED

BLOOD AGENTS

Blood agents are absorbed into the body primarily by breathing. They affect bodily functions through action on the enzyme cytochrome-oxidase, thus preventing the normal transfer of oxygen from the blood to body tissue. They are generally volatile and their utility lies in their quick effect and quick dispersal afterwards.

VERIFIED

Arsine (SA)

Chemical name: Arsenic trihydride, arsine.
Rate of hydrolysis: Rapid, but an equilibrium condition is reached quickly. (Under certain conditions, SA forms a solid product with water which decomposes at 30°C).
Odour: Mild, garlic-like.
No skin or eye toxicity.
SA interferes with functioning of the blood, damaging the liver and kidneys. Slight exposure causes headache and uneasiness. Increased exposure causes chills, nausea and vomiting. Severe exposure damages blood, causing anaemia.
Duration of effectiveness: Short.

VERIFIED

The reality of chemical warfare – British troops about to evacuate a casualty in a casualty bag

Cyanogen Chloride (CK)

Chemical name: Cyanogen chloride.
Rate of hydrolysis: Very low.
Odour: Its irritating and lacrymatory properties are so great that the odour can go unnoticed.
Skin and eye toxicity:, highly irritating to eyes and mucous membranes.
Rate of action: It is assumed that the effect of CK arises from its conversion to AC in the body. In general, CK may be considered a rapid-acting chemical agent.
The general action of CK is similar to that of AC — it interferes with utilisation of oxygen by the body tissues. However, it differs from AC in that it has a choking effect, a strong irritating effect and causes a slow breathing rate.
Duration of effectiveness: Short. Vapour may persist in jungle and forest for some time under suitable weather conditions.
CK can be used for filter element penetration.

VERIFIED

Hydrogen Cyanide (AC, also known as HCN)

Chemical name: Hydrogen cyanide or hydrocyanic acid.
Rate of hydrolysis: Low under field conditions.
Odour: Similar to peach kernels.
Skin and eye toxicity: Moderate.
Rate of action: Very rapid. Death occurs within 15 minutes after a lethal dosage has been received.
AC interferes with utilisation of oxygen by the body tissues by inhibition of the enzyme cytochro neoxidase. AC causes a marked stimulation of the breathing rate. A variation of AC in powdered form, known as Zyklon B, was used in the German extermination camps prior to 1945.
Duration of effectiveness: Short. The agent is highly volatile and in the gaseous state it is lighter than air.
AC can be used to generally clog respirator and other filters.

VERIFIED

CHOKING AGENTS

Choking (pulmonary) agents attack the lining of the lungs by damaging the membranes and allowing leakage of fluid from the delicate alveoli inside the lungs. Fatalities of this type have been referred to as 'dry-land drownings'.

The choking agent originally employed as a war gas, Chlorine gas, is now unlikely to be considered for use as a military chemical agent but may still be encountered as a result of industrial accidents.

VERIFIED

Chlorine

Chemical name: Chlorine.
Rate of hydrolysis: Slow.
Odour: Pungent and unmistakable, like bleaching powder.
Skin and eye toxicity: Irritates eyes.
Powerful irritant, first on upper and then on lower respiratory tract.
Duration of effectiveness: Short.
Chlorine is mentioned here for historical purposes as it is now unlikely to be considered as a military chemical agent and is thus rarely encountered in a military context, although it may be encountered as a result of industrial accidents or as a secondary result of destructive military activities.

VERIFIED

Diphosgene (DP)

As a shell filling, DP has the advantage of a high boiling point which permits filling in the field, whereas CG, with a low boiling point, must be kept refrigerated during filling operations under summer conditions. However, DP does have certain disadvantages. Since it is slightly lacrymatory, troops are not as easily surprised as with CG. Furthermore, its lower volatility adds to the difficulty of setting up an effective surprise concentration. (DP is converted to CG in the body and exerts its effect after this conversion.)
Chemical name: Trichloromethyl chloroformate.
Rate of hydrolysis: Slow at ordinary temperatures.
Odour: New mown hay or grass; green corn.
Skin and eye toxicity: No effect on skin; slight lacrymatory effect.
Rate of action: Delayed. Although symptoms may follow exposure to a high concentration of DP, a delay of 3 hours or more may elapse before exposure to a low concentration causes any ill effects.
Since DP is converted to CG in the body, the physiological action is the same for both agents. The casualty effect is exerted solely on the lungs and results in damage to capillaries.
Delayed or immediate action casualty agent, depending upon dosage rate.

VERIFIED

PFIB

PFIB produces pulmonary oedema in humans and animals, even in small concentrations and has been associated with the condition known as 'polymer fume fever'. It is produced when polytetrafluoroethene (PTFE) is pyrolised. Very little information has been released regarding this agent.

Chemical name: Perfluoroisobutene.

Odour: None.

Colour: None.

No known eye or skin toxicity.

PFIB exerts its effects solely on the lungs. After exposure to even small amounts (measured in parts per million) victims develop headache, coughs, sub-sternal pain, dyspnoea and fever within the first hour and the symptoms become worse at 6 to 8 hours. The pulmonary oedema can cause death 8 to 48 hours after exposure. Even minute concentrations inhaled may lead to influenza-type disabling ailments.

Possible delayed action casualty agent.

VERIFIED

Phosgene (CG)

CG, normally a chemical agent with a short duration of effectiveness, was used extensively in the First World War. More than 80 per cent of the First World War chemical agent fatalities were caused by CG.

Chemical name: Carbonyl chloride.

Not readily hydrolised under usual field conditions; however, rain destroys its effectiveness and heavy vegetation, jungle and forests cause considerable loss of CG by hydrolysis on leaf surfaces.

Odour: New mown hay or grass; green corn.

Rate of action: Delayed. Although immediate symptoms may follow exposure to a high concentration of CG, a delay of 3 hours or more may elapse before exposure to a low concentration causes any ill effects.

No skin or eye toxicity.

CG exerts its effect solely on the lungs and results in damage to the capillaries. It causes seepage of watery fluid into the air sacs. When a lethal amount of CG is received, the air sacs become so flooded that air is excluded and the victim dies of anoxia (oxygen deficiency). If the amount of CG is less than lethal and proper care is provided, the watery fluid is reabsorbed, the air cell walls heal and the patient recovers. However, respiratory problems of one form or another may remain chronic for years. The severity of poisoning cannot be estimated from the immediate symptoms, since the full effect is not usually apparent until 3 or 4 hours after exposure. Most deaths occur within 24 hours.

Duration of effectiveness is short. However, vapour may persist for some time in low places under calm or light winds and stable atmospheric conditions (inversion).

Delayed action casualty agent.

VERIFIED

NERVE AGENTS

Nerve agents perhaps are the most commonly considered of all chemical agents. While different varieties of nerve agents differ chemically, they all act in the same way. They interfere with the balance between the sympathetic (adrenergic) and parasympathetic (cholinergic) nervous systems which together form the autonomic nervous system. Normally in the body there is controlled discharge within the nervous system due to destruction by cholinesterase of acetylcholine, a product of nerve cell metabolism. The nerve agents were found by German scientists to react with the cholinesterase in an irreversible reaction in tissue fluid, permitting accumulation of acetylcholine and continual stimulation of the nervous system.

Over the years, researchers have refined the types and effectiveness of different nerve agents. Toxicity has been significantly increased, principally with the emergence of VX (see separate entry in this section), and persistence can be altered according to the effect required by the aggressor.

Rapid use of so-called autonomic blocking agents which act directly on the effector nerve cell will nullify the effect of acetylcholine. No apparent chemical reaction seems to occur between these autonomic blocking agents and acetylcholine. Atropine salts are the most commonly used autonomic blocking agents although research is bringing other antidotes to light. See under Protection (individual) – medical countermeasures.

VERIFIED

Sarin (GB)

Sarin (GB) is a colourless liquid; its vapour is also colourless.

Chemical name: Isopropylmethylphosphono fluoride.

Very rapidly hydrolysed in alkaline solutions.

Almost no odour in pure state.

Skin and eye toxicity: Twice as high as GA; LD50 is 0.28 mg/kg by mouth and 0.05 mg/kg by eye.

(a) Eye effect. Very high toxicity; much greater through eye than through skin. Vapour causes pupil of eye to contract; vision difficult in dim light.

(b) Skin effect. Liquid does not injure skin but penetrates it rapidly. Immediate decontamination of the smallest drop is essential. Vapour also penetrates the skin.

Rate of action: Very rapid – death usually occurs within 15 minutes after fatal dosage absorbed.

Individuals poisoned by GB display approximately the same sequence of symptoms regardless of the route by which the poison enters the body (whether by inhalation, absorption or ingestion). These symptoms, in normal order of appearance are: running nose, tightness of chest, dimness of vision and pinpointing of the eye pupils, difficulty in breathing, drooling and excessive sweating, nausea, vomiting, cramps, involuntary defaecation and urination, twitching, jerking and staggering, headache, confusion, drowsiness, coma and convulsion. These symptoms are followed by cessation of breathing and death. Symptoms appear much more slowly from skin dosage than from respiratory dosages. Although skin absorption great enough to cause death may occur in one to two minutes, death may be delayed for one to two hours. Respiratory lethal dosages kill in 1 to 10 minutes and liquid in the eye kills nearly as rapidly. The number and severity of symptoms which appear are dependent on the quantity and rate of entry of the nerve agent which is introduced into the body. (Very small skin dosages sometimes cause local sweating and tremors with little other effect).

GB evaporates at approximately the same rate as water. Duration of effectiveness depends upon the munitions used and the weather.

VERIFIED

Soman (GD)

Soman (GD) is a colourless liquid which gives off a colourless vapour.

Chemical name: Pinacolylmethylphosphono
Fluoridate.

Odour: Fruity; camphor with impurities.

Skin and eye toxicity: Three times that of GA.

(a) Eye effect. Very high toxicity; vapour causes pupil of eye to contract, resulting in difficulty in seeing in dim light and generally distorted vision. Toxicity much greater through eye than through skin.

(b) Skin effect. Extremely toxic by skin absorption. Liquid does not injure the skin but penetrates it rapidly. Immediate decontamination of the smallest drop is essential.

Rate of action: Very rapid. Death usually occurs within 15 minutes after fatal dosage is absorbed.

Individuals poisoned by GD display approximately the same sequence of symptoms regardless of the route by which the poison enters the body (whether by inhalation, absorption or ingestion). These symptoms, in normal order of appearance are: running nose, tightness of chest, dimness of vision and pin-pointing of the eye pupils, difficulty in breathing, drooling and excessive sweating, nausea, vomiting, cramps, involuntary defaecation and urination, twitching, jerking and staggering, headache, confusion, drowsiness, coma and convulsion. These symptoms are followed by cessation of breathing and death. Symptoms appear much more slowly from skin dosage than from respiratory dosages. Although skin absorption great enough to cause death may occur in 1 to 2 minutes, death may be delayed for 1 to 2 hours. Respiratory lethal dosages kill in 1 to 10 minutes and liquid in the eye kills nearly as rapidly. The number and severity of symptoms which appear are dependent on the quantity and rate of entry of the nerve agent which is introduced into the body. Very small skin dosages sometimes cause local sweating and tremors with little other effect.

Duration of effectiveness depends upon the munitions used and the weather. Heavily splashed liquid persists one to two days under average weather conditions.

VERIFIED

Tabun (GA)

First discovered in 1936, Tabun is a colourless to brownish liquid giving a colourless vapour.

Chemical name: Ethyl-n-dimethylphosphoroamide-cyanidate.

Reacts slowly with water but fairly rapidly with strong acids or alkalies.

Odour: Faintly fruity; none when pure.

Skin and eye toxicity: Stated to be 20 times more toxic than CG.

(a) Eye effect. Very high toxicity; much greater through eye than through skin. Vapour causes the pupil of the eye to contract, resulting in difficulty in seeing in dim light, or generally distorted vision.

(b) Skin effect. Liquid decontamination of the smallest drop is essential. Vapour penetrates skin readily.

Rate of action: Very rapid.

Individuals poisoned by GA display approximately the same sequence of symptoms regardless of the route by which the agent enters the body (whether by inhalation, absorption or ingestion). These symptoms, in normal order of appearance are: running nose, tightness of chest, dimness of vision and pin-pointing of the eye pupils, difficulty in breathing, drooling and excessive sweating, nausea, vomiting, cramps, involuntary defaecation and urination, twitching, jerking and staggering, headache, confusion, drowsiness, coma and convulsion. These symptoms are followed by cessation of breathing and death. Symptoms appear much more slowly from skin dosage than from respiratory dosage. Although skin absorption great enough to cause death may occur in 1 to 2 minutes, death may be delayed for 1 to 2 hours. Respiratory lethal dosages kill in 1 to 10 minutes and liquid in the eye kills nearly as rapidly. The number and severity of symptoms which appear are dependent on the quantity and rate of entry of the nerve agent which is introduced into the body. (Very small skin dosages sometimes cause local sweating and tremors with little other effect).

Duration of effectiveness depends upon munitions used and the weather. Heavily splashed liquid persists one to two days under average weather conditions.

VERIFIED

V-Agents

Included in the category of nerve agents are the V-agents. The standard V-agent is VX while others include VE, VG and VS. They are generally colourless and odourless liquids which do not evaporate rapidly or freeze at normal temperatures. As they have low volatility, their vapour effect is limited and the duration of their effectiveness is increased. V-agents are absorbed by vegetation.

In liquid or aerosol form, V-agents affect the body in a manner similar to that of the G-agents. They are usually disseminated as liquid droplets which produce casualties when absorbed through the skin. Since liquid G-agents evaporate quickly from the skin, the dosage required to produce casualties by that route is high. The appearance of casualties is correspondingly short compared with the much less volatile V-agents. If evaporation is excluded, the appearance of casualties would be roughly similar at the same dose level with both V- and G-agents. Aerosol V-agents produce casualties by inhalation and absorption of agent droplets through the skin.

The chemical name for VX is ethyl S-2 diisopropylaminoethyl-methylphosphonothioate.

VERIFIED

CMPF (GF)

GF is a cyclohexyl organophosphate ester whose full chemical name is Cyclohexyl methyl phosphonofluoridate. It has been manufactured by Iraq and is similar in many ways to GB although less volatile and more persistant. Its vapour is less toxic than GB (LCt_{50}: 75-120 mg min m^{-3}) but percutaneous toxicity is about the same. It is a colourless liquid, said to have a faint fruity odour The boiling point is 239°C.

VERIFIED

TEAR AGENTS

Tear agents are used operationally for riot control but are often also used to train personnel in the use of individual and collective protection procedures and drills. Although often considered relatively modern, they were in fact first used in war by the French against German forces in 1914 but with little effect. Their impact lies in their ability to cause intense irritation to affected parts of the body and to overstimulate the tear ducts to produce fluid. They have little more than nuisance value in war in view of the effectiveness of modern respiratory and skin protection, although they have been trialled in conjunction with other agents to break down protection discipline.

VERIFIED

Bromobenzylcyanide (BBC)

Rate of hydrolysis: Very slow.
Odour: Like soured fruit but not unpleasant.
Skin and eye toxicity: Irritating; not toxic.
Rate of action: Instantaneous.

BBC produces a burning sensation of the mucous membranes and severe irritation and lacrymation of the eyes with acute pain in the forehead. BBC is considerably less toxic than CG.

Duration of effectiveness depends upon the weather and the munitions used. Heavily splashed liquid persists one to two days under average weather conditions.

VERIFIED

Chloroacetophenone (CN)

Not readily hydrolised.
Odour: Fragrant; similar to apple blossom.
Skin and eye toxicity: Irritating; not toxic in concentrations likely to be encountered in the field.
Rate of action: Practically instantaneous.

In addition to powerful lacrymatory effects, CN is an irritant to the upper respiratory passages. In higher concentrations, it is irritating to the skin and causes a burning and itching sensation, especially on moist parts of the body. High concentrations can cause blisters. The effects are similar to those of sunburn, are entirely harmless and disappear in a few hours. Certain individuals may experience nausea following exposure to CN.

Duration of effectiveness is short, because the agent is disseminated as an aerosol.

VERIFIED

CNB

CNB was adopted in 1920 and remained in use until it was replaced by CNS. The advantage claimed for CNB was that its lower chloroacetophenone content made it more satisfactory than CNC for training purposes. Actually, the same result can be obtained with CNC merely by using a lower concentration.

No chemical name. Solution of chloroacetophenone in benzene and carbon tetrachloride.
Odour: Like benzene.
Skin and eye toxicity: Irritating; not toxic.
Powerfully lacrymatory.
Duration of effectiveness is short.

VERIFIED

CNC

No chemical name; solution of chloroacetophenone in chloroform.
Odour: Similar to chloroform.
Skin and eye toxicity: Irritating; not toxic.
Rate of action: Instantaneous.
Causes the flow of tears, irritates respiratory system and causes stinging of skin.
Duration of effectiveness is short, because the agent is disseminated as an aerosol.

VERIFIED

CNS

No chemical name; mixture of chloroacetophenone, chloropicrin and chloroform.
Odour: Like flypaper.
Skin and eye toxicity: Irritating; not toxic.
Rate of action: Instantaneous.

In addition to having effects described under CN, CNS also has the effects of chloropicrin (PS) which acts as a vomiting agent, a choking agent and a tear agent. CNS may cause lung effects similar to those of CG and may also cause nausea, vomiting, colic and diarrhoea which may persist for weeks. The lacrymatory effects of PS are less marked than those of CN and are relatively unimportant in CNS, as shown by the fact that these effects are no greater than with CNC which contains no PS.

Rate of detoxification: The effects of chloropicrin are long-lasting and cumulative and may cause the effects of CNS to be prolonged for weeks. Such a prolonged effect may be highly undesirable in training and riot control so CNS is now rarely used.

Duration of effectiveness is short.

VERIFIED

O-chlorobenzalmalononitrile (CS)

CS is a white crystalline powder. It has a minimum purity of 96 per cent, is insoluble in water and ethanol, but is soluble in methylene chloride.
Rate of hydrolysis: Unknown.
Odour: Pepper-like.
Skin and eye toxicity: Highly irritating but not toxic.
Rate of action: Immediate.

CS produces immediate effects even in low concentrations. The onset of incapacitation is 20 to 60 seconds and the duration of effects is 5 to 10 minutes after the affected individual is removed to fresh air. During this time affected individuals are incapable of effective concerted action. The physiological effects include extreme burning of the eyes accompanied by a copious flow of tears, coughing, difficulty in breathing and chest tightness, involuntary closing of the eyes, stinging sensation of moist skin, running nose and dizziness or swimming of the head. Heavy concentrations will cause nausea and vomiting in addition to the above effects.

Duration of effectiveness is short.
CS has now largely replaced CN and other tear agents as a riot control agent.

VERIFIED

INCAPACITATING AGENTS

Incapacitating agents are designed to produce physiological or mental effects which interfere with the victim's capacity to operate effectively for a significant period. Effects are reversible and recovery is generally complete.

The types include those which produce temporary physical disability such as paralysis, blindness or deafness and those that produce temporary mental dysfunction.

Experiments by the United States sought to identify a group of agents that could be used to deliver incapacitation amongst opposing forces in order to gain a military goal, without the need to deal with large numbers of casualties or the disposal of the deceased. However, the effects are these agents were unpredictable. It may seem surprising, with the opprobrium that lethal agents attract, that more modern research has not been conducted to identify agents which can incapacitate people for limited periods. Although the ethics of the use of such agents are complex and questionable such a capability may have a role in peacekeeping and riot control.

VERIFIED

BZ

In the 1950s, the United States experimented with non-lethal agents that were designed to alter the decision-making process of the opponent. One such agent was BZ. Sometimes referred to as 'Buzz', it was designed to meet these requirements. However, trials revealed that BZ and similar agents, such as Lysergic Acid Derivative (LSD), were both unpredictable and unreliable. BZ was withdrawn from use and existing American stocks of BZ have all been destroyed.

BZ (Benactyzine) is a crystalline solid at normal temperatures and is sufficiently stable to be disseminated as smoke from a pyrotechnic device. Other members of this glycolic acid ester series may be liquid at usual temperatures and absorbed through the skin. Once in the body, BZ acts by blocking the activity of cholinergic synapses in a manner similar to that of atropine but, unlike atropine, BZ produces predominantly central rather than peripheral effects.

General symptoms from agent BZ are: interference with ordinary activity, dry, flushed skin, tachycardia (that is, abnormally rapid beating of the heart), urinary retention, constipation, slowing of mental and physical activity, headache, giddiness, disorientation, delusions, hallucinations, drowsiness, sometimes maniacal behaviour and an increase in body temperature. For some individuals these symptoms may persist for several days. The symptoms may also vary widely from individual to individual.

Some reports credit the Yugoslav National Army with the use of BZ against Muslim civilians at Srebrenica in 1994. These reports recur from time to time but they remain unsubstantiated. It seems highly unlikely that such an agent would have been chosen by these forces as a weapon of war since, as pointed out earlier, the effect of BZ is so unpredictable.

VERIFIED

Swiss troops in full NBC kit wearing respirators fitted with Protector Sabre filter canisters

VOMITING AGENTS

A vomiting agent is unlikely to be a weapon of choice by an aggressor. However, the unpleasant effects these agents cause could be utilised in conjunction with other agents to break down the discipline of personal and collective protection. Normally solid at room temperature, they vaporise when heated and then condense to form toxic aerosols. Under field conditions, vomiting agents cause great discomfort. When released in enclosed spaces, the high concentrations can cause serious illness or death to vulnerable victims.

VERIFIED

Operating a Chemical Agent Monitor (CAM) on a naval vessel

Adamsite (DM)

Chemical name: Diphenylaminochloroarsine (also phenarsazine chloride).
Rate of hydrolysis: Quite rapid when in aerosol form.
No pronounced odour.
Skin and eye toxicity: Irritating; relatively non-toxic.
Rate of action: Very high. Only about 1 minute is required for temporary incapacitation.

General symptoms in progressive order: irritation of the eyes and mucous membranes; viscous discharge from the nose similar to that caused by a cold; sneezing and coughing; severe headache; acute pain and tightness in the chest; and nausea and vomiting. The effects develop more slowly than with DA and, for moderate concentrations, last about 30 minutes after an individual leaves the contaminated atmosphere. At higher concentrations, the effects may last up to 3 hours.

Duration of effectiveness is short because the agent is disseminated as an aerosol.

VERIFIED

Diphenylchloroarsine (DA)

Rate of hydrolysis: Slow in mass but rapid when finely divided.
No pronounced odour.
Skin and eye toxicity: Irritating; not toxic.
Rate of action: Very rapid; within 2 or 3 minutes after 1 minute of exposure.

General symptoms in progressive order: irritation of the eyes and mucous membranes; viscous discharge from the nose similar to that caused by a cold; sneezing and coughing; severe headache; acute pain and tightness in the chest; and nausea and vomiting. For moderate concentrations the effects last about 30 minutes after an individual leaves the contaminated atmosphere. At higher concentrations, the effects may last up to several hours.

Duration of effectiveness is short because the agent is disseminated as an aerosol.

VERIFIED

Diphenylcyanoarsine (DC)

Rate of hydrolysis: Very slow.
   Odour: Similar to a mixture of garlic and bitter almonds.
   Skin and eye toxicity: Irritating; not toxic.
   Rate of action: Very rapid. Higher concentrations are intolerable in about 30 seconds.
   General symptoms in progressive order: irritation of the eyes and mucous membranes; viscous discharge from the nose similar to that caused by a cold; sneezing and coughing; severe headache; acute pain and tightness in the chest; and nausea and vomiting. DC is more toxic than DA. For moderate concentrations the effects last about 30 minutes after an individual leaves the contaminated atmosphere. At a higher concentration, the effects may last up to several hours.
   Duration of effectiveness is short, because the agent is disseminated as an aerosol.

VERIFIED

PRECURSORS

Elements and compounds necessary for the production of CW agents, but which also have widespread and essential use in industry, are known as precursors.

It is this aspect of CW that poses one of the biggest proliferation headaches for arms controllers. Before the CWC, earlier international agreements had been supplemented by the work of the Australia Group, a group of nations that became active in the mid-1980s to address the CW and BW proliferation issues through the exercise of tougher export controls. Both the Australia Group and the CWC provide comprehensive lists of materials whose use should draw scrutiny. The CWC, with its more formal framework of enforcement, lists these key precursors in the Chemical Annex (B: Schedule of Chemicals). It also gives criteria for the inclusion or exemption of materials by the signatories in the future.

   Control of precursors is clearly a contentious issue. The trickiest part is to decide whether a particular order for a scheduled material by a particular end-user satisfies the criteria of the Convention. If the nation or organisation concerned is not a signatory, there is little that can be done. Verification methods need to work at two levels on the supervision of precursors. At one level, the diplomatic work must ensure that the rules are apparently being complied with. At another level, there needs to be sufficient covert policing to ensure that evidence of non-compliance can be gathered and presented to OPCW and to the UN in a form which makes it irrefutable and unambiguous — not an easy task.

   The CWC text, together with its schedules, can be found at URL: http://www.opcw.nl. The list in Table 1 is not exhaustive. It will be maintained under review in future editions.

Table 1 – Precursors

Precursor chemical	Industrial uses	Agent production
2-Chloroethanol	organic synthesis, chemical manufacture, insecticides, solvent	HD, HN-1, Q
3-Hydroxy-1-methylpiperidine	poss pharmaceuticals, other specific uses not identified	poss BZ
3-Quinuclidinol and 3-Quinuclidone	hypotensive agent, poss pharmaceuticals	poss BZ
Ammonium bifluoride	ceramics, disinfectant for food equipment, electroplating, etching glass	GB, GD, GF
Arsenic trichloride	organic synthesis pharmaceuticals, insecticides, ceramics	SA, L, DM, DA
Benzilic acid	organic synthesis	BZ
Diethyl ethylphosphonate	heavy metal extraction, petrol additive, anti-foam agent, plasticiser	ethyl sarin (GE)
Diethylaminoethanol	organic synthesis, anti-corrosion compositions, pharmaceuticals, textile softeners	VX, VM
Diethyl methylphosphonate	organic synthesis	VX
Diethyl N. N-dimethyl phosphoramidate	organic synthesis, other specific uses not identified	GA
Diethylphosphite	organic synthesis, paint solvent, lubricative additive	VG, GB, GD, GF
Diisopropylamine	specific uses not identified	VX
Dimethyl ethylphosphonate	organic synthesis	GE
Dimethyl methylphosphane (DMMP)	flame retardants	GB, GD, GF
Dimethylamine	organic synthesis, pharmaceuticals, detergents, pesticides, petrol additive, missile fuels, rubber vulcanisation	GA
Dimethylamine HCl	organic synthesis, pharmaceuticals, surfactants, pesticides, petrol additives	GA
Dimethylphosphate	organic synthesis, lubricant additive	GB, GD, GF
Ethylphosphonous dichloride	organic synthesis, other specific uses not identified	VE, VS, GE
Ethylphosphonous difluoride	organic synthesis	VE, GE
Ethylphosphonyl dichloride	organic synthesis, other specific uses not identified	ethyl sarin (GE)
Ethylphosphonyl difluoride	organic synthesis, other specific uses not identified	ethyl sarin (GE)
Hydrogen fluoride	chemical reactions catalyst, liquid rocket fuel additive, uranium refining	GB, GD, GE, GF
Methyl benzilate	organic synthesis, tranquillisers	BZ
Methylphosphonous dichloride	organic synthesis	VX, VM, GB, GD, GF
Methylphosphonyl dichloride	organic synthesis, other specific uses not identified	GB, GD, GF
Methylphosphonyl difluoride	organic synthesis, other specific uses not identified	GB, GD, GF
N. N.diisopropyl aminoethanethiol	organic synthesis	VX, VS
N. N.-diisopropyl-(beta)- aminoethanol	organic synthesis, other specific uses not identified	VX

Table 1 – Precursors continued

Precursor chemical	Industrial uses	Agent production
N. N.diisopropyl-(beta)-aminoethyl chloride	organic synthesis	VX, VS
O-ethyl.2-diisopropyl aminoethyl methyl-phosphonate (QL)	specific uses not identified	VX
Phosphorous oxychloride	organic synthesis, plasticisers, petrol additives, hydraulic fluids, insecticides, dopant for semi-conductors grade silicon, flame retardants	GA
Phosphorous pentachloride	organic synthesis, pesticides, plastics	GA
Phosphorous pentasulfide	organic synthesis, insecticide, mitocides, lubricant oil additives, pyrotechnics	VX, VG
Phosphorous trichloride	organic synthesis, insecticides, petrol additives, plasticisers, surfactants, dyestuffs	VG, GB*, GD*, GF*
Pinacolone	specific uses not identified	GD
Pinacolyl alcohol	specific uses not identified	GD
Potassium bifluoride	fluorine production catalyst, coal treatment, fluid in silver solder	GB, GD, GF
Potassium cyanide	extraction of gold/silver, pesticide	GA
	fumigant, electroplating	HC
Potassium fluoride	fluorination of organic compounds, cleaning, disinfecting of dairy, brewing, food equipment, glass and porcelain manufacture	GB, GD, GF
Sodium bifluoride	antiseptic, laundry neutraliser, tin plate production	GB, GD, GF
Sodium cyanide	extraction of gold/silver, fumigant, dye/pigment manufacture, metal hardening, nylon production	GA, HC, CK
Sodium fluoride	pesticide, disinfectant, dental prophylaxis, glass and steel manufacture	GB, GD, GF
Sodium sulfide	paper manufacturing, rubber manufacturing, metal refining, dye manufacture	HD
Thiodiglycol	organic synthesis, carrier for dyes, lubricant additives, plastics manufacture	HD
Thionyl chloride	organic synthesis	GB, GD, GF, HD
Thionyl chloride	chlorinating agent, catalyst	Sesqui mustard
Thionyl chloride	pesticides	HN-2
Thionyl chloride	engineering plastics	HN-3
Triethanolamine	organic synthesis, detergents, cosmetics, corrosion inhibitor, plasticiser, rubber accelerator	HN-3
Triethyl phosphite	organic synthesis, plasticisers, lubricant additives	VG
Trimethyl phosphite	organic synthesis, insecticides	used to make DMMP

(*   processing stages)

VERIFIED

TECHNICAL DEVELOPMENTS

TECHNICAL DEVELOPMENTS

INTRODUCTION

This highlights key technical developments in each area of NBC defence: detection, protection, contamination control (decontamination), demilitarisation, training and simulation. Science continues to provide answers. The skill of the developer is to spot the 'star' technologies early enough to get ahead of the field.

The procurement process

The procurement authority has a more difficult task. It has to convince sceptical defence and civil emergency budget-holders that the science is sufficiently advanced to provide durable solutions. In addition, it has to be sensitive to fashion. In other words, if NBC defence is out of favour, the authority's requirements in the annual funding competition may lose out against demands in other key areas. It may need to hold back. A danger to which many advisers fall prey is that of over dramatising the problem – alarming senior staffs who, whilst receptive to the needs of NBC defence at the outset, realise later that the requirement has been overplayed. After all, they have to balance the overall needs of the organisation. As a result, future bids are likely to be treated with extreme scepticism or, ignored.

Driving forces

There has been a sharp focus on technical development this year, given impetus by the events of 11 September 2001 and afterwards. However, evidence from the many conferences and exhibitions points towards an industry more prepared to relaunch established technology than to invest heavily in new ideas. The technical developments shown here are specifically NBC-related. They deal with the detection of radiation, pathogens and toxic chemicals. They address protection for people against the toxic hazard, prevention of damage to equipment by toxic agents and the training and maintenance of NBC standards. They do not address the many other areas which may serve to reduce the NBC challenge by other means. It should not be forgotten by the NBC defender that defending his own forces starts in the enemy's HQ. The best form of defence against an aircraft CW attack is to destroy the enemy's airbase. NBC defence is not just about masks and suits.

UPDATED

DETECTION

Radiological (nuclear) detection

The basic science in radiological detection has altered little over the years. Ever since the Chernobyl incident focussed the world's attention on the risks from older reactors (especially those of the Chernobyl type – RBMK), authorities have realised the need for better monitoring and warning at low levels of radiation. With the instability which followed the break-up of the Soviet Union, this has become a live defence and civil emergency planning issue.

Much of the instrumentation designed in the 1950s and 1960s was specialised towards monitoring the rate at which injurious levels of radiation were being accumulated (and the total personal dose). Equally, 'survey meters' were aimed at measuring highly contaminated post-burst situations. These were instruments for use in the highly active fallout environment following a nuclear detonation where people were going to get ill. It was a matter of commanders spreading the misery as equally as possible to complete essential military tasks. They needed instrumentation which worked in these ranges. Today, justified fears over nuclear terrorism and the vast increase in the number of peacekeeping operations in areas of potentially high nuclear risk, have raised the need for new equipment to detect radiation in the lower ranges as well.

Legal pressures have encouraged this trend. Litigation has become popular and citizens discover they can bring actions against governments more easily. Therefore ministries have begun to address personnel liability. Armed forces can be placed in areas where nuclear power plants or laboratories introduce new casualties of war or accident. There is therefore, a new need to monitor low-intensity radiological risks in ways which were previously unique to the nuclear power industry ('health physics').

The biggest development in all areas of detection has been in signal processing and presentation. There are some good examples to review in this year's edition, including the Siemens EPD2 electronic dosemeter, which measures neutron as well as γ-radiation. The French nuclear industry has a strong presence in this market and has acquired a large range of both North American and European-based commercial detection capability under groups such as COGEMA and Saint-Gobain. New personal devices are about 50 times more sensitive that the TLD badge and about 200 times more so than a film badge. Also, the ability to download the data directly into a network enables much faster presentation of information to the command. This benefit of improved signal processing and computer telephony integration applies equally to chemical and biological detection.

Detection of neutrons is a challenge as, being uncharged, they cannot be measured directly. One technique (Hitachi) uses compounds containing hydrogen and granular Boron (B^{10}) in a substrate, acting as a proton radiator. Fast or thermal neutrons generate recoil protons allowing the impacts to be measured. Similar solid-state techniques are being developed to detect α and β radiation using ultra-thin (<1 μ) layers of silicon. This technology may eventually replace traditional GM or scintillator techniques.

Developments in superconductor technology, continue to produce systems which operate at higher temperatures than in the past. A new compound designated BiPbSrCaCuO and developed by DREV Canada, offers a wide spectrum of detection (from visible to millimetre wavelengths) previously only available to superconductors maintained at close to absolute zero. This capability is offered at the temperature of liquid nitrogen and is ripe for the development of small-scale portable radiation detectors operating at sensitivities, response speeds and ranges previously unavailable.

BW agent detection

The 2001 anthrax events in the USA have given tremendous impetus to the biodetection industry. Even before this, it had become a top priority for most nations with developed NBC defence capability, based on the very real fear of BW use by Iraq during the 1990-1991 Gulf War. The twin challenges of real-time detection and long-range stand-off detection have yet to be overcome. The density of pathogenic bugs in the air can be extremely small. The bugs only reveal their true identity when persuaded to react with their target. Therefore, detection systems have to sample massive quantities of the surrounding atmosphere. This requires sophisticated airflow management and collector devices which strip away a sufficiently large concentration of pathogenic material to allow it to be identified. Even then identification is difficult. The raw approach is to analyse the chemical constituents of the material to see if it is living matter or not (does it contain ATP for example) or whether it matches a pattern typical of a group of bacteria or viruses. Gas Liquefied Chromatography (GLC) can profile the long-chain fatty acids in cell membranes and identify organisms down to strain level but it requires large sample quantities, making it a technique more useful in verification than in tactical detection. Few techniques identify the material with the precision required to administer correct antidotes. None of them are quick as, only by forcing the pathogen-antibody reaction can the characteristic enzymes unique to that organism be generated. By tagging the enzymes (with radioactive, coloured or fluorescent molecules for example), an alarm can be made to trigger and the precise strain identified. Immunoassay and bioassay techniques which speed up this process (ELISA for example) are commonly used in all current systems. PCR and Restriction Fragment Length Polymorphism methods (RFLP – often known as 'genetic fingerprinting') are accurate but still require reactions to take place. Detection times are coming down (the order of 5 minutes or less is now common). In the future, mass spectrometry and laser ionisation techniques in combination may offer a solution and work in the USA and Germany is proceeding to develop a technology called MALDI-TOF (Matrix-Assisted Laser Desorption Ionisation – Time

This new MAV design from Singapore Technologies may be suitable to carry an NBC detection device to probe a suspect area for contamination (Singapore Technologies)   2002/0137935

MCAD – The Man-portable Chemical Agent Detector shows how small size remains a key goal for detection (Graseby Dynamics)   2002/0137937

Of Flight mass spectrometry). This appears to deliver the high ion transmission rate needed to give accurate discrimination over the enormous mass range of harmful organisms. The equipment is large and there is pressure to reduce its size sufficiently to make it viable for portable field use. It can, however, contribute to detection where space, weight and size are less important, in warships or reconnaissance vehicles for example.

The UK continues to evaluate a Prototype Biological Detection System (PBDS) which integrates together many of the technologies described above to give near-realtime detection and identification of BW agents. Similar programmes are being fielded in the US (BIDS for example), Russia, Germany and France.

Long-range detection is also being explored. IR, Doppler radar and laser techniques are being studied in the USA, Russia and Europe and this technology, mounted on an airborne platform, could provide a general alert to force commanders of a hostile BW event. Point-detectors could then be deployed further towards the source. The latter would use existing capture and sampling techniques to confirm or negate the IR alert. These techniques will continue to take some time to refine, although the detection-to-result times are constantly reducing. Trials at Dugway Proving Ground in the US, using a LIDAR system developed by Fibertek Inc, have produced some encouraging results. This equipment is designed to autonomously discriminate between BW and non-BW aerosols. The system detects the laser-induced fluorescence caused by UV excitation of the constituents of the BW aerosol cloud.

CW agent detection
CW detection is a much more mature technology but, like BW, it still requires samples of water, soil or the surrounding air. Whilst similar long-range techniques are being explored, the interim solution has been to mount existing technology detector systems on vehicles capable of deployment down the threat axis or upwind of friendly forces. An example of this is the Chemsonde, developed by BAE Systems. This is a SAW detector which can be parachute dropped into the target area where it radio-links its detection data to HQ. UAVs offer a significant opportunity to deploy current detection technology up threat and upwind. A number of countries are examining winged, balloon-based or rotorcraft UAV platforms as a means of gaining longer range warning and the ability to carry out area surveys. Again, miniaturisation is important here and the US Defence Advanced Research Projects Agency (DARPA) has released contracts for a number of studies into the creation of Micro Air Vehicles (MAVs) – airborne remote surveillance devices which are hand-sized or smaller.

Technical progress in miniaturisation has allowed much greater portability in CW detection, with SAW devices such as JCAD and IMS devices such as the Lightweight Chemical Detector offering real-time individual CW detection. Additionally, accumulated experience with CBW agent identification algorithms has allowed the software in these devices to significantly reduce false alarm rates and improve both reliability and maintainability.

Power
The portability of detection methods has improved as a result of advances in battery technology. Commercial demand for longer life and lightweight power in the portable computer and mobile telephone industries has applied considerable pressure. Lithium-ion technology (Li-Ion) appears to offer energy densities up to 130 Wh/kg (Watt hours per kilogram), compared with up to 80 Wh/kg for current Ni/Cd or Ni/MH batteries. An interesting development comes from a new company called PowerPaper in Israel. By discarding the conventional casing to reduce weight and volume, the company discovers that the essential components can be impregnated into ordinary paper. The chemical inks act as the anode, cathode and electrolyte and can be produced in any arrangement. There are many technical challenges to overcome but the outlook is promising.

C³I and reconnaissance
C³I continues to benefit from improved computer modelling but there are two areas which would significantly enhance decision support. One is the need for real-time CW and BW footprint prediction. Current flat-mapped systems cannot provide the answer on land, although results at sea, unaffected by topography, could be based on existing models.

Integration of real-time meteorological profiles with the topographical data will offer better information to commanders. On land, 3-D polygonal databases, while costly to develop, appear the only path to better predictive accuracy for the soldier. Developments in the US have been aimed at predictive tracking of epidemiological data following a BW incident, giving much earlier and clearer indications of whether a suspect outbreak of disease is, in fact, a BW incident, or not.

The other area worthy of attention is the integration of NBC information into existing command system data highways both on land and at sea. The data exchange or refresh rate would not be high. In comparison with the high weapon and sensor control data load, it would be negligible. In the 1990-1991 Gulf War, half the command area was sea. Integrated real-time BW C³I data could have been needed. Fortunately it was not.

COLLECTIVE PROTECTION

COLPRO
COLPRO presents a significant design challenge. Its requirements vary. The permanent underground command centre can benefit from heavyweight physical protection and from high-capacity filtered and conditioned intake air. At the other in the scale, fast mobile land-based forces require lightweight, flexible mobile COLPRO solutions. All the elements of COLPRO are heavy. The CW-resistant tentage materials are generally heavier than standard fabric and the other essential

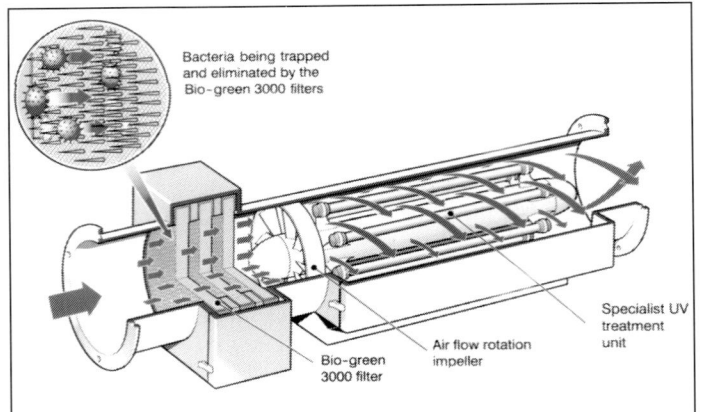

The Microgenix air purification unit is a scalable solution for destroying airborne pathogens (Microgenix Limited)   2002/0059318

elements include self-powered filtration and air-conditioning plant, airlocks, and cleansing stations. To sustain injury-free operations, additional requirements include: COLPRO monitoring systems, disposal facilities for contaminated materials and clothing, and DECON facilities for the whole range of operational equipment from aircraft to personal weapons.

The technological focus is on reducing the weight of each element that comprises the COLPRO system. This includes smaller, lighter prime-movers for filtration and air-conditioning plant, the use of lighter materials for COLPRO tentage and the use of modular construction to allow short-notice changes to the COLPRO requirement in response to a rapidly changing operational environment.

On the air management front, there are two developments which show great promise. The first, regenerable filtration, obviates the need to change and disposal of filter elements. This technology utilises two separate filter units. During a continuous controlled cycle, one unit filters whilst the other is purged of contaminants. This is achieved through management of the pressure or the temperature within the system, or both. The second development, whilst only addressing BW COLPRO requirements, uses a novel surface coating which physically damages over 90 per cent of the pathogens with which it is challenged. In a second phase of the cycle, the remaining pathogens within the incoming air-mass are swirled past a high-intensity UV source, ensuring their destruction. There are illustrations of these techniques in the COLPRO section.

UPDATED

INDIVIDUAL PROTECTION

Impermeable materials
Impermeable suits are made from a wide variety of materials. In an NBC environment, they are likely to be used by decontamination teams or by those involved in DEMIL operations. In addition to a requirement for prolonged and full resistance to CW agents, the materials must have as low a thermal load as possible and be highly resistant to tearing, abrasion or puncture. These materials are also commonly used for the manufacture of *ad hoc* COLPRO tentage and as tarpaulins for the protection of exposed stores and equipment from liquid CW damage (CARM for example).

MultiLaminar Film (MLF) materials
Examples of materials suitable for the internal layers include polyvinylidene chloride, polyamide and polyester. Outer layers are frequently polyolefine-based, modified to enhance adhesion within the matrix.

Inexpensive materials such as Amilon, Riloten and Mylothene are adequate for CW protection where high durability is not an issue. These materials are vulnerable to puncture damage.

In the Swiss ROLAMIT material, the laminated layers are bi-axially orientated. This matrix of 100 μm thick multiple films has a weight per unit area of between 100 and 150 g/m², offering high resistance to H agent penetration (> 1 day); much improved mechanical properties and good resistance to abrasion and puncture. In a toughness test of the material, servicemen marched on a gravel road wearing ROLAMIT overboots, made from three layers of 100 μm film, for 45 minutes without the integrity of the overboots being compromised. Other materials, such as the DuPont Tyvek® and Tychem® series, offer improved resistance and are widely used.

Coated fabric materials
The earliest gas-proof fabric comprised a rubberised canvas or linen material. Such materials were designed to offer very long-term protection against gross liquid contamination. Whilst these materials have reached weights in the range 250 to 500 g/m², modern developments seek to reduce significantly the weight per unit area whilst maintaining good long-term resistance. Nevertheless, the thermal load on the wearer is high and these materials require the wearer to be provided with a cooling system.

Permeable materials
In designing individual protection for military personnel, some compromises have to be made in balancing protection against sustainability. It is a principle of permeable protection that liquid agent falling on the outer layer is absorbed and

Woodland and desert camouflage cotton outer layers cover the SARATOGA™ filter layer underneath in these designs for the US Marine Corps (Blücher)
2002/0102790

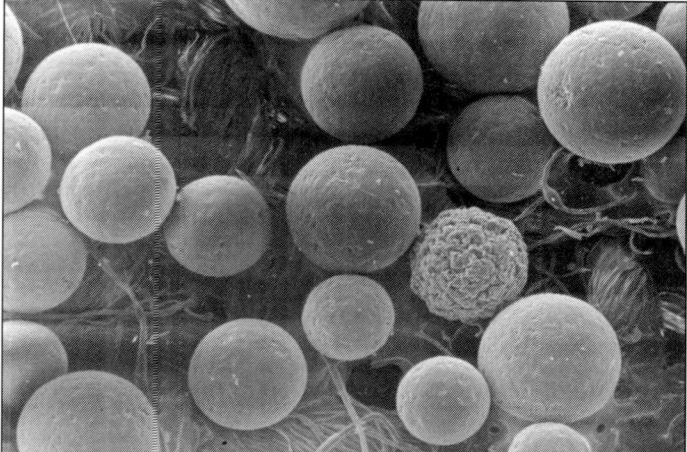

Advanced charcoal spherical absorber (×40 magnification) (Alfred Kärcher GmbH)
0050708

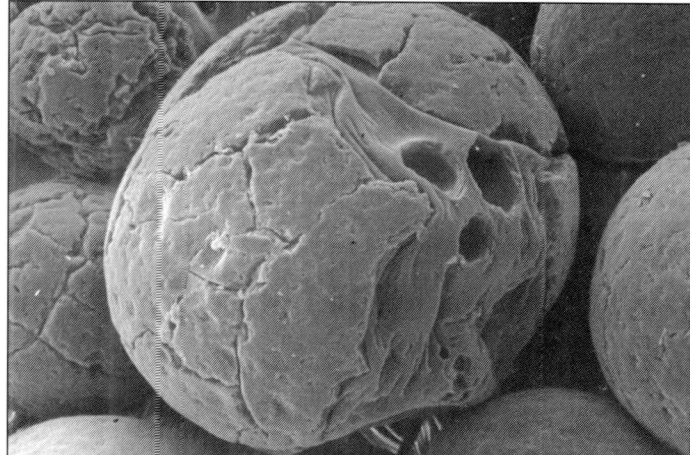

Advanced charcoal spherical absorber (×100 magnification) (Alfred Kärcher GmbH)
0050709

spread within the thickness of the layer. This 'wicking' effect increases the area over which a droplet can evaporate. Clearly, evaporation can take place in both directions. The movement of the wearer encourages this evaporation process. Underneath, there is normally a woven intermediate layer. Bonded to this, on the inside, is a layer of activated carbon whose purpose is to adsorb agent vapour, preventing it passing through to the skin of the wearer. Additionally, normal working clothes are worn underneath.

This principle creates a high-level of protection against the intensity of the liquid CW agent environment likely to be found on the battlefield. Whilst the integrity of the suit against gross liquid contamination may be limited to a matter of hours, it not only offers adequate protection for military personnel but also, unlike impermeable materials, allows prolonged wear without significant heat stress. However, heat stress becomes an increasing limitation as the workload and the ambient temperature increase. Most modern research focuses on improving the adsorptive properties of the activated carbon layer and on improving the durability and wearability of the suit.

Carbon powder
The activated carbon layer can be bonded to a variety of material types. These include low-density flexible polyurethane foam, cotton flannel or non-woven fibre. The latter appears to the modern material of choice. The active layer can be a fine carbon powder, 96 per cent of whose particles have a dimension of less than 25 µm. With the right variety of pore sizes, this layer offers very high adsorption properties, minimising the effect of pore blocking and enhancing the shelf-life of the suit product.

Spherical adsorbers
A second technique, developed by Blücher in Germany as SARATOGA™, uses activated carbon spheres 0.5 to 1 µm in diameter that are point-bonded in a predetermined pattern on to a fabric layer. This layer can be knitted polyamide, polyester, woven cotton or single jersey cotton. This gives a layer of adsorbing carbon approximately 1 mm thick.

In contrast to powder systems, the spherical adsorbers offer about 85 per cent of their surface area to the vapour challenge (the remaining 15 per cent are glue-covered for attachment to the fabric). Carbon in either powder or spherical adsorber form adsorbs vapour at the same rate. However, the latter, with three

times the amount of charcoal per unit area, can adsorb about three times as much vapour. The total capacity of the bonded sphere type system for chemical agents is about three times the level of the impregnated non-woven fabric system. The SARATOGA™ suit material weighs 265 to 310 g/m² and the outer cover of the suit is constructed of a heavy fire-resistant cotton fabric treated with an oil and water-repellent finish. The latter feature prevents much more of the liquid agent from 'wicking' into the fabric. Wear characteristics are described as good and a higher protection level is gained at the expense of some increase in both cost and heat stress.

Durability
Activated carbon is susceptible to sweat poisoning and performance degrades after prolonged use n an environment of high temperature and high humidity.

Considerable attention has therefore been given over the years to the durability of the permeable suit. Clearly it has to be durable enough to provide successful protection for a period of several hours without degradation of the protective layer. It must also be tough enough, as an item of working clothing, to cope with arduous activity regimes, contamination by common-range battlefield contaminants such as POL and be at least no more physiologically stressful than normal battle wear. Risk analysis by NBC defence staffs will define the precise design targets but there are general features of the permeable suit worth mentioning. In most cases, washing the suit will cause degradation of the carbon lining – up to 30 per cent in the non-woven case. Suits designed using the spherical adsorber concept claim to be much more durable and can survive repeated washing (10 to 20 times). Whether this is useful depends on how the suit is used. Earlier design concepts required suits to be discarded after use, especially if contaminated. Today, there is more interest in durability, driven by a need to conserve stocks in training, where the NBC protection factor s only vital if simulants are being used. However, in peacetime it is easy to give the durability and ease of wear attributes more attention at the expense of the suit's ability to deliver assured protection under liquid hazard conditions in war. In summary, the choice is between a low-stress light discardable suit with a low storage volume and a slightly bulkier suit with a higher storage volume but offering greater durability. The protection factor is the first priority.

UPDATED

INDIVIDUAL PROTECTION

Masks
The general design of respirator systems has changed little over the last few years. The material of choice, to provide the greatest resistance against CW agents is chlorobutyl rubber. The biggest design challenges involve providing an adequate

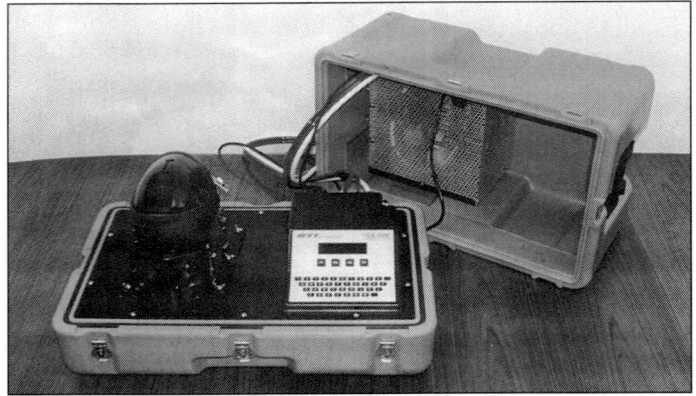

TDA-99M dual-purpose field mask leakage tester (Air Techniques International USA)   0011447

airtight seal with the face and in ensuring the highest degree of comfort. An Armed Forces replacement programme for individual respirators is expensive, as the production volume is enormous. The unit cost of earlier designs of mask, such as the UK S6 respirator of the 1960s, was relatively high and defence procurement agencies face continual downward pressure on costs. The S6 achieved an effective seal on the face with an annular air-filled chamber. Today, computer aided design and manufacture (CAD/CAM) techniques allow integrally moulded dam seals to be made, which are equally effective. This ability to create intricate and accurately machined mouldings has only become available in recent years.

Today, the general issue lung-powered respirator and filter canister assembly is designed to allow wearers to maintain their military or other tasks for extended periods without causing discomfort. In high ambient temperatures, perspiration not only fogs the eye-pieces or visor but also causes the mask to move or slip on the face, risking face-seal leakage. Modern mask design ensures that the interior shape of the mask provides an internal airflow pattern effective enough to keep the user's face as perspiration-free as possible and minimise misting of the eye pieces.

Eyesight correction remains a constant challenge to the respirator designer. Solutions here vary. The use of specially designed spectacles within the respirator is now being superseded by the use of snap-on inserts or 'outserts' added to the eye pieces lenses. Additionally, the simple shape of a round eye-piece probably provides the best seal with the rubber face-piece. Other solutions use a flexible panoramic visor such as the French ARFA mask. Here separate eyesight correction has to be provided.

The area below the eye-pieces/visor is a designer's battleground. The speech enhancement device, the drinking device, and the respirator canister attachment point all need, ideally, to be both as close to the wearer's nose and mouth as possible and centrally located to avoid off-centre loading. To provide better sustainability, the filter needs to provide assured protection for longer than in the past and, with current filter design, this inevitably leads to heavier canisters. In recent years, the trend has been to place the respirator canister on one side of the face-piece. The corresponding point on the other side is fitted either with a blank or with further speech enhancement, such as a microphone attachment point. This arrangement offers the great advantage of choice. The canister can be attached on either side to allow both left and right-handed personnel to operate personal weapons without obstruction. Alternatively, two canisters can be attached either side to give enhanced protection and lowered breathing resistance. In other designs, the canister attachment point is located vertically below the speech enhancement device. This avoids off-centre loading but can severely restrict downward head movement. One novel solution to this comes from Austria. The Canister Swivel Connector SMK (J Blaschke Wehrtechnik GmbH) allows a centrally mounted canister to be swung to either side (see under PROTECTION (individual) — Filters).

The design of the filter canister has changed little and, classically, comprises a particulate (HEPA) filter made from coated glass fibre or other materials. Downstream, there is normally an activated carbon filter bed designed to adsorb the known CBW agents. The enormous surface area of activated carbon ($> 700 \text{ m}^2$ gm^{-1}) is impregnated with other materials to cope with those toxic vapours that undergo only weak physical absorption at ambient temperature. For example, impregnation with copper provides sites which will absorb HCN and related compounds. Silver is used to deal with arsenical compounds. This type of filter can never offer 100 percent respiratory protection to the user and is affected both by age and exposure to water vapour in the air. The level of challenge to the filter canister also reduces its life. Therefore, in order to guarantee full respiratory protection under high concentrations of harmful vapour over extended periods, self-contained breathing apparatus (SCBA) is necessary. Additionally, organisations have become more aware of the need to test the respirators their issue to their personnel, for the effectiveness of both the seal and the filter canister. Test sets are available with varying degrees of sophistication. An example is the TDA-99 M dual-purpose field mask leakage tester. This system not only mimics the face seal for bench testing but can be used to test the respirator whilst being worn.

There have been significant improvements in the design of speech-enhancement devices. The majority of service respirators are fitted with 'compressed horn' type speech transmitters. They additionally carrying receptacles which allow the attachment of a microphone or other speech transmission device. 'Snap-on' speech enhancement and communications devices are also available, designed to fit a range of modern respirators.

UPDATED

CONTAMINATION CONTROL

Whilst technology can help DECON in three main areas — after the event — equipment designers can do a great deal to both minimise the impact of a liquid CW attack and allow quick, easy and effective decontamination. They need to appreciate two important factors. Firstly, the design principles are exactly the same as those which render equipment, vehicles, aircraft and ships less prone to general corrosion and easier to clean and maintain. Secondly, the principles of design for 'stealth' are also those for good contamination avoidance. The effective constituent of most CW agents is an organic solvent with a low surface energy, capable of finding its way into nooks and crannies, screw-threads, bolts and joints. The most difficult task lies in convincing designers that the incorporation of good contamination avoidance principles at the earliest possible stage in the design, also attracts the lowest possible through-life cost. Addressing the issue late in the design process becomes extremely costly and, therefore, likely to be ignored.

Preventing penetration by NBC agents is the key preoccupation of scientists and a variety of new surface coatings are being researched. There are conflicting requirements to address. For example, military vehicles and aircraft often require repainting in different camouflage when they are tasked to operate in new environments. Therefore, any coatings ideally need to be water-based and capable of application by relatively unskilled personnel in the field. Secondly, exterior surface coatings need to deliver the minimum visual or IR signature and minimise the reflectivity to searching radars and other detection systems. One area which shows promise is that of sacrificial coatings whereby the underlying surface is tough and highly resistant to penetration by NBC agents but the visible top-coatings can absorb and retain the liquid agents within its structure. The latter quality keeps the agent where it lands until DECON can be achieved. An example of this is TCC produced by the Le Havre-based company, International Celomer (owned by Akzo Nobel). Secondly, research is focused on improving the hardness and durability of the underlying permanent coating scheme.

The second main area of technological development is in the avoidance of injury to personnel. Here, the pharmaceutical industry and defence research organisations are developing novel methods both of combating the effects of agents on personnel and providing prophylaxis in media which are easy to administer and have the minimum of side-effects. Minimising the effects of contamination on the skin, personal equipment, and clothing is currently achieved with a range of absorbent powders and barrier preparations. The Canadian preparation RSDL shows considerable promise in this area.

The third area of development is in DECON itself. Here, lessons have been learnt from the civil emergency services who, constantly facing incidents involving spillage of hazardous chemicals, demand equipment which is not only highly efficient and effective but also minimises the risk to themselves. New and more effective shower units and delivery systems for decontaminants are being developed. The contaminants themselves, are under review. Modern polymer-based materials are being researched, designed to minimise the damage to the underlying surface and reduce the toxicity of the decontamination residue. Considerable research and development work has been carried out in Germany to improve the facilities available for decontamination both in the field and at base. The decontamination of sensitive material, including electronic components in enclosed spaces such as aircraft cockpits, is achieved through combining vacuum systems, steam delivery and extraction systems with traditional methods. All the systems are made easy to use by personnel with the minimum of training. See examples like the 'Decont Shuttle' and the vacuum decontamination system for sensitive material (both from Alfred Kärcher GmbH) under Decontamination).

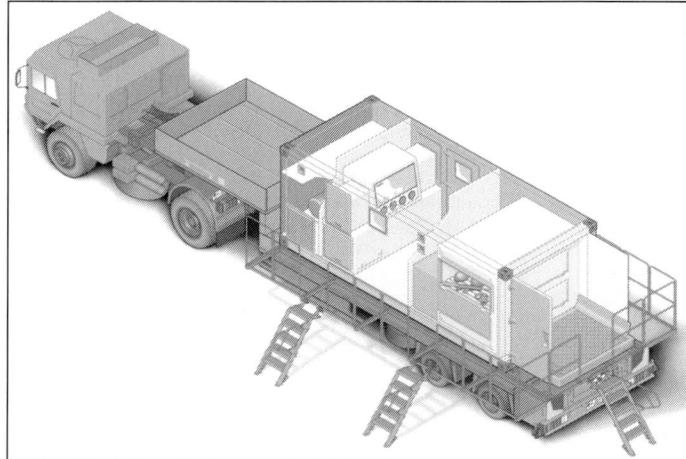

The EURODECONT trailer-mounted system is a joint German-French project involving Giat and Kärcher. It is designed for the field decontamination of a wide range of sensitive material, such as electronic systems, personal weapons, optical devices and so on, either in situ or passed through the container-based unit itself (Alfred Kärcher GmbH)   2002/0098350

DEMILITARISATION

The agreements made under the terms of the Chemical Weapons Convention have to be enacted by the states parties within 10 years of accession to the treaty. Whilst this may at first appear a generous timescale, it will be difficult to achieve for some

This NBC DEMIL facility, developed jointly by Aircontrol Technologies Limited and DSTL CBS Porton Down was used during Operation ABBOTT, a task which involved extensive searching and retrieval of ACWs from a site close to the Defence NBC Centre, Salisbury, UK. The air filtration units can be seen (centre right) (Aircontrol Technologies)   **2000/0050936**

Good simulation `s the key to successful training. The CAMSIM001 system is a CAM emulation which uses an electronic technique to deliver realism to training exercises (Argon Electronics)   **2000/0085454**

nations. The remote location of SCW storage sites, the location of ACW on the sites of other nations and sheer lack of resources will all make compliance a tough issue.

DEMIL processes

The processes fall into two broad types: those that are effective in ambient conditions and those that require a harsh environment to complete the task. The first two in the following list fall into the former category.

Chemical neutralisation uses chemicals appropriate to the type of agent and a mechanism developed through the defence decontamination and industrial waste-disposal industries. Of course, the neutralised waste has to be treated and this can be dealt with on-site using a further downstream chemical or biodegradation process. An alternative solution is to transport the neutralised waste off-site to another licensed or approved specialist facility. The Russian KUASI system has reportedly successfully destroyed 4,000 leaking munitions containing 200 tons of nerve agent. KUASI is a two-step process involving chemical neutralisation of the agent with an organic material and incineration of the waste products.

Electrochemical oxidation involves injection of the stream of agent into a continuous batch reactor containing a solution of silver nitrate dissolved in concentrated nitric acid. The process operates at 90°C at normal pressure and the oxidation products include CO_2, CO and water, together with other inorganic residues. The process chemicals are recycled to a large extent and the waste is in liquid form.

With gas-phase chemical reduction, the agent is drawn into a vessel containing hydrogen at a temperature of 800°C or higher where it is split into its constituents. Agents such as HD are reduced to methane and hydrochloric acid and non-chlorinated species such as VX reduce to methane and minor quantities of light hydrocarbons. A commercial scale plant is currently operating in Australia, processing up to 20 tons of fluid daily, with the object of destroying stocks of obsolete or prohibited pesticides.

Molten metal catalytic conversion evolved from the steel industry. This system passes the agent stream through a 1,500°C bath of molten metal where the agent becomes dissociated into its constituents. Parallel systems are used for VX and HD and the residue gas generated is recycled on-site for use as fuel for the process.

JACADS is the US 'baseline' technology against which alternatives were measured in the 1996 study. It involves incineration of the agent and safe disposal of the residue. Based at Johnson Island, 800 miles west of Hawaii, it consists of four incinerator plants working in parallel. Whilst it raised one or two alarm bells amongst the environmental lobby regarding its suitability for transfer to the continental USA, the JACADS utility has successfully disposed of the majority of the US stockpile of H agent.

There are other solutions worth examining, but the design of these technologies has matured significantly in the five years since the US study and further work would identify how these processes could be tuned individually or combined to offer the range of solutions needed by the world community to support the aims of the CWC.

TRAINING AND SIMULATION

NBC training is generally seen as a chore. There is a perception that, during a long career in the armed forces, the chances of having to operate under an NBC hazard are so slim as to be worth ignoring. Soldiers know that, going into battle, they will inevitably face bullets but, only very rarely, gas. The fear of having to work in a toxic environment during the 1990-91 Gulf War focused renewed attention on training.

Memories are short and NBC trainers therefore continue to face an uphill battle in ensuring that all training, from basic drills to major exercises are as realistic as possible. In NBC, more than any other topic, there is a very thin window of opportunity between conducting indifferent drills and creating an over dramatic scenario that lacks credibility. People need to believe that it is real.

The most successful approach appears to depend on invariably weaving an NBC dimension into every single exercise, even if just to consider the impact of NBC on a particular aspect of an operation. The worst approach simply appends an NBC aspect as a separate phase at the end of an exercise.

Realism is the key. NBC is simply another limiting environment in which warfare has to be practised. After all, flying has to continue at night. Underway replenishments have to take place in rough weather and trenches have to be dug in the rain. These are regularly practised. This is how simulating the NBC environment needs to be approached. New methods of creating realism are constantly emerging.

There are two types of aid for simulating the NBC hazard. One is designed to create the toxic environment for 'live' drills. This type includes pyrotechnics, simulation of NBC injuries and instrument performance simulators. The second is designed to assist commanders to improve decision making under an NBC environment. This includes team monitoring simulators, hazard prediction simulation and command team training.

IT-based training is gaining ground, with its ability to create more realistic NBC scenarios in both incividual and team instruction and exercises. The advent of low-cost PC-based multimedia now allows familiar TV and film-type treatment of NBC topics either on a single PC or in full-mission simulation.

Several new simulation products have emerged using updated and more precise prediction models whereby they are able to offer better decision support data to commanders (See under DETECTION (C³I Systems).

Practical training, especially in the difficult task of effective decontamination, has also improved dramatically as technology allows better solutions to emerge.

SUMMARY

There is significant development on all fronts, driven largely in recent months by the events in the USA. However, NBC generally draws a low priority for funding. It is vital to keep the research community well directed and adequately funded in order to ensure that NBC countermeasures remain in ahead of the game. The materials selected must stand up to the subtle pressure of NBC agents.

Manufacturers of key source products are now included under the relevant entry sections in this edition.

Procurement staffs need to think laterally in order to ensure that non-NBC solutions to the NBC hazard are not ignored. Project teams need to take account of the NBC hardening aspects of design and both testing of candidate equipment and training of personnel need to be realistic. In individual protection, the right balance has to be struck between the durability and comfort of equipment and clothing and its ability to deliver adequate protection for sustained operations.

UPDATED

DETECTION (sensor systems)

Nuclear
Biological
Chemical

DETECTION (sensor systems) - Nuclear

AUSTRIA

Air reconnaissance system for radiation detection

Description
The Austrian Research Centre Seibersdorf has long-standing experience in the measurement and assessment of radioactive contamination and the design and deployment of associated measuring equipment. In accordance with this tradition, Seibersdorf has developed a simple airborne reconnaissance system which allows rapid deployment.

Additional requirements have been defined as fully autonomous operation, high efficiency with respect to scanning speed, simple set up and operation and minimum maintenance. The system is based on the radiation protection survey meter SSM-1 (see separate entry this section), with an additional probe and ancillary equipment such as Global Position System (GPS).

The Seibersdorf air reconnaissance system is in use with Austrian defence forces and civil emergency services. Hungarian and Slovenian military authorities have acquired demonstration units.

Status
Available.

Manufacturer
Austrian Research Centre Seibersdorf.

VERIFIED

Radiation early warning system

Description
The Austrian Research Centre Seibersdorf has developed a radiation early warning system based on the HYDRODAT-S (S - Stationary) data logger and the SSM-1S radiation protection survey meter (see separate entry this section). Linked to sensors for measuring meteorological data, the two units can be combined to deliver early warning of a radiation incident over a wide area network.

A sensor station comprises the meteorological sensors, one SSM-1S survey meter and a HYDRODAT-S data logger. The stations can be sited at strategic locations, where a solar panel can be used for power generation (a separate power source is required for sensor heating).

Each station measures air temperature, relative humidity, precipitation, wind vector and radioactivity. The data logger is based on HYDRODAT stations installed in Austria, Slovakia and Bulgaria. It carries out the acquisition and processing of data measured by the SSM-1S and the other sensors, data archiving and transmission to the host computer and various other functions. Network communications between data loggers and the host computer can be by land-line or RF transmission media.

Status
Available.

Manufacturer
Austrian Research Centre Seibersdorf.

VERIFIED

A typical Seibersdorf radiation early warning system station

Radiation protection survey meter SSM-1

Description
The radiation protection survey meter SSM-1 is a versatile radiation measuring system for γ dose rate measurement and for α /β /γ contamination monitoring. It was developed by the Austrian Research Centre Seibersdorf and is based on third-generation microprocessor circuitry. Auto-ranging from 0.5 μSv/h to 5 Sv/h, it covers a wide operating range from natural background to very high dose rates.

The SSM-1 was designed according to ergonomic principles ensuring safe and easy handling even when wearing NBC protective equipment. A carrying belt allows free hand movement and ensures optimum visibility of the display, even when wearing a respirator. The shape of the lightweight unit permits unencumbered movement when in severe environments, while the display and controls are located on the protected side facing the user. The α /β /γ contamination probe is housed in the instrument case and cables are contained in the carrying belt.

The SSM-1 may be used as a portable or static (SSM-1S) system and has applications during the rapid survey of large area contamination and other emergency situations. A selection of different probes and detectors, as well as a standard V.24 interface for connection to printers and personal computers, allows a variety of further applications.

An SSM-1S is used as part of the Seibersdorf radiation early warning system (see separate entry in this section).

Status
Available.

Manufacturer
Austrian Research Centre Seibersdorf.

VERIFIED

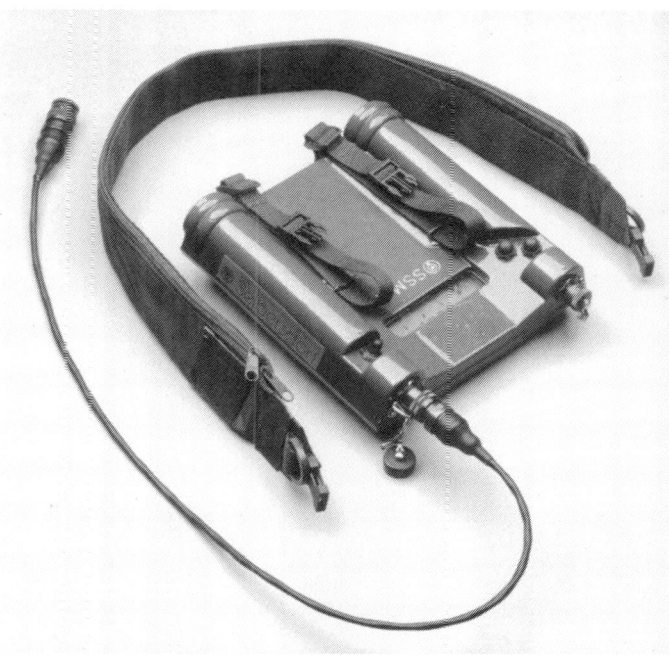

Radiation protection survey meter SSM-1

BULGARIA

ASR-3 automatic radiation detector alarm

Description
The ASR-3 automatic radiation detector alarm is intended for static and vehicle installation, providing a constant monitoring of local nuclear radiation levels and also visual and audible alarm signals when radiation readings rise above a selected preset level. Alarms can be initiated at levels of either 0.05, 0.1, 0.25, 1, 5, 10 or 50 R/h.

A complete ASR-3 system consists of a control unit, detector probe, connecting cable, spare parts and an accessories kit. Power required for operation is 12 V and may be taken from a vehicle electrical system. The weight of a single ASR-3 is 2 kg.

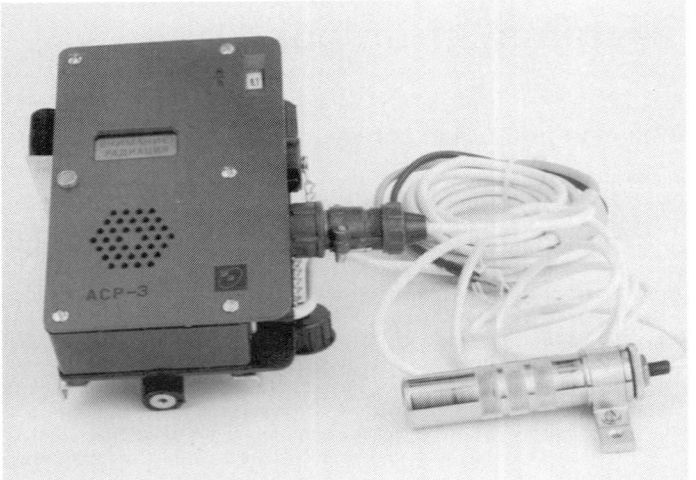

ASR-3 automatic radiation detector alarm

Status
Believed to be still available.

Marketing agency
Kintex.

UPDATED

ID-1 dosimeter kit

Description
The ID-1 dosimeter kit consists of 10 individual dosimeters and a piezoceramic recharging unit. The dosimeters are of the fountain pen, quartz fibre type and are carried and stored in a plastic case. Each dosimeter measures from 20 to 500 rads and weighs 80 g.

There is also an ID-02 dosimeter kit which is basically similar to the ID-1 other than the revised flat case packing and that each of the 10 dosimeters in the kit measures from 0 to 200 rads. The operational temperature range for both kits is –50 to +50°C.

Status
Believed to be still available.

Marketing agency
Kintex.

UPDATED

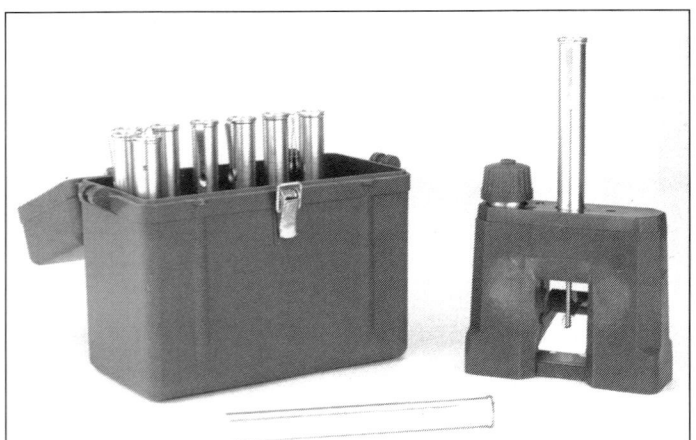

A complete ID-1 dosimeter kit

RBM vehicle-mounted radiation detector

Description
The RBM vehicle-mounted radiation detector is used for mounting on various NBC reconnaissance and tactical vehicles and is intended to measure the ambient level of γ radiation in the surrounding terrain. The instrument may be set to indicate various levels of contamination and may also be used to introduce audible alarms into vehicle intercom systems.

The RBM can be used to measure dose rate readings between 1 and 10,000 R/h and may be used to provide warnings and indications over the same range. The accuracy error rate is given as 20 per cent.

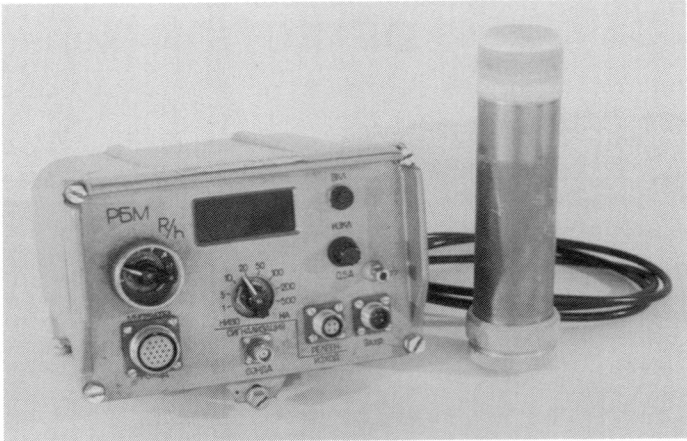

RBM vehicle-mounted radiation detector

The complete RBM consists of a control/indicator unit, detector probe, connecting cable, spare parts and an accessories kit. A voltage supply of 12 to 24 V is required for operation. A complete RBM packed in a cardboard box weighs 8 kg.

Status
Believed to be still available.

Marketing agency
Kintex.

UPDATED

VDP-90 combined arms survey meter system

Description
The VDP-90 is a multipurpose radiation measurement system intended for portable, mobile and static use by defence forces. It comprises a survey meter, with a control/indicator, a series of power supply modules with associated components, various detector probes, interconnecting cables, screening plates and sampling kits.

The VDP-90 may be used to measure γ radiation using three ranges, or β radiation using two ranges. A standard RS-232C terminal is provided to interface the system with automated data processing systems. If required, the VDP-90 can be powered directly from 220 V main power supplies.

The VDP-90 is delivered in two containers which carry the unit itself and the accessories. The combined weight of the two packed containers is 16 kg.

Status
Believed to be still available.

Marketing agency
Kintex.

UPDATED

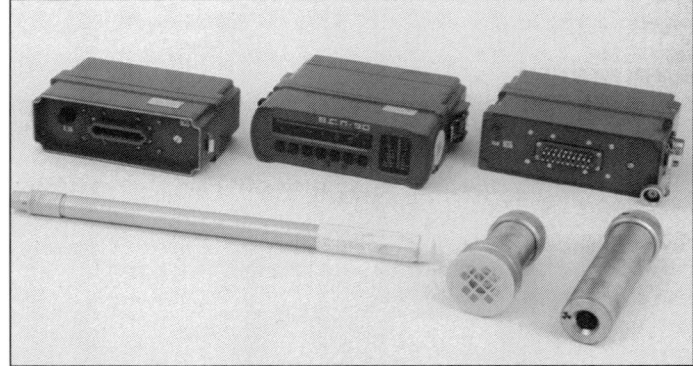

Various components from the VDP-90 combined arms survey meter system

CHINA, PEOPLE'S REPUBLIC

M-78 radiac meter

Description
The M-78 radiac meter is a simple portable meter with a radiation detection probe carried on a telescopic rod. The meter can be used to detect either γ radiation or, with a cover over the end of the sampling probe open, α and β radiation. Readings

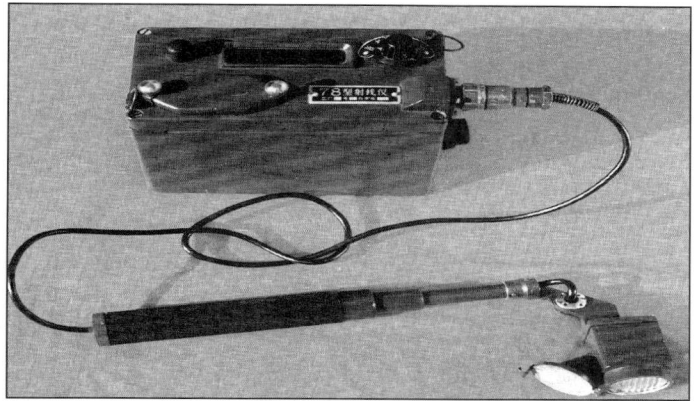

M-78 radiac meter (Jianan Instrument Factory)   **1999**

are given in digital form on a display located on top of the main meter unit. Also on the unit is a single control knob.

The measuring range for α and β radiation is quoted as 5 to 5 × 10⁴ Bq/cm and 1.29 × 10⁻⁷ to 0.129 Ckg/h for gamma radiation.

The M-78 weighs 2.5 kg and the main meter unit measures 195 × 100 × 160 mm. The telescopic rod can vary in length from 205 to 545 mm.

Status
In service with the Chinese armed forces.

Development agency
Research Institute for Chemical Defence.

Manufacturer
Jianan Instrument Factory.

UPDATED

M-83 dosimeter system

Description
The M-83 dosimeter system consists of a shoulder-slung plastic box containing 10 individual dosimeters, a dosimeter charger/reader unit and carrying case and a set of instructions. The measuring range is quoted as 0 to 0.2 C/kg.

M-83 dosimeter system (Research Institute for Chemical Defence)

Each fountain pen type individual dosimeter weighs 50 g, is 120 mm long having a diameter of 16 mm. The complete boxed system weighs 2 kg and the carrying case measures 225 × 145 × 110 mm.

Status
In service with the Chinese armed forces.

Development agency
Research Institute for Chemical Defence.

UPDATED

M-84 γ ray monitor

Description
The M-84 gamma ray monitor is a light and portable monitoring device that appears to indicate the presence of low levels of γ radiation by a lamp on the monitor flashing when radiation is present. The measuring range of the instrument is given as 2.58 × 10⁻⁵ to 5.16 × 10⁻² C/kg. The instrument weighs 500 g and measures 140 × 70 × 40 mm.

Status
In service with the Chinese armed forces.

Development agency
Research Institute for Chemical Defence.

Manufacturer
Jianan Instrument Factory.

UPDATED

M-84 γ ray monitor (Jianan Instrument Factory)

TLD Thermoluminescence Dosimetry (TLD) system

Description
The Thermoluminescence Dosimetry (TLD) system consists of two main components, the dosimeters and the reader. It was designed to measure the personal and environmental accumulative radiation dose.

The dosimeter is a badge known as the TLD 469 containing two or three detectors. The basic detector, known as the GR-100-M, consists of LiF chips (LiF is an alloy of magnesium, copper and phosphorus) sensitised with combined ultraviolet and thermal annealing. The GR-200 detector has a high signal-to-noise ratio while the GR-200-F uses LiF film.

There are two models of the TLD reader, the RGD-89 and the RGD-3. Both can measure a dose range of 10⁻⁸ to 10 Gy and display the result on a four-digit display. Output interfaces provided include an RS-232 C port for a computer, a 16-column microprinter and an X-Y recorder terminal. The reader has memories for the storage of measuring parameters and results and all heating control and data processing is carried out by a microcomputer.

The system is built to full military and nuclear emergency standards. The reader weighs 35 kg and measures 480 × 40 × 220 mm.

Reader for the Thermoluminescence Dosimetry (TLD) system (Research Institute for Chemical Defence)

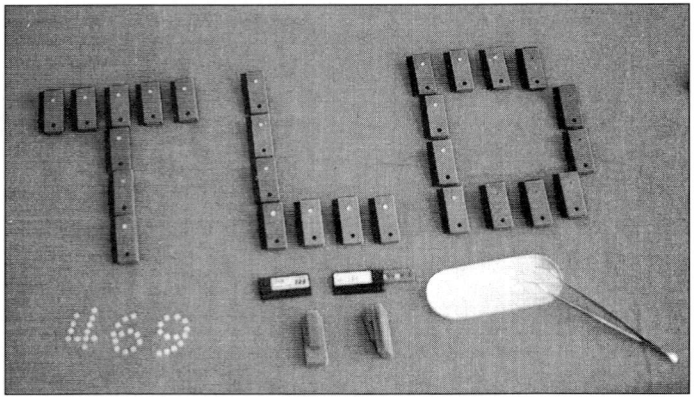

Badges and detector chips used with the Thermoluminescence Dosimetry (TLD) system (Research Institute for Chemical Defence)

Status
In service with the Chinese armed forces.

Development agency
Research Institute for Chemical Defence.

UPDATED

CZECH REPUBLIC AND SLOVAKIA

DP-95 dose rate meter

Description
The DP-95 dose rate meter is intended for radiation reconnaissance and monitoring and can be integrated into vehicular mobile NBC reconnaissance systems. It is a part of the RUDA NBC reconnaissance system. The equipment comprises an intelligent dosimetric probe connected to a control and display unit. Both components benefit from microprocessor operation. The intelligent probe can communicate with a computer through a standard RS-232C interface and with the control and display unit through an RS-485 interface.

Two Geiger-Müller detectors are employed for γ radiation dose rate measurement inside the NBC reconnaissance vehicle. The measurement range is from 1 Gy/h to 10 Gy/h, with a measurement time of 5 seconds, offering a statistical accuracy of 10%. The intelligent probe performs multiplication of the

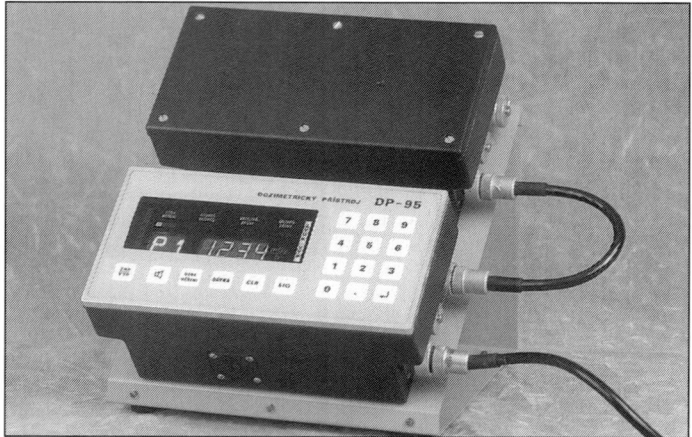

DP-95 doserate meter

measured value by an attenuation factor and integrates the dose. The control and display unit shows values of measured dose rate and dose, provides warning levels for both doserate and dose and also triggers audible and visual alarms. The equipment has integral setting-up, remote control and self-calibration functions and can operate in a simulation mode (connected to a computer) for training purposes.

The weight of the intelligent probe is 2.65 kg. The control and display unit weighs 2.35 kg. Operating temperature range: -30° to +50°C.

Status
Available.

Manufacturer
B.O.I.S. Engineering Praha sro.

VERIFIED

Model DK-62 tactical dosimeter

Description
The Model DK-62 dosimeter operates on the thermoluminescence principle using a single-crystal phosphor glass as the energy-storing component. The dosimeter is contained in a round black plastic case 40 mm in diameter and 20 mm thick. The energy component is 24 × 18 × 11.5 mm and weighs 30 g. The case can be worn around the neck or in a uniform pocket with the dosimeter itself held inside the case on a piece of white plastic foam. For reading, the case is opened and the dosimeter is placed inside a VDK-62 dosimeter reader, where it is heated. Heating reduces the energy level of the dosimeter to zero and the energy produced by the passage of radiation is released in the form of light which is optically matched to a colour scale to determine the level of radiation. The Model VDK-62 dosimeter reader is powered by a 12 V power source such as a vehicle battery and measures 200 × 150 × 120 mm.

Status
In service with the Czech and Slovak armed forces.

Manufacturer
(Unknown).

VERIFIED

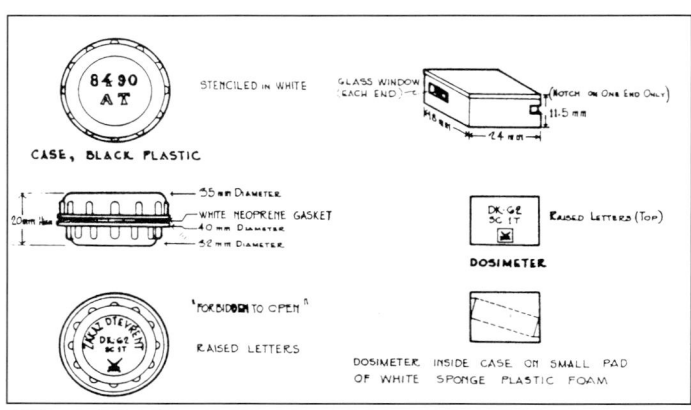

DK-62 tactical dosimeter

FINLAND

RADOS AAM-90 area monitoring system

Description
The RADOS AAM-90 area monitoring system is designed for automatic detection and monitoring of γ and X-ray radiation. It can be used for environmental monitoring, in research establishments, nuclear power stations and hospitals or wherever radiation may be encountered. The system can be used at varying levels covering areas from as small as a single room to entire cities or a country.

The basic AAM-90 measurement station includes a personal computer, detector, associated equipment and a local display which can be either a survey meter or a personal computer. The system also includes the software required for gathering and collating data at the system central computer.

The system involves the RD-02 detector which has two Geiger-Müller tubes operating over a range from background to full disaster levels. Power for the detector is passed through the measurement cable and standby batteries allow long-term continuous operation. The detector also has continuous self-monitoring with automatic error notification, but measurement continues if one of the detector tubes develops a fault.

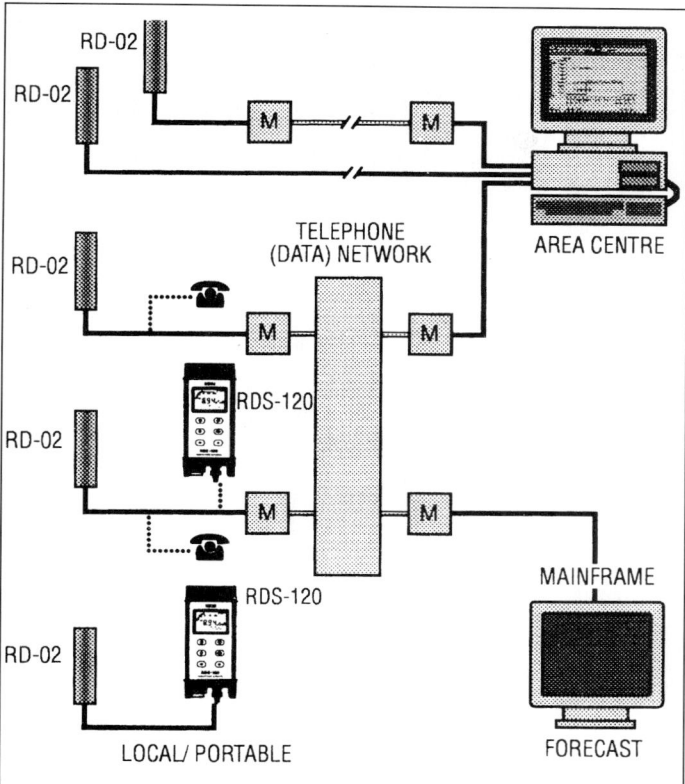

Block diagram of various applications of the RADOS AAM-90 gamma monitoring system
2001/0089992

If further measurement points are required they can be connected using existing cables (for example, a telephone line) and the relevant data is inserted into the computer. A total of more than 1,000 measurement points may be used. Over long distances modems or radiotelephone links may be used. No manned measurement points are involved as system control is completely automatic and the computer can diagnose any fault.

All aspects of operation for each measurement point may be checked and results provided when required. The personal computer uses a 'Windows®' type of programme. It is used for setting measurement and reporting intervals as well as setting data transmission parameters, detector identification and for setting up to nine different alarm levels. The computer is also used for calculations, storage of results and averages and for providing alarms and error messages.

The RD-02 detectors can function completely independently. Even without a personal computer results are recorded in memory and average results calculated. If the memory becomes full, results are compressed by integrating stored values. Normally the RD-02 operates as a passive detector giving information on request with immediate notification of alarms.

The AAM-90 system can be used for remote monitoring with single or multiple detectors connected directly to an area centre using a direct telephone connection, a modem or radiotelephone, or for remote local monitoring with an RDS-120 survey meter display.

AAM-95 uses Windows® software and has otherwise similar performance characteristics to the AAM-90. AAM-95 is used with the RADOS AAM-50 portable area monitoring monitor with the Global Positioning System (GPS).

Status
In service with the environmental gamma monitoring networks of the MoI, Finland and the SCPRI, France.

Manufacturer
RADOS Technology Oy.

Marketing agency
BNFL plc, Instruments.

UPDATED

RADOS RD-10 universal survey meter

Description
The RD-10 survey meter was designed for operation under military and other demanding conditions to the extent that the usual moving coil meter display has been replaced by a line of 40 LEDs that are easy to read under a wide range of user environments. The RD-10 uses solid-state circuitry housed in a rugged polycarbonate spray-tight case and is able to withstand hard knocks, including falls on to a hard floor.

There are two γ and X-ray measurement ranges: from 0.03 to 300 mrad/h and from 0.03 to 300 rad/h. The ranges are selected using a two-way switch; the only

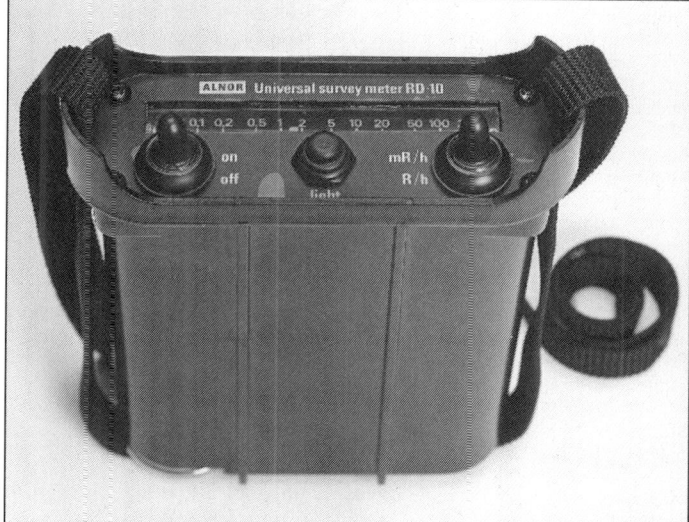

RADOS RD-10 universal survey meter (RADOS Technology Oy)

other controls are an on/off switch and a scale illumination switch. An overload display is provided for readings up to 5,000 rad/h. The radiation detectors are two halogen quenched, energy compensated Geiger-Müller tubes. Power is produced by two standard 1.5 V dry cells located in their own sealed compartment.

The RD-10 can be supplied with a carrying strap and the instrument may be connected to a loudspeaker to produce an audible pulse rate signal. A variant, the RD-10 B, can be supplied together with an external detector for the full range of α /β /γ and X-ray detection.

The RD-10 measures 210 × 155 × 56 mm and weighs 1.1 kg complete with batteries.

Status
Production complete. In service with the Finnish defence forces.

Manufacturer
RADOS Technology Oy.

UPDATED

RADOS RDS-100 radiation survey meter

Description
The RADOS RDS-100 radiation survey meter is a hand-held instrument designed to meet the requirements of civil defence and paramilitary personnel. The energy range of the instrument is from 50 keV to 1.25 MeV over a dose range of 0.1 to

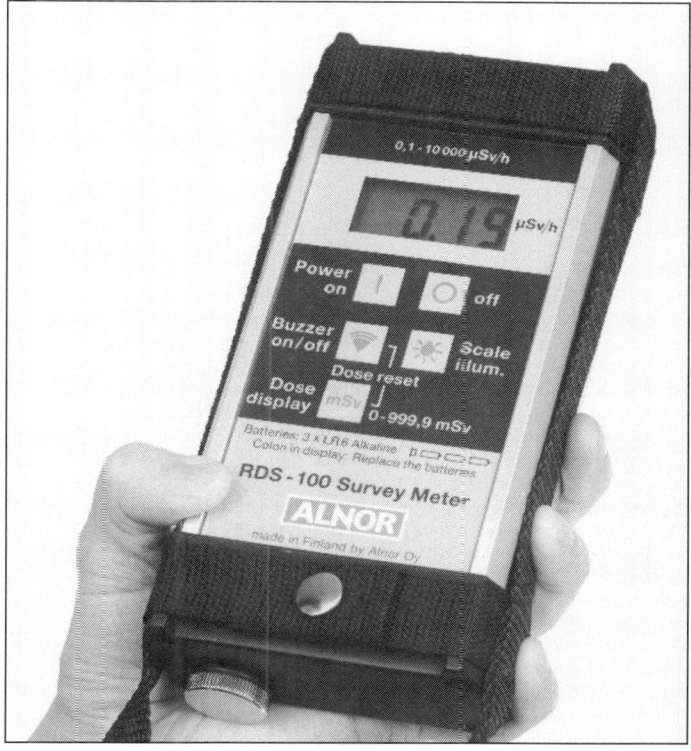

RADOS RDS-100 radiation survey meter (RADOS Technology Oy)

999 μSv/h and a dose value range of 0.001 to 999.9 mSv. One halogen quenched and energy compensated Geiger-Müller tube provides the detection method; γ and X-rays can be detected.

A microprocessor inside the RDS-100 enables a number of user-friendly functions to be incorporated. These include checking all segments of the four-digit LCD display when first switching on, a non-volatile RAM memory, automatic indication of overload (a blinking display) and continuous function status checking. A defect display if the detector tube provides too few pulses in a specified period is also included. Other features include a two-stage alarm for low battery condition and a 'two keys' philosophy to reset to zero and avoid accidental reset. A buzzer can be switched on to provide audible information of each pulse provided by the detector tube. A scale illumination feature allows the instrument to be read in low-light conditions.

The RDS-100 has a robust splashproof aluminium case and is powered by three AA cells. The instrument's dimensions are 88 × 196 × 42 mm and it weighs 650 g including the three IEC LR6 alkaline batteries (570 g without). The operational temperature range is from −25 to +55°C.

Status
In service with the Finnish defence forces.

Manufacturer
RADOS Technology Oy.

UPDATED

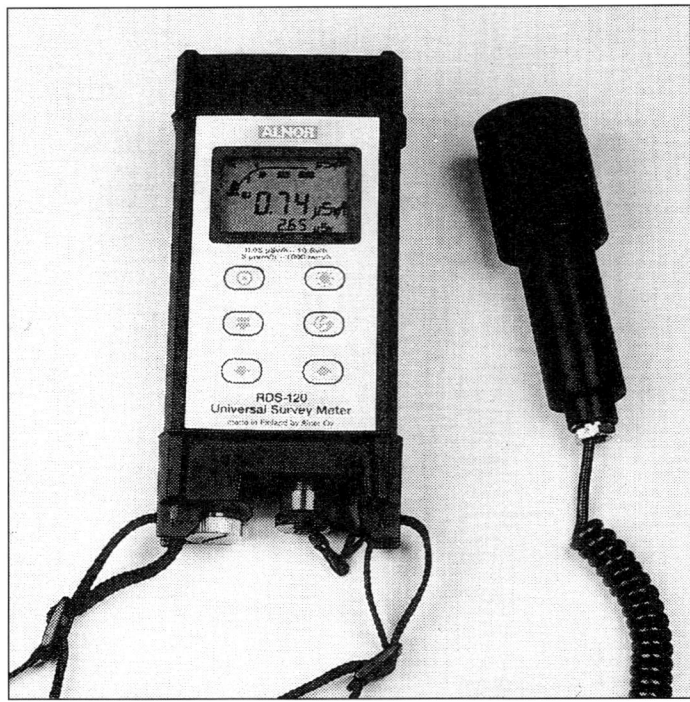

RADOS RDS-120 universal radiation survey meter (RADOS Technology Oy)

RADOS RDS-110 multipurpose survey meter

Description
The RADOS RDS-110 multipurpose survey meter has been designed for a wide range of applications and uses a rugged and splashproof aluminium construction. Functions comprise dose and dose rate measurements with an extended upper limit (100 mSv/h) and can be connected to an end window for Geiger-Müller tube contamination monitoring. It can also be used as a fixed γ alarm monitor with an external gamma probe and separate alarm device. Both the dose rate and dose alarm levels can be selected from a menu.

Audible and visual alarm levels are operator settable for accumulated dose and dose rate. There is also an automatic indication for the low battery state and for possible meter defects. With three standard alkaline batteries, the useful operating time is up to 200 hours in normal background radiation conditions.

The RDS-110 dimensions are 88 × 185 × 42 mm and it weighs 650 g including the three IEC LR6 alkaline batteries (570 g without). The operational temperature range is from −25 to +55°C.

Status
In production.

Manufacturer
RADOS Technology Oy.

Marketing Agency
BNFL plc, Instruments.

UPDATED

RADOS RDS-120 universal radiation survey meter

Description
The RADOS RDS-120 universal survey meter is a multipurpose radiation meter designed for military, civil defence, industrial, hospital and laboratory applications. Two halogen quenched and energy compensated Geiger-Müller tubes combined with microprocessor technology provide an energy range from 50 keV to 3 MeV over a dose rate range of 0.05 μSv/h to 10 Sv/h and a dose range of 0.01 μSv to 10 Sv with a fast and reliable response even in low background radiation fields.

The linearity of the RDS-120 is accomplished by the Time Interval Method developed by RADOS in the early 1980s. Using this method the dose rate is calculated from the time intervals measured between pulses, thus cancelling the need for taking into account the dead time of the GM tubes. Another advantage is that there is no need for multipoint calibration of the meter.

The RDS-120 microprocessor performs continuous self-diagnostic checking of the unit components and operations, for example calibration, memory operations and Geiger-Müller tubes giving error messages followed by an audible signal. A two-stage alarm procedure for low battery conditions is also included.

The measured dose rate is displayed in analogue and numeric form with the possibility to select Sv/h or rem/h units. Pulse indication and scale illumination are available. Audible and visible alarms are given when exceeding dose and dose rate alarm levels, which are pre-programmed by the user and stored in the meter memory. Up to 999 manually or automatically (pre-set time interval) collected dose rate readings can also be stored in the memory and transferred later via an interface to a personal computer or printer.

An external β or γ probe can be connected to the RDS-120. The RDS-120 can also be used as a local display for the RD-02 detector of the RADOS AAM-90 Area Radiation Monitoring System (see separate entry this section).

The RDS-120 weighs 700 g including the three IEC LR6 alkaline batteries and measures 92 × 199 × 44 mm. The operational temperature range is from −30 to +55°C. The dust and waterproof case is both EMP and RF-shielded.

Status
Production complete. In service with the Finnish Defence Forces.

Manufacturer
RADOS Technology Oy.

UPDATED

RADOS RDS-200 Advanced Survey Meter

Description
The RDS-200 utilises the same design principles and similar electronic components to the RDS-120 (see separate entry within this section). It differs from the latter in being designed for applications where accurate dose rate measurement at low dose rate levels is important. In addition to the instant dose rate display, the instrument continuously calculates a rolling average of dose rate values for the previous 5 minutes (or other user-selectable periods). This can be displayed at the touch of a button. The average values are stored into the internal memory (up to 864 values) and can be downloaded from the RDS-200 to a PC via a serial port. The telephone modem data transmission protocol makes RDS-200 meters particularly suitable for integration within an area nuclear monitoring system. The protocol is also compatible with the proprietary RADOS AAM-system software.

Optionally, a range of GM tube-based external probes is available for γ and β contamination measurement. These probes simply plug into the base of the unit and the RDS-200 automatically sets up the instrument for the relevant measuring range and selected probe.

Status
Customised model. In service with the Finnish Defence Forces.

> **Specifications**
> **Dose rate:** 0.05 m Sv/h to 10 Sv/h
> **Dose:** 0.01 m Sv to 10 Sv
> **Interface:** external detectors, PC and printer
> **Enclosure:** IP67

Manufacturer
RADOS Technology Oy.

UPDATED

FRANCE

APA particulate sampler

Description
The APA particulate sampler is a high flow rate instrument intended for the rapid sampling of atmospheric contamination. It can be linked to a DOM DOR 309 radiation meter (see separate entry in this section) equipped with a 125 mm² α probe and can operate positioned both horizontally and vertically.

Flow rates can be varied by three different types of filter. A pink filter provides a flow rate of 650 litres/min, a yellow filter 1,050 litres/min and a white filter 630 litres/min. Sampling times are factory preset at 3 or 10 minutes, selected by a three-way switch.

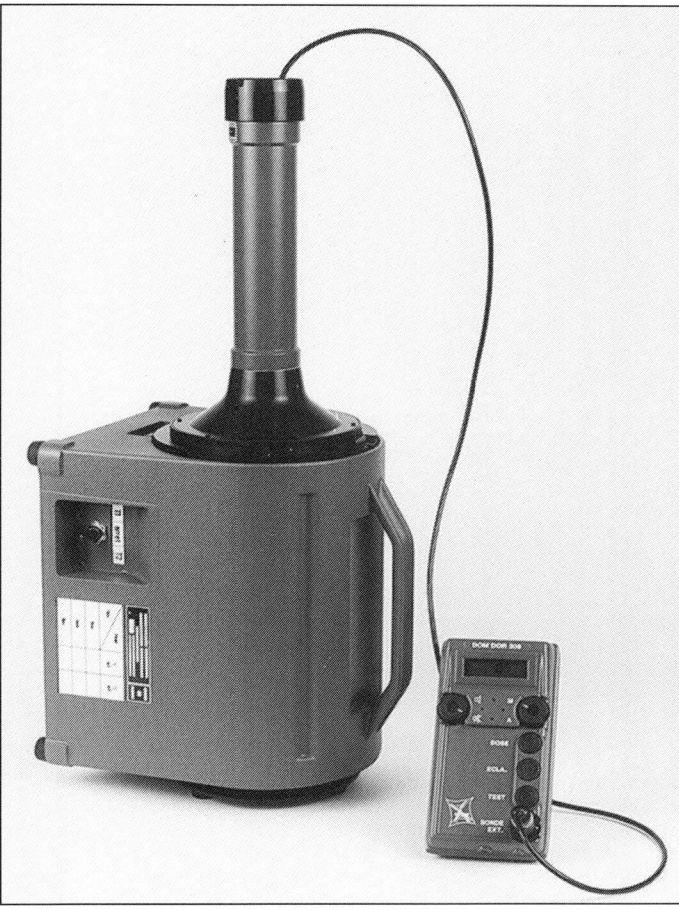

APA particulate sampler with an alpha probe unit

APA particulate sampler

The unit is housed in a reinforced polyurethane housing measuring 270 × 210 × 240 mm; weight is 6.8 kg. Power is provided from a 230 V mains supply although other voltages are available on request.

Status
In product on. Adopted by the French and Belgian armies.

Manufacturer
MGP Instruments.

VERIFIED

BEFIC individual dosimeters

Description
The BEFIC individual dosimeter is termed a *Stylodosimetre* as it resembles a fountain pen in size and construction. The dosimetre is 122 mm long, 13 mm in diameter and weighs 45 g. The dosimeter measures the irradiation by photons or electrons with energies higher than 60 keV; a version for levels of 10 keV is also available. The dosimeter is read by looking through the instrument towards a source of light.

Six versions are available measuring radiation doses up to 0.1, 0.2, 0.5, 1, 5, 100 and 200 rad (other ranges are available on request). The dosimeters are normally issued in multiples in canvas carrying bags that also contain the dosimeter charging unit.

Status
In service with the French armed forces and civil defence and some other European nations.

Manufacturer
BEFIC.

VERIFIED

DOM DOR 309 - X probe

Description
The lead for the X probe for the DOM DOR 309 meter is attached to the external probe receptacle on the front of the meter (see separate entry this section). This probe allows monitoring of α contamination by X-radiation with automatic subtraction of the ambient γ background level. The scanning rod on the probe allows a constant monitoring distance from the target to be maintained and measured using the attached gauge. The DOM DOR 309 meter displays 0-9999 c/s with this probe attached, measuring X-radiation in the range of 10 to 30 keV. It is sensitive to Plutonium [239] at 185 kBq/m² at a distance of 30 cm. In addition to the light and audio indicator on the meter, the probe also has an indicator light. The unit weighs 688 g and measures 246 mm by 47.8 mm (probe diameter).

Status
In production. Adopted by the French Army and Italian armies.

Manufacturer
MGP Instruments.

VERIFIED

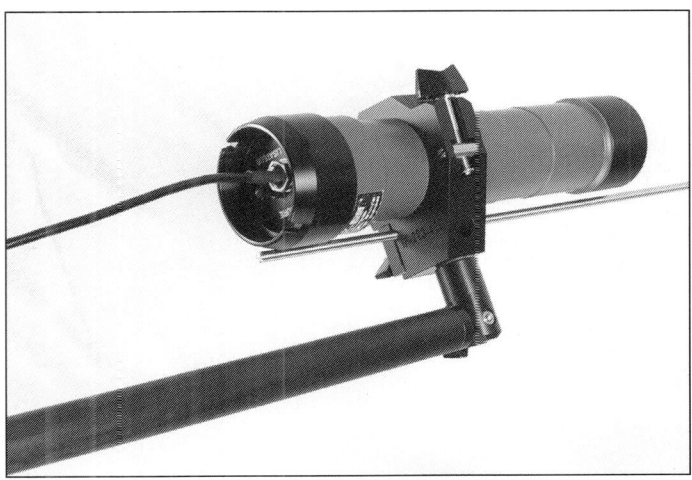

DOM DOR 309 - X probe

DOM DOR 309 - TGS γ probe

Description
The lead for the TGS γ probe for the DOM DOR 309 meter is attached to the external probe receptacle on the front of the meter (see illustration). This probe allows extremely rapid detection of contamination and is aimed at locating large areas of surface contamination and monitoring concentrations. It can be selected between 'gross' or 'compensated' modes, the latter allowing the level of residual contamination to be isolated from the background level. The DOM DOR meter displays 0-9999 c/s with this probe attached, measuring in the range 0.05 to 5 MeV. In addition to the light and audio indicator on the meter, the probe also has an indicator light. Its efficiency is 1,000 c/s/μGy/h and the unit weighs 1,220 g and measures 209 mm by 155 mm (probe diameter).

Status
In production. Adopted by the French and Italian armies.

Manufacturer
MGP Instruments.

VERIFIED

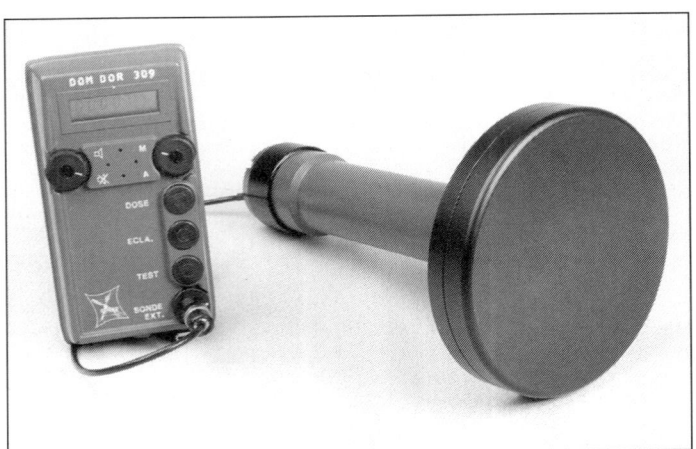

DOM DOR 309 - TGS γ probe   0052786

DOM DOR 309 - α probe '125'

Description
The α probe '125' for the DOM DOR 309 meter is designed for monitoring personnel and equipment. It can also be used with the APA particulate sampler (see separate entry this section) for monitoring contamination in the air. It is attached to the external probe receptacle on the front of the meter (see illustration). The DOM DOR meter detects α emissions in the range 2 to 6 MeV at up to 9999 c/s. In addition to the light and audio indicator on the meter, the probe itself also has an indicator light. The probe unit weighs 800 g, measures 300 mm in length and has a detection area of 125 mm^2 diameter.

Status
In production. Adopted by the French, Belgian and Italian armies.

Manufacturer
MGP Instruments.

VERIFIED

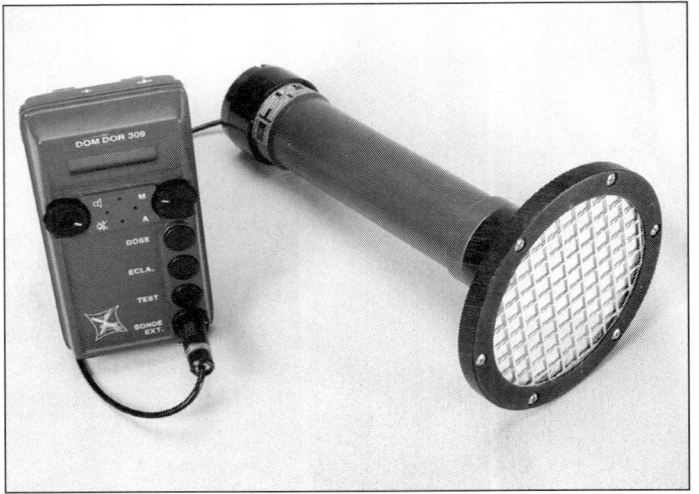

DOM DOR 309 - α probe '125'   0052789

DOM DOR 309 - α 'crayon' probe

Description
The α 'crayon' probe for the DOM DOR 309 meter is designed for wound monitoring by medical personnel. It is attached to the external probe receptacle on the front of the meter (see illustration) and consists of a very small silicon detector inside a stainless steel casing which allows very precise location of the source of contamination. The meter detects α emissions in the range 2 to 6 MeV at up to 9999 c/s. In addition to the light and audio indicator on the meter, the probe also has an indicator light. The unit weighs 200 g and measures 160 mm by 22 mm (probe diameter).

Status
In production. Adopted by the French, Belgian, Italian and Finnish armies.

Manufacturer
MGP Instruments.

VERIFIED

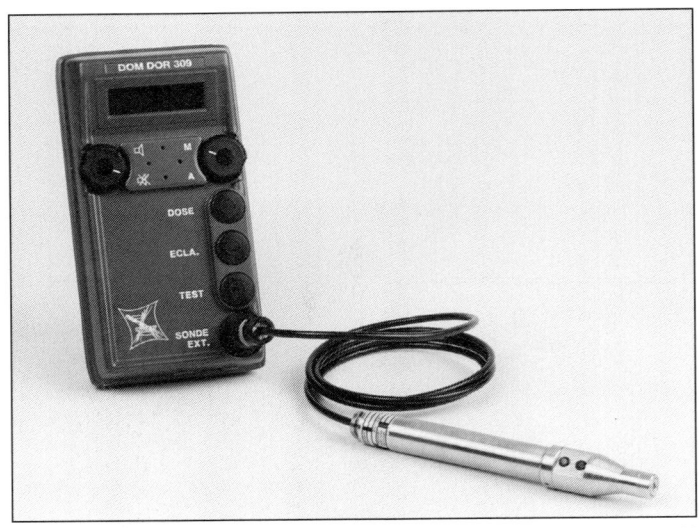

DOM DOR 309 - α 'crayon' probe   0052788

DOM DOR 309 - β/γ probe

Description
The lead for the β/γ probe for the DOM DOR 309 meter is attached to the 'sonde ext' receptacle on the front of the meter (see entry this section). This probe allows monitoring of personnel, vehicles and radioactive residue for β (in the range 0.25 to 5 MeV) or γ activity (in the range 0.1 to 5 MeV). The removable shield allows for differentiation between the types of emission (shield on for γ). Probe efficiency is approximately 65,000 c/s/cGy/h and the unit weighs 480 g and measures 52 mm by 280 mm with the shield on.

Status
In production. Adopted by the French, Italian, Belgian, UK, US and Canadian armies.

Manufacturer
MGP Instruments.

VERIFIED

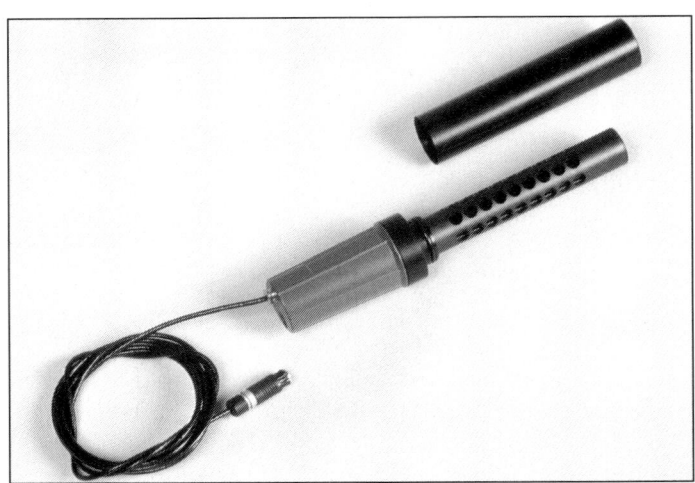

DOM DOR 309 - β/γ probe   0052790

DOM DOR 309 portable radiation meter

Description

The DOM DOR 309 radiation meter is a lightweight portable self-contained instrument designed for measuring γ radiation dose rates and accumulated dose over a given time period in a contaminated area. Its light weight and size allow it to be worn on the wrist, belt, or carried in a pocket, making it easy to use and compatible with most NBC IPE. It has an LCD readout panel which is illuminated on activation and time-delayed to off. The meter generates audible and visual alarms when a given dose is exceeded (selectable on or off). It can be used very flexibly, either as a simple hand-held detector on its own or connected to an external supply. In the latter mode, it can be mounted in a vehicle or fixed facility, drawing data from external probes. The large range of waterproof (splash proof) external probes to which it can be attached are shown in separate entries in this section. The meter has a comprehensive self-check routine.

The DOM DOR meter detects γ radiation dose in the 80 keV to 3 MeV range and can display dose rate in the range 0.001 to 999.9 mGy/h. Accumulated or 'mission' dose is calculated and displayed in the range 0.001 to 999.9 mGy. A trend indicator shows the increase or decrease in dose rate. The meter is sealed in a tough waterproof case and can be powered by four LR6 1.5 V batteries, giving it an operational life of more than 70 hours, or from an external supply. Its dimensions are 171 × 91 × 45 mm and it weighs 580 g. The operational temperature range is from -25 to +50°C.

Status

In production. Adopted by the French Army and the Belgian Air Force. Sold in France, Belgium, Italy, USA, Canada, Finland and UK.

Manufacturer

MGP Instruments.

VERIFIED

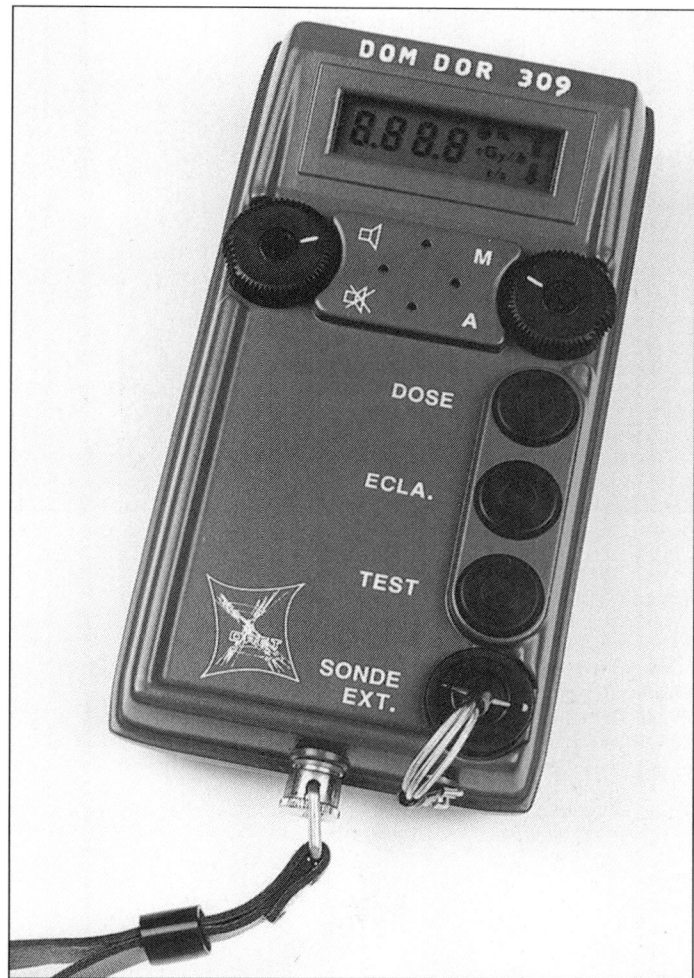

DOM DOR 309 radiation meter

DOSICARD

Description

The DOSICARD is a credit-card sized personal γ-dosimeter with an electronic memory. It measures dose and dose rate and has a programmable alarm. The energy compensated silicon detector has a sensitivity of 100 c/µSv. The equivalent dose - HP (10) - is in compliance with the ICRUM 39 standard and the card has an energy range of 50 keV to 2 MeV. Dose is displayed from 1 µSv to 10 Sv and dose

The EURISYS MESURES DOSICARD personal dosimeter (CANBERR EURISYS SA)   *1998*/0011415

rate from 1 µSv/h to 1 Sv/h. The card has a data capacity of one year (8 hour/day). There are 3 keys to select the operating mode and to display the values of current dose and dose rate on the LCD screen. Integrated doses per day/month/quarter/year/5-year can also be shown.

Status

Available.

Manufacturer

CANBERRA EURISYS SA.

UPDATED

JUK 450 airborne radiation reconnaissance system

Description

The JUK 450 airborne radiation reconnaissance system was designed for installation on slow- and low-flying aircraft such as helicopters. It is used to survey areas exposed to radiation and to determine safe routes through such areas. The JUK 450 is used by the armed forces of France, Germany and Italy.

The system includes an ionisation chamber to be mounted externally on the carrier aircraft, a data processing unit and a control/display unit. The ionisation chamber can measure a dose rate from 0.02 to 250 rad/h. The data processing unit uses microprocessor techniques under the control unit which also has a display panel and controls for a paper printer using paper tape to record the required outputs. These are delivered in a series of figure codes as follows: two figures for position indication; three figures or symbols for indication of printing rate or time in minutes; three figures for altitude indication; six figures for indication of dose rate, plus two symbols for overload (one for measurement out of range and the other for hot-spot marking) and one letter to indicate the selected mode. Printouts can be made automatically every 5, 10 or 20 seconds or under operator control. The total recording time is 3 hours.

The JUK 450 system can be used to indicate the dose rate at 1 m above ground level.

The total weight of the JUK 450 is approximately 15 kg. The main processor unit is normally contained in a case measuring 380 × 200 × 200 mm. Power is derived from the aircraft's normal 19 to 32 V DC supply and the operational temperature range is from -30 to +60°C.

Status

In service with the French, German and Italian armies. A few low-range detector units are in service with the French civil defence.

Manufacturer

BEFIC.

VERIFIED

Preproduction version of the JUK 450 airborne radiation reconnaissance system with ionisation chamber (left), data processing unit (centre) and control/display unit (right)

Model DOK 420 radiation meter

Description

Designed as a portable radiation meter, the Model DOK 420 can detect and indicate both γ and X-ray radiation. It is a single unit with a single scale indicating from 0 to 500 rad/h. The power is supplied by two 1.5 V type BA58 cells which can power the meter for up to 150 hours of use. Accuracy is stated to be ±30 per cent. The unit measures 120 × 75 × 35 mm and weighs 350 g.

It is suspected that, although this product probably remains in service in some units, it will have been progressively replaced by more modern systems such as the DOM/DOR 309 (see entry within this section).

Status

In service with the French armed forces.

Manufacturer

Originally Nardeux SA. Company believed to have been absorbed by Eurysis Mesures (see under Contractors).

VERIFIED

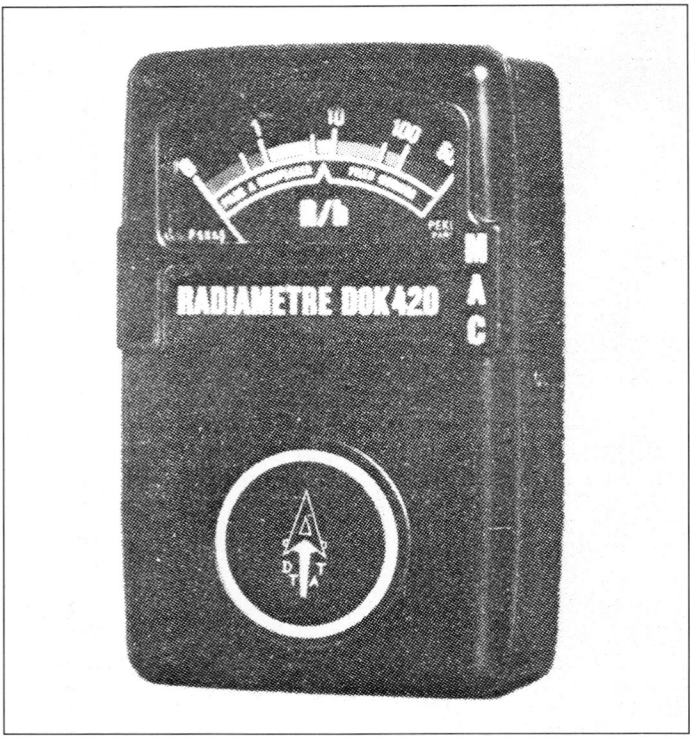

DOK 420 radiation meter

Model DOK 803 radiation meter and DUK 807 collective unit

Description

The basic DOK 803 radiation meter is an instrument contained in a lightweight rectangular moulded plastic case which can be easily carried and used in one hand. Using a Geiger-Müller tube as a detector it can monitor gamma radiation. It has two indicators. One is a small lamp that flashes when radiation present is less than 0.01. At rates from 0.01 to 120 rad/h the dose rate is indicated in digital form on a display panel. The only control is a single on/off switch and power is produced by two internally housed 1.5 V BA30 cells. These provide power for the instrument to be used continuously for 20 hours and intermittently for 40 hours.

The DOK 803 can be used by itself and may be carried and used in a stout fabric case with clear panels for observation of the display panel and the lamp. It may also be removed from this case and used in what is described as a 'Collective Unit' which converts the DOK 803 from a portable detection unit into a monitoring and alarm unit for static use at various locations. In this arrangement the unit forms part of the DUK 807, which, apart from the DOK 803, has two radiation probes and an electronics unit into which the DOK 803 is plugged. One of the probes is a low-level unit used from 10 to 120 rad/h, the other is for use at higher dose rate levels from 10 to 1,200 rad/h. These probes are plugged via cables into the side of the electronics unit as appropriate. When the DOK 803 is plugged into the electronics unit it provides the readout for whatever dose rate the probe in use detects.

The electronics unit may be used to provide an aural warning whenever the dose rate level reaches 10 mrad/h (when the low-level probe is in use) or 10 rad/h (when the high-level probe is in use). The alarm may be switched off using a switch on the electronics unit. This has its own internal battery supply and when the voltage level drops too low for continued use the lamp on the DOK 803 will flash. The DUK 807 may be powered by a mains voltage.

Model DOK 803 radiation meter

DUK 807 collective unit with DOK 803 inserted and showing probes

Specifications
DOK 803 only
Weight: 450 g
Length: 160 mm
Width: 75 mm
Height: 41 mm
Power supply: 2 × 1.5 V BA30 cells

Each of the probe units is 150 mm high and has a diameter of 65 mm. They may be connected to the electronics unit by cables up to 50 m long.

Status

In use with the French civil defence authorities.

Manufacturer

Originally Saphymo-Physiotechnie. Company believed to have been absorbed by BEFIC (see under Contractors).

VERIFIED

Model DOM 410 radiation meter and accessories

Description

The basic unit of this equipment is the Model DOM 410 survey meter to which other parts can be fitted to undertake a variety of tasks. The basic Model DOM 410 is a survey meter containing a Geiger-Müller counter tube. The unit is portable and

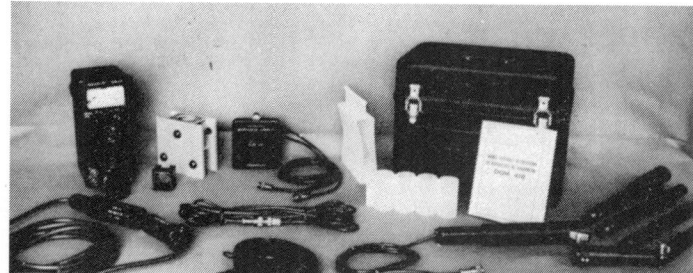

Model DOM 410 radiation meter and accessories

waterproof and may be fitted to vehicles or fixed installations. The meter has six scales: 0 to 10, 100 and 1,000 mrad/h and 0 to 10, 100 and 1,000 rad/h.

Power comes from two 1.5 V type BA30 dry cells in the meter carrying handle, but if required, an external 110/220 V AC supply can be connected. The 1.5 V cells can keep the meter in use for about 16 hours. The meter weighs 2.5 kg and is 230 mm high, 100 mm wide and 190 mm deep. Issued with the meter are a carrying strap, hermetically sealed carrying case, extension cable for a probe, set of instructions and a small radiation test source.

Two other units can be fitted to the basic Model DOM 410 meter, a visual alarm device and a decontamination probe which can be used to detect beta radiation. If required both devices can be fitted to the basic meter at the same time.

For training purposes the Model ROK 410 radiation simulator has been designed consisting of a transmitter, various cables and batteries and 10 receivers. The receivers are all identical to the normal Model DOM 410 meters but their interiors have been altered to pick up the variable signals emitted by the equipment's transmitter which has a range up to 8 km and can be fitted with either directional or omnidirectional aerials. Operating on a frequency of between 3,250 and 3,400 kc/s the transmitter can emit signals that simulate radiation readings on all the receivers in range. The transmitter weighs 33 kg and each receiver weighs 2.6 kg.

Status

In production. In service with the armed forces of France and several other countries.

Manufacturer

Originally Saphymo-Physiotechnie. Company believed to have been absorbed by BEFIC (see under Contractors).

VERIFIED

Model DUK DUR 430 airborne radiation meter/dosimeter

Description

Although this equipment is intended primarily for use in aircraft, it can also be fitted to land vehicles and maritime craft. It consists of a single unit which can be mounted on to an instrument panel and incorporates a radiation meter capable of indicating up to 500 rad/h and a dosimeter which automatically indicates the dose absorbed up to 999.9 rad. The unit requires a power source of 27.5 V AC and measures 80 × 80 × 188 mm. The weight of the complete unit is 1.2 kg.

Status

In service with the French Army.

Manufacturer

Originally Saphymo-Physiotechnie. Company believed to have been absorbed by BEFIC (see under Contractors).

VERIFIED

Model DUK DUR 430 airborne radiation meter/dosimeter

Model RA 73 control and alarm radiation meter

Description

The Model RA 73 is a small radiation meter designed to be carried on the user's belt and capable of emitting an audible alarm once a preset level of radiation exposure has been reached. The alarm threshold is set at the manufacturing stage and may be from 10 to 100 mrad/h. Apart from the alarm, the meter can measure from 0 to 1,000 rad/h on three logarithmic scales. The power comes from a single 4.5 V battery which can supply the meter for up to 100 hours. The meter measures 165 × 75 × 50 mm and weighs 700 g.

The RA 73 was produced under licence by the Swiss firm Autophon which manufactured 25,000 examples for the Swiss armed forces and a similar number for the Swiss civil defence organisations.

Status

In service with the Swiss armed forces and civil defence organisations.

Manufacturers

Originally Saphymo-Physiotechnie. Company believed to have been absorbed by BEFIC (see under Contractors).

Licenced production by Autophon AG.

VERIFIED

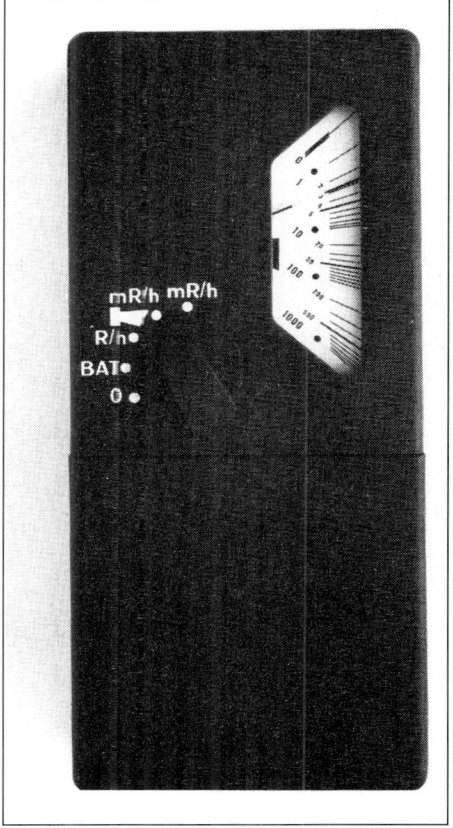

Model RA 73 control and alarm radiation meter

SOR tactical dosimetry system

Description

The SOR tactical dosimetry system comprises two types of dosimeter and three 3 types of reader unit, supported by a suite of bench testing and support equipment. Training software is also included.

The SOR/R dosimeter reads γ radiation, especially at low levels, while the SOR/T reads γ and neutron radiation over a much wider range. Both readers are supported by the XOM series of reader/recorders. A portable version, suitable for field use, the XOM/T is derived from the XOM, a fixed reader designed for dose management and access control. The XOM/M variant is a higher-capability unit which can be set up at unit level for wider resource management of dosimeter readings.

All the units are EMP, EMC and TREE-resistant, waterproof and blast/shock proof. All units satisfy current NATO and FINABEL standards.

Specifications

Both dosimeters incorporate HP(10) dose equivalent measurements, and store their data by EEPROM (>10 year life - without battery). The meters operate for one year on a standard battery. The alarm thresholds for dose and dose rate are user-selectable and both units benefit from comprehensive and regular auto-rechecked BITE algorithms. Historical records are retained for periods of 10 secs, 1 min, 10 mins, 1 hour or 24 hours. The 6-digit displays are selectable between cGy and cGy/h, mSv and mSv/h, mrem and mrem/h. Teledosimetry transmission facilities are included and saturation indicators operate above 10 Gy/h.

Specifications:		SOR/R	SOR/T
Dosimeters			
Flash γ dose:		-	5-10 cGy
Flash neutron dose:		-	5-10 cGy
Relative flash measurement error:		-	±30% over measurement range
Ambient γ dose measurement:		1µGy -10Gy	1µGy -10Gy
γ dose rate measurement:		Background to 10 Gy/h	Background to 10 Gy/h
Relative ambient error measurement:		<±20% over range	<±20% over range
Energy response (in the range 60 keV to 2 MeV):		<±20% in range 60 keV to 2 Mev	<±20% in range 60 keV to 2 Mev
Environmental range (normal):		-20 to +50°C	-20 to +50°C
Environmental range (extended):		-40 to +50°C	-40 to +50°C
Dimensions:		80.4 × 48 × 9 mm	80.4 × 48 × 9 mm
Weight:		55 g	55 g

The XOM series of readers can output digital data and are equipped with LCD display panels:

Reader units	XOM/R	XOM/T	XOM/M
Role:	Fixed unit. Designed for area access control , dose management and wearer localisation.	Portable. Designed for field management of individual or unit dose/dose rate.	Transportable. Extended features over XOM/T. Designed for unit level use.
Capacity:	Simultaneous reading of several dosimeters.	Simultaneous reading of several dosimeters.	Simultaneous reading of several dosimeters.
Power supply:	110-220 V AC (Option for 24 V DC)	Battery or external. 12-28 V DC or 110-230 V AC	220 V AC
Environmental range:	-10 to +50°C	-40 to +50°C	-10 to +60°C
Dimensions:	235 × 297 × 97 mm	260 × 190 × 80 mm	400 × 320 × 130 mm
Weight:	3.5 kg	2 kg	4.9 kg

SOR/T dosimeter

0052794

SOR/R dosimeter

0055056

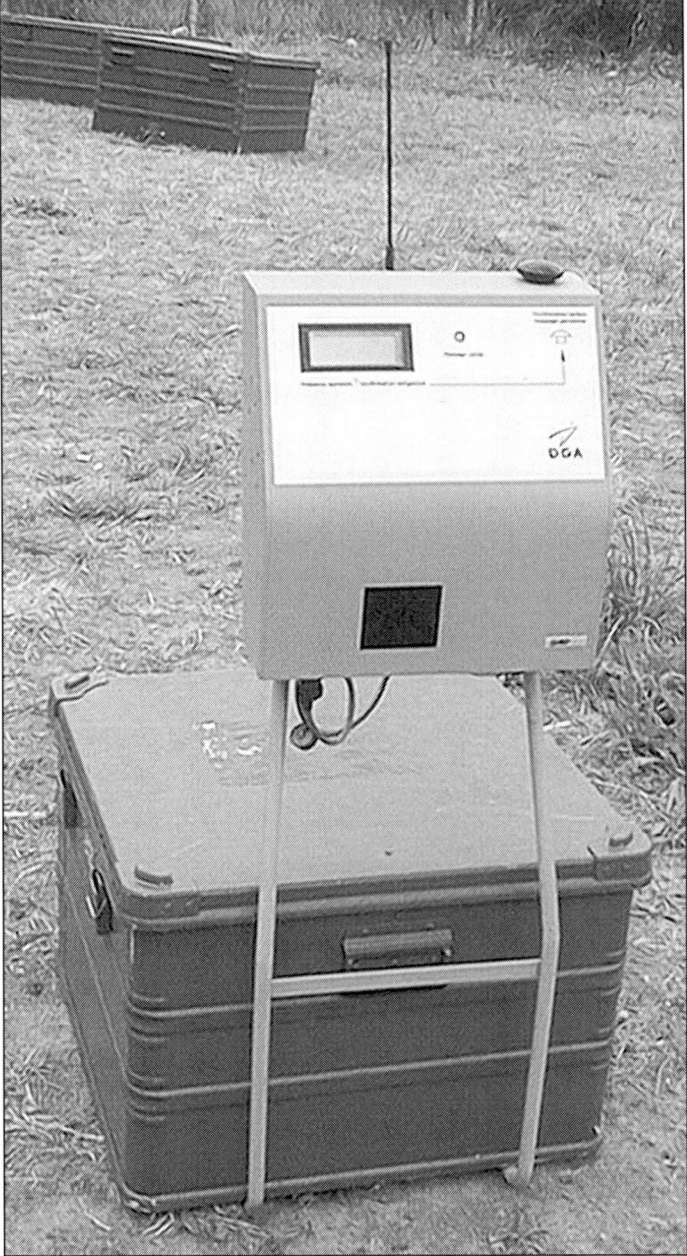

XOM/M unit level transportable dosimetry reader

0055057

XOM fixed dosimetry management unit 0052796

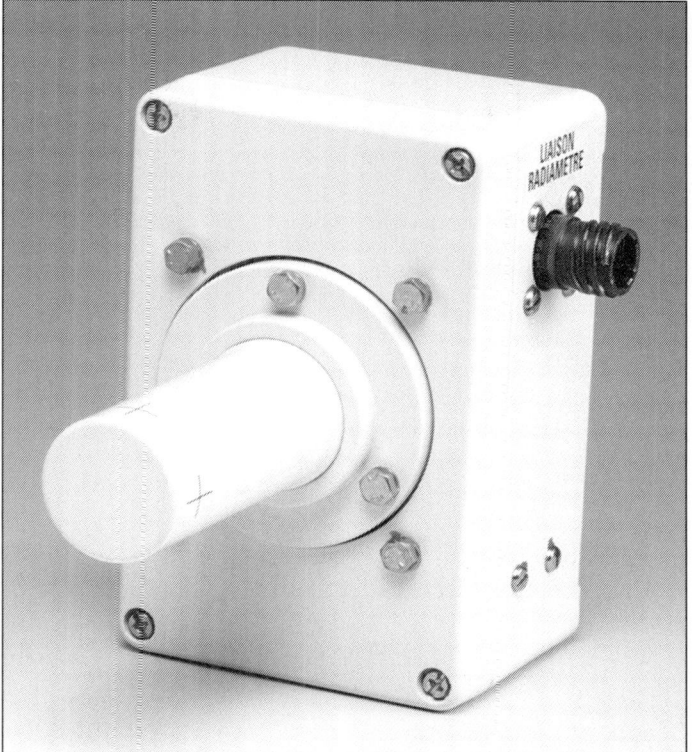

SUK SUR radiation metering unit 0052793

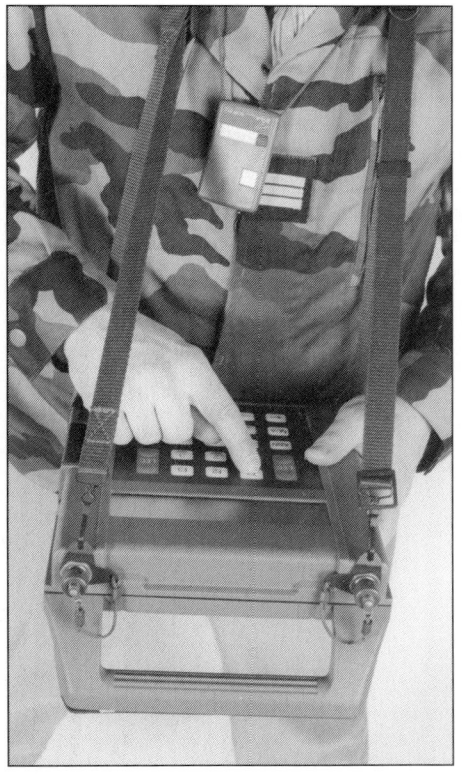

XOM/T portable dosimetry reader - for field use 0052795

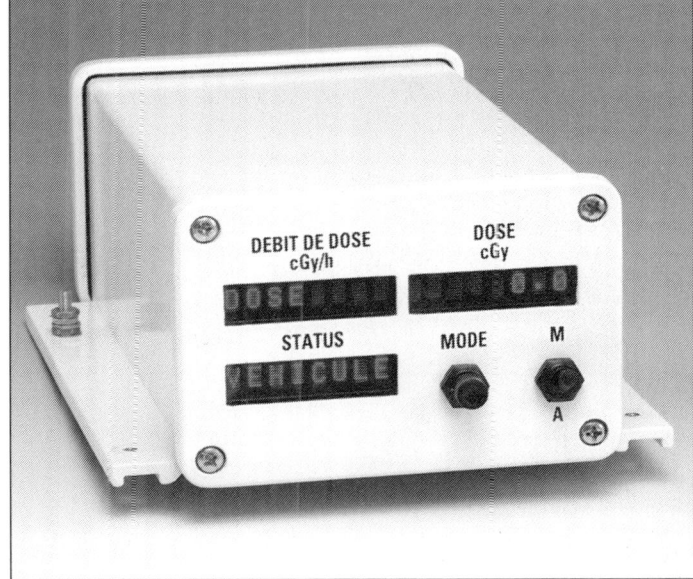

SUK SUR radiation probe 0052792

Status
In production. Adopted by the French Army and the Belgian Air Force. Sold in France, Belgium, Netherlands, Norway, Italy, Sweden, Germany and Canada.

Manufacturer
MGP Instruments.

VERIFIED

SUK SUR radiation meter system

Description
The SUK SUR vehicle-mounted radiation meter system enables γ ground and area contamination to be determined by measuring the absorbed dose. It currently equips the Thales VABRECO armoured reconnaissance vehicle (see *Jane's Armour and Artillery* for details of the vehicle). It comprises three identical and interchangeable silicon detector probes. One is mounted on the left of the vehicle, one on the right and one internally. All can be connected to a dedicated internal display unit (shown) or, as digital RS422 data, to a computer. The two external detectors provide ground contamination data while the third provides environmental monitoring inside the crew compartment. The system can also be used in shelters or as part of a monitoring network at a base location.

The SUK SUR radiation detection system is fitted to VABRÉCO armoured reconnaissance vehicle 0052785

The dedicated display can present the average dose rate of the three detector probes, the total accumulated dose for a mission and the operational status of the system, including test functions.

The system operates in the 80 keV to 3 MeV energy range. Internal or external dose rates are measured from 0.1 to 999.9 cGy/h; the 'mission' or cumulative dose is measured from 0.1 to 9999 cGy. Response time is less than 3 seconds. Audio alarm threshold levels are programmable for all measurement ranges and there is a comprehensive autotest function. The vehicle version has alarms for external dose rate and cumulative mission total dose. Data output can be read by French PERVENCHE bench test equipment and for training (French type RUK RUR 440).

The system is water, shock and vibration resistant. Power is provided from the host vehicle's DC electrical system (18 to 32 V DC). The operational temperature range is from –30 to +50°C.

Radiation metering unit dimensions are (227 × 144 × 77 mm and the total weight (3 detectors and display unit) is 1,350 kg. Each probe measures 133 × 80 × 125 mm and weighs 850 g.

Status

In production. Adopted by the French Army.

Manufacturer

MGP Instruments.

UPDATED

GERMANY

Graetz personal electronic dosimeter EDW150

Description

This dosimeter, which equips the German Fuchs vehicle, detects and stores individual total radiation dose and dose rate. It uses GM tube technology and has a 7-digit LCD display for γ and X-rays. Four pre-set dose and dose rate alarm thresholds are user-selectable. The unit is programmed via the EDAG 02 read-out unit and data is presented on a PC screen via the MS Windows®-based Graetz software. The dosimeter runs on three 1.5 V LR1 batteries, giving an operating life of approximately 12 months at background radiation level. A spare battery is included.

Status

In production. In service with the NBC Reconnaissance Fox/Fuchs vehicles (see separate entries under Detection - Reconnaissance Systems).

Distributor

Graetz Strahlungsmeßtechnik GmbH.

NEW ENTRY

Specifications
Dose indication range: 0.1 μSv - 10 Sv
Dose rate indication range: 0.1 μSv/h - 1.5 Sv/h
Energy range H$_p$(10): 50 keV - 2.0 MeV
Dose alarm thresholds: 4 pre-set values in the range of 1 μSv to 10 Sv
Dose rate alarm thresholds: 4 pre-set values in the range of 1 μSv/h to 1.5 Sv/h
Dimensions: 136 × 40 × 17 mm
Weight: approx 160 g
Temperature range: –20 up to +60°C

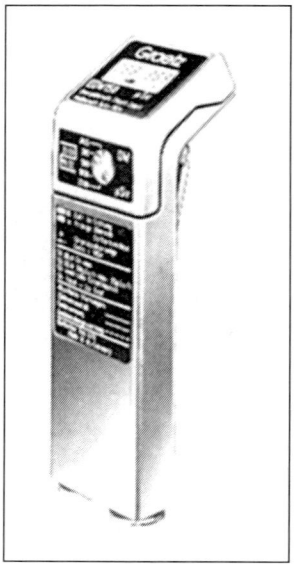

The Graetz EDW150 personal dosimeter forms part of the German Fuchs vehicle's nuclear detection suite (Graetz Strahlungsmeßtechnik GmbH)
***2002**/0137265*

RADOS H13422 MicroCont nuclear contamination monitor

Description

The MicroCont is a portable, hand-held monitor for measuring α/β/γ contamination. It is easy to use, has a large multifunctional display and can cover a large detection area. The unit is designed to be mounted on a docking station and can connect to the vehicle's main electrical supply. Apart from sample measurement, a special bogie arrangement also allows it to be moved at a set height for ground contamination measurement. The set includes an α/β test source and the instrument is protected by a rugged ABS case.

Status

In production. In service with the NBC Reconnaissance Fox/Fuchs vehicles (see separate entries under Detection - Reconnaissance Systems).

Manufacturer

RADOS Technology GmbH.

NEW ENTRY

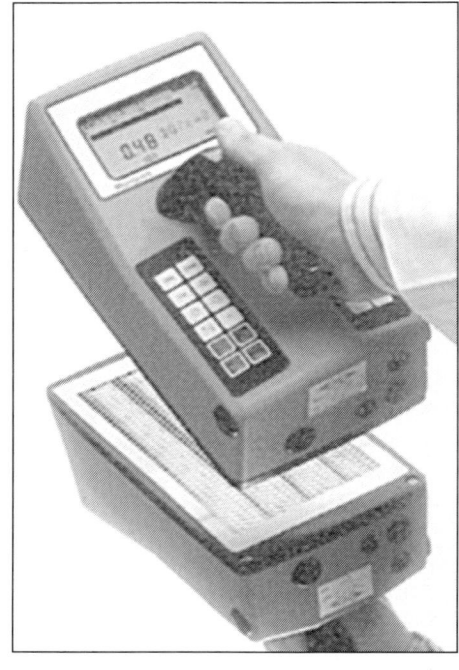

H13422 MicroCont nuclear contamination monitor (RADOS Technology GmbH)
***2002**/0137264*

Rheinmetall Landsysteme NBC field laboratory

Description

The Rheinmetall Landsysteme NBC field laboratory is a comprehensively equipped mobile detection and analysis facility. The integrated equipment suite is sheltered in standard ISO containers mounted on military five-ton trucks. Ancillary equipment is carried on trailers, allowing rapid deployment by land, sea or air. Each air-conditioned shelter is equipped with comprehensive VHF, digital telephony and satellite communications. The unit is capable of collecting, detecting and analysing

View inside the radiation and HazMat facility   0084042

Specifications
(Radiation element)

Collection	Detection	Analysis
Aerosol	Low level γ: 2 germanium detectors - one portable, one lead-shielded against background radiation inside the container	Mass spectrometer with nine interchangeable GC modules, two injectors and an air and water-sampling sensor
Air samples (personnel monitoring)	Portable γ detector	Portable IMS system
Dust sample collector	Portable monitor for α, β, and γ radiation	Air, dust, water, gas
Water sampling unit	Neutron meter	Gas analyser for toxic industrial compounds
	Portable detector for uranium and plutonium detection	Fresh/waste water analysis set
	Dose meter	Portable X-ray fluorescence spectrometer
	HF radiation meter	

Rheinmetall Landsysteme NBC field laboratory   0050715

NBC field laboratory facilities   0050716

background and residual radiation from an incident at current occupational, environmental protection or military levels. Material can be analysed from the integral collection facilities or from samples collected by the Fuchs or other NBC reconnaissance vehicle (see under DETECTION (C³I systems)).

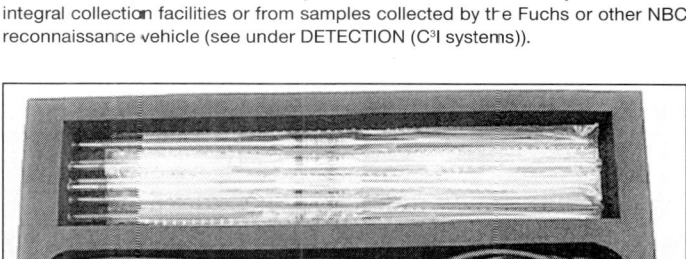
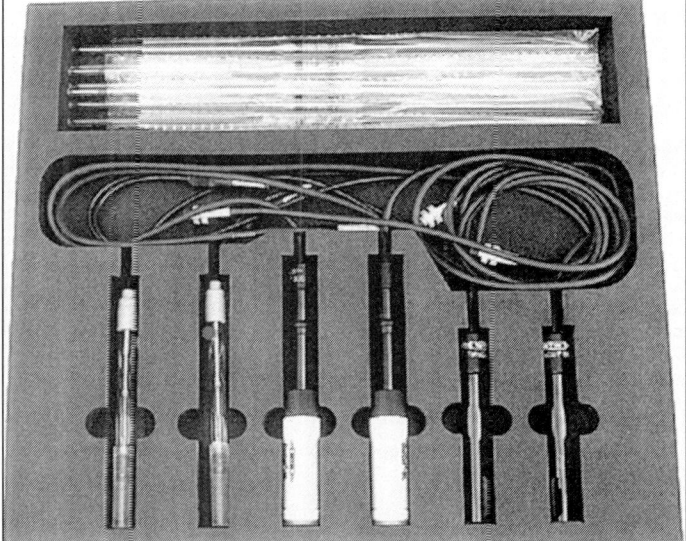

γ probes and sample collectors - NBC field laboratory   0050717

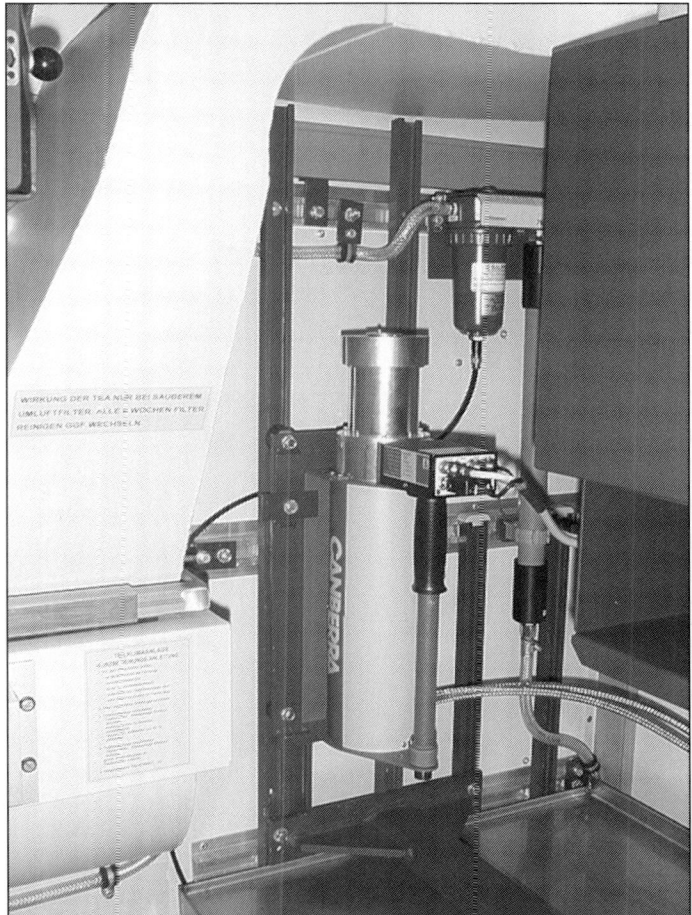

Uranium and plutonium detector   0084043

The NBC Field Laboratory consists of the following subsystems:
- 1 radiation and HazMat (Hazardous Materials) analysis lab shelter
- 2 biological analysis lab shelters (see under DETECTION (sensor systems) Biological)
- 1 chemical analysis lab shelter (see under DETECTION (sensor systems) Chemical)
- 1 command vehicle. Also used for sample collection and transport.

Each subsystem has personnel services and an integral power supply to allow autonomous operation for up to 48 hours in any climatic condition. The output from the laboratory is designed to enhance command decision support in managing an NBC incident, an industrial accident or in preparing for a military operation under NBC hazard. It also has an important role in the arms control arena in the collection of evidence of agent development, storage or use contrary to current conventions.

Status
Available.

Manufacturer
Rheinmetall Landsysteme GmbH.

UPDATED

MVM 1000 radiation detection system

Description
The MVM 1000 radiation detection system provides continuous measuring and recording of γ radiation in a radioactive environment.

The system consists of a central processing unit with a display control, monitoring and alarm components. There is also a built-in simulation system. The central unit can be supplied with data from up to five sensors, each of which can be up to 150 m away (10 km if a repeater device is used), and a recorder. The complete equipment may be installed in a vehicle or a static location. Normally the recorder is located close to the main unit, but a second can be located at a distance with the aid of a special cable and can serve as a remote display.

Specifications
Main Unit
Weight: 35 kg
Width: 330 mm
Height: 490 mm
Depth: 390 mm

Sensor
Weight: 1 kg
Diameter: 110 mm
Height: 160 mm

Recorder
Weight: 10 kg
Width: 220 mm
Height: 230 mm
Depth: 360 mm
Supply voltage: 115 V, 60 Hz
Power consumption: 100 VA
Measurement range: 0.1 mrad/h to 1,000 rad/h or 0.0001 to 1,000 cGy/h
Simulation: microprocessor-controlled, 10 to 90 min in 10 steps
Recorder: 2-channel compensation recorder
Sensor: watertight to a pressure of 70 kp/cm^2

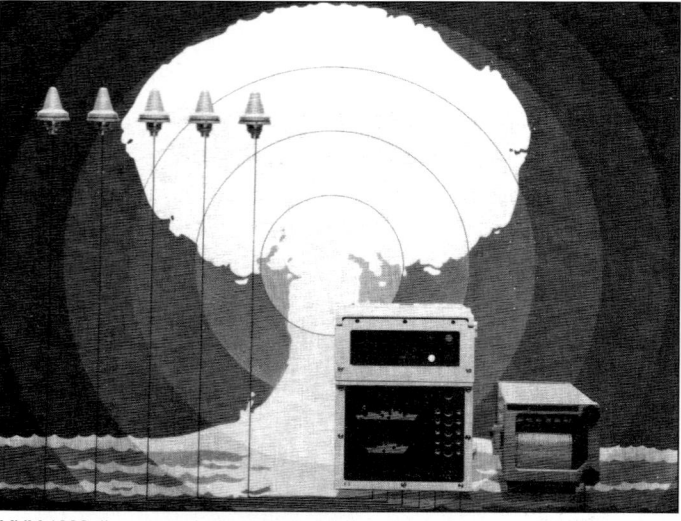

MVM 1000 diagrammatic illustration of radiation detection system with five sensors and recorder

Status
In service with the German Navy and other navies.

Manufacturer
Philips GmbH.

UPDATED

SVG 2 Hand-held Radiation Detector

The SVG 2 is a lightweight hardened radiation detector. This hand-held instrument is microprocessor-controlled and its design is based on state-of-the-art semiconductor technology. The SVG 2 comprises a base instrument, with integrated energy-compensated MOSFET and PiN-diode sensors for γ–and neutron radiation detection. An external personal dosimeter and an external α/β/γ probe are included. Initial and cumulative doses of neutron and γ radiation are monitored even when the instrument is switched off. Both current γ-dose rate and integrated γ-dose over specified time can be displayed. The basic instrument is easy to handle with a low power consumption.

The internal memory allows storage of up to 330 data groups consisting of time, geographic location, and current dose rate (complies with NATO NBC-4-NUC standards). Up to 280 of these data groups can be automatically stored allowing a further 50 values to be recorded when required. For mobile monitoring, two external γ probes can be fixed to reconnaissance vehicles, feeding information to the SVG 2. This allows measurement of surface activity and contamination of liquids, including separate and combined measurements of α/β/γ radiation. The SVG 2 can be detached from the vehicle for hand-held, dismounted missions.

Alarm thresholds for all the modes are adjustable. The alarm is either optical (red flashing LED) or acoustic (a horn (>90 dBA) or both.

Base unit specifications
Dimensions (W × H × L): 196 × 96 × 122 mm
Weight (inc batteries): 1.7 kg
Power supply (batteries): 3 × 1.5 V (>90 hours of operation)
Power supply (vehicle): 10-32 V DC

Measurement ranges (base unit and external vehicle γ-probes)
γ **energy:** 70 keV-1.5 MeV
γ **dose rate:** 0.5 μGy/h-20.0 Gy/h
γ **dose:** 0-20.0 Gy/h

Internal dosimeter measurement ranges
γ **energy:** 0.1-3 MeV
γ **dose:** 0-20.0 Gy/h
This value continues to be measured even when the instrument is switched off.

External α/β/γ-probe measurement ranges
α/β **radiation:** 0-500 kCps (0.5-800 kBq/cm^2)
γ **energy:** 70 keV - 1.5 MeV
γ **dose rate:** 0.5 μGy/h - 2.0 Gy/h

Status
Available.

Manufacturer
Bruker Daltonics®.

Marketing Agency
Jasmin Simtec Limited (UK).

UPDATED

SVG 2 Hand-held Radiation Detector (Bruker Daltonics®)   ***2001**/0077506*

HUNGARY

BNS-97 Local Radiation Monitor

Description

The BNS-97 Local Radiation Monitor comprises a high sensitivity γ radiation meter and an alarm unit in a rugged airtight case. The range of γ radiation detection is very large and depends on a unique patented device (see entry: IH-95 Universal radiation meter in this section). As an approved radiation protection device, the BNS-97 monitor benefits from certification according to the UK NAMAS standard and that of the Hungarian National Measurement Office.

The BNS-97 is a flexible monitoring solution. Distributed around the perimeter of a fixed installation, detectors offer the possibility of continuous site monitoring. The units have local audio and visual alarms and data from individual detectors can be integrated to offer a sophisticated site warning and monitoring system.

Events from two fixed alarm threshold levels can be downloaded to a computer and the device includes a third dynamic alarm level which is calculated from detection data to indicate the trend of radiation activity over time. Local LCD displays show the radiation levels and the unit can trigger an alarm at three user preset threshold levels.

Specifications
Measurement range: 50 nGy/h to 200 nGy/h (±10%)
Indication range: 200 mGy/h to 10 Gy/h (±30%)
Nuclear parameters: IEC 1017-1:1995
Energy range: 60 keV to 1.5 MeV
Set up time: 4 s
Environmental range: −30° to +50°C
Data: RS-485
Network limit: 32 units on the same cable
Power: +12 V local or remote supply
RF compatibility: IEC 801 (resistance), IEC 55011 (emission)

Status
Available.

Manufacturer
Gamma Technical Corporation.

VERIFIED

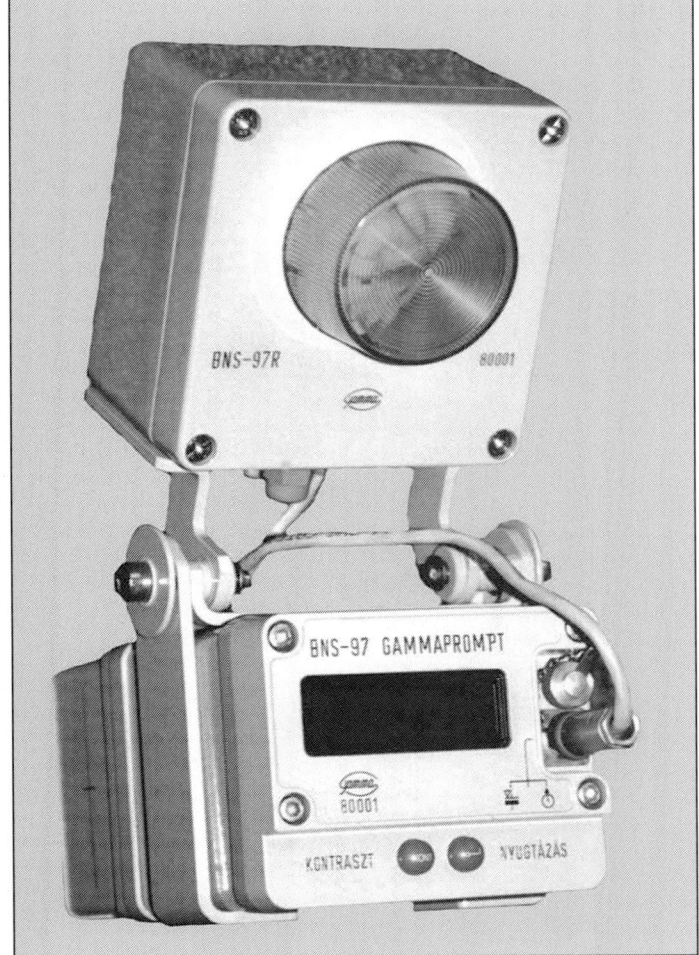

BNS-97 Local Radiation Monitor   **2000**/0084049

BNS-98 remote dose-rate transmitter

Description

The BNS-98 is a highly sensitive γ-radiation meter with an extremely wide measurement range. The device was designed to operate even at extreme meteorological conditions. Its case is sealed and there are no external control buttons. The remote dose-rate meter and transmitter can be fitted in vehicles, aircraft or fixed sites. The BNS-98 is the radiation detector for the TVS-3 environment monitoring station. The device contains a special algorithm which alarms at a significant increase in radiation level over a user-preset timescale. The BNS-98 communicates via a bidirectional line can be connected to a computer.

Specifications
Measurement range: 50 nGy/h to 500 nGy/h (±10%)
Indication range: 500 mGy/h to 100 Gy/h (±30%)
Nuclear parameters: IEC 1017-1:
Energy range: 60 keV to 1.5 MeV
Set up time: 4 s
Environmental range: −30 to +50°C
Data: RS-485
Network limit: 32 units on the same cable
Power: +12 V local or remote supply
RF compatibility: IEC 801 (resistance), IEC 55011 (emission)

Status
Available.

Manufacturer
Gamma Technical Corporation.

VERIFIED

IH-95 Universal Radiation Meter

Description

The IH-95 is a general-purpose high-sensitivity intelligent portable device for measuring α, β and γ radiation. It combines the capabilities of a dosimeter and a contamination meter and its design is optimised for military use. While enclosed in its case (see illustration) it operates as a γ detector, sensing through the dosimetric filter built into the case itself. Removal from this shielding allows it to also detect α and β radiation through the large sensor on the rear of the instrument. The instrument has a user-selectable LCD display and threshold levels can also be

Specifications
IH-95
γ dose rate: 50 nGy/h to 0.5 Gy/h
γ dose: 1 nGy to 1 Gy
Energy range: 60 keV to 1.5 MeV
Standard: satisfies IEC 532
Audio alarm threshold: selectable
Surface α contamination: 2 Bq/cm^2 to 3 MBq/cm^2
Surface β contamination: 0.2 Bq/cm^2 to 300 kBq/cm^2
β radioactive contamination: 2 kBq/l to 3 GBq/l
Measurement time: 2 s to 2 min (automatic)
Environmental range: −20 to +50°C
Dimensions: 350 × 350 × 250 mm
Weight: (not reported)

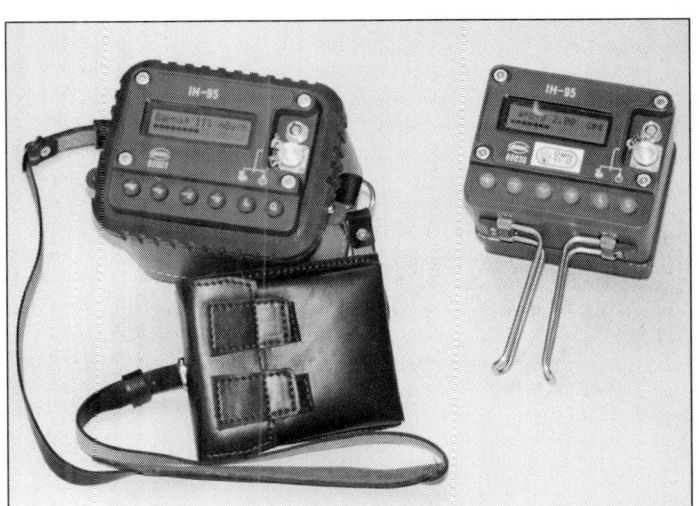

IH-95 general purpose radiation meter   0052783

selected. Data can be viewed directly from the display or via output to a PC. The meter automatically displays the menu appropriate to its function.

Uniquely, an award-winning device incorporated in the design allows the IH-95 a much wider measurement range (9 decades) than the industry standard (2.5 decades). Additionally, the device limits the GM tube to 1000 cps, thus preventing saturation and increasing significantly the life of the tube.

Status
Available. In service with the Hungarian armed forces and civil defence.

Manufacturer
Gamma Technical Corporation.

VERIFIED

TVS-1 nuclear and meteorological sampling station

Description
The TVS-1 station can be erected at unit level and, although primarily designed to monitor nuclear and meterological data, its data acquisition module allows for other digital input, such as from CW or BW detectors. The TVS-1 comprises a data acquisition module and several high-quality meteorological detectors for measuring barometric pressure, relative humidity, wind direction and velocity. It includes two thermometers separated in height above ground and one soil thermometer. The data acquisition unit is equipped with five inputs for 4/20 mA transmitters, an output computer connection and two RS-232 receptacles, allowing the connection of other intelligent input devices. The data acquisition module calculates contamination levels and the integrated data can be presented on a graphical display or sent to a remote terminal up to 2 km away. TVS-1 units can be integrated into a network for wide area monitoring.

The construction allows for a variety of detectors to be attached in addition to the basic configuration. Options include a rainfall sensor, a water-level sensor and a radiation meter (BNS-98). CW detection is achieved from any modern detector with a suitable connection.

> **Specifications**
> **Dose rate:** 50 nGy/h to10 Gy/h
> **Relative humidity range:** 0 to 99.9 at ±2%
> **Temperature range:** -40 to +60°C at ±0.2°C
> **Pressure range:** 960 to 1080 mbar at ±1%
> **Wind velocity range:** 0.7 to 30.0 m/s at ±2%
> **Data transmission facility:** RS 232 over 2 km

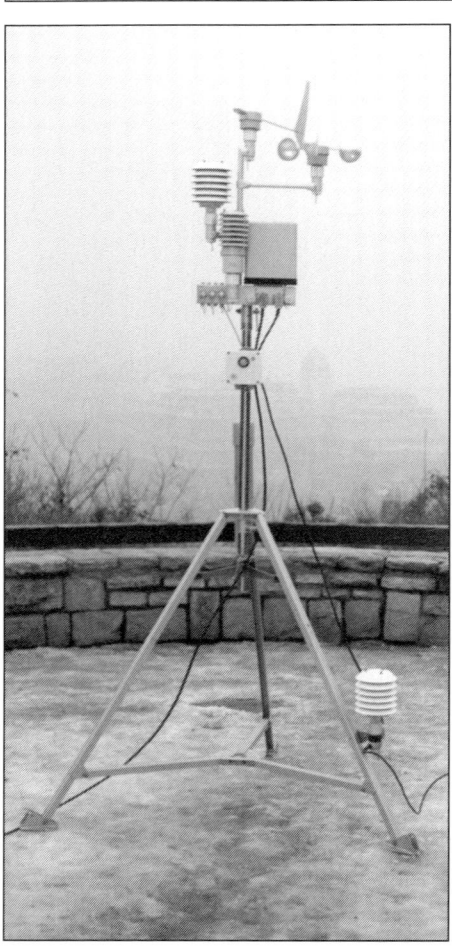

TVS-1 nuclear and meteorological sampling station
0052782

PC-processed data includes average data values (previous 30 minutes). In addition to data storage, the computer triggers the alarm if a parameter exceeds a preset value, calculates minimum, maximum and average values for the previous 24 hour or 30 minute periods. The data can be viewed graphically and printed out.

The TVS-1 is designed to be used as a continuous environmental monitor at the scene of any NBC incident.

Status
Available. In service with the Hungarian armed forces and civil defence units.

Manufacturer
Gamma Technical Corporation.

VERIFIED

TVS-3 environment monitoring station

Description
The TVS-3 mobile environment monitoring station evolved from the TVS-1 (see this section) and was developed to improve the measurement of radiation,

TVS-3 mobile nuclear and meteorological sampling station (Gamma Technical Corporation)
2002/0125534

TVS-3 mobile nuclear and meteorological sampling station - close up of detector array (Gamma Technical Corporation)
2002/0125533

meteorological data and the vapour from CW agents, TICs and TIMs. The TVS-3 is fully automatic and up to 256 detectors can be deployed over a 2 km radius.

Status
Available.

Manufacturer
Gamma Technical Corporation.

UPDATED

....

VSMF Contamination sampling set

Description
VSMF is a comprehensive and versatile set of tools for contaminated sample collection. It includes receptacles for solid-, liquid- or gas-phase materials.

The set is suitable for inspection teams in the gathering, secure preservation and safe transportation of samples for analysis. The array of equipment is carried in a purpose-designed shock-resistant plastic and aluminium suitcase to prevent damage or loss (see illustration).

An awl, shovel, dagger, scalpel, scissors and forceps are included for solid material sampling (soil, food, plants, etc). Liquid samples can be taken at the surface and at depth from exposed water or from deep wells, for example. A specially designed power tool helps retrieve sediment samples. An electronically controlled, battery-powered air-sampling device allows airborne particles to be collected onto filter pads (supplied).

Samples are stored in separate sealed plastic bags on which the collection data can be recorded. The sample bags are stored in the carrying case or in a secure cool box if necessary. Tubes are included for food sampling.

Toolkit:
Battery-driven air pump
Powered sediment sampler
Aerosol filters
Sampling tubes
Sealed plastic bags
Powered water pump
Water sampling ladle
Awl, shovel, dagger, scalpel, scissors, forceps
Cool box

Status
Available. In service with the Hungarian armed forces and civil defence units.

Manufacturer
Gamma Technical Corporation.

VERIFIED

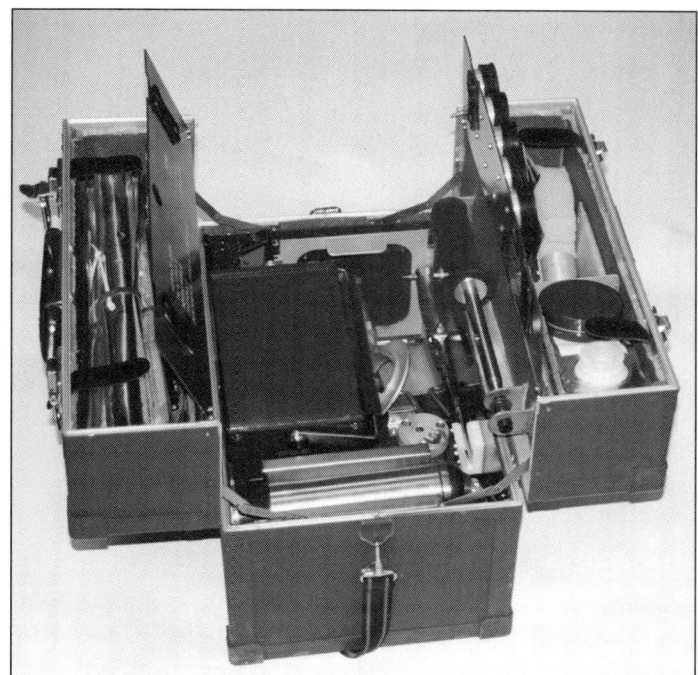

VSMF nuclear and chemical sampling set　　　　0052781

ROMANIA

DET-2 individual dosimeter

Description
This instrument is entirely Romanian designed and produced and is used for detecting and measuring γ and X radiation. It can detect and measure dose rates from 0.01 to 600 rad/h and calculate the flux if the time from the nuclear incident is known. An alphanumeric display is augmented by visual and acoustic warnings if the dose rate increases above a designated danger level. There is a facility for automatic data transfer. The battery life is approximately 50 hours.

DET-2S
The DET-2S variant is designed for stationary measurement of dose and dose rate at fixed points at a permanent base.

DET-2P
This variant comprises an individual DET-2 detector, an external probe with mounting bracket, an integral power supply and data transmission unit. It is designed to be mobile and can be fitted to land vehicles or to aircraft.

Specifications		
Type	**DET-2P**	**DET-2S**
Detector	2 Geiger-Muller tubes with energy compensation	4 Geiger-Muller tubes with energy compensation
External probe cable length	150 m	
Radiation discharge range	0.01 mR/h to 600 R/h	
Dose range	0-600 R	
Error	±30%	
Environmental range	−30 to +50°C	0 to +50°C
Power supply	10-30 V DC at 2.5 A	3 × 3.6 V Li batteries
Dimensions (main unit)	69 × 196 × 175 mm	(not recorded)
Dimensions (external probe)	60 mm diameter × 280 mm	(not recorded)
Weight	(not recorded)	
DET-2 Unit		0.8 kg
External probe with mounting support		0.8 kg
PSU with cables		2 kg
External probe cable		20 kg

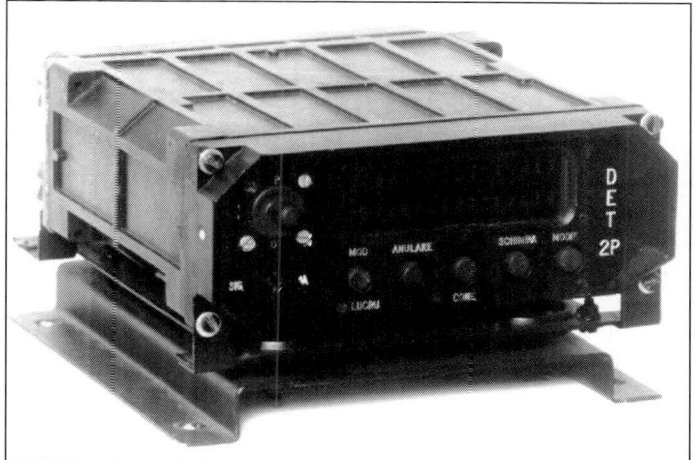

DET-2P dosimeter for vehicles and other mobile units (ROMTEHNICA)
***2002**/0121886*

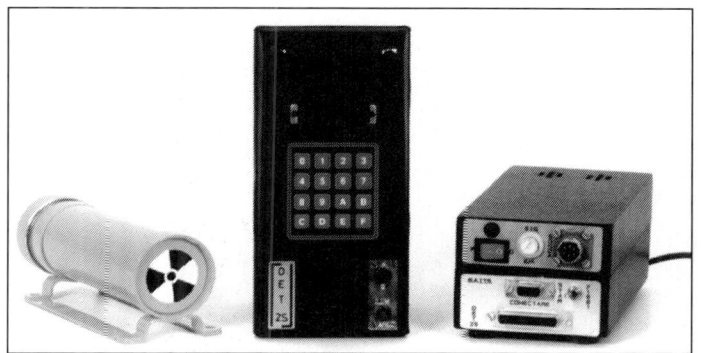

DET-2S components, showing the probe (left), dosimeter unit and 'BAITS' power supply/interface/transmission unit (right) (ROMTEHNICA)　　***2002**/0121887*

Detection data is automatically corrected for the vehicle environment and for height when fitted to aircraft.

Status
In service with the Romanian armed forces and possibly with Romanian civil defence organisations.

Marketing agency
ROMTEHNICA.

UPDATED

Mini radiometer-roengenometer

Description
The mini radiometer-roentgenometer is designed to measure γ radiation and contamination of terrain, liquids and equipment. It consists of a control and display unit offering a dose rate scale of 0.1 m rad/h to 600 rad/h with a maximum error of ±20 per cent. It is powered by four 1.5 V batteries (R14) giving an operational time of 40 hours. The instrument weighs 0.88 kg and measures 113 × 62 × 105 mm.

Status
Available. Offered for export sales.

Marketing agency
ROMTEHNICA.

VERIFIED

RBCA airborne radiation detector

Description
The RBCA airborne radiation detector was designed to be used on helicopters to conduct radiation surveys when flying at altitudes between 50 and 500 m. The single RBCA unit is secured to a panel in the helicopter and is connected to external radiation sensors by cables. Readings are shown on a lamp display and the 27 V power supply is taken from the helicopter's electrical system.

The RBCA measures γ radiation at ground level between 5 and 3,000 rad/h and at flight altitudes between 30 mrad/h and 30 rad/h. Response time is 1 second. The system weighs 5.5 kg.

Status
In service with the Romanian armed forces.

Marketing agency
ROMTEHNICA.

VERIFIED

RUSSIAN FEDERATION AND ASSOCIATED STATES (CIS)

DP-5V dose rate meter

Description
The DP-5V dose rate meter is a portable area survey meter intended for measuring levels of γ radiation and contamination; the instrument may also be used to detect beta radiation. A complete DP-5V comprises the main instrument carried in a case

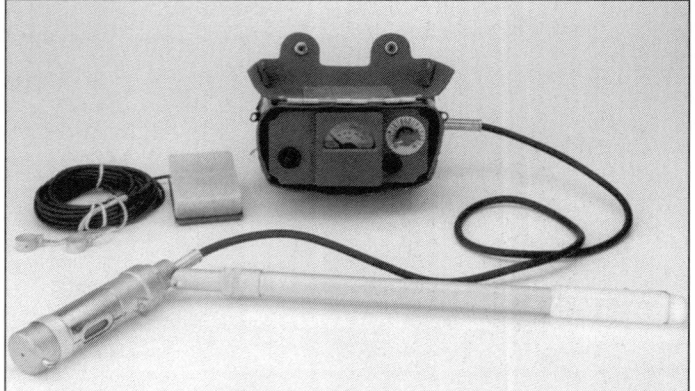

A complete DP-5V dose rate meter

slung around the neck, connecting cable, detector probe unit and a telescopic handle to carry the probe. Also included are an accessories kit and some spare parts.

The DP-5V measures gamma radiation over a range of 0.05 to 200 rad/h with an error range of 30 per cent. The instrument may be powered by an internal 4.5 V battery or a 12 V vehicle or other external supply. The operating temperature is from -50 to +50°C. The combined weight of the instrument and probe is 1 kg.

The DP-5V is also manufactured in Bulgaria where it is marketed by Kintex.

Status
Available from Russian state factories before privatisation. In service with members of the former Warsaw Pact armed forces.

Manufacturer
Unknown.

UPDATED

IMD-2 dose rate meter

Description
This series of meters is designed to measure the γ radiation dose rate. Instruments are available for fixed installations (IMD-2S) and vehicles (IMD-2B). There is also a man-portable version (IMD-2N).

Status
In production and available for export.

Specifications
Measurement range (rad/h): from 1×10^{-5} to 1×10^{3}
Indication error: ±30%
Energy dependence over energy range 0.08 to 1.25 MeV: ±25%
Time to settle: 1 min
Available after: 3 min
Time to measure: 2 s (subrange to 1×10^{3} rad/h), 4 s (up to 1×10^{3} mrad/h), 40 s (up to 5×10^{2} mrad/h)
Power supply: 4 × A-343 batteries or 12 and 27 V DC or 220 V AC direct supply.
Service life: 10,000 h (min)
Weight: 1.6 kg

Manufacturer
Unknown.

Marketing agency
Rosoboronexport.

VERIFIED

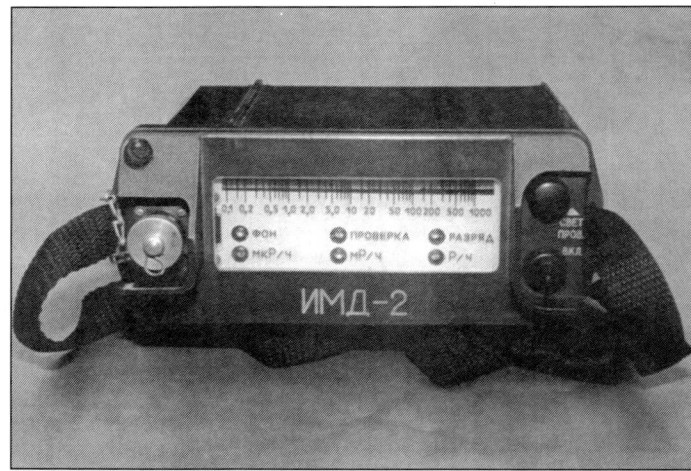

IMD-2 dose rate meter   0019375

IMD-5 dose rate meter

Description
The IMD-5 dose rate meter is a portable area survey meter intended for measuring levels of γ radiation. The instrument may also be used to detect β radiation. A complete IMD-5 comprises the main instrument carried in a stowage box. The meter unit is carried in a strapped pouch round the neck, with its connecting cable, detector probe unit and a telescopic handle to carry the probe. Headphones are provided and the case includes an accessories kit and some spare parts.

IMD-5 dose rate meter with probes, headphones and stowage box   0055209

Specifications
γ dose rate (energy range
   0.084-1.25 MeV): 5 × 10⁻⁵ to 200 cGy/h 30%
β flux density detection range: 50 to 5 × 10⁴ particles/min/cm²
γ dose measurement error: ±30% over measurement range
Measurement time: 45 s
Environmental range: -50 to +50°C
Battery operational life (type A-343): 100 min
Dimensions: 142 × 262 × 402 mm
Weight: 3.5 kg (meter) 9 kg (full equipment)

Status
In service with members of the former Warsaw Pact armed forces.

Marketing Agency
Rosoboronexport.

VERIFIED

IMD-21 series of γ dose meters

Description
This range of instruments is designed to measure γ radiation. It offers selectable ranges between 1 and 10,000 rad/h (error ±25 per cent) and the detected dose rate is displayed on an LED panel 10 seconds after detection. The instruments take about 5 minutes to warm up and a switch allows threshold selection between 1.5, 10.5 and 100 rad/h. Presented in tough, vibration-resistant, dust and damp-proof cases, measuring 191 × 247 × 113 mm, the IMD-21 series of instruments is capable of operating in the temperature range of -50 to +65°C. The probes measure 68 mm in diameter and are 316 mm long. Service life is 25,000 hours and a useful life of 11.5 years is claimed.

The instruments are designed for permanent installation in vehicles, aircraft and protected ground installations. There are five variants: the variants are IMD-21, B, S8A and SA-R, each designed for a particular system installation.

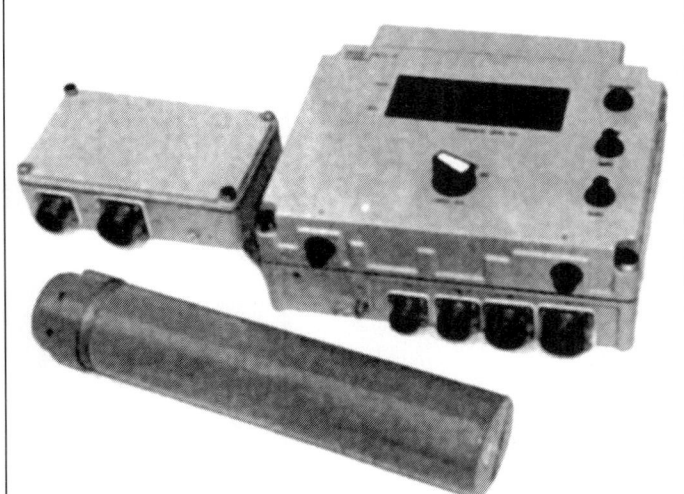

IMD-21SA-R γ dose rate meter   0019373

IMD-21B γ dose rate meter   0019374

Status
In production and available for export.

Manufacturer
Unknown.

Marketing agency
Rosoboronexport.

VERIFIED

Model DP-5A radiation meter

Description
The Model DP-5A radiation meter is used to monitor personnel, food and water for low levels of radiation and has only a very limited application in area monitoring and reconnaissance. The meter is usually worn slung on the chest but the sensing elements are contained in a rod-like probe which can reach into otherwise inaccessible areas. Headphones are supplied for the aural detection of beta radiation.

The meter has six separate ranges from 0.05 to 200 mrad/h. The power is supplied either from three 1.6 V batteries or any external source providing between 3.5 and 12 V DC. The meter, which is contained in a glass fibre reinforced plastic case, weighs approximately 2.1 kg. The whole equipment, including the probe, carrying case, headphones and so on, weighs approximately 7.6 kg.

Status
In service with RFAS and former Warsaw Pact armed forces.

Manufacturer
State factories.

VERIFIED

Model DP-12 contamination survey meter

Description
The Model DP-12 is a highly sensitive contamination meter which is primarily intended to monitor the radioactive contamination of personnel, vehicles and equipment, but can also be used in aircraft or helicopters for rapid area surveys. It consists of two main components, the control case and the probe, but a headset for the aural detection of radiation is also provided. The probe unit is connected to the control case by a flexible cable and consists of a shaft, to the end of which, is secured a Geiger-Müller tube sensitive to both γ and β radiation; the β radiation can be shielded by a rotating collar over the tube. The shaft itself is telescopic.

When in use, the Model DP-12 is carried on the user's chest and the probe unit is held about 25 mm away from the monitored surface and, as the probe is waterproof, it may also be used to monitor the contamination of fluids.

Status
In service with RFAS and former Warsaw Pact armed forces.

Manufacturer
State factories.

VERIFIED

Model DP-23 individual dosimeter set

Description

This dosimeter set is contained in a carrying case and consists of a charger/reader unit, 150 Model DS-50 dosimeters packed into three cases each with 50 dosimeters and 50 Model DKP-50A dosimeters. Each case also contains a torch. The Model DP-23 charger/reader unit is used to charge both models of dosimeter, but of the two models, only the Model DKP-50A can be read directly. It is issued to officers and senior NCOs. The Model DS-50 dosimeters can be read only in the charger-reader unit, that is, under supervision and are issued to the rank and file soldiers. Both models read from 0 to 50 rad in 2 rad steps and both weigh approximately 30 g. The Model DS-50 is 120 mm long and 14 mm in diameter and the Model DKP-50A is 130 mm long and 13 mm in diameter. Both can operate in temperatures between -40 and +50°C. The charger/reader unit, which is battery-powered, weighs 3.5 kg and is contained in an aluminium case. The reader function is carried out in increments of 2.5 rad on a meter scale.

Status

In service with RFAS and some of the former Warsaw Pact armed forces.

Manufacturer

State factories.

VERIFIED

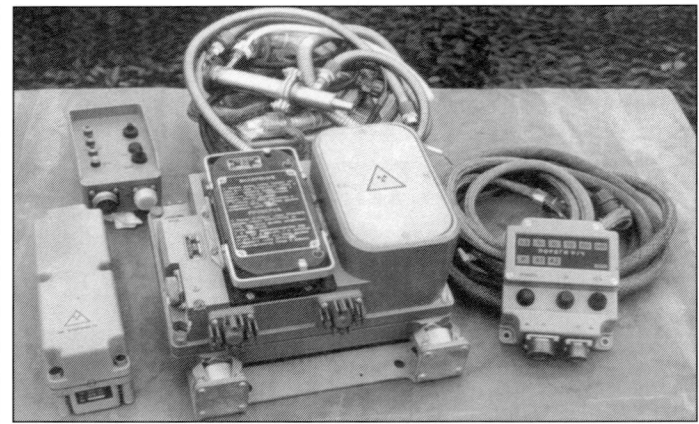
PKUZ-1 vehicle NBC monitoring and control system   0019376

and the levels on the contaminated terrain it moves over (channel R). CW agents are detected in channel C.

The threshold for nuclear detection is 4 rad/h for γ radiation with energy of 1.26 MeV direct radiation (Channel N). For gamma radiation with energy of at least 0.66 MeV (channel R) caused by radioactive contamination of terrain, the threshold is 0.5 rad/h (max).

For CW agent detection, the threshold concentration to trigger the system is 1×10^{-5} mg/litre (first range) and 2×10^{-4} mg/litre (second range). Control signal transmission times through the channels are: 0.1 seconds (Channel N), 0.5 seconds (Channel R) and 30 seconds (Channel C). The system is fully operational 20 minutes from switch-on (maximum) and requires 27 V from the vehicle supply, drawing 200 W of power. Installed, the PKUZ-1 weighs 18 kg (maximum) and has a minimum operating life of 2,000 h.

Status

In production and available for export.

Manufacturer

Unknown.

Marketing agency

Rosoboronexport.

VERIFIED

Model DP-63A lightweight area survey meter

Description

The Model DP-63A lightweight area survey meter can detect and measure γ and β radiation and is widely used by all units of the former Warsaw Pact forces. Using the ion chamber principle, it is powered by two dry cell batteries with an operational life of about 50 hours.

Normally the instrument is carried in a leather carrying case (weight 400 g) with a shoulder strap. The instrument has two logarithmic ranges selected by push-buttons. The bottom scale measures from 0.1 to 1.5 rad/h and the other 1.5 to 50 rad/h. β radiation is measured on the bottom scale only using a β radiation window. Both scales can be illuminated for night use. The Model DP-63A weighs 800 g and the case measures 165 × 115 × 90 mm.

Status

In service with RFAS and former Warsaw Pact armed forces.

Manufacturer

State factories.

VERIFIED

Model DP-70M chemical dosimeter and Model PK-56M field colorimeter

Description

The DP-70M and PK-56M are used in together. Detection is achieved on the colorimeter principle in which any exposure to radiation of a chemical solution will cause it to change colour. This system has the advantage that the dosimeter can be issued without prior charging or any other preparation, with each dosimeter containing a single ampoule of the detecting solution. The field colorimeter Model PK-56M is used to compare the colour of the exposed sample, via a beam of light, with a standard unexposed ampoule, to give the level of radiation absorbed by the Model DP-70M dosimeter. A scale allows comparison of the colour differences to determine radiation doses from 0 to 800 rad in 50 rad steps.

The Model PK-56M is issued with a carrying case and weighs about 1.5 kg.

Status

Available from Russian state factories before privatisation. In service with RFAS and former Warsaw Pact armed forces. Current manufacturer unknown.

Manufacturer

Unknown.

VERIFIED

PRKhR radiation and chemical agent detector system (Model GO-27)

Description

The PRKhR radiation and chemical agent detector system, also described as Model GO-27, is intended for installation in armoured vehicles and provides continuous monitoring for three NBC hazards. They are: the exposure to strong γ -type radiation such as that produced by the detonation of a nuclear device; the reaching of a preset dose rate of gamma radiation such as that produced over terrain contaminated by nuclear fallout; and the presence of chemical agent vapours in the atmosphere close to the vehicle. The three sensor units involved (Channel N, R and C respectively) are interconnected by coaxial cable and connected to a central control unit placed close to the vehicle commander's position. Visual and audible warnings are initiated by the control unit and, under its control, the vehicle engine will stop when a NBC hazard is detected. The system will also automatically initiate the closing of vehicle and ventilation intake ports and switch on the vehicle's internal air circulation, overpressure and filtration systems.

The PRKhR system utilises a through-the-system air flow of 3 litres/min. It was designed to operate at ambient temperatures of −40 to +50°C and incorporates a heating system for use at low temperatures. Preparation time is 20 minutes. Power is provided by the carrier vehicle's 24 V DC supply.

PKUZ-1 vehicle NBC monitoring and control system

Description

The PKUZ-1 is an ingenious device designed to both detect NBC agents and activate vehicle COLPRO automatically when pre-assigned threshold levels are reached. It is optimised for use in armoured vehicles and the detection and activation data is transferred through three separate channels. In the event of nuclear operations, it registers the γ radiation caused by the detonation (channel N)

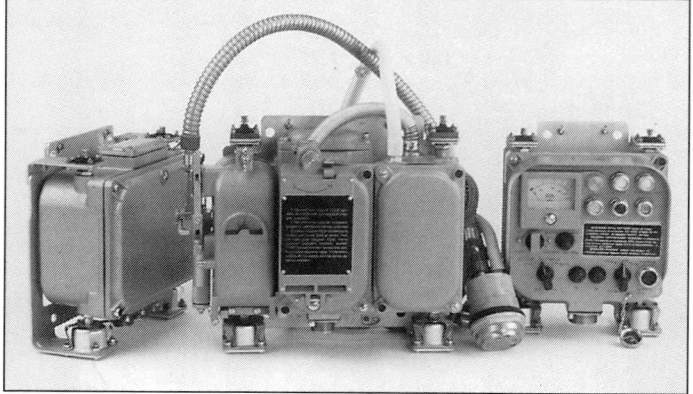
Main components of the PRKhR radiation and chemical agent detector system

The PRKhR can be set to operate when radiation levels of between 0.2 and 150 rad/h are reached. Channel N (immediate radiation) can provide a response time of 0.1 second, Channel R (fallout radiation) in a maximum of 10 seconds and Channel C (chemical agents) in a maximum of 40 seconds.

A complete system weighs 28 kg when in its transport case. The case contents include some spare parts and an accessories kit.

Status
In production. Offered for export sales.

Manufacturer
Signal Instrument Making Plant JSC.

Marketing agency
Rosoboronexport.

UPDATED

SWITZERLAND

IMS 2000 γ detector

Description
The γ detector module is an enhancement to the IMS 2000 CW agent detector (see under *DETECTION (sensor systems) - Chemical*). A removable unit installed in the IMS 2000 runs independently of the CW detection module even when the latter is switched off. The γ sensor, which measures dose and dose rate, is connected via a cable to the IMS 2000 unit and derives its power from a Lithium battery with a life of one year. The internal data back-up battery has a life of 10 years for data storage.

Status
In service with the Swiss Army and civil defence units.

Manufacturer
Bruker AG.

VERIFIED

IMS 2000 γ detector   0050769

UNITED KINGDOM

Harwell Instruments radiation detection and measurement systems

Description
Harwell Instruments Ltd (a subsidiary of Canberra Industries Inc) supplies a complete range of instruments for the detection and measurement of nuclear radiation (γ and neutrons). The range includes personnel dose and dose rate meters, surface contamination monitors, airborne radioactivity monitors and installed systems where separate instruments can be networked to cover particular buildings or local areas. Instruments are also designed to meet specific customer requirements.

Equipment is also available for non-destructive materials analysis based on neutron interrogation methods.

Status
Available. In widespread service.

Manufacturer
Harwell Instruments Ltd.

UPDATED

Local Area Radiation Monitoring System (LARMS)

Description
This system was originally designed for monitoring radiation in and around dockyards where nuclear-powered vessels were likely to be present. It can equally be used to monitor radiation levels in buildings, bunkers and other defence or nuclear-capable sites such as airfields. Each system can manage up to 9 locations and provide a full range of alarm and recording facilities.

LARMS uses the same type of γ radiation detector as SIRS (see separate entry in this section). The system is installed at all nuclear-powered submarine berths in the UK as well as some overseas locations.

There is a transportable version of LARMS which fulfils the same monitoring task as the fixed system but is designed for rapid deployment at temporary locations. A feature of this system is a remote display and alarm facility which can be deployed at a distant location.

Status
In service with the UK MoD and other governments.

Manufacturer
Siemens Environmental Systems Limited.

UPDATED

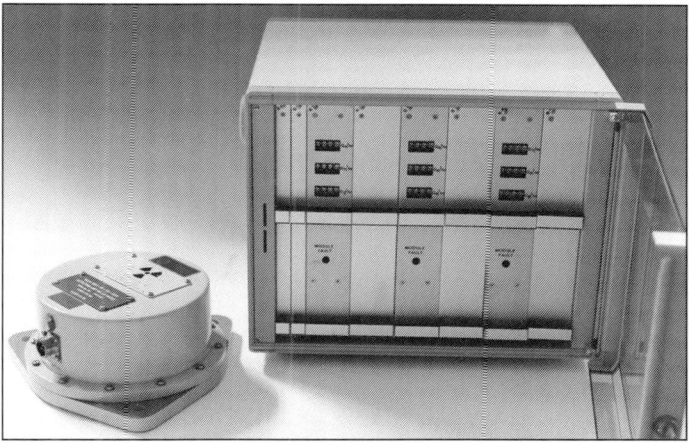

Control cabinet Local Area Radiation Monitoring System (LARMS) (Siemens)
0052801

MD1 and MD2 radiac set

Description
The radiac sets, MD1 and MD2, are portable instruments comprising a control unit, hand-held probe and a screened interconnection cable. The equipment measures and indicates on a meter in the probe, γ radiation over the range of 0.1 to 1,000 rad/h (MD1) or 0.1 to 1,000 mrad/h (MD2). It will also indicate the presence of β particles provided the level of γ radiation is not excessive.

The control unit contains self-testing facilities and all power requirements of the equipment are derived from three 1.5 V primary cells. The equipment is controlled by a four-position switch on the side of the unit; the functions of this switch are:

OFF - batteries disconnected
BATT - battery test
TEST - control unit test
PROBE - set fully operational

A Strontium 90 radioactive test sample is included in the stowage and transit case for testing the complete equipment.

The detector stages comprise two Geiger-Müller tubes, their outputs being paralleled at the input to the head amplifier. When subjected to a radiation field, the tubes generate pulses which are amplified and used to trigger a monostable

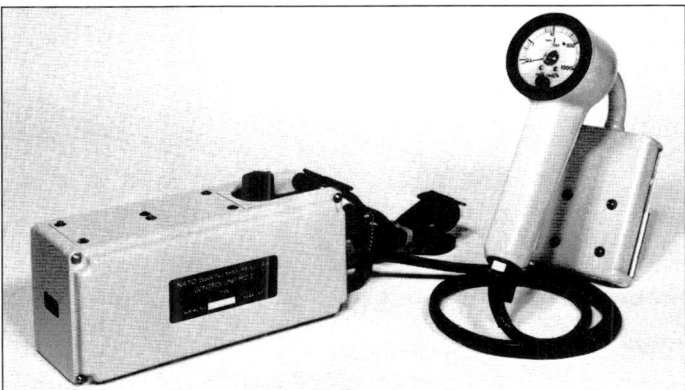

MD2 radiac set (Jasmin Simtec Limited)

Specifications
Weight: 2.7 kg
Dimensions:
(control unit) 208 × 89 × 102 mm
(probe) 218 × 77 × 140 mm
(connecting cable length) 920 mm

circuit. The monostable output is shaped and amplitude-stabilised and then fed to a diode pump circuit which, connected to the meter, provides a current proportional to the logarithm of the radiation level. The meter is calibrated in terms of radiation level. When γ radiation measurements are being made, a sliding screen covers the two windows in the probe face. When the screen is moved to expose the windows, the low-energy γ response is enhanced and the equipment will also detect β -emitting contaminants.

The probe unit is housed in a sealed casing and the measuring head can swivel through an angle of approximately 270°. A sliding screen is fitted to the probe face. The control unit is housed in a sealed case with a carrying strap at one end.

Status
In service with the Royal Navy.

Manufacturer
Jasmin Simtec Limited.

UPDATED

SAIC PD-2 and PD-12i γ detectors

Description
The SAIC PD-2 uses recorded data to inform personnel when their level of exposure to radiation has exceeded a safe limit. The PD-2 is worn clipped to an external clothing pocket and used in conjunction with an Automated Radiological Control System (ARCS), a personal computer-based console able to hold current information of up to 25,000 personnel, including their cumulative individual exposures to radiation. When a user logs into the ARCS with a PD-2, past radiation exposure is downloaded to the portable meter. The software in the ARCS allows the PD-2 to provide features such as total dose measurement, dose rate measurement, dose alarm, rate alarm and sonic chirp. When leaving a sensitive area the user plugs the PD-2 back into the ARCS and the recorded data updates the database.

The PD-2 has a range of 0.02 μSv to 50 Sv and a resolution of 0.02 μSv. Visual and aural alarms can be set across a range of values for dose, pre-dose (to warn when the safe exposure limit will soon be exceeded), dose rate, stay time and pre-stay time. The PD-12i is a further development of the PD-2.

The PD-2 is powered by a single standard AA-type battery providing up to 1,000 hours of use. Weight with the battery is 90 g and dimensions are 48 × 72 × 17 mm.

Status
In production.

Manufacturer
SAIC (UK) Limited.

VERIFIED

SAIC PD-12i personal dosimeter   0011442

SAIC PPM-100A portable portal monitor

Description
The SAIC PPM-100A portable portal monitor is claimed to be capable of monitoring radiation for eight times more personnel per hour than a conventional Geiger counter. It is intended for use by emergency services and the nuclear industry; it also has military applications. Up to 240 personnel can be monitored by the PPM-100A in 1 hour.

The PPM-100A portal can be installed by unskilled personnel in about 2 minutes. When a person steps into the PPM-100A they are detected by an electronic device and the unit starts to count. The count is displayed on a liquid crystal display panel and an alarm will sound if a preset safe dosage level is exceeded. Low levels of contamination can be detected in less than 10 seconds. As the person steps out of the portal the unit automatically reverts to counting the background level.

The PPM-100A consists of a four-piece frame and control box which are stored in two suitcase-type containers. The two side pieces of the frame, which fold in half, each contain two plastic scintillation detectors connected by cables to a control box. The design is modular and can be expanded to monitor large objects such as vehicles and livestock.

Status
In production.

Manufacturer
SAIC (UK) Limited.

VERIFIED

SAIC PPM-100A portable portal monitor in use

Ship Installed Radiac Systems (SIRS) Mk22NRS and Mk23NRS

Description
The Ship Installed Radiac Systems (SIRS) are high-integrity multipoint monitoring systems which measure the g dose rate over the range of 0 to 1,000 cGy/h and provide total dose information. There are two types of system. The Mk22NRS is designed for large ships including destroyers and frigates. The Mk23NRS is available for small ships such as MCMVs, OPVs and submarines. Both systems are designed to UK MoD specifications.

Mk22NRS
The Mk22NRS system measures γ radiation in the atmosphere and below the ship's waterline. This information is displayed at a central control position and also, if required, in the operations room or elsewhere in the ship. Alarm signals warn when the dose rate exceeds 100 mGy/h. The Mk22NRS is also suitable for use in on shore installations.

The detector units are grouped into warning channels and control channels, using Mk28NH (low level) and Mk29NH (high level) detectors. The warning channels provide low level indication of water dose rates from 0 to 999 mGy/h, giving early warning of radiation levels both above and below the waterline. The control channels provide high level indications of dose rates from 0 to 999cGy/h. The total dose is displayed separately for each control channel on counters which the user can reset.

The detectors are strategically located around the ship or boat, offering the NBC Protection Officer a continuous picture of the overall radiation dose levels in

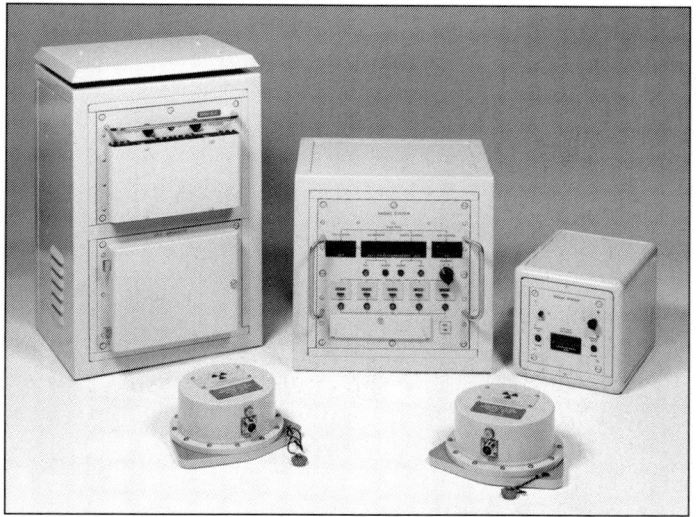

Components of the Mk22NRS version of the Ship Installed Radiac System (SIRS) (Siemens) 0052799

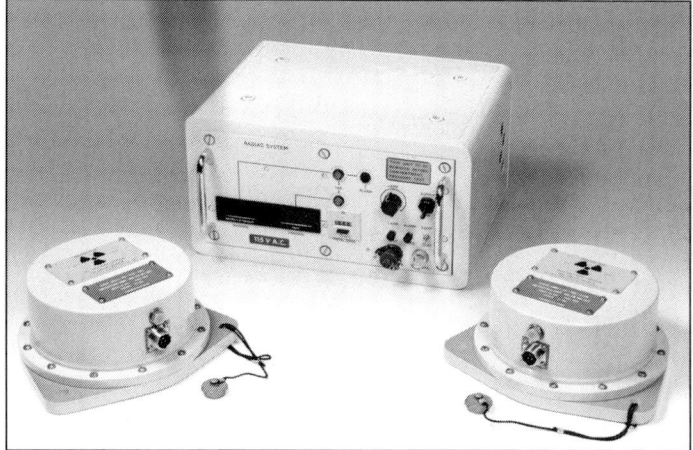

Components of the Mk23NRS version of the Ship Installed Radiac System (SIRS) (Siemens) 0052800

strategic positions. This enables him to advise the command and plan for the minimum impact of radiation on individuals in maintaining essential long-term watchkeeping operations. It also allows him to plan decontamination and to advise on the location and timing of labour intensive activities such as RAS.

For training, a fallout profile simulator is supplied with the Mk22NRS system. The programmes vary from 10 minutes to 2 hours to suit the training need.

Mk23NRS

Although smaller and lighter than the more comprehensive Mk22NRS, the Mk23NRS is derived from it, sharing many common components. It is equally suitable ashore or for installation in small ships.

It comprises an indicator unit, one Mk28NH low-level and one Mk29NH high level detector. The entire signal processing function, power supply management, dose and dose rate displays, alarms and fault indicators are housed in the indicator unit.

A simple, manually operated training module is also available.

Status

In service with the Royal Navy, European and other navies.

Manufacturer

Siemens Environmental Systems Limited.

UPDATED

....

Siemens Electronic Personal Dosemeter (EPD)

Description

The Siemens Electronic Personal Dosemeter (EPD) is a small lightweight dosemeter for measuring β and γ dose and dose rate. With immediate readout of both superficial and deep doses on a small liquid crystal display with optional back lighting, the EPD provides instant and accurate information on radiation hazards.

Running on standard AA batteries, the EPD logs data and holds it in memory for record purposes. Recorded information can be downloaded into a management computer system via a reader and transmitted over command information systems.

Alarm levels, display configurations and identity details can be set in the EPD via a simple IrDA communications link with a PC.

EPD shown clipped to standard webbing (Siemens) 0052797

Siemens Electronic Personal Dosemeter (EPD) (Siemens) 0052798

Specifications

Accuracy (full details from manufacturer):	Dose:	±10% (20 keV to 1.5 MeV)
	Dose rate:	±10% - 0.5 Sv/h penetrating
	Dose rate:	±20% - 1 Sv/h penetrating
	Dose rate:	±20% - 1 Sv/h superficial
Energy response linearity (full details from manufacturer):	Photons, 20 keV to 1.5 MeV	±30%
	Photons, 1.5 MeV to 10 MeV	±50%
Angular response:	±20% up to ±60° off axis over specified energy ranges	
Environmental range:	In use:	−10 to +40°C
	Storage:	−25 to +70°C
	Relative humidity:	20-90% (non-condensing)
Power supply:	Single AA Li battery offering 5 months continuous use (at an average dose rate of <5 mSv/h and alarm activation at <5 h total. A single AA alkaline battery gives 55 days operational use	
Dimensions:	85 × 63 × 19 mm	
Weight:	95 g (including battery)	
Casing:	High-impact ABS/ polycarbonate blend	

The EPD is rugged but extremely sensitive. Ideal for low-level radiation monitoring, it measures from background to full radiac levels. It is the only electronic dosimeter in the world to have been approved for measuring the radiation levels for health and safety 'legal' dose record purposes.

The EPD is calibrated for life but a simple irradiator for checking and confidence testing is available.

The EPD can be supplied in NATO green or grey.

Status
In service with NATO forces and other overseas armed forces.

Manufacturer
Siemens Environmental Systems Limited.

UPDATED

Siemens Neutron Electronic Personal Dosemeter (EPD-N)

Description
The Siemens Neutron Electronic Personal Dosemeter (EPD-N) is part of the same family of dosemeter products as the EPD. It measures γ and neutron dose and dose rate. It has similar features to the EPD (see separate entry within this section).

Siemens Neutron Electronic Personal Dosemeter (EPD-N) (Siemens) **2001**/0098704

EPD-N close-up (Siemens)   **2001**/0098703

The EPD-N is powered by standard AA batteries and is compatible with the EPD, its readers and dose management software. Dose information is displayed to the wearer on a backlit LCD panel which ensures clarity even in conditions of low light.

Status
Available.

Manufacturer
Siemens Environmental Systems Limited.

UPDATED

Specifications

Dose rate alarms:	γ:	10 µSv/h to >1 Sv/h (1 mrem/h to >100 rem/h)
	N:	100 µSv/h to 1 Sv/h (10 mrem/h to 100 rem/h)
Energy response - Photon, $H_0(10)$ (referred to Cs^{137}):	Range 50 keV to 1.5 MeV:	±25%
	Range 1.5 MeV to 6 MeV:	±30%
	Range 6 MeV to 10 MeV:	±50%
Energy response - Neutron:	Test data available from manufacturer[1].	
Angular response - $H_0(10)$ (referred to Cs^{137}):		±20% up to ±75°
Accuracy - $H_0(10)$ (referred to Cs^{137}):		±20% (further details from manufacturer)
Dose rate linearity - $H_0(10)$ γ:	Range <0.5 Sv/h (<50 rem/h):	±10%
	Range 0.5 Sv/h to 1 Sv/h (50 to 100 rem/h):	±20%
	Range 1 Sv/h to 2 Sv/h (100 rem to 200 rem/h):	±30%
	Range 2 Sv/h to 4 Sv/h (200 to 400 rem/h):	±50%
	Range 4 Sv/h to 50 Sv/:	Continues to accumulate dose at a rate >4 Sv/h
Test regime	On issue and at configurable time intervals	
Environmental Range	In use	−10 to +40°C
	Storage	−25 to 70°C
	RH:	20-90% (non-condensing)
Power supply:	Single 1.5 V alkaline AA battery:	Typically 50 days continuous operation[2].
	3.5 V lithium battery:	Typically 5 months continuous operation[2].
Weight		110 g
Dimensions:		85 × 63 × 19 mm
Case material:		High-impact polycarbonate/ABS blend[3]

Notes:
1. The 'neutron response factor' can be adjusted for calibration in the field (by reference to a standard dosimeter or by analysis of neutron flux spectrum).
2. Selecting the standby state extends battery life considerably.
3. Specifications quoted apply under standard conditions of 20°C

UNITED STATES OF AMERICA

ADM-300 survey meter

Description
The auto-range ADM-300 survey meter is a microprocessor-based, battery-operated, digital dose ratemeter and contamination monitor for radiological surveys and monitoring of personnel. An external γ probe is also available.

Specialised microprocessor technology provides auto-ranging and linear operation over the entire operating range. Statistically reliable dose rate measurements, even at very low intensities, are achieved without compromising any rapid step function response to quickly rising radiation fields. A simultaneous analogue/digital display is provided with automatic change of scale and calibration for the detachable external α , β , γ and neutron probe.

The ADM-300 was designed to measure γ dose rates from 1 μrad/h to 10,000 rad/h as well as γ/β radiation contamination levels. The unit can be provided with an optional lithium battery, RAM for storage of historical peak dose rates and accumulated dose over specific time intervals. The survey mode allows collection of survey data at up to 100 locations.

Other optional equipment includes a pancake Geiger-Müller tube β probe, γ probe for remote area monitoring, large area α probe, neutron probe, head set and carrying case and an external power converter.

Status
In production. In service with the US Air Force.

Marketing agency
Tradeways Limited.

VERIFIED

ADM-300 in use   0052803

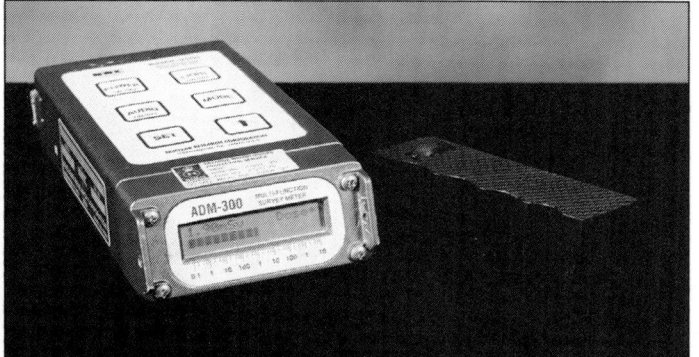

ADM-300 survey meter   0052804

AN/ADR-6(V) Aerial Radiac System (ARS)

Description
Developed by the US Army Electronics Research and Development Command (ERADCOM), the AN/ADR-6(V) Aerial Radiac System (ARS) is designed to be carried in any US Army helicopter and allows the rapid survey of large areas of terrain contaminated by radioactive material. The ARS uses sensors external to the helicopter to detect the radiation and then combines the detected radiation signal with an AN/APN-209(V) radar altimeter signal to compute the dose rate on the ground. The result is displayed on the instrument's dose rate meter and recorded on a 'hard' paper copy. The reading may also be transmitted to a ground station or another aircraft. The sensors used are of the compound scintillator type and can be used at altitudes of from 1 to 305 m.

Main components of the AN/ADR-6(V) Aerial Radiac System (ARS)

The unit may also be used in fixed-wing aircraft and Remotely Piloted Vehicles (RPVs). The power required is 28 V DC with a current requirement of 5 A at normal operating temperatures and 7 A below 0°C. The complete installation weighs less than 33 kg. The maximum aircraft dose rate measurement is from 0.04 to 400 rad/h. This will indicate a ground dose rate measurement up to 1,000 rad/h.

Status
All units delivered to US Army.

Agency
Tradeways Ltd.

UPDATED

AN/PDR-77 radiac set

Description
The AN/PDR-77 radiac set is identical in construction and circuitry to the AN/VDR-2 radiac instrument (see separate entry) with the added capability to measure α and X-rays as well as β and γ rays. It consists of a radiac meter to which can be connected either of three probes (α , β/γ or X-ray) for measuring a particular type of radiation. The three probes are part of the overall set.

Also included in the set are a carrying pouch with a shoulder strap, a headset and volume control assembly and some maintenance items. There is built-in test operating on a continuous basis and a built-in automatic calibration capability. The display is a backlit three-digit LCD with floating decimal point and unit indicator. Increasing or decreasing radiation levels are indicated by two front panel trend indicator (up-down) displays. It is possible to select a visual or audible alarm level over the entire dynamic range. An RS-232 data port is available with a plug-in module.

Range dose rates are as follows: 0 to 999,000 cpm; 0 to 999,000 mr/h; 0 to 999K dpm/100 cm² and 0 to 999K μc/m².

Component parts of the AN/PDR-77 radiac set   0011411

Power is provided by three 9 V batteries providing 100 hours of continuous operation with the β /γ probe or 50 hours with the α or X-ray probes.

The complete equipment weighs approximately 5.4 kg.

Status

In production. In service with US Armed Forces.

Marketing agency

Tradeways Limited.

VERIFIED

AN/UDR-13 radiac set

Description

The AN/UDR-13 radiac set is a compact, hand-held or pocket-carried device capable of measuring a γ /neutron dose from a nuclear event and dose rate. It therefore combines the functions of a dosimeter and a dose rate meter. The dose rate meter portion of the instrument measures dose rates from 0 to 999 cGy/h and the dosimeter records 0 to 999 cGy of neutron and/or γ dose.

The AN/UDR-13 radiac set has a LCD display with a push button pad for mode selection, functional control and the setting of audible and visual alarm thresholds for both dose rate and mission dose. The display can be illuminated for use at night. A system sleep mode with an automatic wake-up function is provided to enhance battery life; the instrument is powered by four 1.5 V AAA dry cell batteries.

The first AN/UDR-13 prototype appeared during Fiscal Year 1994. Delivery commenced to US armed forces in 1998 at the rate of one to each platoon or equivalent sized unit.

The AN/UDR-13 radiac set weighs 227 g and measures approximately 107 × 66.3 × 25 mm. The unit is ruggedised and is intended for use over a wide range of temperatures (−46 to +49°C), climatic conditions (95 per cent humidity) and altitudes (up to 4,500 m).

The AN/UDR-13 now includes an RS-232-type communications facility, enabling the equipment to be controlled as well as 'read' by a computer or similar device. The communications channel operates via an optical, infra-red communications port. This capability has two purposes.

Firstly, to reduce life-cycle costs. The semi-automatic, computer-controlled self-calibration facility obviates the need for traditional calibration routines thereby reducing life-cycle costs dramatically.

Secondly, optically-coupled probes (α , β , γ , X-ray and so on) can be added to the radiac set by making use of the communications feature. This effectively provides long-term 'future-proofing' of this equipment.

Status

Available. Current production run of 16,000 for the US Army.

Marketing agency

Tradeways Limited.

VERIFIED

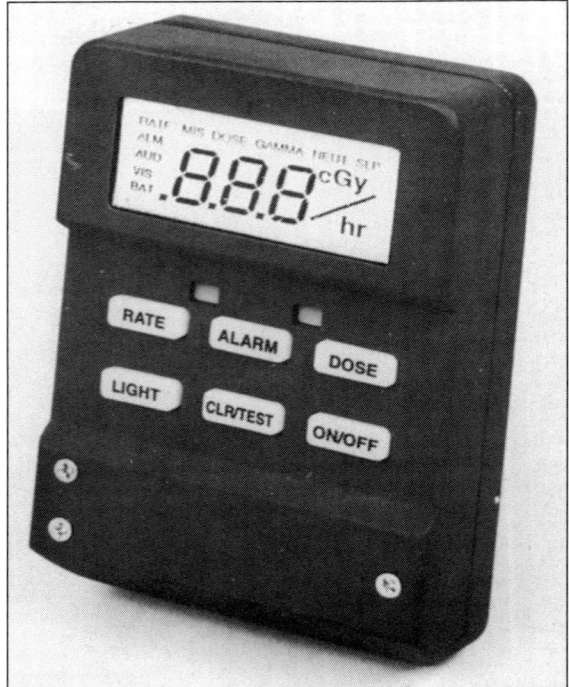

*AN/UDR-13
radiac set*

AN/VDR-2 radiac instrument

Description

The AN/VDR-2 radiac instrument is a lightweight auto-ranging survey meter and dosimeter for radio logical surveys and personnel monitoring. It is microprocessor-based and is designed for measuring γ dose rates from 0.01 mrad/h to 10,000 rad/h as well as β dose rates from 0.01 mrad/h to 4 rad/h.

The instrument functions simultaneously as a dose rate meter and dose meter with independent adjustable alarms that can be set to any level over the entire range of each function. Also provided is measurement of the total integrated dose up to 1,000 rad. Dose data is independently stored in a non-destructive memory for display on command and may be retained when the unit is switched off. Another feature is a self-test programme which is activated automatically when the unit is switched on. The test sequence includes a calibration check on all ranges, processor and memory test, battery condition test and a detector check.

The AN/VDR-2 comprises a radiac meter and a probe. The front panel of the radiac meter, which is contained within a sealed and gasketed aluminium case, contains a liquid crystal display and all operating controls (power on/off, clear/test, rate, dose, light, attenuation, alarm audible/visual). The probe, which is connector-coupled to the radiac meter via a 0.914 m coil cord, contains two Geiger-Müller tube detectors. A low-range tube is housed in the interior of the probe and is active over the γ range 0.01 mrad/h to 5 rad/h. When the probe window is open the tube mica window is exposed to enable the detection of β radiation over an equivalent range. The high-range Geiger-Müller tube is located in a plug-in housing which rides piggy-back on the probe and covers the range from 5 to 9,999 rad/h. The high-range tube does not have a β detection capability.

When the AN/VDR-2 is used in the area survey mode by dismounted personnel, the radiacmeter can be carried in a haversack on a belt or shoulder strap and with preset alarms preselected for total dose and dose rate levels. The instrument can be used for personnel, food, water and equipment monitoring. Experimental work is being conducted to include the use of the AN/VDR-2 as a communicable monitor used on a Remotely Piloted Vehicle (RPV). Special carrying racks are available to allow the AN/VDR-2 to be fitted to a number of vehicles including the M577 carrier, command post; M113 series armoured personnel carriers; M151A2 Jeep; M60A1, M60A3 and M1 Abrams tanks; M2 Bradley IFV; and M880 series (4 × 4) trucks.

The AN/VDR-2 is capable of being used as a fixed station monitor with the detection element either remotely located (up to 1,524 m away) or used in the vicinity of the readout. Provision has been made for operation from normal line power and for the mounting of up to 16 equipments to enable widespread coverage of an area or multiple locations. When used in this mode, an RS-232 interface unit can be used with the instrument to permit expanded capabilities and operation using computer control, display and recording of radiation information.

Specifications		
Weight:		<1.8 kg
Length:	(instrument)	162 mm
	(probe)	159 mm
Width:	(instrument)	102 mm
	(probe)	38 mm
Height:	(instrument)	44 mm
	(probe)	38 mm
Radiation measured:	(γ)	0.01 mrad/h-10,000 rad/h
	(β)	0.01 mrad/h-4 rad/h
Modes of operation:	(ratemeter)	0.01 mrad/h-10,000 rad/h
	(dosimeter)	to 1,000 rad
Response time:	(above 1 rad/h)	2 s
	(below 1 rad/h)	5 s
Power supply:		3 × 9 V batteries
Battery life:		in excess of 100 h
Operating temperature range:		−20 to +50°C
Storage temperature range:		−40 to +60°C

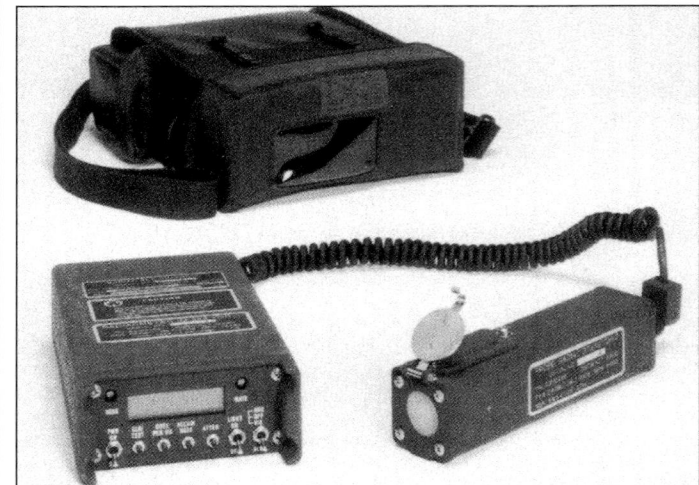

AN/VDR-2 radiac meter and probe unit (foreground) with carrying haversack

The RS-232 unit output may be transmitted over long distances using conventional telephone modem units.

Measurement of underwater radiation to depths of 30.5 m may be conducted using the AN/VDR-2 and a special underwater probe. The probe is constructed of stainless steel and includes a stainless steel support cable and an easily decontaminated non-porous signal cable. The AN/VDR-2 may also be used for aerial surveys with continuing altitude correction data from aircraft altimeters being applied to the AN/VDR-2 and radiation measurements corrected to the 0.914 m level (that is, ground level intensities) will be provided by the instrument readout.

Other available accessories include a telescopic extender probe up to 3.05 m long. Probe extensions 30.5 and 61 m long are available as standard lengths and additional lengths up to 1,524 m are available on request.

Status
Available. In service with US, European and Middle East armed forces.

Marketing agency
Tradeways Limited.

VERIFIED

....

DART® E-METER Automated U-235 Enrichment Monitor

Description
The DART® E-METER is a complete monitoring system, comprising the following components: the E-METER detector, the DART® MCA multichannel analyser, a sub notebook computer and the EM-B32 E-METER application software. Any of these components can be ordered separately. The detector (see specifications below) uses the U^{235} 185 keV g ray to determine the degree of enrichment of the uranium sample. To simplify the calculations, the algorithm assumes infinitely thick samples of relatively low atomic number and a uniform and homogenous distribution of uranium in the sample. The 'infinitely thick' definition implies the attenuation of the 185 keV g ray. For UO_2 this is 3.2 mm. Normal fuel pellets are thick enough to be considered infinite. The system is calibrated by measuring standards of known enrichment. The assumption is not valid if the matrix contains high atomic number material in addition to uranium, and will cause the meter to over-read. Inhomogenous uranium mixtures can also cause errors because the detector only counts the part of the sample closest to the detector. If the uranium is not uniformly distributed in the sample it can similarly cause errors. For this reason, the meter is not recommended for testing materials of low uranium content such as waste.

The DART® multichannel analyser claims to be the leading nuclear MCA of performance, weight, and battery life. See below for specifications.

The DART® E-METER application software has been designed for ease of use. Each E-METER detector has unique characteristics so that the software needs to be told the identity of the detector to which it connected. Detection and visual presentations on the screen are automatic and the software offers the user an

The DART® E-METER System   *2001*/0100722

The DART® E-METER can collect field data (stored in memory) for later analysis by the MAESTRO-32 software   *2001*/0100723

Specifications
E-METER - system components
- 51 BS 12.7/2-E3-T-Am-X
- 51 mm diameter ×12.7 mm thick crystal, coupled to 51 mm diameter PMT, EMI 9266
- Integral voltage divider/preamp to IAEA specification
- 7-pin connector LEMO ERAS 3S 707
- Light guide mounted RODAN 05DB102 thermistor for temperature measurement.
- Am^{241} activated source. 1,000 cps ±20% g equivalent energy 2.5-2.8 MeV
- Stainless steel case
- FWHM 662 keV: ≤7.5%
- 122 keV: ≤15%

DART® multichannel analyzer
- Lightweight: <2.38 kg
- Long life: >7 h with a typical (Safeguards type) HPGe detector. Longer with E-METER NaI detector
- High resolution (up to 8k channel resolution)
- High intrinsic stability, plus digital stabiliser
- Automatic pole-zero adjustment
- Full computer control
- High-speed data transfer to computer via parallel interface
- Front-panel ratemeter
- 100 kHz multichannel scaling mode
- Supports HPGe, CZT and NaI(Tl) detector types
- Reads temperature sensor in E-METER detector

intuitive interface and comprehensive help support. Results presented on screen are saved to a database which is compatible with MS Access*. The software can run on any PC which supports the Windows* 9× operating system and occupies at least 50 MB of disk space.
(*Registered trademarks of the Microsoft Corporation)

Status
Available.

Manufacturer
PerkinElmer (formerly EC&G Ortec).

VERIFIED

DigiDART™ HPGe Spectrometer

Description
The DigiDART™ is a lightweight hand-held monitoring device and its technology is derived from the successful DART HPGe-grade Micro-Channel Analyser. Ruggedly designed for field measurement, it has a built-in display and numeric keypad allowing a much greater degree of processing and analysis within the unit itself and without the need to attach it to a computer. The live spectral display shows region of interest, peak information and online activity calculations. Library calibration information can be downloaded from PerkinElmer's proprietary Maestro software via a PC to the DigiDART™ and, although this can also be changed in the field to adapt to changed needs, it allows activity to be calculated for a list of up to nine nuclides. The spectral data can be saved and later re-analysed in more detail using a more sophisticated PC-based analysis package such as GammaVision.

Protection of data, chain-of-custody integrity and intact audit trail are all vital aspects in the recording of nuclear activity. PerkinElmer 's new SMART-1 detector technology allows significant improvements in detector management practice and data control.

Status
Available.

Manufacturer
PerkinElmer (formerly EC&G ORTEC).

VERIFIED

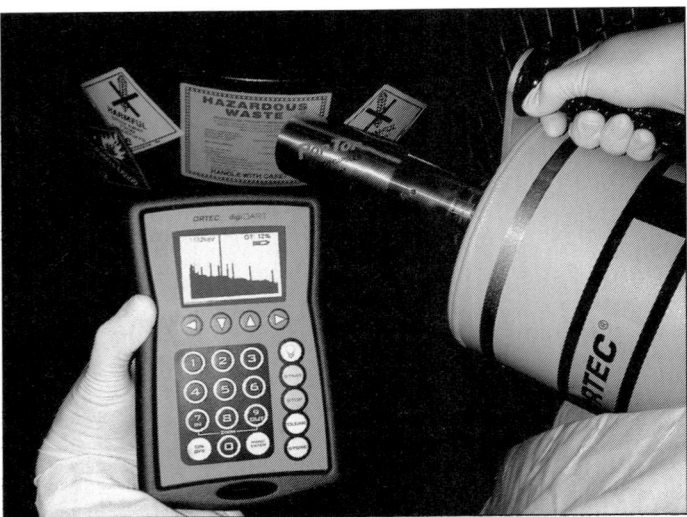

The DigiDART™ portable HPGe spectrometer with its probe (right) monitoring a waste canister for nuclides   **2001**/0059316

Inovision Model 451B survey meter

Description
The Model 451B ion chamber survey meter is a hand-held battery operated unit designed for use in both rugged and normal environments. The Model 451B measures α (above 4 MeV) β (above 100 keV) and γ (above 7 keV) and X-radiation, measuring rate and dose simultaneously. It uses microprocessor and LCD technology and features a rugged ionisation chamber with a Mylar window and protective steel mesh. An integral β shield serves as an equilibrium thickness for photon measurements. The display features an analogue bar graph, two half-digit digital readout, low battery and freeze mode indicators. User controls comprise an on/off button and a mode button. The unit is auto-zeroing and auto-ranging. The display features circuitry that automatically activates the backlight in low ambient light conditions. A tripod mount for stationary, area monitor applications is available. The power comes from two 9 V batteries which can supply the meter for up to 200 hours.

An RS-232 communications interface, allows the 451B to display data on a computer via an add-in for Microsoft Excel®, enhancing the functionality of the instrument. The software allows for real-time data logging, data retrieval

Specifications

Operating ranges	Response time (seconds)
0-50 µSv/h	8
0-500 µSv/h	2.5
0-5 mSv/h	2
0-50 mSv/h	2
0-500 mSv/h	2

Dimensions: 10 × 20 × 15 cm
Weight: 1.11 kg
Temperature range: −40 to +70°C

(up to 2,700 data sets stored in unit can be downloaded), user parameter selection, and provides a virtual instrument display with audible (requires sound card) and visual alarm indication. The software may be customised by the user for specific applications.

A pressurised ion chamber model is also available (model 451P) , working in the µR/h range.

Status
In service in a wide range of medical, health and physics applications in the US. Regularly used by emergency response and HAZMAT teams, nuclear medicine laboratories, hospital radiation safety officers and nuclear power industries.

Manufacturer
Inovision Radiation Measurements.

NEW ENTRY

Naval Ship Radiation Monitoring System (NASRAMS)

Description
The Naval Ship Radiation Monitoring System (NASRAMS) is a continuous real-time monitoring system for nuclear hazard identification, monitoring for both external airborne and seawater contaminants and for internal munition stores, particulate and water contaminants; monitoring radiations levels (dose rate and accumulated dose) of the environment at selected areas throughout the ship and providing local and/or remote visual and audible alarms. It also takes the necessary remedial action in the case of an event by controlling ventilation air flows and filters, establishing areas to be evacuated and indicating necessary course changes. NASRAMS therefore provides an overall awareness of environmental contamination levels and permits the mapping of affected areas.

NASRAMS employs the AN/VDR-2 microprocessor based radiac set, utilising the time-to-count principle of radiation measurement. AN/VDR-2s are networked to provide up to 16 channels of radiological information to an onboard PC-compatible computer.

The AN/VDR-2 β/γ probe is available in three types of housing:
DT-616P: hand-held or shockmounted
DT-616W: for underwater installation
DT-616B: bulkhead mounting.
Each probe is operated through its local AN/VDR-2 radiac meter equipped with a RS-232/422 communication port powered by the ship's power supply.

A typical NASRAMS installation provides local digital readouts at each detector location and also provides local, visual and audible alarms. In addition the complete system status is presented on a CRT display at the ship's Damage Control Centre and/or any other selected location or locations.

Computer software programs offer various optional data and colour graphics. User friendly menu-directed programs require minimum operator training to analyse trends and current conditions at each location. System self-diagnostics and troubleshooting can also be exercised and observed on the Control Centre CRT screen.

Customised software can be provided to incorporate an ionisation chamber chemical point detector into NASRAMS. Programs can be expanded to accommodate up to 64 channels of information from radiological, chemical, meteorological or other instrument inputs.

Status
In production.

Marketing agency
Tradeways Limited.

VERIFIED

OS1700 Tritium collector

Description
The Discriminating Tritium Collector, Model OS1700, from EG&G ORTEC uses a two-stage collection process - one for oxides of tritium and one for elemental tritium and tritiated carbon compounds. The OS1700 is a stand-alone room-air and environmental sampler; with options, it can be a stack sampler.

Because of different release limits for the two forms of tritium due to different metabolisation in biological organisms, a two-stage collector is required. Most bubblers or collectors have a single stage that collects only tritium oxides.

The OS1700, with its built-in palladium-catalysed oxidisation oven, oxidises gaseous tritium and also cracks and oxidises tritiated carbon compounds, thus enabling collection of those chemical forms in the second stage. Samples are counted in the user's liquid scintillation counter to measure activity at tritium-in-air concentration levels as low as -0.37 Bq/m (10 mCi/ml). The OS1700's built-in microprocessor and mass flow meter record the total flow, thereby allowing the user to record the averaged activity concentration of the sample air. A 60 ml, as well as a 20 ml, sample vial is available to accommodate longer sample times or

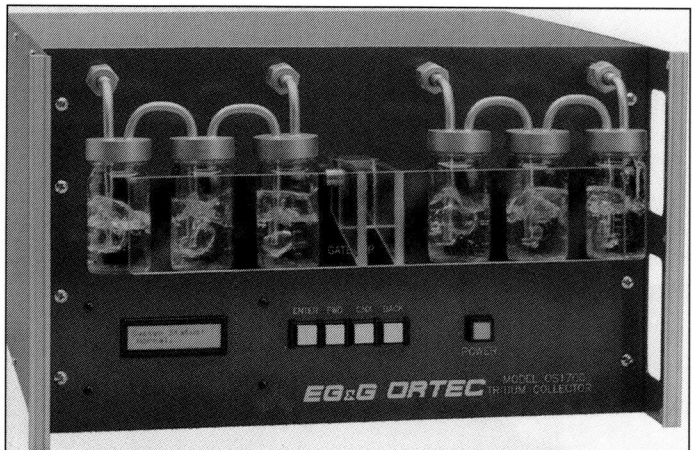

OS1700 Tritium collector   0011444

high-humidity areas. The OS1700 is provided in a bench-top enclosure, which may be mounted in a 19 in rack. The OS1700 replaces ORTEC's previous tritium product, the EL700.

Status
In production.

Manufacturer
PerkinElmer (formerly EG&G ORTEC).

UPDATED

SE Digilert 50 nuclear radiation monitor

Description
The SE Digilert 50 nuclear radiation monitor is a personal instrument intended for use by civil and military personnel. It replaces the earlier Digilert model. It can be used to alert the user to high radiation levels, monitor changes in radiation levels and detect radiation leaks. It is based on a halogen-quenched Geiger-Müller detector with a mica end window (LND712) and can be used to detect and monitor α, β, γ and X-rays. It has an operating range of 0.001-50,000 mrad/h and 0 to 60,000 accumulated counts or 0 to 50,000 counts/min (cpm).

The Digilert 50 provides an audible alert once a user-set level of counts or cpm has been reached. In addition, a red LED flashes and a beeper sounds with each count (the beeper can be muted). Readings are provided on a clear LCD. The instrument has a dual miniature jack output which can be used either to provide counts to a computer, a data-logging device or provide an alert signal to an external warning device. A sub-mini jack provides audio output to an external earpiece, amplifier, or tape recorder.

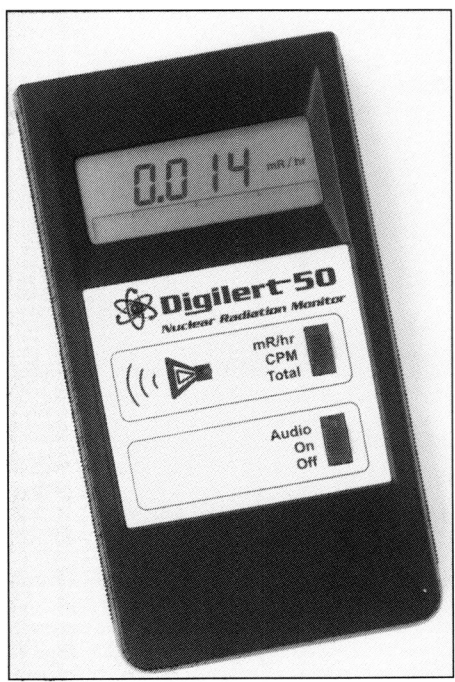

SE Digilert 50 nuclear radiation monitor (SE International Inc)
2000/0084056

The Digilert is powered by a 9 V alkaline battery which provides a typical 2,000 hours of use at a normal background. The instrument measures 150 × 80 × 30 mm and weighs 270 g including the battery.

Status
In production.

Manufacturer
SE International Inc.

VERIFIED

SE Gateman multifunction survey meter

Description
The SE Gateman is a multifunction analyser/survey meter designed for ease of use, with the minimum of training. It is a hand-held computer-based instrument used with an external scintillation probe having two outputs including an RS-232 port for connection to a computer or external warning device. Lighted indicators show the count rate and exceeded alarm levels. The sensitivity is optimised to the selected radionuclide. It offers a high background alarm with automatic background count, automatic or manual alarm adjust function and is of rugged and weather proof construction. The Gateman is powered by a rechargeable battery.

Status
In production.

Manufacturer
SE International Inc.

VERIFIED

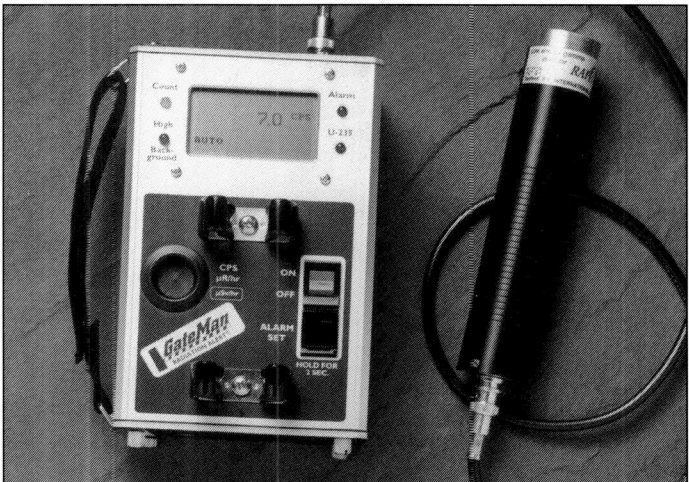

Gateman multifunction analyser/survey meter (SE International Inc)   0011441

SE Inspector EXP survey meter

Description
The SE Inspector EXP survey meter is optimised for use by emergency services personnel. It is similar in technical specification to the SE Inspector (see separate entry in this section), offering small size and ease of use. The Inspector's built-in detector is replaced with an external probe (see illustration). The instrument, stowed in its rugged vinyl carrying case, can be belt-mounted, allowing one-handed

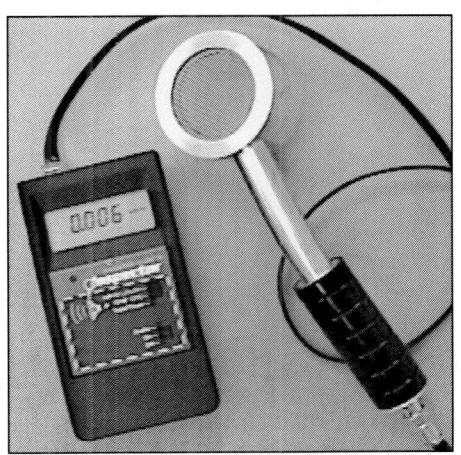

SE Inspector EXP survey meter with external probe (SE International Inc)
2000/0084059

operation of the probe by the operator. The probe has a protective cover which can be stowed clear one-handed as well.

Status
In production.

Manufacturer
SE International Inc.

VERIFIED

SE Inspector survey meter

Description
The SE Inspector survey meter is a microprocessor controlled measuring instrument with a sensitivity to low levels of α, β, g and X-rays. The four-digit readout is displayed with a red count light and an audible bleep to provide constant indications of the radiation level. Other features include an adjustable timer and external calibration controls.

The detector is a halogen-quenched uncompensated Geiger-Müller tube with thin mica windows and an effective diameter of 45 mm. Displays update every 3 seconds. At low background levels the update is the moving average for the past 30 second time period. The time period for the moving average decreases as the radiation level increases.

The operating range in counts per minute is from 0-5,000 or 0-300,000. The range can be from 0.001 to 100 mrad/h or an optional 0.01 to 1,000 μSv/h. Readout will hold at full scale in fields as high as 100 times the maximum reading.

The SE Inspector is powered by one 9 V alkaline battery, weighs 301.4 g and measures 150 × 80 × 30 mm.

Status
In production.

Manufacturer
SE International Inc.

VERIFIED

SE Inspector survey meter (SE International Inc)

SE Radiation Alert CHARGER

Description
The patented hand-held SE Radiation Alert CHARGER can zero a variety of quartz or carbon fibre pencil dosimeters including the SE PEN200 (see following entry). Powered by a piezoelectric generator, its operation does not require batteries.

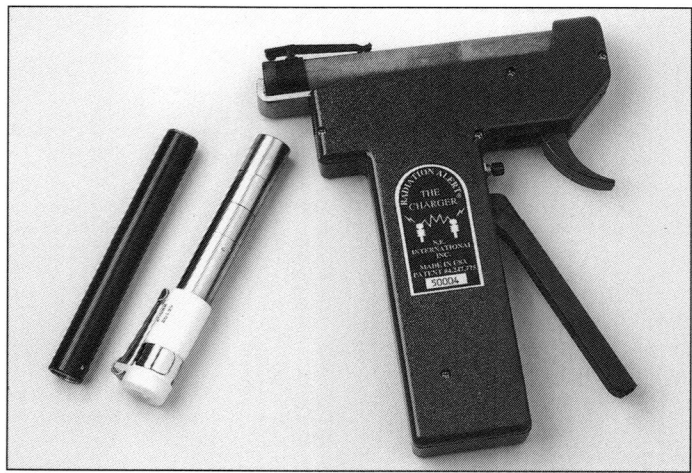

SE Radiation Alert CHARGER (SE International Inc)

Pulling a trigger allows a dosimeter to be easily placed in or removed from the CHARGER unit while a clamping action automatically holds the dosimeter.

Weight is 245 g. Operating temperature range is −20 to +50°C.

Status
In production.

Manufacturer
SE International Inc.

VERIFIED

SE Radiation Alert MC1K survey meter

Description
The SE Radiation Alert MC1K is a compact, hand-held radiation survey meter capable of detecting g and X-rays up to 1,000 mrad/h. It has an internal energy-compensated, halogen-quenched Geiger-Müller tube with a dual-scale analogue meter that reads four ranges, from 0.01 to 1,000 mrad/h and 0.001 to 10 mSv/h. Radiation activity is also displayed by a red count light and an internal beeper.

One 9 V alkaline battery provides 2,000 hours of operation at normal background levels. A carrying case is provided with each unit.

The MC1K measures 145 × 72 × 38 mm and weighs 188 g without the battery. Operating temperature range is from −25 to +40°C.

Status
In production.

Manufacturer
SE International Inc.

VERIFIED

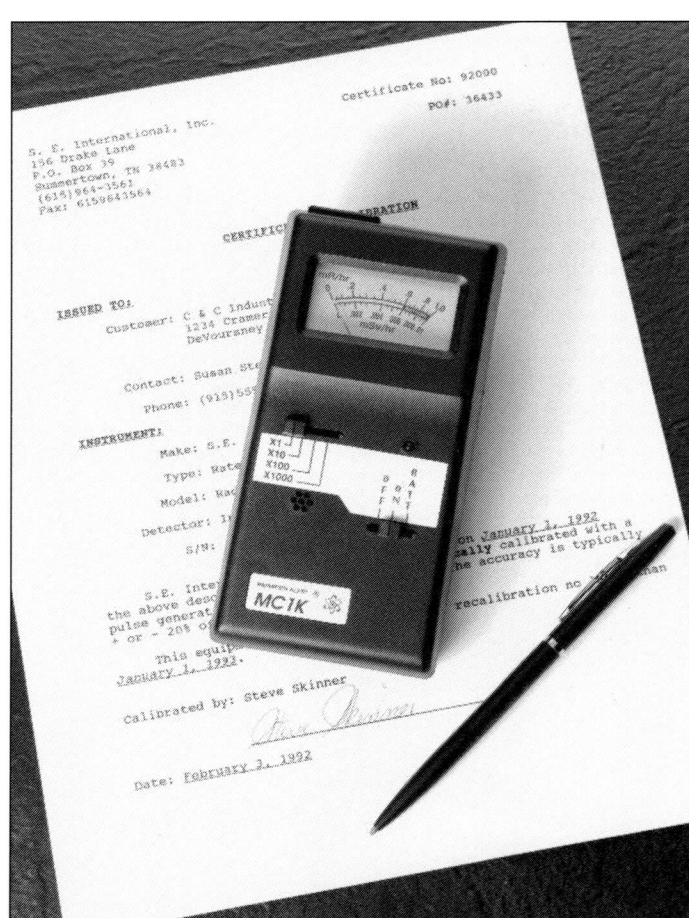

SE Radiation Alert MC1K survey meter (SE International Inc)

SE Radiation Alert Monitor 4 survey meter

Description
The SE Radiation Alert Monitor 4 survey meter is a general purpose Geiger-Müller radiation α, β, γ and X-ray radiation.

When using the Radiation Alert Monitor 4 the presence of radiation is displayed by a flashing count light, an audible bleeper that can be switched off for silent operation and an easy-to-read dual-scale analogue meter. The meter has three

SE Radiation Alert Monitor 4 survey meter (left) with Monitor 4EC (right) (SE International Inc)

selectable scales (×1, ×10 and ×100). An internal Geiger-Müller tube eliminates external probes and wires for ease of handling. A single 9 V alkaline battery can provide up to 2,000 hours of operation at normal background radiation levels.

An anti-saturation circuit forces the meter to read full-scale in radiation fields as high as 100 times the maximum reading in the highest range. The dual-scale meter reads from 0 to 500 µSv/h and 0 to 50 mR/h.

The Monitor 4EC is an energy-compensated version of the Monitor 4 that provides a more accurate mrad/h reading.

An instruction manual is provided with each instrument. Also provided is a padded vinyl carrying case with a wrist strap and belt clip. The instrument measures 145 × 72 × 38 mm and weighs 198 g without the battery. Operating temperature range is from −20 to +55°C.

Status
In production.

Manufacturer
SE International Inc.

VERIFIED

SE Radiation Alert Monitor 5 survey meter

Description
The SE Radiation Alert Monitor 5 is a Geiger-Müller survey meter with a built-in detector capable of α, β, γ and X-rays. It can be used for the checking of clothing, equipment and working areas and for locating small sources of radiation. It can also be used as a personal monitoring and warning instrument.

The presence of radiation is provided by either a flashing red light, a beeper that can be switched off for silent operation, or an analogue meter reading. The meter

SE Radiation Alert Monitor 5 survey meter (SE International Inc)

reads from 0 to 50,000 cpm over three selectable ranges. Each range is independently calibrated. An optional tri-scale reads 0 to 1,250 cpm, 0 to 50 mrad/h and 0 to 500 µSv/h.

The Monitor 5 is powered by one 9 V alkaline battery with a life up to 2,000 hours at normal background levels. Each instrument is provided with a padded vinyl bag, a wrist strap and a belt clip. The Monitor 5 measures 145 × 72 × 38 mm and weighs 240.5 g without the battery. Operating temperature range is from −20 to +55°C.

Status
In production.

Manufacturer
SE International Inc.

VERIFIED

SE Radiation Alert PEN direct reading dosimeters

Description
The SE PEN direct reading dosimeters are a range of carbon fibre direct reading instruments for γ and X-rays offering ranges of 0-200 mrad, 0-2 rad, 0-5 rad and 0-2 mSv. It is claimed that the advanced design imparts improved shock resistance and reliability over other dosimeters of its type. The dosimeters are hermetically sealed, immersion proof. A high-strength metal clip attaches the instruments to clothing or objects.

The PEN200 meets US military specification for RADIAC METER IM-264/pd and ANSI N322.

Both dosimeters have a diameter of 15 mm and weigh 19 g.

Status
In production.

Manufacturer
SE International Inc.

VERIFIED

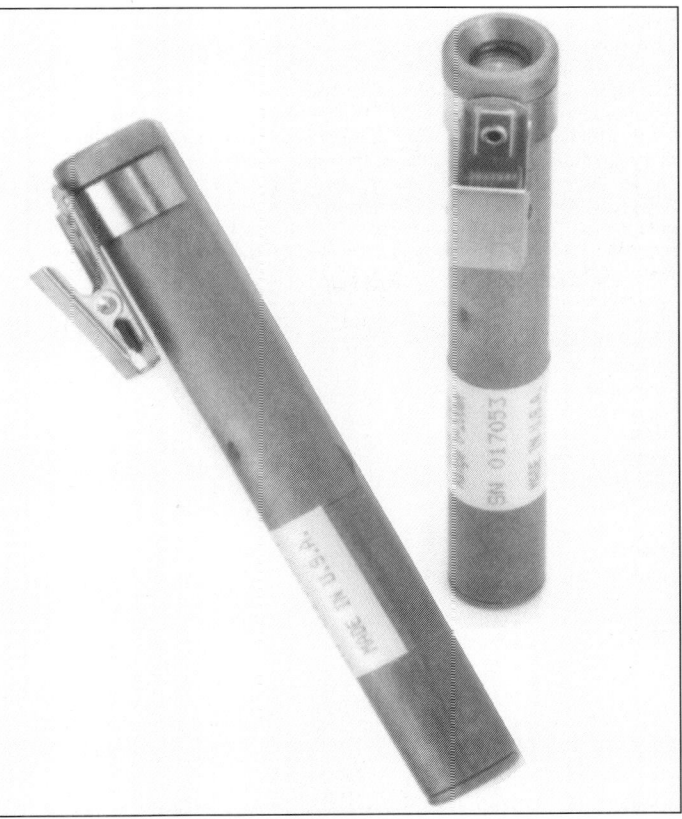

SE Radiation Alert PEN200 direct reading dosimeters (SE International Inc)
***2000**/0084058*

Smartbadge intelligent personal radiation monitor

Description
The Smartbadge intelligent personal radiation monitor was developed to allow personnel working in a radiation environment to automatically make a number of safety and other calculations relating to radiation exposure and provide a warning when certain predetermined limits have been exceeded.

Smartbadge intelligent personal radiation monitor

Smartbadge is about the size of a hand-held electronic calculator and is carried in a belt pouch. The black plastic case is sealed, shock- and waterproof and is EMP-resistant. The keyboard has large keys which incorporate a tactile feedback for use when wearing NBC gloves. A 32-character liquid crystal display prompts the user to enter basic information on dose restraints. User-friendly software then provides data in response to all likely radiation calculations. For instance, Smartbadge can determine the waiting time necessary before entering a currently too-radioactive area; calculate remaining allowable work time in a radioactive area; predict future exposures, including automatic calculation of radioactive decay; and calculate the residual dose. The instrument continually measures and displays the radiation rate, dose, time remaining until a user-designated dose limit is reached and provides an audible and visual warning of dangerous conditions.

Smartbadge uses a microprocessor-controlled energy-compensated Geiger-Müller tube to measure γ radiation. The response time is 4 seconds from 0.1 to 1,000 rad/h and the display updates every second. In an audible 'chirper' mode, chirps are produced in proportion to the sensed radiation rate.

Power is provided by a small power pack containing four disposable or rechargeable AA batteries. The display warns when the batteries need changing and there is a built-in self-test routine every time the power pack is installed. Dose information, alarm level settings, user identity and other data is stored in a continuous memory.

Specifications
Weight: (with batteries) 553 g
Dimensions: 162 × 102 × 43 mm
Dose rate range: 0.01-1,000 rad/h
Dose range: 1-1,000 rad
Response time: 4 s
Operating temperature range: 0 to +50°C

Status
In production.

Manufacturer
System Planning Corporation.

VERIFIED

Inovision Model 190 survey and count-rate meter

Description
The Inovision (formerly 'Victoreen') Model 190 survey and count-rate meter is compatible with GM detectors, neutron probes, proportional counters, and scintillation probes operating between 300 and 1,300 V. Depending on probe selection, the Model 190 detects α, β, γ, neutron or X-ray radiation within an operating range of 1 μR/h or 1 to 1,000,000 counts per minute. The unit is available with either MHV or BNC connector to provide the user with versatility in probe selection. Visual indication of selected parameters, as well as measured values, are displayed on the analogue/digital display. The unit weighs (without the probe) 709 g and measures 234 × 92 × 50 mm.

The 190 series comprises the 190I (time integrator included), 190N (Neutron), 190EX (included extended 'pancake' probe).

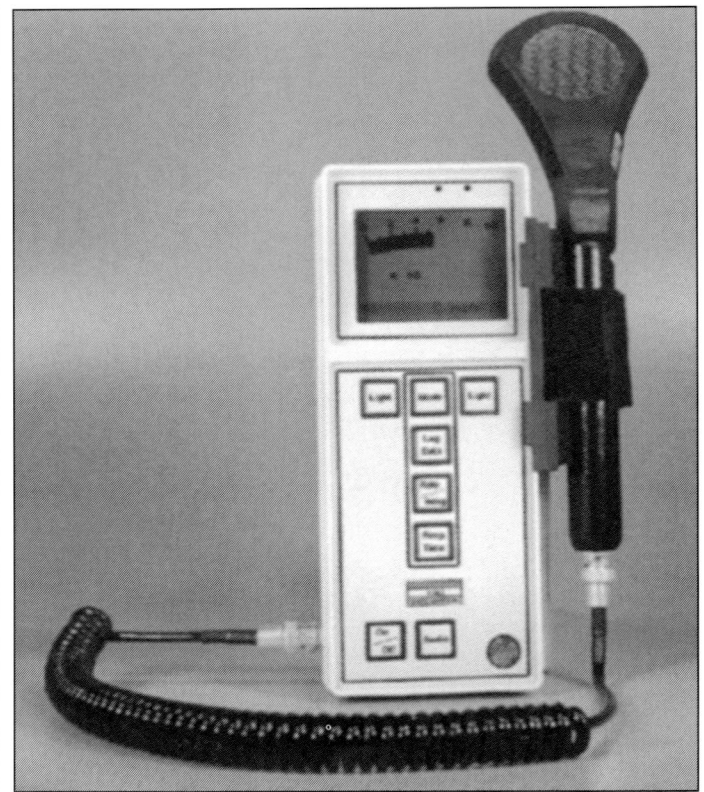

Inovision Model 190 (Inovision Radiation Measurements)   **2001**/0109584

Status
In production.

Manufacturer
Inovision Radiation Measurements.

UPDATED

YUGOSLAVIA, FEDERAL REPUBLIC

ASZO collective protection warning system for installations

Description
The ASZO collective protection warning system is used to monitor the atmosphere, preventing the penetration of radioactive or chemical agents into a static installation and also to provide warnings for an installation following a nuclear blast. In addition, audible and visual warnings are provided when such events take place, prompting the occupants to take the necessary closing down procedures.

The ASZO system can detect a nuclear explosion and/or flash at a range of 18 km. It can also monitor γ radiation dose rate over a range from 0.01 to 999 cGy/h. Doses are measured over a range from 0.1 to 999.9 cGy. Accuracy is ñ20 per cent and the warning level is 0.05 cGy/h.

The ASZO system can detect Sarin (GB), Soman (GD), VX, Phosgene (CG) and Hydrogen Cyanide (AC). Response time is 60 seconds. The system can operate for 8 hours on one electrolyte charge. The system is ready for use within 15 minutes at temperatures above +15° C, or 40 minutes between 0 and +15° C. If the temperature is from –30 to 0°, then 90 minutes of preparation time will be required.

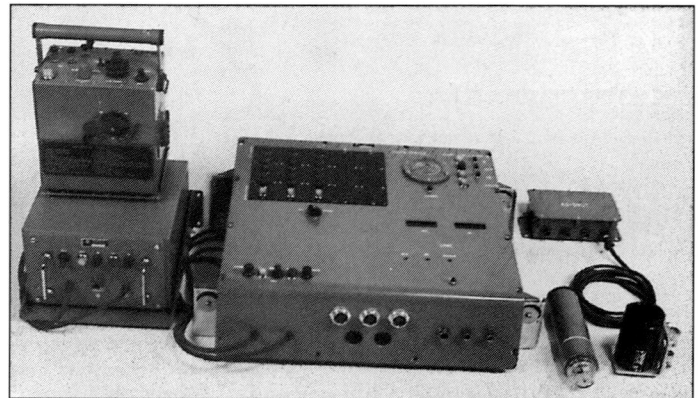

Main components of the ASZO collective protection warning system for installations

For both the nuclear and chemical agent detection modes the sensor may be placed up to 200 m from the rest of the equipment. Power is provided by a 220 V 50 Hz or 24 V DC supply.

A complete ASZO system consists of a central control unit, radiological measuring probe S-13N, DAH-M1 automatic chemical detector, blast wave detector, power supply unit, KS-PS and KS-VUT connecting boxes, spare parts, accessories and connecting cables.

Status
In service with the Yugoslav Army.

Contractor
Yugoimport SDPR.

VERIFIED

DRBIV M81 radiation detector for combat vehicles

Description
The DRBIV M81 radiation detector is intended for use in combat and reconnaissance vehicles. It can be used to measure the γ radiation dose and dose rate and can transmit data through the vehicle's radio network.

The DRBIV M81 consists of an external sensor unit and an internal display unit together with the associated cables. The sealed sensor unit is mounted in an appropriate position on the exterior of the user vehicle. The wall-mounted internal display unit uses two LED arrays to display the dose and dose rate. The dose range is from 0.1 to 999.9 cGy and the dose rate from 0.01 to 999 cGy/h. A socket on the display unit may be used to connect the unit to the vehicle radio to transmit data to a central monitoring point.

The DRBIV M81 operates off the vehicle DC supply which may be between 10.8 and 30 V DC. The complete equipment weighs less than 3 kg with the sensor unit measuring 58 × 200 mm; the display unit measures 199 × 194 × 51 mm. The operational temperature range is from -25 to +55°C.

Status
In service with the Yugoslav Army. Unable to trace current manufacturer or supplier of service and parts.

Contractor
Yugoimport SDPR.

VERIFIED

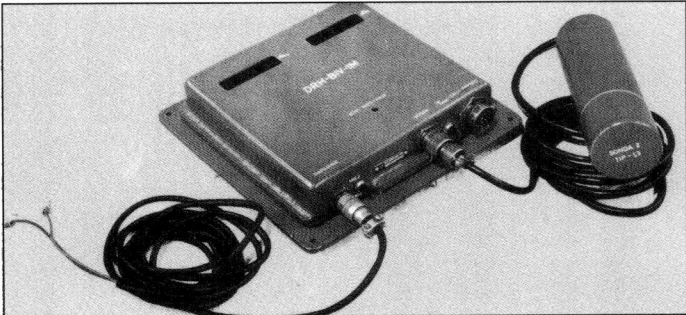

DRBIV M81 radiation detector for combat vehicles

DRHT radiation and chemical detection equipment

Description
The DRHT radiation and chemical detection equipment is intended for use in armoured fighting vehicles such as tanks. It is used for the detection of nuclear radiation, for the measuring of nuclear radiation dose rates from the surrounding terrain, warning the vehicle crew when a preselected dose rate is exceeded either from the local terrain or within the vehicle and detecting nerve agents and other chemical agents in the surrounding atmosphere and warning the vehicle crew. The DRHT may also be used to automatically transmit alarm signals to other locations via a radio link.

The complete equipment weighs less than 30 kg and has a power consumption of less than 7 A when operating off the vehicle's 24 V DC power supply. There are five main units. They are the radiation detector, an automatic chemical agent detector, a measuring and control unit, a compressor unit and a cyclotron. There is also a consumables kit and various tools and accessories including the interconnecting cables.

The DRHT operates over a temperature range of −15 to +55°C. The radiation dose rate measuring range is from 0.05 cGy/h and the radiation dose from 0.1 to 999.9 cGy. The vehicle internal dose rate sensitivity is such that an indication is given when primary γ radiation is detected in excess of 5 cGy/h. Agents such as nerve gases can be detected in concentrations of less than 6 × 10⁶ mg/dm³.

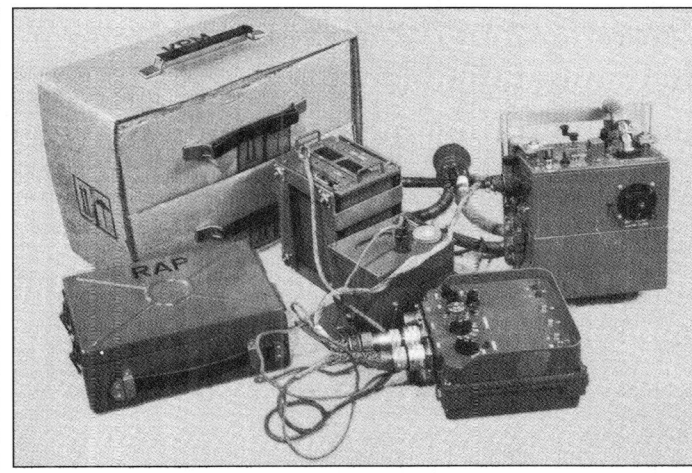

DRHT radiation and chemical detection equipment complete with consumables kit in cardboard case

Nuclear dose rate indications can be provided after a period of 8 seconds, while chemical agent indications are provided after 90 seconds.

Status
In service with the Yugoslav armed forces.

Contractor
Yugoimport SDPR.

VERIFIED

DRZON general purpose radiological detector

Description
The DRZON general purpose radiological detector is intended for radiological detection and contamination control with tactical field units, for civil defence and in some industrial applications. It is a compact, rugged, easy to operate unit involving microprocessor data processing.

The DRZON may be used as a dose rate meter and as a dosimeter. The dose rate measuring range is from 5 μGy/h to 999 cGy/h. The dose measuring range is from 1 μGy to 999 cGy. An audio alarm is triggered when the dose rate exceeds 0.5 cGy/h. Response time is less than 4 seconds while accuracy is less than ±20 per cent.

The DRZON is powered by a single R6 battery which enables the instrument to remain operational for more than 100 hours. Weight is 770 g and the instrument measures 185 × 75 × 50 mm. Its operational temperature range is from -25 to +55°C.

Status
In service with the Yugoslav Army. Unable to trace current manufacturer or supplier of service and parts.

Contractor
Yugoimport SDPR.

VERIFIED

The general purpose radiological detector DRZON in use

M81 radiation detector and alarm

Description
The M81 radiation detector and alarm is a fully automatic system that can monitor radiation and provide an alarm when either the radiation dose or absorbed dose rate reaches a preset level. It is intended for installation at military and other special locations and provides real-time detection and monitoring of hazards from γ radiation. It can also be used to automatically switch on filter and other ventilation systems when necessary.

The system consists of a measuring unit, up to three Type MS2 sensor probes, connection box Type SK2, interconnecting cables and maintenance accessories. The three Type MS2 sensor probes may be located up to 200 m away from the central measuring unit.

The system can measure a radiation dose from 0.1 to 999.9 cGy and an absorbed dose rate of from 0.01 to 999 cGy/h. The alarm threshold level can be preset within the absorbed dose rate range so that when a sensor probe detects contamination at that preset level, visual and aural alarms are activated and filter and ventilation systems may be switched on. LED displays are used to provide information on radiation levels at each sensor probe.

Status
In service with the Yugoslav Army. Unable to trace current manufacturer or supplier of service and parts.

Contractor
Yugoimport SDPR.

VERIFIED

M81 radiation detector and alarm ready for installation

MRK-M87 radiation contamination measuring monitor

Description
The MRK-M87 radiation contamination measuring monitor is a hand-held measuring instrument intended for measuring radiation contamination on

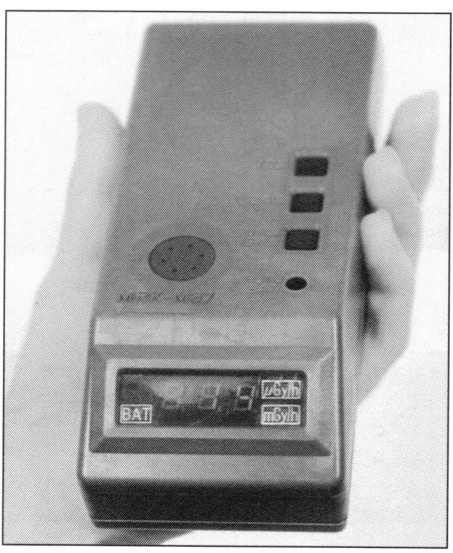

The MRK-M87 radiation contamination measuring monitor

personnel, water, food, weapons and equipment and vehicles. It is compact, rugged, easy to operate and easy to decontaminate.

The dose rate measurement range of the MRK-M87 is 0.1 µGy/h to 15 mGy/h, with a response time of 4 seconds for a dose rate over 45 µGy/h.

The MRK-M87 is powered by four R6 batteries or a 6 V vehicle battery. Weight is 770 g and the instrument measures 178 × 75 × 47 mm. Its operational temperature range is from -25 to +55°C.

Status
In service with the Yugoslav Army. Unable to trace current manufacturer or supplier of service and parts.

Contractor
Yugoimport SDPR.

VERIFIED

MZ-10 and MZ-100 γ radiation alarm monitors

Description
The MZ-10 and MZ-100 γ radiation alarm monitors are portable instruments designed to provide visual and audible warnings when the ambient level of γ radiation rises above a preset level. They are intended for use in a variety of environments from nuclear laboratories and installations to nuclear shelters and may also be used in vehicles.

The MZ-10 and MZ-100 both weigh 2.5 kg and are contained in a waterproof case. The monitors have their own integral sensors but each may be provided with an optional detector probe that can be used up to 100 m from the instrument.

The MZ-10 may be preset within a measurement range of 10 to 9,999 pC/kg/s which corresponds to 14 to 140 mrad/h; other ranges can be produced on request. The MZ-100 may be preset within a measurement range of 1 to 999 pC/kg/s which corresponds to 1.4 to 14 mrad/h; other ranges can be produced on request. The preset alarm level may be set at selected fixed intervals within this measurement range and when it is exceeded the four-digit LED display will start to flash. A small loudspeaker on the instrument will also begin to emit sound pulses, as will the device on the optional probe, if used.

The MZ-10 and MZ-100 can be powered by a normal 220 V 50 Hz mains supply or by a 12 V battery. There is an automatic overload protection device for use when the monitors enter a high-level radiation field.

Status
In service with the Yugoslav armed forces. Unable to trace current manufacturer or supplier of service and parts.

Contractor
Yugoimport SDPR.

VERIFIED

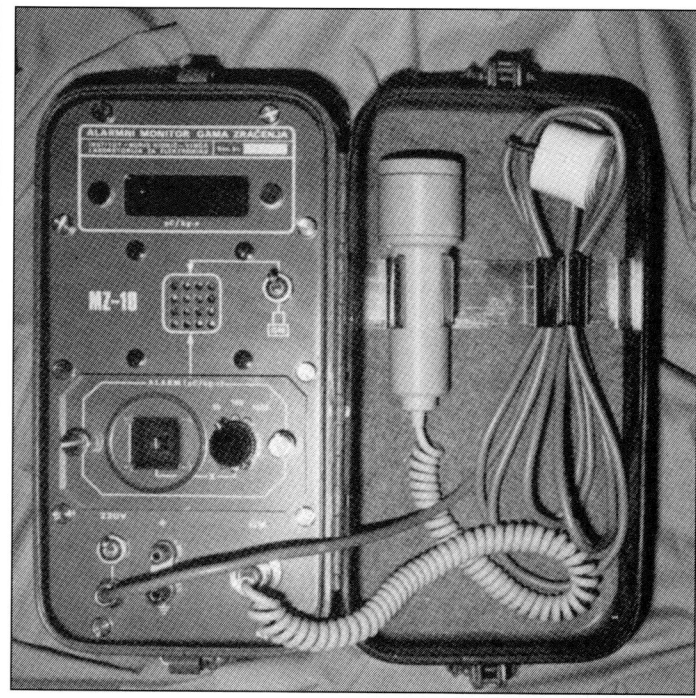

MZ-10 γ radiation alarm monitor; the MZ-100 is virtually identical

UPNE nuclear burst detection and analysing equipment

Description
The UPNE nuclear burst detection and analysing equipment is intended to provide a definite confirmation that a nuclear detonation has taken place. It provides as much relevant data as possible relating to that burst for subsequent planning purposes.

The equipment consists of two main components plus interconnecting cables. The UPNE sensor consists of a 360° azimuth sensor (elevation detection limits are -15 and +30°) under a hardened transparent dome supported on a metal tripod. The related computer is based upon a microprocessor and uses data from the sensor unit to display to the operator the real time and magnitude of a nuclear burst, its azimuth and elevation from the sensor unit and also the distance of the centre of the fireball from the sensor. Nuclear bursts with powers from 0.01 kT to 1 MT can be analysed with a detection range of over 20,000 m for a 0.1 kT burst. The computer, contained in a hardened case, has an automatic data store for up to 10 nuclear bursts and the resolution between successive bursts is less than 10 seconds for bursts of less than 100 kT.

The UPNE takes two people less than 20 minutes to install and may be powered by a 220 V 50 Hz supply or by two 12 V batteries. The integral real-time clock is powered by an independent lithium battery with an operational life of one year. The UPNE has a built-in automatic test feature for all functions.

The UPNE sensor unit weighs approximately 80 kg together with its transport packaging; the computer weighs approximately 25 kg. The operating temperature range is -25 to +60°C.

Status
In service with the Yugoslav armed forces. Unable to confirm current manufacturer.

Contractor
Yugoimport SDPR.

VERIFIED

UPNE nuclear burst detection and analysing equipment sensor unit

UPNE nuclear burst detection and analysing equipment computer

DETECTION (sensor systems) - Biological

CANADA

4WARN Urban

Description
The 4WARN Urban BW agent detection system is the latest in the 4WARN series of detection systems designed by Computing Devices Canada (see separate entry within this section). It comprises a fully integrated cost-effective response vehicle aimed at domestic preparedness and protection/migration applications. It can be mounted in a standard sport utility vehicle with minimum vehicle modifications and is able to reliably detect low concentrations of BW agent within seconds, giving near-realtime monitoring of ambient air. No consumables or wet chemistry techniques are involved. 4WARN Urban has a range of integrated sensors. Signal processing software allows for near-realtime detection, alarm and messaging functions. The components include:
- Fluorescence aerosol biodetector
- Particle concentrator
- Embedded computer
- GPS receiver
- Meteorological station
- Secure satellite communications
- Interior workstation with laptop computer
- Custom roof-mounted sportspack enclosing sampling equipment and antennae.

Computing Devices Canada is the prime contractor for the Canadian Integrated Biological Agent Detection System (CIBADS) (see separate entry within this section). CIBADS and its subsequent 4WARN family of detectors have been successfully field trialled and deployed operationally.

4WARN Urban Additional Capacity
The 4WARN Urban current configuration for biological agent response is expandable to include nuclear and chemical detection, as well as other NBC functions. Optional features include:
- Automated Ticket Reader (ATR) Identification
- Additional detectors: including chemical and radiological
- Alternate communications: including radio transceiver, Ethernet, Serial (RS 232, RS 422, E1A-48S)
- Test aerosol delivery capability.

Status
Available.

Manufacturer
General Dynamics Canada.   **NEW ENTRY**

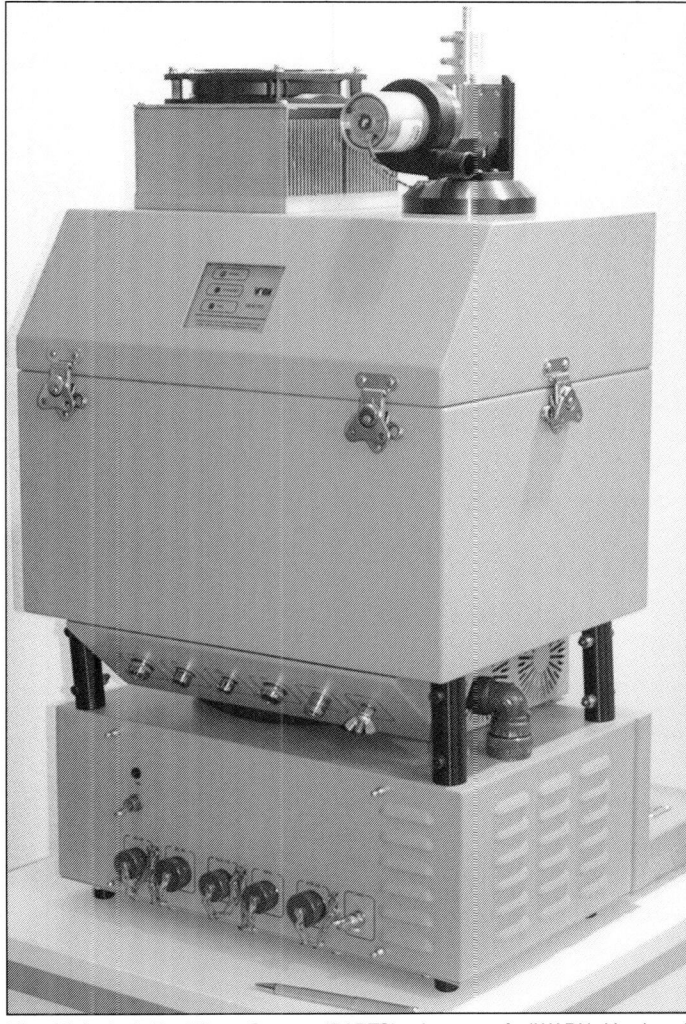

The Biological Real-Time Sensor (BARTS) element of 4WARN Version 2 (John Eldridge)
2002/0137903

CIBADS and 4WARN

Description
4WARN is a range of detection and identification systems which integrates COTS BW and/or CW agent sensors with other technologies to offer reliable near-realtime detection. The units are robust, automated and designed for ease and safety of deployment. The modular design facilitates incorporation of new sensors or other upgrades as they emerge in future. Computing Devices Canada led an integrated project team, working with the Canadian Defense Research Establishment Suffield (DRES) and other organisations including the University of Alberta, the Canadian Forces, SIL, Dycor and TSI Inc to develop a strong BW agent capability which became the Canadian Integrated BioChemical Agent Detection System (CIBADS). The CIBADS Advanced Development Model followed on from this and formed the basis of the current 4WARN family.

CIBADS aimed to address the following:
- Detection of biological and chemical agents in time to warn and protect personnel in real time (seconds rather than minutes)

Specifications

Generic detection:	Fluorescence particle detection, Biological Real Time Sensor (BARTS).
Identification:	Antibody assay strips and automated reader. Self-Contained Automated Multi-Assay Reader (SCAMAR). See illustration.
Concentrators:	Particle impingers. 4WARN V.2 MesoSystems MicroVic Concentrator (integral with BARTS) 28 l/min.
Sample Collectors:	Triggered by fluorescence particle detector 400 l/min MicroVic with impingement module. Dry samples for verification sample.
Additional Sensor:	Port for ion mobility agent sensors. Port for optional radiation sensor.
System Processor:	PC 104.
Software:	Fully automated/no user intervention and remote operating software.

Physical and performance data

Set up time (no assembly required)	5 minutes
Sensitivity	10 ACPLA for *bacillus globigii* with 20 second response time
First test results	15-20 seconds
User selectable (default 3 seconds) sampling time with generic detection and alarm	20-22 minutes
Liquid sample collection and identification	
Consecutive results (detection independent of identification)	20 minutes
Total cycle time detection and identification	
Dimensions (2 boxes)	12 × 41 × 56 cm
Weight	45 kg per box
Power	400 W normal
Battery capability	115/22 V, 12/24 V battery

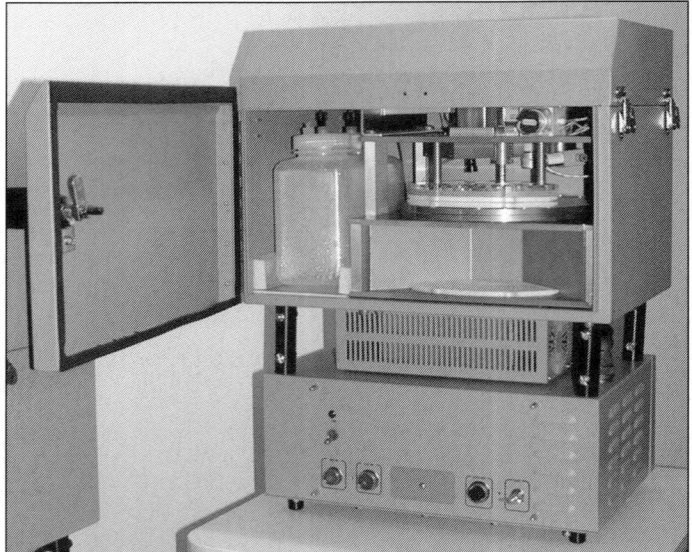

The Self-Contained Automated Multi-Assay Reader (SCAMAR) of 4WARN Version 2 identifies the agents from the sampler unit (John Eldridge) **2002**/0137904

- Identify biological and chemical agent in time to treat casualties (chemical agents under 1 minute, biological agents in under 15 minutes)
- Collect samples for off-site verification and absolute proof of agent use
- Monitor levels of contamination
- Allow integration with command and control functions.

4WARN Version 2 (V.2)
4WARN V.2 is a fully automated third-generation BW agent detection and identification system based on fluorescence for real-time detection of BW agents and antibody based assays for agent identification. It has multiple assay capability and can identify 11 agents simultaneously using arrays of assay strips.

Status
Available.

Manufacturers
General Dynamics Canada.
Dycor (XMX series collectors/concentrators, CBNet software).
Mesosystems USA (particle concentrator).
TSI Inc USA (FL/APS, Airborne Particle Sizer®).

NEW ENTRY

FRANCE

BIOWARD 1

Description
BIOWARD 1 is a prototype BW agent detection system, developed as part of EUCLID RTP 13.7 - a European project involving both state (DGA-CEB/Dutch MoD) and industrial (TNO-PML and Giat Industries/NBC-Sys) organisations.

The unit comprises a two-wheeled transport case containing the three elements of this prototype system: the bio collector, the sensor system and the user interface.

The biocollector
The miniature wetted-wall cyclonic biocollector concentrates the particles contained in the sample aerosol by sucking in the ambient air at a flow rate of 225 1/min. The particles entering the vortex tube are thrown out onto the walls whilst the air continues and is drawn out at the top of the tube. The particles become trapped in the fluid medium (buffer solution) on the walls of the tube are collected in a tank from which they are re-injected again. This recirculation amplifies the particle to enhance detection and identification of the BW agent. Operationally, the operator deploys the biocollector from its stowage inside the carrying case and connects its electrical supply and sample flow tube. The bio collector stores up to 600 ml of buffer solution, enough for 30 measuring cycles. The detection element of the system is based on the antigen/antibody immunoreaction.

The sensor system
The sensor system, which is thermally isolated from the outside and from the other compartments in the case, incorporates three detector units currently targeted at

Specifications

Biocollector

Type	Wetted-wall cyclone with recirculation loop
Air collection rate	To 225 l/min
Sample concentration ratio	Up to 100,000
Embedded buffer solution	300 ml
Concentration cycle duration	5 min
Dimensions	142 × 200 × 300 mm
Weight	2.8 kg

Sensor system

Principle	SPR. Low volume flow cells.
Reagent solutions	15 ml (per analyte)
Biochemical interface capacity	20 measurement/regeneration cycles
Duration (single measurement/ regeneration cycle)	8 mins

Unit (packed)

Weight	55 kg
Dimensions	900 × 525 × 360 mm
Power supply	115/230 V AC or 18-32 V DC

Environmental

Operating temperature range	+5 to 45°C
Management	Cooling/heating system with PID regulation

three different agents: *Staphyloccocal Enterotoxin B*, MS2 and *Erwinia Herbicola*. Others, including *Escherichia Coli*, are also being developed. The miniature sensors use Surface Plasmon Resonance (SPR) as the detection principle (see *Technical Developments*). The Texas Instruments ™ Spreeta® sensors measure the refractive index of the liquid in contact with a sensitive surface. This built-in and rugged system also comprises an LED, a sensitive gold electrode surface and an array of photodiodes to detect the light reflections. SPR technology was selected over volume acoustic wave technology because it appears to offer the advantages of increased measuring stability, sensitivity, better reproducibility and shorter response times.

The user interface
A touch-sensitive TFT screen allows the user to read the output from the three sensor units. The user activates the sampling and detection cycle, which starts the biocollector, from the screen. There is a visual alarm which displays when the pre-set threshold is exceeded, indicating a high concentration of agent present in the sampled airflow. Menu selection allows the user access to two additional screens to check the correct operation of the biocollector and the sensors.

Operation
BIOWARD 1 is delivered in a wheeled IP 64 sealed carrying case, transportable by two persons. Opening the top cover reveals the front panel, which houses the thermoregulation system's external fans, the user-interface touch-sensitive screen and the energy conversion electronics fan. The on/off switch, the indicator lights and the fuses are also arranged on the front panel. Operationally, the user selects the on/off button and the menu-driven interface appears on the screen.
A handle lifts the hinge-mounted user interface panel to reveal three compartments underneath: the biocollector storage housing, the control and checking compartment (containing the microcomputer) and the detection system.

A single trained operator can set up and run the system. Setup requires the user to arrange the sensors in their flow cells, fill the nine measuring cylinders with solutions and connect the biocollector. No special tools are required. The biocollection/detection cycle is initiated via the interface and the system is operational after three minutes. Detection information is supplied eight minutes after the beginning of the first concentration cycle. As the biocollector operates concurrently with the measuring cycle, the result of the second detection cycle can be delivered three minutes after the first

Status
Under trial for the EUCLID RTP 13.7 project by France and the Netherlands.

Development Agency
DGA-CEB (France) / Netherlands MoD.

Manufacturers
Giat Industries (France).
TNO-PML (Netherlands).

NEW ENTRY

GERMANY

Chemical Biological Mass Spectrometer (CBMS) Block III

Description
This combined CW and BW detection system uses pyrolysis mass spectrometry. (See separate entry in DETECTION (sensor systems) — Chemical).

Status
In production. Standard CBMS in service with the US Army.

Manufacturer
Bruker Daltonik GmbH.

VERIFIED

··

Rheinmetall Landsysteme NBC field laboratory

Description
The Rheinmetall Landsysteme field laboratory is a comprehensively equipped mobile detection and analysis facility contained in three half-sized air-mobile customised ISO containers. A fourth container contains crew support facilities. Mounted on two standard German Army lorry-trailer units, the entire laboratory can be air-lifted or driven to unit HQ areas to provide NBC agent detection and analysis. Co-ordinated with data retrieved from deployed Fuchs NBC reconnaissance vehicles (see entry under DETECTION (reconnaissance systems)), the facility offers rapid command decision support or on-site verification. NBC data is integrated in a suite of high-performance IT based data management and communications equipment. The BW suite comprises the equipment listed below.

Toxic samples are introduced via the microbial isolator (see illustration) and the pressure inside the laboratory can be maintained at below ambient pressure to avoid escape of pathogenic material to the exterior. Biohazardous materials can be isolated and grown under safe and controlled conditions for both on-site analysis

Rheinmetall Landsysteme NBC field laboratory BW analysis suite   0050718

Specifications
Spectrometry: Gas chromatography. Mass spectrometry. Automatic sampler. Micro High Performance Liquid Chromatography (HPLC) system.
Culture management: Microscopes (fluorescence, inverse, stereo). Culture growth media, centrifuges, vibrators, stirring units, sterilisation units. Photometer with micro titre plate. Safety level L3 microbiological isolator.
Analysis: PCR-assisted fluorescence. Enzyme-linked immunosorbent assay.
Sample collection: Air sample, dust, water, gas.
Communications: VHF, digital telephony (including ISDN), SatCom.

The micro-biological isolator   *2000*/0084045

and for safe transport to research laboratories off-site. See under DETECTION (sensor systems) nuclear and chemical for details of the nuclear and chemical suites.

Status
Available.

Manufacturer
Rheinmetall Landsysteme GmbH.

UPDATED

View into the Biological Analysis Laboratory   *2000*/0084044

UNITED KINGDOM

Prototype Biological Detection System (PBDS)

Development

The UK MoD took delivery of the first of several prototype systems in December 1998. The current PBDS, as it develops, capitalises on the earlier success of the nine Biological Detection Systems (BDS) fielded successfully during the 1990-1991 Gulf War. In the latter, the BW detection process was achieved by passing batches of air samples through a variety of different stages. The first is the collection process. Secondly, airborne particles in the 2 to 10 µm range are scrutinised using a light-scattering technique. Subsequently, a luminometer surveys the collected particles and determines the presence of biological material by searching for traces of adenosine triphosphate (ATP) which is present in all living matter. The technique closely follows that used in the US XM31 BIDS (qv) in that mixing the particles with luciferase allows a luminometer to detect and measure the amount of light emitted. A third stage involved the use of antibody reactions to determine the exact nature of the sample involved.

Further developments to PBDS aim to streamline the processes, moving closer to the achievement of a comprehensive near-realtime detection capability for a large range of BW threat agents. The Aerosol Shape Analyser (ASAS) integrates two of the processes, analysing the light-scatter data to deduce the proportion, shape and size of particles in the sample. A rapid variance from the background triggers the alarm. Further refinements to the overall facility, each element of which is currently a man-in-the-loop system, aims for much greater automation and will allow the system to tell whether an organism is one of nature's or has been artificially developed.

The PBDS is operated by the Joint NBC Regiment, formed in 1999. The detection suite is housed in a standard army 4 ton truck chassis.

Integrated Biological Detection System (IBDS)

In March 1999, the PBDS consortium won a £50M contract from the UK MoD for full development and production of mobile Integrated Biological Detection Systems (IBDS). Based on the experience of PBDS, the new system will provide a unique BW agent detection capability for the UK armed forces. New technologies are also expected to be explored for an unattended biological sensor system (UBS).

Integrated Naval Biological Detection System (INBDS)

This system remains under parallel development at DSTL (Porton Down) for trials at sea.

The onboard installation of INBDS   **2002**/0127267

Integrated Biological Detection System (IBDS) shown deployed   **2000**/0084046

Prototype Biological Detection System (PBDS)   0050701

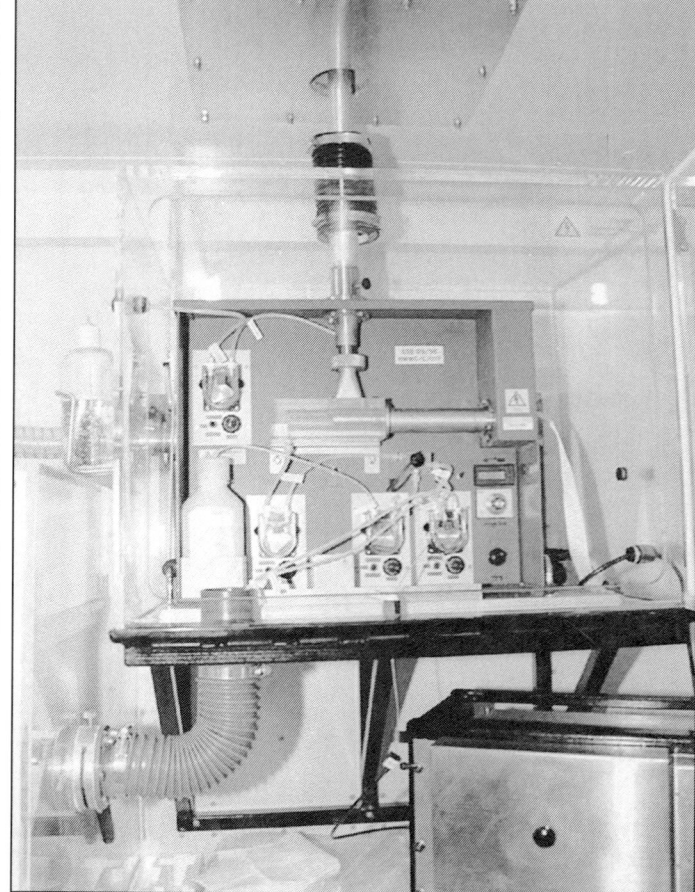

The air sampler unit inside PBDS   **2002**/0127264

Status
Seven prototypes delivered for trial. Four systems deployed to Kuwait December 1998 in support of Operation BOLTON.

Development agency
DSTL.

Manufacturers
INSYS Limited (system integration).
Graseby Dynamics (detection technology).
EDS Defence (part of consortium).
Integrated Photomatrix Limited (external detection sensor).

UPDATED

UNITED STATES OF AMERICA

BioCapture™ BT-500 and BT-550 portable samplers

Description
This novel hand-held bio-detection device comprises several modules. At the top front of the detector is the air intake which draws the particle-laden air across a rotating arm impactor which separates the biological particles from the air stream, throwing them into the path of a continuously flowing film of fluid. The coating on the impeller arms and the inner walls of the housing are designed to increase collection efficiency, resulting in more concentrated liquid samples and therefore potentially shorter sampling times. The system is designed to work in wet or dry conditions in a wide range of temperatures. In trials conducted at the US Dugway Proving Ground and the wind tunnel at MesoSystems Technology, the performance of the BioCapture™ system was assessed against two other types of collector: the All Glass Impinger (AGI) and the Slit Sampler. The relative collection efficiency was measured against the total number of Colony Forming Units (CFU) of single cell, 1 m diameter, aerosolised *bacillus globiggii* (BG) per litre of sampled air. The results are shown in the histogram below.

The concentrated samples are collected in cartridges (see below). Maintenance and diagnostic cartridges are also included.

Both versions are presented in a carrying case and can be fitted with two carrying straps. A sampling hose and adapter allows sampling of confined spaces and a charger is included for the four lead-acid rechargeable batteries.

BT-550 offers on-site sampling as well as sample collection
The BT-550 version is additionally equipped to take the Tetracore BTA™ Test Strip (see separate entry this section), allowing on-scene collection and analysis of the biohazard.

Specifications
Particle collection size range: 0.5 to 10 m
Operating flow rates: 150 litres per min
Battery life: 12 V DC, providing up to one hour of continuous run time
Dimensions: 5 × 2 × 2.5 cm
Weight: 4.5 kg

Status
Available. In use with several First Responder units in the USA, including those in Seattle, San Diego, Los Angeles, El Pas, New York City, Montomery County, Maryland and the 9th WMD Civil Support Team in California.

Manufacturer
MesoSystems Technology Inc.

NEW ENTRY

Bio Detector (BD)

Description
The Bio Detector (BD) has been developed as an integral component of the P3I Biological Integrated Detection System (BIDS) under contract to the US Army CBDCOM. The BD will be used to perform fully automated immunoassays of liquid samples to detect and identify the presence of biological warfare agents. The BD is an on-demand system capable of operating continually during a 14-hour mission. When a threat is detected, the BD generates both an audible and visual alarm and provides specific agent identification as well as concentration data.

The BD employs a rugged and automated application of an immunoassay technique known as Light Addressable Potentiometric Sensor (LAPS). LAPS technology provides the BD with flexibility to simultaneously detect and identify eight biological warfare agents, including bacteria, viruses and toxins, within 15 minutes.

The BD includes sample input, fluidic system, sensor module, electronics and consumables in a single rugged housing. A liquid sample with known biological

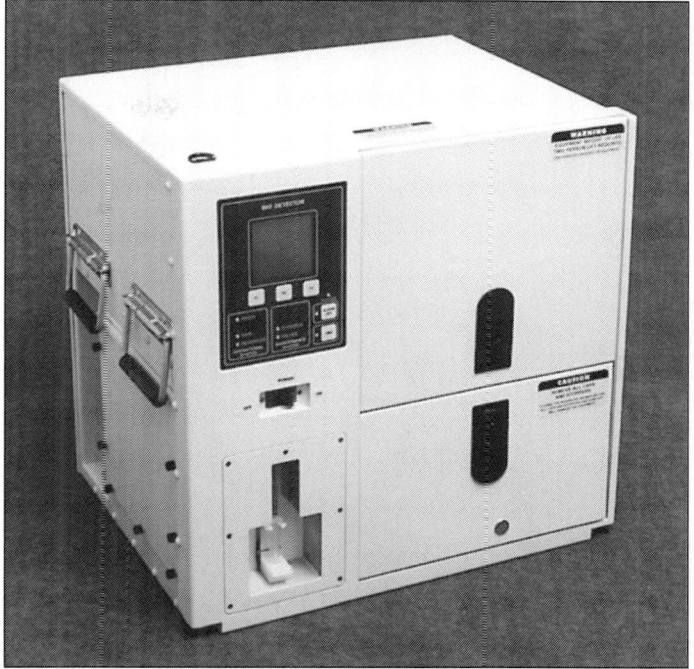

Bio Detector (BD)

Specifications
Mission length: 14 h continuous operation, 40 detection requests
Dimensions: 55.9 × 60.7 × 45.7 cm
Weight: 61.2 kg
Power: 110 V AC (50-60 Hz)/28 V DC.
Operational environmental range
 −19 to +63°C (initial start-up, using shelter temperature controls)
 +10 to +27°C (maintained operational temperature)
Instrument storage temperature
 −46 to +71°C
Instrument storage life: 5 years
Reagent storage life: 2 years at +4 to +8°C
Self-test: BITE - over 170 failure conditions

---

materials of interest is captured by specific antibodies and detected with a light-addressable potentiometric silicon sensor. Up to eight different analytes can be monitored simultaneously on the same silicon chip.

Extensive laboratory testing with a bacterial simulant, *Bacillus subtilus var. niger*, has been performed, demonstrating a limit of detection better than 20,000 CFU/ml. Toxins (for example, Botulinum Type A, Staphylococcal enterotoxin B) have a demonstrated limit of detection better than 10 ng/ml. Additional capabilities have been added for several biological warfare agents of interest, including *Bacillus anthracis* (Anthrax), *Yersina pestis* (Plague), *Ricin*, *Brucella* and *Francisella tularensis*. The required reagents and consumables developed and produced by ETG allow the BD the capability of performing fully automated biological warfare agent identification[2].

Status
Development – see also Biological Integrated Detection System (BIDS) entry in this section.

Manufacturer
Environmental Technologies Group Inc (ETG).

VERIFIED

Biological Aerosol Warning System (BAWS)

Description
BAWS (Tier 1) is an integrated network offering area BW detection. It consists of a base station with a display and control console allowing an operator to monitor up to 150 remote stations within a 10 km zone. It integrates incoming BW detection information with positional information from a GPS receiver and real-time meteorological data. The integrated detection data determines and triggers an alarm condition for the area covered. Outgoing information can be passed to a PC workstation for analysis, or to other networked units by UHF radio or landline using standard ATP1B NBC1 and NBC4 message formats.

Each Remote Station is a tripod-mounted detector unit with sensors for BW aerosol detection (using an airborne particle counter linked to a bio-fluorescence detector), a chemical agent detector (either the M88 ACADA or LSCAD systems - see entry under DETECTION (sensor systems) - Chemical), an anemometer a wind direction vane and an electronic compass. There is a UHF telemetry link to the base

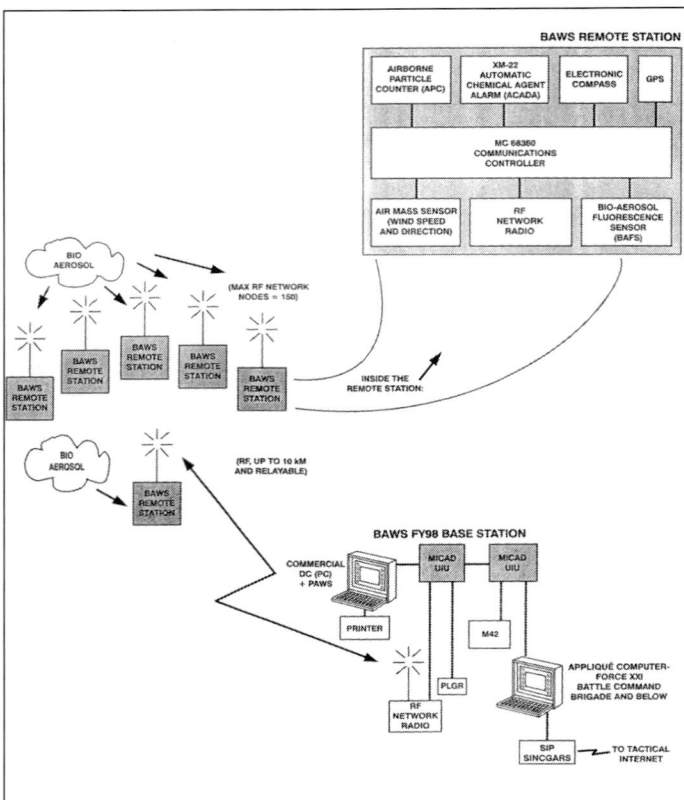

BAWS Remote and Base Stations - arrangements   0050920

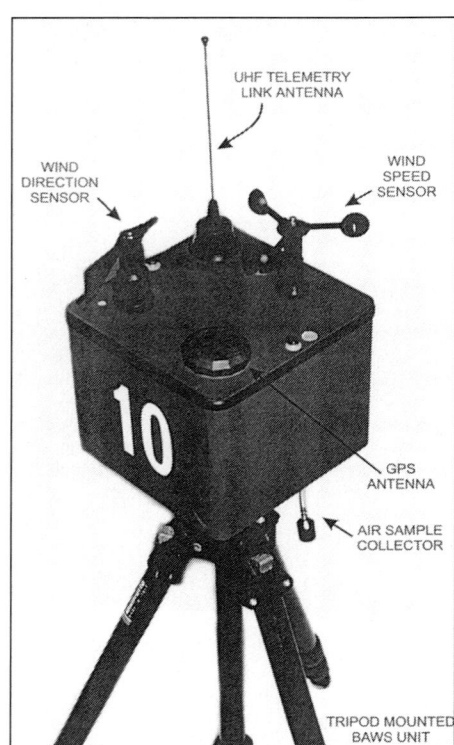

BAWS Remote Station showing sensor array
0050921

station and a GPS receiver. The units can be powered by battery or from a combat vehicle power supply. The Remote Stations can act as relays to other, more remote units to take account of the line-of-sight range limitation imposed by UHF.

A BAWS version suitable for internal zone monitoring, of a large building or facility for example, is also available.

Status
Available.

Development agency
Joint Program Office for Biological Defense.

Manufacturer
Lockheed Martin.

UPDATED

Guardian BTA™ System

Description
The Guardian BTA™ System is designed for the rapid collection, detection and identification of BW agents. Being portable and easy to operate, it is specifically designed for field use and delivers a fast, 15 minute reaction in screening for biological agents. Currently available tests include anthrax, ricin, botulinum toxin SEB and plague. Tests are being developed for other potential BW diseases. At the heart of the Guardian System are the BTA™ test strips, from Tetracore Inc. Based on innovative, patented technology, the BTA™ test strip employs agent-specific antibodies to positively identify the potential threat. During the sample evaluation, the BTA™ strip also performs an internal quality control test to assure the validity of the test.

On suspecting contact with a BW agent, the suspect material, solid or liquid, is mixed with an aqueous solution, the BTA™ Sample Buffer. This prepares the sample for testing. Five drops of the liquid mix are added to the sample port of the test strip. The sample interacts with the reagents and moves along the test material, inside the test strip's plastic case. Lateral flow chromatography then provides the results.

Screening results are produced in 15 minutes. Two solid bands, one in the control area and one in the sample area, indicates a positive result. One solid band (in the control area) indicates negative results. Any other combination of bands indicates an invalid result, in which case the test should be re-run. The results may be read visually or, for greater accuracy, the Guardian BTA™ Test Strip Reader can be used. The test strip reader is designed to accept and analyse BTA™ test strips. The reader offers greater accuracy, as its optical technology can recognise positive results that might be missed by the human eye, due to faint positives or poor ambient lighting. Guiding the operator through the evaluation procedure, the reader provides a print-out of the test results and date. Embedded radio frequency identification (RFID) technology ensures that the chain of custody is documented for each individual BTA™ strip test; documentation crucial in later phases of the investigation.

Status
In production.

Manufacturers
Alexeter Technologies LCC.
Tetracore Inc (test strips).

NEW ENTRY

The Guardian BTA test strip, showing a positive test result in the small window (the left hand red stripe, under 'S' for Sample and the control result under 'C') (Alexeter Technologies LCC)   *2002*/0137375

Hand-held microluminometer

Description
As part of a collaborative development programme with Edgewood Research, Development and Engineering Center (ERDEC), a hand-held microluminometer has been produced which allows quantitative analysis of BW agents. The requirement is to enable forces to detect the presence of BW agents in the field in real time.

The meter works by using the chemiluminescent reaction to track the presence of ATP - a constituent common to all living organisms except viruses. The instrument ingests a 10 ml sample of the ambient atmosphere and the LED screen indicates the presence or absence of a possible agent after approximately 10 seconds.

It is planned to make the developed system available to the first response community.

Status
Available.

Manufacturer
New Horizons Diagnostic Corporation.

Agency
Edgewood Research, Development and Engineering Center.

VERIFIED

Joint Biological Point Detection System (JBPDS)

Description
In April 1997, Lockheed Martin Librascope, a business unit of Lockheed Martin Corporation, was awarded a 33 month, US$32 million Engineering and Manufacturing Development (EMD) contract to develop a JBPDS by the US Army's Chemical and Biological Defense Command (CBDCOM). The JBPDS design target is to evolve 28 systems in nine different configurations to suit most types of defence platform, providing each service with the capability to operate and survive in a BW environment. The nine different system configurations are designed around a single basic detection module. This is matched to each platform-unique environment by appropriate interface units, an approach which greatly simplifies the complex logistics problems normally expected in a US joint service application.

The JBPDS provides the user with rapid warning and identification of BW hazards, being capable of detecting point source releases as well as sea and ground level releases or line aerial sources. It provides local and remote warning capability; and formats threat information for distribution over standard military communication systems. The JBPDS collects, contains and provides suspect samples, for laboratory conformation. The system is housed in a ruggedly designed enclosure for day or night operation in a battlefield environment. The interface kits allow the JBPDS to be used in a wide variety of battlefield applications such as S788 Shelters, ½ ton trucks, stands or poles for perimeter monitoring systems, deck-mounted shipboard installations and NBC reconnaissance vehicles such as HMMWV, S788 and Light Armoured Vehicles (LAV). A more compact version of the detection module is designed for man-portability.

The key features of the JBPDS include automatic triggering, collection, agent detection and identification. The system is to take 30 minutes or less to set up, initialise (using built-in test capability) and operate. Agents will be identified in 15 minutes or less with a false negative detection rate of <0.1 per cent (false positive: <2.0 per cent). Facilities will be provided for the collection of 25 and 50 ml samples for independent analysis. Maximum use is to be made of COTS components, including software. Interactive technical manuals are to be available on electronic media and the system will offer IT-based embedded training.

JBPDS is scheduled to replace both the Navy Interim Biological Agent Detector (IBAD) and the Army Biological Integrated Detection System (BIDS) (see separate entries in this section).

Status
Engineering and manufacturing development. Planned in-service date of 2005.

Development agency
Joint Program Office for Biological Defense.

Manufacturer
Lockheed Martin.

UPDATED

Long-Range Biological Standoff Detection System (LR-BSDS)

Description
The Long-Range Biological Standoff Detection System (LR-BSDS) is intended to provide the earliest possible standoff warning of a biological attack. It will be an airborne system carried by a UH-60 helicopter to detect manmade aerosol clouds containing biological and chemical agents at long range.

A Non-Developmental Item (NDI), LR-BSDS with a detection range of 30 km or more, was fielded in June 1997. An objective system, known as the Counterproliferation (CP) LR-BSDS, which has a detection range of 50 km or more, is currently being developed.

The LR-BSDS consists of an infra-red laser, receiver and detector with an information processor. The system will provide a long-range and large area aerosol detection and tracking capability. The early warning provided will allow troops to reach or assume adequate protection prior to agent exposure and cue tactical point detection assets such as the M31 Biological Integrated Detection System (BIDS - see separate entry).

In May 1995 Schwartz Electro-Optics Inc, of Orlando, Florida, were awarded a contract for development of the CP LR-BSDS. Also involved in the contract were Fibertek Inc of Herndon, Virginia. A major Biological Agent Detection program is the Counter Proliferation Long Range - Biological Standoff Detection System (LR-BSDS). This program developed a high power, high resolution LIDAR that scans

out to 50 km or more to detect the presence of man made aerosol clouds which may present the threat of biological warfare.

The laser is a pulsed, diode-pumped, Nd:YAG design that is frequency shifted by an Optical Parametric Oscillator to the wavelength of 1.54C um. In order to achieve long range performance and eye safe operation the transmitter beam is expanded to 12 inches in diameter. The backscatter return laser energy is collected by 24 inch optics and concentrated on a sensitive photon level detector system. The transmitter and receiver telescopes are mounted on a stabilized gimbal.

The processed signal is integrated with feedback from the onboard inertial navigation system in order to calculate and display the size, relative intensity, exact geographic location and drift of the aerosol cloud. System operation is controlled through an operator's console that provides display of all system parameters. An embedded training system permits full simulation of all functions and performance. The complete system is mounted on a single pallet which enables the unit to be quickly installed into an unmodified UH-60 Blackhawk Helicopter. The Biological Agent Detection was produced under U.S. Army Contract No. DAAM01-95-C-0041. Several very successful field tests were performed in early and mid-year 2000, however, due to a change in government requirements. this system will not be fielded. Sales of this technology for airborne or ground based systems is available for FMS or Commercial use.

Status
The CP LR-BSDS was scheduled for fielding in October 1997.

Development agency
Joint Program Office for Biological Defense.

Manufacturers
Fibertek Inc.
Schwartz Electro-Optics Inc.

UPDATED

M31 Biological Integrated Detection System (BIDS) and the Mobile Biological Agent Detection System (MBADS)

Description
The M31 Biological Integrated Detection System (BIDS) is a mobile detection and identification system for BW agents deployed with US land forces. The main part of the system, the detection and analysis equipment, is located in an S-788 shelter containing the M31 BIDS. The shelter is carried on a M1097 heavy HMMWV vehicle which also tows a PU-801 trailer-mounted 15 kW generator.

The S-788 shelter is normally carried on the HMMWV but can be dismounted for fixed-site operations or transport by helicopters. The shelter is collectively protected and environmentally controlled. The shelter is provided with navigation aids and meteorological sensors.

From within the shelter, continuous monitoring of the atmosphere is undertaken by an Aerodynamic Particle Sizer® (APS). An increase in particle density or change in size distribution exceeding the set threshold values will trigger an alarm that alerts the operator to begin collecting samples.

There are two sample collectors within BIDS, one to support remote analysis and the other for local operations. The biological sampler concentrates aerosol particles in the 2 to 10 μm range for transport to a laboratory. The Liquid Sampler (LS) provides the samples needed for limited analysis within the BIDS.

Four devices allow BIDS to perform local analysis: a bioluminometer, a flow cytometer, a threshold device and a range of smart tickets (see separate entry within this section).

The bioluminometer operates on the principle that there is a relatively constant amount of Adenosine TriphosPhate (ATP) present in the cells of all living

M1097 heavy HMMWV carrying the S-788 shelter used as part of the XM31 Biological Integrated Detection System (BIDS) (T J Gander)

organisms. The operator can assess the concentration of living organisms such as bacteria and compare it to the normal atmospheric background values for the area.

The Flow CytoMeter (FCM) breaks down particles into their components. The FCM provides data on cell size, shape and fluorescence that, using pattern recognition techniques, can differentiate bacteria from natural airborne matter such as pollens and mould spores.

Any warnings provided by the two devices mentioned above trigger the BIDS operator to perform more specific tests. One involves a threshold device which takes a sample from the LS, mixes it with antibodies against specific BW agents and filters it through a specially coated ticket. It provides simultaneous detection of four different agents within 10 to 12 minutes.

Smart tickets serve as back up to the threshold device. Each ticket is specifically produced to detect a single agent. If a specific agent is present, a bright red spot will appear on the ticket within 20 minutes.

Status

In service. It will eventually be replaced by the JBPDS system (see separate entry within this section). In 2000 and 2001, BIDS evolved further, taking advantage of maturing technologies, and the Mobile Biological Agent Detection System (MBADS) emerged.

Development agencies

EAI Corporation.
Joint Program Office for Biological Defense.
SBCCOM.

Marketing agency

Tradeways.

UPDATED

RAPID™ (Ruggedized Advanced Pathogen Identification Device)

Description

RAPID™ is a portable BW detection device incorporating a Web-based surveillance system. Developed by Idaho Technology Inc and the US Air Force, RAPID™ is based on Polymerase Chain Reaction (PCR) technology. Up to 32 prepared test samples can be amplified and analysed in less than 30 minutes. Fluorescent dyes in RAPID™'s reagent mixture attach to DNA and make it glow, allowing for fast and quantitative analysis. The device can survive extreme temperatures, vibrations and is airtight and watertight. It is capable of withstanding a 1 metre drop in transport mode and a 15 cm drop while operating. The Web-based application allows quick exchange of information about suspected biowarfare and disease outbreaks.

The software is designed for ease of use and is divided into basic and advanced modes. Laboratory personnel can design, create and test analysis protocols using advanced mode, making them available to less technically trained field users as simple push-button tests (basic mode). Field personnel prepare test samples,

---

Specifications
Total weight: 22.5 kg
Instrument and case: 13.5 kg
Laptop computer: 6.5 kg
Mini-centrifuge: 1 kg
Backpack: 1 kg
Size: 493 × 363 × 267 mm
Fluorescence optics modules: 3 colour optics modules
Sample rate: 6 seconds for 32 samples
Modes: Continuous; once per cycle: continuous scanning mode
Programming interface: Windows NT program with dual interface. Simplified operation with automatic results analysis for field use
Temperature range
Accuracy:
  Room temperature to 120°C
  Within and between sample: ±0.2°C

---

place them in the instrument, and push the button. RAPID™ runs the appropriate reaction, analyses the fluorescent change in the samples and displays the formatted results on the integral laptop screen. Test results can be printed on commercially available printers.

Status

Available. In service with the US Air Force as part of LEADERS (Lightweight Epidemiology Advanced Detection and Emergency Response System).

Manufacturer

Idaho Technology Inc.

NEW ENTRY

Sensitive Membrane Antigen Rapid Test identification tickets (SMART)

Description

The SMART identification tickets are self-contained, colorimetric, solid phase immunofiltration assays designed to be used in conjunction with a liquid interface. Two types of SMART devices have been developed. One kit is capable of detecting endospore-forming bacteria. The other kit is capable of detecting proteinaceous toxins or soluble antigens including bacteria. The SMART devices utilise a colloidal gold particle concentration immunoassay to effect sensitive and selective detection of biological materials. Antibodies specific to the agent of interest are conjugated to colloidal gold particles. When concentrated on solid surfaces, these particles can be visualised by the naked eye. Labelled antibodies can easily be lyophilised and reconstituted without losing activity or specificity.

The presence or absence of the target antigen is indicated colorimetrically. A small red dot appears on the ticket which the user compares with a colour chart.

SMART tickets to detect anthrax and botulinum toxin were issued to military personnel during operation Desert Storm in 1991. The technology is further developed and incorporated in a variety of the BW detection devices included in this section.

Other collection, screening and identification assay systems for potential BW agents are available. The **SWIPE** collection system includes media for surface, powder, liquid or air sample collection and, combined with a luminescence ATP system, allows screening for bacteria or spores. The SMART immunoassay can then be applied to allow field identification of the agent.

The entire system: collection, screening, and ID has been incorporated into numerous HazMat team protocols for responding to a BW event. Assay systems for anthrax, SEB, ricin, cholera, plague, botulinum toxin, glanders, brucella and others have been developed. Technologies utilised include colloidal gold, luminescence and fluorescence.

Status

Available. SMART anthrax and botulinum toxin media have been used by several countries since the early 1990s. The systems was first issued to military personnel during operation Desert Storm in 1991. The latest version is the **SMART II** lateral flow assay.

Manufacturer

New Horizons Diagnostic Corporation.

UPDATED

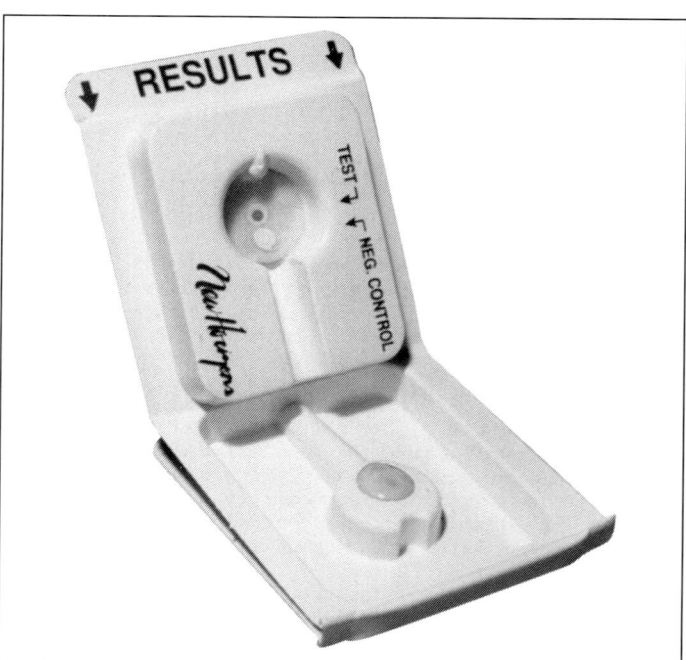

The RAPID™ system (Idaho Technology Inc)   *2002*/0137266

SMART identification ticket (New Horizons)   0050919

Short-Range Biological Standoff Detection System (S-RBSDS)

Description

The Short-Range Biological Standoff Detection System (S-RBSDS) is intended to provide a standoff warning of an imminent biological attack. It will be able to detect man-made aerosol clouds to ranges of 3 km and determine if they contain aerosol agents. In November 1998, Fibertek and the JPOBD completed a series of trials with S-RBSDS at Dugway Proving Ground, Utah. The trials verified the viability of the concept and demonstrated the ability to auto-detect and auto-track BW simulant aerosol clouds whilst, at the same time, discriminating between biological and non-biological aerosol. The trials included several disseminations in a matrix of ranges between 1 and 3 km, a variety of BW simulant types, some mixed with known interferents and several dissemination techniques.

The S-RBSDS is an eye-safe multiwavelength lidar system capable of standoff detection of biological agent aerosol and has successfully demonstrated the ability to autonomously detect and tract a biological aerosol cloud while discriminating between biological and non-biological aerosols, background pollens and hard targets. The S-RBSDS is an active standoff detection system with both IR and UV capability. The IR wavelength provides cloud detection, acquisition and tracking capability in a rapid single pulse laser firing manner. The UV wavelength provides near-realtime detection and ranging of a particulate cloud along with demonstrated discrimination capability. This provides the ability to achieve early warning, tracking and discrimination of biological agent clouds combined with the capability to reduce false detections from battlefield interferents and improve sensivity of detection in high aerosol background conditions.

The S-RBSDS is configured for an unmodified HMMWV installation and is designed to be set-up and operated by a two-person military crew with minimal training. Also, the S-RBSDS provides a 'detect-to-warn' capability allowing personnel to take protective measures in advance of arrival of a biological warfare cloud. The S-RBSDS was developed by Fibertek Inc of Herndon, Virginia in conjunction with the US Army Soldier and Biological Chemical Command (SBCCOM) at Aberdeen Proving Ground, Maryland.

Key Features

- Autonomous Operation (no dedicated operator required)
- Set-up in less than 45 minutes by two persons
- Safe for operation at air bases, ports, assembly areas
- Automatically searches in azimuth and elevation
- Sensor network integration
- Rugged design, HMMWV compatible
- Military generator or HMMWV power
- Eye-safe per US Army Center for Health Promotion and Preventive Medicine
- Developed Training, Operators and Training Manuals.

Status

Completed development, field testing and performance and military utility assessments.

Development agencies

Joint Program Office for Biological Defense.
US Army Soldier and Biological Chemical Command.

Manufacturer

Fibertek Inc.

UPDATED

Short-Range Biological Standoff Detection System (S-RBSDS) (Fibertek Inc)
0055058

Transportable Emergency Response Monitoring Module (TERMM)

Description

The Transportable Emergency Response Monitoring Module (TERMM) is produced by Engineering Computer Optecnomics Inc (ECO) of Annapolis, Maryland, for the US Departments of State, Defense and Energy. The TERMM is a transportable shelter module which was produced to determine several factors following an NBC incident.

The first factor is to determine whether agents have been released. The TERMM can be used to determine exactly what agents are involved, where the agents are concentrated, where they are likely to spread and monitor and confirm when decontamination or dissipation of the agents is complete.

The TERMM is a specially prepared analytical laboratory within a shelter. The shelter may be used either mounted on a semi-trailer transporter or on the ground and acts as a self-contained unit for extended periods. The basic crew is two men but for extended shift operations a five-man team working shifts can be accommocated and catered for.

A TERMM can be rapidly deployed (including by helicopter - deployment preparation takes less than 2 hours) and has all the necessary transport connectiors. It has an air and overpressurisation system, complete with a personnel airlock for entry and exit. There is also a glove box into which samples can be passed directly from the outside. Internal support equipment includes secure verbal and data communication systems, a hardened computer served by fibre optic cables, an electrical generator, toilet facilities complete with a shower unit and heating, ventilation and air conditioning systems. There are integral tanks for fuel, drinking water and for liquid wastes, plus a separate water system for an onboard personnel decontamination system. Optiona extras include ballistic protection and satellite communication systems.

The equipment used with TERMM can identify and monitor all known radiological, chemical and biological agents. It has the potential to detect agent developments, such as micro-encapsulated or bio-engineered agents, that are undetectable by current methods. Specialised equipment contained in a TERMM include radiation detectors, analysers and monitors (fixed and portable), aquatic biomonitoring systems, chemical analysis equipment, a refrigerator/freezer, glove box and hood and acid/solvent cabinets.

Being modular, the TERMM can be produced to one of three configurations. The basic TERMM (TERMM I) can be enhanced to an intermediate (TERMM II) or full capability (TERMM III) system and housed in shelters with standard ISO dimensions of up to 53 ft.

Status

Available.

Manufacturer

Engineering Computer Optecnomics (ECO) Inc.

VERIFIED

A Transportable Emergency Response Monitoring Module (TERMM)

UVAPS BW agent detector

Description

The UltraViolet Aerodynamic Particle Sizer® spectrometer (UVAPS) reliably detects the presence of airborne particles that fit a defined profile of specified BW agents and electronically triggers downstream analysis instruments to look out for the sample. UVAPS is deployed on the US Army's BIDS system (see related entry within this section) and forms part of other national BW response systems. UVAPS

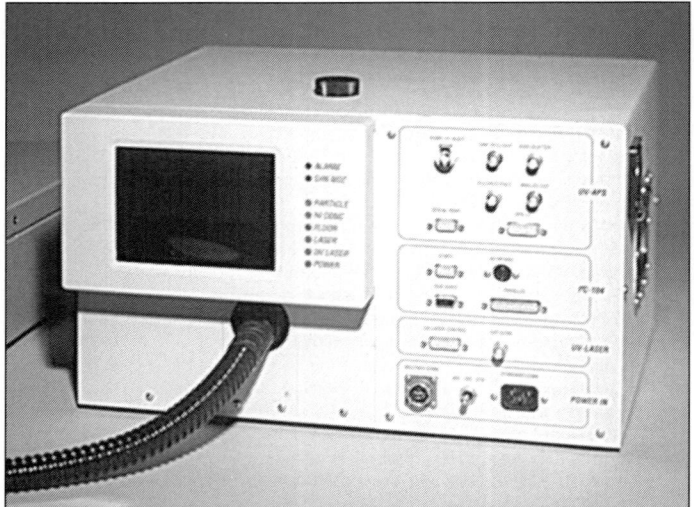

The UVAPS BW agent analyser uses TSI's APS® technology to categorise the type of agent and point it at BW analysis instruments. It forms part of the US DoD's BIDS system (TSI Inc)
2002/0137854

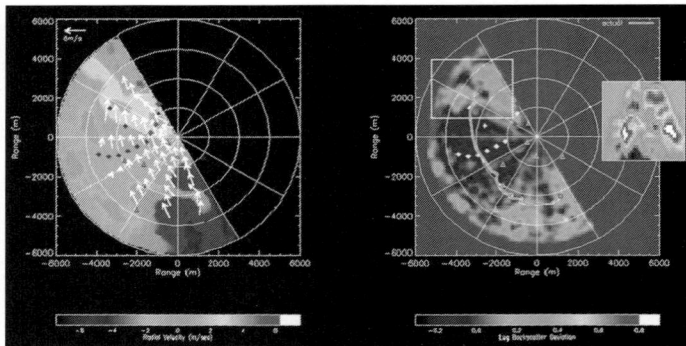

WindTracer® bio-aerosol standoff detection   **2001**/0109598

Specifications
Typical range: 80 km (2 > 15 km)
Minimal range: 400 m
Range resolution: 50 - 100 m
Velocity resolution: 0.5 m per second
Detection sensitivity: 100 - 1,000 particles per litre
Weight: 130 kg
Volume: 1 m³

---

examines both the size and size spectrum of the aerosol. A UV laser pulse irradiates particles in the range of interest (defined as possible BW agents) and the system quantifies the fluorescence emissions. Software algorithms analyse the size, time history and fluorescence data to determine if the particle fits the criteria for a range of BW agents. Each individual particle is scrutinised (not just batches of particles) giving UVAPS exceptional sensitivity and discrimination. Further refinements to UVAPS technology are being developed for both military and civilian BW environments.

UVAPS applications include continuous monitoring of indoor and outdoor air around critical buildings, facilities and process points (such as mail sorting systems). Suspected BW agent contamination can be gathered, concentrated and passed to automated portable detection systems.

Status
Available and in production. In service with the US Armed Forces.

Manufacturer and developer
TSI Inc.

NEW ENTRY

....

WindTracer® bio-aerosol stand-off detection

Description
This BW agent aerosol stand-off detection system comprises a Doppler radar, signal processing software, and a communications suite, mounted on a lightweight single-axle trailer. The WindTracer® infra-red Doppler radar detects and tracks aerosol levels above normal conditions. For high concentrations, detection ranges in excess of 15 km and detection sensitivity of a few hundred to a few thousand particles per litre is achievable. Wind data is simultaneously and directly monitored using Doppler techniques and early detection of wind shifts result in improved downwind warning. Integration with plume dispersion models provides accurate forecasting of the plume track. The resultant early threat warning and exposure assessment reduces contamination of critical resources.

Typical applications include:
• Air, ground and maritime defence surveillance
• Exposure mapping for post event mitigation
• BW plume detection and assessment following a suspected or actual incident.
The Doppler radar system with its controlling signal processing software allows real-time direct plume detection and prediction as well as remote sensor control. High-resolution meteorological information is incorporated into the model and vector or raster map underlays can be provided improving incident management decision support.

Status
Available.

Manufacturer
CLR Photonics Inc.

VERIFIED

DETECTION (sensor systems) - Chemical

AUSTRIA

NBC sampling set

Description

The Blaschke NBC sampling set is issued in an easily decontaminated aluminium carrying case containing all the various items included in the set. The set involves the use of 10 flexible storage containers with a spring-clamped self-sealing system which can keep the contents safely isolated for periods up to 2,000 hours without problems. Once samples have been placed inside the container the seal is automatically applied and the neck of the container can be further sealed with adhesive tape. The container can then be rolled up and kept closed with an integral strap arrangement.

Also included in the sampling set and separated by a series of disposable foam panels are the following: 10 alcohol spray bottles, a funnel, an expendable shoulder bag, a roll of adhesive tape, 15 markers, an alcohol bottle, 15 plastic and 15 aluminium sample scoops, 20 sampling pouches, 15 sets of tweezers, 15 knives, five sample containers, 30 sample wiping cloths and at least two spare respirator filters and documentation.

Blaschke NBC sampling set 0052130

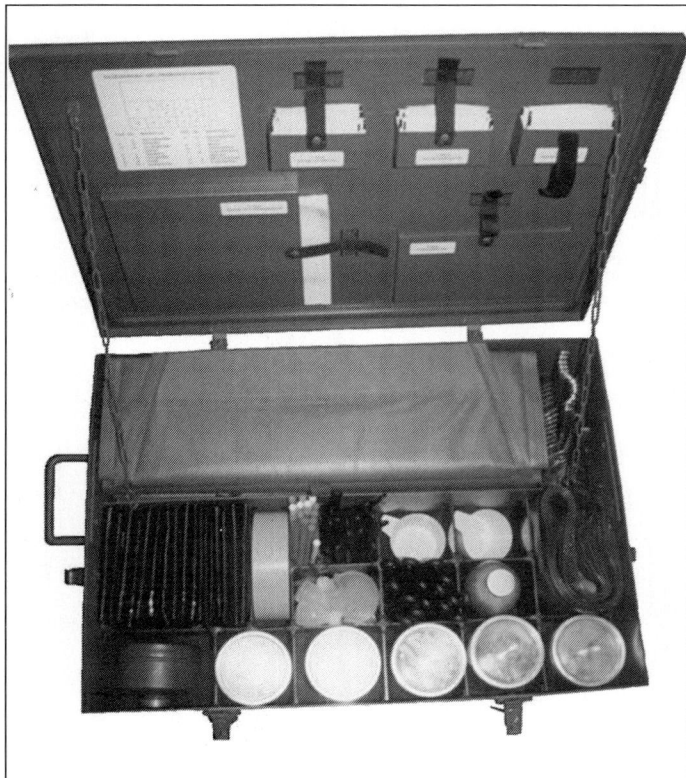

Blaschke NBC sampling set showing contents 0052131

Status

In production. In service with the Austrian Army.

Manufacturer

J Blaschke Wehrtechnik GmbH.

VERIFIED

CANADA

Chemical agent liquid detector papers - 3-way, M-8 and M-9

Description

These chemical agent liquid detector papers were designed to meet the need for a simple and rapid method of detecting and differentiating between the three major groups of liquid chemical warfare agents and can detect G, V and H agents. The material used is basically a dye-impregnated paper sensitive to liquid chemical agents. They are produced either in booklet form (3-way with backing adhesive or M-8 without) or in a dispenser roll (M-9). Colour comparison charts and instructions are provided with the booklets.

To use, a piece of the paper is detached from a booklet or dispenser roll. The paper is exposed to a surface suspected of contamination by a liquid chemical agent, either by wiping the suspected surface with the paper or by attaching it to a surface or piece of equipment which may be exposed at some point. If the paper comes into contact with a liquid chemical agent, it will change colour. The 3-way and M-8 papers exhibit one of three colours to signify the presence of G (shades from yellow to orange), V (very dark blue-green to light blue-green) or H agents (dark red). The M-9 paper develops coloured spots in the presence of any of the agents.

The 3-way papers are issued in booklets containing 12 sheets of paper each measuring 100 × 65 mm. The M-8 papers have similar dimensions but are issued in booklets containing 25 sheets. The M-9 papers are issued in a roll 10 m long contained in a dispenser box measuring 75 × 75 × 60 mm, this has a serrated feedout edge to cut lengths of paper as required and is issued in a sealed bag.

Status

In production. In service with the Canadian and US armed forces.

Manufacturer

Anachemia Canada Inc.

UPDATED

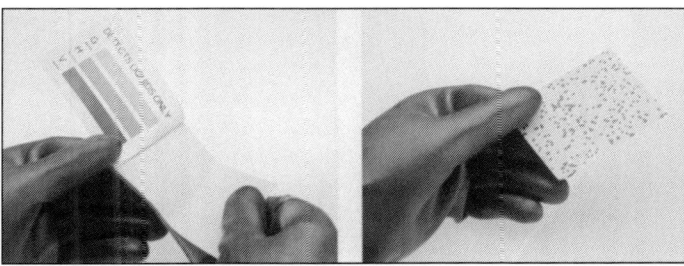

Chemical agent liquid detector 3-way (Anachemia Canada Inc) 0050695

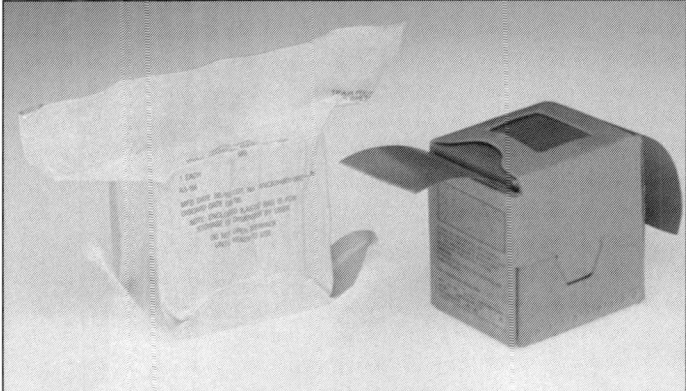

Chemical agent liquid detector M9 (Anachemia Canada Inc) 0050696

Chemical Agent Vapour Detector

Description

Developed by the Canadian Forces, this simple detector is intended for issue down to individual level and is intended to indicate the presence of nerve agent, supporting command decisions to maintain protection or remove protective masks and IPE after a nerve agent attack. The detector is in two parts, both using clear plastic bases. On the main body is a disc of enzyme-impregnated test paper, while a holder has a small sample of a chemically impregnated test paper. In use the paper on the body is moistened and the paper on the holder is pressed against it. If the test paper on the body changes colour to blue or green no nerve agent is present, but if the test paper remains unchanged, nerve agent is present.

Each detector measures 55 × 25 × 2 mm and is packaged in an airtight foil wrap together with an instruction sheet and a silica-gel air dryer pack. The individual units are packed 40 to an airtight, moisture-proof container measuring 95 × 95 × 51 mm.

Status

In production. In service with the Canadian Forces.

Manufacturer

Anachemia Canada Inc.

UPDATED

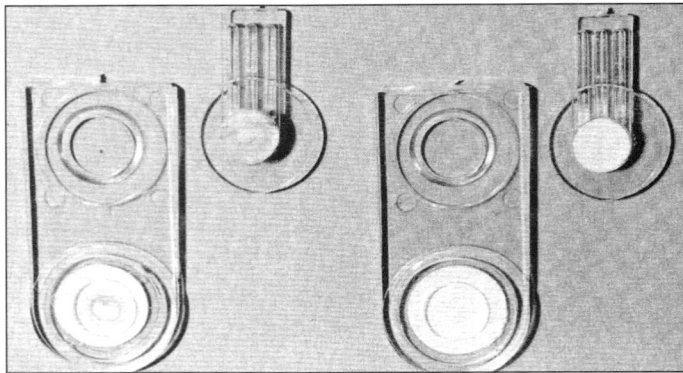

Chemical Agent Vapour Detector (Anachemia Canada Inc)

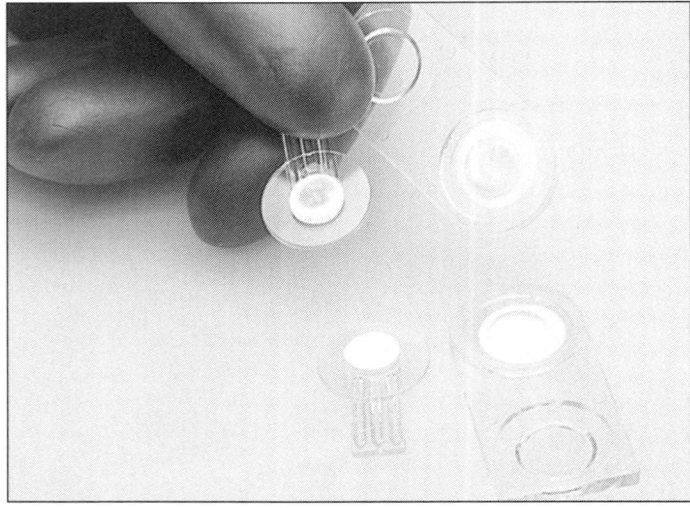

Chemical Agent Vapour Detector - test papers (Anachemia Canada Inc)
2000/0056273

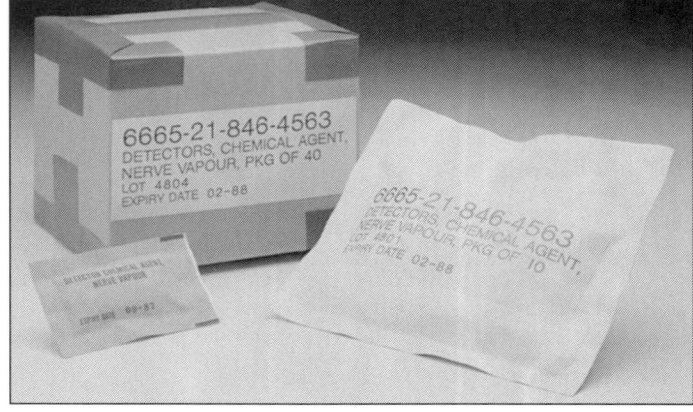

Chemical Agent Vapour Detector - plastic detector paper holders showing no nerve agent present (Anachemia Canada Inc)
2000/0056274

Detector kit, chemical agent (C-2)

Description

This kit is contained in a vinyl-coated carrying case and may be used by one person after only a minimum of training. It has a variety of operational roles including determining the presence or absence of chemical agents after an attack, identifying chemical agents, collecting samples for later analysis, identifying when it is safe to unmask for long or short periods, testing for area contamination, monitoring the expected arrival of a vapour hazard and testing for the presence of agents after decontamination operations. The kit can be used to detect and identify Tabun (GA), Sarin (GB), Soman (GD), V agent (VX), Mustard (H, HN and T), Phosgene oxime (CX), Hydrogen Cyanide (AC), Cyanogen chloride (CK) and Phosgene (CG).

The kit contains: a booklet of chemical agent liquid detector paper – 3-way (see separate entry); 20 detectors; chemical agent; nerve vapour (see separate entry); 30 plain detector tubes; 20 detector tubes with white bands for sampling; three bottles of chemical reagents; an air sampling pump; a container intended to hold 10 detector tubes; an instruction card set; a pencil; water bottle; some anti-freeze solution; report cards and envelopes. The items are all stored in arranged compartments and pockets and the kit is normally carried slung from a shoulder.

The kit is intended for issue down to small unit level.

Specifications
Weight: 1.4 kg
Length: 230 mm
Width: 70 mm
Height: 150 mm

Status

In production. In service with the Canadian armed forces.

Manufacturer

Anachemia Canada Inc.

UPDATED

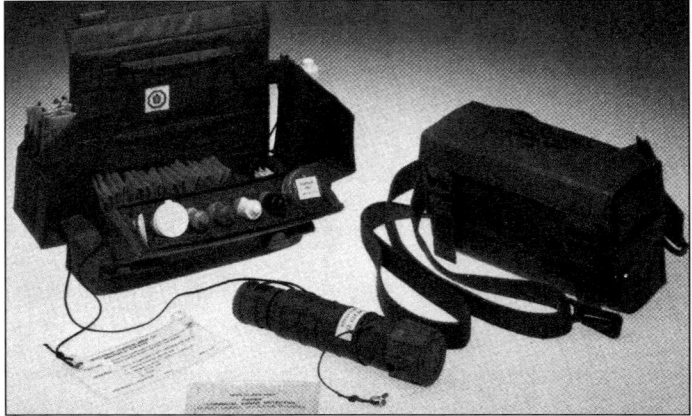

Detector kit, chemical agent (C-2) (Anachemia Canada Inc)

Detector kit, chemical agent, M256A1

Description

The M256A1 chemical agent detector kit replaces the earlier ABC-M15A2 and ABC-M18A2 kits for most field use applications. It is designed to allow troops in the field to determine within 10 minutes whether it is safe to remove protective masks

M256A1 chemical agent detector kit (Anachemia Canada Inc)   *2000*/0056272

and clothing. The kit contains 12 pocket-sized sampler-detectors, a book of M-8 chemical detection papers and instruction cards, all carried in a rugged case.

Each sampler-detector contains an impregnated test spot for blister agents, a circular test spot for blood agents, a star test spot for nerve agents and a Lewisite (L) pellet and marking pad. There are eight glass ampoules, six containing reagents for testing and two in an attached chemical heater. In use, the reagent ampoules are crushed by hand and preformed channels in the plastic sheet of the sampler-detector direct the flow from the ampoules to the appropriate test spot. Simple safe/danger indicators are printed on the back of each sampler-detector and show the colour that each test spot develops if a chemical agent is present. The foil packet enclosing each sampler-detector also contains instructions printed on its outer surface.

The M256A1 kit measures 180 × 130 × 75 mm and weighs 450 g. It requires about 4 hours training to use it with skill. A Training Kit (Simulator) for the M256A1 is available.

Status
In production. In service with the Canadian and US armed forces.

Manufacturer
Anachemia Canada Inc.

UPDATED

A typical cable-connected CADS II sampling station complete with solar cell (Scientific Instrumentation Limited)

DRES Chemical Agent Detection System (CADS)

Description
During the deployment of Canadian Forces to the Persian Gulf in 1990-1991, a requirement for the remote detection of chemical warfare vapours became apparent. At the Defense Research Establishment Suffield (DRES) in Alberta, an initiative was undertaken to develop and produce a system based on the Chemical Agent Monitor (CAM™ - see separate entry under UK in this section for details) already in service with the Canadian Forces.

DRES developed, assembled and field-tested three Chemical Agent Detection Systems (CADS) for deployment in the Gulf. Each system was controlled by a computer capable of monitoring four sampling stations at distances up to 1,000 m. Each sampling station, connected to the computer by cable, contained two CAM units to facilitate simultaneous real-time monitoring of both mustard and nerve agent vapours. Three CADS systems were deployed at Canadian Forces installations in Qatar and Bahrain during October 1990 and a fourth system was deployed at a US airbase in Qatar.

The CADS control station consists of a ruggedised 386 computer, a power supply, a switchable audio alarm and an interface panel, all housed in a shock-resistant aluminium case. The control station power supply sends DC voltage to the sampling stations, thus negating the use of batteries in the CAM units. The computer also takes data from each CAM unit, via an interface card in the computer and displays both the identity of the agent detected and the threat level in a bar display similar to that used on the CAM.

Each of the four sampling stations contains a mounting pole, an interface module, a sun shade and three long stakes. The poles were erected with two CAM units mounted on an adjustable bracket using Velcro straps a short distance below the sun shade. The interface module, connected to the CAM units by a CAM Y-cable, regulates power to the CAM units and transmits serial data from the units to the control station.

The original CADS stations were developed, assembled and field-tested on a tight schedule that precluded the incorporation of technical improvements. When in use it became apparent that CADS would benefit from improvements to reduce the cost of the system while increasing the range of possible uses. In addition, a requirement to support maritime operations was also identified. These factors led

to the joint development and production of a second generation of CADS, designated CADS II, by Scientific Instrumentation Limited of Saskatoon, Saskatchewan.

The CADS II central control unit is a rugged portable unit incorporating a single-board computer running EPROM embedded applications software and a switchable high-level audio alarm. The power supply is capable of operating from 90 to 130 and 200 to 260 V AC (50/60 Hz) and an internal battery allows uninterrupted operation in the event of a power failure. The unit was designed to be positioned remotely at distances up to 4,000 m from the sampling stations. The identity of detected chemical warfare agents and threat level data from up to eight sampling stations, each containing two CAM units, are updated on the real-time display every 4 to 8 seconds in a bar format similar to that used on the CAM.

CADS II has three extra functions controlled by the operator through a three-button keypad. Any of the sampling stations can be turned on and off by the operator. The operator can also set the alarm threshold level of any CAM unit and initiate the operation of new or existing sampling stations.

The CADS II sampling station contains a telescopic mounting pole, an interface module, a low-profile sun shade, three long stakes and an optional radio frequency transceiver. Two CAM units are mounted below the sun shade on a bracket with the interface module and transceiver.

Sampling stations may be connected to the CADS II in any combination of three possible configurations:

1. The sampling station may be connected to the central control unit with up to 1,000 m of cable, as with the original CADS; power is supplied to the CAM units from the central control unit.

2. The sampling station may be connected to the central station with up to 3,000 m of light cable for serial data transmission; power to the CAM units is provided by a solar cell with a rechargeable battery back-up.

3. The sampling station may communicate serial data using radio frequency transmission over distances up to 4,000 m; power is provided by a solar cell with a rechargeable battery back-up.

A marine sensor station to protect the CAM units from salt spray is also available.

Status
In service with the Canadian Forces.

Development agency
Defense Research Establishment Suffield.

Manufacturer
Scientific Instrumentation Limited.

UPDATED

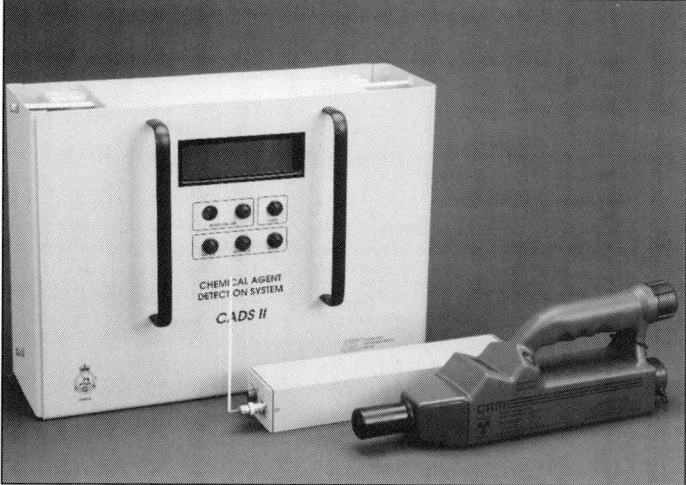

A CADS II control unit with radio frequency receiver (centre) and Chemical Agent Monitor (CAM) on the right

CHINA, PEOPLE'S REPUBLIC

Chemical agent detector M02A

Description
The M02A detector kit is a portable field instrument which samples the ambient air to detect CW vapour. The technology is suspected to be of first generation, similar to the UK NAIAD (see separate entry this section), whereby an air sample is drawn over a reagent which causes an alarm to sound when a nerve agent reacts with it. The chemistry emulates the way nerve agents act on cholinesterase. The output drives a moving coil meter and an audio and visual alarm. The operational time between reagent pack changes.

Status
In production. In service with the Chinese armed forces.

Chemical agent detector M02A (Research Institute of Chemical Defence) 0050693

Specifications
Sample flow rate: 7800-1200 ml/m
Purging airflow: 0.25-0.16 of sample airflow rate
Sensitivity (GB): 1.5 mg/m³. <5 s response time
Sensitivity (VX): 3-4 mg/m³. 5-10 s response time
Operating temperature: −20 to +40°C
Interferents: Herb-burning smokes. Explosion products, engine exhaust, or anthracene smoke are reported as not likely to cause false alarms.
Power supply: 6 type UM/SUM-1, R20 batteries (Chinese designation)
Endurance: >4 h at STP
Weight: <1.6 kg
Size: 200 × 90 × 158 mm

Manufacturer
Research Institute of Chemical Defence.

UPDATED

Chemical agent detector paper, 3-way liquid, adhesive-backed

Description
These chemical warfare agent detector papers are issued in booklet form packed into clear plastic envelopes. They are highly sensitive to droplets of chemical agents and coloured stains appear when agents are present. Indicator strips on the inside cover of the booklet can be used to determine the type of agent involved: G-agents produce an orange to light yellow colour, V-agents produce dark green and H-agents produce red.

These adhesive-backed papers are issued in booklets each containing 10 sheets; each sheet has a protective backing which is removed before use. Each booklet measures 100 × 65 mm.

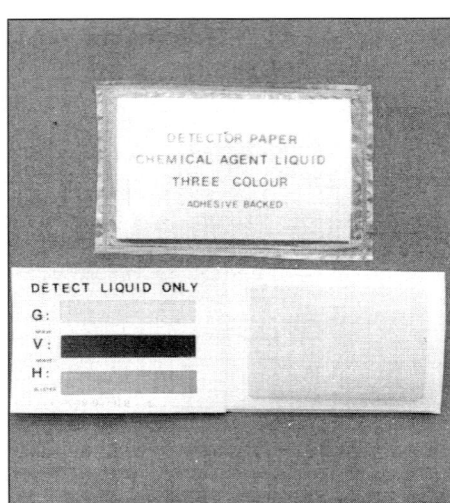

Chemical agent detector paper, 3-way liquid, adhesive-backed (Research Institute of Chemical Defence)

Status
In service with the Chinese armed forces.

Development agency
Research Institute of Chemical Defence.

UPDATED

Chemical warfare agent identification kit, M-75

Description
The M-75 chemical warfare agent identification kit was designed for use by special troops to detect all known chemical warfare agents. Contained in a carrying box equipped with a shoulder strap are a small air pump, racks of glass ampoules filled with various reagents and, some other items including a set of instructions inside the lid that acts as a working tray when opened. In use, an ampoule is placed into the air pump after one end has been broken off. As air is pumped through the reagent in the ampoule, colour changes will take place if chemical warfare agents are present - this change of colour can be used to identify the agent present. The kit contains ampoules to identify nerve agents (V and G agents), mustard (H), Lewisite (L), nitrogen mustard (HN), hydrogen cyanide (AC), cyanogen chloride (CK), phosgene (CG), diphosgene (DP), chloro-acetonephenone (CN) and adamsite (DM).

The M-75 kit weighs 2.6 kg complete. When packed it is 245 mm long, 105 mm wide and 143 mm high.

Status
In service with the Chinese armed forces.

Development agency
Research Institute of Chemical Defence.

Manufacturer
Great Wall Instrument Factory.

UPDATED

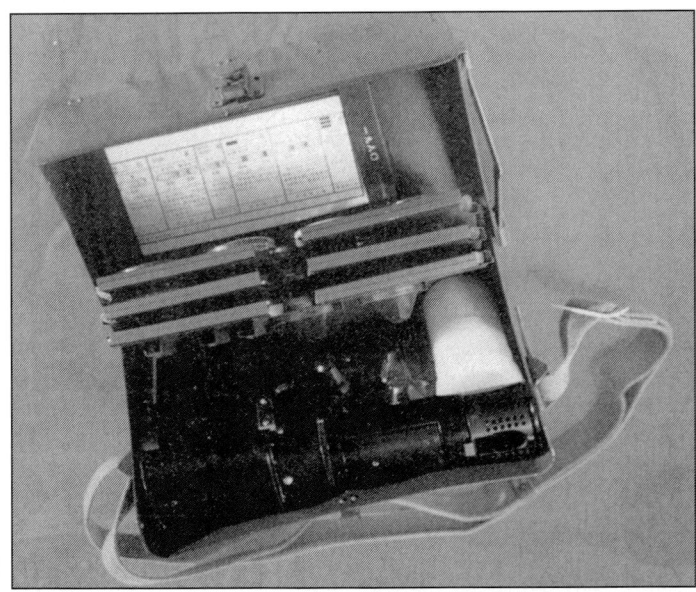

Chemical warfare agent identification kit, M-75 (Great Wall Instrument Factory)

Chemical warfare agent vapour detector kit, M-86

Description
This kit is used at sub-unit level to detect the presence of chemical warfare agent vapours and determine if it is safe to remove protective respirators and clothing after a chemical warfare attack. The M-86 kit uses the puffer principle, in which a small reagent ampoule is placed in the puffer/sampler and the puffer is actuated several times. If the reagent changes to certain colours a chemical warfare agent is present. The exact colour change will provide an indication of the type of agent present.

The kit will detect 0.02 mg/l of nerve agents (VX, GB and GD), 2 mg/l of mustard agents (H and HD), 40 mg/l of hydrogen cyanide (AC) and 20 mg/l of cyanogen chloride (CK).

Each kit is packed into a flat canvas wallet with Velcro-type closures and contains one puffer/sampler, three or four reagent ampoules, chemical agent detector papers and a set of instructions.

The M-86 kit weighs 500 g and measures 160 × 130 × 60 mm when packed.

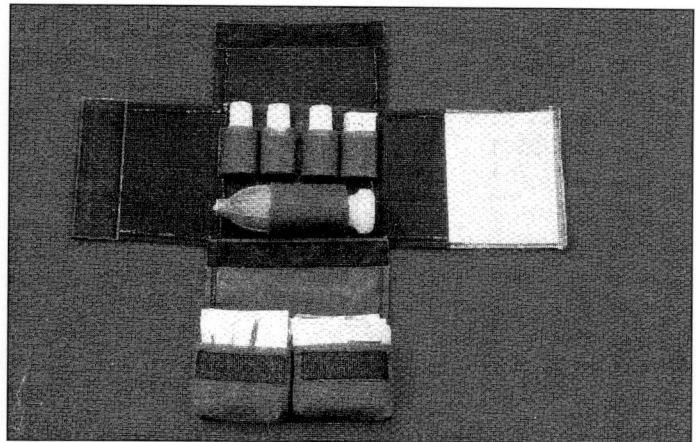

M-86 chemical warfare agent vapour detector kit (Great Wall Instrument Factory)

Status

In service with the Chinese armed forces.

Development agency

Research Institute of Chemical Defence.

Manufacturer

Great Wall Instrument Factory. *UPDATED*

Detector paper booklet, chemical warfare agent, liquid

Description

These chemical warfare agent detector papers are issued in booklet form packed into clear plastic envelopes. Each sheet of paper has an adhesive backing. Two types of booklet are produced. In one, the Type X-1, the paper is coloured dark blue and in the other, the Type X-3, the paper is deep red. Exposure of either type of paper to mustard or nerve agents will result in a noticeable colour change.

Status

In service with the Chinese armed forces.

Development agency

Research Institute of Chemical Defence.

UPDATED

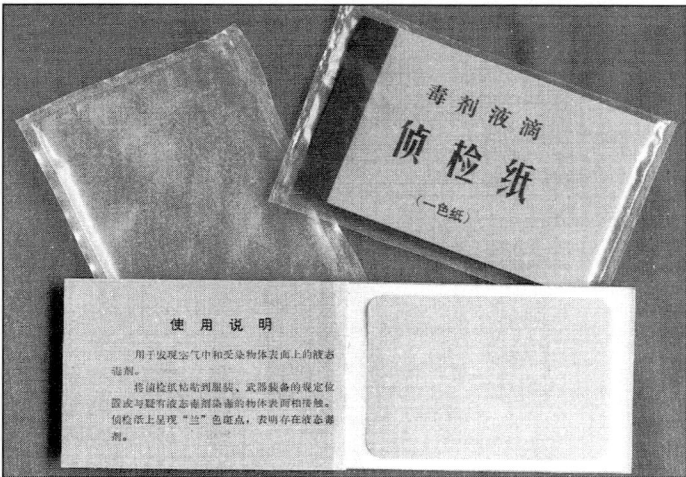

Detector paper booklet, chemical warfare agent, liquid, Type X-1 (dark blue) (Research Institute of Chemical Defence)

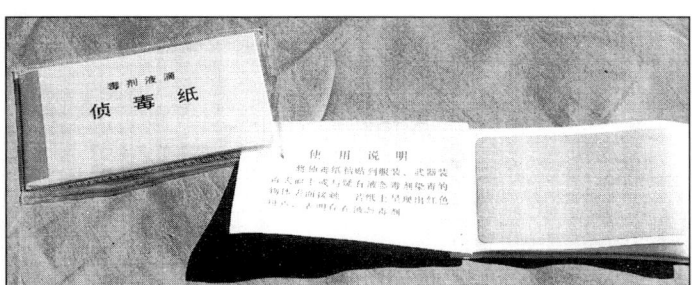

Detector paper booklet, chemical warfare agent, liquid, Type X-3 (deep red) (Research Institute of Chemical Defence)

CZECH REPUBLIC

CALID-3 chemical agent liquid detector papers

Description

CALID-3 chemical agent liquid detector papers are designed to meet the need for a simple and rapid method of detecting and differentiating between the three major groups of CW agents (G, V or H) in liquid form.

The detectors consist of dye impregnated paper which is sensitised to the type of liquid CW agents. They are provided in booklets with adhesive backed sheets and colour comparison charts and instructions.

To use CALID-3 a piece of the paper is detached from the booklet and exposed to a surface suspected to be contaminated by a liquid chemical agent, or to the agent itself by wiping the paper against a suspect surface. If required the paper may be attached to an exposed surface or to the sleeve of an IPE suit, for example. If the paper comes in contact with an agent it changes colour according to the type of liquid agent present.

The booklets are polyethylene sealed. Each booklet contains 12 sheets, measuring 100 × 65 mm.

ORITEST Limited also produce CW agent detector tubes and other detection kits (see separate entries within this section).

Status

In production. In service with the Czech armed forces and some other armed forces.

Manufacturer

ORITEST Limited.

UPDATED

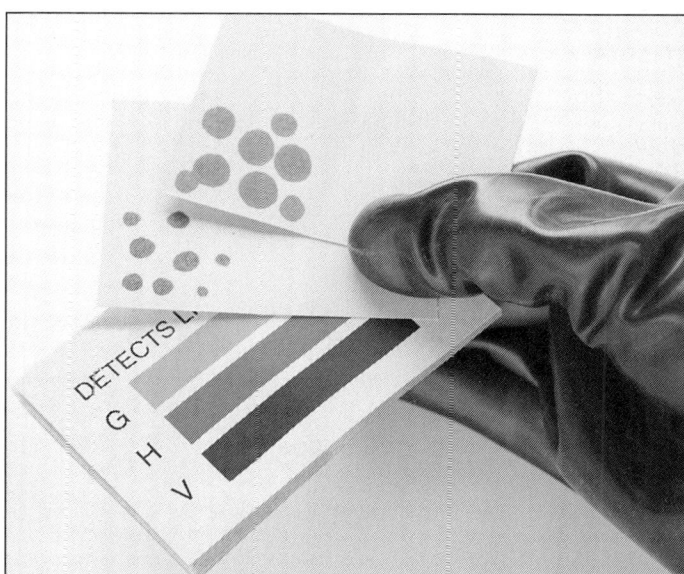

CALID-3 chemical agent liquid detector papers (ORITEST Limited)   0098765

DETEHIT nerve agent detector

Description

DETEHIT was developed to provide a simple means of detecting nerve agents and similar cholinesterase inhibitors. The DETEHIT principle relies on the biochemical reaction of acetylcholinsterase from bovine brain tissues immobilised on cellulose cloth. Nerve agents and associated chemical products reduce the activity of the acetylcholinsterase. The degree of reduction activity depends on the toxicity, dose and time of influence of the inhibitors.

Specifications

Nerve Agent type	Detection limits			
	In air (mg/l)		In water (mg/l) and on surfaces (gm/m²)	
	After exposure			
	2 min	20 min	5 min	30 min
GB	1.10^{-5}	4.10^{-7}	1.10^{-3}	1.10^{-3}
GD	8.10^{-6}	2.10^{-7}	5.10^{-3}	4.10^{-4}
VX	5.10^{-5}	5.10^{-7}	3.10^{-3}	3.10^{-4}
GP	5.10^{-6}	2.10^{-7}	5.10^{-3}	2.10^{-4}
GA	8.10^{-5}	1.10^{-6}	6.10^{-3}	2.10^{-3}
GF	3.10^{-6}	1.10^{-7}	2.10^{-3}	1.10^{-4}

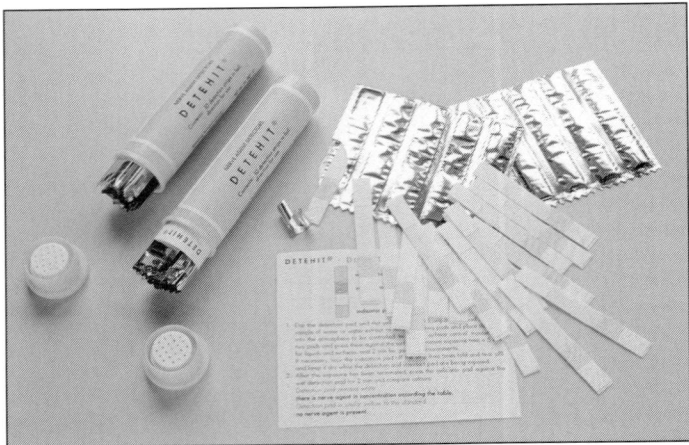

DETEHIT nerve agent detectors and packs (ORITEST Limited)   **2001**/0098764

DETEHIT detectors are issued in strip form, mounted in 3, 5 or 10-strip moisture-proof metal foil packs. They are also available in 10-strip plastic tubes.

Each detector strip has three sections comprising a white detection cloth, a yellow standard section and an indicator paper. In use the detection cloth area and the yellow standard section are either moistened or dipped into water to be sampled. If an air sample is to be taken the strip is held in the air for 1 minute. For surface sampling the detection cloth area is pressed against the target surface. If required, it is possible to tear off the indicator paper section across the perforation area to keep it dry. After exposure the pads may be re-moistened if they have dried off and the detection and indication areas are pressed together for 2 minutes. At the end of that period, the detection cloth area will remain white if nerve agents are present. It will turn yellow if none are present; the yellow standard area of the strip is provided for comparisons. Instructions are provided.

DETEHIT is marketed in Austria by J Blaschke Wehrtechnik GmbH.

Status
In production. In service with the Czech armed forces and some other armed forces.

Manufacturer
ORITEST Limited.

Marketing Agency
J Blaschke Wehrtechnik GmbH (Austria).

UPDATED

ORI-217 detector kit for CW agents

Description
The ORI-217 detector kit for CW agents is a portable field laboratory for the simultaneous detection of all types of CW agent. The kit comprises both the DETEHIT set and a range of CW agent specific tubes for vapour detection as well as a hand-held bellows pump and a battery-powered electrical pump for air sampling.

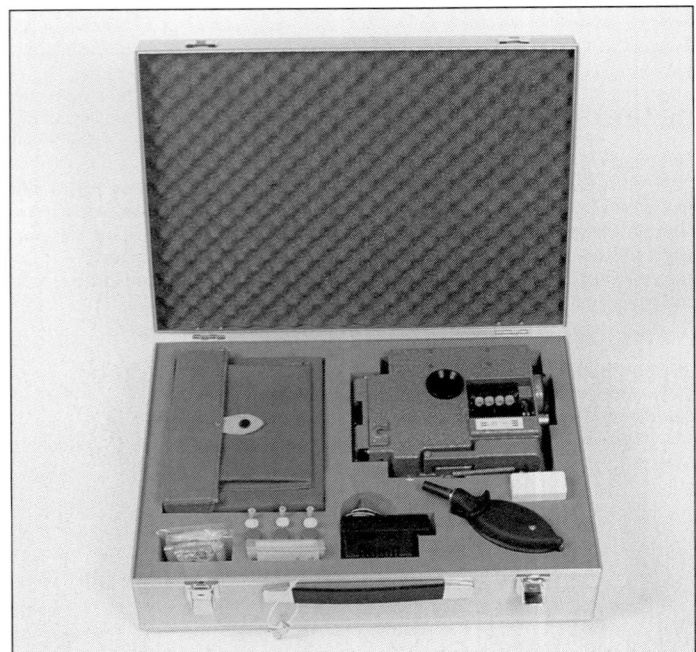

ORI-217 detector kit for CW agents (ORITEST Limited)   **2002**/0098763

CALID-3 booklets, for liquid agent sampling, are also included (see separate entries for all these equipments in this section). All the elements of this kit are housed in a tough transit case.

Status
Production as required. In service with the Czech and Slovak armed forces.

Manufacturer
ORITEST Limited.

UPDATED

ORITEST range of detection tubes

Description
This range of detection tubes is designed to be used with a wide range of commonly issued hand or electrically-driven vacuum pumps. The range offers tubes to detect the full range of known important CW agents. Each tube has a diameter of 6 mm and a length of 93 to 105 mm.
Advantages:
- Wide choice of tubes (BZ available)
- Environmentally friendly - no Hg and Os used in the construction
- Shelf life: 5 years
- Suitable for most types of pump (Draeger, AUER, Yugoslavian types and others)
- Simulant tubes and simulant agents are available for training.

Specifications	
Agent Detected	**Sensitivity (mg/m³)**
G, V	0.01
H, HD, T, Q, HN	0.5
HD	1.0
HN	1.0
L	1.0
CG, CK, AC	5.0
AC, CK	0.05/0.5
CK	0.5
CG, DP, H	3/30
BZ	1.0
CN	0.5
CS	1.0

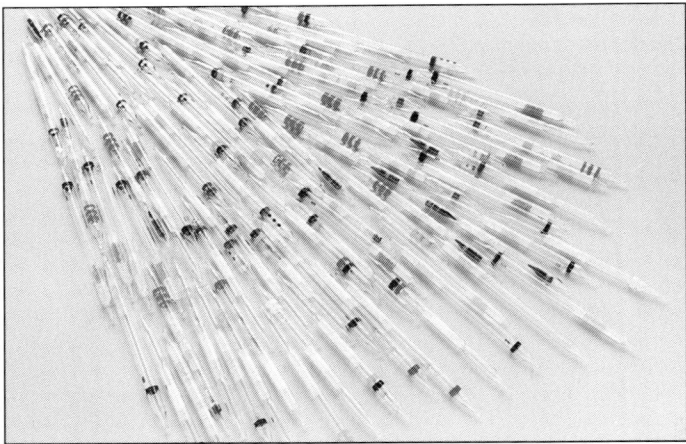

ORITEST CW agent detector tubes (ORITEST Limited)   **2002**/0098766

Status
In production. In service with the Czech armed forces and some other armed forces.

Manufacturer
ORITEST Limited.

UPDATED

DENMARK

Innova 1312 Photoacoustic Multi-gas Monitor

Description
The 1312 Photoacoustic Multi-gas Monitor uses the photoacoustic spectroscopy principle. By choosing and installing up to five appropriate optical filters on an internal filter carousel arrangement, the 1312 can selectively measure up to five component gases and water vapour in any air sample. This offers the facility of being able to detect and measure mustard (H) and nerve (G) agents

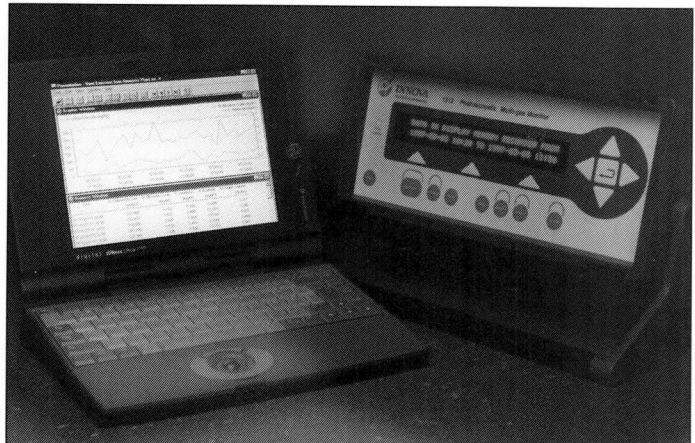

1312 Photoacoustic Multi-gas Monitor (Innova)   **1998**/0011463

simultaneously, in military and civilian locations, including shelters, submarines, command posts and laboratories. The detection threshold of the 1312 is gas-dependent but is typically below one part per billion.

The casing used for the 1312 Photoacoustic Multi-gas Monitor is not weatherproof but it can be mounted in a weatherproof box in external locations. For long-term monitoring, air samples are drawn into the instrument through Teflon tubing from points up to 50 m away. Used with one or two 1303 multipoint dose and sampler units, it can monitor air samples collected from 6 or 12 different locations. The 1312 includes standard PC-based Windows 95™ software which can be used to set up all the parameters, including selection of the gas to measure and the sample integration time. Data collected can be presented in graphic or table formats and the results analysed statistically. The monitor can be used on its own or with a PC.

The 1312 is controlled by push-buttons on the front panel and is user-friendly. Each time a button is pressed a short self-explanatory text appears on a fluorescent display screen to guide the user through each operating procedure. Therefore, it requires very little formal training to operate successfully.

Users can programme the monitor to perform almost any type of monitoring task. Results are stored in the 1312's memory and can be downloaded to the PC storage, display and printout.

The 1312/1303 system can also be used for tracer gas studies such as the measurement of filter breakthrough, air change rates and air movement.

Reliability of measurement results is ensured by regular self-tests which the 1312 performs, to check that it is functioning correctly, as well as by the 1312's ability to compensate any measurement for the interference caused by the presence of water vapour and other known interferents. Calibration and changing the filter papers in the air filtration units are usually required no more than twice a year.

The 1312 Photoacoustic Multi-gas Monitor is 155 mm high, 355 mm wide and 300 mm deep. Weight is 9 kg.

Status
In production.

Manufacturer
Innova AirTech Instruments.

UPDATED

FINLAND

ChemPro100 Hand-held Chemical Detector

Description
The ChemPro 100 is a new hand-held detection and identification system. It is small enough to be used as a personal detector, a monitor for surveying after an event, or a fixed installation detector. It provides continuous operation without the need of expendable desiccant cartridges and is designed for low-life cycle and operating costs. It is the next-generation sensor from Environics, based on the patented Open Loop Ion Mobility Spectrometry (IMS) technology. The ChemPro 100 uses an improved Ion Mobility Cell™, which provides improved selectivity and sensitivity. It is designed to detect all known CW agents as well as Toxic Industrial Compounds/Materials (TICs/TIMs).

The ChemPro 100 weighs less than 1 kg and can be powered by a rechargeable battery pack or 'AA' batteries. The system has an easy to use operator interface, which can be operated single-handed. The user display provides the operator with battery life indicator, concentration bar display, agent class, agent ID, relative time based dose, horn volume level, date and time. The ChemPro 100 stores agent alarm information for retrieval at a later time to provide a historical log of events. It has a memory capability for 50 different chemical libraries with each library consisting of up to 100 chemicals. The user can easily choose and change the library even during the mission.

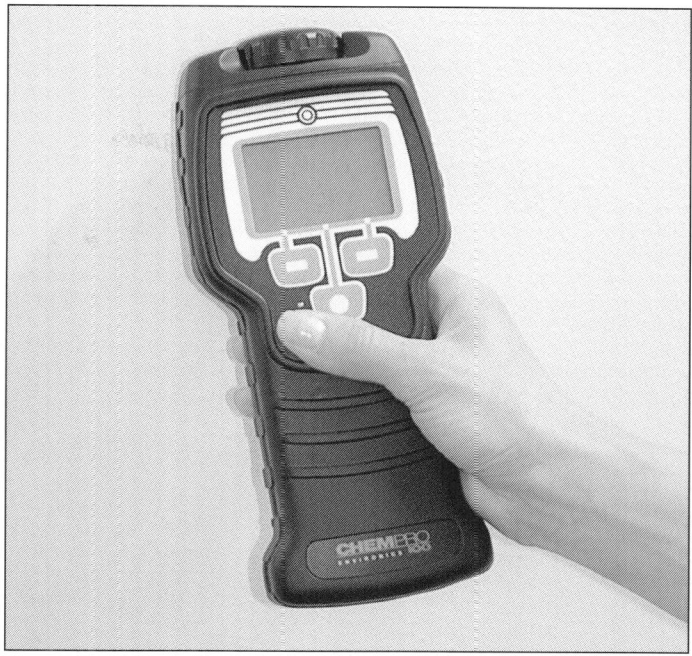

ChemPro 100 Hand-held Chemical Detector   **2002**/0137938

Status
Production commenced first quarter of 2002. Available.

Manufacturer
Environics Oy.

NEW ENTRY

M90 CW agent detection system

Description
The M90 (version M90-D1-C) is a rugged CW agent field detector providing intelligent analysis and flexibility. With a low false-alarm rate, it can recognise all types of nerve, blister, blood and choking agents. Detection is fully programmable, giving the operator complete freedom to decide what to monitor. The M90 uses a combination of improved sensor technology, a patented application of Ion Mobility Spectrometry (IMS), and a new Advanced Signal Pattern Recognition Method (ASPRM). A simplified sensor structure claims to deliver improved discrimination, increased sensitivity and shorter than normal response and recovery times.

The M90 can be used independently as a portable detection unit, as an element in a monitoring network, as a personal detector, a vehicle-mounted detector and for contamination monitoring using intake tubes to pick up the target agent. The same detector may also be used in fixed installations in civil defence and military shelters.

The M90 alarms in less than 10 seconds at all concentrations. Agent concentration is indicated in three (user adjustable) levels for each group of CW agents. RS-232, RS-422 and RS-485 ports allow self-diagnosis of the detector unit, fault logging, updating of CW agent library data and performance adjustments. Data can be downloaded for PC-based analysis or integration with other data as part of a C³I system.

Whilst operational, the M90 detector continuously monitors the ambient atmosphere. Measurements are collected from the IMS sensor and analysed by ASPRM using mathematical algorithms and microprocessor technology.

Operating a chemical agent detector M90 in the field (Environics Oy)   **2002**

Processed measurements are continuously compared with an internal data library of agents. The detection algorithm is modified heuristically.

Besides the standard accessories for field use, including a carrying bag, a Ni/Cd battery and a battery charger, various field network accessories are available. For cable connections up to 1 km, an M90-RH2 Alarm Centre is used for single detectors or an M90-RAC2 Alarm Centre is available to concentrate alarms from four detectors. A flexible local RF network can be set up using the M90-TM transmitter alongside the detector and several M90-PA personal alarms for deployed troops. Accessories for fixed installations include mounting racks, cabinets and control units. The detector can be unshipped from the rack for portable use and shipped just as easily. The control unit provides the same information as the detector. Computerised control centres are also available. Integrated multisensor environmental monitoring systems have been tailored for shelters, CW DEMIL plants and for other purposes.

Standard batteries, mains power unit and vehicle power unit supplies are also available in addition to the rechargeable Ni/Cd battery. A full mission radio-based

training system consisting of a trainer unit and one or more training detectors is available.

The improved M90-C version, released in 1999, offers design improvements including a modified cell (type SC) which increases operational time to 3,000 hours, an upgraded air pump, a new programmable 60-agent library and a much reduced maintenance load. Rate of change of agent concentration level can now be displayed to allow trend analysis, and both sensitivity and selectivity have been increased. The alarm trigger level can be adjusted and the audio alarm muted by the user. Dual-polarity operation has increased the stability of the IMS cell.

Specifications
Weight: 4.7 kg
Length: 280 mm
Width: 105 mm
Height: 280 mm

Status
The M90 and previous models have been successfully used in 15 countries and is in service with the US Armed Forces. The latest model is the M90-D1-C (2001).

Manufacturer
Environics Oy.

UPDATED

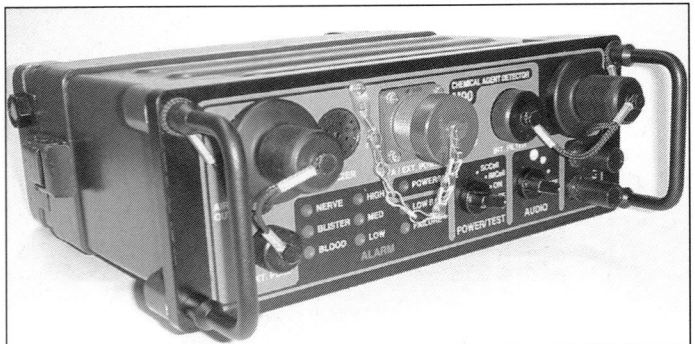

Chemical agent detector M90 (Environics Oy) **2002**/0102796

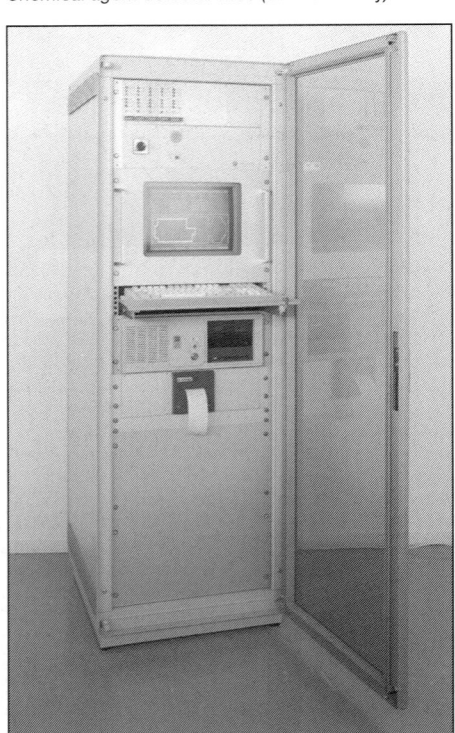

M90-COC personal computer-based control centre for M90 fixed installations (Environics Oy) **2002**

M90-AK mounting cabinet for M90 fixed installations (Environics Oy) **2000**

FRANCE

Giat detection kit for toxic chemicals KDTC

Description
The Giat detection kit for toxic chemicals KDTC is a portable equipment intended for field use to detect toxic agents in the air or on equipment, identify the nature of the toxic agent(s) involved and evaluate the concentration of those agents.

The kit is contained in a hermetically sealed case provided with a bleed valve for storage. The kit includes a manual pump with an integral counter which samples contaminated air, sampling tickets and specific reagents for each toxic substance to assist detection. A sealed lamp is provided for operations at night. The kit also includes various accessories and instructions. A case of spare consumables is available as is an instruction kit containing simulated toxic agents.

The kit can be used to detect agents A, CG, CK, GA and GB.

The KDTC kit enables the following minimum concentration detection levels to be made:

A - 350 µg/m³
CG - 2,000 µg/m³
CK - 2,000 µg/m³
GA - 1µg/m³
GB - 1µg/m³

Status
In production.

Manufacturer
Giat Industries.

VERIFIED

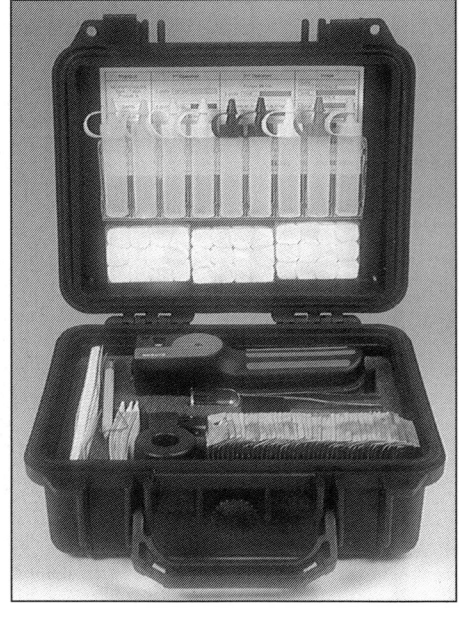

Giat detection kit for toxic chemicals KDTC

Giat NBC adhesive detector paper PDF1

Description
Adhesive detector paper PDF1 is adhesive on one side and can stick to any military equipment surface. The paper is issued in booklets sealed in transparent polyethylene wrapping and in two sizes, 90 × 140 mm and 90 × 25 mm. As with other detector papers, the presence of chemical agents is indicated by the formation of coloured spots. Blister agents will produce red to violet spots, nerve agents yellow to orange spots and blood agents blue-green to black spots.

Status
In service with the French armed forces.

Manufacturer
Giat Industries.

VERIFIED

Giat NBC DET INDIV mle F1 individual nerve agent detector

Description
The DET INDIV mle F1 individual nerve agent detector depends on the inhibition biochemical reaction of cholinesterase principle to detect nerve agents. Each DET INDIV mle F1 unit consists of an 80 × 25 × 8 mm plastic wafer which includes a glass container filled with a buffer solution, a first-paper tablet (white) impregnated with enzymes, a tube allowing the liquid to flow towards the enzyme wafer and another impregnated paper tablet (pink).

In use the detector is removed from its sealed aluminium/polythene sachet and a protective film is removed. The glass container is broken and the first-paper tablet is moistened. The wafer is then exposed to the atmosphere for 5 minutes, after which the first (white) paper is pressed against the second (pink) for 20 seconds. After a further 5 minutes a blue colour on the wafer will denote that no nerve agents are present; a white colour will denote that nerve agents are present.

Within 5 minutes the DET INDIV mle F1 can detect Sarin (GB) at a concentration of 10 µg/m^3, Soman (GD) at 20 µg/m^3 and Tabun (GA) at 50 µg/m^3.

DET INDIV mle F1 units are supplied in sets of five or in packs of 275 or 550. Storage life is three years.

Status
In production. In service with the French Army.

Manufacturer
Giat Industries.

VERIFIED

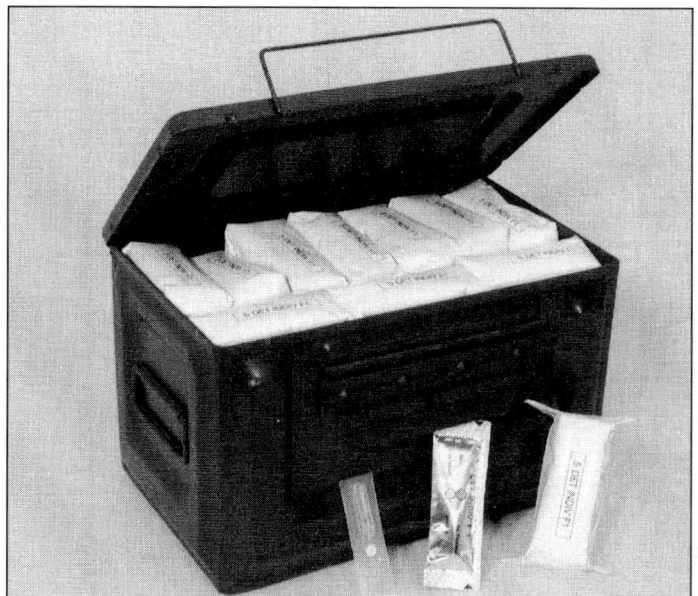

Pack of 550 DET INDIV mle F1 individual nerve agent detectors

Giat NBC toxic agent detection and identification kit

Description
This toxic agent detection and identification kit was designed to detect toxic agents in the atmosphere or on materials, in soil, food, fatty substances and meat.

The kit contains various chemicals, equipment and implements which can be used by NBC reconnaissance or medical units to detect toxic and chemical agents.

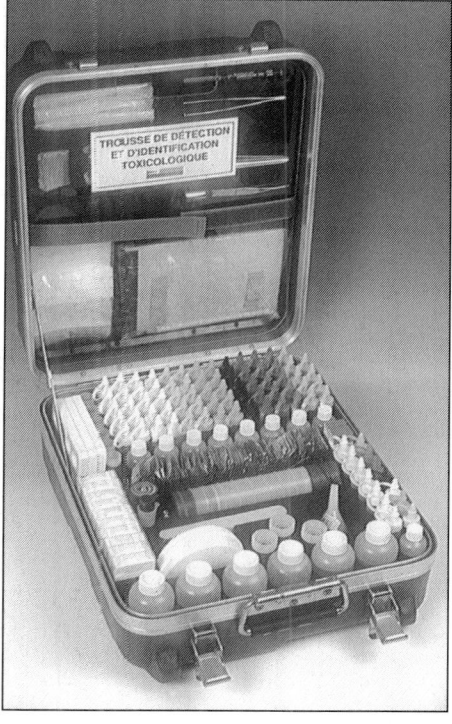

Giat NBC toxic agent detection and identification kit

The complete kit weighs 5 kg and measures 400 × 400 × 150 mm and the contents may vary according to requirements. A training version containing simulants is available.

Status
In production.

Manufacturer
Giat Industries.

VERIFIED

MS/MS DAXEL Analyser

Description
The DAXEL chemical and biological analyser from MGP Instruments is a cabinet-sized CW and BW agent detection and identification system. It can analyse gaseous, liquid or particulate species collected and passed to it by appropriate collection devices. By using double mass spectrometry (MS-MS), it avoids the need for sample separation before analysis, thereby saving valuable seconds in sample processing. It also makes DAXEL highly selective and able to avoid false positive results. Identification is achieved by comparing the fragmentation patterns of the

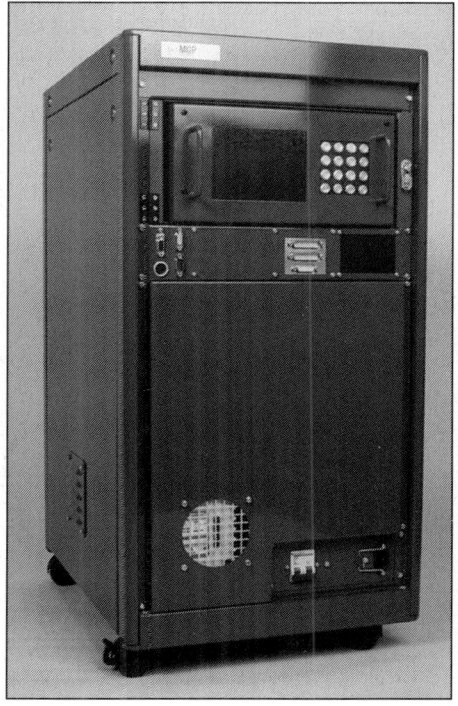

MS-MS DAXEL analyser
00⁻1464

Specifications
Source: Easy-to-change high-luminosity filament electron impact source
Analyser type: MS-MS. Wien filter and 90° energy analyser. High-energy (8 keV) collision system. Daly type high-dynamic detector
Mass range: 12-400 Da
Weight: 85 kg
Dimensions: 610 × 440 × 860 mm
Power supply: 220 V AC. 50 Hz
Power consumption: 600 W (basic version)

collected sample heuristically with comprehensive agent data held in an electronic library. Library data is regularly updated.

Operation is simple and intuitive, allowing users to manage data using the 16-button keyboard and integrated colour screen. The comprehensive screen data presented includes the nature and risks of the detected agent. Data can also be output for external use.

There are a number of options available including a variety of inlet and sampling devices, uninterruptible power supply, transport container and data libraries.

Whilst optimised for defence, DAXEL can be used for environmental, internal security and counter-terrorist surveillance of public places (for example sites used for illegal drugs operations, subways and polluted areas).

Status
Development in progress.

Manufacturer
MGP Instruments.

VERIFIED

PROENGIN chemical agent detector for fixed or mobile use AP2C-V

Description
The AP2C-V version of the PROENGIN chemical agent detector for fixed or mobile use was developed under a DRET-CEB contract with the co-operation of the Centre du Bouchet. It can detect GA, GB, GD, GF, VX and HD agents.

The detector is based on flame spectrophotometry techniques and thus couples rapid detection with a fast return to zero, even after detecting a strong concentration. It is suitable for equipping reconnaissance vehicles which have to provide continuous readings of current concentrations, regardless of the travelling speed and the concentrations encountered. It may be used on any type of armoured or light vehicle and ships. It may also be fitted to shelters and other structures.

The detector is housed in a light-alloy casing and weighs 4 kg. The dimensions are 300 × 147 × 106 mm and the operating temperature range is from –32 to +55°C.

Sensitivity for GA, GB, GD, GF and VX is 5μg/m³ and 400 μg/m³ for HD. Response time is 2 seconds. A detector can operate independently under remote control for 24 hours powered from a vehicle electrical supply, or by using some form of power adapter, with the unit status transmitted via a computer link.

Status
In production. Adopted by the French armed forces. Under evaluation by several other armed forces.

Manufacturer
PROENGIN SA.

VERIFIED

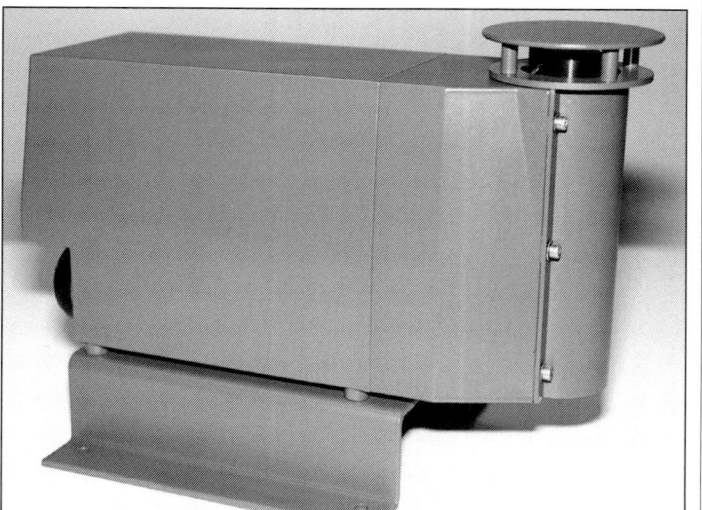

PROENGIN APC-V CW agent detector for fixed or mobile use   **2001**/0097409

PROENGIN chemical detection unit for fixed installations (ADLIF)

Description
Developed by PROENGIN under a DRET-CEB contract with the co-operation of the Centre du Bouchet, the PROENGIN local chemical detection unit for fixed installations (ADLIF) is intended for the detection of GA, GB, GD, VX and HD agents at fixed positions in locations such as shelters, ventilation ducts, outdoor locations and on warships.

ADLIF units operate using flame spectrometry principles and are fed with hydrogen from an electrolysing unit. It operates continuously and automatically, requiring no more care than topping up with 1.2 litres of distilled water every month. A remote alarm is housed in a box to trigger an alarm for a predetermined concentration of agents in the atmosphere. The alarm is triggered according to the type of agent detected (G, V or H). A remote-control unit can be used to check the operation of the ADLIF unit by automatically injecting simulants into the detector.

An ADLIF unit weighs 45 kg and is housed in a watertight cast aluminium housing measuring 850 × 490 × 410 mm. It operates using a 19 to 32 V DC electrical supply and can operate over a temperature range from –25 to +55°C. Sensitivity for GA, GB, GD and VX is 5 µg/m³, and 400 µg/m³ for HD. Response time is 2 seconds.

The ADLIF has an autocheck system and is claimed to have a Minimum Time Between Failure (MTBF) of 10,000 hours.

Status
Adopted by the French Navy. In service with the naval forces of several countries.

Manufacturer
PROENGIN SA.

VERIFIED

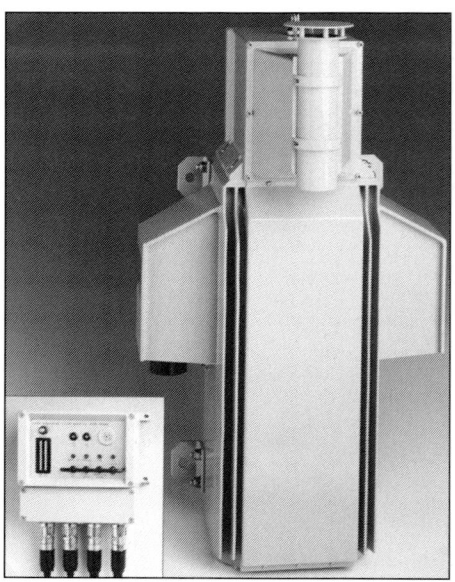

PROENGIN chemical detection unit for fixed installations (ADLIF)

PROENGIN portable chemical contamination monitor AP2C and AP2Ce

Description
The PROENGIN portable chemical contamination monitor AP2C *(appareil portable pour le contrôle de la contamination chimique)* was developed by PROENGIN under a DRET-CEB contract with the co-operation of the Centre du Bouchet. It is a self-contained instrument which operates on the flame photometer principle to detect a wide range of vapours produced by chemical agents, including nerve agents, with a detection response time of less than 1 second for phosphor-based agents. An ultra-fast return to normal sensitivity is claimed, even after the detection of a high agent concentration.

The AP2Ce is a version optimised for use in an explosive atmosphere.

The instrument is contained within a watertight aluminium housing and can be held and operated by one hand. Power is supplied by lithium batteries or an auxiliary electrical supply. A liquid crystal display panel shows simultaneously the total phosphor- and sulphur-based concentrations of any agents present. A device for taking samples allows the detection of persistent agents over the complete operating temperature range. Involved in the detection process is a hydrogen supply carried in a rechargeable metal cartridge. When using the internal batteries and hydrogen cartridge the AP2C has an operating autonomy of 12 hours. Two plug-in points at the rear of the instrument body can be used to plug in an optional auxiliary electrical supply and a set of headphones for aural warnings and indications.

A diagnostic module forms part of the AP2C support and maintenance equipment. It allows internal self-tests to be cross-checked while also checking the electrical parameters, which can be accessed using the diagnostic socket in the

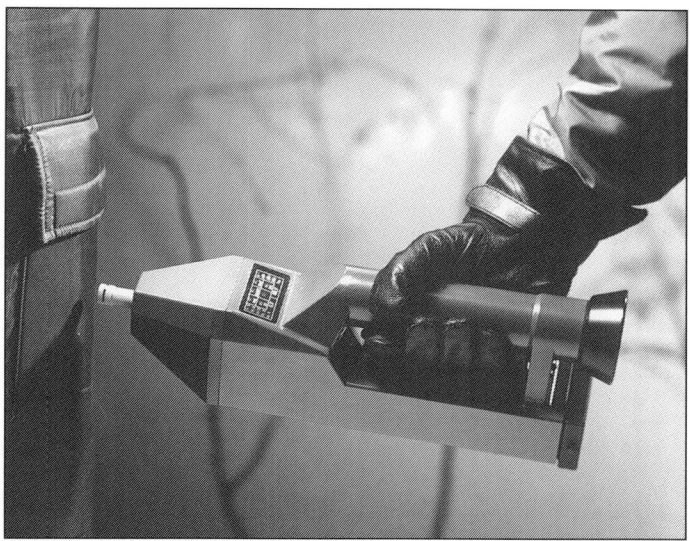

PROENGIN portable chemical contamination monitor AP2C configured for the vapour detection role

AP2C portable chemical contamination monitor configured for the persistent liquid agent role

Specifications
Weight: 2 kg
Length: 350 mm
Width: 85 mm
Height: 125 mm
Operating temperature: –32 to +55°C
Storage temperature: –39 to +71°C
Operating autonomy: 12 h
Response time: <2 s
Sensitivity:
In vapour form:
  (GD) 10 µg/m³
  (HD) 400 µg/m³
In liquid form:
  (VX, GD) 10 µg/m²/15 mn
  (H) 80 µg/m²/15 mn

AP2C battery compartment. When running test sequences, the diagnostic module places the AP2C under various operational configurations and checks the response and adaptation of the unit to these configurations. The diagnostics module is inserted into the AP2C battery compartment, replacing the battery pack.

Status
In production. Adopted by the French armed forces and several other armed forces.

Manufacturer
PROENGIN SA.

Marketing agency
MSA Defense Products.

VERIFIED

PROENGIN portable chemical control and alarm apparatus (APACC)

Description
The PROENGIN portable chemical control and alarm apparatus (APACC) was developed by PROENGIN under a DRET-CEB contract with the co-operation of the Centre du Bouchet. The APACC is an alarm detector for GA, GB, GD, VX and HD agents.

An alarm box (previously called ADAC) rapidly converts the AP2C chemical contamination monitor into an APACC. The alarm box fits into the AP2C battery compartment to form a complete APACC unit. The alarm operates on the given dose flow rate principle (C × T).

The APACC can be used with AP2C batteries or, on a vehicle, with a 28 V DC converter. At any time, the basic AP2C can be reconfigured as a contamination monitor.

An APACC alarm threshold default setting is at a concentration of 200 µg/m³ for G and VX agents for 1 minute and/or 10 µg/m³ of HD for 1 minute. The thresholds are user selectable. The APACC can be reset using a push-button on the independent version and from a control panel on the vehicle version.

The APACC can be deployed clear of the ground on a purpose-designed tripod. An alarm box (rear of the APACC) allows alarm signals to be transmitted by wire to units up to 400 m away.

Alternatively, a remote alarm unit can replace the alarm box, allowing up to 10 AP2C units to feed, by wire, into a network. In this way, all units can be interrogated by the software of a surveillance station pre-installed on the hard disk of a laptop computer.

Status
In production. Adopted by the French armed forces.

Manufacturer
PROENGIN SA.

VERIFIED

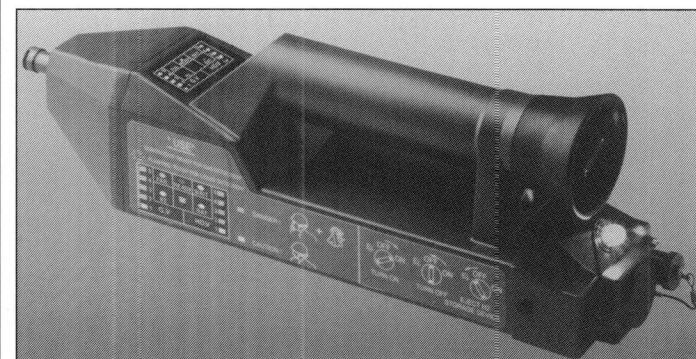

PROENGIN portable chemical control and alarm apparatus APACC

SYDERAL chemical threat warning detection and identification system

Description
The SYDERAL chemical threat warning detection and identification system is intended for the remote detection and identification of nerve and blister agents in vapour or aerosol form. It is a laser-based system intended for autonomous stand-alone applications or for mounting on lightweight vehicles.

SYDERAL uses a frequency-agile carbon dioxide laser which provides detection and identification of agent concentrations by the spectrally dependent absorption of the agents involved. The system can operate by day or night and involves the automatic acquisition of usable reflective topographic targets in the field. Automatic 360° surveillance is then achieved by a series of measurements on all targets using several wavelengths. When an agent is detected the system undertakes a confirmation routine before audible and visual warnings are provided for the operator. In addition, an automatic report is provided through an NBC communications network. If required, a manual operation mode is available.

SYDERAL has a vapour detection range of between 3 and 5 km; the range for aerosols is 1 to 2 km. The detection probability claimed for both cases is 0.95 and the false alarm rate is 1 in 120 hours. The system utilises a modular design for easy maintenance and there are built-in test facilities.

Status
Under development.

Manufacturer
Thales.

VERIFIED

GERMANY

Chemical Biological Mass Spectrometer (CBMS) Block III

Description

The CBMS Block III is a highly reliable detector for CBW agents. Both types of agent can be monitored concurrently in the air and CW agents can be monitored on surfaces. The CBMS can be operated manually or automatically and the data management system allows agents of biological origin to be detected as well as known CW agents. Unknown CW agents can be analysed and their data signatures included in the database. No liquid consumables are required as the system uses pyrolysis mass spectrometry. The CBMS Block III features a rugged design, making it suitable for mobile use in the field, in compliance with US Army MIL-STDs. A special inlet and vacuum system shortens start-up times: to 20 minutes from −19°C, for example. The system allows the minimum of false alarms. The operator interface is easy to use, comprising a touch screen, and the system can be integrated into sensor networks at fixed installations or onboard vehicles and ships. In the CBW agent air-monitoring mode, the system is capable of unattended operation.

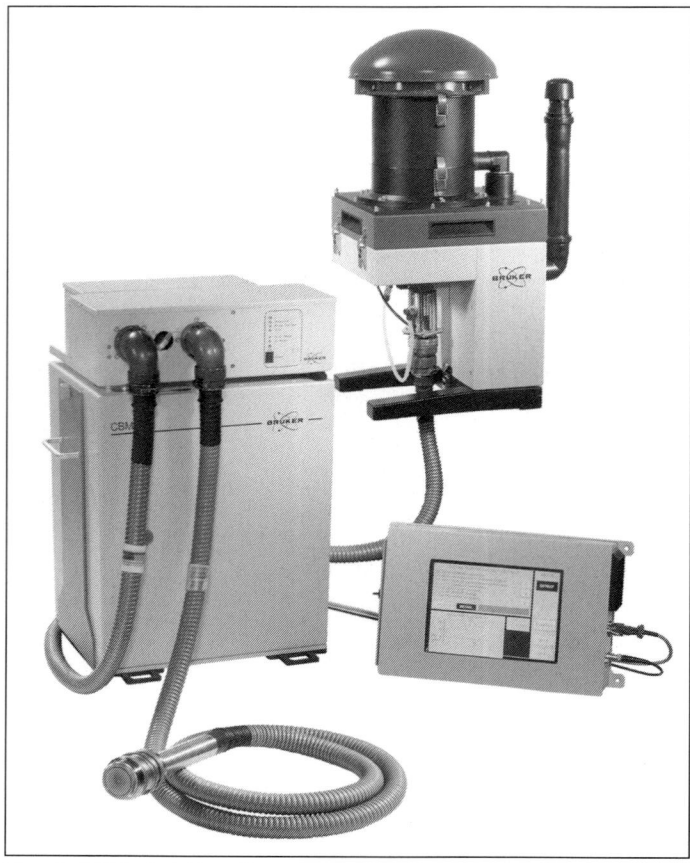

Chemical Biological Mass Spectrometer (CBMS) Block III (Bruker Daltonics®)
***2002**/0137482*

Screen snap from the Chemical Biological Mass Spectrometer (CBMS) Block III software (Bruker Daltonics®)
***2002**/0137483*

Specifications

Dimensions: 650 × 470 × 350 mm
Weight: 65 kg
Power supply: 24 V DC at 25 A (40 A max)
Mass range:
    10-450 amu (manual mode)
    45-250 amu (automatic mode)
Scan speed: 28,000 amu/s
Ion getter pump: 20 ions/s
Max pyrolysis temp: 600°C

Detection of:

Agent type	Detection limits (mg/m³)
GB	0.04
GD	0.04
VX	0.02
HD	0.07
L	1.4
Bacteria spores	100 ng (manual mode)
	1 μg (automatic mode)

CBMS control software

The CBMS control GUI displays all information relevant to the operator in one screen. Fast access to all operating modes is provided through the six function keys on the right of the display (see illustration). Upon detection of a CW agent, the CMBS Block III immediately alarms and displays the result of the identification. For BW agents the system displays a status message every three minutes. The graphics display can be selected to show the chromatogram or line spectrum view, giving a broad indication of system performance. Error self-checking is continuous and the operator is alerted to faults and suggested remedies on screen.

Status

In production. Standard CBMS in service with the US Army. Incorporated into the FOX NBCRS programme (see separate entry under Detection (Reconnaissance systems)).

Manufacturer

Bruker Daltonics®.

UPDATED

EM 640 mobile mass spectrometer

Description

The EMS 640 is a mobile analysis facility which uses a ruggedised one-piece quadrupole directly coupled GC/MS system. It has a large dynamic range (see below), is designed to be readily transportable, being contained in four tough transit cases. The system is ready for use within 30 minutes of arrival at a site. Mounted in vehicle, ship or aircraft, it can maintain continuous monitoring whilst in motion. The GC/MS unit outputs data to a ruggedised PC running specialised analysis software which is 'blinded' or proofed against tampering. This ensures

EM 640 mobile mass spectrometer (Bruker Daltonics®)   ***2002**/0137484*

high confidence in the use of EM 640 data for verification purposes by OPCW or other inspectors. Other features of EM640 include:

- Specially designed mounting unit allows continued running without external power supply and during transfer of EM640 between different modes of transport: for example from land vehicle to helicopter.
- Flexible and quick re-configuration for newly assigned analysis task. Modular peripherals and easily changed GC ovens, primary modules and sample introduction systems (secondary modules).
- Simple software GUI allows confident use by non-technical personnel.

System components
- Membrane inlet system
- Ruggedised one-piece quadrupole
- Two filaments
- Ion getter pump
- Ultra high-vacuum system
- Instrument can be left switched off for lengthy periods
- Large dynamic range
- Mass range m/z 1-64
- Temperature range 0-50°C
- 24 V DC, max 650 W
- Shock tested 1g; 1-100 Hz.

Data system
- Pentium processor ruggedised PC
- Multitasking operating system, simultaneous data acquisition and identification
- NIST and other customer specific libraries
- Bruker proprietary operational analysis software designed for simple accurate use by non-technical personnel.

Status
In production. In service with OPCW inspection teams.

Manufacturer
Bruker Daltonics®.

UPDATED

··

MM-1 mobile mass spectrometer

Description
The Mobile Mass Spectrometer MM1 is a ruggedised GC/MS-System designed for field use. More than 600 systems are deployed in units such as the Fuchs (Fox) NBC Reconnaissance Vehicle (US, UK, Germany, Saudi Arabia), the French VAB RECO and the South Korean K216 and K316 NBCRS (see separate entries under Detection (Reconnaissance Systems)).

Different samplers can be user-exchanged, allowing the system to respond to a range of different hazards in samples of contaminated air, soil or water. The sensitivity and range of hazardous agents capable of detection by the MM1 were both increased in 1996. The following improvements and additions have also been made, aimed at integration into the improved German FUCHS NBC reconnaissance vehicle.

- GC Sampler. The Gas Chromatographic Sampler (GC Sampler) is equipped with a 12 m DB5 capillary column. This allows better separation of mixtures of organic chemical substances by use of an automatic temperature program. The

compounds are absorbed onto sampling tubes via the Air Sampler or the Surface Sampler and afterwards desorbed in the injector of the GC Sampler. The GC Sampler is fixed in a special mounting on the left panel of the sensor unit. It can easily be exchanged for the Air/Surface Sampler by the operator.

- **Air Sampler.** The Air Sampler comprises a time-controlled electrical suction pump and is mounted on the inside of the NBC pack at the rear of the vehicle. Contaminated air continuously streams through the inlet system of the Air Sampler while the pump is running. The chemicals are absorbed onto special tubes which are fixed into the tube holder. Afterwards the tube holder together with the tube is inserted into the injector of the GC Sampler.
- **Surface Sampler with remote control.** This works in a similar way to the Air Sampler but in conjunction with the Air/Surface Sampler (probe). The probe head is heated up to 260°C and pushed onto contaminated surfaces. Chemicals are vaporised and sucked in through the heated probe line by a pump. At the end of the probe all organic compounds are absorbed on special tubes which are inserted into the tube holder of the Surface Sampler. Afterwards the tube holder together with tube is fixed into the injector of the GC Sampler. The remote control can be mounted anywhere inside the vehicle.
- **Transputer interface.** The Transputer interface, which is mounted in the electronics cabinet, allows the MM1 to be remotely computer-controlled. Spectra can be continuously acquired and transferred to the computer. Realistic software simulation programmes can be utilised for operator training in the vehicle. Different values for the transmission rate, up to a maximum of 115 kbaud, can be programmed or set up through the hardware.
- **Central Data Station.** The Transputer Interface allows MM1 system data to be integrated with other systems in the vehicle such as the ASG 1, SVG 2, AN/VDR2, GPS and so on. The MM1 workstation table allows a standard computer keyboard to be fitted. Training can also be conducted using notebook computers fixed into a docking station. The picture shows the original Data station and monitor of the improved German Fuchs system.
- **Sampling tubes.** Commercial sampling tubes made of glass (such as Dräger tubes) are considered unsuitable for military use and Bruker has developed universal sampling tubes, filled with a special absorbent material and charcoal. The sampling tubes are used to collect and 'fix' chemicals with the Air Sampler or the Surface Sampler. Special tubes pre-filled with a mixture of simulants are provided for testing the condition of the GC Sampler. The tubes are stored in airtight cartridges which can also be used to safely retain samples for later laboratory analysis.

The improved MM1
The improved version of the Mobile Mass Spectrometer MM1 incorporates further enhancements to the performance of the detection system, including a ten-fold increase in sensitivity. A central data station is connected to all of the detection units in the vehicle as shown in the picture.

Status
In production. In service with France, Germany, South Korea, Thailand, UK and the US Army.

Manufacturer
Bruker Daltonics®.

UPDATED

··

RAID series of CW agent detectors
(Rapid Alarm and Identification Device)

Description
RAID comprises a range of user-friendly, reliable fixed and portable CW agent detectors which operate on the IMS principle. Equipped with automatic alarms, there are variants designed for both personnel and contaminated area CW agent monitoring. The detection library includes TICs and the instruments are ruggedised for field operations in vehicles, ships, fixed shelters and for hand use.

General design features
- Design and materials conform to relevant DEFSTANs
- Compact single-tube design with automatic polarity switching
- Fully microprocessor-controlled
- Remote display and control option (Remote unit/PC).

Detection performance
- Continuous autonomous air monitoring for the presence of CW materials. Visual and audible alarm
- Identifies type and concentration level of both 'classic' CW agents and selected TICs
- Sensitive: low detection limit
- Short response and recovery times
- Low false alarm rate
- Prevention of saturation
- Re-programmable for newly discovered CW hazards
- Equipped for simulant operator training.

RAID-1 versatile hand-held CW Agent Detector
Automatic polarity switching allows the RAID-1 to detect continuously all militarily significant CW agents. A built-in microprocessor evaluates the recorded ion mobility spectra. The data on agent identity and calculated concentrations are

The MM-1 mass spectrometer, showing the various components of this rugged, MS vehicle-mounted system   **2002**/0137481

Specifications

	RAID-1	RAID-S
Dimensions	75 × 165 × 180 mm	405 × 500 × 130 mm
Weight	2.6 kg	12 kg
Power consumption	10 to 32 V PC	6 to 32 V PC
Data interfaces	RS 485, RS 232, RS 422 (optional)	
Temperature range	−50 to +75° C (storage)	
	−25 to +55° C (operation)	
Humidity	all conditions	
Drying filter lifetime	approx 800 hrs of operation	approx 9000 hrs of operation
MTBF	>1,000 hrs	
MTTR	<10 mins (operator level)	<10 mins (operator level)
	<45 mins (direct support)	<45 mins (direct support)
Training (to proficient operational level)	2 hrs	

	RAID-M	RAID-E
Dimensions	400 × 115 × 165 mm	Details supplied when available
Weight	2.9 kg	
Power consumption	12 to 24 V PC	
Substances detectable		
CW agents	VX, GA, GD, GF, GB, HD, HN, L, AC	
Test substances	GS (DPM), HS (MS)	
TICs	CL$_2$, CG	

shown on the LCD display. Sophisticated software algorithms prevent false alarms from the typical battlefield environment and from decontaminants or other commonly used chemicals. Built-in protection against detector overload at extremely high concentrations is included.

All known threat agents are covered in the detection library and additional agent parameters (TICs, for example) can be added on request. User updates for the RAID-1 library are available by download from a secure site. The new spectral data are simply programmed into the detector unit via a PC. The RAID-1 can be operated in two different modes:

- Continuous operation with automatic alarm function
- Intermittent operation, which allows higher sensitivity measurements.

RAID-1 can be used by troops in the field and by monitors inside COLPRO or, attached to vehicles, as a deployable area CW monitoring facility. The system includes realistic training via the instrument's ability to respond to non-toxic simulants.

Status
Available.

RAID-S (Rapid Alarm and Identification Device - Stationary) fixed trace-gas detector for continuous monitoring operations
The RAID-S is a trace-gas detector for the detection of CW agents and other toxic substances. The detection unit of this instrument is based on the RAID-1 IMS drift tube. It is aimed at long-term unattended operations (maintenance interval one year). It can be used independently or linked in series to a data highway, providing remote coverage for a larger site via a management and alarm system located in a central control room. Roles for RAID-S include:

- Environmental air monitoring of ambient air for hazardous compounds in buildings, shelters, warehouses or chemical plants
- Mobile CW agent detection in vehicles, ships or aircraft
- COLPRO monitoring. Providing an alerting capability to any breakdown in COLPRO integrity.

Status
Available.

RAID-M (Rapid Alarm and Identification Device - Monitor) hand-held CW agent monitor
The RAID-M is a CW agent detector based on the principle of IMS. It is lightweight and designed to be operated whilst being held with one hand. It comes with a carrying sling, but it can also be installed on a vehicle mounting. It is used to monitor the contamination of personnel or equipment in the field and within collective protection facilities. Other applications of the detector include anti-terrorism and counter-proliferation measures as well as real time monitoring of toxic and hazardous industrial chemicals in the environment. The RAID-M is able to detect, classify, quantify and continuously monitor concentration levels of the CW agents specified. The identity of substances detected is either indicated by class "G" or "H", or the specific agent or simulate identity is displayed. Hazard levels are indicated by an incremental hazard level display with five or eight increments. The detector sends out an audible alarm when identifying any hazardous agent. The RAID-M provides very low detection limits and short response times. Due to the state of the art measuring cell and highly sophisticated software, nerve agents can be detected at miosis level. Short recovery times are achieved by an automatic purge mode.

Status
Available.

RAID-E (Rapid Alarm and Identification Device - Enhanced) man-portable detector
The RAID-E is a portable version, designed to offer unattended enhanced detection sensitivity in the field where extremely low concentrations of agent can affect performance. Trace levels of CW agents and selected TICs can be detected. This is particularly important in the case of nerve agents where trace levels of agent can significantly affect performance of intricate tasks by causing miosis. The system can also be mounted on vehicles.

Status
Available 2002.

Manufacturer
Bruker Daltonics®.

UPDATED

RAPID stand-off detector

Description
RAPID is a highly reliable infra-red stand-off detector for CW agent clouds. All known CW agents and important TICs are automatically monitored.

The lightweight system can be mounted on vehicles, ships and helicopters and performs real-time field screening whilst under way. It can also be installed as a stand-alone device. The sensor, scanner, electronics and control unit are integrated in one compact housing.

The RAPID is based on the proven Bruker RockSolid™ flex-pivot interferometer and is resistant to mechanical shocks, vibrations, humidity and extreme temperature differences. It is hardened for field operations in harsh and rugged environments. Features of the RAPID include:

- Easy handling
- Robust and compact design
- Low weight
- Minimal power consumption

RAPID stand-off detector showing resilient supports - for use with vehicle mounting (Bruker Daltonics®)
2002/0137486

Specifications

Sensor module: FTIR (patented RockSolid™ interferometer)
Dimensions: 500 × 331 × 386 mm
Weight: 27 kg
Power consumption: 80 W
Environmental range
  −50 to +75°C (storage)
  −30 to +55°C (operation)
Humidity range: 0 to 95%
Shock and vibration: approved to MIL-STD 810D
Spectral range: 700 to 1,300 cm⁻¹
Spectral resolution: 4.0 cm⁻¹
Scan rate: 15 spectrals at 4 cm⁻¹ spectral resolution
NEDT (Noise Equivalent Delta Temperature): Better than 0.15 K for one scan with a resolution of 4 cm⁻¹ and a mirror speed of 160 kHz, depending on the detector
Maximum speed of rotation
  Azimuth: 120°/sec⁻¹
  Elevation: 20°/sec⁻¹
Field of view: 90 mrad
Search field
  Azimuth: 360°
  Elevation: −10 to +50°
Vehicle mounting
  Dimensions: 789.5 ×285 × 445 mm
  Weight: approx 9 kg
Remote Display and Control Unit
  Dimensions: 174 × 185 × 70.5 mm
  Weight: 1.2 kg
Data interfaces: RS 422 or Ethernet

- Low detection limits
- Fast measurement and alarm
- Continuous monitoring whilst in motion.

The elements of the RAPID remote and display unit are identical to those integrated into the housing of the RAPID. The unit is connected via an RS 422 interface.

RAPID Control Software

The Windows NT™ based GUI displays all relevant information in one window. Fast access to operating modes is provided through function keys. The RAPID gives an alarm immediately after detection of any target compound. Additionally, the PC software graphically visualises the cloud's position, extension and direction. The database includes all known CW agents and important TICs. New chemical agents can be added to the detection library. NBC reports according to ATP 45 can be automatically generated and sent via data interface if the system is connected to a GPS.

Status

In production.

Manufacturer

Bruker Daltonics®.

UPDATED

Rheinmetall Landsysteme NBC field laboratory

Description

The Rheinmetall Landsysteme field laboratory is a comprehensively equipped mobile detection and analysis facility contained in three half-sized air-mobile customised ISO containers. A fourth container contains crew support facilities.

Specifications

Detection/analysis: Gas chromatography. Mass spectrometry (capillary chromatographic apparatus, ion trap. Spectral photometer (water analysis). X-ray fluorescence spectrometer. Fourier transform IR spectrometer. High performance ion meter
Sample collection: Air sample, dust, water, gas
Communications: VHF, digital telephony (including ISDN), SatCom

RAPID with tripod
(Bruker Daltonics®)
***2002**/0137487*

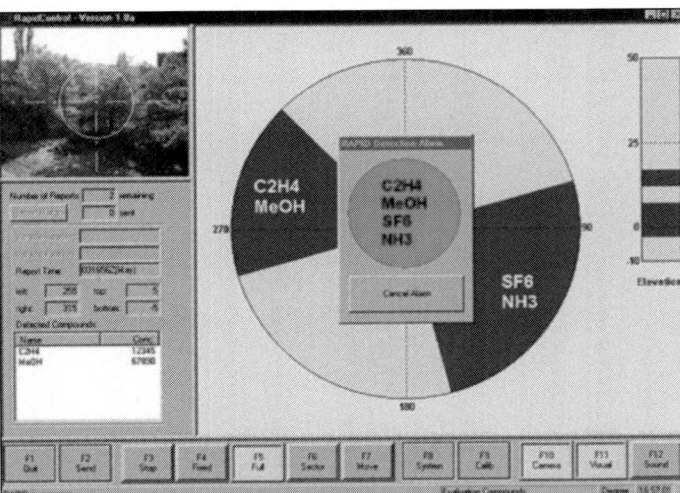

Screen snap showing direction and concentrations of the hazard to the user via the RAPID management software (Bruker Daltonics®)   ***2002**/0137488*

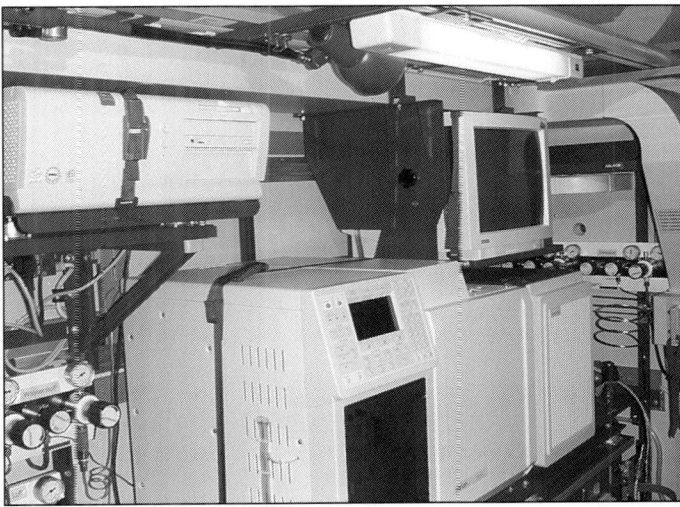

View into the Chemical Analysis Laboratory   ***2000**/0084047*

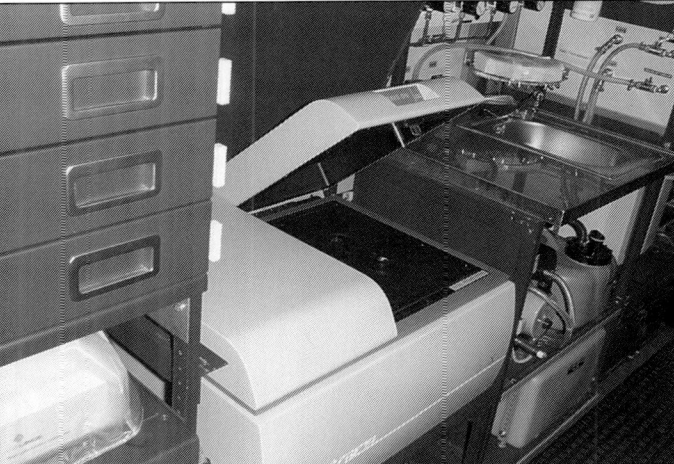

The X-ray fluorescence spectrometer   ***2000**/0084048*

Mounted on two standard German army 5-ton lorry-trailer units, the entire laboratory can be airlifted or driven to unit HQ areas to provide NBC agent detection and analysis. Co-ordinated with data retrieved from deployed Fuchs NBC reconnaissance vehicles (see entry under Detection (reconnaissance systems)), the facility offers rapid command decision support or on-site verification. NBC data is integrated in a suite of high-performance IT based data management and communications equipment. The CW suite comprises the equipment listed below. See under Detection (sensor systems) - nuclear and biological for details of the N and B suites.

Status
Available.

Manufacturer
Rheinmetall Landsysteme GmbH.

UPDATED

HUNGARY

Chemical agent sensor GVJ-2

Description
The chemical agent sensor GVJ-2 is a portable hand-held instrument able to detect nerve and blister chemical warfare agents and various toxic industrial materials. The device automatically monitors the air for the presence of agents with air analysis being carried out in an ionisation cell under the control of a microprocessor. Ion mobility spectroscopy techniques are employed.

The GVJ-2 has a sensitivity of 0.00005 mg/litre for GB, GD and VX and 0.002 mg/litre for HD. The response time in both instances is 5 seconds.

The GVJ-2 has only two operator controls, an on/off push-button and a mode change button. The two operating modes employed are a single measurement mode in which only one sample of air is pumped into the system, analysed and the results are then displayed, while in the periodic measurement mode the same sequence is repeated at preselected time periods.

Results are displayed on a liquid crystal display panel using an array of eight blocks. For instance, a heavy concentration of nerve agent will result in all eight blocks being displayed, lesser concentrations will produce fewer blocks. The display can also be used to provide a battery low indication (BAT. LOW.) and an indication that the usual self-test after switching on has resulted in a component failure (TEST. FAIL).

The GVJ-2 is 400 mm long over its longest point and weighs approximately 1.8 kg with four Ni/Cd batteries, which provide at least 6 hours of continuous use and (typically) 10 hours of more normal use. Its operating temperature range is from -25 to +55°C with a relative humidity up to 95 per cent.

Status
Under development by Hungarian state factories before privatisation. Current manufacturer unknown.

Manufacturer
MoD Institute of Military Technology.

VERIFIED

Chemical agent sensor GVJ-2

Chemical contamination detector VSJ-1

Description
The chemical contamination detector VSJ-1 is used to detect any chemical contamination of vehicles, terrain and other military equipment and during effectiveness checks following decontamination operations. The device uses fluorescence techniques and consists of three sub-units: a reagent system, an electronic module and a sensor head. The reagent system contains the reagent solution and a pump. The electronic module includes the lamp housing, the detector unit and the signal processing unit. A common fibre light is used to guide the excitation and detection light. All supply cables are linked into the sensor head which contains a nozzle for the reagent spray.

The system weighs 35 kg and has a response time of 10 seconds. Power is taken from a 24 V supply. The operating temperature range is from −10 to +50°C.

Status
Available from Hungarian state factories before privatisation. Current manufacturer unknown.

Manufacturer
MoD Institute of Military Technology.

VERIFIED

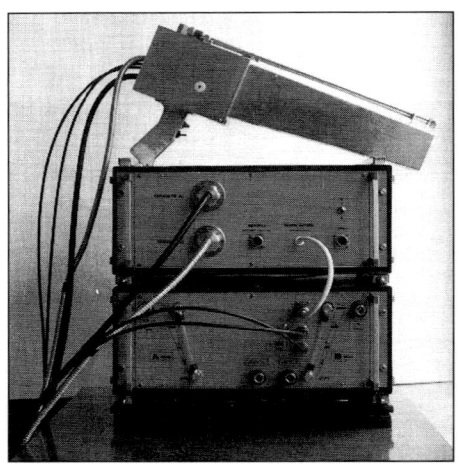

Chemical contamination detector VSJ-1

Double way chemical agent indicator patch

Description
The double way chemical agent indicator patch is intended for use in a number of ways to act as a monitoring device to indicate the presence of chemical agents. Described as a 'stripe', the patch has a self-adhesive backing that allows it to be secured to uniforms, protective clothing, vehicle windscreens or metal and other surfaces. The patch has two separately treated indicator fields that can indicate the presence of V, G, H and L agents by a colour change. The exact colour change will indicate the nature of the agent which can be indicated by reference to a colour chart provided in the centre of the patch. By using two separately treated indicator

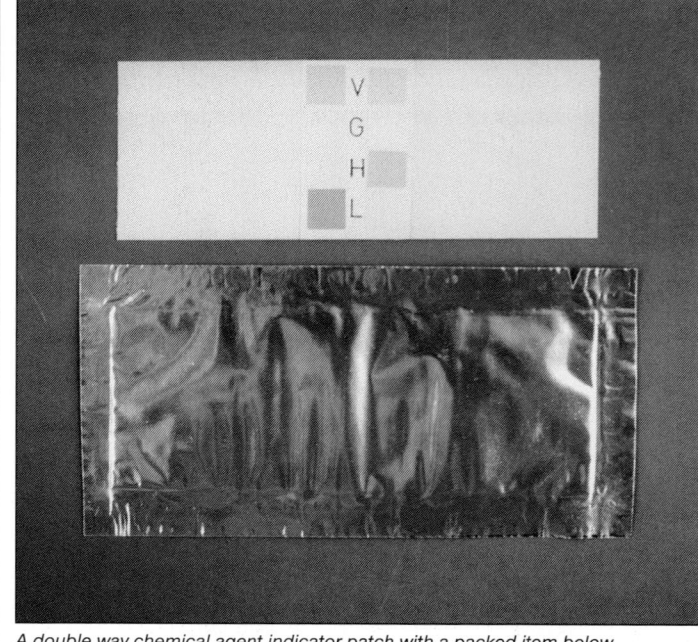

A double way chemical agent indicator patch with a packed item below

areas spurious indications such as those caused by fuels or explosives can be largely avoided.

For issue each patch is packed into an individual aluminium foil bag.

Status
Available from Hungarian state factories before privatisation. Current manufacturer unknown.

Manufacturer
MoD Institute of Military Technology.

VERIFIED

Portable field chemical laboratory TVL-63

Description
The portable field chemical laboratory TVL-63 is used for the detection and identification of numerous chemical substances in soil, water and other materials. The laboratory is carried in a suitcase-type container carried by one person and measuring 800 × 600 × 300 mm; weight is 27 kg. The suitcase holds 235 items of laboratory glassware and other items, a 50 g chemical balance and an alcohol burner as well as 83 different chemicals. The equipment can be used for filtration, distillation, extraction and other chemical processes to detect different chemical agents, provide quantative analysis of samples, detect alkaloids and verify decontamination processes.

The shelf-life of the laboratory is five years.

Status
Available from Hungarian state factories before privatisation. Current manufacturer unknown.

Manufacturer
MoD Institute of Military Technology.

VERIFIED

Remote chemical agent sensor VTB-1

Description
The remote chemical agent sensor VTB-1 is a field-deployable laser radar which remotely detects chemical agents.

The measurement principle employed is differential absorption spectroscopy. The system has two continuous wave radio frequency excited CO_2 waveguide lasers as the power source. The lasers are tunable to approximately 40 lines in the wavelength region between 9 and 11 µm. Each element in the atmosphere has a characteristic adsorption spectrum so the laser wavelengths may be chosen, in order that the wavelengths emitted are in coincidence with the features of this spectrum as high and low adsorption levels. Thus differential adsorption is measured to identify atmospheric pollutants.

The system is double ended, in that it measures the average or integrated concentration between the lasers mounted in an air conditioned truck and a 1 m² tripod-mounted sandblasted aluminium retro-reflector positioned up to several kilometres away. It is possible to monitor large facilities by erecting an optical 'fence' around them using multiple reflectors. For instance an industrial area could be monitored using several separate paths and traversing the laser system through 360° horizontally and 15 to 20° vertically.

Optical heterodyne detection is applied for the measurement of backscatter signals. The minimum measurable light intensity is around 10 to 20 W, with a bandwidth of only 1 Hz. One laser can be the source for both the local oscillator and

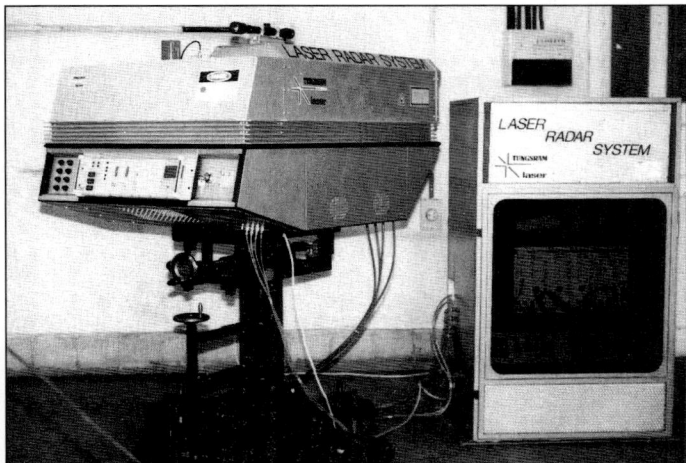

Remote chemical agent sensor VTB-1

the transmitter for one channel so that the optical path is self-controlled. The required frequency shift is produced by FM modulation.

The optical arrangements are Michelson-based, having great sensitivity with normal alignment procedures. A cassegrain-type telescope is used for both the transmitter and receiver parts of the system, which consists of very few optical components for ease of use and reliable operation.

The processing electronics consist of a state-of-the-art low-noise preamplifier followed by a band pass filter. The output laser beam is chopped which allows further processing of the lidar echo using lock-in techniques. Full automatic data acquisition is used for both real-time interpretation or storage for later use.

The system can detect all G and V vapours with concentrations of 130 mg/m² at ranges between 250 and 5,000 m.

Status
Under development at Hungarian state factories before privatisation. Current manufacturer unknown.

Manufacturer
MoD Institute of Military Technology.

VERIFIED

Type 66-M chemical reconnaissance set

Description
The Type 66-M chemical reconnaissance set is the Hungarian version of the RFAS and former Warsaw Pact PKhR series of chemical agent and identification kits and is carried and used in much the same way. While the Type 66-M is similar to the PKhR series in many ways and uses an almost identical hand pump to that used with the PKhR-54, the internal layout and contents of the set differ.

The Type 66-M is operated by one person and can be used to detect up to five different chemical agents simultaneously. Detection time for each agent is 1 to 2 minutes. A heater is provided for the detector tubes when the temperature falls below +10°C and a filter disc is used when detecting chemical agents in smoke samples.

The Type 66-M carrying case contains a torch and a silica gel container to protect the contents. The weight is 3.2 kg. Also provided is a carrying strap. A supplementary kit is available for the replenishment of consumable items.

The detector tubes used with the Type 66-M can detect VX, CN, DM, L and BZ. Detector tubes containing simulants are produced for training purposes.

Status
Available from Hungarian state factories before privatisation. In service with Hungarian armed forces. Current manufacturer unknown.

Manufacturer
MoD Institute of Military Technology.

VERIFIED

Type 66-M chemical reconnaissance set

Type FVJ chemical agent detector

Description
The Type FVJ chemical agent detector is intended for use on vehicles and uses an internal electrically driven air pump to continually draw air through up to four chemical agent detector tubes. The presence of a chemical agent will be indicated

A selection of detector tubes used with the Type FVJ chemical agent detector

Type FVJ chemical agent detector with carrying case

by a change in colour of one or more of the detector tubes. The colour involved will indicate the type of agent involved. At temperatures below +10°C the detector tubes may be heated. A drawer at the bottom of the unit can hold spare tubes. The detector tubes available can detect HD, CG, AC, CS, VX, CN, DM, L and BZ. Those containing simulants are available for training purposes.

The Type FVJ requires 10 minutes of preparation before use and thereafter can be left unattended, other than viewing the detector tubes through a transparent panel on the unit access door. It is powered from a 24 V DC battery or a vehicle 24 V power supply. Maximum power consumption is 2 A. The detector weighs 6.5 kg and measures 346 × 240 × 135 mm. When not in use it is stored in a wooden carrying case.

Status
Available from Hungarian state factories before privatisation. In service with Hungarian armed forces. Current manufacturer unknown.

Manufacturer
MoD Institute of Military Technology.

VERIFIED

ISRAEL

CDK chemical detection kit

Description
The CDK chemical detection kit is a simple, light and reliable chemical detection kit, which was designed for ease of handling by non-specifically trained personnel after only a few minutes of instruction. It employs two separate samplers which provide identification of nerve or blister agents by specific colour reactions.

The CDK consists of two plastic cases, each containing four seal-wrapped detection samplers for either nerve or blister agents respectively. In addition there is a plastic adaptor for mounting on a respirator filter canister. Operating instructions are included in each kit case. The samplers are carried on the canister and are removed to test for the presence of either nerve or blister agents, in either a vapour or aerosol state, by pressing the sampler ampoules and releasing reagent onto an impregnated pad. An entire test cycle takes from 6 to 10 minutes (depending on temperature) and positive detection of agents is provided by clearly visible discolourations of the paper.

Nerve agent detection limits are 2 to 5 µg/litre of air for GB and GD, 6 to 12 µg/litre for GA and 8 to 15 µg/litre for VX. The sensitivity limits for HD are 2 to 4 µg/litre of air.

CDK chemical detection kit

A CDK nerve agent sampler is 78 mm long, 39 mm wide and 11 mm high. Weight of a sampler is 11.1 g or 33.4 g with the adaptor.

A CDK mustard agent sampler is 80 mm long, 48 mm wide and 11 mm high. Weight of a sampler is 12.5 g or 34.8 g with the adaptor.

All CDK test reagents have a guaranteed shelf-life of five years at +37°C.

Status
Available. Approved for use by the Israeli Defence Forces.

Manufacturer
Israel Institute for Biological Research (IIBR).

VERIFIED

ROMANIA

Chemical agent detector kit

Description
This chemical agent detector kit relies on the same hand-operated air pump and glass indicator tube principles as other similar kits used by the former Warsaw Pact (and other) nations. The kit may be used to detect virtually all known chemical warfare agents (including BZ) by placing up to six indicator tubes inside a hand pump assembly and pumping air through. A maximum of 12 pumping strokes is usually sufficient to produce a colour change in one or more of the indicator tubes to denote the chemical agent(s) involved. The kit may also be used to detect the presence of agents on food, soil and in water.

The kit contains a small heater for the indicator tubes as the kit is intended for use over a temperature range of from –30 to +40°C. Detection papers are also included in the kit.

The kit weighs 2.5 kg complete with its carrying case.

Status
Available.

Marketing agency
ROMTEHNICA.

VERIFIED

Nerve agent alarm ASTN-2

Description
The nerve agent alarm ASTN-2 automatically provides optical and acoustic signals following the detection of nerve agents in the surrounding atmosphere. The signals are provided only when a predetermined nerve agent concentration has been exceeded. Warning signals can be provided from a remote location via a cable if required and the device can be mounted in a vehicle, in which instance an anti-vibration base is used.

The ASTN-2 employs a sampling air flow of approximately 1 litre/min and an alarm can be provided within a maximum of 20 seconds from detection. The normal autonomous operating cycle is 24 hours. Power is supplied from a battery or from a vehicle power supply between 12 and 24 V. Maximum power consumption under normal climatic conditions is 0.8 A in the sampling mode and 1 A in the alarm mode.

*Nerve agent alarm
ASTN-2*

Total weight of the system is approximately 8 kg of which 2 kg is the anti-vibration base. Dimensions, with the base, are 245 × 210 × 225 mm.

Status
In production. Has been offered for export sales.

Marketing agency
ROMTEHNICA.

VERIFIED

RUSSIAN FEDERATION AND ASSOCIATED STATES (CIS)

Chemical agent detector GSA-1

Description
Few details are available regarding the chemical agent detector GSA-1 other than that it is designed to detect the nerve agents GA, GB and VX. It appears to employ some form of ion mobility spectrometry, although this has yet to be confirmed.

Status
In service with RFAS armed forces. Offered for export sales.

Marketing agency
Rosoboronexport.

VERIFIED

*Chemical agent
detector GSA-1*

GO-27 radiation and chemical agent detector system

Description
The GO-27 radiation and chemical agent detector system (PRKhR) is intended for installation in armoured vehicles and provides continuous monitoring for the three NBC hazards (see under Detection (sensor systems) - nuclear).

Status
In production. Offered for export sales.

Manufacturer
Signal Instrument Making Plant JSC.

Marketing agencies
Kintex.
Rosoboronexport.

VERIFIED

Model GSP-11 automatic nerve agent detector alarm

Description
The Model GSP-11 differs mainly from the Model GSP-1 in that it can detect G-type and V-type nerve agents; the radiation detection element has been removed. The same basic tape and reagent system as that used in the Model GSP-1 is retained but the reagent uses an enzyme-inhibition reaction using two reagent solutions. This enzyme-inhibition reaction is temperature sensitive so the interior of the Model GSP-11 is maintained at between 28 and 38°C by a thermostatically controlled heater operated by external batteries in a battery case.

The GSP-11 has two sensitivity ranges, E1 and E2. The E1 is the most sensitive and operates at intervals of 60 to 80 seconds. This range allows the GSP-11 to remain in operation for 2 hours on one reagent filling and each operating cycle lasts between 22 and 26 seconds. The air flow through the system using the E1 range is between 0.7 and 1 litre/min. On the E2 sensitivity range the display interval lasts between 5 and 8 minutes. This allows the unit to remain operational on one reagent filling for 10 to 12 hours. Each operating cycle lasts between 1.5 and 2.5 minutes and the air flow is reduced to 0.5 to 0.7 litre/min.

Specifications
Weights:
  (control unit) 12 kg
  (alarm unit) 0.5 kg
  (battery case) 15 kg
  (indicator set) 2 kg
Total weight: 29.5 kg
Warm-up times:
  (0°C) 1 h
  (−40°C) 3 h
Time taken to refill reagent: 10 min
Operating temperature: −40 to +40°C
Refills in indicator set: 3
Voltage: 12 V
Heat output: 120 to 140 W
Number of battery sets: 2
Battery life:
  (+20°C) 6 h
  (−40°C) 1 h

Status
Available from Russian state factories before privatisation. In service with RFAS and some former Warsaw Pact armed forces in 1995. Current manufacturer unknown.

Manufacturer
Unknown.

VERIFIED

Model GSP-11 automatic nerve agent detector alarm

Models GSP-1 and GSP-1M automatic chemical agent and radiation detector alarms

Description
These detectors can be used to detect G-type nerve agents and nuclear radiation and to provide visible and aural alarms in the presence of both. The GSP-1 and GSP-1M, which are basically similar, provide detection and alarms only and do not determine the nature of the chemical agent or the intensity of the radiation.

The nerve agent detection system consists of a tape transport system, a reagent dropper system, an air system and a photodetector system. The tape transport system and the reagent dropper system are interconnected and operate in 5 minute cycles separated by a 5 to 8 second cycle change, both controlled by a clockwork mechanism. At the start of each cycle, reagent is dropped on to the tape which is then moved on so that the fresh drop lies under the photodetector system. An electric pump draws air into the detector system via a screening filter and over the drop on the tape. The reagent changes colour if there is any G-type nerve agent in the air and the photodetector system compares the drop colour with a reference formed from a reference beam. This is formed in different ways on the two models: the Model GSP-1 is derived from splitting the source beam and shining part of it on to an unmoistened part of the tape; and on the Model GSP-1M, the split light source is passed through a system of filters. In both cases the intensity of the two beams is compared and any change from the norm initiates the alarm systems. Air is drawn into the detector at 1.5 litres/min and on the Model GSP-1 is then passed directly out of the system whereas on the GSP-1M it is used for interior cooling before being passed out through a grille. The detector alarm contains enough tape and reagent to run for up to 8 hours. Electrical power is provided by two rechargeable batteries and the system is designed to operate at temperatures from –30 to +40°C.

The nuclear radiation detector system uses a halide tube detection unit. It initiates the alarm system as soon as the radiation level reaches in excess of 0.1 r/h. The alarm horn and a flashing lamp are both contained inside the case and operate both from nerve agent and radiation detection.

Without the batteries both models weigh 10 kg; with them, the weight is 18 kg. The dimensions are approximately 450 × 300 × 150 mm.

Status
Obsolescent. Available from Russian state factories before privatisation. In service (1995) with RFAS and some former Warsaw Pact and Middle Eastern armed forces. Current manufacturer unknown.

Manufacturer
Unknown.

VERIFIED

Preparing a GSP-1M automatic chemical agent and radiation detector alarm for use

Models MPXR and MPXL field chemical laboratory

Description
The Model MPXR field laboratory can be used to detect and identify numerous chemical substances in soil, water and other materials and to analyse water and other similar substances. The equipment in the laboratory can be used for numerous chemical procedures including filtration, distillation, extractions and the like. The laboratory is contained in a hinged box which opens into compartmented sections. The box holds higher quantities of the items found in the Model PKhR series of kits (see separate entry in this section), with the addition of numerous items of laboratory glassware and other hardware including a chemical balance, an alcohol burner and writing materials.

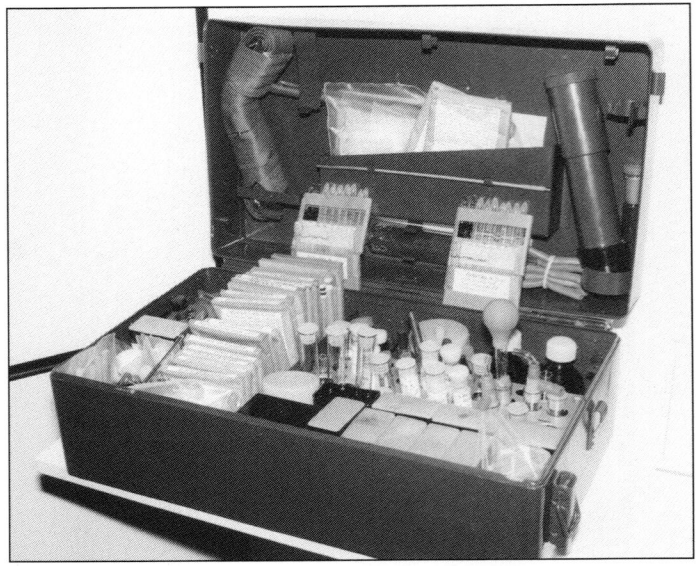

MPXL field chemical laboratory, the smaller version of the MPXR (T J Gander)

The MPXR container dimensions are 564 × 282 × 459 mm. It weighs 35 kg fully loaded.

A smaller kit is known as the MPXL. This is generally similar to the MPXR but contains fewer items to suit it for its more portable field role. The MPXL weighs 7.5 kg and its container case measures 427 × 242 × 162 mm.

Status
Available from Russian state factories before privatisation. In service with RFAS armed forces. Has been offered for export sales. Current manufacturer unknown.

Manufacturer
Unknown.

VERIFIED

Models PPKhR and PPChR semi-automatic chemical agent detection and identification kit

Description
This kit is carried on various types of radiological and chemical reconnaissance vehicles such as the BRDM-1 RKh. Basically it is a Model PKhR series kit with an electrically driven rotary air pump in place of the cylindrical hand pump and an electric heater replacing the indicator tube heater. Both these components are driven by the vehicle battery. The East German version was the Model PPChR, a modified Model PChR-54U kit.

Status
Obsolescent. Programme (1995 onwards) to replace with later models such as PKhR (see separate entry this section). Available from Russian state factories before privatisation. In service with the RFAS and former Warsaw Pact armed forces. Current manufacturer unknown.

Manufacturer
Unknown.

VERIFIED

PKhR series of chemical agent detection and identification kits

Description
This series of chemical agent identification and detection kits has been under development since the late 1940s and has undergone gradual changes over the years. The kits are in service in several forms, all using the same basic operating processes. Some were also produced in the former East Germany while recent versions are still in production in Bulgaria. Some of the earlier kits are now obsolete but are still in use for training and by some former Warsaw Pact nations. The following list summarises the main types:

RFAS designation	Former East German designation	Comments
PKhR-50	n/a	Obsolete. Used by Bulgaria
PKhR-51	n/a	Obsolete. Used by Bulgaria
PKhR-54	KA-54, PChR-54	
PKhR-63	PChR-54 (improved)	
None	PChR-54U	Former East German version of PKhR-63
PKhR	WPChR-64	

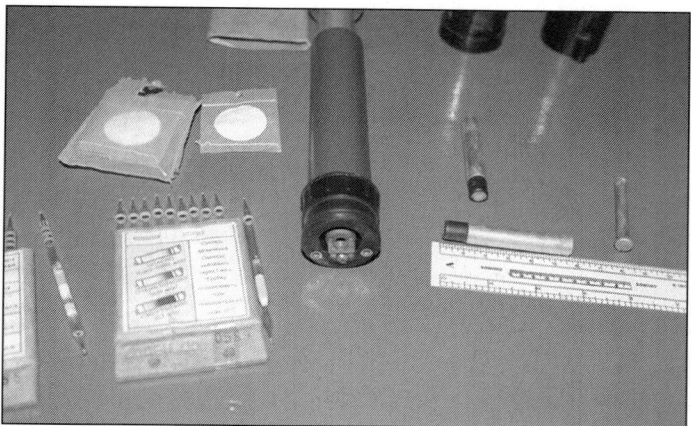

VPKhR contents tubes and colour comparison charts   **2002**/0137542

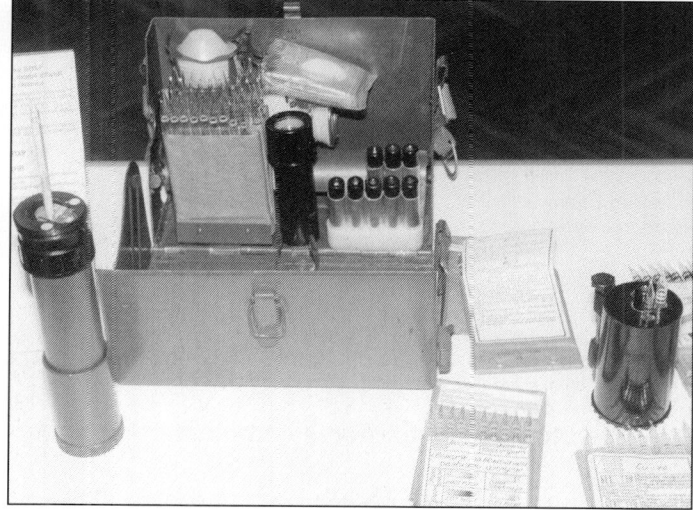

VPKhR chemical agent detection and identification kit (T J Gander)

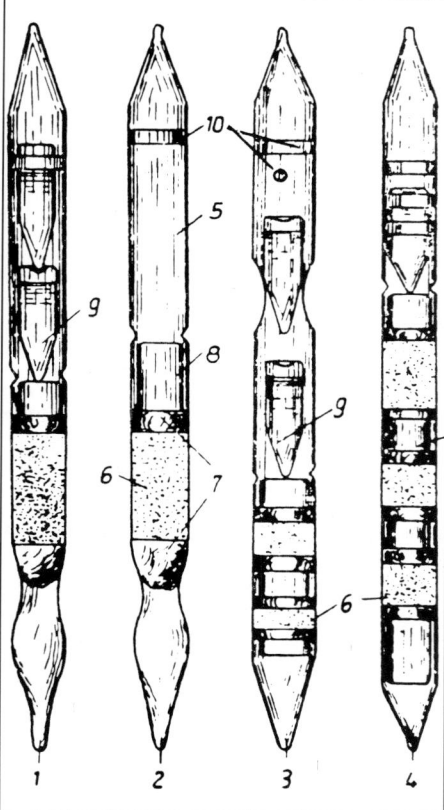

Indicator tubes **(1)** *1 red ring tube* **(2)** *1 yellow ring tube* **(3)** *1 red ring with 1 red dot tube* **(4)** *3 green ring tube* **(5)** *transparent glass tube* **(6)** *inert filter* **(7)** *cotton plug* **(8)** *flow-through canal* **(9)** *reagent ampoule* **(10)** *colour-code marking*

If the Model PKhR-54 is taken to be typical of the above series the kit is contained in a metal case fitted with waist and shoulder straps. The case lid opens forward to form a working tray. The kit box contains a hand-operated cylindrical air pump, 9 or 10 different types of indicator tubes, a pump attachment, smoke filters, perforated plastic caps, a spatula, two sample jars, tape for marking affected areas, a torch, gloves, report forms and an instruction sheet. The kit is capable of detecting and identifying mustard, nitrogen mustard, Lewisite, hydrogen cyanide, cyanogen chloride, phosgene, disphosgene, chloropicrin, adamsite, chloroacetophenone and G-type nerve agents. V-type nerve agents can also be dealt with by the addition of suitable indicator tubes. Each type of tube can normally identify and detect one type of agent, but some of the more recent indicator tubes can detect up to four types of agent. The tubes are 7 mm in diameter and 110 mm long. They are packed in tens and are colour-coded as follows:

Tube marking	Agent detected
1 yellow ring	mustard (H)
2 yellow rings	nitrogen mustard (HN)
3 yellow rings	Lewisite (L)
1 black ring	hydrogen cyanide (HC)
1 green ring	chloropicrin
2 green rings	cyanogen chloride (CK)
3 green rings	hydrogen cyanide (HC)
	cyanogen chloride (CK)
	phosgene (CG)
	diphosgene (DP)
1 white ring	chloroacetophenone (CN)
2 white rings	adamsite (DM)
1 red ring	G-type nerve agents
1 red ring with 1 red dot	G-type and V-type nerve agents

Each tube contains a measured quantity of reagent in a sealed ampoule. The end of the tube is broken off by inserting it in a hole on the end of the pump handle and snapping it. The ampoule is then pierced by a stiff wire contained in the pump handle and the tube placed in a socket in the head of the air pump. Hand pumping draws air through the tube and the ampoule will change colour according to the type of agent present. The pump can hold several tubes at once and the degree of agent concentration can be determined by reference to a calibrated colour scale. Fitting the air pump attachment enables sampling of soil, fabric or surfaces and a smoke filter can be used to screen smoke fumes which might affect detection. The sample jars can be used to take samples for further laboratory analysis. The kits from the Model PKhR-54U onwards contain an indicator tube heater as an extra in place of one of the sample jars. This heater is a hollow plastic cylinder closed at both ends and filled with cotton wadding. The top has four wells to hold three indicator tubes and a heating ampoule. When this heating ampoule is pierced using one of the wires in the pump handle, an exothermic reaction heats the tubes which are placed in the heater for 1 minute. The heater is used only when the temperature is below 15°C.

The current RFAS kit is the Model VPKhR which is smaller and lighter than the previous models, although with a more limited detection capability. Only three types of indicator tube are carried, but by using multi-detection tubes the kit can identify and detect mustard, phosgene, diphosgene, hydrogen cyanide, cyanogen chloride and G-type and V-type nerve agents. The sample jars have been replaced by plastic bags and a tube holder is also included. This kit is produced in Bulgaria as the VPHR and is offered for export sales by Kintex. The Hungarian Type 66-M

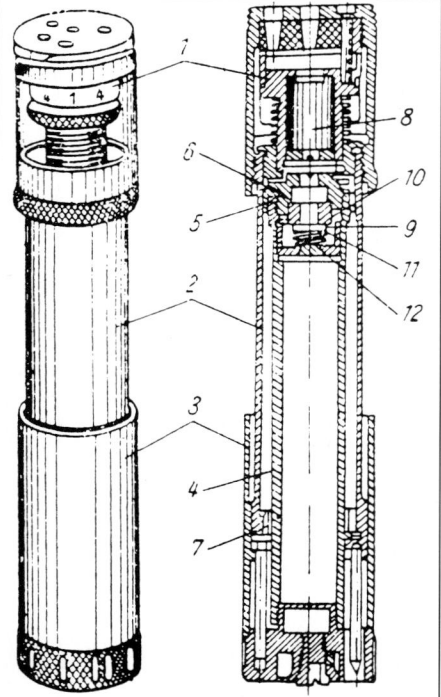

Piston-type air pump PKhR-54 **(1)** *collector* **(2)** *pump housing* **(3)** *handle with ampoule opener* **(4)** *piston rod* **(5)** *sleeve* **(6)** *seat* **(7)** *spacer ring* **(8)** *protective cartridge* **(9)** *valve* **(10)** *valve seat* **(11)** *spring* **(12)** *spring stop*

Specifications		
Model	**PKhR-54**	**VPKhR**
Length	240 mm	205 mm
Width	100 mm	100 mm
Height	140 mm	140 mm
Weight	2.7 kg	2.3 kg

chemical reconnaissance set (qv) is essentially similar to the PKhR-54 kit. Romania also produces a similar kit.

Status
Available from Russian state factories before privatisation. In service with the RFAS and former Warsaw Pact forces, with the Model VPKhR being the current version. Current manufacturer unknown.

Manufacturer
Rosoboronexport.

UPDATED

SWEDEN

CW detection device for nerve and mustard agents

Description
These Swedish CW detection devices are produced by Åkers Krutbruk Protection AB who refer to them as detection tickets. They are produced in two forms, one for detecting nerve agents and the other for mustard. Both forms are basically similar in operation and differ mainly in the reagents involved.

The nerve agent detection ticket is based on enzymatic reaction and contains all the elements necessary to perform operations in both air and water. Sensitivity is high and can be increased by using a pump which will also decrease exposure time. The system works over a temperature range from −20 to +40°C, while the tickets can be stored in a frozen condition which will increase the shelf-life from 5 to 20 years.

The device is activated by removing the ticket from its envelope and folding the flap backwards. The ampoule is crushed by thumb pressure. Liquid enters the enzyme impregnated paper and the ticket is exposed to the air for 2 minutes or inserted in a pump for 10 strokes. The flap is folded forward again and pressed closed for 2 minutes before opening and reading the result. A clearly visible blue mark will appear when no harmful concentrations of nerve agents are present.

The nerve agent detection ticket, with packaging, weighs approximately 10 g and measures 80 × 28 x 10 mm. The detection limit in air is 0.1 mg/m³ without a pump whilst the detecting limit in water is 0.1 mg/dm³. Shelf-life is five years at +25°C or 20 years at −18°C.

The mustard agent detection ticket contains all the elements necessary to perform detection operations. Sensitivity is high and requires a pump to reduce the reactive exposure time, or an adapter mounted on the respirator filter inlet. The system works over a temperature range from −10 to +40°C. Shelf-life is 20 years.

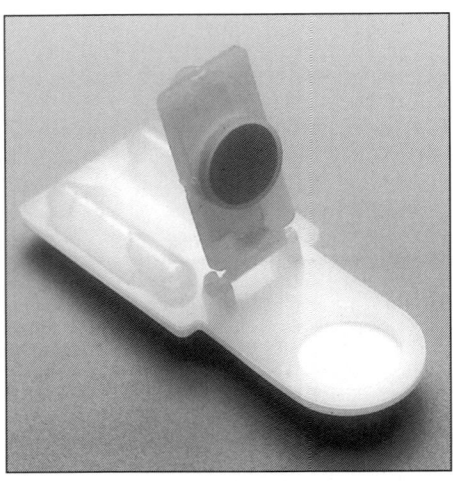

Åkers Krutbruk CW detection ticket for mustard agents

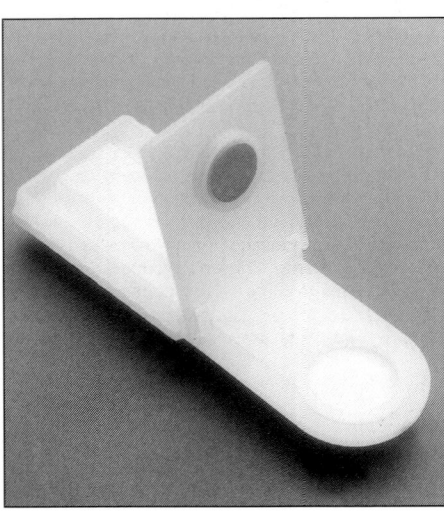

Åkers Krutbruk CW detection ticket for nerve agents
0050689

For use the ticket is removed from its envelope and the ticket flap is folded backwards to crush ampoule No 1. The liquid soaks the paper and the ticket is exposed to air by using a pump for 10 strokes (or after 10 breaths when attached to the adapter on the respirator filter). The flap is folded forward again so that applying pressure on top of the flap will break ampoule No 2 to start a heating stage. Heating will take place for 2 minutes before folding the flap backwards and breaking ampoule No 3 and reading the result. A clearly visible violet mark will appear when harmful concentration of mustard agents are present.

The mustard agent detection ticket, with packaging, weighs approximately 10 g and measures 80 × 33 × 18 mm. The detection limit in air is 1 mg/m³ using the pump.

Status
Nerve agent detector ticket in series production. Mustard agent detector ticket in series production.

Manufacturer
Åkers Krutbruk Protection AB.

UPDATED

SWITZERLAND

IMS 2000 CW agent detector

Description
The IMS 2000 CW agent detector is a hand-held, battery-powered unit which works on the IMS principle, using a low-level reactant source. The tough alloy casing houses the IMS drift tube assembly, battery, detector nozzle, LCD display and removable tracer gas cartridge. Two buttons allow the user to switch the unit on or off and to select between continuous monitoring and 'all clear' modes. The latter is the highest sensitivity mode, mainly used for monitoring during DECON or cleansing operations. There is a comprehensive start-up BITE routine during the 2-minute warm-up time and continuous operational BITE monitoring. The clear LCD display simultaneously shows agent identification and the level of H or G agent detected at the nozzle. The system is equipped with a self-protection algorithm and, if the IMS system becomes overwhelmed by excess agent challenge, it can be purged to recover within minutes (automatic backflush).

The modular design allows easy field replacement of consumables (molecular sieve pack, carbon filters, dopant cartridge) and the IMS 2000 offers a variety of target agent applications, including TICs in addition to most threat CW agents.

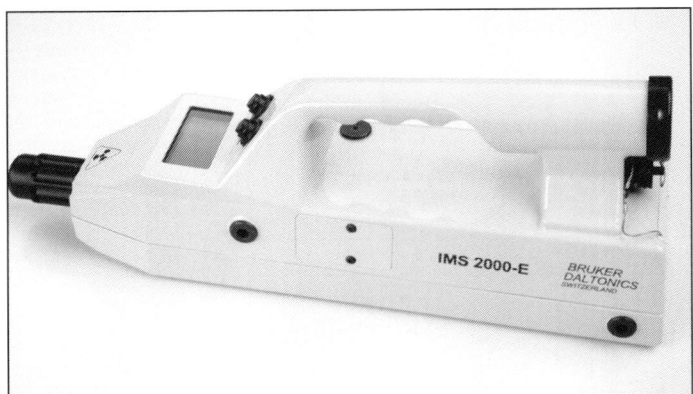

The IMS 2000 'COLPRO CAM' variant of the monitor is designed for monitoring inside collective protection facilities (Bruker Biospin AG)   ***2002***/0134216

IMS 2000 CW agent detector (Bruker Biospin AG)   ***2002***/0134215

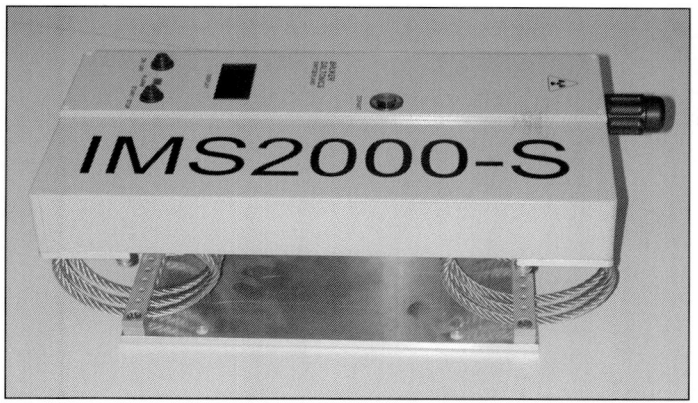

This shock-mounted version of IMS 2000 is designed to be fitted to mobile vehicles for site detection and monitoring (Bruker Biospin AG)   ***2002**/0134217*

Specifications

Sensitivity:	Nerve agent (20 mg/m³)
	Blister agent (200 mg/m³)
Dynamic range:	20 mg/ m³ to 5 mg/m³
Switch-over time to backflush:	<3 s
Warm-up time:	1 minute at 20°C
Alarm:	Flash LED, power buzzer
Weight:	2.2 kg

The IMS2000 has an RS232 port for the export of digital data to a standard PC and a port for external alarm activation. PC-based field analysis allows more detailed data to be provided at unit level for command decision support.

The unit is shock protected, making it an effective option for mounting in vehicles or aircraft. Built-in pressure compensation allows the unit to be used at altitude in helicopters or fixed-wing aircraft.

Powered internally from a LiMgO$_2$ battery, the unit delivers 24 hours of intermittent use or 16 hours of permanent use. External power supply: 9-24 V DC.

Status

Available. Delivery of IMS 2000 to the UK MoD under way from mid-2001 for the detection and monitoring of CW agent levels within COLPRO facilities (COLPRO CAM).

Manufacturer

Bruker Biospin AG.

UPDATED

UNITED KINGDOM

Chemical Agent Monitor (CAM)

Development

The Chemical Agent Monitor (CAM) is a hand-held instrument used to warn of the presence of CW agent vapours. The first CAMs were produced in the mid-1980s since which time a continuous programme of product improvements has taken place. Several variants of CAM are now in service and the monitor has been deployed with the armed forces of 28 countries, including the UK and USA. In excess of 57,000 CAMs have been produced, serving as a key part of UNSCOM's detection and monitoring capability and with various national forces operating in support of the UN in the former Yugoslavia.

Description

CAM responds to nerve, mustard, blood and choking agents and is able to produce a quantitative readout. The abilities of CAM are invaluable for monitoring personnel, ground equipment, vehicles and aircraft for contamination to determine whether it is safe for personnel to relax chemical protective measures. CAM is also a vital instrument in monitoring whether personnel entering collective protection areas are contamination-free, to check battlefield casualties for contamination and to confirm the effectiveness of training procedures, using simulated agents.

Principles of Operations

CAM makes use of ion mobility spectrometry, (IMS), principles to respond selectively and accurately to toxic chemical agent vapours. Air is drawn into the unit and is ionised. The molecules of certain types of agent vapours are characterised by their ability to form low-mobility ionic clusters. These ionic clusters are then classified according to their mobility relative to an onboard dopant vapour source.

IMS enables specific agents to be monitored at very low levels, whether found at isolated ground locations or present in the general atmosphere. It also discriminates against other vapours likely to be found in battlefield situations. IMS has gained a reputation as the best technique to detect very low levels of toxic vapours on the battlefield. It provides many possibilities for analysis in varied fields,

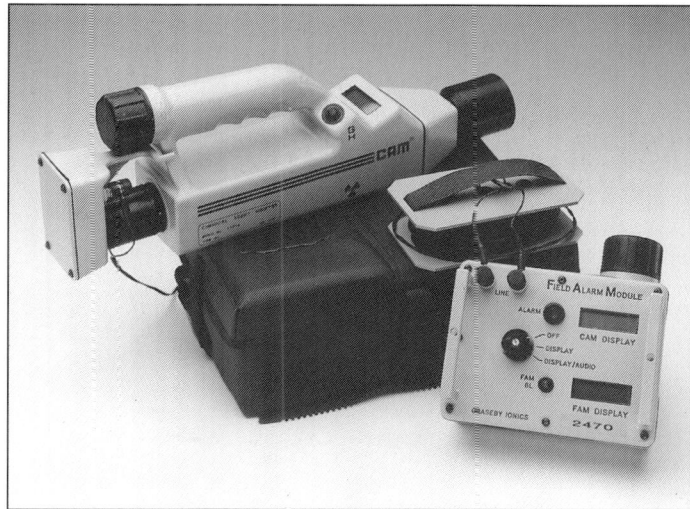

Field Alarm Module (FAM) for CAM

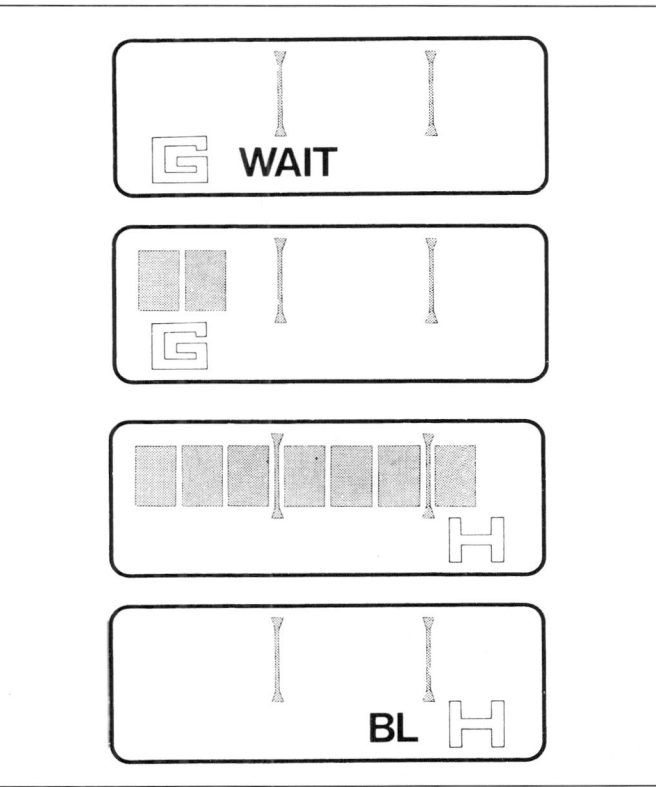

Examples of CAM display from top: nerve agent mode (G) selected, instrument in warm-up phase; nerve agent mode (G) selected with relatively low concentration indicated (two blocks visible); blister agent mode (H) selected with relatively high concentration indicated (seven blocks visible); battery low indication (BL)

in particular, for continuous emission monitoring and ambient air monitoring of hazardous air pollutants.

It has been designed to be used by personnel in full NBC equipment, whatever the climate or weather. Operator controls are simple, consisting of an on/off push button and a mode change switch. One-hand operation is normal and no chemicals are required. It is easily decontaminated, highly rugged and fully compatible with other NATO equipment. Battlefield confidence is assured via a confidence sample that tests the CAM for operational viability.

Variants

ECAM

In 1999, Graseby commenced the full scale production of a further generation of CAM for the Swiss Defence Procurement Agency. This unit has many advanced hardware features, specified by the customer. Features include a removable dopant source for improved long term storage, automatic sample overload protection, simultaneous detection of nerve and blister interactive display and simplified operations.

CAM plus

CAM plus is the latest improved software standard now being incorporated in all new production CAM detectors. It allows detection of a greater range of CW agents and also includes a capability against a specified range of TICs. This upgrade is available for retrofit.

FAM being set up for use (Graseby Dynamics)   **2002**/0137741

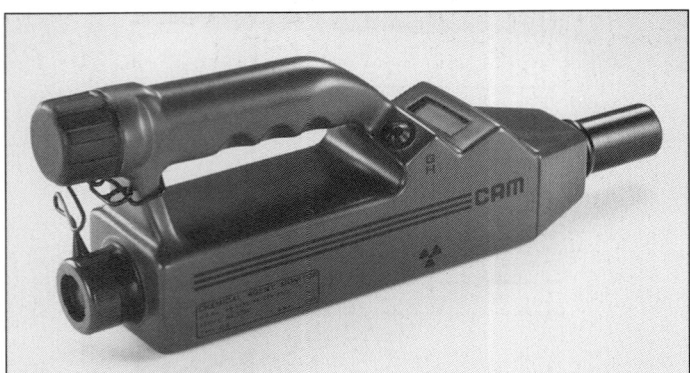

Chemical Agent Monitor (CAM)

CAM in use

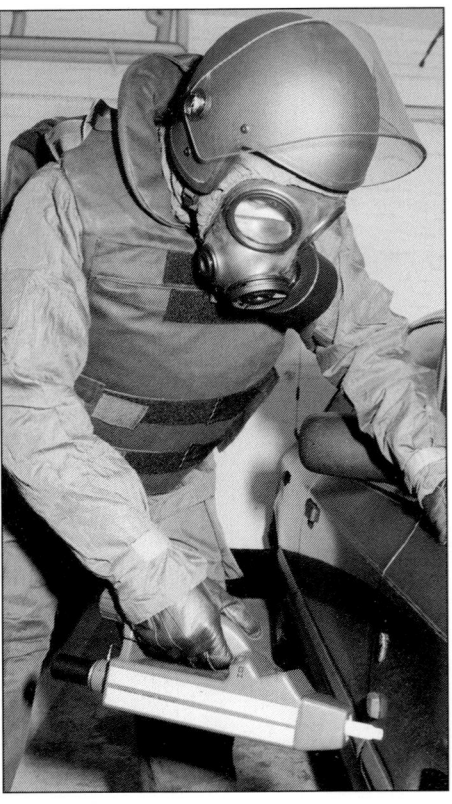

CAM in the internal security role - detecting explosives (Graseby Dynamics)
0050688

Specifications

Weight:	(hand unit) 1.5 kg
Length:	(over longest dimension) 380 mm
Battery life:	(minimum): 6 h continuous
	(typical): 15 h
Temperature range:	(operating): −30 to +55°C
	(storage): −55 to +70°C
Power supply:	single 6 V battery
Sensitivity:	to present NATO requirements
Durability:	to DEF STAN 07-55

Field Alarm Module (FAM)

To address the need for a local sentry alarm, Graseby developed the Field Alarm Module (FAM) which extends the role of CAM to provide a local detection capability whilst still allowing its use as a chemical agent monitor. FAM increases the operational flexibility of CAM by offering a remote display/alarm feature, automatically switching CAM between the two detection modes.

FAM saw operational service during the 1990-91 Gulf War, when it was used by the Netherlands armed forces to protect their Patriot batteries and by the British Army when fitted to the Fuchs. The use of FAM on the Fuchs allowed the CAMs to detect the outside atmosphere while the crew could remain within the vehicle's NBC collective protection system and observe the CAM display on the FAM units. The American M93 Fox NBC reconnaissance vehicle also includes CAM as part of its sensor suite.

The FAM is operational with the British Army, Royal Air Force, Australian Army, Belgian Air Force, Netherlands armed forces and a further seven nations.

Otto Fuel Monitor (OFM)

The Otto Fuel Monitor (OFM) was developed from CAM for use in Royal Navy submarines and, potentially, on surface vessels and armament depots where Otto fuel (iso-propyleneglycol dinitrate) is used in torpedoes. The OFM is a fixed-mode device with one on/off button. Powered by a 6 Volt rechargeable battery. The OFM is now supplied to an OFM-2 standard common to CAM, offering improved life-cycle costs and ease of maintenance. The OFM is in service with the Royal Navy and the navies of eight nations including Australia, Canada, Netherlands and Turkey.

Status

In production. Over 57,000 in service with 26 countries (13 from NATO) including Belgium (Army and Air Force), Canada, Denmark, Italy, Luxembourg, NATO (HQ AFCENT), Netherlands, Norway, Portugal, Spain, Turkey, UK, USA (Army, USAF and USMC). Australia, New Zealand, Sweden, the GCC and several Asian countries have CAM in service. Some 5,600 in production for Switzerland. Under evaluation by many others.

250 ECAMs were ordered by the Austrian ministry of defence for delivery in 2001.

Manufacturer

Graseby Dynamics Limited.

UPDATED

Detector kit, chemical agent, residual vapour, No 1 Mark 1 (RVD)

Description

Known to UK service personnel as the Residual Vapour Detector or RVD, this CW agent detector kit is carried by monitoring teams as a simple but effective means of detecting the presence of H or G-type agents. The kit comprises a set of plastic bottles containing agent-specific reagent, a small rubber-bulb air pump, plastic sample ticket holders, weatherproofed instruction sets and comparison charts, all contained in a 200 × 200 mm stiff canvas pack.

Each bottle contains a binary chemical reagent product (a sealed capsule within a liquid) which is activated by squeezing the bottle to crush the capsule, allowing the contents to mix. A ticket, moistened with a drop of the reagent liquid, is positioned in the plastic frame on the end of the air-pump. By squeezing the bulb to draw air through the ticket, the monitor can see whether agent is present by observing a change in ticket colour.

The kit has a shelf-life of at least 10 years.

Spares (reagents and tickets) and training sets (nerve agents) are available.

Although largely being replaced with hand-held electronic IMS or GC/MS systems the RVD continues to offer a simple, lightweight and reliable detection capability, requiring no power supply and little training.

Status

In production. In service with the UK armed forces.

Supplier

Richmond Packaging (UK) Limited.

UPDATED

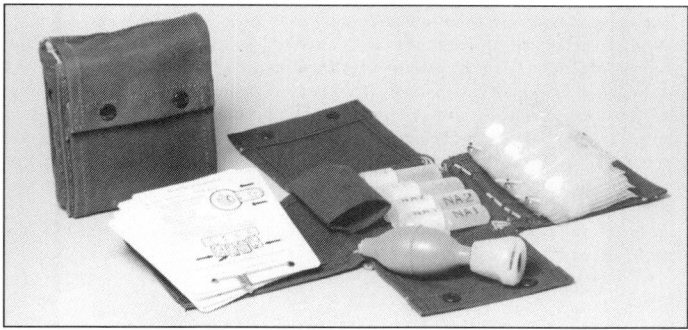

No 1 Mark 1 detector kit (Richmond Packaging (UK) Limited)   **2000**

Eclox™-M Military Water Test Kits (Biological and Chemical) (WTK (B&C))

Description

The Eclox™ -M Water Test Kits (Biological and Chemical (WTK (B&C)) have been developed by Severn Trent Services in conjunction with the UK Defence Procurement Agency NBC Integrated Project Team. This new kit replaces the

The Eclox™ Water Test Kit (Biological and Chemical) (Stella-Meta)   **2002**/0137751

Water Test Kit (Poisons) which has been in service with the British Forces since the 1960s.

The WTK ensures that both raw and treated water is free of a wide range of chemical and other contaminants and is fit to drink. It uses Eclox™ technology, which measures the light generated from the reaction when chemicals are added to a water sample. After a few minutes the WTK operator can tell how pure the water supply is and what additional treatments may be needed to make the water fit to drink. The water is then tested again and, if safe, can start being used immediately. Among other contaminants, the WTK can test for pesticides, arsenic and mustard gas. Full results are available within 30 minutes, significantly quicker than the initial requirement of 45 minutes.

The WTK is battery-powered, hand-portable, shock and vibration proof. It has been extensively tested for use anywhere in the world and in any climate, from sub-zero to +50°C. The WTK can be stored complete with reagents for two years at ambient temperature without deterioration. No refrigeration is required for the reagents. The new kits are easier to use than their predecessors and operators need less training than previously.

Status

Delivery of production models began in March 2001. In use with the British Army, Navy and Air Force.

Manufacturer

Stella-Meta Filtration Systems.

NEW ENTRY

GID-2A fixed chemical agent detector

Description

The GID-2A is a fixed detector which offers a continuous real-time detection and assessment capability against nerve, blister, blood and choking agents. The GID-2A is designed to be integrated into a wide range of collective protection facilities, both at sea and on land.

Development

The GID-2A was jointly developed with the UK Ministry of Defence to meet specific requirements for internal and external ship citadel or COLPRO CW detection. Primarily designed for maritime use, GID-2A is equally suited to CW detection in command and control bunkers and hardened aircraft shelters. The system provides a visible and audible alarm in the event of a chemical attack, both locally or at a remote point through Network Control Units (NCU).

Principles of operation

The GID-2A detector operates using a patented twin drift tube Ion Mobility Spectrometry (IMS) technology. This technology has been developed to target warfare agents extremely accurately and thus ensure a very low false alarm rate. The GID-2A responds simultaneously to nerve, blister, blood and choking agents in real time and is capable of being reprogrammed to meet future threats.

The GID-2A detector incorporates an automatic self-flushing device, which is activated in very high vapour concentrations. This enables the equipment to operate continuously throughout a chemical attack, giving real-time information on the rise and fall of the threat levels. This equipment also indicates which type of

GID-2A Ship-Installed Chemical Detector (SICS Mark 10) (©Graseby Dynamics)   **2001**/0106863

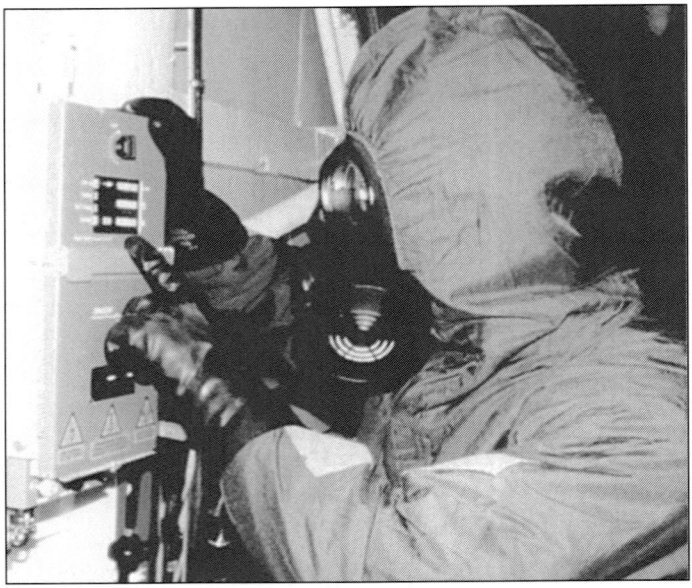

GID-2A (SICS Mark 10) (©Graseby Dynamics)   **2001**/0106882

agent has been deployed. GID-2A offers both audio and visual warnings and may be interfaced to an external remote alarm.

Status

In service with the Royal Navy and 7 others, including the Royal Swedish and Royal Netherlands navies. In 2000, the UK Royal Navy purchased 250 GID-2A (RN designation is SICS Mk 10), supplementing the 120 units already in service.

Manufacturer

Graseby Dynamics Limited.

VERIFIED

GID-3 point chemical agent detector

Description

GID-3 was developed as a result of lessons learnt during the 1990-91 Gulf War to overcome perceived weaknesses in existing CW detectors. The development incorporated design features to improve agent detection capability, reduce false alarms, allow better discrimination between, and identification of, agents and improved ease of use by deployed troops.

Following initial design, GID-3 was extensively tested in a variety of vehicle applications to check its use in both hard wired and VETRONIC armoured fighting vehicles. Predominant amongst these tests were the GEC Marconi Centaur Weapon Control System and the UK DERA VERDI (Vehicle Electronics Research Defence Initiative) 1 and 2 vehicle electronics demonstrators (this DERA activity now subsumed by the new DSTL organisation). In this latter vehicle, GID-3 was subjected to some 14 weeks of user troop trials and the lessons learnt were incorporated into the later versions.

Graseby GID-3 point chemical agent detector (Graseby Dynamics)   **2001**/0106864

Experience gained proved successful when the GID-3 was accepted into service following competitive trials in Canada, UK and the USA. In the USA, GID-3 won the competition and was accepted into service as the M22 ACADA detector.

The GID-3 has a full range of accessories and fitting plans available for different vehicles.

Status

In production. In service with Canada, Kuwait, UK and USA. In March 2000, the Canadian Armed Forces purchased an additional 40 GID-3 units to the 39 already in service, fitted to the LAV armoured personnel carrier (reconnaissance version).

Manufacturer

Graseby Dynamics Limited.

UPDATED

Lightweight Chemical Detector (LCD)

Description

The Lightweight Chemical Detector (LCD) is a hand-held device based on the successful IMS technology used in CAM (see separate entry this section). A need for a small, personal CW agent detector was identified during the Gulf war (UK designation: Operation GRANBY). The LCD is designed to offer extremely fast and sensitive, simultaneous individual detection and monitoring of nerve, blister, blood and choking agents in real time. It can also be used to detect TICs and TIMs in a military or domestic response environment.

When switched on, the unit samples continuously through the nozzle, shown (covered) top left of the illustration. The LCD offers a low false alarm rate as well as fast detection response and recovery following a high-concentration attack. All the IMS detection, sampling and processing features have been miniaturised and improved to give lighten the detector and reduce its power consumption. The unit can operate for 40 hours on four AA type batteries and does not use a radioactive source for ion production. Other features include an audio and/or visual alarm and a communication port.

As an individual detector, the LCD can be deployed unobtrusively by military, law enforcement or disaster response personnel.

Status

Graseby's LCD was runner-up for the US Joint Chemical Agent Detector (JCAD) requirement and was extensively tested by both the Soldier Biological-Chemical Command (SBCCOM) and the Special Operations Command (SOCOM) for the competition. A variant of LCD won the UK MoD Lightweight Chemical Agent Detector (LCAD) contract in December 2000. A purchase of approximately 10,000 units is planned, for delivery in 2004/2005. The Japanese Police has purchased 60 LCDs (June 2002). A further variant, LCD-3 is also available (see illustration).

Manufacturer

Graseby Dynamics Limited.

UPDATED

The LCD-3 variant of the Lightweight Chemical Detector (Graseby Dynamics)
2002/0137546

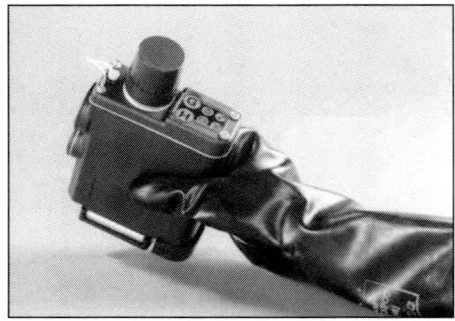

The Lightweight Chemical Detector (LCD) (Graseby Dynamics)
2002/0137545

NAIAD

Description

NAIAD stands for Nerve Agent Immobilised enzyme Alarm and Detector. It is an automatic alarm system which continually monitors the surrounding atmosphere to provide audible or visual warnings of the presence of nerve agents in either vapour or aerosol forms. The equipment consists of a sampling detector which can provide its own warning plus up to a further three remote alarm units which can be up to 500 m away.

Its operation depends on the action of an enzyme, cholinesterase, which occurs naturally in the human body and reacts to nerve agents. Inside NAIAD a small sample of this enzyme in a plastic holder is continually irrigated by a supply of an organic ester, butyrylthiochlorine. The two substances combine by hydrolysis to form thiocholine. When the cholinesterase reacts to a nerve agent the resultant thiocholine shows changes in its electrochemical reactions which are detected by a graphite-measuring electrode which initiates the alarms. NAIAD can be used to detect any nerve agent, even those that are at present unknown, by using natural body chemistry. It can also be used to detect hydrogen cyanide in attack concentrations.

The mechanics of the system depends on electrical power derived from a standard 3.3 Ah Clansman radio battery which supplies the alarm circuits, reagent liquid pump, incoming air pump and the air heater. The reagent pack has the same non-stop service life as the battery: both are changed every 12 hours. The reagent pack is sealed, as is the enzyme pad holder, so there is no danger of contaminating the system by handling and the whole system has a rapid response with a low false alarm rate. The detector is nuclear-hardened and can operate even when tipped on its side or inverted. The alarm units have flashing lights, a clearly audible alarm signal and built-in test equipment automatically tests the cable connections between the alarm and the detector.

The detector is normally carried in a backpack which also contains spare batteries and reagent packs, but it is possible to fit it into vehicles or fixed installations. A special rack for carrying the detector on the outside of vehicles has been developed. Two test units are available, a base workshop test set and a field test kit.

The Simtec NAIAD is used in a ship chemical alarm system known as Ship Installed Chemical Alarm System (SICAS). This uses the existing hardware and through-bulkhead units of the Royal Navy's Ship Installed Chemical System

NAIAD interior showing sealed units
(Henry Dodds)

Specifications
Detector unit:
  (size) 251 × 209 × 475 mm overall
  (weight) approx 12.5 kg with battery
Alarm unit:
  (size) 232 × 177 × 99 mm
  (weight) approx 2.5 kg with battery
Climatic range: −31 to +52°C, 0-100% relative humidity
Min detectable concentration: 0.005 to 0.05 mg/m³ dependent on agent

(SICS - see following entry) and involves the SICAS detector, through-bulkhead units, a remote status panel and alarm unit.

The Airfield Chemical Agent Detection System (ACAS) is a proposed system using NAIAD for airfield use. It involves the placing of a number of detector units around an airfield perimeter with more at selected positions. Remote status panel and alarm units are involved.

Status

Production complete. Remains in service with the armed forces of Portugal, Spain and UK.

Manufacturer

Jasmin Simtec Limited.

UPDATED

NAIAD detector unit
(Jasmin Simtec Limited)

UNITED STATES OF AMERICA

Agilent/Dynatherm Agent Monitor (A/DAM)

Description

The Agilent/Dynatherm Agent Monitor (A/DAM) is a transportable laboratory for CW and environmental monitoring assessment. It is designed for site remediation or CW demilitarisation programs. Agents detected include GA, GB, HD, VX direct or G-analog, common simulants, precursors and manufacturing by-products. The A/DAM has two main components: the IACEM 980 air sampler and the Agilent 6852A Gas Chromotograph (GC). The detection process combines simultaneous sampling, separation, identification and measurement.

The A/DAM can also be used for domestic preparedness and CW counter terrorism response. It has been used by the UNSCOM post-Gulf War Iraqi CW assessment and by the Chem/Bio Response team at the Atlanta Olympic Games.

Sensitivity
GB	0.0001 mg/m³	4 min
VX	0.00001 mg/m³	5 min
HD	0.003 mg/m³	4 min

IACEM 980

The IACEM 980 air sampler is designed to quantitatively collect and release chemical agents from air on solid sorbent cartridges and transfer the analytes to the Agilent 6852A gas chromatograph for separation, identification and

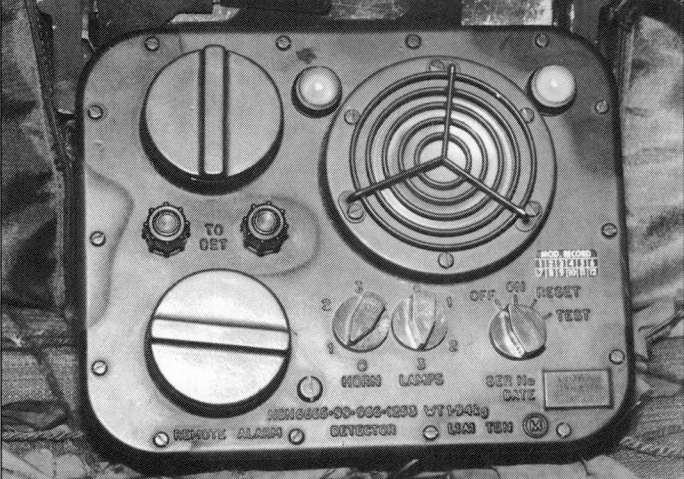

Operating panel of NAIAD remote alarm unit (Henry Dodds)

measurement. The IACEM's dual-collection tubes operate in alternating cycles, providing continuous, uninterrupted sample concentration. The integrated sampler and GC provide sensitivity through sample collection efficiency and GC provide sensitivity through sample collection efficiency and state of the art chromatographic resolution, while the alternating collection scheme provides faster turn-around times.

- Sample collection flow: controlled by integral mass flow controller, max 2.5 litres/min. Accuracy ±1% full scale.
- Times events: 5 sequence positions, adjustable from 0-99.9 minutes in 0.1 minute increments.

Specifications
Heating rate °C/minute
 Sample tube: 1,000
 Focusing trap: 900
Dimensions: 3,000 × 340 mm
Weight: 19 kg
Power requirements: 115 V AC single phase (110 to 125) 8A max 50/60 HZ

Agilent 6852A gas chromatograph
The 6852A GC is equipped with two flame photometric detectors, available in either single wavelength sulphur mode or phosphorous mode. All sample introduction is via the IACEM sampler. A 6852A GC software driver and ChemStation software provide control and data handling for 6852A GC's only. The 6852A software driver is included with the GC. The ChemStation software for the GC is sold separately.

The photometric detectors share an electronic pneumatics control module set for either pressure or flow control. Pressure setpoint increments 0.01 psi. Temperature and pressure sensors compensate for ambient variation and altitude. There are 3 channels for detector gases (pressure/flow) and 3 channels for auxiliary gases (pressure only). The maximum operating temperature is 250°C.

Specifications
Dimensions: 283 × 605 × 568 mm
Weight: 29 kg
Power requirements: 120 V (+10%, −10%) approx 1,440 VA (max) at 120 V, approx 2,000 VA at 230 V

Status
Available. In service with the US Army in mobile laboratories monitoring chemical weapons stockpile storage sites and chemical weapons small burial sites.

Manufacturer
CDS Analytical Inc.

NEW ENTRY

....

Alarm, chemical agent, automatic: portable, M8 and M10 to M18

Description
This alarm system has 10 variations all based on the use of two basic modules, the M43A1 detector and the M42 alarm. Each detector unit can activate up to five alarm units. By the use of various components and associated equipment these two basic units can form the basis for the following alarm systems:

Alarm, chemical agent, automatic: portable, manpack, M8 or M8A1;

Alarm, chemical agent, automatic: portable, fixed emplacement, M10;

Alarm, chemical agent, automatic: portable, fort truck, utility, 1 ton, M11;

Alarm, chemical agent, automatic: portable, for truck, 2½ ton, M12 (no longer in production);

Alarm, chemical agent, automatic: portable, for truck, 2½ ton, M13;

Alarm, chemical agent, automatic: portable, for full-tracked APCs and ARVs, M14;

Alarm, chemical agent, automatic: portable, for carrier, command and reconnaissance, M15;

Basic alarm, chemical agent, automatic: portable, M8A1 showing (left) high-profile rack with space for a power pack, (centre background) M43A1 detector unit, (centre foreground) M42 remote alarm and (right) Test Set M140

Alarm, chemical agent, automatic: portable, w/power supply for truck, utility, 1 ton, M16;

Alarm, chemical agent, automatic: portable, w/power supply for truck, utility, 2½ ton, M17 (no longer in service);

Alarm, chemical agent, automatic: portable, w/power supply for truck, 2½ ton, M18.

These systems differ mainly in the type of mounting used, but common to all is the M253 winterisation kit which has two special batteries and a cable assembly. This kit was not purchased by the US Army.

The M43 detector operates by passing air through an oxime solution surrounding a silver analytical electrode and a platinum reference electrode. If a chemical agent is present, a reaction occurs in the solution which increases the potential between the two electrodes. Any such potential change is amplified and the alarm signal is triggered.

The M43A1 chemical agent detector, developed at the US Army Armament Research and Development Command's Chemical Systems Laboratory, CB Detection and Alarms Division, serves as a direct replacement for the earlier M43. The M43A1 is designed to detect low concentrations of nerve agents in less than 1 minute. The M43A1 is structurally and functionally similar to the M43 but it differs mainly in its internal sensor characteristics and the way in which it operates. For operation, air is drawn continuously through the internal sensor by a pump. The air consists of its normal constituents plus any chemical agents which may be present. As the molecules are drawn past a radioactive source, a small percentage are ionised by α-rays. Through multiple collisions some of the ions cluster with agent molecules which then become very stable. As the air and agent ions are drawn through the baffle sections of the cell, the lighter and less stable air ions diffuse to the walls and are neutralised more quickly than the heavier and more stable ones. As a result, the collector senses a greater current when nerve agents are present compared to the current when only clean air is sampled. An electronic module monitors the current produced by the sensor and triggers the alarm when a critical threshold is reached. The alarm has a variable pitch with a maximum range of 90 dB. The M43A1 detector incorporates a nuclear compensation circuit to assure continued proper detector response after exposure to the neutron flux resulting from a nuclear event.

The Test Set M140 maintains detector and alarm modules in a state of readiness. All elements of the system, including power, pressure and electronics can be monitored in seconds to assure response to chemical agents.

Brunswick Defense (now Intellitec) produced a Vesicant Agent Detector (VAD) module which can be attached to a standard M43A1 and is M42 alarm compatible. The VAD is a stand-alone Mustard/Lewisite detector unit which weighs 1.6 kg and measures 165 × 140 × 89 mm.

The Israeli Chemical Agents Tracer (CAT) appears to be based on the M43A1. The KM43A1 is produced in South Korea.

The detector unit, simulator automatic chemical agent alarm, XM81, has been developed for use with the M8 system and when fitted is attached to the M43 detector unit (see Chemical training and simulation section).

Status
In service with the US Army and German forces (M8A1). 26,000 portable manpack M8 sets were ordered for the US Army from 1975 to 1982. More than 35,000 M43A1 detectors were delivered to the US Army from 1983 to 1986.

Manufacturer
General Dynamics Armament and Technical Products.

UPDATED

Specifications Component	M43 and M43A1 detector	M42 alarm kit	M229 Refill Supply	BA3517 battery kit	M10 power kit	M228 mounting	M182 mounting	M10A1 power Supply
Length	165 mm	180 mm	330 mm	160 mm	150 mm	191 mm	191 mm	190 mm
Width	140 mm	97 mm	264 mm	196 mm	163 mm	254 mm	254 mm	165 mm
Height	275 mm	60 mm	229 mm	127 mm	188 mm	318 mm	203 mm	81 mm
Weight	3.4 kg	1.9 kg	6.3 kg	3.4 kg	8.2 kg	7.3 kg	6.8 kg	2.9 kg
Max distance of detector from alarm	400 m	400 m	400 m	400 m	400 m	400 m	400 m	400 m
Life on one refill (continuous use)	15 days	n/a	15 days	n/a	n/a	n/a	n/a	n/a

Akorn Cyanide Antidote Package

Description
This kit is aimed primarily at the medical element of first responder organisations to deal with the extreme speed needed to prevent deaths from the blood agent HCN. The Akorn Cyanide Antidote Package contains the following key elements:

- 2 ampoules of sodium nitrite injection (USP). 300 mg in 10 ml of water.
- 2 vials of sodium thiosulphate injection (USP). 12.5 g in 50 ml water.
- 12 ampoules of amyl nitrite inhalant (USP). 0.3 ml.
- 10 ml and one 60 ml sterile disposable syringes (one each).
- Stomach tube.
- Sterile disposable needle.
- Tourniquet.
- Non-sterile 60 ml syringe.
- Comprehensive instructions for use.

According to analysed results from canine trials, the kit is capable of detoxifying 20 lethal doses (based on sodium cyanate in dogs).

Specifications

Antidote	Advantage	Limitation
Sodium nitrite:	Removes cyanide ions from various tissues	Excess dosage can induce dangerous methemoglobinemia
Sodium thiosulphate:	Converts poisonous cyanide ions to thiocyanate	-
Amyl nitrite:	Assists in detoxification of CN ions.	Excess dosage can induce dangerous methemoglobinemia. Continuous use may inhibit oxygenation of the blood.

Status
Available. Used by medical and first responder organisations in the USA.

Manufacturer
Akorn Inc.

NEW ENTRY

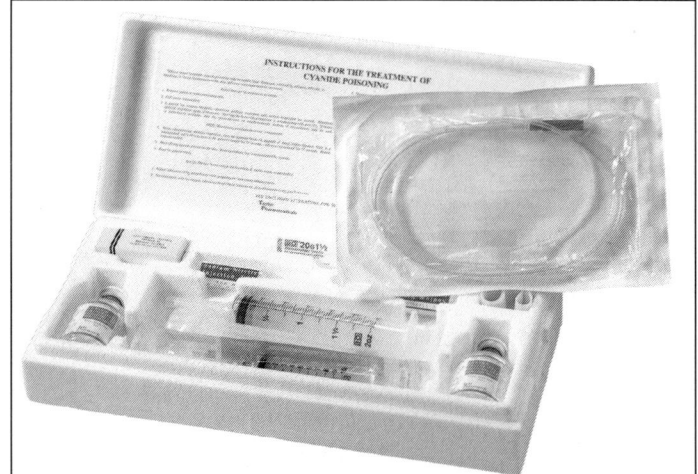

Akorn Cyanide Antidote Package (Akorn Inc)   **2002**/0137855

AN/KAS-1 Chemical Warfare Directional Detector (CWDD)

Description
The AN/KAS-1 Chemical Warfare Directional Detector (CWDD) is a common module, two fields of view, forward-looking infra-red system utilised by the US Navy for the standoff detection of nerve agents. A selectable spectral filter (one of four available) enables an operator to interrogate potential threat. By comparing images produced by the filter's three available spectral bands they can distinguish the presence of nerve agent accumulations. The unit has a secondary function by providing thermal imaging for night surveillance.

Late production versions of the AN/KAS-1 include a remote monitoring feature which enables the central command station to view the scene at the same time as the operator.

The system comprises a sensor unit and a power conditioning unit. The sensor has 60 detector elements and is cryogenically cooled. Two fields of view are available: wide, 3.4 × 6.8° and narrow, 1.1 × 2.2°. Total weight of the equipment is 20.9 kg and it is fully submersible.

The AN/KAS-1 (latest variant AN/KAS-1A) entered production in 1985. A total of 555 units were ordered by the US Navy and delivered. An additional contract for 362 units was awarded in 1991.

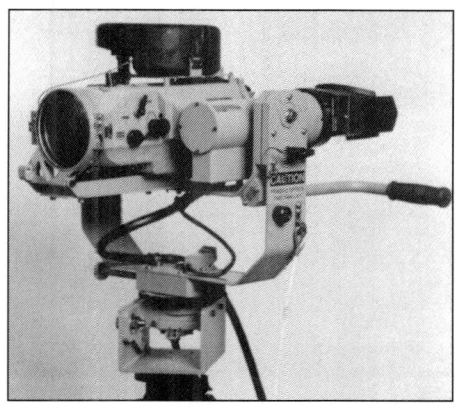

AN/KAS-1 Chemical Warfare Directional Detector (CWDD)

Status
In service with the US Navy.

Manufacturer
General Dynamics Armament and Technical Products.

UPDATED

AN/PSR-2 Automatic Liquid Agent Detector (ALAD)

Description
The AN/PSR-2 Automatic Liquid Agent Detector (ALAD) was developed and produced by the Calspan Corporation under contract to the US Air Force, based on the joint requirements of the US Army and Air Force. The ALAD was designed to a specification requiring a lightweight, low-maintenance, durable unit and sensor for neat and solid GD, VX, HD, L and thickened agent detection and warning down to 200 μm diameter across a temperature range of −35 to +52°C. In January 1993, Calspan completed the delivery of 2,100 AN/PSR-2 systems, together with 210 BZ-90/PSR-2 auxiliary alarms.

The AN/PSR-2's sensor technology and detection algorithms are derivatives of the US Army XM85/XM86 ALAD system. Development of the XM85/XM86 ALAD was completed in 1982, when the system commenced a series of qualification tests.

The AN/PSR-2 ALAD detects liquid droplets of the chemical agent directly. It uses a dry 136 mm diameter sensor with a rigid nylon grooved substrate containing the agent sensitive material at the bottom of the grooves. Detection response time

AN/PSR-2 Automatic Liquid Agent Detector (ALAD) detector unit and auxiliary alarm unit

AN/PSR-2 Automatic Liquid Agent Detector (ALAD) detector unit

is within 60 seconds. The sensor is located on a Detector Unit which may be used to provide local or remote warnings and may be networked with other chemical agent detectors via wire, fibre optics or radio. The ALAD Detector Unit can provide continuous unattended operation for up to one month when operating from BA 5588 lithium batteries in a sealed internal chamber. As an alternative, the unit may be powered by 110 or 220 V AC mains power. In the latter mode the unit may be used at temperatures of –30°C by using an integral heater.

The ALAD Detector Unit has a built-in self-test function which enables an operator to determine the status of the AN/PSR-2 from the end of the link to the unit. Installing the unit takes less than 1 minute while monthly maintenance consists of changing the battery and replacing the sensor.

For local warnings the Detector Unit has an integral horn and alarm lamp, each of which may be independently disabled. If required, an M42 sound alarm may be connected. As an alternative to the M42 an ALAD Auxiliary Alarm can be provided to give stronger aural alarm signals than the M42.

For remote warnings the Detector Unit can be connected to a communications network. The unit can provide three signals to a central warning site, namely an 'operating and OK' signal, a 'needs service' low battery signal and an 'Alarm' signal.

The AN/PSR-2 Detector Unit measures 315 × 213 × 113 mm and weighs 4.5 kg.

Status
In service in the US Air Force.

Manufacturer
Veridian Corporation.

UPDATED

APD 2000 CW agent detector

Description
The APD 2000 CW agent detector is hand-held device designed not only for defence use but for environmental protection and domestic preparedness as well. It is adaptable and simple to operate. It can be powered by commercial alkaline or rechargeable batteries or by AC vehicle power. The clear LCD screen is operated using eight clearly-marked buttons on a touch-pad at the rear of the instrument. With sophisticated software algorithms, it can simultaneously detect for H, G or V agents and for domestic threats such as pepper spray or mace.

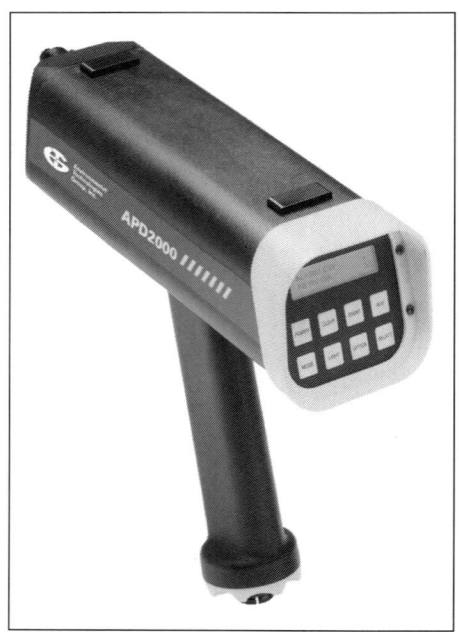

APD 2000 CW agent
detector
0055060

Additional options include an integrated radiation dosimeter, data logging of all detection and monitoring events, sample preservation and remote alarm delivery by RF.

Status
Available. Widely used by first responder organisations in the USA.

Manufacturer
Environmental Technologies Inc.

Agency
Barringer Technologies Inc (also marketed by other organisations).

UPDATED

Chemical agent detector kit ABC- M18A3

Description
The ABC- M18A3 replaces the earlier M18A2 version CW agent detector kit and comprises the following performance improvements:
- Individual detector tickets (40 per kit) are now foil-sealed.
- Elimination of the aerosol substrate dispenser (required to activate the detector tickets in the earlier version).

The M18A3 is aimed at the collection and identification of CW agents, TICs and TIMs. The kit contains sealable sampling tubes for the safe transport of unidentified but suspect samples to approved laboratories. It is designed for use at unit level by NBC specialist personnel. In addition to its use in the field to establish contamination extent, perform CW reconnaissance or check DECON operations, it is also suitable for the formal collection of samples for forensic analysis in the treaty verification role.

Agents detectable by ABC-M18A3:	
Agent	**Designation**
Cyanogen Chloride	CK
Mustard	H, HD, HN, HT
Phosgene oxime	CX
Hydrocyanic acid	AC
Phosgene	CG
Lewisite	L
Ethyl dichloroarsine	ED
Methyl dichloroarsine	MD
Nerve agents (all types)	V and G agents

Status
In production.

Agency
Tradeways Limited.

UPDATED

Chemical agent detector kit M256A1

Description
The M256A1 is a reagent-based chemical detection and verification kit. Presented in a lightweight (0.5 kg) plastic carrying case, it contains a sealed booklet of test papers (designated M-8), instruction cards and 12 pouches containing the chemical agent detector samplers. The samples are trapped onto the test discs for testing. The samplers are contained in crushable ampoules which the operator crushes to release the reagents. Exposure to air reveals a colour change appropriate to the agent. The instruction cards show colour swatches against which the test paper is compared.

Specifications

Target agents:	GA, GB, GD, GF, VX, HD, L, pepper spray, mace	
Sensitivity:	V agent at 4 ppb:	30 s (10 s in high concentrations)
	G agent at 15 ppb:	30 s (10 s in high concentrations)
	H agent at 300 ppb:	15 s (10 s in high concentrations)
	L agent at 200 ppb:	15 s (10 s in high concentrations)
Self test:	BIT for electronic, pneumatic and power conditions	
Operator service:	5 mins per 24 operational hours	
Power requirements:	AC (unspecified), 9-18 V DC from 6 standard 'C' batteries	
Data port:	RS232	
Environment:	–30°C to +52°C (operational)	
	–62°C to +71°C (storage)	
Reliability:	>2,000 h MTBF	
Dimensions:	100 x 90 x 280 mm	
Weight:	<3 kg	

The container is 180 × 130 × 75 mm and contains 12 detector sets, instruction cards and the M-8 booklet of detector papers. The kit is designed to test for G and V nerve agents, AC, CK, CX, H (and HD) and L.

Status

In production. Manufactured in Canada. See separate entry this section for further details and illustration.

Agency

Tradeways Limited.

VERIFIED

Chemical agent water testing kit M272

Description

This kit contains reagent sets for determining whether water has been contaminated with nerve (GA, GB, VX), blood (A) or blister (H, L) agents. The test tickets provided with the kit are first wetted with the suspected nerve agent, then left to soak for a specified time before being pressed against a reagent patch for test. The colour indicates the presence or absence (blue) of nerve agent. Water samples suspected of being affected by other agent types are treated with reagent fluids, introduced from reagent tubes provided in the kit. Samples of suspected H agent are heated before the reaction. Again, a colour change indicates the type and degree of contamination.

The kit is delivered in a convenient and robust carrying case.

Status

In production.

Development agency

Tradeways Limited.

VERIFIED

CW Sentry

Description

CW Sentry is an automated system for detecting trace levels of CW agent. It is designed for 24 hour monitoring of fixed installations, performing sample analysis every 60 seconds. Alarm concentrations are reported by means of simple switch closures or an optional RS232C serial communication line. The system uses a pair of SAW microsensors, similar to the SAW MINICAD Mk II (see separate entry this section).

Status

Developed in 1998 for issue. Current status unknown.

Developer

Microsensor Systems Inc.

UPDATED

Specifications

Alarm level: 0.2-1 mg/m³
Target agents: GA, GB, GD, GF, HD
Response time: 60 s
Alarm: Audible and LED
Warm-up time: 2 mins
Display: 8-point LED, including alarms and system performance. Reset and test buttons
Data output: RS232C (optional)
Operational environment: +5 to +40°C. RH 0-95% (non-condensing)
Power supply: 24 V DC at 0.3 A
Dimensions: 230 × 400 × 180 mm
Weight: 11.4 kg

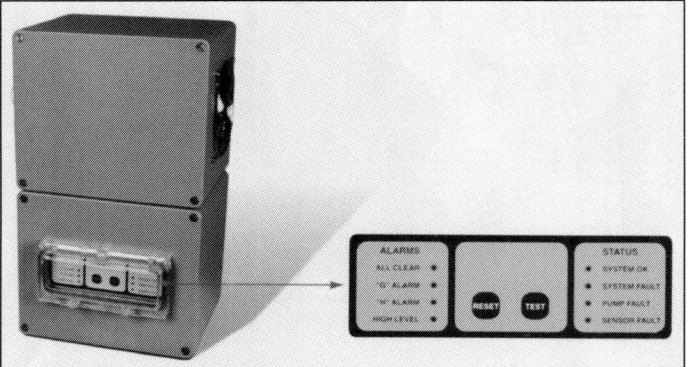

CW Sentry   0050700

ETG Advanced Portable Detector (APD) family

Description

Environmental Technologies Group Inc (ETG) is producing the Advanced Portable Detector (APD) family for the detection of chemical warfare agents for battlefield and environmental clean-up operations and for treaty monitoring and verification purposes. The APD can also perform the chemical agent monitoring role using a carry strap and user-selected 'Monitor Mode' which defers backflush until a high concentration is identified.

The family of detectors includes the APD Improved Chemical Agent Monitor, the APD-2100 and several application specific variants. The family features built-in tests for electronic, pneumatic and power conditions, RS-232/RS-422/RS-485 communications interfaces, the ability to remain fully operational in collective protection (ammonia) environments, on-the-move vehicle detection and low maintenance costs.

The APD, the successor to the ICAM-D, features the direct integration of the CAM/CAM2/ICAM monitor assembly or ETG's Sensor 2100 with ETG's advanced electronic platform to permit simultaneous detection of both nerve and blister vapour and aerosol agents.

The APD features an onboard visual alarm that can be programmed in any desired language, local and remote alarms, automatic cleardown and reset following a detection and power miser circuitry that delivers 60 hours of operation at +20°C with a standard set of four BA-5847 LiSo₂ batteries.

The advanced electronics contain: an improved detection algorithm that provides superior battlefield interference rejection; a remote communications capability allowing spectra and internal data to be monitored via modem from anywhere in the world; a programme download capability via the communications connector allowing upgrade for specific threats; a troubleshooting software package that provides self-diagnostics for every function.

The APD is currently configured to detect GA, GB, GD, VX, HD, HN and L. It can be reprogrammed to meet new agent threats or to detect any of the organic compounds or inorganic and acid gases typically detected with ion mobility spectrometry.

Sensitivity performances are as follows:
G - 0.1 mg/m³ <30 s; 1 mg/m³ <10 s
H/L - 0.1 mg/m³ <30 s; 5 mg/m³ <10 s
Weapons grade VX - 0.04 mg/m³ <10 s
The ADP weighs less than 5.5 kg. Accessories available include a remote alarm, a vehicle mounting and a hard-sided transit case.

Status

APD is in production and available in the USA and other licensed countries.

Manufacturer

Environmental Technologies Group Inc (ETG).

VERIFIED

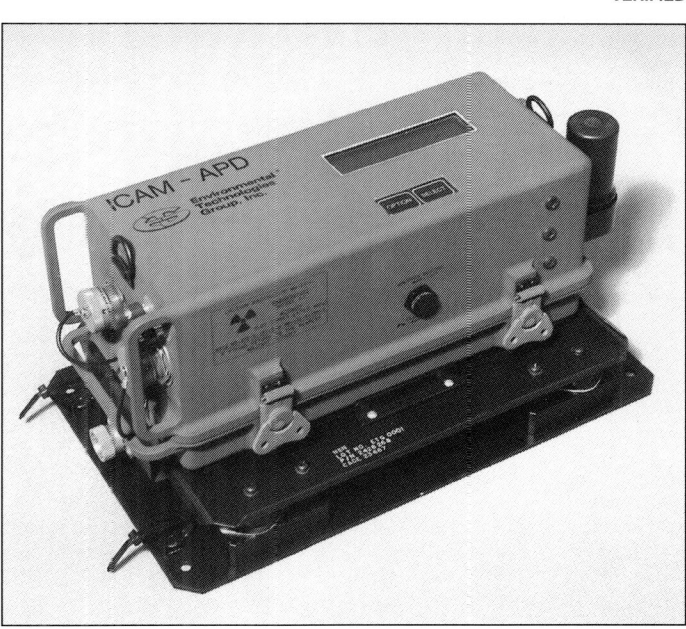

Advanced Portable Detector (APD)

Fixed Site/Remote Chemical Agent Detector

Description

The Fixed Site/Remote Chemical Agent Detector is a tripod-mounted unit. It offers a flexible choice of sensors including radiation, meteorological conditions and GPS reception as well as CW agent detection. It can be connected to a central management system at a base station, through a radio frequency, satellite, RS232 or land-line link. Its mode of operation is similar to BAWS (see under Detection (sensor systems) – Biological)). Of lightweight rugged construction, units can be

Fixed Site/Remote Chemical Agent Detector
0055059

Network components of a remote/mobile chemical agent detector network involving ICAD

linked together and moved as required to offer trend data during decontamination operations.

Status
Available.

Manufacturer
Environmental Technologies Group Inc.

VERIFIED

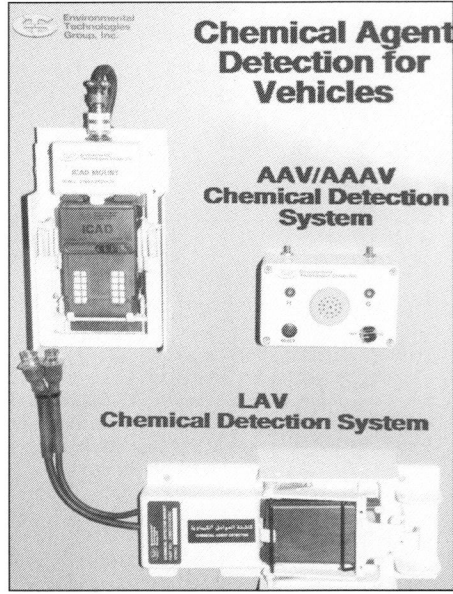

The main components for the ETG chemical agent detection system for vehicles
(T J Gander)

GI-CAD

Description
GI-CAD is an ultra-sensitive CW agent detector combining gas chromatography with IMS and proprietary two-dimensional detection algorithms to achieve rapid detection and identification. It is designed to be fitted in vehicles or to be used as a unit-level portable detector. The RS232 port allows digital data to be distributed over a data management network such as MICAD (see under Detection (C³I systems)).

Specifications
Target agents: GA, GB, GD, GF, VX, HD. L
Detection time: <2 mins
Power requirements: 28 V DC 100 W (peak)
Operating life: 200 h
Data port: RS232
Shock/vibration: To US MIL-STD-810
Operating temperature range: –30°C to +60°C
Reliability: >2,000 h MTBF
Dimensions: 305 × 380 × 200 mm
Weight: 11.3 kg

Status
Design candidate for US DoD CW detection suite in the Fox NBC reconnaissance vehicle (see entry under Detection (reconnaissance systems), the Lightweight NBC Reconnaissance System (LNBCRS) and Biological Integrated Detection System (BIDS) (see separate entry under Detection (sensor systems) - Biological).

Manufacturer
Orbital Science Corporation.

VERIFIED

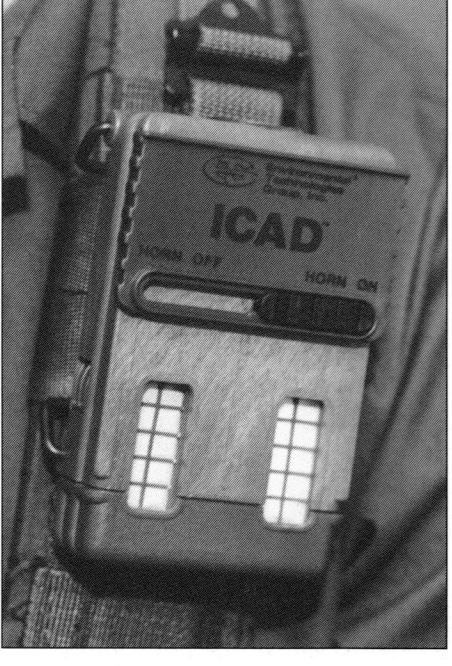

ICAD miniature chemical agent detector

ICAD miniature chemical agent detector

Description
Environmental Technologies Group Inc (ETG) developed the ICAD miniature chemical agent detector to address the chemical agent threat of the 1990s. The ICAD is in production and over 8,600 units have been delivered to the US Marine Corps and US Army.

The ICAD is a low-cost, miniature detector utilising an electrochemical method of detection. The ICAD is compact and lightweight, measuring 110 × 66 × 20 mm and weighing 215 g.

The system consists of an electronics module and a replaceable sensor module. The electronics module, consisting of the detector processor and an audible alarm and warning light, is reusable. The sensor module, which contains the battery power source and sensor cells, is disposable and provides four months of continuous use. The sensor contains three independent interference-rejection techniques and provides an alarm response to nerve, blister, blood and choking agents. After an initial alarm the sensor automatically resets and is capable of accepting subsequent chemical agent challenges.

Applications for the ICAD include a point sensor for squad level formations, a chemical warfare detection network for fixed and mobile installations, a detector for installation on board combat vehicles and a filter monitoring device for collective protection equipment. The ICAD has been dropped from a remotely piloted vehicle in a remote detection role.

The ICAD Chemical Warfare Agent Detector Network was developed to remotely detect the presence of chemical agent threats and transmit near-realtime warnings to decision makers over battlefield distances up to 5 km. The network identifies and provides visual and audio alarms to nerve, blood, blister and choking agent threats, individually and in combinations.

The network consists of up to 30 deployed field units, a base station and an installation kit. These components combine to produce a computer-generated display showing the exact location of each up to 30 fixed or moving detectors and, identify the class of chemical agent encountered. Field unit locations are identified using the Global Positioning System (GPS) included in the installation kit.

ETG developed a basic vehicle mount which can be adapted and customised for any vehicle. The system is capable of operating within the vibration, temperature, EMI/RFI and power supply environments. ETG is supplying 1,145 vehicle chemical detection systems for installation on LAVs. In addition, preproduction vehicle mount systems have been delivered to the US Marine Corps for installation on AAVs.

Status

The US Marine Corps purchased a large number of ICADs for use during Operation Desert Storm. ICADs were also deployed with US Army, US Navy and NATO forces operating in the Persian Gulf area. ICADs are also being purchased by other countries for use as point sensors and vehicle-mounted detectors.

Manufacturer

Environmental Technologies Group Inc (ETG).

VERIFIED

....

Improved Chemical Agent Monitor (ICAM)

Description

ICAM is one of the standard CW agent monitors in service with the US Armed Forces and the National Guard. The Improved Chemical Agent Monitor (ICAM) electronics assembly was type-classified by the US Army in August 1993 as a standard-issue item. It was designed as a low-cost, non-proprietary replacement for current CAM and ICAM electronics. ICAM is a hand-held battery-powered monitor for nerve and mustard agents which works on the IMS principle. Other features include an RS232 interface for IT-based data analysis, reduced power consumption, built-in diagnostics and the ability to reprogram the CAM/ICAM through the device's external connector. Built-in reporting features permit remote monitoring of detected agents with detailed description of agent detected and

Improved Chemical Agent Monitor (ICAM) (Intellitec)

ICAM improved electronics (Intellitec)   *2001*/0106865

concentration level. It uses special timing and microprocessor techniques to reject interference and false alarms. The ICAM upgrades the CAM (see separate entry within this section) by significantly improving reliability and maintainability.

Status

There were 3,716 units produced in FY00 (US Army: 2,934, US Navy: 390, National Guard 342). Final procurement target includes a further 3,100 ICAMs (US Army: 3,003, National Guard: 97).

Manufacturer

Intellitec Division, Advanced Technical Products Inc.

UPDATED

....

Improved (chemical agent) Point Detection System (IPDS)

Description

IPDS is a shipboard CW agent detection system which links distributed point sensors to a display unit in the Damage Control Center (DCC). The sensors are elongated IMS drift tubes designed to reduce the false alarm rate, especially from interferants commonly found on board. The integral data processing units use a software-based detection algorithm and library data. Developed to a stringent MilSpec, IPDS has been successfully tested against both the high shipboard RF/EM environment and the harsh atmosphere of the maritime environment. The following components comprise IPDS:

- External Air Sampling Unit (EASU). 2 per DD/FF-sized platform: one either side, mounted internally and sampling the atmosphere via a weather-proof airtight seal
- Detection Unit assembly (DU). These twin IMS cell assemblies are mounted close to each EASU to achieve detection. Contain B TE
- Control Display Unit (CDU). The whole system can be managed from the CDU in the DCC. The system can be checked and tested as well as displaying status, alarms, detection data and error messages.
- Remote Display Unit (RDU). Normally located in the alternate DCC (the bridge). Mimics information on the CDU.

The Improved Point Detector System extends the capability of the currently fitted CW detection installed on US Navy surface platforms.

Specifications		
Sensitivity	G and V nerve agents:	0.1 mg m⁻³
	H agent:	10 mg m³ +
Response time:		< 60 Sec
Power supply:		115 V AC (± 10%), 60 Hz (± 5%)
MTBF:		1,440 hrs
Environmental:	Operational:	0 to +50°C
	Storage:	-40° to +70°C

Status

Available. A production contract was awarded in October 1996 and initial testing was completed in December 1998. Fleet deliveries began in August 1999, with a fleet-wide distribution target of 235 systems. Six IPDS units had been installed by October 1999 and, at a rate of 60 systems per year, deployment is due for completion in FY2003. A new system called Ship ACADA (see separate entry this section) is due to phase replace IPDS over time.

Manufacturer

Powertronic Systems Inc.

Agency

Tradeways.

NEW ENTRY

....

Joint Chemical Agent Detector (JCAD)

Description

JCAD is a multimission CW agent point detection system currently in development for the US armed forces, where it is designed to fulfil the following missions:

- Local detection and warning. JCAD is small and light enough to be hand-held, or worn in pouches that attach to general issue webbing. JCAD units can also be more permanently installed in vehicles, aircraft, maritime platforms and fixed sites.
- **Area warning.** JCAD units can be placed singly or interfaced as part of a network around the perimeter of an air base or key installation.
- **Cumulative dose measurement.** JCAD will have the ability to accumulate and report cumulative concentrations (at nerve agent miosis levels) whilst, at the same time, providing continuous rapid alarm response to high concentration exposure from other types of agent. JCAD will store up to 72 hours of cumulative dosage data.

Specifications

Sensor type: Surface Acoustic Wave (SAW)
Operating temperature: –32 to +49°C
Operating altitude: 0 - 7,620 m
Volume: >655 cc
MTBF: >2,400 hours

Internal power: Stock BA-5800 LiSO$_2$ cells or rechargeable BA-380 NiMH cells
External power: 12 to 28 V DC or 110 to 240 V AC (at 50 - 400 Hz)
User interface: LCD display brackets operate in direct sunlight and is BG compatible; LED alert; Audio alert
External interface: RS 232

US DoD JCAD specification

Agent	Threshold detection level (mg/m-1)	Detection Response Time (Sec)	Temperature Range (°C)
VX	1	<10	–10 to 49
	0.1	<30	–10 to 49
	0.04	<90	–10 to 49
G-series nerve agents	1	<10	–30 to +49
	0.1	<30	–30 to +49
H-agent and Lewisite	50	<10	–18 to +49
AC blood agent	2,500	<10	–32 to +49
	22	<60	–32 to +49
CK blood agent	20	<60	–32 to +49

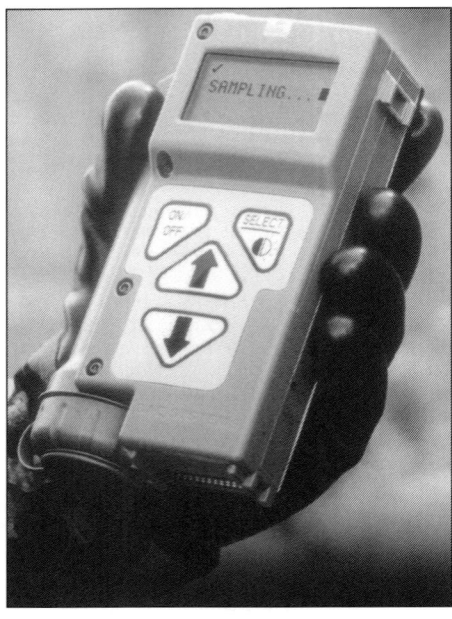

The hand-held SAW-based JCAD detector (BAE Systems)
2002/0132279

The unit can detect, identify quantify and report the presence of H, G or AC agents and TICs. Data is displayed to the user through an LED flat panel and a selectable audio or LED alarm. External data can be downloaded to a network for warning and reporting, or for later analysis though a RS 232 port. The communication protocol is compliant with the US Joint Technical Architecture (JTA) and the Joint Warning And Reporting Network (JWARN).

JCAD will operate on internal battery or external mains power. It is ergonomically designed for ease of use whilst IPE is worn and in low-ambient light conditions.

Status
Available. The US DoD is understood to be planning the procurement of over 270,000 JCAD units beginning in 2002. Full production is planned for FY04. The DoD JCAD programme is intended to provide replacement detection capability for all other current US military CW point detection systems.

Manufacturer
BAE Systems.

UPDATED

Joint Service Lightweight Standoff Chemical Agent Detector (JSLSCAD)

Description
The JSLSCAD is a modular, lightweight, passive, standoff chemical agent detector. It is capable of providing up to 360° on-the-move vapour detection from a variety of tactical and reconnaissance platforms at distances up to 5 km. The JSLSCAD is a second-generation chemical agent vapour detector and improves on the capabilities of the M21 Remote Sensing Chemical Agent Alarm (RSCAAL) first-generation system (see separate entry within in this section). Warfighter protection and manoeuvre unit combat capabilities will be increased with the JSLSCAD. When avoidance is not possible, JSLSCAD will provide extra time for personnel to don full protective equipment.

Objectives for FY01 include completion of Engineering Design Test and incorporate test results into system design; complete integration into JSLNBCRS, CH-53 and C-130; and to fabricate 40 Production Qualification Testing/Initial Operational Test & Evaluation (PQT/IOT&E) test article and conduct PQT/IOT&E.

Objectives for FY02 include procurement of 40 JSLSCAD for First Article Test (Army 30; Marines 10), and to refurbish 30 PQT/IOT&E units (Army 15; Marines 15).

Status
Production is planned during FY02 to FY07.

Manufacturer
General Dynamics.

UPDATED

Lightweight Standoff Chemical Agent Detector (LSCAD)

Description
The Lightweight Standoff Chemical Agent Detector (LSCAD) is intended for use on a variety of platforms including unmanned aerial vehicles, rotary- and fixed-wing aircraft, tracked and wheeled vehicles, amphibious vehicles and fixed emplacements. It scans the surrounding atmosphere for chemical agent vapours and so provides a passive standoff detection capability for both contamination avoidance and chemical reconnaissance. On ground vehicles LSCAD is equipped with a 360 × 60° scanner that provides wide area surveillance out to 5,000 m. Aerial operation can be co-aligned with video.

LSCAD is a passive infra-red detection system that detects the presence or absence of chemical warfare agents in the 800 and 1,200 wave number region of the electromagnetic spectrum by monitoring the ambient background infra-red radiation. It can operate when stationary or moving. LSCAD signal processing hardware discriminates between chemical targets and other non-toxic substances in a battlefield environment. If a cloud or vapour is cooler than the background an emission feature is observed. The spectral location and intensity of these features indicate the material or substance detected.

LSCAD consists of an interferometer block, servo electronics, timing and reference laser, a combined detector, cooler and preamplifier, a power supply, optics, and processing hardware and software. It is intended that the prototype will weigh less than 10 kg. Power is provided by either the carrier vehicle supply or a lithium battery.

The AAI Corporation are responsible for the gimbal mounting involved when LSCAD is carried by the Pioneer UAV, and the SSG Corporation will provide the 360 × 60° scanner.

Status
Development.

Development agency
Joint Program Office for Biological Defense.
Edgewood Chemical Biological Center.

UPDATED

M21 Remote Sensing Chemical Agent Alarm (RSCAAL)

Description
The M21 Remote Sensing Chemical Agent Alarm (RSCAAL) is a unit which can detect nerve and blister agent clouds at ranges up to 5 km. Development began in 1980 for the US Army's Chemical Research Development and Engineering Center at Edgewood, Maryland.

The M21 detection system utilises a passive infra-red sensing system which detects the background radiation in the field of view and determines if chemical agents are present. Chemical agents are identified by their infra-red spectral absorption signature. The M21 has a horizontal field of view of 60° which is

*M21 Remote Sensing
Chemical Agent Alarm -
RSCAAL*
(Advanced Technical
Products Inc)

scanned in less than 60 seconds. The detector has an all-weather capability and is fully automatic in operation, being capable of unattended operation for 24 hours. The system can be utilised in a tripod-mounted, fixed site or vehicle-mounted configuration.

The detector design is based on a Michelson interferometer and the use of an MC68000 microprocessor. The complete equipment consists of the detector which weighs about 23.6 kg (dimensions are approximately 508 × 432 × 254 mm), a tripod weighing 6.8 kg, an M42 remote alarm (used with the existing M8 alarm system), a power cable and a transit case. Two people can carry the complete equipment and it takes 10 minutes to prepare for unattended use.

The sensitivity of the M21 for nerve agents is 90 mg/m². For Lewisite (L) the sensitivity is 500 mg/m² and for Mustard (HD) it is 2,300 mg/m². The operational temperature range is from −32 to +48°C.

The M21 is powered by batteries or by standard military generators delivering 120 W of start-up power, normal power consumption is 80 W.

The M21 detector is constructed from nine major hardware components. These are the base assembly, the spatial scanner, the interferometer, an infra-red detector/cooler, analogue signal electronics, a signal processor, a power supply, a control panel and a sighting device. Using the sighting device on top of the detector, the M21 is normally placed looking into the wind to view a scene which consists of the background and the air path along its line of sight to the background. The M21 measures and stores a background spectrum which consists of the ambient energy contained in the scene. When an agent cloud enters the line of sight, the scene is spectrally altered by the absorption/emission characteristics of the cloud. The M21 software detects the changes and provides an alarm. The M42 alarm may be more than 400 m from the detector. The M21 is provided with an RS-232 data port for external command and control.

It has been proposed that detection devices based on the M21 could be mounted on reconnaissance vehicles, helicopters and remotely piloted drones. When fitted to the M93A1 Fox NBC reconnaissance vehicle the M21 is mounted on an extending telescopic mast, with the detector head controlled in azimuth and elevation by a joystick from inside the vehicle.

A total of 205 units have been delivered to the US Army. A further 125 systems have been delivered to the US Marine Corps, where the M21 is used in the tripod mode only. A number of systems were successfully deployed during Operation Desert Storm. The unit was Type Classified in March 1995.

Status
In production for the US Army and Marine Corps.

Manufacturer
General Dynamics Armament and Technical Products.

UPDATED

M22 Automatic Chemical Agent Detector and Alarm (ACADA)

Description
The US ACADA system is based on the Graseby Dynamics GID-3 (see separate entry within this section). An automatic chemical agent alarm system capable of detecting, warning and identifying standard blister and nerve agents simultaneously, ACADA is man-portable, operates independently after system start-up, and provides an audible and visual alarm. It is fitted with a communications interface for automatic battlefield warning and reporting, can operate with the

M279 Surface Sampler and is compatible with MICAD (see separate entry within Detection (C³I Systems). Improvements over the M8A1 Automatic Chemical Agent Alarm System include increased sensitivity, decreased false alarm response to interferants, ability to operate inside COLPRO, and suitability for fixed operation on and inside mobile vehicles.

The ACADA was selected by the US Armed Forces after extensive trials conducted in competition with two other detector manufacturers. These trials included environmental testing in the Arizona desert, Panamanian jungle and Alaska.

Production for land-based ACADAs runs from FY00 to FY02. An ACADA system has been designed for ships, based on a different product, manufactured by STR Inc (see separate entry this section). M22 production is planned to be complete by FY01.

Status
Available US procurements for FY00 were 4,655 Army; 398 National Guard; 235 Ship ACADAs. FY01 objectives are to procure a further 6,903 ADADAs and 300 M279 Surface Samplers for the Army.

Manufacturer
Graseby Dynamics Limited (UK).

Agency
SBCCOM.

NEW ENTRY

MINICAMS®

Description
The CMS series 3000 MINICAMS® is an automatic, continuous air monitoring system using gas chromatography and selected detectors and samplers to monitor for the presence of CW agents, simulants and related military compounds. MINICAMS detects CW agents at the US Surgeon General's, 8 hour time-weighted average concentrations. It is a fieldable, expendable system developed specifically to detect CW agents and simulants including GA, GB, GD, HD, VX, DF, DFP, DIMP, half mustard, MES, DEM and OSDMP.

The unit is compact, lightweight (less than 20 lb), and highly portable. It measures approximately 1 ft³ in size. It has been ruggedised for field use and its modular design enhances configurability and maintenance. The field MINICAMS automatically collects an air sample, performs an analysis and reports the results. Many operating parameters can be changed conveniently by the operator to meet site-specific requirements. Once the parameters are changed, the system continues to operate automatically until operator intervenes.

MINICAMS using the solid-sorbent sampling configuration (Module SFI-100) is a sophisticated analytical and alarm system based on the concentration of agents and simulants from a large volume of air using solid-sorbent collection. Components in the sample are separated using temperature-programmed, capillary gas chromatography and detected using systems specified by the client. Examples include: Flame-Photometric Detector (FPD), Flame-Ionization Detector (FID), Photo-ionization Detector (PID) or a Halogen Selective Detector (HSD). The solid-sorbent sample collection provides the sensitivity necessary to determine the chemical agents at Time-Weighted Average (TWA) concentrations, which range from 0.9 parts per trillion (pptv) (0.00001 mg/m3) for VX to about 450 pptv (0.003 mg/m³) for agent HD. Applications using the solid-sorbent configuration include TWA-level monitoring to protect workers from exposure to chemical agents

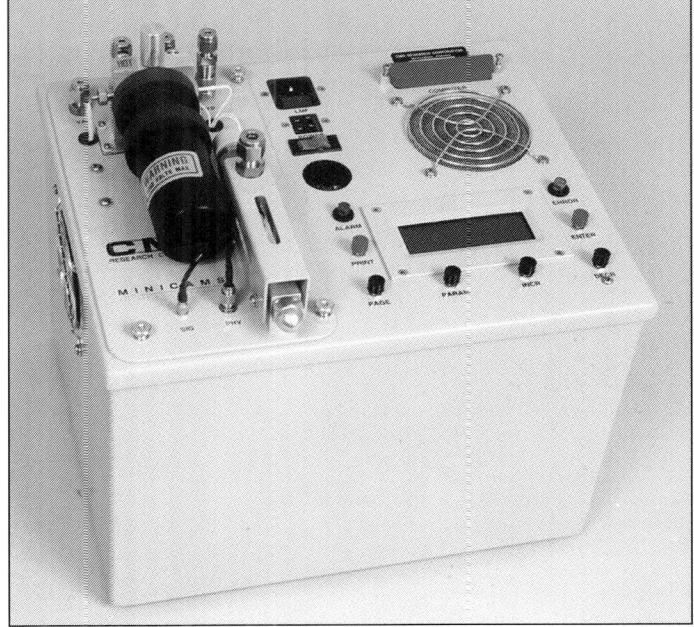

MINICAMS® (OI Analytical)   0011431

and other monitoring tasks requiring measurement of agents, simulants and related compounds at extremely low concentrations.

MINICAMS using sample-loop collection (Module LFI- 100) is identical to the system above except that its sensitivity range is extended to higher concentrations. A small volume of air is collected in the sample loop and injected on the gas chromatographic column for separation. Typical sensitivity for agents and simulants ranges from 0.1 to several hundred parts per million by volume (ppmv) or 0.1 to 10,000 mg/m³. This sampling configuration is useful in laboratory testing and other evaluations in which large concentrations of agents or simulants must be measured.

CMS offers a choice of seven detector options for use with the MINICAMS. Although, a FPD is normally used for chemical agent monitoring, other detector combinations are available for the determination of other classes of compounds. The FPD detects phosphorus and sulphur-containing compounds such as GA, GB, GD, VX, and HD at TWA levels. The FID detects simulants and other compounds that do not necessarily contain sulphur or phosphorus such as MES and DEM. The FPD/FID includes two detectors operating simultaneously from one flame. The PID detects aromatics and other VOCs such as MES. The FID/PID detects VOCs, saturated hydrocarbons, MES and aromatics. Two new MINICAMS detectors are the XSD which provides real time monitoring for lewisite, mustard, chloroform, chloropicrin, phosgene, cyanogen chloride and nitrogen chloride mustards; and the Pulsed Flame Photometric Detector (PFPD) for the monitoring of G-nerve agents and VX. All detectors or combinations of detectors described above are completely interchangeable and take less than 10 minutes to change. The MINICAMS system is designed with a degree of flexibility that allows rapid changeout of detectors/detector combinations without requiring changes in other elements of hardware or software. The CMS MINICAMS system includes options and accessories for remote, fixed-site and laboratory operations including multipoint sampling; Depot Area Air Monitoring System (DAAMS) for remote vapour and liquid sampling, MINI-LINK and MINI-NET systems for local and remote data reporting, storage and communications; heated sample lines, customised fly-away cases for mobile operations. The MINICAMS is being used for air monitoring at US CW demilitarisation sites and facilities and for air monitoring at US CW depots and storage sites. Laboratory versions of MINICAMS are used in CW laboratory studies and for the permeation testing of CW protective fabrics. MINICAMS also have been installed on vehicles for mobile CW monitoring.

Status
CMS has sold over 450 MINICAMS to US government and international customers.

Manufacturer
OI Analytical/CMS Field Products.

UPDATED

Portable Continuous ppb VOC Detector Monitor (ppb Rae)

Description
The ppb RAE is a hand-held Photo Ionization Detector (PID) with internal pump. It measures VOCs and other ionizable compounds with a 3 second response time and is accurate enough to detect continuously at levels down to 1 ppb. The user is

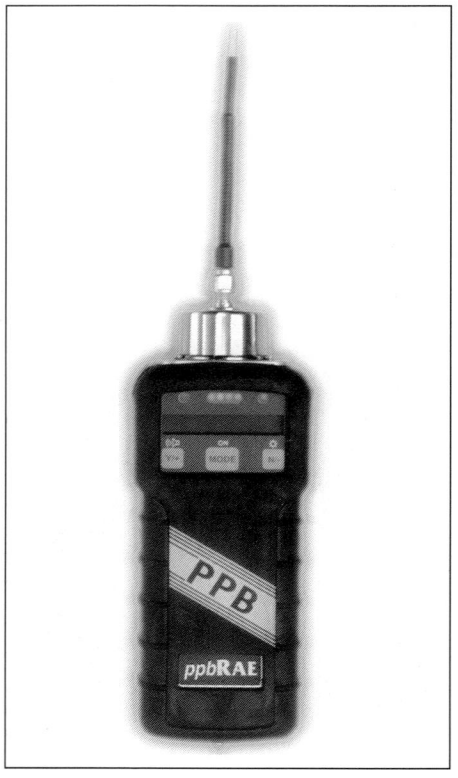

ppb RAE
2001/0106866

able to control the sensitivity and to 'zero out' the background VOC level. Default baseline readings can be restored without affecting the calibration. ppb RAE is designed to measure highly toxic compounds with low vapour pressures. A 'bagging port' enables sample capture of unknown compounds for laboratory analysis. Computer-based data logging, working in the MS Windows™ environment shows concentration against time. The ppb RAE has a high-capacity internal memory and is powered by rechargeable NiH batteries, giving an endurance of over 10 hours (continuous if connected to wall charger). Thus, VOCs concentrations can be monitored for up to 7 days and later analysed on computer.

Status
In service with the US Marine Corps (CBIRF), FBI and Secret Service and US National Guard Civil Support Teams (CSTs previously called 'RAID' teams).

Manufacturer
Rae Systems.

VERIFIED

Rae Systems range of Photo Ionisation Detectors (PID)

Description
Photo Ionisation Detectors (PID) produce a measurable electric current, proportional to the concentration of a sample of ionisable gas passing by a UV lamp. For ionisation to take place, the Ionisation Potential (IP) of the gas must be lower than the light energy of the ionising lamp (measured in eV in this context). With a range of lamps, the VOC component from a wide selection of compounds can be measured and, by using electrodeless lamps, Rae Systems claims significant performance advantages in longer life, lower power consumption and the avoidance of RF or EM interference. Both TICs and CW agent breakdown products have discrete measurable VOC components and, according to research by Rae Systems, the thresholds of particular components can produce results which tie them to particular compounds in particular agents. The PID in fact detects the natural breakdown products of the agent and there is a close correlation between the threshold levels of components in a gas passing over a PID and the

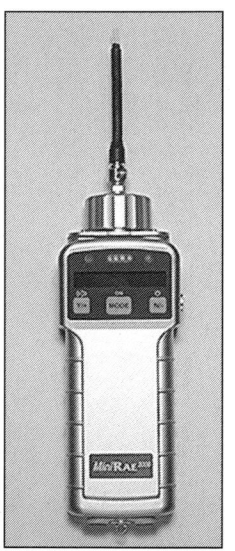

The MiniRae 2000 portable VOC monitor
0055635

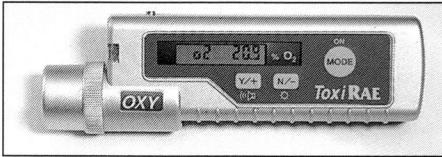

The ToxiRae PLUS personal gas monitor
0055120

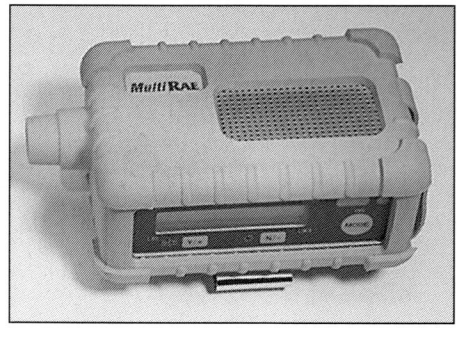

The MultiRae PLUS gas monitor with VOC detection
0055119

composition of a CW agent. A significant advantage of PID technology is that it measures continuously. Current lightweight CW agent detection and identification technology, IMS for example, measures a 'snapshot' of the inspired air sample at the nozzle at the moment it is ionised. Whilst not currently agent specific, PIDs measure the concentration of the discrete volatile component continuously. With further development, the technology appears to combine the advantages of continuous monitoring with high-confidence agent identification - a significant development. Additionally, PID measurement ranges down to the order of parts-per-billion would further widen the range of cover.

A further advantage of PID technology is its compactness. Primarily designed to measure TIC levels in the atmosphere or percentage oxygen levels in a confined space, the Rae series of adaptable VOC meters (which includes the MiniRae 2000, MultiRae PLUS and ToxiRae PLUS systems) is extremely compact, especially when compared with existing hand-held IMS or GC/MS systems. As an example, the MiniRae 2000 offers a 0-10,000 ppm range, 10 hours continuous data logging, 15,000 point data storage, easily downloaded to a PC and 102 built-in correction factors from a list of 250 chemicals. The lamps and the battery are easy to change and the device is ease to use whilst wearing NBC IPE.

The range of gas monitors produced by Rae Systems is widely in service in the US, being issued to national security staff at high level as well as to environmental monitoring authorities and organisations.

Status
In service with US HazMat specialists, including US Marine Corps (CBIRF), FBI and Secret Service.

Manufacturer
Rae Systems.

VERIFIED

SABRE 2000

Description
Sabre 2000 is a hand-held, battery-operated chemical detector designed for ease of effective operation by non-technical personnel. Aimed primarily at the first responder market, it docks on a quick-release base station which continually charges the battery and maintains full functionality of the unit. SABRE 2000 is a dual-mode instrument. Its detection system allows it to be targeted against classic CW agents or TICs. Alternatively, it can be used to look for explosive materials in its general security role. Its detection library covers more than 30 CW agents including Tabun, Sarin, Soman, Cyclosarin, VX and V, Nitrogen Mustard 3 and Mustard gas. It can analyse both vapour and trace particulate samples. The explosive detection range includes RDS, PETN, TNT, Semtex, NG and Ammonium Nitrate. In this role,

---

Specifications
CW agent range: V and G nerve agents including Tabun, Sarin, Soman, Cyclosarin, VX and Vx, Nitrogen Mustard 3 and Sulphur Mustard
Explosives range: RDX, PETN, TNT, Semtex, NG, Ammonium Nitrate and others
Drugs range: Cocaine, Heroin, THC (Cannabis) Methamphetamine, and others
Analysis time: 15-20 seconds
Warm up time: Less than 10 minutes
Size: 33 × 11.5 × 13 cm
Weight: under 2.6 kg (with battery)
Power supply: 12 V DC, 110/220 V AC; 2-hour battery

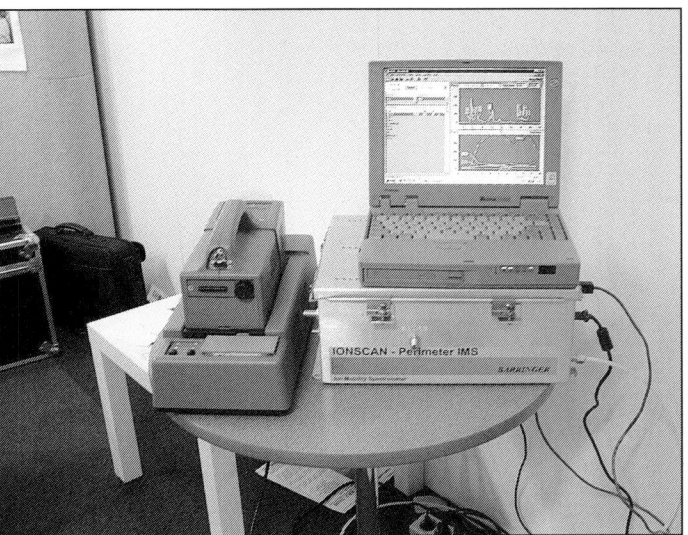

SABRE 2000 shown in its battery charger/docking station (left). The notebook computer shows full analysis of the detected substance through a MS Windows®-based user interface (John Eldridge)   ***2002**/0137270*

SABRE 2000 CW agent and explosives detector showing indicator panel, detector nozzle (at left) and detector ticket in use (top, in front of LED panel) (John Eldridge)   ***2002**/0137271*

SABRE 2000 is used to screen suspicious items received or abandoned in public places, conduct routine checking of vehicles, or of people, at entrances to facilities or during times of high security alert. SABRE 2000 uses the proven IMS-based IONSCAN® technology used in other Barringer security products.

Status
Available. In use with many security organisations.

Manufacturer
Barringer Instruments Inc.

NEW ENTRY

Sampling kit, CB agent, M34 and M34A1

Description
The M34 kit can be used to sample soil, water and other items for the presence of various chemical and biological warfare agents. The kit is issued in a fibreboard box which weighs just over 2 kg and consists of two soil sampling kits, a glass phial container and two pairs of gloves. The kit can be used for preliminary soil testing. The original M34 is superseded by the M34A1 which replaces the glass components with Teflon-coated jars. There are four sections in the carrying case to allow easy location of each essential component required for the collection process. Soil, liquid, and other surface samples can be collected for analysis. Together with the Teflon containers, there are discardable spatulas and other materials for taking several of each type of sample. The jars can store contaminated samples at a range of temperatures up to 120 °F. Other components in the kit include gloves, tweezers, syringes, scoops, and M8 Detector Paper (see separate entry within this section). Sample labelling materials are also included.

Status
In service with the US Army, Navy and Air Force.

Manufacturer
Edgewood CB Center.

UPDATED

SAW MINICAD Mk II

Description
The MINICAD Mk II is a highly selective, lightweight, solid state, personal CW agent detector. Powered by four standard Li batteries or a rechargeable battery pack, it simultaneously detects trace levels of H, G or V agent, using a pair of surface acoustic wave (SAW) microsensors. SAW sensors are piezoelectric crystals that

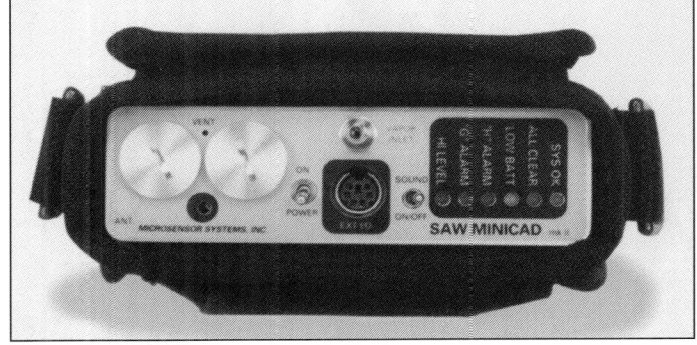

SAW MINICAD Mk II   0050699

Specifications

Alarm level:	0.2 - 1 mg/m^3
Target agents:	GA, GB, GD, GF, HD
Response time:	60 s
Alarm:	Audible and LED
Warm-up time:	2 min
Display:	6 LEDs
Data output:	RS232, 2400 Baud (6-pin mini-DIN) or optional low power 50 MHz AM radio beacon
Shelf life:	5 years (Limit: RAM battery)
Operational environment:	, +5°C to +40°C. RH 0-95% (non-condensing)
Operating life:	2-4 h (internal battery). 6-8 h (external)
Dimensions:	32 × 110 × 130 mm
Weight:	0.5 kg (including batteries)

detect the mass of chemical vapours absorbed into chemically selective coatings on the sensor surface. This absorption causes a change in the resonant frequency of the sensor. The internal microcomputer measures these changes and uses them to determine the presence and concentration of CW agent. The SAW sensor coatings have unique physical properties which allow a reversible absorption of chemical vapours.

Status

In service with a wide variety of US government and city authorities including the Defense Protective Service, the Federal Emergency Management Agency and New York City HazMat teams.

Developer

Microsensor Systems Inc.

Manufacturers

Analytix Inc.
Triquint Semiconductor (Sawtek Division).

Distributors

Environmental Safety Group (ESG).

UPDATED

Shipboard ACADA MK27 MOD 0

Description

Science & Technology Research Inc produces the US Navy's version of the Automatic Chemical Agent Detector and Alarm (Shipboard ACADA) system at its production facility in Fredericksburg, Virginia. See separate entry in this section for land-based ACADA. The Shipboard ACADA is a portable, battery-powered detector, alarm and measuring system with a low false alarm rate in multi-interferent environments. Vapour analysis is achieved by the use of two Ion-Mobility Spectroscopy (IMS) cells; a radioactive source, sealed inside each cell, is used as an ionizer.

Operational performance of the Shipboard ACADA system has been successfully tested in the field and laboratories against live agents and against various interferents present in shipboard environments. Laboratory tested agents include GA, HD, GB, GD and VX. The system is easily upgradeable for new and novel agents.

The Shipboard ACADA can operate for approximately three hours on its rechargeable battery box and indefinitely on its 110 V AC power cord. Typically two battery boxes are provided with each detector unit. The detector is equipped with a wand on a five-foot long hose for monitoring small spaces. The system also can be configured to monitor ventilation lines.

Specifications

Response time: typically less than 3 seconds
Interferents: rejects most, 1 to 3% false alarm rate in shipboard environment
Start-up time: 30 minutes (from cold start)
Alarm: audio and visual
Size: 152 × 152 × 381 mm high
Weight: 11.5 kg including battery
Power: battery for mobile use, AC for fixed
Communications: RS 232 cable port

Status

There were 235 Ship ACADA units procured for the US Navy. Production completed mid-2001.

Manufacturer

Science & Technology Research Inc.

NEW ENTRY

Shipboard Automatic Liquid Agent Detector (SALAD)

Description

The Shipboard Automatic Liquid Agent Detector (SALAD) is a US Navy programme being undertaken in recognition of the chemical warfare liquid agent threat to maritime forces in littoral operating areas. The programme is sponsored by the Chief of Naval Operations (CNO) Surface Ship Survivability Directorate (N86D) and managed by the US Naval Sea Systems Command Program Manager for Chemical and Biological Defense located in Arlington, Virginia. Also involved as an industrial partner is the Science & Technology Corporation.

SALAD is an automatic, externally mounted, liquid agent point detection system using off-the-shelf technology that will recognise blister and nerve agents. It consists of a detector package housing chemically treated paper, optical scanners, a central processing unit which recognises the changes in paper optical qualities when exposed to agents and alarm devices in key locations, such as the damage control station and the bridge, in a ship. The system can be operated and maintained by personnel wearing chemical protective clothing.

Preliminary specifications include a weight of 22.7 kg and a volume of 0.11 m^3.

Status

Development.

Manufacturer

Science & Technology Corporation.

Sponsor/agency

CNO/N86D.

VERIFIED

WADI mobile NBC field laboratory

Description

The WADI is a compact, completely equipped, self-contained mobile NBC laboratory designed for field service. The rugged, custom-built steel and aluminium body is available for use on 4 × 2 or 4 × 4 drive chassis or trailers. The interior is designed to permit the conduct of many laboratory tests and is both air conditioned and heated. It has its own 12.5 kW petrol-engined power source. Liquid propane bottled gas is piped to various outlets and around the laboratory area and storage is provided for all equipment. The WADI can be designed for use as a NBC, medical, environmental pollution or general field laboratory.

The basic laboratory equipment includes the following: a liquid propane gas system with outlets, bunsen burners and a stove; a refrigerator, incubator and fume hood; various instruments including scales, oven, glassware, balance, forceps, tongs, filters, mortars and pestles; and a centrifuge, water bath and microscope.

Specialised equipment can provide the WADI with the following NBC defence capabilities: air sampling from battlefield environments and disseminating surfaces; chemical agent detection; airborne collection of bacteria and fungi; medical diagnosis; water testing; radiation detection and measuring; toxicity detection and measurement; and residual chemical warfare agent detection.

The approximate interior dimensions of a WADI field laboratory are 4.11 × 2.29 × 2.13 m.

Status

Available.

Marketing agency

Tradeways Limited.

VERIFIED

A typical WADI mobile NBC field laboratory

YUGOSLAVIA, FEDERAL REPUBLIC

AHD automatic chemical detector

Description
The AHD automatic chemical detector is a backpack detector designed for alternative installation in combat and reconnaissance vehicles and static installations. It is an automatic detector which continually monitors the atmosphere to detect vapours and aerosols of chemical warfare agents including GB, GD, VX, CG and AC. If any of these agents are detected, audio and visible warnings are provided. Response time is a maximum of 60 seconds.

The complete equipment comprises the following: automatic detector unit (ADJ); remote alarm unit (DAJ); disposable material set (KPM); power supply unit 12/24 V DC (DNP-83); power supply unit 110/220 V AC (ANP-84); carrying and installation kit; and spare tools and accessories (RAP).

The detector can operate for 12 hours without an electrolyte change; the KPM set provides sufficient consumable for 360 hours of operation. Preparation time for temperatures higher than +15°C is 10 minutes. For temperatures between –30 to +15°C the preparation time is 60 minutes. The operating temperature range is from –30 to +55°C.

Dimensions of the automatic detector unit are 230 × 150 × 320 mm; weight is 8 kg. Dimensions for the remote alarm unit are 230 × 164 × 85 mm.

Status
In service with the Yugoslav armed forces.

Contractor
Yugoimport SDPR.

VERIFIED

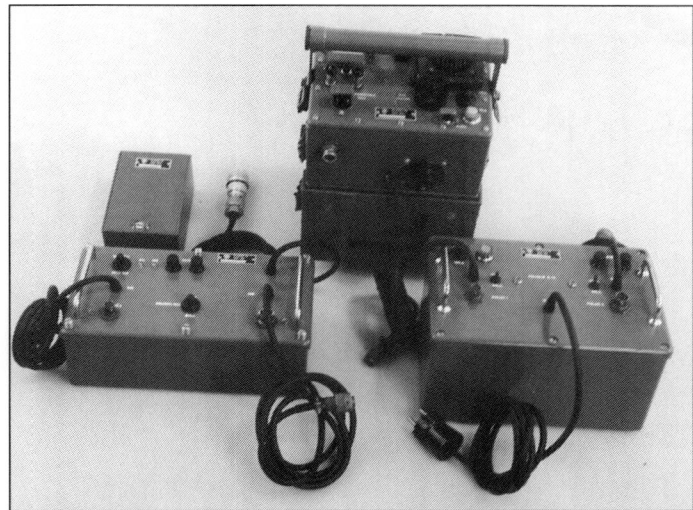
AHD automatic chemical detector

DHM-11B chemical detection kit

Description
The DHM-11B chemical detection kit is used for taking and identifying detection samples from chemical agents in the air, on the ground, on equipment, in loose material and in granular forms of food. It uses the pump and indicator tube method of sampling involving colour changes of chemicals. Agents that can be detected include the nerve gases, hydrocyanic acid, cyanogen chloride, phosgene, diphosgene, mustard agents and the nitrogen mustards.

The DHM-11B consists of a shaped sheet steel carrying case with a shoulder or belt strap, a manual piston pump with an adaptor, an indicator tubes holder, indicator tubes, an accessories holder and various accessories and spare parts. In use various indicator tubes are placed in the end of the pump and air is hand pumped through them for about 5 to 10 pump actions. The indicator tubes are carried in a double-ended and tubular aluminium holder with provision for 20 indicator tubes at each end. Before use each indicator tube has both ends cut off by using a cutting device on the pump. The tubes are then inserted into the pump for the sampling process. Various colour codings on the tubes denote their detection capabilities, as follows:

Tube with yellow and green ring with yellow point: phosgene, diphosgene, mustard, nitrogen mustard;

Tube with blue ring and blue point: cyanogen chloride, hydrocyanic acid, tabun, CN group;

Tube with red ring: nerve agents.

The normal operating procedure is to use the tube with the red ring first, then the tube with the blue ring and blue point and finally the tube with the yellow and green ring and yellow point. Preparation time is between 1.5 and 3 minutes; each sampling takes between 1.5 and 5 minutes.

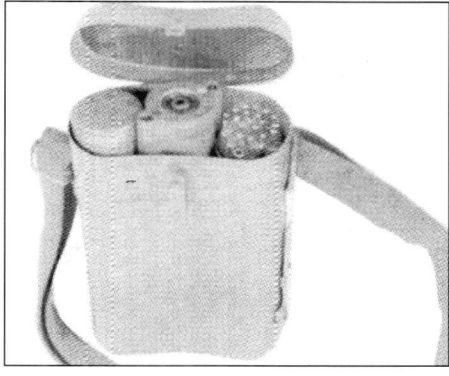
DHM-11B chemical detection kit

There is also a tube filled with silica gel for the sampling of aerosols and another filled with cotton wool for sampling unknown fumes. After the pumping action has been completed these tubes are resealed at both ends with plastic plugs and sent to a mobile chemical laboratory for further analysis. The DHM-11B kit also contains various colour-change detection papers and, a small sampling spoon is carried on the outside of the case for taking specimens to be placed into jars or sample tubes. A torch is normally carried with the kit. The pump adaptor is used to take agent samples from equipment.

Extra indicator tubes are delivered in sealed packs, each pack holding five tubes.

Status
In service with the Yugoslav armed forces.

Contractor
Yugoimport SDPR.

VERIFIED

DRHT radiation and chemical detection equipment

Description
This equipment is designed to offer nuclear and CW agent warning to armoured vehicle crews. For further details, see under Detection (sensor systems) - Nuclear.

Status
In service with the Yugoslav armed forces.

Contractor
Yugoimport SDPR.

VERIFIED

KDBOT personal detector kit for chemical agents

Description
The KDBOT personal detector kit for chemical agents is used to take samples from the atmosphere and equipment surfaces to determine if it is safe to remove personal protective equipment. The kit contains 12 personal detectors, a booklet of chemically-treated detection paper SDP-1 and an instruction manual and list of parts.

Each personal detector has a test spot for nerve agents (G and V), blood agents (AC and CK), mustards (H and HD), phosgene (CG) and lewisite (L). If an agent is present each test spot develops a colour. Observations on safe/danger conditions are printed on the back of the detector.

SDP-1 detection paper detects liquid nerve and blister agents by colour changes. It has a self-adhesive back to allow it to be fixed to clothing, equipment or weapons. A comparison chart is printed on the booklet cover.

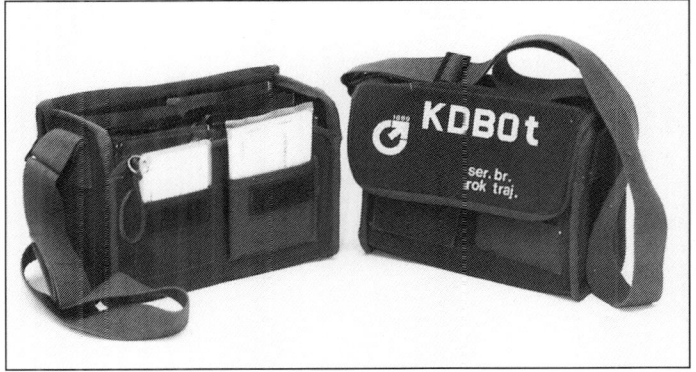
KDBOT personal detector kit for chemical agents

The kit requires about 15 minutes for a complete analysis of vapours, aerosols and chemical agent droplets. The kit measures 235 × 115 × 175 mm and weighs 0.8 kg.

Status
In service with the Yugoslav armed forces.

Contractor
Yugoimport SDPR.

VERIFIED

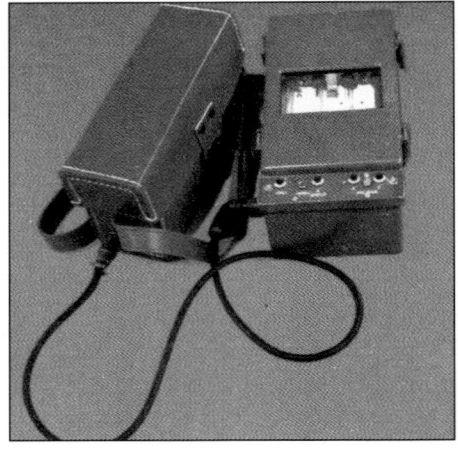

PHD semi-automatic chemical agent detector

PHD semi-automatic chemical agent detector

Description
The PHD semi-automatic chemical agent detector is a portable unit connected by a cable to a power supply contained in a separate case provided with a shoulder strap. A second case is required to carry consumables, spare parts and accessories. It is intended both for continuous and occasional monitoring of the presence of chemical agents in gaseous and aerosol form. The detection response time for agents such as mustard (H) and phosgene (CG) does not exceed 10 to 20 seconds for battlefield agent concentrations but may be of the order of 3 to 5 minutes for small agent concentrations. Alarms are provided by flashing lights on the detector unit.

The detector is powered by two 6 V batteries or from a vehicle's 12 V supply; power consumption is less than 700 mA. Weight of the complete equipment is 5 kg.

Status
In service with the Yugoslav armed forces.

Contractor
Yugoimport SDPR.

VERIFIED

DETECTION (C³I systems)

DENMARK

NBC-ANALYSIS

Description

NBC-ANALYSIS is a risk management software system designed to deliver decision support to commanders faced with hazards arising from the use of Nuclear, Biological and Chemical (NBC) weapons and Releases Other Than Attack (ROTA) of radiological material and Toxic Industrial Materials (TIM). It is available as a Commercial Off The Shelf (COTS) product, available to military or civil emergency authorities and, in its stand-alone version, includes extensive mapping functions, message handling, a communications module and the capability to display unit positions. Integration into C⁴I systems allows NBC-ANALYSIS to connect with external data sources including cartographic data (in either vector or raster format), unit databases and external communications.

NBC-ANALYSIS is in service with the majority of NATO commands, members and other nations. The system can be customised to meet national requirements and both the USA and UK use adapted versions (NBC-ANALYSIS for JWARN and BRACIS NT respectively. See separate entries within this section).

NBC Capability

Reports of an attack or toxic release can be entered into NBC-ANALYSIS automatically or manually, using data received from detectors via a range of communications channels. The software system automatically calculates the expected hazard area and displays it on a map of the region of interest. The same data is used to highlight units at risk. ROTA predictions are calculated for the following types of releases:

- Nuclear material released into the atmosphere from reactors, nuclear fuel reprocessing or production facilities.
- Releases from nuclear waste or radiological material storage facilities.
- Intentional radiological release.
- Releases from stockpiled BW agents, bunkers or production facilities.
- Releases from stockpiled CW munitions.
- Toxic industrial material released during transport whilst in storage.

NBC-ANALYSIS integrates prevailing meteorological information with hazard data to calculate the current and predicted hazard zones, allowing clear command evaluation of a NBC or ROTA event. Unit locations and deployment areas can be displayed on the map using NATO standard symbology (APP-6A - in turn based on the US MIL-STD-2525A). The system complies with NATO ATP-45(B) and AEP-45 (STANAG 2497) and NBC formatted messages are automatically generated in accordance ADatP-3. Nation-specific message formats have also been implemented. For example, the US Army has specified its Joint Variable Message Format (JVMF). Although designed to the NATO standard, the systems are considered Commercial Off The Shelf (COTS) products. In addition to the NATO standard warning areas, national diffusion models can also be incorporated. No software development is required to interface with models that comply with the ATP-45(B) message format. An example of such a model is the HAPPIE-RIOT model produced by the TNO Prins Maurits Laboratory in the Netherlands. Models that do not have an ATP-45(B) interface can also be integrated but may require some development to link to the model interface. An example of this approach is the integration of VLSTRACK for the JWARN project in the USA.

An additional EOD (Explosive Ordnance Disposal) module is available for calculating hazards arising from unexploded munitions.

NBC-ANALYSIS uses NBC 1-6 NUC, BIO, CHEM and ROTA Reports, NBC STRIKWARN, NBC SITREP as well as NBC BWR (Basic Wind Report), NBC EDR (Effective Downwind Report) and NBC CDR (Chemical Downwind Report) for the reporting of source data and prediction results.

The map module can utilise vector or raster maps and aerial photographs. Maps are normally sourced from national mapping institutes/agencies and can be imported using NATO or US mapping standards, such as VPF, ASRP, CADRG, ADRG or CRP.

The data communication module handles the transmission of NBC reports. Communication can be made over military or public telephone lines, secure military radio communications or mobile phones. Reports can also be passed via, cable, Local Area Networks (LAN) or the existing communications utility in MS Windows®. E-mail and fax can also be used.

Detector interfaces

Bruhn NewTech first demonstrated integration of detection results into NBC-ANALYSIS in 1995 by linking the system to GPS, radiological, chemical and meteorological sensors. Most modern detector systems have built-in digital interfaces that allow data to be analysed and forwarded instantly and accurately up the chain of command. Driver software for the specific detector is essential and several are available for common detectors, including CAM, ACADA and GID-3 from Graseby Dynamics and the EPD dosimeter from Siemens Environmental Systems. New driver development is a low-cost service.

Platforms

NBC-ANALYSIS supports MS Windows 98, Windows NT Windows 2000, Windows Millennium (ME) and Windows XP. In addition, integrated versions of NBC-ANALYSIS also support UNIX and LINUX.

Maintenance

NBC and ROTA hazard prediction is a growth area and NATO standards continue to evolve or, sometimes, change radically. Therefore Software Update and Maintenance Agreements ensure that the system takes account of both changing NATO standards and developments in functionality. The R&D load is shared across the wide customer base for NBC-ANALYSIS to reduce costs to the customer.

Status

Operational with the defence and civil emergency services of Belgium, Canada, Czech Republic, Denmark, Finland, France, Germany, Hungary, Italy, Luxembourg, NATO, Netherlands, Norway, Poland, Slovak Republic, Spain, Sweden, Turkey, UK and the USA.

Manufacturer

Bruhn NewTech.

UPDATED

....

NBC-ANALYSIS for UNIX and LINUX

Description

Variants of the NBC-ANALYSIS system are available to run on a UNIX or LINUX platform, offering access to its features over larger scale networks. NBC-ANALYSIS for UNIX is fully integrated into the US Marine Corps (USMC) Global Command and Control System (GCCS). Current platforms are Sun SPARC terminals running Sun OS 4.1.3 or Solaris 2.x. It is planned also to offer a version to run on HP machines

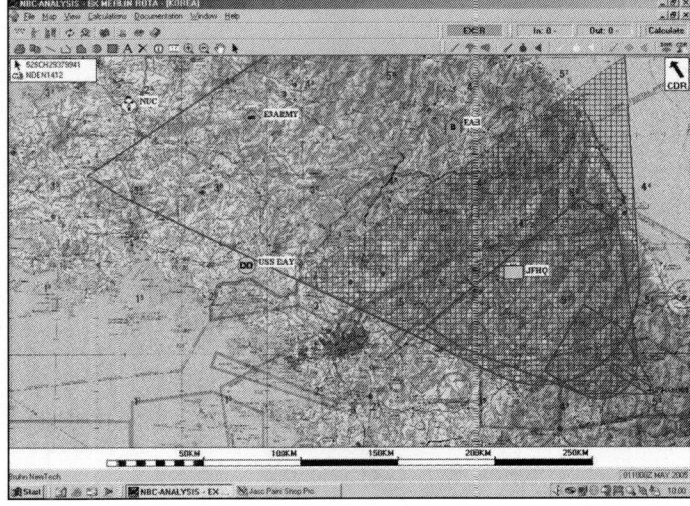

NBC-ANALYSIS screen snap showing predicted hazard data overlaid on an aerial map, using NATO standard notation (Bruhn NewTech) **2002**/0134232

NBC-ANALYSIS for UNIX forms the core of the US DoD's JWARN system (Bruhn NewTech) **2002**/0134231

running HP-UX. NBC-ANALYSIS for UNIX forms the core of the US DoD's JWARN system (see separate entry within this section).

Status
NBC-ANALYSIS for UNIX version 1.0 available and tested with the US Marine Corps Tactical Systems Support Activity (MCTSSA) during the Joint Warrior Interoperability Demonstration (JWID) at Camp Pendleton, California, USA, in 1995 and 1996. The application was used during 1997 and 1998 (Exercise Roving Sands) running on MCS Beta. NBC-ANALYSIS for UNIX has been ported to the US Army's Maneuver and Control System (MCS Block IV) as part of the JWARN contract signed in 1998 with the US Government.

Manufacturer
Bruhn NewTech.

UPDATED

FRANCE

NABOUCO hazard prediction system

Description
The NABOUCO system is an integrated NBC command and data management system contained in an air-transportable container. With a comprehensive array of sensors for NBC, GPS and meteorological data collection, unit operators can provide integrated command decision support data and deliver ATP45 formatted messages over a variety of communications routes.

NBC data can be integrated into existing command data and used for forecasting and localisation of the threat.

Status
Available. Under evaluation by the French Air Force.

Manufacturer
MGP Instruments.

VERIFIED

Standard ISO containers house the NABOUCO NBC hazard prediction system
***2000**/0081482*

The NABOUCO NBC system attached to a field shelter (see also Bachmann NBC shelters under PROTECTION (Collective) ***2000**/0081483*

GERMANY

Hazmat response detection and command vehicle

Description
This vehicle is designed to offer on-site analysis of materials released during a toxic incident and to provide command and control facilities for hazardous incident commanders at the scene. Based on the Mercedes Sprinter chassis, the vehicle is manned by a driver and assistant with two analysts, including the incident commander. There are planning facilities and workstations to allow hazard monitoring and prediction. The vehicle is fitted with full toxic protection consisting of air filtration and air conditioning plant. The interior can be closed down very quickly if a hazard is anticipated.

Detection and analysis is achieved by external detectors and analysed with an infra-red gas analyser. Results are compared with database information. The vehicle has two computer workstations which operate systems to identify and make predictions about hazard spread. Real-time meteorological data is also supplied from external sensors and integrated with the hazard data to aid decision making.

There is a generator on board to provide power when the vehicle engine is stopped.

The Henschel Hazardous Material Reconnaissance Vehicle has a similar suite of detection equipment. See entry under Detection (reconnaissance systems).

Status
Available.

Manufacturer
Odenwald-Werke Rittersbach GmbH.

VERIFIED

Exterior of Hazmat response detection and command vehicle 0011466

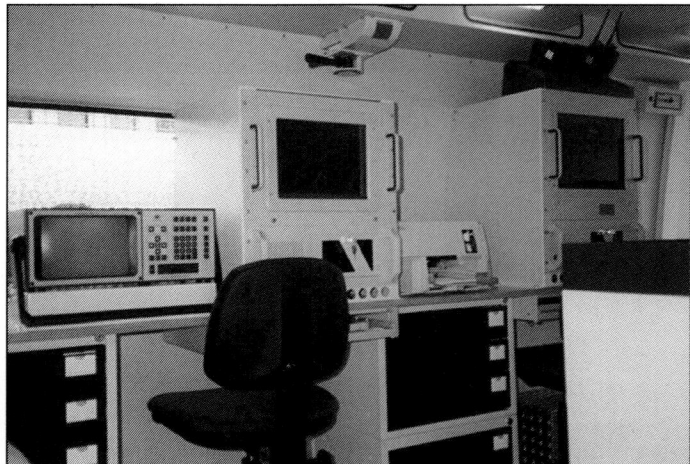

Hazmat response detection and command vehicle - interior of cabin 0011467

SNOOPER 2.0: IT-based NBC management tool

Description
SNOOPER is a PC-based software application designed to offer NBC data analysis and decision support to field commanders. Initially aimed at the German Fuchs NBC reconnaissance vehicle (see under Detection (reconnaissance systems)),

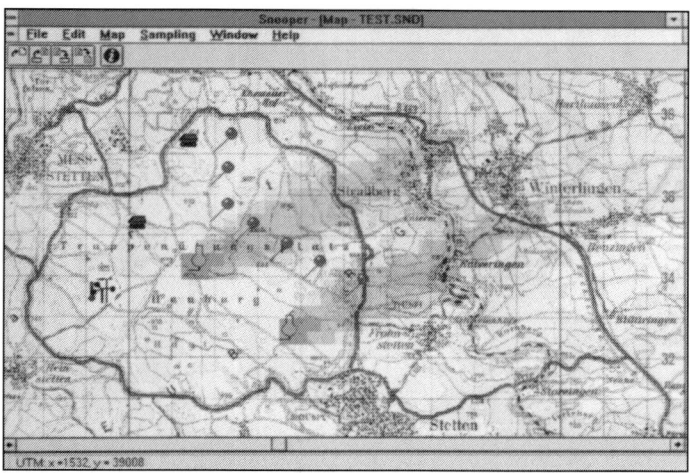

SNOOPER NBC management application - digital map with overlay   0050712

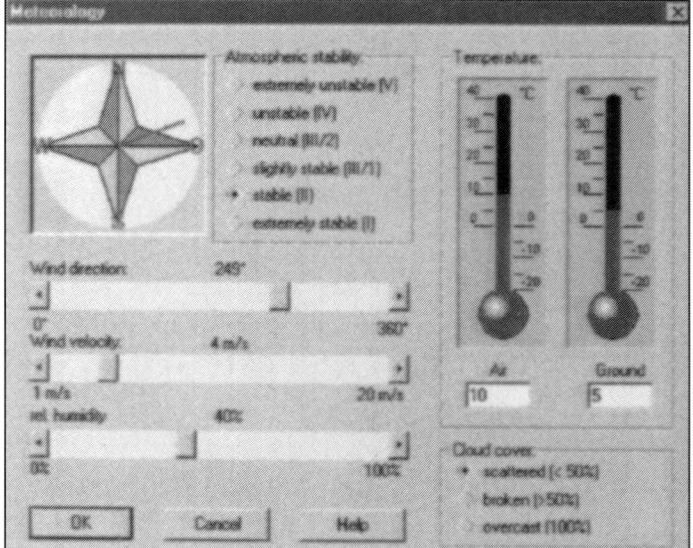

SNOOPER NBC management application - meteorology window   0050713

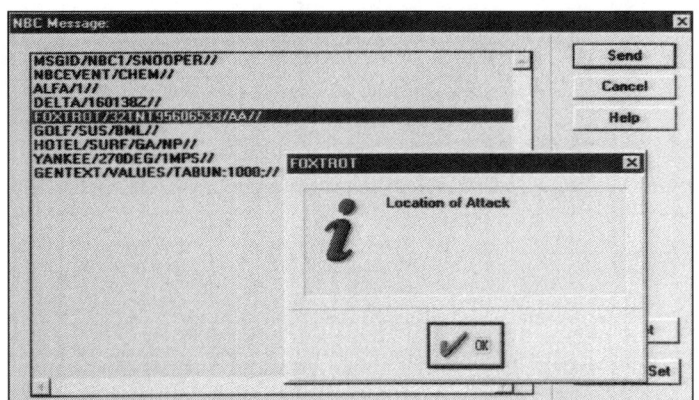

SNOOPER NBC management application - NBC message window   0050714

SNOOPER's functionality allows it to be used in a variety of NBC command and control situations including domestic response to an NBC terrorist incident.

Version 2.0 of this application runs in the MS Windows® PC-based environment - in Windows 95, 98 or, networked, in Windows NT. Windows multitasking allows several situations to be managed simultaneously. The main window uses a digital map (Digital Terrain Elevation Data) on which symbols for mobile NBC units and contamination incidents are overlaid. The command post window offers data analysis and control, access to the CW agent information database and NBC message handling in ATP-45 standard formats. The database includes information about decontamination measures.

In the measurement window, detected concentration levels are listed, together with associated MET data. The meteorology window allows manual input and editing of MET data.

The application is flexible enough to take direct digital sensor input or manual input via the keyboard. Calculated fields show likely locations of sources of contamination from arrays of peripheral detection data and met data. Using Gaussian and Lagrange risk analysis calculation models, dispersal plumes can be

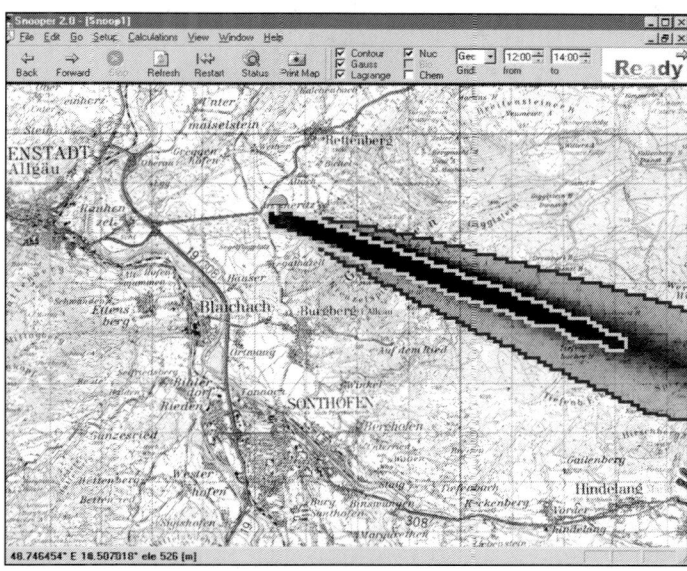

SNOOPER 2.0: Results of Gauss and Lagrange prediction algorithms   **2000**/0088182

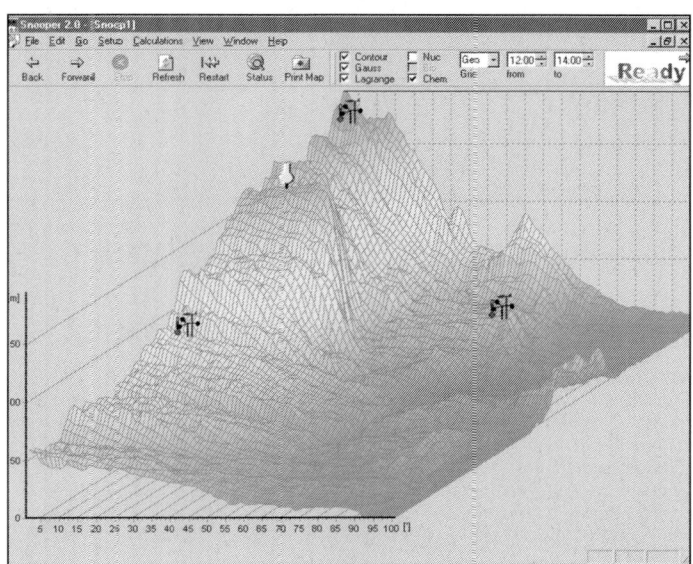

SNOOPER 2 0: View of topography with NBC event symbols   **2000**/0088183

accurately plotted, taking account of weathering rate, environmental and topographical characteristics. The latter feature alone makes this a powerful NBC management tool.

SNOOPER has been upgraded to version 2.0. This version is optimised for the domestic arena and its data management and command support features make it equally suitable for use in an NBC terrorist incident, a radioactive leak or an incident at a chemical facility.

Status
Available.

Manufacturer
Rheinmetall Landsysteme GmbH.

UPDATED

UNITED KINGDOM

BRACIS NT

Description
The Biological, Radiological And Chemical Information System - BRACIS NT is a version of NBC-ANALYSIS developed for the UK MoD to run on MS Windows NT. The UK has been operating BRACIS since February 1995 and this upgrade is in operational use by all three armed services of the UK. An installation at the Defence NBC Centre at Winterbourne Gunner is used for training. Versions are also available to run on LINUX and BRACIS NT supports all versions of Windows including ME and XP.

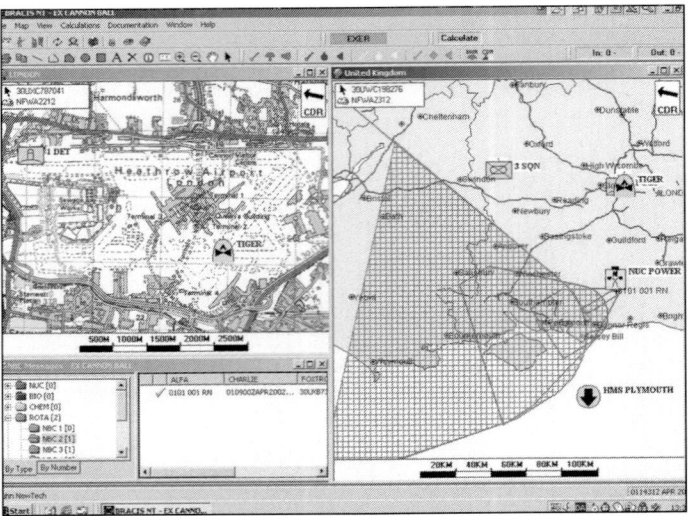

The UK MoD Tri-service Biological, Radiological and Chemical Information System - BRACIS NT showing a nuclear incident together with the location of units (Bruhn NewTech)
2002/0134230

Status
BRACIS NT is available as a Commercial-Off-The-Shelf (COTS) system.

Manufacturer
Bruhn NewTech.

UPDATED

UNITED STATES OF AMERICA

Joint Warning And Reporting Network (JWARN)

Description
The JWARN is a command and control system designed to offer commanders an integrated picture of the NBC aspects of the battlefield. It will be installed in major command and control centres and will offer analysed data for decision-making and automate warnings for dissemination to the lowest level of unit. JWARN data will be automatically retrieved from deployed NBC detectors and other sensors, providing commanders and C⁴I systems with the clearest possible NBC picture. It is intended that JWARN will eventually interface with Artemis (see separate entry within this section).

JWARN is a three-phase program:
- Block I: Interim Standardisation (IS). Initial procurement and deployment of Commercial-Off-The-Shelf (COTS) and Government-Off-The-Shelf (GOTS) software to standardise NBC warning and reporting for the US Armed Forces.
- Block Ia: COTS (Bruhn Newtech) NBC Analysis software (see separate entry within this section) for DOS-based platforms and GOTS hazard prediction models software.
- Block Ib: NBC Analysis software integration with Automated Nuclear, Biological and Chemical Information System (ANBACIS) Battlefield Management functionality for the US Army Manoeuvre Control System/Phoenix.
- Block Ic: NBC Analysis software integration with ANBACIS Battlefield Management functionality for Windows 32-bit environment and GOTS hazard prediction models software.
- Block II: Block Upgrade (BU) provides the total JWARN capability by integrating NBC detector systems, NBC Warning and Reporting Software Modules and NBC Battlefield Management software modules into the Armed Forces C⁴I systems.
- Block III: Product Improvement Proposal/Program (PIP).

Status
Engineering and Manufacturing Development (EMD) contracts were awarded during FY00 for:
- Block II integration of NBC legacy and future detector systems.
- Developed NBC warning and reporting modules.
- Battlespace management modules for use by Joint Services C⁴I systems.

Manufacturer
Bruhn NewTech A/S (US Office: Bruhn NewTech Inc).

NEW ENTRY

MIDAS-AT

Description
This Windows-compatible, networked management tool is designed to assist defence and domestic managers to predict and manage NBC incidents on the battlefield or in a domestic toxic incident such as a terrorist attack. The Meteorological Information and Dispersion Assessment System Anti-Terrorism (MIDAS-AT) is available in versions P1 through P4 in increasing functionality. It allows the effects of a nuclear, biological or chemical attack to be plotted on a vector or raster digital map overlay. Menus allow detailed information on the nature and concentration effects to be viewed, as well as the manual or real-time digital input of meteorological data. Using complex risk analysis models and taking

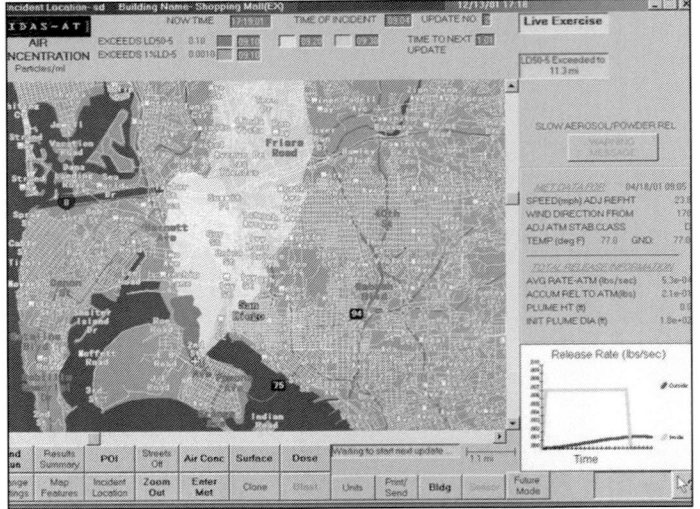

MIDAS-AT NBC incident management system (PLG Inc)   **2002**/0137746

MIDAS-AT - urban incident example (PLG Inc)   **2002**/0137748

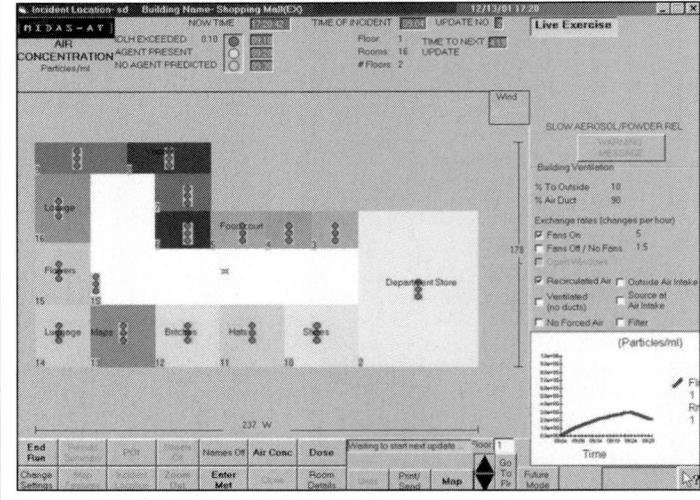

MIDAS-AT - internal building incident example (PLG Inc)   **2002**/0137747

account of topography, the application allows incident managers to brief and control response teams to achieve the minimal risk of injury.

The inside-building model allows the incident controller to explore the dispersion of a toxic agent inside a building, subway or other enclosed site.

The application is adaptable to user requirements and allows realistic NBC incident training.

Status
Avaialble.

Manufacturer
PLG Inc.

UPDATED

Multipurpose Integrated Chemical Agent Alarm (MICAD)

Description

In April 1993, Lockheed Martin Librascope, a division of the Lockheed Martin Tactical Systems Sector, was awarded a US$28 million engineering and manufacturing development contract to provide an integrated NBC detection and warning system for battlefield equipment and personnel. The 48 month contract, awarded by the US Army Chemical and Biological Defense Agency, called for Lockheed Martin Librascope to complete advanced developmental design, produce test units and support formal testing of the Multipurpose Integrated Chemical Agent Alarm (MICAD). Under the contract Lockheed Martin will design three new system configurations (see below) to provide combat, armoured and tactical vehicles with MICAD detection systems, as well as vans and shelters already equipped with positive pressure collective protection equipment.

For the MICAD programme, Lockheed Martin Librascope is teamed with Calspan of Buffalo, New York. The user component of the team is the US Army Chemical Center and School (USACCS) and the developer is Chemical and Biological Defense Command (CBDCOM).

MICAD provides automatic warning of NBC attacks throughout the battlefield and automatically provides NBC-1 and NBC-4 reports to the chain of command over existing tactical communication systems (for example, SINCGARS and MSE). It is intended to operate with existing sensors such as the M43A1, XM21, M22 (ACADA) or the AN/VDR-2 radiac set and any future NBC detectors. The system samples internal and external environments for chemical contamination, interface with vehicle navigation systems and collective protection equipment and automates the NBC report preparation and the transmission process from platoon to battalion level.

MICAD is compatible with the Army Tactical Command and Control System (ATCCS) and the Automated Nuclear Biological And Chemical Information System (ANBACIS). The system design is flexible, allowing it to be used in an area warning role via a telemetry link. It is also designed for NBC survivability and ease of decontamination.

There will be three MICAD Alarm Monitor Groups (AMGs), as follows:

XM26, NBC Detectors, SINCGARS Radios/MSE, M42 Alarm, Collective Protective Equipment and Position Location Device will be mounted in a tactical van or shelter, interconnected and provided with connections to appropriate power sources;

The XM27, NBC Detectors, SINCGARS Radios/MSE, M42 Alarm, Collective Protective Equipment and Position Location Device will be mounted in a tactical vehicle, interconnected and connected to the vehicle power;

The XM28, NBC Detectors, SINCGARS Radios, Voice Intercom, M42 Alarm, Collective Protective Equipment and Position Location Device will be mounted in an armoured vehicle, interconnected and connected to the vehicle power.

The AMGs perform the functions of pre-operational checks and maintenance, monitoring, warning and activating and transmitting data. MICAD furnishes information to the battlefield command structure (ATCCS and ANBACIS) through SINCGARS radios and MSE. Up-to-date locations of each host system and platforms are derived from the Global Positioning System (GPS) or other position locating devices. Interface to vehicle Collective Protection Equipment (CPE), when present, is automatic. Detection and alarm systems with a Sample Transfer System (STS) are mounted on armoured vehicles, vans and shelters to complete the

MICAD hardware suite around the AMGs. System support packages in the form of software and software development tools, maintenance items and packaging are also included in delivered materials. The soldier-machine interface is designed for use when wearing cold weather clothing and NBC protective clothing.

MICAD incorporates a number of system components, as follows:

Display/Control (DC): This operator/system interface is an integrated hardware and firmware component that allows the operator to configure, monitor and control the MICAD system. There are no external knobs or switches on the front and it can be mounted in any position. The design features a MC68332 microprocessor, 8 MB RAM, 1 MB ROM, Built-In Testing (BITE), a power down mode and a touch panel display.

Sample Transfer System (STS): STS units for combat and armoured vehicles or tactical vans and shelters are of one basic design and provide loss-free transfer of chemical agents from the external environment to NBC detectors. The STS also allows sampling of interior air for contamination. It has a built-in interface to the Interface Architecture (IA) bus for reporting and controlling functions.

Universal Interface Unit (UIU): All non-communications device interfaces are provided by UIUs. They contain all the circuitry necessary for two-channel interface with NBC detectors, GPS, collective protection equipment, telemetry links, alarms and voice intercom systems. They connect in tandem on the IA bus and are linked to the DC.

Two channels are provided by a Communication Interface Unit (CIUs). It contains two modem/processors that link tactical radios or switches to the DC via the IA bus. Each modem provides analogue or direct digital message processing.

Telemetry Link: This consists of a pair of identical small receivers/transmitters (Telemetry Link Radios - TLR) for relaying alarm data from a remote detector to the DC via a UIU or CIU. The TLR also functions as the activating transmitter for the Alert Device.

Alert Device: This is a commercial type personal paging unit which will be issued to soldiers to warn them of a chemical attack.

Specific MICAD installations have been developed for the following platforms: Fox NBC Reconnaissance System (see section Detection (reconnaissance systems)), M1 Abrams tank, M2 Bradley FV, Standard Integrated Command Post System (SICPS), HMMWV and the Biological Aerosol Warning System (BAWS). Installation in 38 Fox systems started in 1998. A further option of 24 is likely to be taken up. The system has successfully passed cold and warm region testing (Alaska and Panama).

Status

Advanced development. System type-classified in June 1997 and in full-scale production in July 1998.

Main contractor

Lockheed Martin.

UPDATED

Palmtop Emergency Action for Chemicals (PEAC)

Description

Palmtop Emergency Action for Chemicals (PEAC) is a palmtop computer equipped with a comprehensive hazard management database software. It allows military or civilian Hazmat incident managers to select safe options during management of

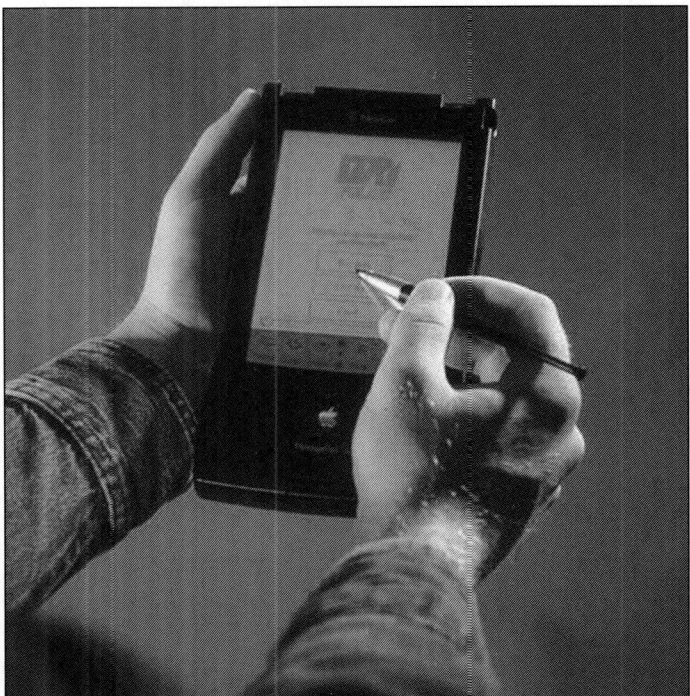

PEAC palmtop Hazmat incident decision support aid **2001**/0077504

MICAD system - layout example 0011462

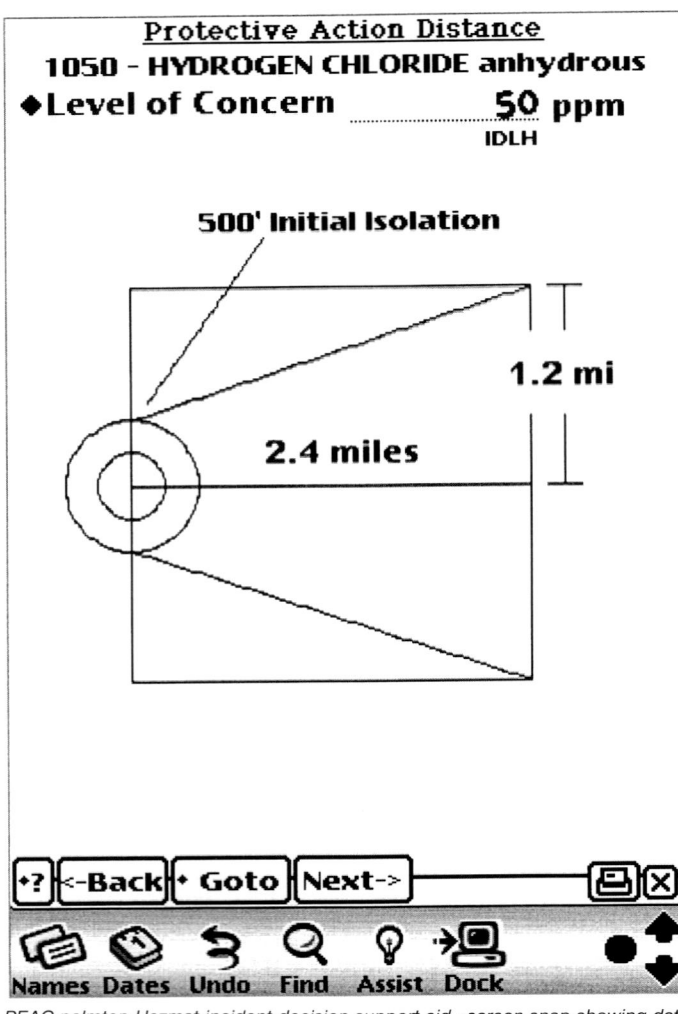

PEAC palmtop Hazmat incident decision support aid - screen snap showing data for HCl
2001/0077505

first response to hazardous materials incidents. Data on 4,000 different chemicals is available, searched by UN number or chemical name. Data on the physical and chemical characteristics of the agent, including toxicity, are offered to the user as well as advice on the required level of protection (personal and collective) and the decontamination techniques. For toxic vapour release, the characteristics of the hazard zone is calculated by PEAC and recommendations offered on the minimum unprotected downwind distances for personnel. The software takes account of meteorological conditions, topography and volume of agent released.

By use of simple on-screen menus and a pen input device, PEAC offers HAZMAT, first response and defence leaders rapid-reaction decision support for the correct and safe management of a battlefield or domestic HAZMAT incident.

PEAC versions are available for both palmtop and PC-based use. Versions include:
- PEAC-WMD 2002 and PEAC-DB 2000 for Windows
- PEAC-WMD 2002 and PEAC-DB 2000 for the Pocket PC.

Status
Available.

Manufacturer
AristaTek Inc.

UPDATED

TRACE

Description
TRACE, now at release 8.2, is an integrated software suite designed for CW risk management. Derived from domestic HAZMAT requirements, the application is equally effective for defence or domestic terrorism management requirements. TRACE can simulate a variety of incidents, including CW agent or TIC release from weapons, tanks or pipes. With its Windows™-based graphical user interface, the software is intuitive and easy to use. It permits data retrieval on the effects of an NBC attack and the location and dispersion of agent over time is overlaid on a vector or raster digital map. Data can be input directly from local or dispersed sensors, or manually by the user. Sophisticated risk analysis models take account of the effects of the physical and climatic environment in developing the dispersion plume, assisting managers in making safe incident management decisions.

The application is adaptable to user requirements and allows realistic NBC incident training.

Release 8.2 includes significant upgrades to the modelling algorithms, especially in multicomponent liquid evaporation. Improved mapping and plume-display capabilities significantly enhance user functionality over previous versions. The new TRACE software is fully networkable and can run under Windows 95/98 and Windows NT 4.0 platforms.

Status
Available. In wide use by authorities concerned with the management of hazardous incidents across the US and the rest of the world.

Manufacturer
SAFER Systems, LLC.

VERIFIED

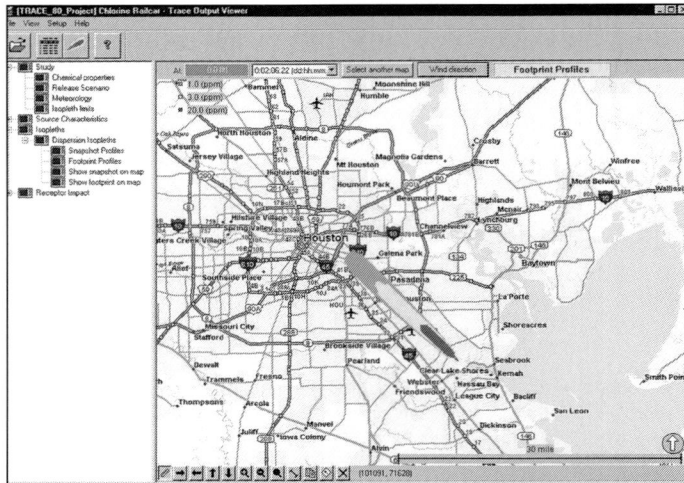

TRACE CW risk management
0055213

DETECTION (reconnaissance systems)

BULGARIA

Maritza NBC reconnaissance system

Description
A complete Maritza NBC reconnaissance system comprises six command vehicles and 36 NBC reconnaissance vehicles, all based on the amphibious MT-LB tracked carrier produced in Bulgaria. The system is designed so that each of the reconnaissance vehicles will automatically transmit coded NBC reconnaissance data, at preset intervals, to a command vehicle located up to 25 km away. Processed NBC data from the command vehicle is then automatically transmitted to higher command levels. Associated messages and data can be manually fed into the system.

Each Maritza NBC reconnaissance vehicle has a crew of four who may dismount from the vehicle to collect data using portable instruments if required. Each vehicle carries contaminated area marker flags and may also provide local NBC warnings by means of warning rockets.

NBC survey speeds can vary between 10 and 40 km/h. Each coded data message transmission lasts 3 seconds with a data transmission speed of 600 bits/s.

Status
Available.

Marketing agency
Kintex.

VERIFIED

Rear view of one of the MT-LB tracked carriers used with the Maritza NBC reconnaissance system

---

CROATIA

LOV-ABK 4 × 4 NBC reconnaissance vehicle

Description
The LOV-ABK 4 × 4 NBC reconnaissance vehicle is a variant of the LOV-OP armoured personnel carrier, itself the Croatian-produced version of the former

Croatian troops in NBC protective clothing standing in front of a LOV-ABK 4 × 4 NBC reconnaissance vehicle

---

Specifications
Crew: 2 or 3
Configuration: 4 × 4
Weight:
 (empty) 7,200 kg
 (loaded) 9,200 kg
Max load: 2,000 kg
Length: 5.89 m
Width: 2.36 m
Height: 1.98 m
Ground clearance: 0.31 m
Track: 1.86 m
Wheelbase: 2.85 m
Angle of approach/departure: 40°/40°
Max speed: 110 km/h
Range: 500-700 km
Max gradient: 65%
Side slope: 35%
Vertical obstacle: 0.5 m
Fording: 1 m
Engine: Torpedo BT 6 L 912S turbo diesel developing 130 hp at 2,650 rpm
Transmission: Z5-35S with 5 forward and 1 reverse gears
Suspension: leaf springs with hydraulic shock-absorbers
Brakes: drum, hydraulic servo-assisted
Turning radius: 13 m
Tyres: 14.5 R 20 MPT80 run-flat
Electrical system: 24 V
Batteries: 2 × 12V, 100 Ah

Yugoslav BOV armoured personnel carrier series. (For details of the latter see *Jane's Armour and Artillery 1998-99*.)

The LOV-ABK has a basic crew of two or three. The vehicle is provided with a filter-ventilation device which draws in air from the surrounding atmosphere and processes it, to detect the presence of NBC agents.

The system involved can detect radioactive radiation over the range from 0.1 μGy/h to 100 Gy/h. It can also detect GA, GB, GD, VX, HD, HN, L, AC and CG agents with a sensitivity of 0/04 mg/m³. The vehicle also carries instruments to gather meteorological data including air and ground temperature, wind direction and speed, relative humidity and air pressure. The LOV-ABK is also provided with a collective protection system with filtration and ventilation equipment.

The armoured hull of the LOV-ABK is proof against 7.62 mm NATO API projectiles fired from a range of 30 m and is also proof against artillery fragments at a range of 40 m. Other protective measures include run-flat tyres with a centralised tyre pressure system, self-recovery winch and an eight-barrel smoke grenade launcher.

Status
Available. In service with the Croatian Army (HVO).

Marketing agency
RH-Alan d.o.o.

VERIFIED

EGYPT

Contaminated area marking system

Description
This area marking system is carried at the rear of light and other vehicles (such as a locally produced YJ-L Jeep), usually in pairs, one unit on each side. Each unit carries up to 20 stakes, each bearing a suitably annotated marker flag. These are driven vertically downwards into the ground in a required sequence by electrically ignited cartridges controlled from a unit carried in the vehicle cab. Power for the system is taken from the vehicle's 12 V power supply. If required the dispensing system can be preprogrammed but may also be directly controlled by the vehicle driver.

The complete system weighs 30 kg.

Status
Available.

Manufacturer
Maadi Company for Engineering Industries (F.54).

VERIFIED

FRANCE

Renault VAB-RECO NBC reconnaissance vehicle

Description
A prototype of the Renault VAB-RECO NBC reconnaissance vehicle was first shown during January 1988 and is a 4 × 4 variant of the Renault VAB armoured personnel carrier (see also *Jane's Armour and Artillery*).

The VAB-RECO was developed for the French Army by the French Ministry of Defence with the initial conversion work being carried out at Roanne Arsenal. Two prototypes of the VAB-RECO were evaluated by the French Army during 1988 and in 1992, it was announced that a team headed by Thales and Giat Industries had been selected to integrate the NBC reconnaissance vehicles and the sensor suites. MGP Instruments supply the nuclear detection suite.

At the end of August 1995, it was announced that a contract for the supply of 40 VAB-RECO vehicles had been placed with Giat Industries on behalf of the French Army General Staff, with deliveries spread during 1997. Giat is responsible for the integration of the rear armoured blister, the mechanical elements and the modifications to the basic VAB vehicle.

It is anticipated that, once in service, the VAB-RECO will operate as part of a reconnaissance platoon in three detachments, each of two vehicles.

The vehicle has a basic crew of four: commander, driver and two system operators, one for sampling and one to operate the analyser.

The vehicle will be used for three main missions; identification and qualification of an NBC threat, threat evaluation and zone marking and information transmission to a commander. In addition the vehicle will provide meteorological reports and take liquid or solid samples for later analysis.

Nuclear detection is accomplished using an automatic radiation detector which triggers an alarm when an adjustable threshold is exceeded and measures the radiation level in a range from 0.01 to 1,000 cGy/h. The system consists of two external probes, in dialectric blisters either side of the vehicle, designed to measure the dose rate received at 1 m from the ground in the absence of the vehicle. There is also an internal probe for checking the dose received by the crew and a processing system to calculate, display and interface with the system's MLX-UR 3000 central unit.

Chemical threat detection and identification is carried out using a Bruker-Franzen MM-1 mass spectrometer coupled to a liquid sampling device (warning detection 1 mg/m³ for vapours and 10 mg/m² for liquids). A sampling system consists of a double silicon wheel arrangement capable of continuously picking up samples at a maximum vehicle speed of 50 km/h and transferring them to the MM-1. A further sampling system is capable of taking soil or other samples and placing them in one of four 100 cc sample containers for laboratory analysis.

An AP2C portable chemical agent monitor is also carried along with a S4PE sampling device. These two monitors also interface with the MLX UR central unit.

Renault VAB-RECO NBC reconnaissance vehicle (T J Gander)

VAB-RECO showing radiation detector (arrowed)   0055096

A GPS land navigation system is integrated with the system, along with an onboard meteorological system.

The centre of the VAB-RECO system is a Thales MLX-UR 3000 onboard computer which centralises and processes data from all system sensors. The computer has four data processing peripherals providing man-machine interfaces and including a high-resolution graphic monitor, keyboard and trackerball and a portable MPC 486/33 MHz computer with a colour screen, used as a remote terminal by the commander. The MLX-UR 3000 is a hardened civil computer using Thales software which carries out mission preparation, carries out various decision-making and operation management functions during a mission, provides a mission report and carries out logistic support operations.

A Thales PR4G transceiver provides radio-interphone links with a rear control post.

The vehicle carries a zone marking system using a set of 40 N or C beacons which can be dropped from the back of the vehicle.

Crew protection is provided by an internal overpressure collective system equipment with NBC filters and a sealed air conditioning system. In addition, a warning is provided for the crew when internal NBC agent levels rise above a predetermined level.

Status
In production for the French Army.

Manufacturer
Thales. Prime contractor (see text for subcontractors).

VERIFIED

GERMANY

Double-wheel sampling unit

Description
This unit consists of two trailing wheel units, operated remotely by the operator of an NBC reconnaissance vehicle. The wheel arm and axle units are jointed at the vehicle and covered by silicon rubber boots. The wheels have silicon tyres, designed to act as the sample collectors, picking up contamination as the vehicle moves forward. The operator, by raising a wheel in front of an NBC detector probe (see first illustration), can introduce agent into a mass spectrometer, where it can be identified and measured. Once contaminated, wheels can easily be replaced remotely by the operator.

The system is currently fitted to the Fuchs NBC reconnaissance vehicle (see separate entry in this section).

Double-wheel sampling unit - wheels deployed (probe at top centre)   0050706

Double-wheel sampling unit - RH wheel at the probe
0050707

Status
Available.

Manufacturer
Odenwald Werke Rittersbach GmbH (OWR).

VERIFIED

··

Rheinmetall Landsysteme Hazardous Material Reconnaissance Vehicle

Description
Based on a commercial transport chassis, the Rheinmetall Landsysteme Hazardous Material Reconnaissance Vehicle (HAZMAT RECON VEHICLE), offers a cost-effective mobile detection and analysis capability to domestic and defence response teams. The mission suite is capable of diagnosing and measuring the NBC and TIC environment during civil disaster relief efforts or military operations. The equipment enables a response team to make a comprehensive on-site assessment of the toxic environment, assisting swift and appropriate action by response personnel.

Detection, analysis and data-processing are conducted on a racked system which can be installed in suitable commercial chassis including the Mercedes Benz Sprinter and Fiat Ducato.

The detection and analysis suite provides PC- managed collection, evaluation, archiving and presentation of NBC data, together with corresponding geographical positional data. The software suite is compatible with GPS and digital map data, allowing accurate tracing of event data. On-line data links also allow remote centres to access the event data. An integrated power supply allows autonomous system performance for up to 4 hours without access to an external power source. External power augmentation can be supplied from a domestic supply (230 V AC) or from the vehicle (12 V DC), supporting the detection facilities and charging the system battery supply at the same time.

The open architecture of the command and control software is designed to allow integration of data from a wide variety of detection devices. Currently, interfaces are provided for the following measuring devices:

Radiological measuring system
Radiological measuring probe with radiometer: FH 4OG display device and NBR probe FHZ 672-2. Accurate measurement of dose, dose rate and count rate. The user can conduct wide area detection of radioactive contamination or focus down on point sources of radioactivity. False alarms are reduced to the minimum by filtering out background emissions. Measurement range for γ dose rate ranges from 0.1 µSv/h to 1 Sv/h. The energy range is from 50 keV to 2 MeV.

Chemical measuring system
IMS system: RAID-1 type (See also entry on RAID-1 under DETECTION (sensor systems) - Chemical). The RAID-1 installation is arranged to monitor the surrounding atmosphere continuously. Concentration levels are shown on a monitor and thresholds are set to gain attention through visual and audio alarms. The following chemicals are detected by the system: SO_2, NH_3, Cl_2, HCN, tetrachloroethene, NO_2, TDI, VX, GA, GB, GD.

Photoionisation detector (PID), AUER TOX-Meter type. This device enables the user to carry out continuous non-specific atmospheric monitoring for unknown chemical substances as well as quantitative analysis of known organic and inorganic gases and vapours, compared against a data library of 50 of the chemical species of greatest interest. The library records ionisation potential and response factors for each species.

The suite of devices can be operated integrally with the vehicle system, giving real-time data on the monitor screen through the DCS (see below). The user can set visual and acoustic alarm levels. Alternatively, it can be operated detached and away from the vehicle whereby data is stored in memory which can later be downloaded to the vehicle-based facility through a standard connection.

Display and control software (DCS)
The DCS is a user-friendly database software (WINDOWS 9x and NT operating systems) for storage, analysis, presentation and warning, based on detection data. The graphic display can show concentration levels by agent against either time or distance run (selectable). Historic data can also be selected and viewed. Being application independent, the ASCII format allows data to be analysed in a variety of common-range software including MS Excel, Access etc. The DCS also allows the user to output detection parameters to the detection devices.

Vehicle navigation unit
The vehicle navigation unit comprises a DGPS (Differential Global Positioning System) and an ALF long-wave receiver, which gives a positional accuracy of ±5 metres. Positional data is time-synchronised at system start and passed to the DCS as the measurement timebase. DGPS data is augmented by dead-reckoned (DR) positional data derived from vehicle start location, and subsequent steered vectors. With DGPS failure or if DGPS data is suspect because of the terrain, the DR data alone is sufficiently accurate for risk-area mapping purposes. Positional data is superimposed on CD-derived raster maps to complete the graphic presentation (German maps are from the Munich-based Land Surveying Office. Scale: 1:50000). Map co-ordinates can be selected between Gauss/Krüger and UTM according to preference.

Central Evaluation Unit (CEU)
The Central Evaluation Unit (CEU) allows a regional HQ to run up to eight NBC measuring systems. Data is integrated at regional HQ from deployed Hazardous Material Reconnaissance Vehicles or other suitable connected detection vehicles via the GSM network. Deployed operators transmit current and historical data to the CEU at intervals. Data is archived for later evaluation if required.

Status
Available. In service with the German Disaster Relief Forces and the Fire Department, NBC-Platoons.

Manufacturer
Rheinmetall Landsysteme GmbH.

UPDATED

··

Marking Set, Contamination: Nuclear, Biological, Chemical (NBC)

Description
The Rapp contamination marking set is contained within a portable slung container which can be easily carried and used by one person. This may be slung on a canvas strap over one shoulder or slung across the chest. It is a light metal frame holding three marking flag dispensers, 13 rolls of yellow marker tape, 48 light mounting stakes and crayons.

In the simplest case, the set may be used to dispense the marker flags for attachment to existing objects such as tree branches or fence posts. The flags are

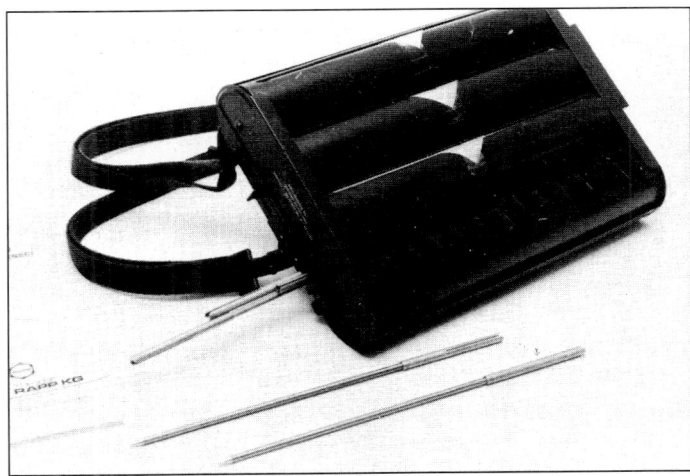

Rapp contamination marking set complete showing stake location

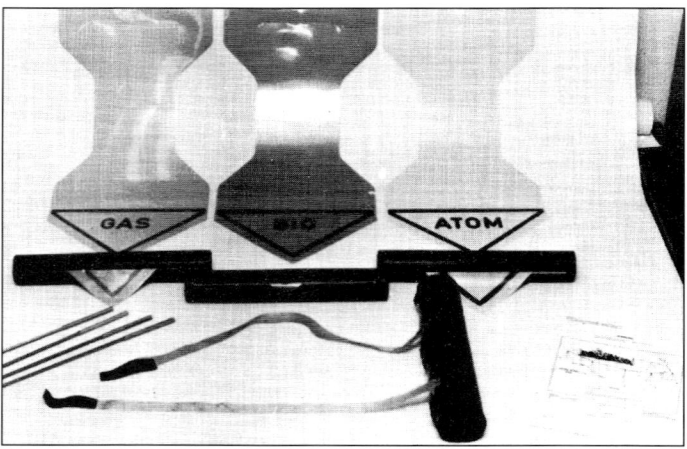

Marker flags and tapes from NBC marking set

Specifications
Container with components
Weight: approx 4.5 kg
Length: 345 mm
Width: 235 mm
Height: 90 mm

Mounting stakes
Length: each 290 mm
Quantity: 48

Marking ribbon
Length of each roll: 20 m
Quantity: 13

Marking flags
Quantity:
  (nuclear) 20 white (marked ATOM)
  (biological) 20 blue (marked BIO)
  (chemical) 20 yellow (marked GAS)

US Army XM93 Fox NBC reconnaissance vehicle (Michael Jerchel)

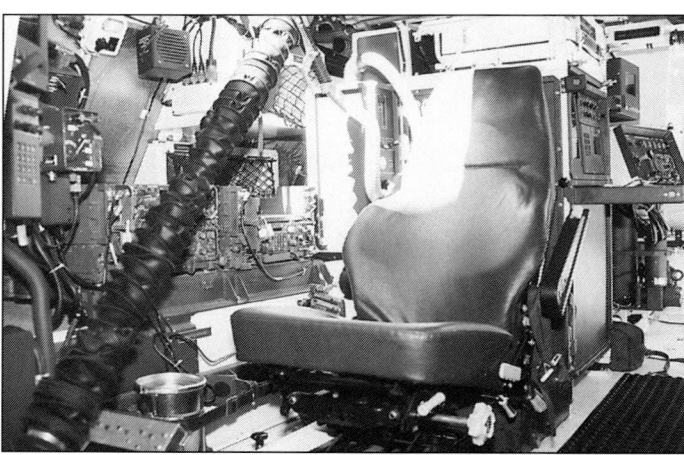

Operator's position inside a US Army M93E1 Fox NBC reconnaissance vehicle (T J Gander)

US Army M93E1 Fox NBC reconnaissance vehicle with M21 RSCAAL sensor mast raised (T J Gander)

simply pulled from one of the three dispensers, a white flag for nuclear contamination, a blue one for biological contamination and a yellow flag for chemical contamination. Using a crayon taken from a clip on the side of the kit container, extra information such as date and extent of contamination can be written onto the flag which can then be fixed to the object by folding the head through a slot in the tail.

When stakes are required, they are removed from the container by opening latches in its side. For most purposes three stakes from the kit are required to make one pole and each stake is provided with a point at one end and a locating collar at the other. Once the first stake has been driven into the ground the others can be fitted into the sockets. The necessary flag can then be fitted onto the top of the pole by impaling it over the top. As before, extra information can be written onto the flag using the crayon from the kit.

At times it will be necessary to isolate an area by fencing it off by poles and tape. The poles are erected in the manner already described and the yellow tape can be taken from the 13 20 m rolls carried on the kit container. The tape can also be used to form pole stabilising guys with extra stakes as the holding peg anchors. The marker flags can then be hung from the tape lengths connecting the poles.

Once used the kit container is disposed of once used and replaced by a fresh kit. No training in using the kit is required other than a simple handbook providing general guidelines and in use, the container is easy to carry and handle. It can be used even when the operators are fully covered by NBC protection clothing and the only safety point to watch, is that the sharp ends of the stakes might cut through protective clothing. The kits are issued ready for use and no form of preparation is required.

Status
No longer in production but support and spares remain available. In service with the US Army.

Manufacturer
Theodor Rapp GmbH & Co KG.

VERIFIED

Transportpanzer 1 'Fuchs' NBC reconnaissance vehicle

Description
This wheeled NBC reconnaissance vehicle is built on the components and hull of the Transportpanzer 1 armoured personnel carrier (see *Jane's Armour and Artillery* for vehicle details).

As part of the German contribution to Coalition forces operating in the Gulf in October 1990, 60 German-configured Fuchs NBC reconnaissance vehicles were released to the US Army by the Bundeswehr. These vehicles were returned to Thyssen Henschel to be 'Americanised' by the inclusion of American communication systems, smoke grenade launchers, air conditioning and other engineering changes. The converted vehicles were despatched to Dharan in Saudi Arabia and issued to US Army units operating in the region. A further 11 units were passed to the British Army and Royal Air Force, eight more were loaned to Israel and two went to Turkey (later returned). In early 1991, Germany ordered a further 45 vehicles from Thyssen Henschel to replace those taken from German Army stocks.

A squadron of British Army Fuchs vehicles now forms part of the UK NBC Defence Regiment which is manned by reserve personnel of the Royal Yeomanry.

For nuclear radiation detection the German-configured Transportpanzer 1 Fuchs carries a set of ASG 1 nuclear tracing equipment with a printout facility that can provide a hard-copy of detection location co-ordinates when coupled to the vehicle's orientation system. For the analysis of soil and other samples, the Fuchs carries an MM-1 mobile mass spectrometer (see separate entry in Detection (sensor systems) - Chemical). This is associated with a double-wheel sampling unit at the rear of the vehicle (see separate entry this section). Two sampling wheels with silicone tyres can be run along the ground surface and lifted up to an external probe connected to the MM-1 mobile mass spectrometer for analysis. This probe is

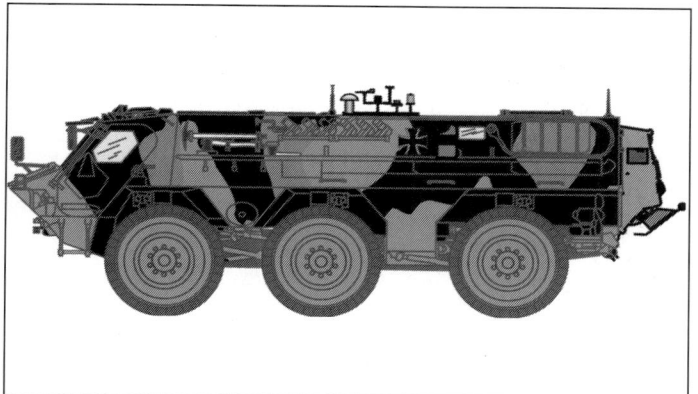

Biological Reconnaissance Vehicle ('Bio-Fox') (Rheinmetall Landsysteme GmbH)
2000/0088180

extendable for use as a ground probe. Also at the rear of the vehicle is a work hatch through which a sampling device can be manually operated to take soil and other samples for later analysis.

Other NBC equipment carried includes a CW agent detection and monitoring unit based on the American M43A1 and NBC marking equipment to mark contaminated areas. The markers are ejected through a work hatch at the rear of the vehicle. The vehicle has an NBC air filtration unit with a capacity of up to 180 m³/h. For operations in the Gulf region during 1990-91, British Fuchs vehicles were equipped with a Chemical Agent Monitor (CAM) mounted above each of the front headlights.

NBC Spürfuchs

Since June 1997, the German armed forces have been taking delivery of the improved NBC Spürfuchs variant. The upgrade embodies improvements to the entire data management system including upgraded Central Data Processing Unit (CDPU), better software, better command data presentation, better integrated internal data highway and improved data interface with other units. In detail:
- Meteorological station with the capability to integrate real-time climatic data with other sensor data
- Automated RF datalink
- GPS and fibre-optic gyro for accurate positional information
- Improved generator and battery packs for more stable and longer-life power
- Improved alarm system
- Improved COLPRO. Refrigerant: R134a. Cooling capacity: d dt =18 k at 49°C and RH = 10%

The Radiation Detection Equipment comprises:
- Radiation Detection Device ASG. This automatic dose rate device measures γ radiation from a range of 10^{-2} to 10^{3}cGy/h
- Dose/Dose Rate Measurement Device MAB 500. Measures γ and X-rays in the low dose range from 0.05 μSv/h to 10 mSv/h. This can be done on the spot or during travelling and it can be operated independently of the vehicle. α and β dose rate meters are carried for environmental monitoring outside the vehicle
- MicroCont Detector (see separate entry in Detection (sensor systems) - Nuclear)
- Graetz personal electronic dosimeters (see separate entry in Detection (sensor systems) - Nuclear)

The CW Agent Detection Equipment comprises:
- MM-1 mobile mass spectrometer, including improved gas chromatographic separation oven (see separate entry in Detection (sensor systems) - Chemical)
- Improved sample collection

Bio-FOX

The Bio-FOX variant is optimised for reconnaissance duty against BW attack. The detection and monitoring suite comprises facilities for continuous air sampling, ground/water sampling, verification operations, safe sample collection and transport. Supplementary armour plating and internal lining improves ballistic and mine protection in this variant. Facilities include:
- Aerodynamic Particle Sizer® (APS) for continuous monitoring;
- Large-volume air sampler;
- Safety level L3/L4 glove box for BW agent testing and verification;
- Probes for ground and water sample collection;
- Freezing and refrigeration plant for viable sample storage and transport;
- Internal storage racks optimised for biological laboratory equipment and supplies.

Status

114 NBC vehicles are in service with the German Army and 11 with the British Army. Eight vehicles loaned to Israel were returned after Operation Desert Storm. The Fuchs is a contender for a 64-vehicle UAE requirement and the Fox M93 and M93A1 variants equip the US Army and the USMC (see separate entry within this section).

Manufacturer

Rheinmetall Landsysteme GmbH.

UPDATED

HUNGARY

K-90 Field NBC reconnaissance system

Description

In order to speed up the overall data collection and processing procedures involved in NBC reconnaissance, the Hungarian NBC industry created the K-90 Field NBC reconnaissance system. The system was designed to merge data from a variety of different sources and is based on 24 armoured NBC reconnaissance vehicles and two vehicle-mounted monitoring centres.

The K-90 system can determine the location of reconnaissance vehicles; measure the characteristics of nuclear detonations; measure meteorological data close to ground level; measure radiation levels; detect chemical attacks; and automatically provide data for decision making. The system gathers data from the reconnaissance vehicles via a controlled radio network. System software processes the data and provides the parameters of nuclear detonations and chemical attacks, levels of radioactive contamination, losses and so on.

The armoured reconnaissance vehicles involved are the wheeled VSBRDM-2. This is a Hungarian conversion of the RFAS BRDM-2 amphibious scout car (see *Jane's Armour and Artillery* for full details of the BRDM-2). The VSBRDM-2 is equipped with a nuclear blast detector AM-2, chemical agent sensor GVJ-2, field meteorological station TMF-2, radiation survey meter H-31K, TNA-3 navigation device, multifunction radiation meter IH-90, microcomputer with data storage unit K-9011 and an R-111 radio set. (Hungary also uses a version of the PSZH-IV armoured personnel carrier known as the FUG for the NBC reconnaissance role but it is not part of the K-90 system - see entry in this section).

The two vehicle-mounted monitoring centres involved with the K-90 system are carried inside command post bodies carried on ZIL-131 (6 × 6) 3,500 kg trucks. The interior of the body contains a digital map table, the system main computer, data input computer (with data acquisition software), data output computer (with graphics software), communications controller, R-111 radio set, high-resolution colour monitor and printer.

Status

Available. Manufactured by Hungarian state factories prior to privatisation but current manufacturer unknown. In service with the Hungarian armed forces.

Manufacturer

Unknown.

VERIFIED

PSZH-IV NBC reconnaissance vehicle

Description

This vehicle is a derivative of the Hungarian FUG (or OT-65) amphibious scout car in armoured personnel carrier form and is deployed and equipped in a similar fashion to the RFAS ERDM-1 RKh NBC vehicles. Lane marking pole racks are located either side of the hull rear and are swung through 90° to be dispensed over the rear of the vehicle. The interior contains the usual former Warsaw Pact suite of radiological and chemical detection equipments (see entry under RFAS for details).

Hungarian Army PSZH-IV NBC reconnaissance vehicles

This vehicle is not associated with the Hungarian K-90 NBC reconnaissance system (see entry in this section). For details of the PSZH-IV armoured personnel carrier, refer to *Jane's Armour and Artillery*.

Status
Previously manufactured by Hungarian state factories prior to privatisation. Current manufacturer or servicing arrangement unknown. In service with the Hungarian armed forces.

Manufacturer
Unknown.

VERIFIED

JAPAN

SU 60 - based NBC reconnaissance vehicle

Description
The Japanese Self-Defence Forces' requirement for an NBC reconnaissance vehicle which began development at a slow pace in 1980 and prototypes were produced for trial. Based on the chassis and hull of the Type SU 60 armoured personnel carrier, they carried a range of NBC detection and monitoring equipment. The SU 60 vehicle is powered by a 220 hp diesel engine and weighs about 11.8 tonnes in action (see *Jane's Armour and Artillery* for more detail).

The NBC equipment carried by these vehicles is unspecified in available source material but it is thought to comprise a range of capabilities similar to other vehicles in this section. Specialised sampling and lane indicator dispensing equipment can be identified at the rear of the vehicle. Equipment is thought to have varied between prototypes as trials progressed.

Status
Prototypes developed 1980-1987. Current status unknown. See other Japanese reconnaissance vehicle types in this section.

Manufacturer
Mitsubishi Heavy Industries.

UPDATED

The second prototype of the Japanese NBC reconnaissance vehicle based on the SU 60 APC (Mitsubishi)

Third prototype of the Japanese NBC reconnaissance vehicle based on the SU 60 APC (Mitsubishi)

Type 82 and Type 87 - based NBC reconnaissance vehicle

Description
The JGSDF is introducing a new series of NBC reconnaissance vehicles into service. Based on the Type 82 command post and the Type 87 reconnaissance vehicles, the system offers a similar capability to the Fuchs and the VAB-RECO vehicles shown in this section. The vehicles are equipped with radiation detectors as well as BW and CW agent detectors. There is a sophisticated onboard agent analysis capability and, at the rear, there are crew operated remote sample collection devices.

Known as Kagaku-bougo-sha, these vehicles are developed and produced by the Komatsu factory and represent a significant advance in capability.

Status
In course of introduction into service with the Japanese Ground Self-Defence Force (JGSDF).

Manufacturer
Mitsubishi Heavy Industries.

VERIFIED

Specifications	
Configuration: 6 × 6	
Weight (loaded): 14,100 kg	
Length: 6.1 m	
Width: 2.5 m	
Height: 2.4 m	
Crew: 4	
Max speed: 95 km/h	
Armament: 1 × 12.7 mm machine gun	

Type 82 (6 × 6) command and communications vehicle from the rear - the basis for the new NBC reconnaissance vehicle (K Nogi)

Type 82 command and communications vehicle (Kensuke Ebata)

KOREA, SOUTH

K216A1 NBC reconnaissance vehicle

Description
The K216A1 NBC reconnaissance vehicle was produced by Daewoo Heavy Industries and is an updated variant of the K200A1 Korean Infantry Fighting Vehicle (KIFV) series which entered production in 1985 (see *Jane's Armour and Artillery* for

Prototype of K216A1 NBC reconnaissance vehicle (T J Gander)

Specifications

Crew: 4
Combat weight: 12,900 kg
Unloaded weight: 11,800 kg
Power-to-weight ratio: 27.8 hp/t
Ground pressure: 0.63 kg/cm²
Length:
  (overall) 5.568 m
  (hull) 5.345 m
Width: 2.67 m
Height:
  (top of hull) 1.93 m
  (top of gunner's shield) 2.198 m
Ground clearance: 0.41 m
Track width: 381 mm
Max speed:
  (land) 70 km/h
  (water) 7 km/h
Fuel capacity: 400 litres
Cruising range: 480 km
Fording: amphibious
Gradient: 60%
Side slope: 30%
Vertical obstacle: 0.63 m
Trench: 1.68 m
Engine: Daewoo D 2848T 14.62 litre V-8 turbocharged diesel developing 350 hp at 2,300 rpm
Transmission: Allison X200-5K with 4 forward and 1 reverse speeds
Suspension: torsion bar
Electrical system: 28 V
Batteries: 2 × 12 V 6TN, 100 Ah
Armament: 1 × 12.7 mm K6 (M2) MG
Ammunition stowage: 2,000 rounds

further information on this series of vehicles). The NBC reconnaissance vehicle has been placed in production to meet a local requirement for 300 vehicles.

The K216A1 NBC reconnaissance vehicle is based on a welded aluminium armour hull which differs from the basic KIFV in that it resembles the US M113 APC. The vehicle lacks any form of roof armament mounting other than a pintle mounting for a 12.7 mm K6 (Browning M2 HB) machine gun located behind an optional armoured shield over the commander's centrally located turret. The driver is located on the front left of the hull with the engine pack to his right. The engine is a MAN turbocharged diesel produced under licence by Daewoo, coupled to an Allison X-200-5K hydrodynamic transmission. There are five main road wheels each side connected to a torsion bar suspension with three shock-absorbers each side.

Apart from the commander and driver, there is provision for a further three crew members. Between them they operate the NBC sensor equipment carried in the vehicle, with one person seated at the rear interior of the vehicle to operate a sampling device which can be lowered to the ground to collect ground samples on one of a pair of small rotating wheels. Each wheel is periodically lowered as the other is raised to offer a wheel to the sensor of an MM-1 mass spectrometer within the vehicle. There are other provisions to collect chemical samples using either a protective glove or a small spade collector with samples being passed through a small NBC-protected hatch to the hull interior. The vehicle carries a KM43A1 chemical alarm system and a AN/VDR-2 radiation detector unit. The sensor suite is completed by a meteorological observation station. Contaminated areas are marked by a marking system; 75 markers are stowed inside the vehicle and emplaced by a pneumatic dispensing system.

Personnel inside the vehicle are protected by a collective overpressure NBC system.

Status

In production. Some 12 vehicles produced during 1996 and 36 in 1997. In service with the Republic of Korea Army.

Manufacturer

Daewoo Heavy Industries Limited.

VERIFIED

KM453 NBC reconnaissance vehicle

Description

Based on the Asian motors KM45 series 4 × 4 truck chassis, the KM453 NBC reconnaissance vehicle is a mobile battlefield laboratory capable of rapid deployment to assess the NBC hazard (see *Jane's Military Vehicles & Logistics* for details of the vehicle). The vehicle is provided with a comprehensive COLPRO system allowing the crew to monitor NBC hazards without having to leave the vehicle. The detection suite comprises four elements:

The weather observation system monitors wind vector, ambient air temperature, pressure and humidity. The data is presented on a built-in monitor and printer.

The radiation detector is portable and can be connected to external sensors on the vehicle or detached and used as a mobile monitor outside the vehicle by the crew. Beta and gamma radiation can be detected.

A mobile mass spectrometer allows precision analysis of air and soil samples for radiation, biological and chemical contamination.

The vehicle carries an onboard launcher which allows flag markers to be placed in the ground to indicate contaminated areas, without the crew having to leave the vehicle.

Status

In production and in service with the Republic of Korea Army.

Manufacturer

Kia.

VERIFIED

Specifications

Crew	4
Length	5.4 m
Width	2.1 m
Height	3.4 m
Gross weight	4,595 kg
Payload	400 kg
Prime mover	6 cylinder 4 litre diesel engine developing 115 bhp at 3,600 rpm
Max speed	80 km/h
Cruising range	600 km

KM453 NBC reconnaissance vehicle is based on the Asia Motors KM45 chassis (T J Gander)   **2001**/0100331

ROMANIA

TAB RC NBC reconnaissance vehicle

Description

The TAB RC NBC reconnaissance vehicle (also referred to as the TABCRH-84) is a variant of the TABC-79 (4 × 4) amphibious armoured personnel carrier. It is a Romanian development of ex-RFAS designs which may incorporate some components of the Romanian TAB-77 (8 × 8) armoured personnel carrier; in its turn the TAB-77 is a Romanian development of the ex-RFAS BTR-70 (see *Jane's Armour and Artillery* for vehicle details).

Romanian TAB RC NBC reconnaissance vehicle

Specifications

Crew: 2 or 3 (seating for up to 7)
Configuration: 4 × 4
Length: 5.65 m
Width: 2.8 m
Height: 2.34 m
Ground clearance: 0.495 m
Wheelbase: 2.9 m
Track: 2.25 m
Max speed:
　(land) 85 km/h
　(water, TAB-CG) 8 km/h
Fording:
　(TAB RC) 0.4 m
　(TAB-CG) amphibious
Gradient: 32°
Side slope: 28°
Vertical obstacle: 0.3 m
Trench: 0.7 m
Engine: Saviem type 798.05 M2 6-cylinder turbocharged diesel developing 154 hp at 2,800 rpm
Transmission: mechanical with 5 forward and 1 reverse speeds
Steering: power-assisted
Turning radius: 10 m
Suspension: torsion bars with hydraulic dampers
Electrical system: 24 V
Armament: 1 × 7.62 mm PKMS MG

The turretless TAB RC has a crew of two or three with a circular roof hatch behind the driver's and commander's roof hatches; there is provision for seating up to seven crew. The circular roof hatch has provision for an externally mounted 7.62 mm PKMS machine gun. Lane marking pole and pennant racks are mounted on both sides of the hull rear. A central tyre pressure regulation system is fitted and the vehicle has a mechanical winch. Night driving equipment is provided.

It is assumed that this vehicle has a similar internally mounted radiological and chemical detection instrument suite as other similar former Warsaw Pact NBC reconnaissance vehicles (see under RFAS for details). The TAB RC is provided with additional shielding against nuclear radiation.

The TAB-CG (CA-SG) is an essentially similar NBC reconnaissance vehicle which differs mainly in having an amphibious capability and lacks the extra radiation shielding of the TAB RC. In the water the TAB-CG is driven by a hydraulic jet pump unit.

Status
In service with the Romanian armed forces. Unable to trace current manufacturer or supplier of service and parts.

Manufacturer
Unknown.

VERIFIED

RUSSIAN FEDERATION AND ASSOCIATED STATES (CIS)

Hind-G NBC reconnaissance helicopter

Description
The Hind-G NBC reconnaissance helicopter lacks many of the undernose electro-optic and RF missile guidance systems associated with most of the MIL Mi-24 Hind assault helicopters. In place of the wingtip weapon stations there are what can be described as 'clutching hand' soil sampling grabs used to take soil samples for

Hind-G NBC reconnaissance helicopter (John W R Taylor)

later analysis. Other adaptations for the NBC reconnaissance role include various sensor stations around the cockpit area. It is not thought that the Hind-G has been produced in quantity.

Status
In service with the RFAS armed forces.

Manufacturer
(Airframe) MIL design bureau.

Marketing Agency
Rosoboronexport.

VERIFIED

KDKhR-1N chemical agent reconnaissance vehicle

Description
The KDKhR-1N chemical agent reconnaissance vehicle is based on the hull and tracked chassis of the ACRV artillery command vehicle and carries a turret-mounted laser-based sensor system intended for the remote detection of nerve agents at ranges of 1 to 3 km. KDKhR-1N translates as Complex for Standoff Chemical Reconnaissance although the vehicle/sensor combination is commonly known as the 'Dal', a term denoting 'farseeing'. Initial testing of the system commenced during 1987 with series production commencing during 1990.

The KDKhR-1N turret can direct its laser system to scan through a 360° azimuth in 60 seconds (elevation range is from −3 to +70°) to detect chemical agent clouds and especially phosphororganic (nerve) agents, using spectral analysis. A computer-controlled system carried inside the hull enables an operator to determine the distance, size and shape of an agent cloud as well as its speed over the ground and direction of movement; the system can determine concentrations as low as hundredths of parts of a milligram of agent per cubic metre of the atmosphere. Results can be transmitted to a central control station via an automated data transmission system, also controlled by the onboard computer.

In addition to its laser detection system the KDKhR-1N carries a full suite of conventional chemical and nuclear reconnaissance sensors and other equipment including the following: IMD-21B dose rate meter; GSA-12 automatic chemical agent detector; PGO-11 semi-automatic chemical agent detector; and a KPO-1 probe unit. Navigation and communication units include a TNA-4-6, a R-123 and a R-171. The crew is three. As far as can be determined no armament is carried.

It has been suggested that in addition to its NBC reconnaissance functions the KDKhR-1N could be used for environmental monitoring purposes.

KDKhR-1N chemical agent reconnaissance vehicle

Status
In service with the RFAS armed forces.

Manufacturer
Volsk Metalworking Plant.

Marketing agency
Rosoboronexport.

VERIFIED

PKhM-4-01 radiation and chemical reconnaissance vehicle

Description
The PKhM-4-01 radiation and chemical reconnaissance vehicle is based on the amphibious BTR-80 armoured personnel carrier (see *Jane's Armour and Artillery* for further details of the BTR-80 vehicle). The crew of three on board the vehicle can carry out a wide range of NBC reconnaissance tasks. Some exterior tasks, such as sample retrieval, are automated to reduce the risk to the crew. The vehicle carries the KZO-2 equipment to deploy marker flags around contaminated zones. The vehicle may also be employed for weather observations.

The range of specialised equipment is shown below. See under *Detection (sensor systems) - chemical* for details of individual equipments. The vehicle is equipped with a night vision device and the TNA-4-4 tank navigation equipment.

PKhM-4-01 radiation and chemical reconnaissance vehicle

KZO-2 warning marker emplacement set carried on rear of PKhM-4-01 radiation and chemical reconnaissance vehicle

Specification	
Equipment	On-board stock
IMD-21B or IMD-1r dose rate meter	1 set
DP-5V dose rate meter	1 set
GSA-13 automatic chemical agent detector	1 set
VPKhR chemical agent detector kit	1 set
PGO-11 semi-automatic chemical agent detector set	1 set
KPO-1 sampler	1 set
MK-3M meteorological kit	1 set
KZO-2 warning markers	6 sets

The communications suite comprises an R-173M or R-123M radio set and an R-174 or R-124 intercom system. The PKhM-4-01 is armed with a 14.5 mm KPVT heavy machine gun and a 7.62 mm PKT machine gun.

There is no EW detection capability.

The weight of the PKhM-4-01 when operating is 13.5 tonnes. Radiation reconnaissance speeds can be up to 30 km/h, while for chemical reconnaissance the speed is up to 10 km/h. Maximum road speed is 80 km/h. The vehicle can remain in operation on one set of consumables for 24 hours.

Status
In service with the RFAS armed forces.

Manufacturer
Tula Plant JSC.

Marketing agency
Rosoboronexport.

VERIFIED

RFAS chemical and radiological reconnaissance vehicles

Description
The RFAS armed forces have numerous vehicles devoted primarily to chemical and radiological reconnaissance. Few of these special vehicles have any drastic modifications for their role, other than the addition of lane-marking equipment or warning marker dispensers (such as the KZO-2) to the rear. However, they nearly all carry a standard equipment kit consisting of one Model GSP-1 or GSP-1M nerve agent and radiation detector-alarm or one Model GSP-11 nerve agent detector-alarm and a separate radiation alarm, one Model VPKhR chemical agent detection and identification kit or the equivalent, one Model PPKhR automatic chemical agent detection and identification kit, one DP-3 area survey meter, one DP-5A radiation meter, one Model KPO-1 biological warfare sampling kit, SKhT signal cartridges, RDG smoke grenades and lane flag markers and emplacers.

UAZ-469-RKh chemical and radiological reconnaissance vehicle carrying lane flags

BRDM-2 RKha NBC reconnaissance vehicle in travelling configuration with 14.5 mm machine gun turret and warning marker dispensers ready for dispensing

RKhM NBC reconnaissance vehicle based on chassis of MT-LB multipurpose armoured vehicle

Vehicles include the UAZ-69-RKh, UAZ-469-RKh, BRDM-1 RKh, BRDM-2 RKhb and RKha, RKhM-2S and the RKhM-4. These have been joined by later vehicles (see separate entries in this section).

Some vehicles have two KZO-2 20-flag marker emplacers carried at the rear, but the UAZ-69-RKh may have either one 10- or 12-flag emplacer, or a flag rack from which the flags are emplaced manually.

The BRDM-2 RKhb carries the GSA-12 automatic chemical indicator and has a special device for firing warning signs into the ground using explosive charges, even when the vehicle is moving at speeds up to 20 km/h. This device has 19 barrels mounted on a plate, rather like a Gatling gun and can be swung through an axis of 270°. The vehicle is also equipped with the TNA-3 inertial navigation device for accurate course and position finding. The combat weight of the BRDM-2 RKhb is 7,090 kg and it is armed with two 7.62 mm machine guns. An essentially similar vehicle, the BRDM-2 RKha, is armed with a single turret-mounted 14.5 mm KPVT machine gun and lacks some of the sensors carried by the BRDM-2 RKhb. The version of the BRDM-2 RKha used by the former East German armed forces was known as the SPW 40 P2 Ch.

A recent addition to the RFAS range of chemical reconnaissance vehicles is the RKhM, based on the chassis of the MT-LB multipurpose armoured vehicle. Modifications to produce the RKhM include repositioning the engine and the provision of a box-like structure on which a machine gun cupola and the commander's cupola are installed. The upper part of the track is covered by a three- or four-part skirting. At the rear is a lane flag dispenser.

During early 1993 it was revealed that there is an NBC reconnaissance vehicle based on the chassis of the tracked GT-MU armoured transporter tractor. Known as the RKhM-2S, this weighs 5.9 tonnes and carries VPKhR, PPKhR, GSA-12, DP-3B and DP-5V equipments as well as an ASP biological reconnaissance set. Also carried are a KPO-1 probe set and a KZO-1 marker set. Maximum speed when carrying out radiological reconnaissance is 30 km/h and 5 km/h for chemical and/or biological reconnaissance. R-123M and R-124 radio equipments are provided. The RKhM-2S is intended for use in areas of terrain with extremely low temperatures.

An NBC reconnaissance version of the BTR-80 wheeled (8 × 8) armoured personnel carrier is known as the RKhM-4-01 (see separate entry in this section).

A chemical agent detection vehicle carrying a laser-based sensor system is known as the KDKhR-1N, (see entry in this section for details.)

Status
All the above are in service with the RFAS armed forces and other former Warsaw Pact forces. These were made available from Russian State factories before privatisation. Unable to trace current manufacturer or supplier of service and parts.

Manufacturer
Unknown.

VERIFIED

UNITED KINGDOM

Lightweight Marking System (LIMAS)

Description
The INSYS Lightweight Marking System (LIMAS) delineates NBC-contaminated areas, minefields and other hazards, route and lane markings and similar operations. It utilises a lightweight picket post with one end slotted to hold tape, wire or light rope; a flanged interface unit secured by a steel pin which is hammered into the ground or tarmac surface; yellow and red tape 16 mm wide; and a simple hand-held dispenser. Marking triangles and signs can be provided. Using this equipment, a team of two can mark perimeters faster than 1 km/h.

Packaging can be arranged to suit customer requirements and satchels are available to enable one person to carry equipment sufficient for 250 m of fencing.

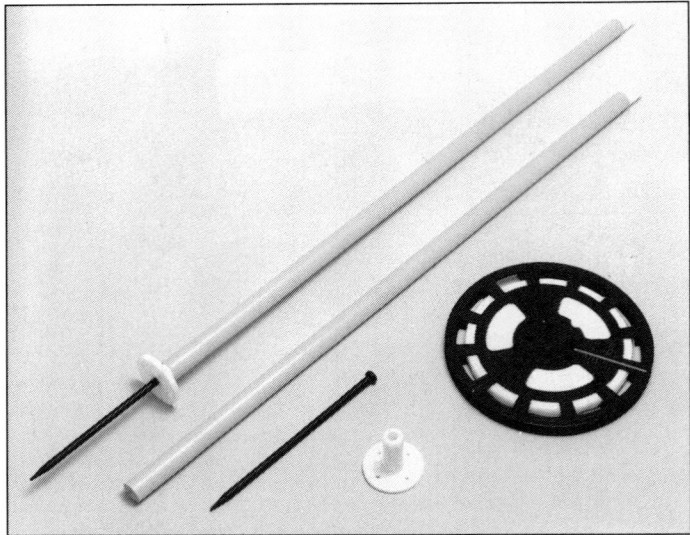

Components of INSYS Ltd lightweight marking system

Status
In production.

Manufacturer
INSYS Limited.

UPDATED

Pearson Pathfinder marking device

Description
To meet the need for an automatic marking device which can be used in adverse conditions, Pearson Engineering developed its Pathfinder marking device. It can be used to indicate the boundaries of areas such as minefields or areas of NBC contamination. The device can also be used to denote the approaches to bridges or similar defiles.

Pathfinder can be attached to any vehicle and operates automatically while the crew remains under cover. It operates by firing lightweight rods into the ground using compressed air. The rods are stored in magazines, each with a capacity of 100 rods, which means that up to 1,500 m of terrain can be marked from a single magazine. The rods are painted Day-Glo white and yellow (or other colours) and can be fitted with lights for use at night. The rod tips are made of stainless steel which ensures penetration into almost all ground conditions. The rods are reusable.

The marker unit is made up of the magazine, an emplacement unit frame, a deployment frame to swing the device 90° over the side of the vehicle to deploy the emplacement unit and a power pack (compressor), sufficient to drive two marker units. A control unit completes the system. The complete unit is fully waterproof and dust protected.

When not in use the Pathfinder is stowed within the width of the carrier vehicle and is deployed to the firing position in approximately 14 seconds, all from under armour. Once deployed, rod spacing can be controlled in either of two modes: by time (5, 10 or 15 seconds), or by full manual operation.

The Pathfinder marking devices are the two columns either side at top rear of the vehicle shown
0050702

Specifications

Emplacement unit
Weight: 210 kg
Length: 1.5 m
Height: 500 mm
Width: 400 mm

Marker rods
Weight: (each) 125 g
Length: 996 mm
Diameter: 12 mm
Soil penetration:
  (sand) 150-200 mm
  (mid-European clay) 70-150 mm
  (hard-baked clay) 30-50 mm

Power pack
Weight: 128 kg
Length: 408 mm
Width: 373 mm
Height: 450 mm

Control box
Weight: 3 kg
Length: 220 mm
Width: 120 mm
Height: 80 mm

Status

In production. In service with the British Army. An updated version is in service with the French Army and two systems are being evaluated by the US Marine Corps.

Manufacturer

Pearson Engineering.

UPDATED

··

Piranha NBC reconnaissance vehicle

Description

The Piranha NBC reconnaissance vehicle is developed by Alvis vehicles in partnership with the Intellitec Division of Advanced Technical Products Inc of the US as a private venture. It is based on the command post variant of the Swiss-designed MOWAG Piranha II (see *Jane's Military Vehicles and Logistics* for technical details of the vehicle). The following NBC detection and protection equipment is fitted:

- A development of the Viking Instruments SpectraTrak CW Agent Detection and Identification System (CADIS) allows continuous air sampling as well as ground sampling from the double-wheel sampling unit (see this section for the OWR example of the latter)
- A cupola on the cabin roof, fitted with the M21 RSCAAL CW agent detector and alarm system (see this section), can be traversed through 360° of bearing and 20° of elevation
- The CAM is carried onboard for both COLPRO and area monitoring
- A new Chemical, Biological Mass Spectrometer (CBMS), under development by Orbital Sciences, is also fitted
- Nuclear radiation levels are monitored with the AN/VDR-2 detector, with exterior and interior probes (see under DETECTION (sensor systems) - Nuclear)
- The TACMET II meteorological system provides real-time wind vector, air temperature, pressure and RH data to a display unit at the vehicle control console

Piranha NBC reconnaissance vehicle   *2001*/0067446

- A LITEF land navigation system uses dead-reckoning with periodic GPS updates integrated with digital map data.

All the detection and met data is integrated using the MICAD system (see under DETECTION (C³I)). The vehicle claims the advantage of having considerably more internal volume than other reconnaissance solutions and a faster response time. The vehicle carries the M13 portable DECON system, a full communications suite and area marking system.

Status

Available.

Manufacturers

Alvis Vehicles (UK).
Advanced Technical Products Inc (US).

VERIFIED

───────────────────────────────

UNITED STATES OF AMERICA

Armoured Light NBC Reconnaissance Vehicle (ALRV) Light NBC Reconnaissance Vehicle (LRV)

Description

These vehicles can act with NBC reconnaissance units to conduct survey activities over a wide area providing warnings of battlefield contamination and designate safe areas for the conduct of operations. This requires the capability to rapidly deploy NBC reconnaissance equipment, detect and characterise contamination (air, ground and water), determine and communicate the location and size of contaminated areas to other units and mark safe lanes and the boundaries of contaminated areas.

The ALRV and LRV are self-contained, highly mobile reconnaissance systems capable of rapid chemical and nuclear detection, identification, marking, mapping and communication. In addition, both vehicles provide the capability for biological sampling.

Both vehicles provide overpressure NBC protection for the crew (on the LRV the driver is provided with a M42 respirator). The ALRV is based on the MOWAG

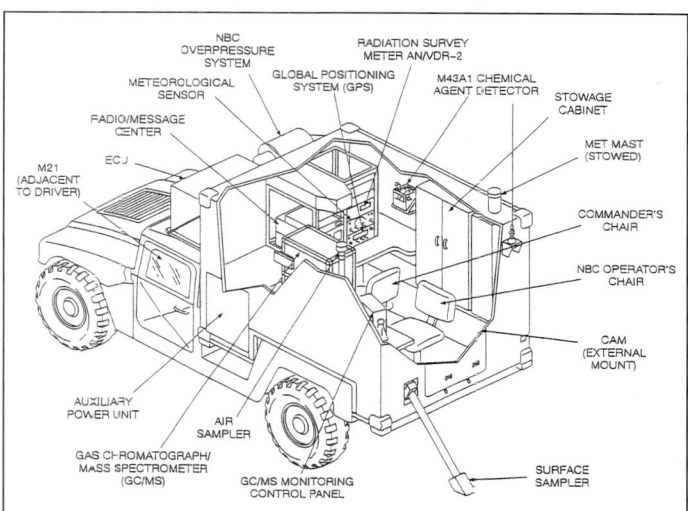

Outline drawing of the layout of a Light NBC Reconnaissance Vehicle (LRV)

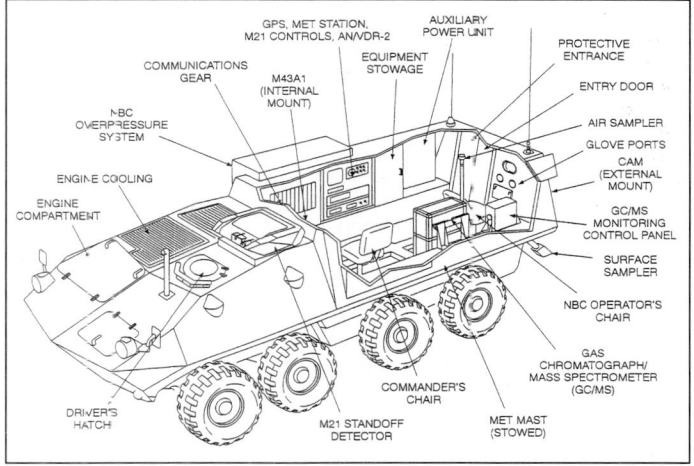

Outline drawing of the layout of an Armoured Light NBC Reconnaissance Vehicle (ALRV)

Piranha/LAV 8 × 8 series armoured personnel carrier and carries a crew of three. The LRV is based in a Standard Integrated Command Post Shelter (SICPS) carried by a HMMWV 4 × 4 multipurpose wheeled vehicle and has a crew of two or three.

Both vehicles can carry a CADIS gas chromatograph/mass spectrometer, two chemical agent point detectors (one CAM, one M43A1), an M21 RSCAAL standoff chemical agent alarm, two chemical agent detection kits (one M272, one M256A1), two radiation detection instruments (one AN/VDR-2, one AN/PDR-77), biological specimen sampling equipment, meteorological sensor, Global Positioning System (GPS) and support and communications equipment. In addition both vehicles have a marking kit for clear lanes and area boundaries and an M13 DAP decontamination apparatus. Both vehicles can carry an Integrated Protective Entrance (IPE) for use when the vehicle is deployed at a static location.

Status
Available.

Manufacturer
General Dynamics Armament and Technical Products.

Marketing agency
Tradeways Limited.

UPDATED

M93A1 Fox NBC Reconnaissance System - BW sensor mast shown raised (General Dynamics Corporation)   0011469

ChemSonde air-deployable CW sensor

Description
ChemSonde (Chemical agent DropSonde) is an aircraft-deployed remote CW agent detection sensor. ChemSonde uses JCAD Surface Acoustic Wave (SAW) sensors to detect, identify, and quantify the presence of CW agent vapour clouds.

ChemSonde provides military and civil defence planners with immediate alert and verification of airborne CW agent presence. ChemSonde transmits real-time data of CW agent type (nerve, blister or blood agent), cloud velocity and direction of travel as well as the upper and lower vapour cloud altitude boundaries.

As the ChemSonde is designed to withstand the shock of pyrotechnic ejection and to fit in the 8.1 × 2 × 2.5 in volume identical to the USAF MJU-10 flare, it can be dispensed from any standard inventory military aircraft chaff/flare countermeasure dispenser system. Chaff/flare dispensers, such as the ALE-47, are currently installed on nearly all military tactical and transport aircraft.

Once ejected from the aircraft, ChemSonde deploys a parachute and antenna before starting to sample the air and transmit data during its controlled descent. Data is transmitted over a VHF/UHF datalink to a ChemSonde data station. Transmitted data includes temperature, pressure, humidity, GPS co-ordinates, GPS velocity, CW agent type and corresponding concentrations.

ChemSondes have been flight-tested and successfully dispensed from military fast jets, fixed-wing aircraft, and UAVs. The current version ChemSonde collects data as it descends with an unguided parachute.

Future BAE SYSTEMS Tactical DropSonde capabilities will include:
- GPS-guided parafoil
- SatCom datalink
- BW agent detection
- Unattended ground operation.

ChemSonde is currently in low-volume production.

Status
Available. ChemSonde is currently in low-volume production.

Manufacturer
BAE Systems.

VERIFIED

Specifications
Max deployable airspeed: 450 kts IAS
Descent rate: 3.2 m per sec at 1,500 m AMSL
Dimensions: 20.3 × 5.0 × 6.3 cm
Operating environment: −10 to +49°C
Operating altitude: Up to 7,600 m AMSL
Weight: <0.7 kg
Battery: 2 SAFT 2/3 C size $LiMnO_2$ +3 V DC
Telemetry omni-directional range: >112 km LOS (max range requires airborne platform)
Detector: SAW (see also JCAD, this section)

Fox NBCRS M93A1 vehicle

Description
The Fox NBCRS M93A1 vehicle is a high-speed, high-mobility armoured carrier wholly dedicated to NBC detection, analysis and warning. Sampling mechanisms allow remote collection of potentially contaminated soil, water and air in its immediate environment. It was used in the Gulf War and at that time was the most

M93A1 Fox NBC Reconnaissance System (ALRV) (General Dynamics Corporation)   0011468

sophisticated, technically complex piece of CW agent detection equipment to be used by US forces. Since then the US has improved Fox doctrine, training and equipment.

The heart of the Fox system is the MM-1 Mobile Mass Spectrometer and, since the Gulf War, the M21 Remote Sensing Chemical Agent Alarm (RSCAAL). See separate entries in Detection (sensor systems) - Chemical. Information from the Fox detection suite, from meteorological sensors and from GPS is integrated by the data management system, allowing automatic transmission of digital NBC warning messages through the Manoeuvre Control System to warn follow-on forces. Planned developments (NBCRS Block II project) will incorporate upgraded CBW detection and analysis capabilities by the installation of the Chemical Biological Mass Spectrometer (CBMS). The Joint Service Lightweight Stand-off Chemical Agent Detector (JSLSCAD) will replace the M21 RSCAAL. See separate entries in Detection (sensor systems) - Chemical.

The first two M93A1 Upgrade Program vehicles were delivered from Henschel Wehrtechnik (now Rheinmetall Landsysteme of Germany) in mid-1997 to General Dynamics Land Systems for validation of technical manuals and the development of troubleshooting procedures and training material. They were delivered to General Dynamics at Anniston, Alabama in November 1997 for final acceptance. Key personnel training commenced at Fort McClellan in December 1997. Army training to support Production Qualification Testing (PQT) began in February 1998. Following PQT, the vehicles returned to the Anniston Army Depot for rework and subsequent fielding. The first equipped unit was fielded in December 1998.

In early 1997, General Dynamics was awarded a US$39 million contract by the US Army Chemical and Biological Defense Command (now the SBCCOM) as part of a US$131.3 million contract to upgrade the system to M93A1 configuration. The upgrade incorporates organic direct support, maintenance and supply support and reduces the number of operators from four to three. The programme (for 62 of the vehicles) required that 59 units would be done at the Anniston Army Depot by a team comprising General Dynamics Land Systems, Henschel Wehrtechnik and Anniston Army Depot personnel.

The basic contract provided for the establishment of a government owned/ contractor operated facility at the Anniston Army Depot where the upgrade activity would take place. Eight vehicles were to be upgraded in conjunction with this basic contract. Upgrade contract options for the remaining vehicles were covered in FY97 (30 systems), FY98 (12 systems) and FY99 (12 systems), FY00 (procured 14 Block I systems of which 7 have been installed. Three training systems were also procured).

Status
The US Armed forces owns 123 Fox vehicles. Of these, a total of 86 are expected to be upgraded to M93A1 standard under the General Dynamics Corporation

contract. By November 2000, 60 vehicles had been delivered to the US Army. In August 2001 a US$26.4 million modification contract was also awarded for the upgrade of 11 vehicles. Work is being performed at Anniston (50 per cent), Germany (35 per cent) and Sterling Heights Michigan (15 per cent) and completion is expected by the end of June 2004.

FY01 objectives include plans for Block II development, test and evaluation and the integration of developmental detectors into the vehicles. Trials are planned for FY04 and the first unit is likely to be equipped in FY05.

Manufacturer
General Dynamics Corporation.

UPDATED

Joint Service Light Nuclear, Biological and Chemical Reconnaissance System (JSLNBCRS)

Description
The US Marine Corps has awarded TRW contract modifications for the engineering and manufacturing development phase of the Joint Service Light Nuclear, Biological and Chemical Reconnaissance System (JSLNBCRS). These modifications total more than US$21 million bringing the total cumulative value of the contract to approximately US$34 million, the largest programme TRW has undertaken for the Marine Corps Systems Command to date. The JSLNBCRS is being developed to detect, identify, mark and report NBC hazards and toxic industrial chemicals on the integrated battlefield. The system represents a seamless integration of sensors and C⁴I capabilities.

Following on from variants delivered during the programme's development risk reduction phase, TRW will refurbish the High Mobility Multipurpose Wheeled Vehicle (HMMWV) variants and then design and fabricate Light Armoured Vehicle (LAV) variants for further developmental testing. Both platform variants will then undergo operational testing between July and October 2003 prior to a production decision and fielding in 2004. Seven HMMWV and two LAV pre-production variants will be tested. Current fielding is estimated at more than 500 HMMWV and 30 LAV variants.

Status
In development.

Manufacturer
TRW.

NEW ENTRY

MicroSTAR Micro Air Vehicle (MAV) for NBC reconnaissance

Description
In June 1998, the US Defense Advanced Research Projects Agency (DARPA) authorised work on the possibility of surveillance using MAVs. Several organisations have developed prototypes for the programme using a variety of design approaches including miniaturised helicopters and ducted propellers as well as conventional mini-fixed-wing aircraft. With NBC sample collection as a possible payload, Sanders developed the MicroSTAR for evaluation. The MicroSTAR is a low-flying miniature air vehicle weighing 85 grams and measuring 15 cm across. The payload will use advanced miniaturisation techniques, including ultralight packaging and the so-called 'Chip-on-Flex' and will represent approximately 20% of the total all-up weight. The vehicle will incorporate an autopilot, INS, the sensor and a data link. Payloads will be fully integrated to accomplish waypoint navigation, provide day and night imagery and real-time data download to the launch or control site.

DARPA expects MAVs to be launched by individuals at platoon level or higher or expended from overflying aircraft. Designed for local area situation awareness, the 30 mph MAVs will be capable of short-duration (20 minutes) excursions at low altitude and offer real-time imagery to over 5 km or of capturing N, BW or CW agent data for download. MAVs could be used for landing area surveillance prior to amphibious operations, for site monitoring by advancing forces, for retrieval of plume data from contamination incidents, for remote surveillance of sites by verification inspectors or for surveying areas too toxic for human safety.

Status
Research project for US DARPA.

Manufacturer
Sanders.

UPDATED

YUGOSLAVIA, FEDERAL REPUBLIC

Contaminated area marking kit KOZ

Description
The contaminated area marking kit KOZ can be used for the day or night marking of NBC contaminated areas and also to provide an indication of the direction of travel to avoid such areas.

The complete KOZ kit comprises a set of 20 flags to mark the boundaries of contaminated areas, a set of two signboards for directions, lighting kit for night use, writing instruments and two bags to carry it all. The carrying bags can be used to hold contaminated signs and flags ready for decontamination.

Status
Available. In service with the Yugoslav Army.

Contractor
Yugoimport SDPR.

VERIFIED

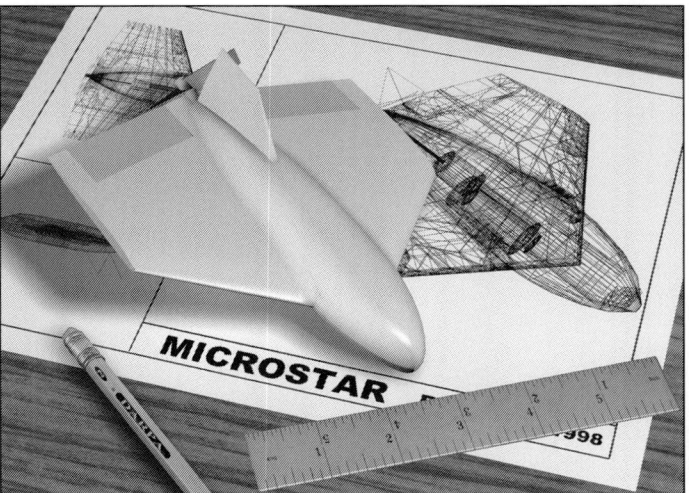

An NBC payload is one option planned for MAVs such as the Sanders MicroSTAR (artist's impression) which forms part of a US DARPA project　　　0055214

Contaminated area marking kit KOZ in use

PROTECTION (individual)

Masks (general issue)
Masks (aircrew)

PROTECTION (individual) - Masks (general issue)

BELGIUM

BEM 4GP general purpose respirator

Description
The BEM 4GP general purpose respirator was developed by Engicom Systems in close collaboration with the Belgian armed forces and various university research centres.

The BEM 4GP uses a facepiece that is stated to be easy to decontaminate and is held in position by a head harness system. The vision device is a one-piece panoramic panel made of coated polycarbonate which is impact- and scratch-resistant and can be supplied as an easily replaceable spare part. A set of spectacles for eyesight correction can be attached inside the respirator by means of a removable fixture. In front of the user's mouth is a monobloc assembly which combines the filter canister interface, the inhalation and exhalation systems, a communication device and a drinking system. There is also a drainage point. The mask is available in three sizes (S, M, L) of which the medium size is designed to fit 86 per cent of male and female users.

The filter canister interface is unusual in that the canister position can be readily altered to suit the wearer. It can be moved through 160° (centre, left or right) to allow the use of weapons or other equipment in various positions. The interface has a locking system and there is no break in protection as the filter canister position is altered.

The speech diaphragm is a straight-through device which is compatible with most types of communication equipment. The drinking system uses a 'drinking straw' arrangement which is guided by a protective sheath. This system allows the intake of high-viscosity liquids and prevents any pollution of the interior of the respirator.

The respirator is carried in a bag that also contains other parts such as a spare filter canister. The bag is designed to be opened in a single movement. This facility, together with the special grip on the respirator head harness, allows the user to achieve the 9-15 sec NATO target for safe donning when under chemical attack. The vision panel, monobloc assembly, speech diaphragm and the inhalation and expiration valves are all available separately as spare parts.

Status
In service with the Belgian armed forces and the state police.

Manufacturer
Engicom Systems NV.

UPDATED

The BEM 4GP general purpose respirator (Engicom Systems NV)

Seyntex NBC facelet

Description
Seyntex produces an NBC protective facelet which can be worn for indefinite periods. It is made from an outer layer of flameproof-treated core-spun polyester/cotton while the interior filter layer consists of a heavy polyurethane foam loaded with active charcoal. The foam is separated from the mouth and nose by a very thin layer of polyester. Four straps are used to hold the facelet on the head.

Status
Available.

Manufacturer
Seyntex NV.

VERIFIED

BULGARIA

PF-90 Combined Arms NBC respirator

Description
The PF-90 Combined Arms NBC respirator consists of a facepiece, one or two filter canisters, a carrying bag and accessories. The filter canister may be worn either on the chin or on the left side of the facepiece. There is also a speech transmitter device, while a drinking device with a flow rate up to 200 mg/min uses a port on the right cheek

The PF-90 is intended to be worn for periods up to 24 hours. Air flow resistance at 30 litres/min is 100 Pa and the weight is 800 g.

Status
In production. Offered for export sales.

Marketing agency
Kintex.

VERIFIED

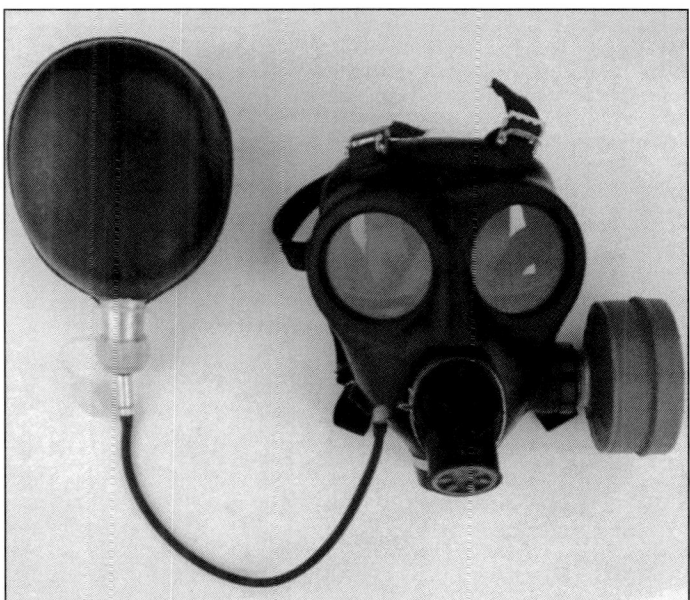

PF-90 Combined Arms NBC respirator with filter canister on left and showing drinking device

PG-1 NBC respirator

Description
The PG-1 NBC respirator is a conventional design with a rubber facepiece held in position by a head harness. Large eyepieces are provided and there is a prominent circular speech unit positioned in front of the mouth. A large filter canister, the EO-18K, is mounted on the chin position. This respirator is provided with a light fabric sling worn around the neck allowing the respirator to be carried on the chest when not on the face. A speech transmitter unit is provided.

Air flow resistance at 30 litres/min is 230 Pa and effective field of view 70 per cent. Weight is 800 g.

Bulgarian P6-1 NBC respirator

Status
In production. Offered for export sales.

Marketing agency
Kintex.

VERIFIED

CANADA

C4 protective mask

Description
The Canadian C4 mask consists of a bromobutyl rubber facepiece with turned-in peripheral face seal. The mask has an internal silicone nose-cup. The one-way valve on the nose-cup deflects incoming air over the eyepieces to eliminate fogging. There are two uniquely shaped, circular, convex eyepieces made of polycarbonate. There are NATO-compliant threaded canister mounts on either side of the facepiece which can take either a canister or a secondary voice emitter. The primary voice emitter is located at the front of the mask just below the eyepieces and has a glass-reinforced nylon protector plate. The black metal outlet valve is below it, underneath a detachable rubber covering to provide dead air space and access to the valve itself. The mask is held in place by a six-point elastic mesh bonnet. After initial fitting, only the two lower clips require adjustment for best seal and greatest working comfort.

The C4 is available 2 colours: green or black. NATO stock number is 42400-221-908-1095 for black and 42400-221-908-1096 for green. The size range comprises extra small, small, medium and large and is designed to cover 94 per cent of the

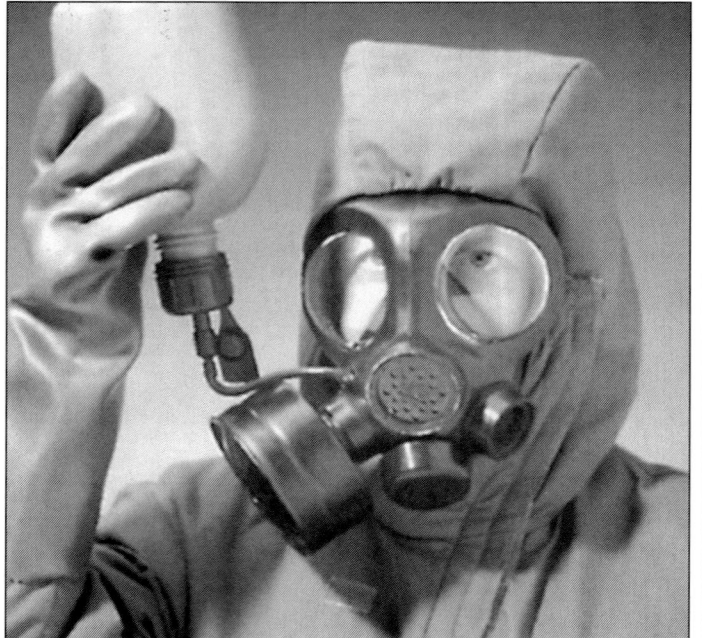

The C4 mask (with canister on wearer's right-hand side) with drinking tube in use (Irvin Aerospace Canada Limited)   ***2001**/0106862*

The C4 mask (canister fitted on left-hand thread) (Irvin Aerospace Canada Limited)
***2001**/0106881*

Specifications	
Total weight	705 g
Canister	265 g (C2 canister*)
Facepiece	440 g
*See under Protection (individual) – filters	

population. The mask meets both Canadian national and NATO standards for protection (NATO Triptych).

This mask also has an integral drinking device (see illustration). This and other minor engineering changes were prompted by field trial testing during Operation Desert Storm, in Canada and in Europe. Eye-correction is supplied in the form of internal wire combat spectacles, currently being developed by DRES.

Modifications and attachments allow the C4 facepiece to be used both in aircraft and vehicle COLPRO (for further details see under Protection (collective) and other entries within this section).

Status
Available. In service with the Canadian armed forces.

Manufacturer
Irvin Aerospace Canada Limited.

VERIFIED

Carleton NBC belt-mounted respiratory system

Description
The Carleton NBC belt-mounted respiratory system was designed for use by personnel obliged to carry out the repair or servicing of equipment, aircraft or vehicles while working in areas contaminated by NBC warfare agents. The system provides the user with clean, filtered air, under pressure to the mask or respirator, preventing the ingress of agents. A standard C2 filter canister is used with the system, or any filter canister with a NATO standard thread and it can be fitted with an optional snap-on prefilter for use in dusty or desert areas.

The user can select one of five air flow speed positions on a control box to suit individual comfort. The system's positive pressure air flow provides the user with

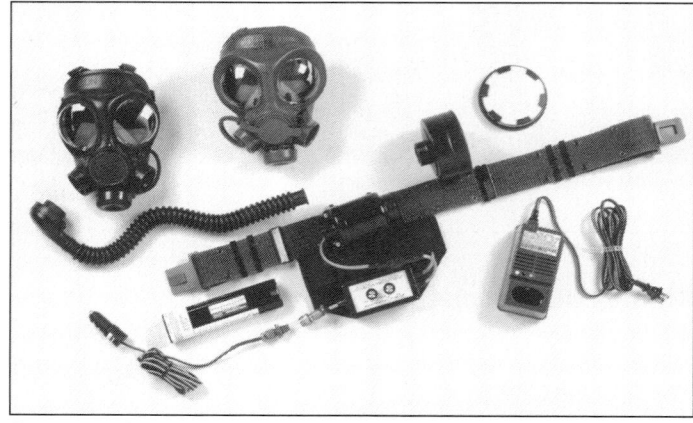

Components for Carleton NBC belt-mounted respiratory system

Carleton NBC belt-mounted respiratory system

cooling and anti-misting of the eyepieces when working in warm temperatures or in extreme circumstances.

The system was designed for use with the C4 mask although other similar masks may be utilised. A rechargeable Ni/Cd battery will operate the system for 2 hours of continuous use at a temperature of 22°C; a battery charger is available. It is possible to connect the system to a suitable vehicle power supply.

Status
Available.

Manufacturer
Carleton Life Support Technologies.

VERIFIED

CHINA, PEOPLE'S REPUBLIC

M-65 protective mask

Description
The M-65 protective mask features a prominent filter housing on the left-hand cheek of the facepiece and a prominent voicemitter at the front. Large lens eyepieces are provided and the mask is held in position by a six-strap head

M-65 protective mask (Shanxi Xinhua Chemical Factory)

harness. The filter has an efficiency of 99.995 per cent and can withstand aerosol droplets down to 0.3 μm in diameter. Air flow resistance for inhalation is less than 176 Pa and less than 98 Pa for exhalation at 30 litres/min.

Weight of the M-65 is 600 g.

Status
Available. In service with the Chinese armed forces.

Development agency
Research Institute for Chemical Defence.

Manufacturer
Shanxi Xinhua Chemical Factory.

UPDATED

M-87 protective mask

Description
The M-87 protective mask appears to have a natural rubber facepiece held in place by a six-strap head harness. A cylindrical filter canister is fitted to the left-hand side of the facepiece and large eye lenses are provided. A prominent voicemitter device is fitted to the front of the facepiece. The filter has an efficiency of 99.995 per cent and can withstand aerosol droplets down to 0.3 μm in diameter. Air flow resistance for the filter canister is less than 147 Pa at 30 litres/min.

Weight of the M-87 is 700 g.

The MF-11 protective mask is a version of the M-87 with a drinking device which has been offered for commercial sales.

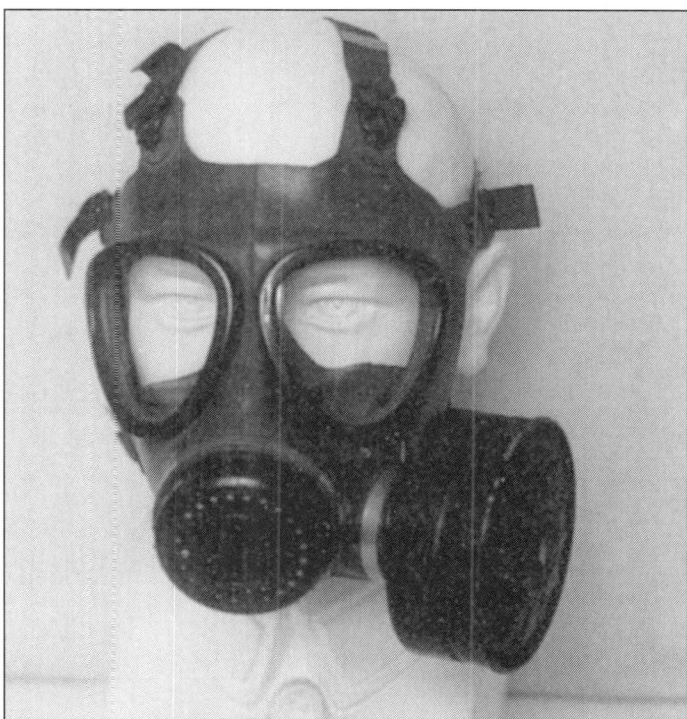

M-87 protective mask (Xinhua Chemical Factory)

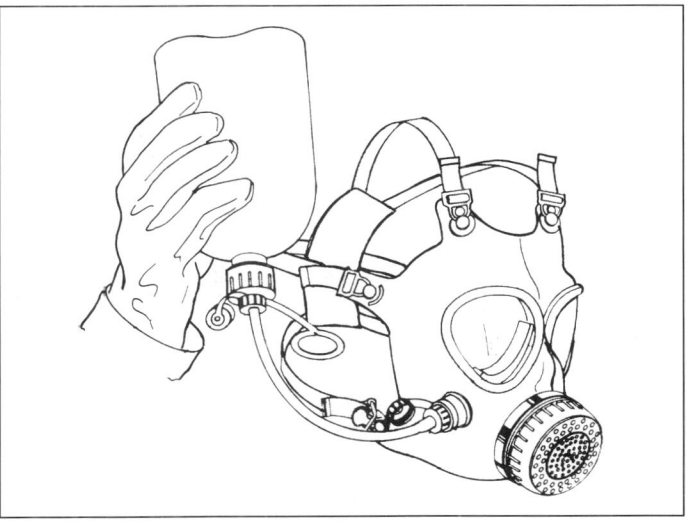

Drawing of MF-11 protective mask drinking device (Xinhua Chemical Factory)

Status
In production. In service with the Chinese armed forces and some other nations.

Development agency
Research Institute for Chemical Defence.

Manufacturer
Xinhua Chemical Factory.

UPDATED

Type 69 protective mask

Description
The Type 69 protective mask is stated to provide the wearer with protection against toxic gases and vapours, smoke, radioactive dusts and bacteria. It consists of a moulded rubber facepiece with wide-vision eyepieces that are probably flexible. A charcoal-based filter canister is located on the side of a large voice transmission device that projects prominently to the front. The voice transmission device is stated to provide 90 per cent articulation at a range of 50 m.

Weight of the mask is 650 g and the filter canister 150 g.

Status
In production. In service with the Chinese armed forces and offered for export.

Manufacturer
China North Industries Corporation.

UPDATED

The driver and crew of an armoured personnel carrier wearing Type 69 protective masks (China North Industries Corporation)

CZECH REPUBLIC AND SLOVAKIA

CM-4 NBC respirator

Description
The CM-4 NBC respirator is optimised for civil defence personnel. It was purchased by the Hungarian civil defence organisation between 1985 and 1990. It is of conventional design, available in three sizes, and has a rubber facepiece with two wide vision eyepieces. The respirator is held on the face by three head straps and one further strap across the upper neck. Drinking and speech enhancement facilities are also fitted The chin-mounted standard filter thread is designed to take a connection to a closed-circuit breathing apparatus or to a standard canister of the MOF-4 type (See separate entry under *Protection (individual) - Filters*).

Specifications
Weight: (average) 400 g
Field of vision: 73%
Breathing resistance (30 l/min)
  Inhalation: 25 Pa (valve chamber)/30 Pa (inner mask)
  Exhalation: 100 Pa
Drinking capacity: 200 ml/min

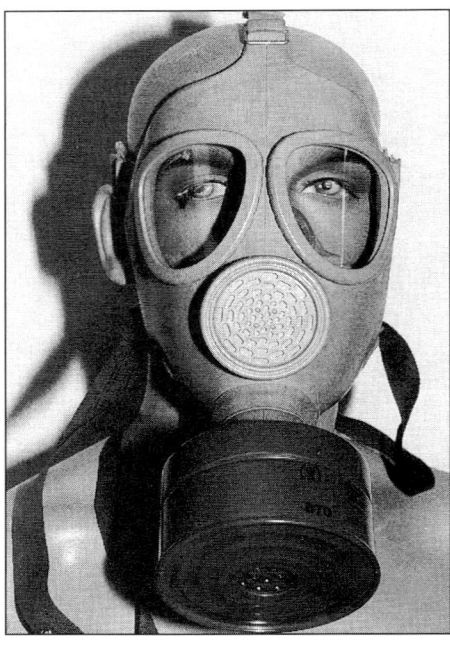

CM-4 NBC respirator
(T J Gander)

Status
Available.

Manufacturer
Gumárny Zubří JSC.

UPDATED

CM-5D and CM-6 respirators

Description
Developed with the experience of the CM-4 respirator, the CM-5D is aimed at first responder organisations. It differs from the CM-4 in offering a panoramic visor facepiece but otherwise has a similar capability. Internal visual correction glasses can be worn. Available in two sizes, it has a five-point face harness, drinking (designated CM-5DM) and speech enhancement facilities. The chin-mounted standard filter thread is designed to take a connection to a closed-circuit breathing apparatus or to a standard lung powered filter canister.

CM-6
The CM-6 is a further development, also designed for use by domestic responders. Available in a single size, it features a redesigned polycarbonate visor, speech

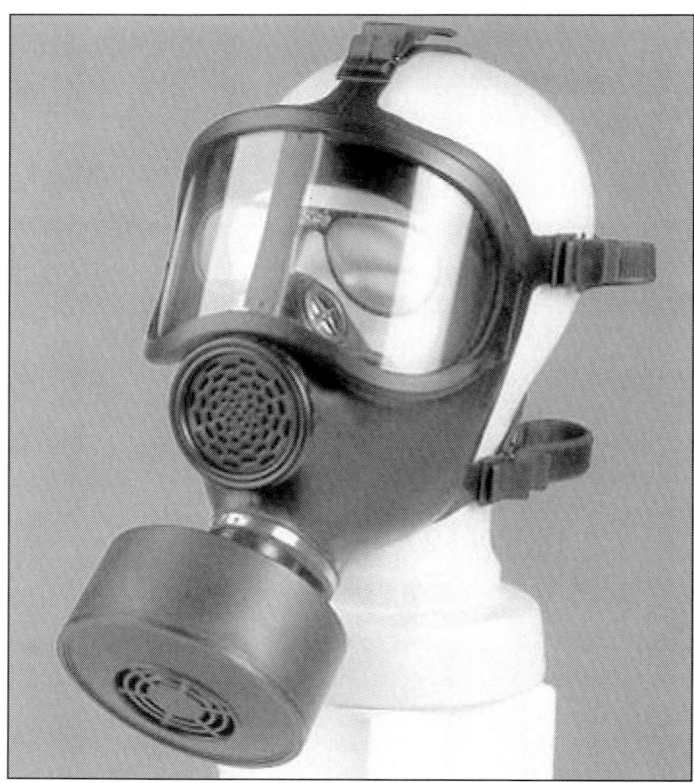

CM-5D respirator with the OM-90 filter canister fitted (Gumárny Zubří JSC)
2002/0121883

CM-6 respirator shown with a standard industrial filter canister fitted (Gumárny Zubří JSC) **2002**/0121882

Specifications

CM-5D
Weight: (average) 500 g
Field of vision: 77%
Breathing resistance (30 l/min):
   Inhalation: 25 Pa (valve chamber) 30 Pa (inner mask)
   Exhalation: 100 Pa

CM-6
Weight: (average) 500 g
Field of vision: 75%
Breathing resistance (30 l/min):
   Inhalation: 25 Pa
   Exhalation: 60 Pa

enhancement device (95 per cent improvement over previous designs) and face seal. The unit has side locations for one or two lung-powered canisters or a connection to a closed-circuit breathing apparatus. Access frees include a drinking device and a communications connection for a radio unit.

Status
Both types available.

Manufacturer
Gumárny Zubří JSC.

NEW ENTRY

Model M-10 protective mask

Description
First issued during 1970, the Model M-10 mask is a copy of the American M17 series (which see). Although it is no longer listed as a current product by this manufacturer, maintenance and support may be provided.

The large eyepieces provide good visibility and the airflow inside the inner mask prevents fogging. The lower part of the mask is very prominent and houses inlet valves on either side of the internally fitted filter system. A single outlet valve at the front below the 'voice emitter' enables the wearer to communicate with relative ease. A head harness holds the mask in place.

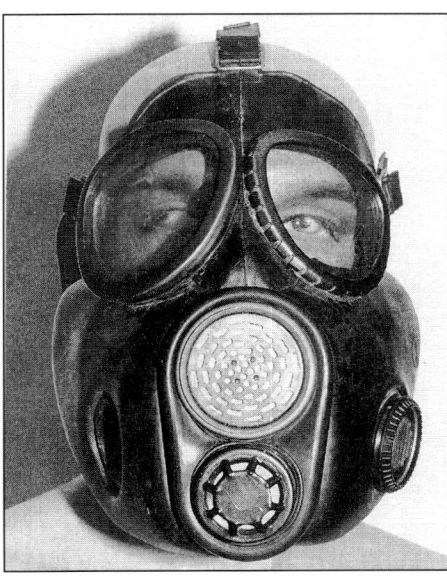

This M-10-type protective mask was manufactured in Bulgaria where it is known as the PDE-1 (T J Gander)

A 'tropical' version with a drinking device has been produced. It is understood that this mode is known as the M-10M.

The M-10 has also been manufactured in Bulgaria as the PDE-1.

Status
Available. In Czech and Slovakian Army service and has been exported. In limited service with the Polish armed forces. Being superseded by later models.

Manufacturer
Gumárny Zubří JSC.

UPDATED

Model OM-90 protective mask

Description
The OM-90 is a modern and efficient face-mask which is designed for military use. It is therefore strongly constructed and lightweight, offers good communications, good optical correction facilities and a drinking facility. There are alternative standard-thread canister mountings on either side of the facepiece. The low-breathing resistance, OF-90 is the recommended filter canister (see separate entry under Protection (individual) - Filters). The round, clear eyepieces are made of hardened glass and are very resistant to impact or scratching. The drinking tube is stowed around the speech diaphragm at the front and a plastic tap above the canister allows fluid to flow from the attached water canteen. The OM-90 is available in three sizes: small, medium and large.

The OM-90 is delivered as part of a kit which includes a single-use protective cape (JP-90).

Status
In service with the Czech and Slovak Armies. Available for export.

Manufacturer
Gumárny Zubří JSC.

UPDATED

Specifications
Weight
OM-90 kit: 2,300 g
Face-blank with NBC canister: 770 g
NBC canister: 255 g
Breathing resistance (30 l/min)
   Inhalation: 150 Pa
   Exhalation: 30 Pa
Face-piece breakthrough time to HD at 30°C: >8 h
JP-90 breakthrough time to HD at 30°C: >20 min
Filter dynamic sorption capacity (30 l/min)
   GB, GD, HD, AC, CK: >2 g
   CG: >6 g
Protective property
   Coefficient of aerosols inward leakage through filter: 1×10^{-4} % (standard turbine oil fog)
   Coefficient of face seal leakage: 5×10^{-3}% (tested with SF)
Voice transmission: 98% minimum
Field of vision: 70% minimum
Drinking capacity: 200 ml/min

OM-90 protective mask (Gumárny Zubří JSC) **2001**/0059322

EGYPT

BSS and CM3 protective masks

Description
These two protective masks are Egyptian-produced versions of the Russian Federation and Associated States (CIS) (RFAS) ShM helmet-type protective head mask (see separate entry in this section). Both are identical to the originals in virtually every respect; they use the same type of filter canister system and are carried in canvas shoulder bags. The BSS is the standard type of ShM mask covering the entire head while the CM3 has a diaphragm voicemitter and a head harness, making it similar to the RFAS Protective Mask, Communication (see separate entry in this section).

Both masks are stated to provide 65 per cent vision and the total weight of both is 1.25 kg.

Production of these masks has apparently ceased although they remain in service with the Egyptian armed forces and others. The only NBC respirator now offered by the Egyptian defence industry is the M2-E, a close derivative of the Yugoslav Protective Mask M-2 (see entry in this section).

Status
Production apparently complete. In service with the Egyptian armed forces and possibly some other Middle Eastern armed forces.

Manufacturer
National Organisation for Military Production.

VERIFIED

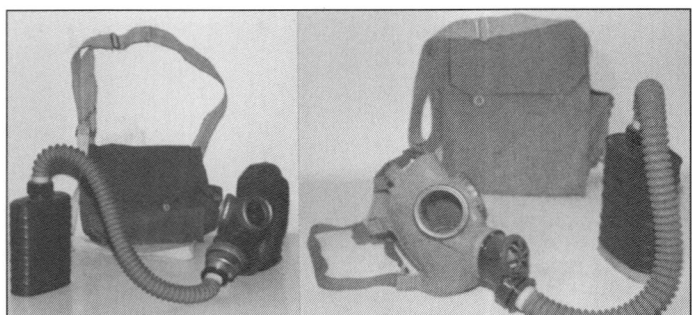

On left is the BSS respirator; on the right is the CM3

M2-E NBC protective mask

Description
The Egyptian M2-E NBC protective mask appears to have affinities with the American M9 mask (qv) although it is a local version of the Yugoslav M-2 (qv). It is stated to be proof against all known chemical warfare agents and is provided together with a canvas carrying bag and a cleaning kit. The associated filter canister has a removable fabric cover and is fitted with a sealing cap.

Total weight of the M2-E is given as 760 g.

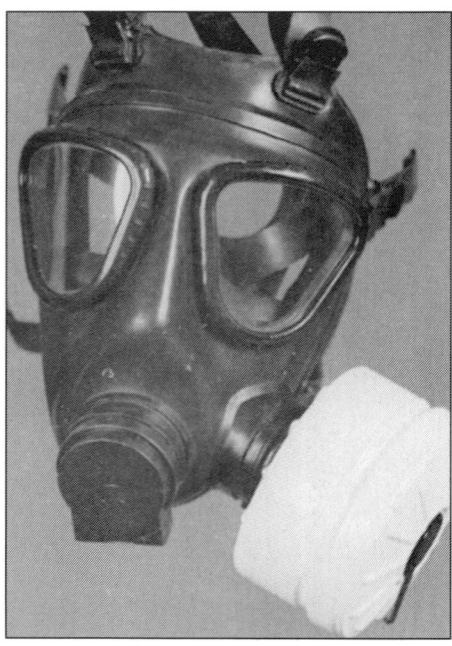

M2-E NBC protective mask

Status
In production. In service with the Egyptian armed forces.

Manufacturer
National Organisation for Military Production.

VERIFIED

FINLAND

Scott Health & Safety CIVIC chemi-hood

Description
This system comprises a hood and visor assembly with an integral filter. It is designed to provide protection against both CB and industrial toxic agents during escape from a hazardous area. The possibility of cost-effective protection for civilian populations following domestic emergencies or a terrorist incident is a key benefit.

The filter is described as offering protection against most hazardous gases except CO. The high-protection factor equates favourably with a conventional NBC respirator mask. Constructed in highly chemical-resistant materials, the unit is designed in one size to fit ages from 12 years to adulthood. The elastomeric collar prevents leakage at the neck. The multipurpose combined filter protects against the majority of hazardous materials.

The Civic hood is prepacked in ready use, quick-opening ABS plastic containers with a shelf-life of 10 years. Two versions of the container allow the hoods to be either wall-mounted or portable by means of a flexible strap-handle. Hoods could therefore be pre-located throughout a high-risk facility or distributed to those forced to escape through a toxic area following an incident.

The material is resistant to HD and other caustic gases such as HCl, Cl_2, HF_2 and other hazardous organic compounds for 24 hours.

The multipurpose combined filter (ABEK P 15) protects against most organic and inorganic gases including industrial hazards. Specifically, protection is good against Cl_2, H_2S, hydrocyanic acid, other acid gases and vapours (SO_2, NH_3, organic ammonia derivatives, radioactive and other toxic particles, bacteria and viruses).

Test gas	ABEK P15 filter	Requirement
Butanon	>50min	>15 min (0.25 Vol %)
Chlorine	>50 min	>15 min
Hydrogen sulphide	>180 min	>15 min
Hydrogen cyanide	>80 min	>15 min
Sulphur dioxide	>24 min	>15 min
Ammonia	>80 min	>15 min
Hydrogen sulphide	>46 min	>5 min (0.1 Vol %)
Particle protection	P3	P2 (EN 143)

Status
In production.

Manufacturer
Scott Health & Safety Oy.

UPDATED

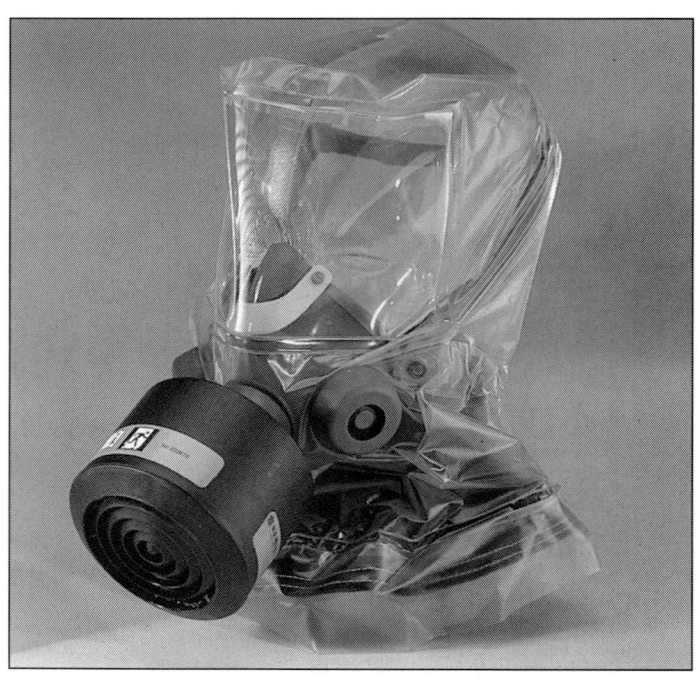

Scott Health & Safety CIVIC chemi-hood   0011446

Scott Health & Safety Safety NBC mask M95

Description

Scott Health & Safety Safety's respirator M95 embodies the highest levels of efficiency and comfort in modern NBC protection and meets the most critical hazards and stresses encountered in combat situations. It is designed and developed using advanced CAD/CAM techniques to ensure a close anatomical fit and user comfort. Total leakage is less than 0.01 per cent. The facepiece is halobutyl rubber. A special silicone inner mask reduces the carbon dioxide content to a minimum. The mask is held in place by a six-point head harness and there is a chin support. The mask can be placed in position in 10 seconds.

There are two high-optical efficiency polyamide lenses, allowing the mask to be used when operating optical devices and weapons. Special sight correction spectacles are available. Two EN 148 threaded locations for the filter canister allow the mask to be worn by left- and right-hand users. A speech diaphragm is provided and there is a hoseless drinking device with an intake rate of 0.25 litres/min. Incoming airflow is routed over the interior of the lenses and a moisture drain is located close to the exhalation channel.

Scott Health & Safety Safety NBC mask M95

The Scott Health & Safety Safety M95 NBC Mask 0011445

Specifications
Dimensions
   Height: 90 mm
   Width: 109 mm
Weight (without canister): 250 g
Gas penetration
   CG at 30 l/min: >40 mins
   Chloropicrin at 30 l/min: >55 mins
   Mustard: >50 mins
Breathing resistance:
   30 l/min: <1.2 mbar
   95 l/min: <4.0 mbar

The M95 mask weighs under 500 g without a filter canister and 720 g with one in place. Inhalation resistance at 30 litres/min is less than 45 Pa and less than 110 Pa at 95 litres/min. Exhalation resistance at 160 litres/min is less than 130 Pa. The mask has a protection factor of more than 10,000 and can provide protection against chemical warfare agents for more than 48 hours.

The M95 can be used over a temperature range of -50 to +70°C. Accessories include spectacles, a drinking bottle with an adaptor and a maintenance tool. Shelf-life is 20 years

Status
Satisfies current NATO standard.

Manufacturer
Scott Health & Safety Oy.

UPDATED

FRANCE

BACANOP leakage tester for respirators

Description

The BACANOP leakage tester is designed to check the overall air-tightness of a respirator in less than 30 seconds. It consists of a compact, lightweight test unit to which an elastomer head is attached. A sealing plug replaces the filter canister and the respirator is attached to the head and depressurised by squeezing gently. The test unit measures the difference between the ambient and internal mask pressures as a function of time. After 10 seconds, the operator presses a button. A green light indicates a pass. The rejection threshold is adjustable. The unit can be powered by mains or battery. Mains requirements are 220 V single-phase AC at 50-60 Hz and battery power can be supplied by a lithium battery (7.3 V) or a 6LR61 commercial battery (9 V)

The elastomer test head can be a multipurpose shape, designed to deal with a variety of common respirator types or tailored to the user's needs. A test head is available specifically to fit the requirements of the French armed forces respirators such as the NMA and ARF-AL1.

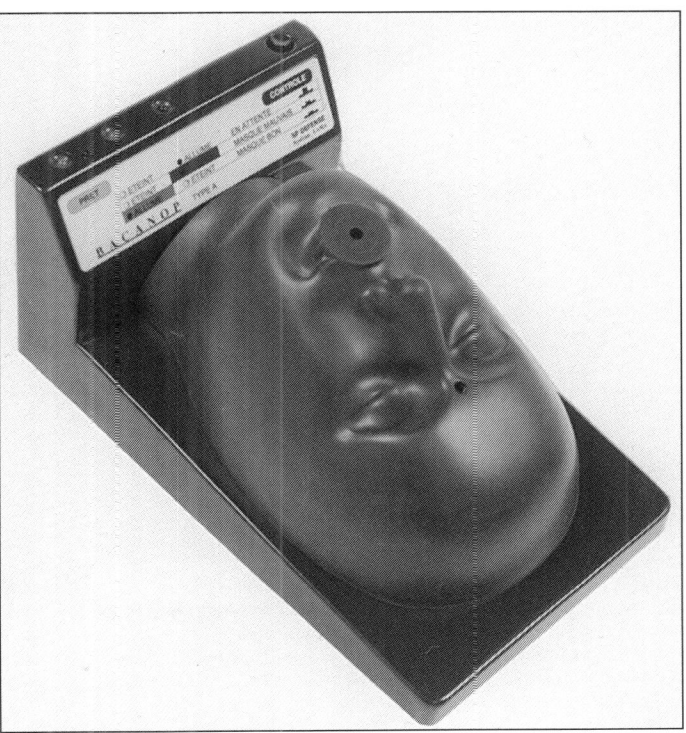

BACANOP respirator leakage tester 0011461

An optional software package and RS232 interface allows leakage test data to be downloaded for archive and analysis on a standard PC.

The BACANOP unit, complete with pressure sensor calibration set, is delivered in a portable case measuring 760 × 745 × 265 mm, weighing 27 kg.

Status
In production. In service with the French armed forces and police. Available for export.

Manufacturer
SP Défense.

VERIFIED

Breathing simulator for mask/filter testing

Description
The breathing simulator is designed to mimic the human breathing process accurately to allow research and testing of masks, filters and breathing support equipment. It faithfully reproduces the human breathing pattern over a wide range of circumstances. Linked to a computer, the simulator provides the following information:
- Respiratory work and pressure
- CO_2 concentration
- O_2 concentration.

It is designed to be used successfully with the minimum of training. The breathing volumes and frequencies can be set over a wide range and are easy to change. In calibrating breathing volume, the deviation is between −2 to +2 per cent. The simulator allows exploration of breathing performance at temperatures, pressures and humidities which would otherwise cause injury to human volunteer subjects.

Status
Available and in production. In service at CEB, with the French Navy and TNO-PML (Netherlands).

Manufacturer
SP Défense.
(The breathing simulator is a joint venture with the French company, Fenzy who make self-contained breathing apparatus for firefighters).

NEW ENTRY

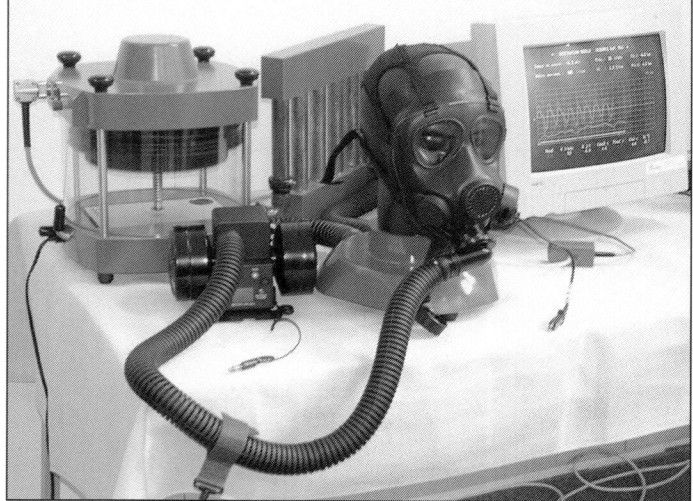

The SP Défense breathing simulator for mask/filter testing (John H Eldridge)
2002/0121879

EVATOX™ NBC hoods for civilians

Description
The EVATOX™ range of NBC hoods are designed to protect civilians of all ages. They protect an adult's face and respiratory tract against chemical agents in gaseous form. Designed for swift donning by untrained personnel, it offers safe escape and survival from a contaminated area where the atmosphere is toxic, following a toxic industrial or terrorist event.

The hood comprises a semi-rigid visor carrying the breathing system. It offers a good field of vision and optical clarity. A low breathing resistance cartridge filters out toxic hazards and it is comfortable to wear.

The childrens' version fits over the upper body and is secured at the waist. In both designs, the filter is out of the way, at the back.

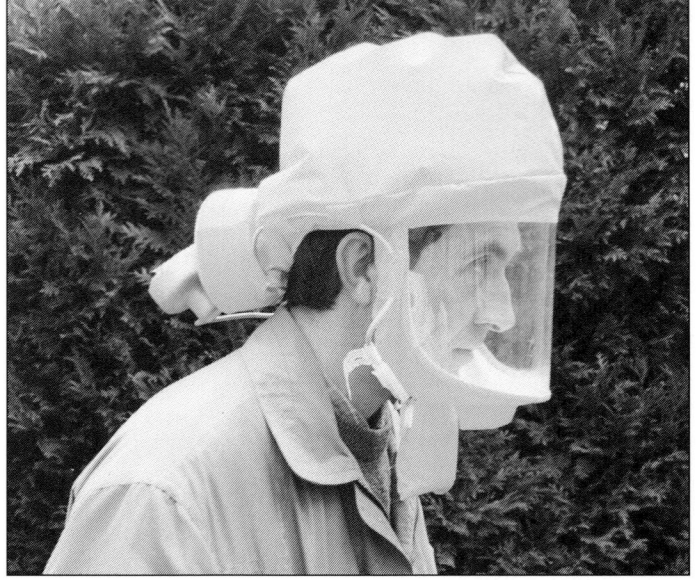

EVATOX™ NBC hood for adults   0052836

EVATOX™ NBC hood for young children   0121889

Status
Serial production. Distributed to French domestic response teams.

Manufacturer
Giat Industries.

UPDATED

Giat NBC ARFA mask

Description
During 1987, Giat produced the first prototypes of a new NBC respirator for military use, known as the *Apparatus respiratoire filtrant des Armées* (ARFA). An initial batch of 200 was delivered for French Army troop trials and a preproduction batch of 15,000 units was manufactured during mid-1990 for the French Air Force. Deliveries to the French Army began during 1992. The ARFA has since been exported to Poland and Hungary, the combined batch totalling 'several thousand'.

The ARFA face mask is made from moulded polyurethane and features an ergonomic face shield with a tough flexible wide-angle panoramic vision visor (also made from polyurethane) integrally moulded with the shield. It is intended that the mask will provide maximum user comfort for at least 24 hours.

Giat NBC ARFA mask

Giat NBC ARFA mask issued as part of collective protection kit for NBC teams

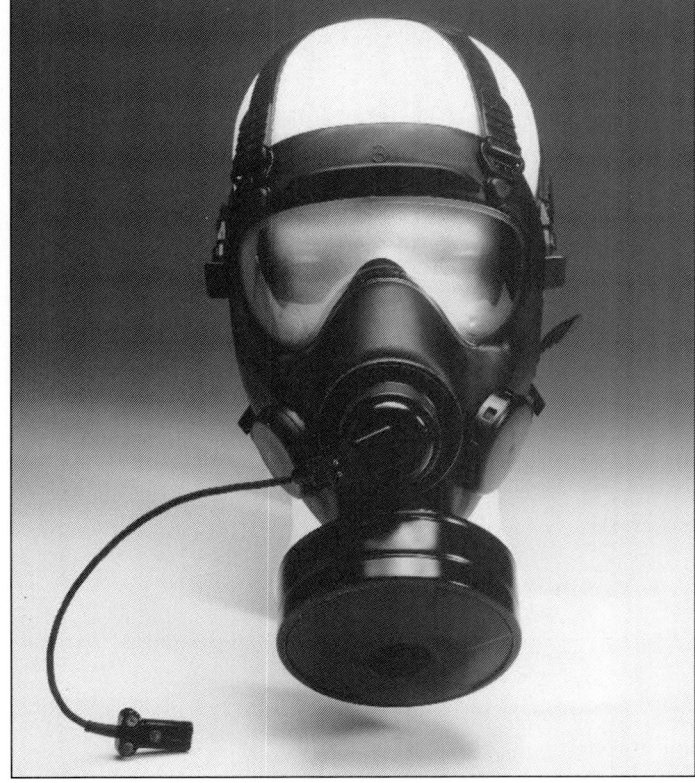

Giat NBC ARFA mask with microphone adaptor

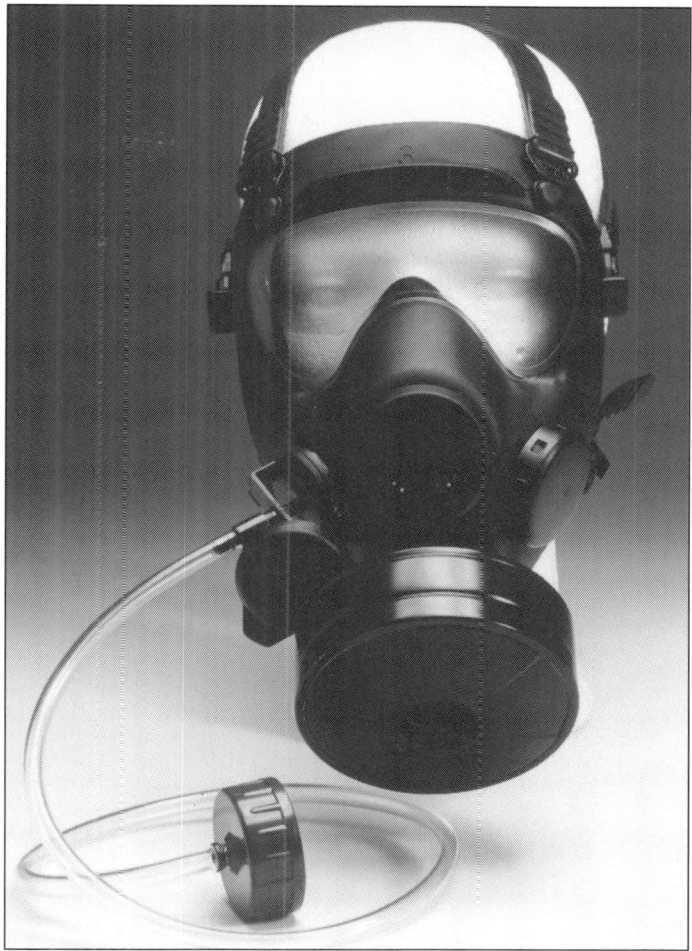

Giat NBC ARFA mask with liquid feed device

Giat NBC ARFA mask in sand-coloured finish

There is an inner mask liner with an anti-mist deflector system to direct incoming air over the visor. The mask is held in position on the head by four hood retainers connected to straps which may be adjustable or preset according to the model. Weight of the complete mask varies between 450 and 495 g according to the size and model; four sizes are available.

The ARFA features a liquid feed device on the right-hand side of the facepiece and an exhalation valve on the left. Features inside the facepiece include a nose cup and nose cup valves and an inhalation valve (not present on all models). The filter canister, which weighs 225 g, is located centrally below the voice transmitter unit in front of the wearer's mouth.

The base model is the ARF OTAN (NATO) version produced to meet full NATO recommendations and with either preset or adjustable hood straps. The ARFA 11 version has been adopted by the French Air Force and has adjustable straps, as it is not intended for issue to any particular individual. This version lacks the inhalation valve and uses a reinforced pressure drop exhalation valve. The ANP VP F1 version, adopted by the French Army, is intended for use by individuals so the straps are adjustable so that, once the straps have been adjusted, they remain in the preset state allowing an individual to put on the mask within seconds. This version lacks the inhalation valve and uses a reinforced pressure drop exhalation valve.

Giat NBC ARFA mask in sand-coloured finish

The ARFC is intended for civil defence use and lacks some features, such as the liquid feed device. The ARF MO is intended for use by paramilitary and other riot control forces. This version is adapted to fit directly on to police helmets in such a way that safety visors can be utilised. Special riot gas filter canisters can be used with this version. Other simplified models may be issued as part of the contents of collective protection kits.

Apart from the liquid feed device, other accessories which may be used with the ARFA series include a belt-mounted carrying bag, a flexible optical insert which can be fitted inside the mask within seconds and a microphone adaptor which may be clipped on to the voice transmitter unit. A film-type sun shield can be clipped into the hood retainers. If required the filter canister can be replaced by a quick-connect swivel coupling allowing the mask to be used with collective protection systems.

Two types of mask cleaning kit are available. One is a disinfected pad while the other consists of a bottle of disinfecting solution and a cleaning pad, both issued in clear plastic bags.

A test bench for checking the facial seals of all the above masks is produced by SP Defense.

Status
In production. In service with the French Army and Air Force and some other armies, including Hungary and Poland.

Manufacturer
Giat Industries.

VERIFIED

Giat NBC ventilated casualty hood

Description
The Giat NBC ventilated casualty hood is designed to protect the face and respiratory tract of an individual with facial or head injuries and who is thus unable to wear a conventional protective mask. The hood is cylindrical in shape and consists of two sections of transparent PVC material with barrier films, proof

Giat NBC ventilated casualty hood

against liquid and gaseous agents. There is a rubber hose distributing clean air from a CASU C420 two-canister filtered air distribution unit (see separate entry in this section for details) to the hood. Also there is a cord with a locking button for adjusting and tightening the hood to the top of an NBC suit or a transport bag. The hood will resist penetration by mustard agents for more than 24 hours.

Status
In production.

Manufacturer
Giat Industries.

VERIFIED

GERMANY

Drägerwerk KARETA M respirator

Description
The KARETA M respirator was developed for civil defence use but also has military and paramilitary applications. It features a mask with a soft lip sealing frame manufactured from an ageing-resistant, non-irritating neoprene/natural rubber compound. The mask is held in position by a five-point head harness with self-locking sliding eyelets for adjustment or rapid removal and donning. If required the mask can be worn ready for use in front of the chest slung from a separate carrying strap.

Two large, cylindrically curved wide-angle transparent visors provide a wide angle of vision in both the vertical and horizontal planes. The visors are made from high-grade laminated glass and are held in place by acetate resin clips for easy changing.

An inner mask guides sound waves in the shape of a funnel to the exhalation valve located at mouth level. An elastic vibratory multiple-seat valve disc transmits the sound waves to the exterior.

Incoming air is deflected in such a way that it flows in close contact with the visor's discs before the airflow passes through two control valves for inhalation into the inner mask which covers the nose and mouth. Most expired air is trapped within the inner mask and removed through the exhalation valve.

Available accessories include carrying bags and a drinking device.

Status
In production.

Manufacturer
Drägerwerk AG Lübeck.

VERIFIED

Drägerwerk KARETA M respirator

Helsatech NBC facelet

Description
The Helsatech NBC facelet was designed to provide interim respiratory protection against chemical agents during stand-by, rest and other periods when a full

The Helsa-Werke NBC facelet (Helsatech)   **2000**

personal protection stance is not deemed necessary. The facelet fits all sizes and is easy to put on and wear, producing a very low breathing resistance and a high level of speech transmission.

Status
Available.

Manufacturer
Helsatech.

NEW ENTRY

....

Kärcher NBC facelet

Description
The Kärcher NBC facelet was designed to provide interim protection against surprise chemical attacks. The facelet is small, lightweight, has a very low breathing resistance, a high level of speech transmission and can be quickly changed to full

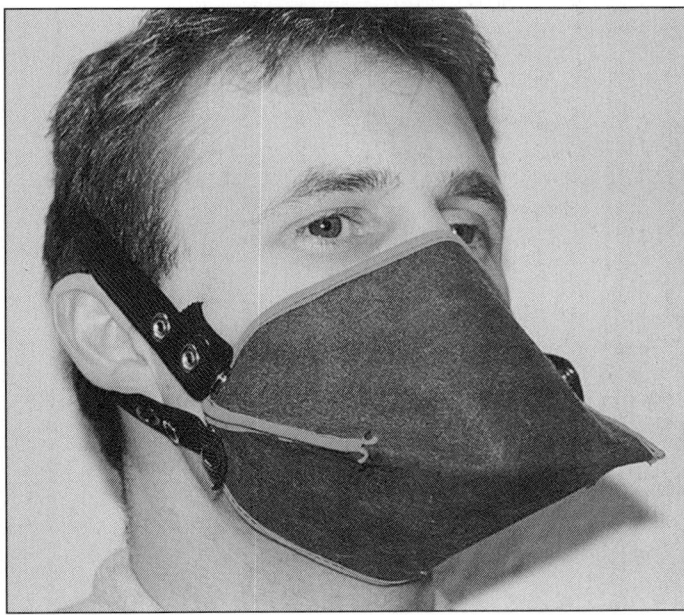

Kärcher NBC facelet

respirator when required. The facelet can be worn for long periods under operational conditions and is supplied with an adjustable harness to fit all sizes.

Status
In production.

Manufacturer
Alfred Kärcher GmbH & Co.

NEW ENTRY

_____

GREECE

COBRA NBC respirator

Description
The COBRA NBC respirator is the latest development of a continuing Hellenic Army General Staff requirement for respiratory protection. The earliest models (the B-3 in 1955) evolved through the B-31 (1965, for police force use), the B-32 (1968, also for police force use) and the B-33 (for industrial use).

The COBRA is a twin visor mask developed to meet NATO requirements as well as Greek and German specifications. It is lightweight, non-magnetic and easy to use and provides both high protection and user comfort. The goggle uses thread Rd 40 x 1/7 in (acc. DIN 3182) and can be adapted to take corrective goggles. The elastic parts of the mask (body, inner mask, head harness, inhalation and exhalation valve discs) are manufactured from a special halogen butyl rubber mixture with a matt black finish. This material ensures good mechanical strength and flexibility as well as high penetration resistance against toxic gases (Mustard penetration time is more than 24 hours), easy decontamination and effective use even under extreme environmental conditions. Airtight sealing of the mask on the face is achieved through an inner rim which seals the frame to the mask body. In addition, a rib on the outer rim of the mask ensures a good fit of the mask inside an NBC clothing hood. The mask is supported on the face by a five-piece head harness which provides a fast fitting time.

The curved eyepieces provide a wide field of vision and are kept free of misting by directing incoming air over their internal surfaces. Exhaled air cannot come into contact with the eyepieces due to an inner mask covering the nose and mouth.

The filter canister is fitted under the high efficiency 'voicemitter' assembly which also contains the exhalation valve. The filter canister used is the SUPERCOMBI NBC with a claimed efficiency of 99.997 per cent.

The COBRA is provided with a drinking device for water and liquid foods. The device is mounted on the right-hand side of the facepiece and consists of an inner mouthpiece, intake valve, external connecting tube and a special canteen top.

The COBRA respirator can be carried in the Type 87 B carrying case. This case has space for the respirator, two filter canisters, spares, an instruction book and NBC first-aid equipment (for example, an atropine self-injector).

Features include:
- Twin visor
- Lightweight and non-magnetic
- Five-point head harness
- Inner mask covers nose and mouth
- Optimum verbal communication
- Optional drinking device

Status
In production.

Manufacturer
BIANA SA.

UPDATED

Two views of the COBRA NBC respirator, with and without a filter canister fitted (T J Gander)

HUNGARY

Type 70 M gas mask set

Description
The Hungarian gas mask set Type 70 M is intended to provide protection for individuals against NBC warfare agents and has three main parts, a face mask, filter canister (described as a filter insert) and a fabric carrier bag.

The face mask (referred to as the EO-18K) is of the helmet type and uses a rubber body stretched over the head to provide the necessary protective seal around the face. Five sizes are available. There are two small flat eyepieces, which permit the use of monocular optical devices, with anti-condensation inserts placed over the inside of the eyepieces. The 70 M filter canister is located on the left-hand side of the face mask and does not provide protection against carbon monoxide. A chin-mounted exhalation valve directs spent air out of the mask while a speech unit is mounted in front of the mouth area.

The 70 M fabric carrying bag is used to carry the face mask and filter and is also used to carry accessories. These include a box containing three pairs of anti-condensation inserts; a separate box contains spare membranes.

Weight of the complete 70 M is 850 g. Air flow resistance at 30 litres/min for inhalation is 210 Pa and 120 Pa for exhalation.

The 70 M has been observed in service with the RFAS armed forces. The designation PMG-2 has been referred to in this context. The 70 M has also been produced in Bulgaria where it is known as the PMG.

Status
Production ceased 1992. Remains in service with the Hungarian and RFAS armed forces and others.

Manufacturer
Respirátor Rt.

VERIFIED

RFAS troops wearing the Type 70 M gas mask set

....

Type 93 M gas mask set

Description
The Type 93 M gas mask set (respirator) is based on the design of the French *Apparatus Respiratoire Filtrant des Armées* (ARFA) (see separate entry this section) and manufactured under licence by Respirátor Rt in Budapest. The facepiece has a standard connection thread for a centrally mounted filter canister and a voice transmitter and drinking device. The mask facepiece and visor are of polyurethane construction, with a ridge above the forehead to locate the IPE hood. The panoramic visor is flexible. Designed to fit a full range of face shapes and

Specifications		
Face mask weight:		550 g
Inhalation resistance	30 l/m:	<50 Pa
	95 l/m:	<150 Pa
Exhalation resistance	95 l/m:	<160 Pa
	160 l/m:	<300 Pa
NBC filter canister weight:		<320 g
Breathing resistance	30 l/m:	<260 Pa
	95 l/m:	<980 Pa
Filter efficiency	liquid particles:	99.999%
	solid particles:	99.95%

Type 93 M gas mask set
0052842

available in 4 sizes, the mask set offers good comfort and full protection against known threat agents for at least 24 hours. The 93 M filter canister is Hungarian and mask accessories include eyesight correction, microphone attachment, sun visor, cleaning kit and carrying case.

Status
In production. In service with the Hungarian and Polish armed forces.

Manufacturer
Respirátor Rt.

VERIFIED

─────────────────────────────────

IRAQ

Iraqi military protective mask

Description
The standard Iraqi military protective mask is a version of the Romanian M-85 mask (qv) produced in Iraq. Quantities of M-74 and M-85 masks were also purchased by Iraq direct from Romania. It is stated to protect the wearer against all toxic gases likely to be encountered during warfare and is provided with a CF4 filter canister. There are two eyepieces that appear to be polycarbonate and a drinking system

Iraqi military protective mask

connected to a water bottle. The mask is said to provide protection against mustard-type agents in aerosol form for a minimum of 6 hours.

The mask weighs 1.02 kg and is supplied together with a fabric carrying satchel. Storage life is two years.

Status
In service with the Iraqi armed forces.

Manufacturer
Iraqi state factories.

VERIFIED

ISRAEL

SHALON-Chemical Industries Inpro Infant Protector

Description
The Inpro Infant Protector is a positive pressure protective hood system intended to protect children aged up to three years from all known NBC agents. It provides complete and effective protection for extended periods with optimum comfort and allowing normal activities for the infant and parent.

The Inpro consists of a head and body covering made of strong impermeable and transparent plastic laminate. The cover has two strong plastic zippers for quick and easy positioning or removal of the infant and an integrated feeding nipple (a feeding bottle is supplied with the system). There is a flexible rubber hose for directing filtered air into the protector and a support system consisting of a soft cushioning material on a rigid plastic carrier. There is also a strap system to secure and carry the infant.

Positive pressure is maintained by a compact and reliable air supply unit which draws in ambient air through a filter canister into the hood. The unit consists of a durable polycarbonate housing enclosing a miniature blower, four lithium batteries and a specially designed filter canister adaptor. The Inpro is supplied in a protective plastic packing case.

Total weight is 1.55 kg. Air flow is 45 litres/min, while the operating time is 14 hours.

Manufacturer
SHALON-Chemical Industries Limited.

VERIFIED

SHALON-Chemical Industries Inpro Infant Protector in use

SHALON-Chemical Industries protective masks

Description
SHALON-Chemical Industries produces a range of NBC protective masks and filter systems. All models are provided with a training attachment that simulates a real filter canister for use during practice drills, thus saving the real filter canister for emergencies. The main models are as follows:

Respirator No 4A1
Produced in one universal size and intended for the use of all adults and children from the age of eight years upwards. It incorporates a drinking system and a voicemitter. The Respirator No 4 omits the drinking system and voicemitter.

SHALON military NBC respirators

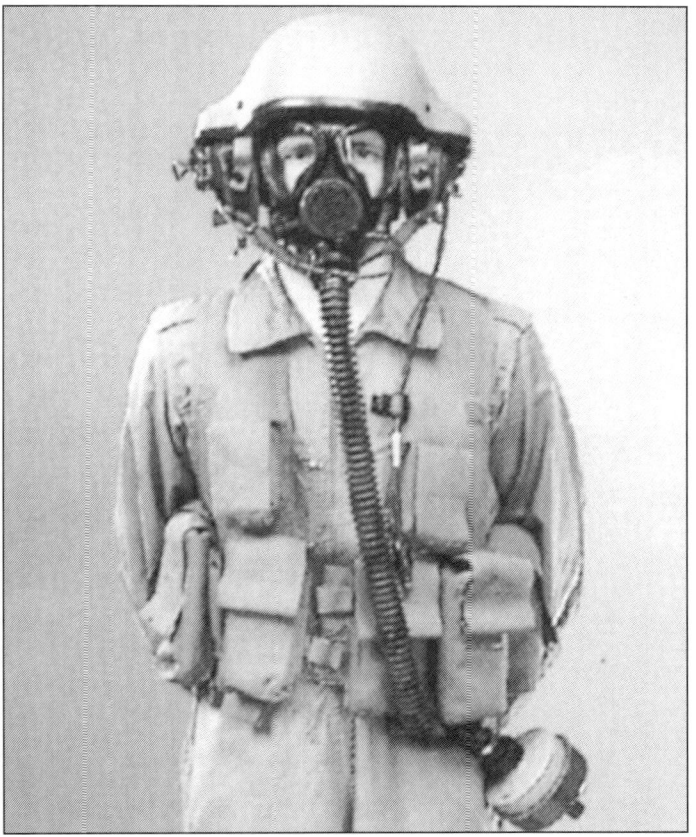

NBC mask No 15-S-80 for tank crewmen

Respirator No 10A1
Intended for use by children of 8 to 16 years. It incorporates a drinking system and a voicemitter. The Respirator No 10 omits the drinking system and voicemitter.

Respirator No 15A1T
This respirator is intended for general military applications and features a drinking system, a voicemitter and a side-mounted speaking diaphragm for use with communications equipment and telephone sets.

Respirator No 15-S-80
Intended for use by the crews of armoured vehicles this respirator can use a flexible rubber hose to connect it to a vehicle air supply system or it can be used with a filter canister attached. The respirator has a drinking system, a voicemitter and a dynamic microphone with an external connector.

Respirator No 30
This is intended for general applications and has a centrally located voicemitter.

PANORAMIC respirator
A PANORAMIC respirator, with a one-piece vision visor and a voice amplifier, is also available. It is intended for use by industrial and laboratory personnel.

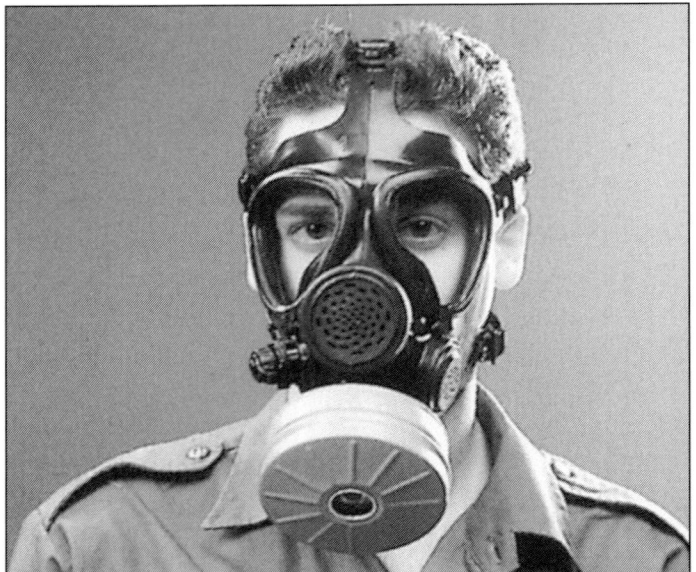

NBC mask No 15A1T

SHALON NBC civilian respirators

Status

In production. In service with the Israeli Defence Forces and police and with several foreign countries. The civilian respirators are the standard approved NBC respirators of the Israeli Civil Defence Authorities.

Manufacturer

SHALON-Chemical Industries Limited.

VERIFIED

SHALON-Chemical Industries Respro protective hood for children

Description

The Respro is a positive pressure protective hood system intended to protect children aged three to ten years from all known NBC agents. It is designed for optimum comfort and safety, providing extended use with minimal interference in normal activities.

The Respro has three main subsystems. They are an impermeable hood made of strong transparent plastic laminate, an air supply unit and a brightly coloured vest acting as an interface between the other two subsystems.

Positive pressure is maintained by a compact and reliable air supply unit which draws in ambient air through a filter canister into the hood. The unit consists of a durable polycarbonate housing enclosing a miniature blower, four lithium batteries and a specially designed filter canister adaptor.

Total weight is 1.2 kg. Air flow is 45 litres/min, with an operating time of 14 hours.

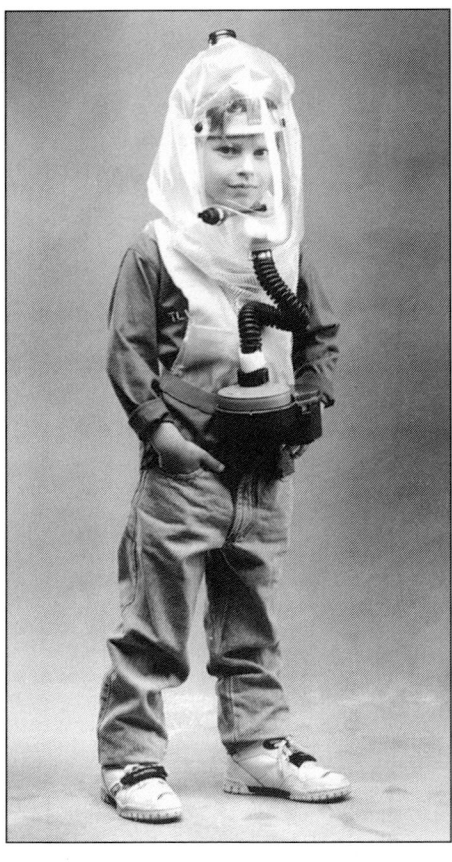

SHALON-Chemical Industries Respro protective hood for children

Status

Available.

Manufacturer

SHALON-Chemical Industries Limited.

VERIFIED

ITALY

M90 NBC protective mask

Description

Development of the M90 NBC protective mask began during the mid-1980s and was directly contracted and partially funded by the Italian Ministry of Defence. Following extensive operational testing it is in service with the Italian armed forces, having largely replaced the earlier M59 and M73 masks.

The M90 respirator consists of the M90 facepiece, the M90 filter canister, a carrying bag and several accessories including a canteen connector, containers for nourishing liquids and other accessories. The entire equipment is designed to provide protection against all known NBC agents over prolonged periods of continuous wear, including when the wearer is asleep. The design is claimed to allow almost continuous wear over a five-day period with minimal discomfort.

The M90 facepiece has an external surface made of bromobutyl rubber and there is a reverse seal made from natural rubber compounds. Both these components are produced during a single manufacturing process and have a

M90 NBC protective mask complete with accessories (Aero Sekur SpA)

smooth external surface for ease of decontamination, using boiling water if necessary. The facepiece is held on the head by a five-strap chloroprene rubber harness, taking nine seconds to put on. A small bead around the periphery of the facepiece ensures an effective seal between the mask and the hood of a protective NBC suit.

The M90 filter canister is located in a downwards-facing central position on the lower front of the facepiece to ensure that the canister does not interfere with the field of vision and to allow the ground area close to the wearer's feet to be easily observed. The central location also assists in lens demisting and allows corrugated hoses for centralised filter systems to be fitted easily.

The M90 combines in a single unit the filter canister connector, the respiratory valves and the single speech diaphragm. Combining these units adds stability to the mask and helps prevent it from swinging and twisting when sudden head movements are made. The unit is constructed using a low-density glass-reinforced polymer with a high impact strength and good resistance to chemical agents. The unit can incorporate a microphone in a protected position by inserting it under the front screen of the unit.

In use, external air is purified by the M90 filter and enters the facepiece interior on both sides between the facepiece and the nose cup. Tissot channels draw the air towards the eyepieces and it is then directed down to reach the wearer's respiratory tract through nose cup check valves. A protrusion on the sides of the nose cup separates the Tissot channels from the check valves area. Exhaled air flows from the mask through the exhalation valve to the front screen of the respiratory unit, cleaning particles from the speech diaphragm as it passes through. For nuclear flash protection the speech diaphragm is protected by a set of three overlapping screens. A patented device included in the facepiece connector allows the safe replacement of a filter canister in NBC contaminated atmospheres. The M90 can accommodate any filter canister with a QSTAG 496 NATO standard 40 mm thread.

The two eyepieces are made from allyl-diglycol carbonate, with the eyepiece profile square to the normal visual axis to allow the use of optical instruments. Purpose-designed corrective spectacles for use within the mask are available.

A nylon coated carrying pouch contains the mask and its accessories.

A drinking facility is located on the right-hand side of the facepiece. The drinking facility is used with a direct connector fitted to a standard canteen and the same facility can be used with sealed and disposable flexible containers for nourishing liquids. The canteen connector uses a self-cleaning needle valve.

The M90 facepiece is available in three sizes: small, medium and large. Weight is approximately 540 g. All parts can be dismantled and reassembled using standard tools such as a screwdriver and a pin wrench. All components can be replaced. Shelf-life is up to 10 years.

Status
In service with the Italian armed forces.

Manufacturer
Aero Sekur SpA.

UPDATED

KOREA, SOUTH

Samgong K-1 and KM9A1 NBC respirators

Description
The Samgong K-1 is a Korean evolution of a design produced under licence. The licence applies to manufacture of the KM9A1 (Samgong designation), a version of the US M9A1 respirator (now no longer manufactured). The KM9A1 is identical to the US original and weighs 1.2 kg complete with its carrier case.

The K-1 design uses a canister mounting adapter of German design. The facepiece has flat wide vision lenses, an efficient voice emitter system, a drinking system assembly and provision for eyesight correction. The inhalation and

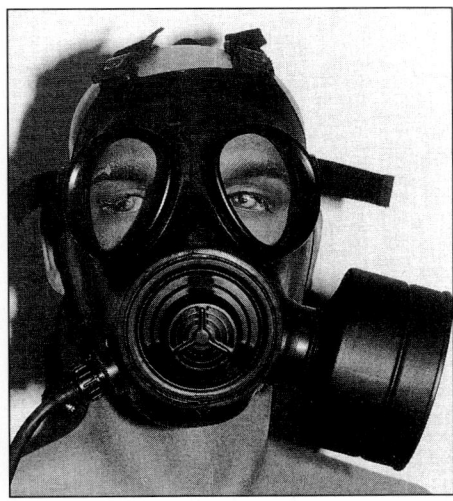

Samgong KM9A1 NBC respirator
(T J Gander)

Samgong K-1 NBC respirator 0052846

exhalation air flow resistance allows distress-free breathing during exertion, for example when the wearer is running.

Weight of the K-1 is 1.29 kg ±20 g.

Status
Both available.

Manufacturer
Samgong Industrial Company Limited.

VERIFIED

Samgong KS M 6685 civil defence hood/mask

Description
The KS M 6885 is designed for NBC protection of the civilian population. It is a simple, low-cost, one-size hood with a visor screen. Inside the hood, a rubber mask covers the nose and mouth. A threaded tube passes out through the front of the mask to which a standard respirator canister is attached. The hood, mask and canister are retained on the head by rubber straps. The visor is treated for anti-fogging (condensation).

This lightweight design has low breathing resistance and is suitable for adults and adolescents.

A range of masks and whole-body ventilators is also available to offer protection to children from birth to teenage.

Status
All available.

Manufacturer
Samgong Industrial Company Limited.

VERIFIED

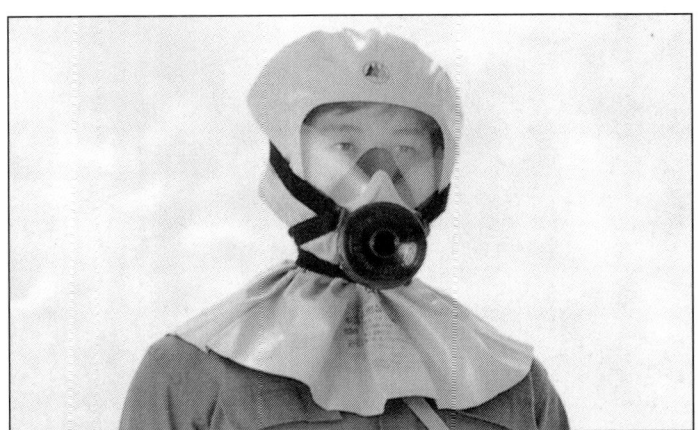

Samgong KS M 6685 civil defence hood/mask 0052845

PAKISTAN

NBC PAK-10 general service respirator

Description
The NBC PAK-10 general service respirator is a variant of the British Avon S-10 and reference should be made to the entry under United Kingdom for details. The NBC PAK-10 appears to be identical to the S-10.

A further variant known as the SERVAIR CD-10 is produced for civil defence and industrial purposes. A range of 40 mm diameter thread filters is available for this respirator.

Status
In production. In service with the Pakistan armed forces.

Manufacturer
Service Industries Limited.

VERIFIED

NBC PAK-10 general service respirator

POLAND

MC-1 NBC respirator

Description
The MC-1 NBC respirator is of conventional design having a rubber facepiece held in position on the head by a six-strap head harness. The facepiece is a close fit to

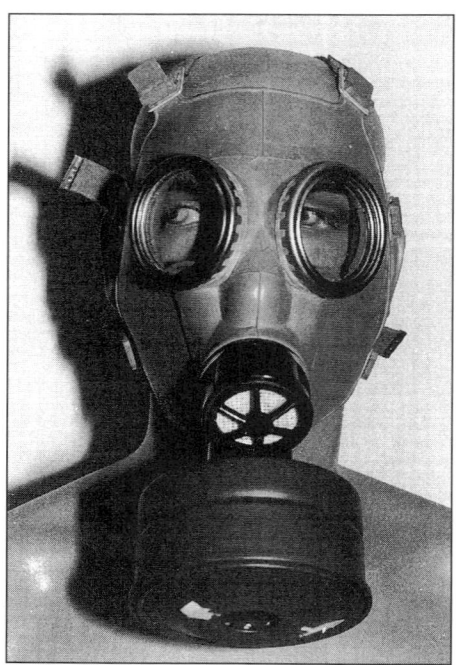

Polish MC-1 NBC respirator
(T J Gander)

the extent that a nose outline is made on the outer contours of the unit. An MS4 filter canister is mounted on the chin.

Status
Manufactured by Polish state factories prior to privatisation. Current manufacturer unknown. In service with some Polish military and civil defence units.

Manufacturer
Unknown.

VERIFIED

ROMANIA

Romanian M-74 and M-85 NBC respirators

Description
The M-74 and M-85 NBC respirators are described by their producers, Romtehnica, as gas masks. The M-74 is held in position on the head by rubber straps, has two polycarbonate eyepieces and a chin-mounted CF4 filter canister. It is stated to be effective for up to 6 hours against chemical warfare agents in the form of droplets, aerosols and vapours. The respirator is intended to be worn for time periods up to 12 hours.

The M-85 respirator is essentially similar but is fitted with a drinking device which can be fitted to a water bottle. It can also be connected to a sound amplifier unit via a cable from a cheek-mounted internal microphone. The sound amplifier unit is intended for use by commanders and weighs 500 g.

Romanian M-74 NBC respirator
(T J Gander)

Romanian M-85 NBC respirator
(T J Gander)

The M-85 respirator and sound amplifier unit can be carried in a special carrying satchel weighing 500 g. The respirator facepiece weighs 600 g and the CF4 filter canister 450 g.

Status
In production. In service with the Romanian armed forces and with Iraq (both M-74 and M-85).

Marketing agency
Romtehnica.

VERIFIED

RUSSIAN FEDERATION AND ASSOCIATED STATES (CIS)

Model GP-4U civilian protective mask

Description
Designed primarily for general civilian use, the Model GP-4U uses a moulded rubber face mask which is secured to the head by three straps fitted to six points on the mask. The round glass eyepieces are held in place by crimped aluminium rings and, as there is no other method provided of keeping the eyepieces clear when worn, an anti-dim kit is issued with the mask. Single inlet and double outlet valves are fitted and air is drawn into the mask via a corrugated rubber hose which is secured to the face mask by wire and tape. A GP-4U canister is used for filtering. When not in use the Model GP-4U mask is carried in a fabric carrier fitted with a shoulder strap and a waist strap. The weight of the complete equipment in its carrier is around 2.5 kg.

The GP-4U filter canister used with this mask is cylindrical and weighs 0.56 kg. It is 150 mm high and 80 mm in diameter.

Status
Available from Russian state factories before privatisation Production complete. Scheduled to be replaced by the Model GP-7. Current manufacturer unknown.

Manufacturer
Unknown.

VERIFIED

Model GP-7 civilian protective mask

Description
It is intended that the Model GP-7 protective mask will eventually replace existing GP-4U, GP-5 and GP-5M masks in use by civilians and some non-combatant military personnel. The GP-7 operates along the same lines as the GP-5 but has some advantages in terms of use and ergonomics and the filter resistance has been reduced. An improved 'floating' facial seal creating less pressure on the face has been introduced along with an improved speech device. Special attention has been placed on making the mask, which is available in three sizes, a good fit for a wide percentage of the user population.

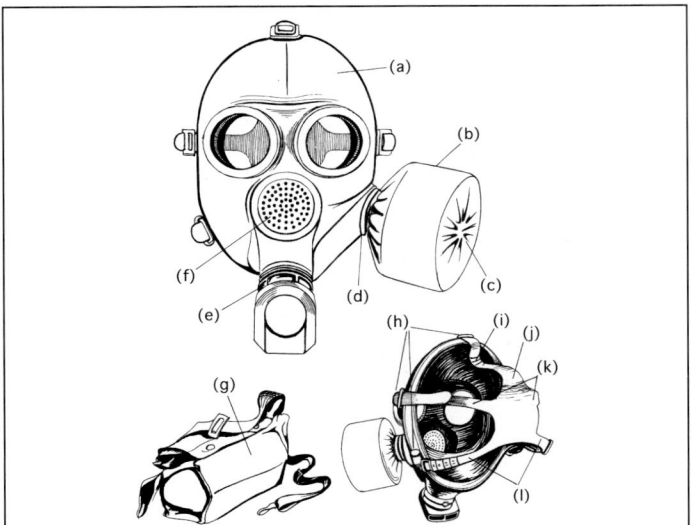

Parts of the GP-7 civilian protective mask: **(a)** *MGP face mask* **(b)** *GP-7K filter unit* **(c)** *filter unit fabric cover* **(d)** *inlet valve location* **(e)** *outlet valve* **(f)** *speech device* **(g)** *holdall* **(h)** *buckles* **(i)** *forehead strap* **(j)** *strap assembly* **(k)** *temple straps* **(l)** *cheek straps*

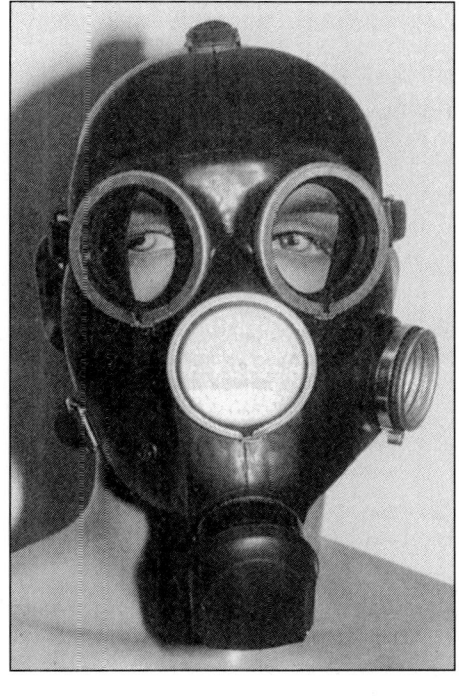

GP-7 civilian protective mask; the filter unit is not fitted
(T J Gander)

The GP-7K filter unit is mounted on the left-hand side of the mask and is covered with fabric. The facepiece is held in position on the head by a rear panel and five straps; the two cheek straps have metal self-tightering buckles. Each mask is provided with a protective fabric cover and a holdall. Also provided with each mask are two thermal inserts and six anti-perspiration layers to absorb facial sweat.

The GP-7 face mask, the MGP, weighs 600 g and the filter unit 250 g.

Status
In production. Available from Russian state factories before privatisation. Current manufacturer unknown.

Manufacturer
Unknown.

VERIFIED

Model GP-7V and Model GP-7VM protective mask

Description
The protective mask, Model GP-7V, related to the civilian protective mask, Model GP-7 (see earlier entry), is intended as a general purpose protective mask fitted with a filter canister providing protection against vapours up to 6 hours and against droplets up to 2 hours. There is provision for mounting the filter canister on either the right- or left-hand side of the facepiece. The Model GP-7VM has provision for a drinking cevice which can be connected to a canteen bottle.

The GF-7V and GP-7VM have a breathing resistance of 180 Pa at 30 litres/min. Each mask weighs 0.8 kg. When stowed in a carrying bag the dimensions are 285 × 250 × 115 mm. The mask is available in three sizes.

A variant of the GP-7V series is known as the PMK (see following entry).

Status
In production.

Marketing agency
NPO 'NEOGANIKA'.

VERIFIED

Model GP-7VM, protective mask with filter canister (right) and canteen bottle (left)
(T J Gander)

Model PGP protective mask and canister leakage tester

Description
This equipment is used to test most types of former Warsaw Pact protective masks and filter canisters for leakage. All the components are contained in a metal box and include two rubber bulb-type air pumps, a manometer, 1 minute sand glass timer, rubber tubing and plugs, valves and various fittings. With both the masks and the filter canisters, the units are sealed and pumped up to pressure. The pressure is measured on the simple manometer and has to be maintained for 1 minute without leakage; the time is determined by the sand glass timer.

Status
In service with RFAS and former Warsaw Pact forces. Available from Russian state factories before privatisation. Current manufacturer unknown.

Manufacturer
Unknown.

VERIFIED

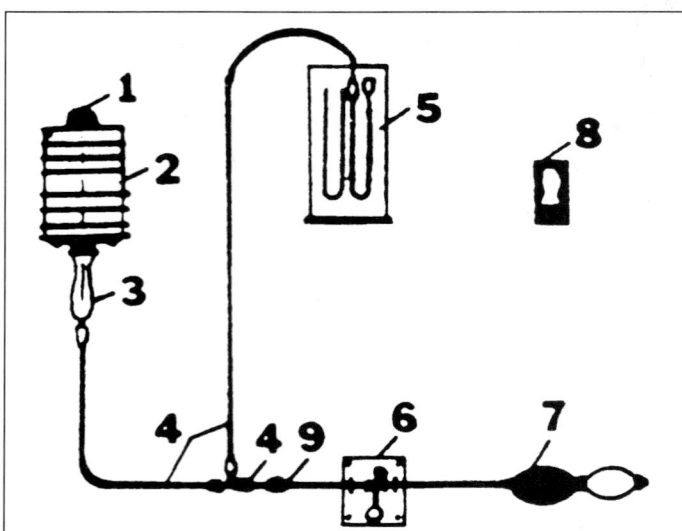

Model PGP protective mask and canister leakage tester canister testing **(1)** *rubber plug* **(2)** *canister* **(3)** *connecting tube* **(4)** *rubber tubing* **(5)** *pressure gauge* **(6)** *wire spring clamp* **(7)** *rubber bulb* **(8)** *one minute sand glass timer* **(9)** *connector*

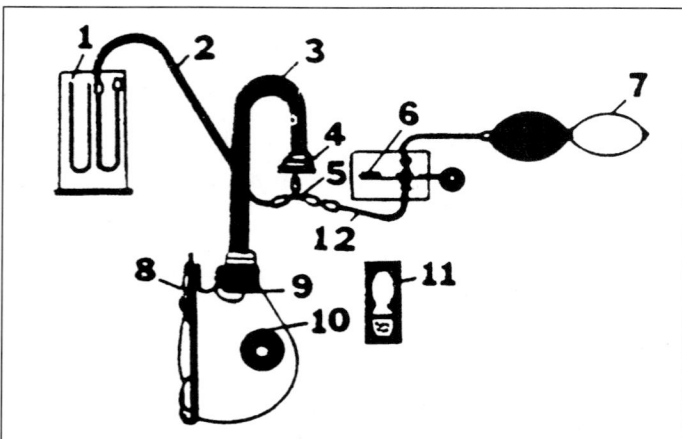

Model PGP protective mask and canister leakage tester mask testing **(1)** *pressure gauge* **(2)** *rubber tubing* **(3)** *protective mask nose* **(4)** *connector* **(5)** *-tube* **(6)** *wire spring clamp* **(7)** *rubber bulb* **(8)** *acepiece clamp* **(9)** *rubber plug* **(10)** *protective mask facepiece* **(11)** *one minute sand glass timer* **(12)** *rubber tubing*

Model PMK protective mask

Description
The Model PMK protective mask may be regarded as the full military version of the Model GP-7V protective mask (see entry in this section). It has a drinking facility, a voicemitter and curved glass eyepieces with a field of view limitation of 70 per cent. The PMK uses a large filter canister mounted on the left-hand side of the facepiece (there is no provision apparent for mounting the filter canister on the right-hand side). The filter canister has a reduced breathing resistance (150 Pa at 30 litres/min) compared to the Model GP-7V so the mask can be worn for periods up to 12 hours at one time.

The mask is provided with a carrying bag. Weight of the mask without the bag is 0.95 kg, of which 0.3 kg is the filter canister.

Model PMK, protective mask

Status
In production.

Manufacturer
Plant KINAP JSC.

VERIFIED

Model ShM helmet-type protective mask

Description
The Model ShM has long been the standard RFAS protective mask and is also used by some other former Eastern Bloc forces. As the mask completely covers the head, it provides good protection against most agents but is uncomfortable to wear for long periods and general visibility is limited when wearing it. The helmet itself, which can be worn under a steel helmet, is made from light grey rubber which turns a beige colour following extended exposure to strong light; black rubber versions have also been observed. Crimped metal rings secure the eyepieces to the face mask and, as there is only a limited anti-fogging system for the mask interior, the eyepieces can be fitted with extra internal gelatine lenses; an anti-dim set is issued with each mask. Incoming air is filtered through an MO-2 or MO-4U canister filter element which can be fitted either directly onto the mask or on to a length of fabric-reinforced corrugated rubber hose. Incoming air is directed across the eyepieces to reduce fogging to some extent.

Despite its general efficiency, the Model ShM is rather heavy and uncomfortable and the external filter canister and hose make it rather awkward to fit and wear. The mask is issued in a fabric carrier which also contains the filter canister, hose and the anti-dim set. The combined weight is about 2 kg.

The Model ShM is produced in Egypt as the BSS (qv).

Air Force technician wearing Helmet-type protective mask, Model ShM

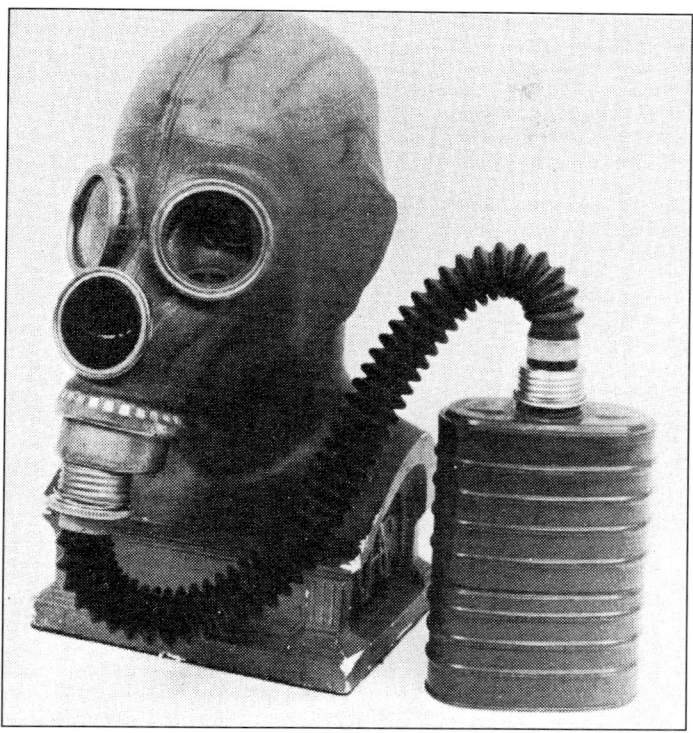

Model ShM mask with outlet valve and hose connection to filter canister

Status
Production complete. In service with RFAS and former Warsaw Pact forces but being replaced by later models. Also in service with Syria. Available from Russian state factories before privatisation. Current manufacturers unknown.

Manufacturer
Unknown.

VERIFIED

Model ShMS special protective mask

Description
The Model ShMS has provision for fitting optically corrective lenses, either spectacle lenses or special lenses to suit binoculars or weapon sights. Normally the Model ShMS is issued with optically flat lenses which are fitted into internal or external grooves, according to the circumstances, the corrective lenses fitting into the other groove. As a measure against dimming, extra gelatine lenses can be fitted inside the internal lenses. The lenses are as small as 40 mm in diameter so the mask's visual efficiency is low but, as it retains the overall protection and filter system of the Model ShM, it is effective against nearly all agents. The mask is made from grey or beige rubber and is issued in five sizes. Weight of the fabric carrier, mask, tube filter canister, hose and anti-dim set is approximately 2 kg.

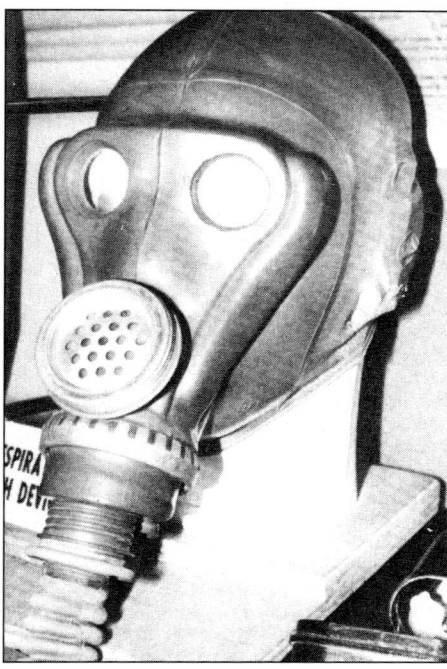

Model ShMS special protective mask, fitted with 'voicemitter' (T J Gander)

This example was found in Afghanistan. Shown with canister (bottom left) and personal DECON material   **2002**/0137541

Status
Production complete. In service with RFAS and former Warsaw Pact forces. Available from Russian state factories before privatisation. Current manufacturers unknown.

Manufacturer
Unknown.

VERIFIED

Models GP-5 and GP-5M civilian protective mask

Description
The Model GP-5 may be regarded as the civilian counterpart of the military Model ShM, as the rubber head-covering facepiece (designated ShM-62) is almost identical. As with the Model ShM, the Model GP-5 uses air deflection across the eyepiece interiors and is issued in five sizes, but the filter canister is fitted directly on to the inlet valve. The filter canister is cylindrical and is fitted via a female connector on the facepiece. A fabric carrier contains the facepiece and filter canister and a set of anti-dim discs for fitting inside the glass eyepieces. Pockets on the carrier contain a small first aid kit and an anti-gas kit.

The Model GP-5M is essentially similar to the Model GP-5 and there are few differences between the two.

Status
Production complete. Scheduled to be replaced by the Model GP-7. Available from Russian state factories before privatisation. Current manufacturer unknown.

Manufacturer
Unknown.

VERIFIED

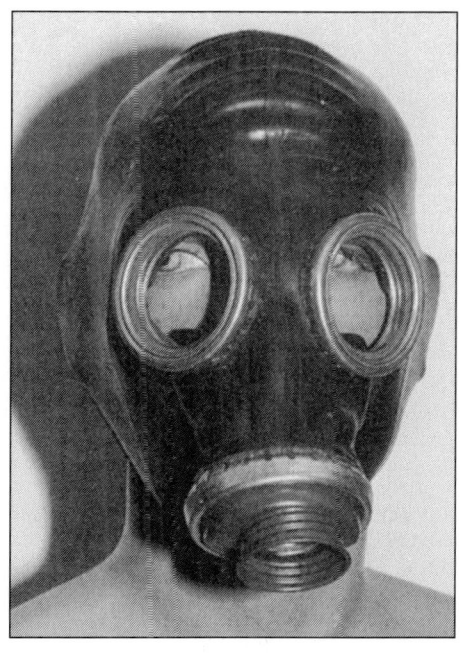

Model GP-5 civilian protective mask; the filter canister is not fitted (T J Gander)

Protective Mask, Communication

Description
Although this mask uses the same external filter system and canister as the Model ShM, it consists of a face mask only, held in place by an adjustable head harness. The ears are exposed which allow headphones to be worn and a diaphragm 'voicemitter' in front of the mouth permits speech to be transmitted more clearly than is possible with the Model ShM. The visual efficiency of this mask is reported to be low.

A similar mask was produced in Egypt as the CM3 (see separate entry in this section).

Status
Production complete. In service with RFAS and former Warsaw Pact forces. Available from Russian state factories before privatisation. Current manufacturer unknown.

Manufacturer
Unknown.

VERIFIED

R-2 protective mask

Description
The protective mask R-2 is intended to provide respiratory protection against radioactive particles and other harmful dust particulates. It covers the nose and mouth only, being formed from three filter layers of a polyurethane sponge, a filtering material and polyethylene film. To assist breathing the mask has inhalation and exhalation valves. The mask can be placed in position on the head in less than 14 seconds.

The R-2 mask weighs 60 g and can be worn continuously for up to 12 hours.

Status
In production.

Manufacturer
Plant KINAP JSC.

VERIFIED

R-2 protective mask

RFAS NBC respirator

Description
First observed during 1990, this NBC respirator is of a novel pattern with a facepiece moulded from two pieces of rubber joined together and held in position over the face by a head harness. There are two relatively small flat-mounted circular eyepieces and a large frontally mounted high-efficiency voicemitter. There does not appear to be any provision for a drinking device.

The main items of note on this respirator are the two relatively bulky cheek assemblies housing an internal filter on each side. The size of the air inlets and cheek assemblies would suggest that the internal filters are quite large and therefore of high efficiency and probably have a low air flow resistance. It would also appear that the mask would have to be removed from the face to change the filters under operational conditions. The size of the cheek assemblies would also appear to rule out the efficient use of aimed weapons such as rifles and most weapon sights, so it would appear that this mask is intended for use by second line support troops rather than front line combat personnel.

RFAS NBC respirator (Jane's Intelligence Review)

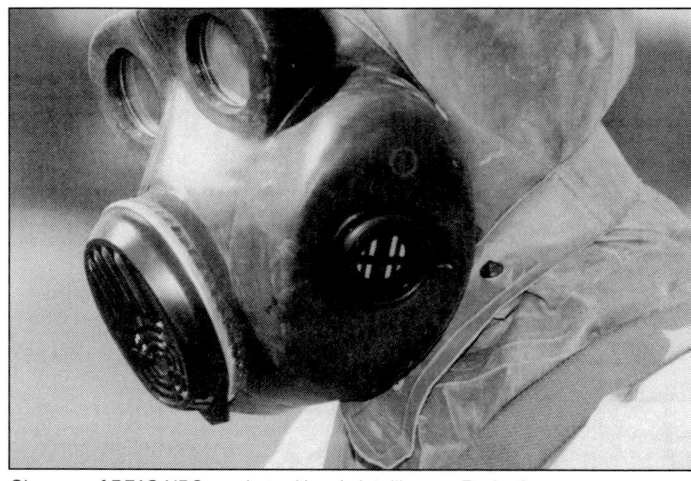

Close-up of RFAS NBC respirator (Jane's Intelligence Review)

Status
In service with RFAS armed forces. Available from Russian state factories before privatisation. Current manufacturer unknown.

Manufacturer
Unknown.

VERIFIED

SWEDEN

Forsheda NBC respirator Type F2

Description
The Forsheda NBC respirator Type F2 was developed to meet a Swedish armed forces requirement for a general purpose military respirator. In April 1994, the Swedish Defence Administration signed an agreement with Forsheda AB for deliveries of approximately 500,000 Type F2 masks (to be known as Protective Mask 90) for use by the Swedish Defence Forces, the Swedish Emergency Services Administration and some other Swedish authorities. There was also an option calling for further quantities to be ordered to replace old masks currently in use. The Type F2 respirators are used in conjunction with the Racal C7 filter canister.

The Type F2 is stated to have a protection factor in excess of 10,000 and is compatible with most weapon systems and optical devices.

The Forsheda NBC respirator Type F2 has a facepiece made from bromobutyl rubber and is made in three sizes; the weight of the medium-sized facepiece is 440 g. The facepiece is held in position on the head by six adjustable straps connected to a fabric mesh panel on the back of the head. There is an exponential speech horn unit and a further speech unit for use with communications equipment is an option; the latter is mounted in the unused filter screw thread connection.

A drinking device with a flow rate of more than 200 ml/min is standard. The eyepieces are made of a substance known as polysulfone which is specially treated to increase scratch resistance. The effective total field of vision is 88 per cent with the effective overlapped field of vision being 61 per cent.

Forsheda NBC respirator Type F2

Forsheda NBC respirators Type F2 with accessories

The respirator is used with a Type F4 filter canister with an aerosol filter made from glass fibre materials and a gas filter containing Whetlerite type activated carbon. Inhalation resistance with the filter canister is 80 Pa at 30 litres/min or 505 Pa at 160 litres/min.

Storage life of the NBC respirator Type F4 is more than 10 years at 30°C or more than 20 years at 20°C.

Status
In production for the Swedish Defence Forces, the Swedish Emergency Services Administration and some other Swedish authorities.

Manufacturer
Forsheda Industrial Polymer AB.

VERIFIED

SWITZERLAND

PM31 and PM33 NBC protective masks

Description
With a claimed protection factor of >10,000, the PM31 and PM33 masks offer similar safety and comfort features to the SM90 and SM3 types, with which they share the major components. The PM31 is an extra-lightweight version with a centrally positioned filter (below the chin). However, like the SM3, the PM33 can accommodate a filter in any of the three positions (below chin or to either side). Both masks offer sizes which claim to cover more than 98 per cent of the adult European population.

The design objective was to create a mask which offered a range of flexible options for users, drawing from a stock of common components to ease the logistic load.

The basic PM31 and PM33 masks have a 6-point rubber harness and hard-coated polycarbonate eyepieces. Upgrade options include:
- Enhanced voice emission, featuring a stainless steel speech diaphragm
- Huber + Suhner triple-safe drinking device with canteen and connector
- Integrated electrostatic microphone
- Hardened mineral glass eyepieces
- Insertable prescription lenses

PM33 mask - basic version   0052840

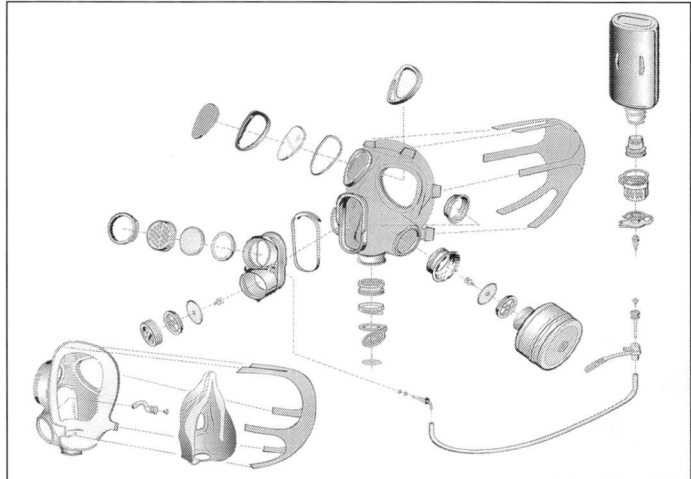

Exploded diagram of PM33 mask assembly   0052841

- Textile mesh/strap harness
- Clip-on anti-glare glasses

Status
Available.

Manufacturer
Huber – Suhner AG.

VERIFIED

SM90 and SM3 NBC protection masks

Description
Following technical and functional tests carried out by the Defence Procurement Agency of the Swiss armed forces it was decided to procure the SM90 NBC protection mask. The SM90 was selected due to its sealing qualities for various head shapes, its high level of wearer comfort and low breathing resistance and for its wide field of vision. Other factors involved were the use of corrective lenses for spectacle wearers, positive guidance of intake air to prevent eyepiece fogging, good communication capability (speaking aid) and a drinking facility.

The SM90 uses a bromobutyl facepiece with twin mineral glass eyepieces. Corrective lenses can either be bonded to the eyepiece or inserted by means of special mounting frames. The filter canister is located on the chin. The speech device is located in front of the mouth. The drinking device is inserted into the speech assembly and can be connected to a standard canteen. A head net arrangement, held in place by six straps, supports the respirator when worn.

When not in use the SM90 is carried in a coated fabric case which also holds spares and accessories. The latter include exterior clip-on anti-glare protection glasses and various adaptors and adaptor systems for special positioning of the filter canister or for connection to a central oxygen supply or filter system. Filter and

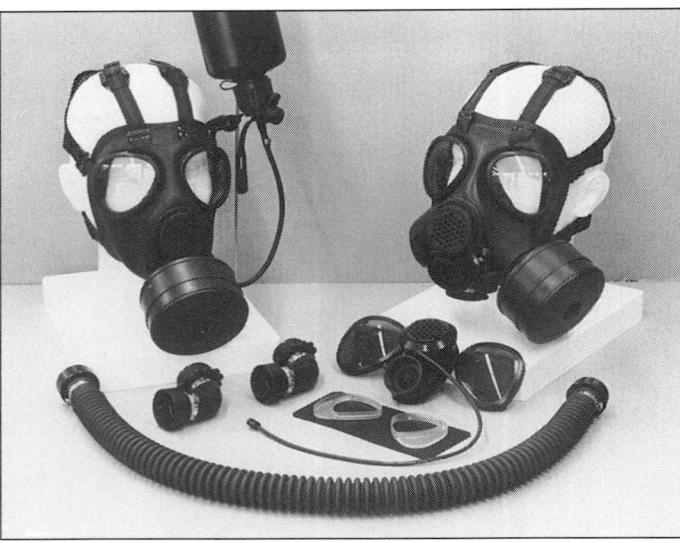

SM90 NBC protective mask (left) with SM3 (right) showing available range of accessories

SM90/SM3 NBC mask maintenance and spare part kit

hose adaptors for use with blower units are available as well as an integrated electrostatic microphone assembly.

The SM3 mask differs from the SM90 in that the filter canister can be mounted in the centre, left- or right-hand position.

The SM90 mask has been produced under licence in South Africa by HAZMAT.

Status
SM90 produced for the Swiss armed forces and police forces. Also in licensed production in South Africa for the South African National Defence Force. A number of other countries have selected the SM90 and/or SM3. Total quantity of SM90/SM3 produced: 600,000.

Manufacturer
Huber + Suhner AG.

VERIFIED

TAIWAN

T3-75 protective mask

Description
The T3-75 protective mask is designed to meet the operational requirements of the Taiwanese armed forces against all known CW agents. The mask is delivered with its own filter canister in a nylon carrying back. Optical corrective lenses can be used and the mask has a speech enhancement device and water drinking device. The mask assembly weighs 586 g and the filter 350 g.

Status
In production. In service with the Taiwanese armed forces.

Manufacturer
Pao-Chang Company.

VERIFIED

T62 protective mask

Description
The T62 protective mask is a locally produced version of the American ABC-M17 mask and differs from it in few respects. It is issued with a canvas carrying satchel which incorporates a cleaning kit. The mask is stated to weigh 410 g and the filter canister 265 g.

Status
In service with Taiwan armed forces and available for export sales.

Manufacturer
Hsing Hua Company Limited.

VERIFIED

TURKEY

Modified NBC SR6 (M) respirator

Description
The Modified NBC SR6 (M) respirator is a slightly modified version of the British Respirator NBC S6 licence-manufactured in Turkey by MKE ELSA AS. The NBC S6 is no longer manufactured in the UK.

The Modified NBC SR6 (M) is basically similar to the British NBC S6 with the facepiece made from high quality natural non-dermatitic rubber. The reflex edge between the mask and the face adapts to the contours of the face making a good seal without undue pressure. Speech transmission is achieved by using a resined nylon diaphragm held under tension. The eyepieces are two optically flat surfaces joined by a parabola. An air guide inside the facepiece separates incoming and outgoing airflows to avoid fogging the eyepieces.

The modified NBC SR6 (M) weighs 830 g and will fit freely into a space measuring 220 × 200 × 150 mm. Shelf-life is in excess of 25 years with a shelf-life of 10 years guaranteed by the company. Three sizes are available; small, medium and large.

The NBC SR6 (M) uses filter canisters designed by Flodins Filter AB of Sweden and manufactured by MKE ELSA SA. They are 110 mm in diameter and 70 mm high.

Status
In production. In service with the Turkish armed forces.

Manufacturer
MKE ELSA AS.

VERIFIED

Modified NBC SR6 (M) respirator and carrying case

UNITED KINGDOM

AFT 500 leak tester

Description
The AFT 500 leak tester has been developed to enable respirators to be tested for leakage after components have been changed by the user. The equipment requires single phase AC 50 Hz mains voltage electricity at 240 V to operate. Equipment needing different voltages for example 110 V USA, can be supplied on request.

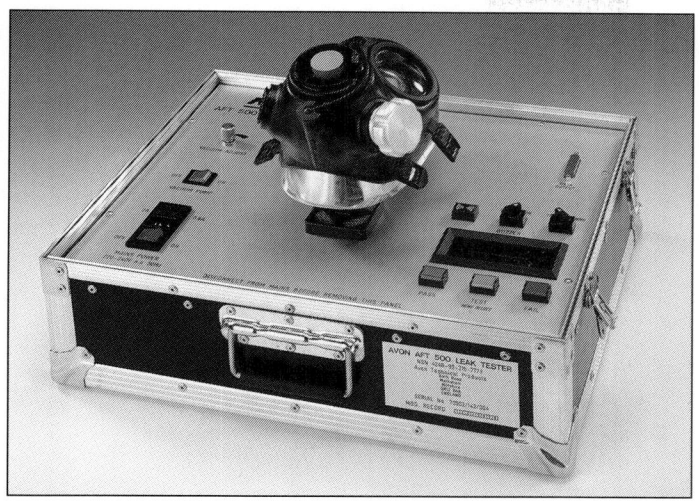

Avon AFT 500 leak tester (Avon Technical Products)   0011432

The tester is contained within a robust case and can be carried by one person. The dimensions of the case overall are 520 × 420 × 295 mm with a weight of 21 kg. In use, the respirator to be tested is sealed on a test head and the pressure within it reduced by approximately 1,000 Pa by an integral vacuum pump.

The degree of vacuum can be adjusted by a flow valve on the panel. The vacuum must be set within ±200 Pa or the computer will not allow the test to proceed. The degree of vacuum is indicated on the LCD display. Once the correct pressure is set, the test sequence is initiated. Following a stabilising period of 15 seconds, the mask is subjected to a test for 30 seconds when a sensitive pressure transducer notes any change of pressure which would indicate a leak. Providing the change in vacuum is not greater than 50 Pa during this period, the respirator is considered satisfactory and a green light illuminates, indicating a pass. In addition, the test results are indicated on the LCD display. Should the change in vacuum be in excess of 50 Pa, a red light indicates failure, again with supporting information on the LCD display.

In addition to testing the respirator, the AFT 500 is also capable of testing the performance of the respirator outlet valve. Again the test conditions for valve testing are programmed in the computer and testing will not proceed unless the prescribed conditions are met. The same green or red lights indicate a pass or fail of the valve. A calibration kit is also available so that the AFT 500 can be checked at regular intervals.

Note that when the AFT 500 is used for checking the FM12 respirator, the two test heads required are contained within the standard case. For the S10, two different test heads are required.

Status
Available.

Manufacturer
Avon Technical Products, Protection Group.

UPDATED

Facelet NBC L1A1

Description
The Facelet NBC L1A1 was developed by the former Chemical Defence Establishment (CDE) at Porton Down, Wiltshire, to provide personnel with interim protection against surprise attack by chemical weapons including nerve agents.

Charcoal cloth Facelet NBC L1A1 in use (DSTL Porton Down)

It entered service with the UK Armed Forces during 1983 but has since been withdrawn from the UK inventory.

The Facelet NBC L1A1 provides oronasal protection and is put on as a precautionary measure when the wearer is on combat alert. It has been designed to exert a minimal physiological load, with the result that it can be worn over long periods and under all operational conditions. Each mask is provided with an adjustable harness which enables one size to fit the whole service population. Other features of the Facelet NBC L1A1 include its very low breathing resistance, high level of speech transmission and compatibility with optical sights and other equipment. It provides a degree of protection at times when the full face respirator is not being worn, including periods of rest and sleep Once a chemical attack is confirmed, the Facelet can be removed quickly and the full face respirator put on. It is small and robust and can be easily carried in a serviceman's combat kit.

The main material used for the Facelet NBC L1A1 is activated charcoal cloth. Facelets are usually issued in vacuum-packed polythene packs with each pack containing three facelets and one harness.

Status
Available. See also similar face facelet devices in this section.

Manufacturers
Charcoal Cloth (International) Limited.
CQC plc
North Safety Products (Siebe Gorman) Ltd.
Remploy Limited.

UPDATED

Respirator NBC FM12

Description
The respirator NBC FM12 was designed and developed to meet the specific operational requirements of a number of European NATO members and has since been taken into service by armed forces around the world. The Dutch Ministry of Defence also selected the FM12 for their future operational requirements and placed an order worth in excess of £8 million in March 1995. Deliveries were to be made over three years, including the provision of filter canisters, water bottles, spares and maintenance kits.

The FM12 may be regarded as a development of the Respirator NBC S10 (see following entry) and is stated to be a high-performance respirator compatible with a wide range of combat and other equipment.

The lightweight facepiece of the FM12 was designed to provide a high level of wearer comfort combined with very low breathing resistance. Scratch- and impact-resistant eyepieces provide a good field of vision and have a non-misting system. Sight correction systems are available for use with the standard mask. An intrinsically safe speech transmitter allows clear direct voice communications, while a secondary speed transmitter for use with communications equipment is an option. A high-flow fail-safe drinking system is provided and it is possible to mount the filter canister on the left or right side. The mask is held in place on the head by a six-strap arrangement connected to a fabric mesh head harness.

The respirator NBC FM12 is stated to be highly robust and easy to decontaminate. Routine maintenance and upgrading can be carried out at unit level using simple tools.

The AMF12 filter canister was designed for use with this respirator.

Respirator NBC FM12 (Avon Technical Products)   0052839

Respirator NBC FM12 - drinking device in use (Avon Technical Products)
0052838

Status

In production. In service with the Danish, Netherlands, Norwegian and Singapore Armed Forces and several other nations.

Manufacturer

Avon Technical Products, Protection Group.

UPDATED

Respirator NBC S10

Description

The respirator NBC S10 was designed and developed by Avon Technical Products for the UK MoD. Following an initial development contract, the production contract was awarded to Avon in September 1987, and the S10 remains in service with UK Armed Forces at the present day. Its advantages include improved levels of protection, communication, vision, comfort, compatibility and maintainability. The S10 also has an integral drinking facility.

For protection the S10 has a high-performance integrally moulded reflex seal. The facepiece is available in four sizes and has a minimum of moulding crevices which improves decontamination efficiency. The canister is supplied in a wrapped condition and it can be quickly replaced without removing the facepiece. The facepiece is held on the head by a fixed buckle harness arrangement that allows the respirator to be repeatedly put on and removed in seconds. Primary communication is via a horn-type voice transmitter and there is also a secondary speech transmitter for use with communications equipment such as handsets,

Respirator NBC S10 (Avon Technical Products)   0052837

SF10 respirator for Special Forces fitted with tinted outsert lenses to protect against flash (Avon Technical Products)

telephones and microphones. Vision is improved by the use of low-profile coated polycarbonate lenses; a corrective lens system for spectacle wearers is available. A facepiece rib ensures a good interface with NBC protective clothing hoods. The drinking device has a double safety valve feature.

The filter canister can be worn on either the left or right of the facepiece. The S10 can also be used with remote filtration units. The filter canister S10 has the STANAG 4155 thread.

The SF10 is a variant of the basic S10 developed for use by Special Forces. It differs from the basic version, in having provision for an internal microphone (for use with a communications harness/radio transmitter and ear defenders) in place of the drinking facility and, a cheek-mounted screw-in plug. This provides access to a second filter canister mount that can accept an air-escape bottle for conditions of oxygen depletion – it also allows the use of twin filter canisters. The eyepieces can also accept tinted outsert lenses to provide protection for the wearer against flash and fragments; they can be quickly removed by hand.

The AR10 is a version produced for use by police forces. It is in service with the majority of police firearm units in the UK.

Status

In production. In service with the Australian, Kuwait, New Zealand and UK Armed Forces and several other nations.

Manufacturer

Avon Technical Products, Protection Group.

UPDATED

SFP Services sodium chloride bench test rig

Description

This bench test rig is used for the production and development testing of filters, filter canisters and respirators for personal protection. It uses the sodium chloride (salt) aerosol test principle in which a sodium chloride aerosol is introduced to one side of a filter and the amount detected on the other side is used to determine the filter's efficiency.

The rig provides a pass/fail system for production testing. A changeover valve enables two test chambers to be used if required and the rig can be used by semi-skilled operators. Respirator filters and materials may be tested at flow rates, ranging from 15 litres/min to 95 litres/min, with a minimum detectable penetration of 0.00005 per cent. The unit can also be used for respirator face seal checks.

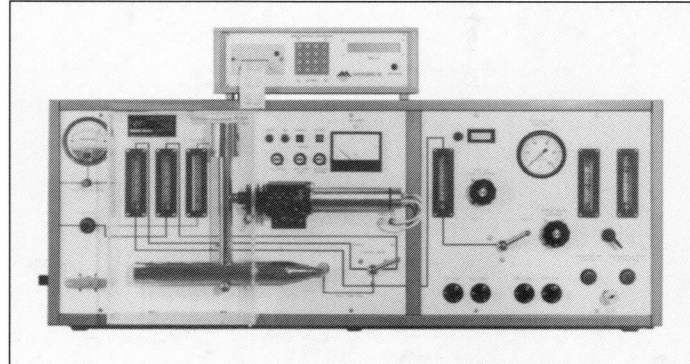

Sodium chloride aerosol bench test rig with direct readout unit

The rig shown in the illustration includes an additional item, a direct readout processor unit. This is a microprocessor-controlled unit used to calculate and display the direct penetration results of tests via an LCD display; it can also produce self-adhesive labels. The labels may be dated and sequenced to satisfy batch testing needs. The unit is available as a retrofit addition to existing bench test rigs.

The bench test rig weighs 90 kg and measures 1,448 × 535 × 560 mm. The direct readout unit weighs 6.5 kg and measures 480 × 330 × 160 mm.

Status
Available.

Manufacturer
SFP Services Limited.

VERIFIED

UNITED STATES OF AMERICA

Advantage® 1000 CBA/RCA mask

Description
The Advantage® 1000 CBA/RCA mask is aimed at internal security forces required to operate in a high RCA environment. Drawing on computer-aided design and anthropometric data from over 7,000 individuals, the mask is designed for a high protection factor and user comfort. Made of soft Hycar rubber, the assembly is similar in form to the Millennium™ and MCU-2/P series of masks. The flexible visor lens is made of urethane and a range of outserts is offered to provide greater physical and glare protection to the eyes. The mask is compatible with spectacle-based eyesight correction, with the ESP communications system and with standard-threaded filter canisters.

See separate entries in this section for the MCU-2P and ESP systems.

Status
Available.

Manufacturer
MSA Defense Products.

UPDATED

AudioPack TriCom voice communications device

Description
The AudioPack TriCom is a three-function device which enhances voice amplification when the user is wearing a protective mask. Secondly, it provides a push-to-talk radio interface designed to reduce voice distortion when an NBC protective mask is worn. Thirdly, it operates as a rugged lapel microphone for use in the field, unmasked.

The TriCom microphone system is weather-proof, easily decontaminated and does not compromise the airtight seal between the user's face and the respirator. Its power is supplied from a single 12 V battery. The outfit is modular in design and user-customisable in the field.

TriCom radio interface and amplifier (Audiopack Technologies Inc)   0134200

Optional accessories include a flexible ear-boom, ear-bud speakers and lapel, finger or arm-operated PTT (Press-To-Transmit) buttons. The latter is paddle-operated by the user's lower arm, enabling hands-free operation where lapel operation is inappropriate (wearing fire-fighting suits for example).

Status
In production. In service with the US Navy Special Operations Forces and with US domestic HAZMAT teams.

Manufacturer
AudioPack Technologies Inc.

UPDATED

BioPak 240

Description
The BioPak 240 is a closed circuit, self-contained breathing apparatus which recirculates a major portion of the user's exhaled gas. A small oxygen cylinder, connected to the breathing chamber, delivers gas into the facepiece. Exhaled breath passes, through a CO_2 absorbent pack, into the breathing chamber where it is mixed with fresh oxygen before becoming available for the next inhalation. A spring-loaded diaphragm in the breathing chamber maintains positive pressure in the system. If the user's inhalation fully depletes the breathing chamber, the demand free-flow valve automatically supplies the additional oxygen required. If the user's exhalation causes the diaphragm to fully expand, excess gas is vented out of the relief valve. A loud, high-pitched, pressure-activated whistle-alarm sounds when approximately 20 to 25 per cent of the service life remains. A manual bypass is provided to override the supply system in an emergency. A 'blue-ice' coolant canister is provided in the inhalation circuit to cool the breathing gas on the way to the mask.

Under normal circumstances, the BioPak 240 delivers at least 4 hours duration. It is simple to use and all controls and displays are easily accessible. The tough outer

> **Specifications**
> **Weight:** 15 75 kg
> **Dimensions (back-pack unit):** 40 × 63 × 19 cm
> **O₂ cylinder:** 0.588 m³
> **Duration:** 240 min
> **Constant flow rate:** 1.78 ±0.13 l/min
> **Materials:**
>   **Shell:** High-impact, fire-retardant NORYL®
>   **Hoses:** Neoprene
> **Environmental:** Approved for use down to −9.44°C

BioPak 240 SCBA in use (BIOMARINE Inc)
***2001**/0121878*

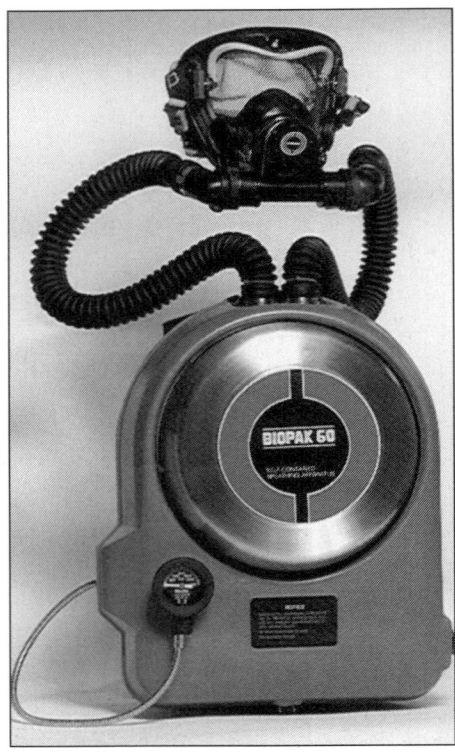

BioPak 60
(BIOMARINE Inc)
***2001**/0121877*

carapace of the backpack protects the unit and the wearer from damage. A fully padded harness with multiple adjustments allows the weight to be borne on the hips, not the shoulders, making the unit extremely comfortable for any sized user to wear for extended periods. The well-fitting facepiece with built-in speaking diaphragm provides a durable, effective seal.

Status
In service with US first responder and HAZMAT teams.

Manufacturer
BIOMARINE Inc.

NEW ENTRY

Interspiro chemical warfare mask for CBWD, EOD and firefighting

Description
The Interspiro chemical warfare mask was jointly developed by Interspiro Inc of Branford, Connecticut and Interspiro AB of Lidingo, Sweden, under contract to the US Air Force Aeronautical Systems Center at Eglin Air Force Base, Florida. Following First Article Testing the mask was approved by the US Air Force for service in September 1994. Under the development contract, the US Air Force has received over 9,000 units.

The Interspiro chemical warfare mask is intended for firefighting applications and provides the capability of selecting a breathing source from either a

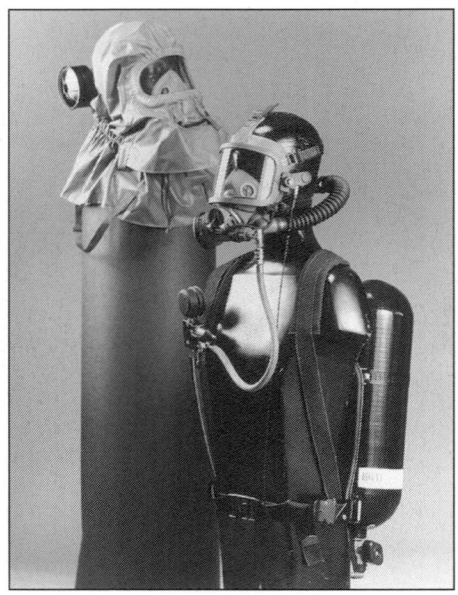

Interspiro chemical
warfare mask for
firefighters

Self-Contained Breathing Apparatus (SCBA) or filtered air through a chemical/biological filter canister.

The CW mask has subsequently been adapted to military EOD requirements as well as civil applications for counter-terrorism protection from weapons of mass destruction.

The mask is provided as part of a kit designed to be used with the Interspiro Spriromatic 9030 SCBA. This SCBA is the standard US Air Force equipment with Interspiro contracted to deliver over 16,000 units. The kit includes a mask, three protective hoods, three C2 filter canisters, a waterproof bag and a canvas carrying case. In a chemical warfare threat environment the firefighter replaces his standard SCBA mask with the chemical warfare mask. If required the mask can be worn without the SCBA. If the SCBA is in use, the fire-fighter can switch from the filter respiration mode to a compressed air supply by engaging the positive pressure selector on the breathing valve. Conversely, the user can switch back to the filter mode by turning off the positive pressure switch.

The system has recently been modified to remove the hood and allow direct interface with the JSLIST style CB clothing.

Status
In production. In service with the US Air Force, Army and Marine Corps.

Manufacturer
Interspiro Inc.

VERIFIED

M40 and M42 Series chemical-biological masks

Description
Following the non-acceptance of the XM30 protective mask (now the MCU/2P (qv)), during 1983, the US Army Chemical Systems Laboratory (later part of the US Army Chemical Research, Development and Engineering Center) produced prototypes of a new mask with a 'low risk' design incorporating the best features of many existing masks. This evolved into the XM40 multipurpose chemical-biological mask which retained many features of the XM30 design plus the hard lenses of the M17 series of masks. The M40 and the M42 (see below) are replacing the M9, M17 and M25 masks as standard issue for the US Army. Most major US commands had been re-equipped with the M40 series by June 1996.

The M40/M42 series of masks feature a rubber silicone facepiece with an internal peripheral seal turned inwards to protect against leakage. Two ballistically hardened hardcoat polycarbonate optically corrected lenses are held in place with metal eye rings to provide broad peripheral and downward vision and covered with outserts. The front voicemitter is located over the mouth while a secondary voicemitter fits into the unused canister sideport. The internal nose cup has two check valves to prevent exhaled air from fogging the lenses. A butyl coated nylon and EDPM rubber hood (known as the Toxic Agent Protection or TAP hood) is provided for chemically contaminated environments.

Other design features include a silicon rubber facepiece and a six-point adjustment head harness. Drinking facilities are provided. The mask is compatible with combat spectacles.

The M40 and M40A1 are used by infantry while the M42, M42A1 and M42A2 are for combat vehicle applications.

M40A1 chemical-biological protective mask with C2 filter canister

M42A1 chemical-biological protective mask with C2 filter canister

M40 chemical-biological protective mask

The M42 mask differs from the M40 and M40A1 in that the filter canister is attached to the mask via a hose with the canister in a carrier. The hose can also be used to attach the mask to vehicle air supply systems. The M42 also features a radio microphone.

Preplanned product improvements introduced on the M40A1 and M42A1 include a 'Second Skin', a covering for the mask face blank material which provides added agent resistance. The material used for the skin is a bromobutyl-EPDM rubber blend. The second skin is designed to be compatible with a quick-removal hood and future overgarments with integral hoods.

In addition to the 'Second Skin' the M42A1 has canister mounting provision for left- and right-hand wearers. The M42A2 has an external microphone assembly for combat vehicle use. This assembly offers the same communication capability as the permanently installed microphone in the M42A1 mask. The assembly consists of a M116 dynamic microphone and cable assembly connected to an adaptor clip for snap-in assembly into the front voicemitter of the mask.

The M40A1 mask is compatible with the MSA ESP® communication system.

Three sizes of M40/M42 series masks are available: small, medium and large.

Status
In production for the US Army and Saudi Arabia.

Manufacturers
ILC Dover Inc.
MSA Defense Products.

Marketing agency
Tradeways.

M41 PATS - Mask fit testing equipment (Portacount® Plus)

Description
The M41 Protection Assessment Test System (PATS) is in service with the US and German Armed Forces to test the quality of the face seal when respirators are worn (TSI market the system as the Portacount® Plus Model 8020). The M41 measures the fit of the respirator whilst it is being worn. NBC training staff use the M41 to identify and document problems. The largest source of leakage is through the facepiece seal and achieving the minimum leakage depends firstly on matching the correct mask size to the wearer's face size and, secondly, on the correct donning procedure by the wearer. Masks, potentially capable of protection factors

Specifications
Fit factor range: 1 to >50,000
Particle concentration range: 0.01 to 500,000 particles/cm³
Particle size range: 0.02 to >1mm
Test duration: 40 s
Sample flow rate: 0.7 litres/m
Alcohol: 8 h operation per charge at 21°C (reagent grade Isopropyl alcohol)
Pass/fail settings: User selectable
RTB calibration interval: 1 year
Size (instrument): 240 × 190 × 140 mm
Size (carrying case): 410 × 380 × 250 mm
Weight (instrument): 1.9 kg
Weight (carrying case): 10 kg

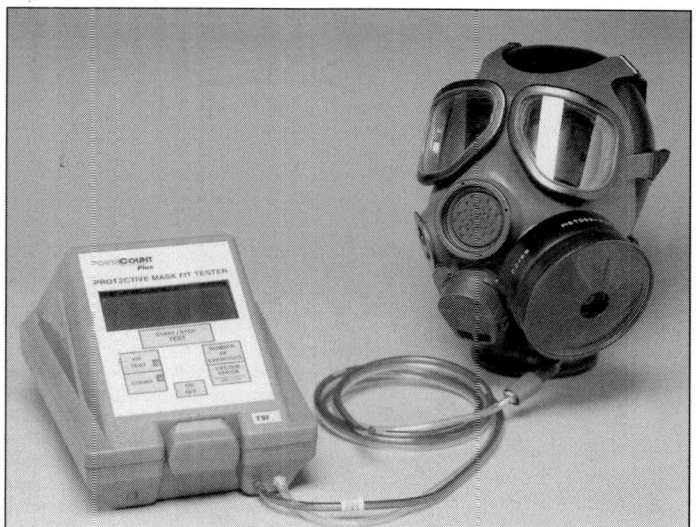

M41 PATS mask fit testing equipment 0052844

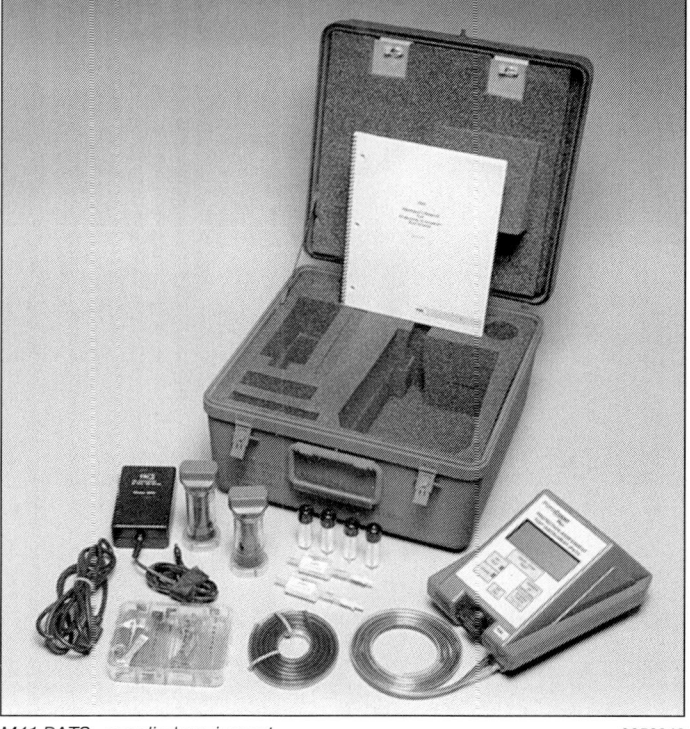

M41 PATS - supplied equipment 0052843

VERIFIED

of over 10,000, may deliver factors of less than 100 if incorrectly sized or fitted. The M41 measures a numerical fit factor as the user conducts specified activities representative of field operations whilst wearing the mask. The numerical fit factors allow the trainer to identify and solve problems with mask fit and donning technique. The equipment is connected to the drinking device, where it compares the ambient level of dust with the level downstream of the filter canister, inside the mask. The particle count is achieved by measuring the light scattered by dust in a laser beam.

The test takes about 4 minutes.

Status
In production.

Manufacturer
TSI Inc.

UPDATED

Mask, Chemical-Biological: Field, M17, M17A1 and M17A2

Description
These masks were the standard issue for all the US armed forces and have been in use since the early 1960s. The M17A1 differs from the M17 by having a self-contained drinking system and a resuscitation capability. The mask has good vision and a voicemitter. The M13A2 filter elements in the cheeks on each side are easily replaced without tools although the mask has to be removed for the changing operation. The masks are issued in an M15 carrier with two filter elements and an M1 CB mask waterproofing bag. The M17A1 mask is issued in an M15A1 carrier which also contains an M1 water canteen cap. The mask, which is available in three sizes, is kept in place by a six-strap head harness. When worn, the mask can be covered by the Hood, Mask, M6A2 which can be held in position with underarm straps. The hood, which is made from lightweight butyl rubber-coated nylon cloth, can be issued in camouflage colour schemes, has openings for the eyepieces and voicemitter and has a zip fastening in the front. Each M6A2 hood weighs 230 g.

A version of the M17A1 mask with a non-essential component omitted is known as the M17A2.

The Czech and Slovakian Model M-10 mask design (*qv*) is a close copy of the basic M17, as is the Bulgarian PDE-1. The Taiwanese T-62 is also identical to the M17.

Status
In service with the US Army, Navy, Air Force and Marine Corps. Has been exported to nations such as Australia and Panama. Being replaced in the US Army by the M40A1/M42A1 (which see) and in the US Navy and US Air Force by the MCU-2A/P (which see).

Manufacturer
MSA Defense Products.

UPDATED

Mask, Chemical-Biological: Field, M17A1

MCU-2/P Series chemical-biological masks

Description
The MCU-2/P Series of NBC masks features an integral moulded polyurethane one-piece panoramic lens bonded to a moulded silicone rubber facepiece. Each mask is issued with an outsert to protect the flexible lens. The mask has left- and

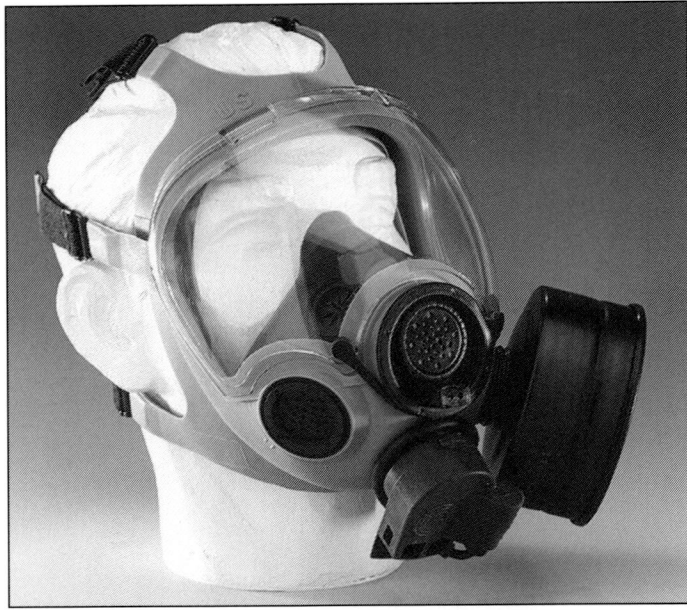

MCU-2/P Series chemical-biological mask with C2 filter cannister

right-hand canister mount options with a side voicemitter inserted into the unused mount. The MCU-2A/P has a primary voicemitter, which provides voice and telephone communications via a M101/AIC microphone, is located over the mouth area. An internal nose cup has two check valves to deflect exhaled air away from the eyepiece to prevent fogging. Drinking capabilities are integrated into the mask.

The MCU-2/P series of masks is available in three sizes; small, medium and large. The mask can accommodate the MSA ESP® communications systems.

Status
In service with the US Navy and Air Force. Also in service with the FBI, US Secret Service, Drug Enforcement Agency, US Coast Guard and the US Department of Energy.

Manufacturer
MSA Defense Products.

UPDATED

Millennium™ CB mask

Description
The Millennium™ CB mask is a development of MSA Defense Product's MCU-2/P series. It is manufactured in Hycar, a nitrile rubber blend that offers an equivalent protection factor and greater comfort than silicone rubber. The Millennium™ mask offers a lower cost solution to high-volume purchasers and maintains compatibility with existing filter canister ranges. The mask assembly is delivered with an optional butyl-coated nylon cloth hood, a low-profile ESP communications system, left- or right- handed canister mount, a drinking device and a range of outserts for additional protection to the main visor. Lens fogging is prevented by the internal

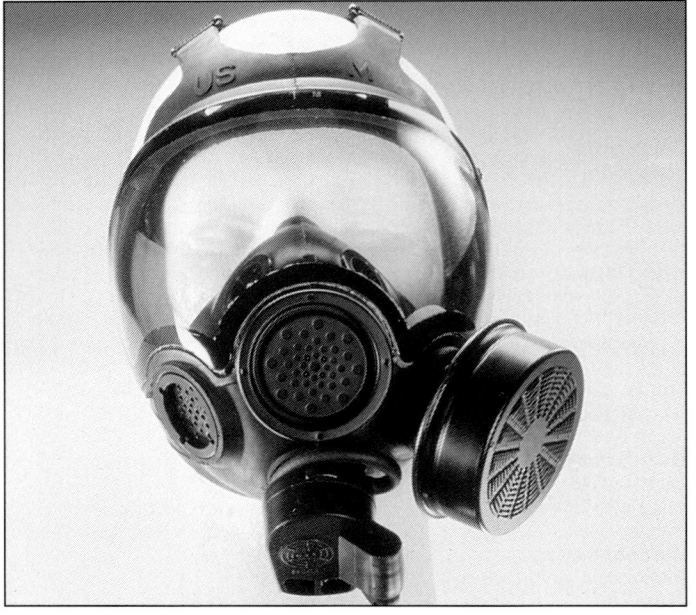

Millennium™ CB Mask   *2001*/0085453

design of the facepiece to route moist air away from the visor. See separate entries in this section for MCU-2P and ESP systems.

Status
Available.

Manufacturer
MSA Defense Products.

VERIFIED

MSA ESP® communications system

Description
The MSA ESP® communications system was designed for use with a wide range of NBC masks, including the M40 series, MCU-2/P, Millennium, Advantage 1000, A-4 Breathing Apparatus and others. Using the system under high-ambient noise conditions, military personnel can communicate clearly with one another with undistorted voice amplification. In addition, the system can be used in conjunction with two-way radios or microphones without experiencing interference or static from radio frequencies.

When used with NBC masks the ESP® system includes a microphone and an amplifier. The system is made from an injection-moulded high-strength polymer and is water and heat resistant. The assembly is lightweight and is equipped with a single on/off switch. The system is powered by a 9 V alkaline battery with a life of 14 to 18 hours.

Status
In production.

Manufacturer
MSA Defense Products.

UPDATED

The Spirotek self-contained breathing apparatus   2001/0079844

fast cylinder changes and a convenient buddy breathing hose which allows others to be connected to the set without interrupting the user's breathing.

Upgrade and maintenance programme was are available for both systems.

Status
In production. In service with Greece, Iceland, Netherlands, Norway, UK (Army and Royal Air Force), US military (Army, Air Force and Sealift vessels). The equipment is also used by numerous police and firefighting forces.

Manufacturer
Interspiro Inc.

UPDATED

TDA-99M dual-purpose field mask leakage tester

Description
The TDA-99M is a multifunction, portable, protective mask testing system. With its full solid-state circuitry, lightweight construction and greater variety of mask testing functions, it replaces the earlier model TDA-99D. The assembly packs into a rugged case measuring 660 × 410 × 460 mm and weighs 27 kg. It requires a power supply of 100 to 250 V at 50 to 60 Hz. The test head is a moulded polycarbonate frame over which any type of western mask can be fitted for test. There is an aerosol generator, vacuum pump, light-scattering chamber, flowmeter and control components which allow the performance results of the mask under test to be displayed to the user. The simulant used is polydispersed oil aerosol (POA) at a particle size of 0.1 to 0.5m. PAO, DOP and DOS are also suitable simulants. The

The MSA ESP® communications system mounted on a M40A1 NBC mask

---

Spirotek and Spiromatic-S self-contained breathing apparatus systems

Description
Both the Spirotek and the Spiromatic-S high-performance breathing sets are designed for use in highly toxic environments. When worn as part of full IPE, both equipments provide the user with the highest possible protection factor.

Both systems are ruggedly designed and simple to operate and maintain. Depending on the cylinder size and pressure, the systems offer up to 60 minutes duration. Other features include a primary low-air alarm comprising an audible whistle and a visual indicator (flashing red LED on the pressure gauge). The systems exceed NFPA requirements and the first-stage regulator delivers high-capacity flow rates to over 1,350 litres per minute. The cartridge-style first-stage regulator and low-air warning alarm is optimised for easy replacement and maintenance. The breathing valve allows automatic positive pressure with airflows exceeding 300 litres per minute. The Spiromatic-S facemask is supplied with both equipments. The harness assembly is robustly designed and uses sturdy quick-release fastenings. Air cylinders can be supplied in carbon, fibreglass or aluminium. The Spiromatic-S system has a new docking mechanism which allows

Specifications
Size: 41 × 51 × 74 cm
Weight: 32 kg
Power supply: 100 - 250 V AC at 50 - 60 Hz
Aerosol simulant: Polydispersed oil aerosol - 0.1 to 0.5 μ particle size
Aerosol detection: Near forward light scattering photometer. Leak detection range: 0.0005 to 100%
Data output: RS 232 serial port

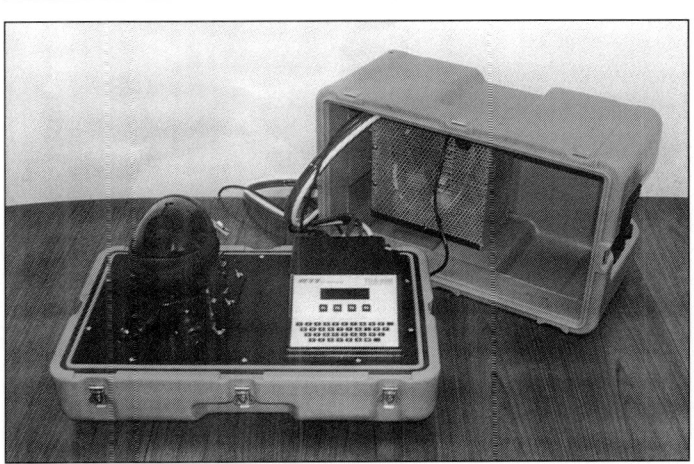

TDA-99M dual-purpose field mask leakage tester   0011447

data logger can record up to 500 complete mask tests, although increased memory capacity is optional. Data can be downloaded to a PC through an RS232 serial port. Leakage from the face seal, lens joints, inhale/exhale valves, drinking tube and voice enhancer can be measured by fitting the mask over the test head. The TDA-99M can also be used to check masks when worn by personnel (face fit test). To perform both tests on personnel and their masks takes about 10 minutes per person. The TDA-99M is a DoD approved device which is set to replace a number of different test sets currently in service. These include the following: M14/TDA-104 leakage tester; Q21, Q179 and Q204 testers; M4A1/TDA-124 and M41 leakage and face-fit testers.

Status
In production. Under evaluation by the US DoD and under test by the US Marine Corps.

Manufacturer
Air Techniques International.

Marketing agency
Technology Marketing.

UPDATED

Voice Projection Units (VPUs)

Description
The Voice Projection Unit (VPU) is a small, lightweight voice enhancement device which can be externally mounted on the respirator without the need for modification. The VPU is mounted over the existing voice diaphragm to project the user's voice to a distance of 75 m. Powered by a single 9 V battery, it is 50% lighter than previous AudioPack units.

The VPU has been type classified and has passed all operational, field and environmental requirements set by the US Armed Forces. It is compatible with the M17, M40 (use model number B-181040A), MCU-2A/P, MCU-2/P (MCU masks use model number B-181050), S10, SF10, SM90, FM-12, C4 and most other NBC respirators.

M45/M40 VPU (Model number B-181045JSP)
This new (2001) VPU is over half the size and weight of previous models. It is powered by two AAA batteries and is also available for MCU-2A/P, OBA, M-17, EAB, MSA Advantage 1000 and MSA Millennium respirators.

OBA VPU (Model B181A)
This VPU is available for the B-181A and other types of SCBA.

Voice Projection Units (VPUs) (AudioPack) **2000**

AudioPack VPUs. The M45/M40 unit is shown on the left (John H Eldridge) **2002**/0121885

AudioPack VPUs shown with a US M40 respirator (John H Eldridge) **2002**/0121884

AudioPack VPUs Model No B-181045JSP (AudioPack) **2002**/0134205

Status
In production. In service with the US Army, Air Force, Navy, Marine Corps, Coast Guard and National Guard.

Manufacturer
AudioPack Technologies Inc.

Marketing agency
Tradeways Limited.

UPDATED

YUGOSLAVIA, FEDERAL REPUBLIC OF

M2 series protective masks

Description
The M2 series of protective masks are intended specifically for efficient protection of the eyes and respiratory system against gaseous, liquid or solid NBC contaminants in atmospheres comprising not less than 17 per cent oxygen by volume. Any appropriate filter can be screwed onto the EN148 compliant connection.

Masks are positioned on the head by a harness of adjustable elastic rubber straps. Whilst aimed primarily at the chemical industry, the M2 series is suitable for defence use and can be safely worn in the temperature range -30 to +50°C. The masks contour comfortably to the face, allowing unrestricted head movement. The eyepieces provide wide-angle visibility, offering a 70 per cent field of vision. Eyepiece transparency is more than 87 per cent.

The M2-F has a speech enhancement device. The M2-FV has, in addition to this, a drinking facility. The speech transmission index is at least 0.85. Water can be consumed at 200 ml/min minimum.

Trayal M2-FV mask - showing drinking facility   0011448

The M2 series masks are made of smooth-shaped natural rubber making them easy to decontaminate. They are available in three sizes (small, medium, large) and can be coloured to customer requirements. The mass is 500 g, (about 800 g with filter). Exhalation resistance through the exhalation valves subassembly is 70 Pa, providing secure protection under all conditions. The carrying case measures 300 × 230 × 150 mm and the total mass of the system when carried is 1,300 g.

Status
Available.

Manufacturer
Trayal Corporation.

VERIFIED

NBC protective mask for children MD-1

Description
The MD-1 NBC protective mask for children is intended for the protection of the respiratory system, eyes and faces of children aged between 6 and 15 years against NBC contamination in the form of droplets, vapours and aerosols and radioactive or biological particles. It is produced in two sizes, small and medium.

The weight of the MD-1 mask is 600 g without its carrying bag.

NBC protective mask for children MD-1   0011449

Status
In production.

Manufacturer
Trayal Corporation.

Contractor
Yugoimport SDPR.

VERIFIED

NBC protective mask for horses M-1

Description
The M-1 NBC protective mask for horses is used to provide respiratory protection against NBC agents for draught, pack and riding horses and mules. The complete mask consists of a facepiece covering the animal's head (covering the nose, eyes and ears) and secured around the neck, two filter canisters, a cleaning kit and a carrying bag. The carrying bag may be carried on a pack or back saddle.

Once fitted, the mask may be placed in one of two conditions, preparatory or protective. The state can be altered by the animal's hand er by altering the position of a lever on the inhalation valve; the process takes about 5 seconds. It is claimed that once the mask is fitted the animal retains 80 per cent vision.

The mask, also known as the MZK M-1, is available in two sizes. The smaller size (S) is for animals with muzzles up to 470 mm in circumference; this version weighs 3.25 kg. The larger size (L) weighs 3.55 kg and is for animals with a muzzle circumference greater than 470 mm.

Status
In service with the former Yugoslav armed forces.

Manufacturer
Trayal Corporation.

Contractor
Yugoimport SDPR.

VERIFIED

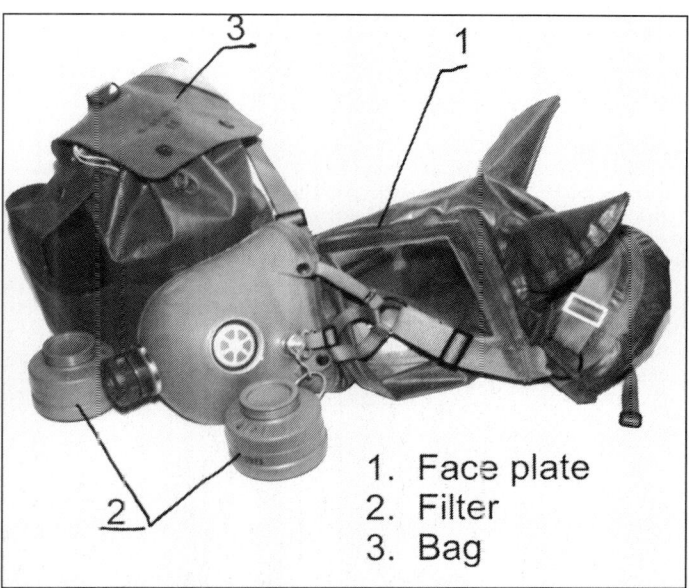

1. Face plate
2. Filter
3. Bag

NBC protective mask for horses M-1

NBC protective mask M-2

Description
The NBC protective mask M-2 evolved from the M-1 version and may be regarded as a progressive development of the earlier model (now largely withdrawn from use).

The M-2 is available in three sizes to accommodate the face shapes and sizes of 99 per cent of the local population. It has a facepiece made from natural rubber with an integrally moulded peripheral seal for a comfortable fit and face seal. A facepiece rio ensures a good interface with protective clothing hoods. The face blank incorporates a simple low-resistance exhalation valve, semi-triangular glass eye lenses and a standard thread connection (40 mm) which enables it to be used with a wide range of optional filters. The filter canister can be replaced without removing the mask from the face. The filter contains a pleated glass paper particulate filter and an activated impregnated charcoal bed. Exhalation resistance through the exhalation valve subassembly is 70 Pa.

NBC protective mask M-2

The M-2 mask weighs approximately 800 g, including the filter canister which weighs 300 g. It will remain operational over a temperature range of from −30 to +50°C. To put on or remove takes under 10 seconds and when not in use the mask is carried in a shoulder-mounted haversack measuring 300 × 230 × 150 mm. The mask was designed for easy cleaning, decontamination and maintenance.

A version of this mask is produced in Egypt as the M2-E (qv).

Status
In production. In service with the Yugoslav armed forces, Iraq and several other foreign countries.

Manufacturer
Trayal Corporation.

Contractor
Yugoimport SDPR.

VERIFIED

....

NBC protective masks M-2F and M-2FV

Description
The NBC protective masks, M-2F and M-2FV, may be regarded as advanced versions of the M-2 mask (see previous entry). Both are produced in three sizes.

The M-2F mask has a combined voicemitter/exhalation valve assembly in place of the former expiratory valve. This provides better speech transmission characteristics with a speech transmission index of at least 0.85.

The M-2FV mask incorporates a drinking system consisting of a drinking tube equipped with two drinking valves and a valve port on the right side of the face blank; the drinking system capacity is over 200 ml/min.

Both the M-2F and M-2FV provide the same levels protection as the M-2 mask. Facepieces have been produced coloured black and olive green.

NBC protective mask M-2F with voicemitter/exhalation valve

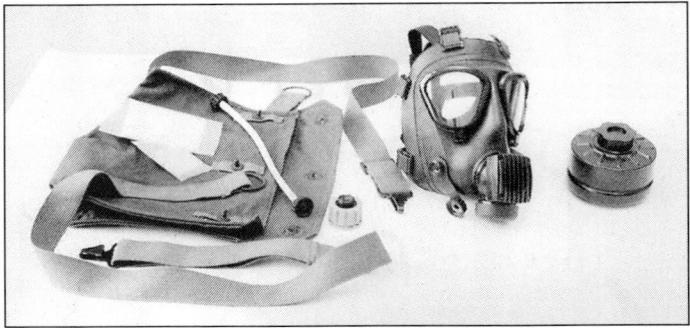

NBC protective mask M-2FV with drinking system

Status
In production for the Yugoslav Army.

Manufacturer
Trayal Corporation.

Contractor
Yugoimport SDPR.

VERIFIED

....

Panoramic NBC protective mask B-2/G

Description
The panoramic NBC protective mask B-2/G uses a natural rubber facepiece with a one-piece vision system to provide a wide field of view. The vision system may be produced using either a triplex material for use in aggressive environments or acrylate for use in more neutral surroundings. Internal dead space is reduced to a minimum to maintain clear vision through the vision system. The position of the facepiece on the head is carried out using adjustable elastic head straps.

The mask is provided with a phonic membrane for voice transmission with a built-in filter connector. Exhalation resistance through the exhalation valve subassembly is 75 Pa.

The mask is available in two sizes, small and medium and can be produced in various colours. Weight of the mask without a filter canister is 650 g. It may be used over a temperature range of −30 to +50°C.

Status
In production.

Manufacturer
Trayal Corporation.

Contractor
Yugoimport SDPR.

VERIFIED

Panoramic NBC protective mask B-2/G

PROTECTION (individual) - Masks (aircrew)

CANADA

Carleton aircrew respirator system

Description
The Carleton aircrew respirator system is used with an NBC protective hood to form part of an NBC aircrew Individual Protective Equipment (IPE). The system was developed for use in helicopter, transport or patrol aircraft and provides protection to the head, neck and respiratory tract against NBC warfare hazards. One of the main assemblies of the system is an AC4 respirator which may be connected to an aircraft oxygen system when required, as all components are oxygen compatible. All metal parts are non-magnetic.

The AC4 respirator assembly is normally used in conjunction with a filter blower and an aircraft power source. If required, the system can be powered by a battery housed in a battery box carried inside a pouch on a belt or can be vest mounted. The respirator has a lung-powered demist function and can operate with or without powered ventilation. When in use, the ventilator (blower) unit (weight 308 g) reduces breathing effort, provides positive pressure for increased protection, provides constant demisting and creates a cooling effect.

The respirator can be used with any filter canister and/or NBC mask assembly with a NATO standard thread.

Status
Available.

Manufacturer
Carleton Life Support Technologies Limited.

VERIFIED

Carleton aircrew respirator system

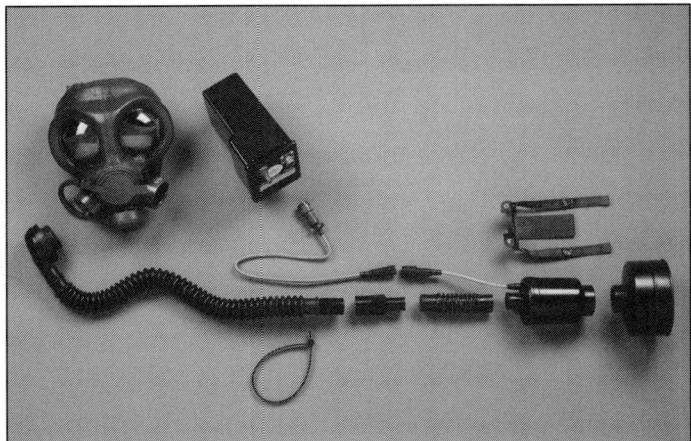

Components of Carleton aircrew respirator system

FRANCE

EPHESE personal breathing protective equipment for aircrew

Description
The EPHESE aircrew respirator is designed to deliver filtered air at positive pressure to all types of aircrew on low-level duty, for example transport and helicopter aircrew, where oxygen is not used. The respirator face-piece is secured on the head by means of a low-profile webbing harness allowing a comfortable fit under the flying helmet. An air-pipe delivers air to the front of the face-piece from filter-pack, temporarily fixed to the aircraft structure for flight use or worn strapped to the waist when on the ground. The pack is energised by a 7.3 V lithium battery which runs the blower for 24 hours and the airflow can be adjusted automatically to personal requirements. A useful LED indicator displays battery life remaining. Air is drawn through two standard NBC filters and the facepiece, harness and air supply are compatible with all flight role equipment such as night-vision aids, flight communications systems and cockpit displays. There is also a separate voice transmitter for ground communications.

The respirator assembly weighs 580 g and is supplied in a choice of three sizes. The blower unit weighs 1,900 g.

Status
Approved by the French MoD and in series production for the French armed forces.

Manufacturer
SP Défense.

VERIFIED

EPHESE aircrew respirator system
0011451

EPHESE system. Note attachment of blower unit to airframe 0052835

Giat NBC protection system for helicopter pilots

Description

Evolved from the EPHESE system (see separate entry in this section), this is a head protection system for the NBC protection of combat helicopter and transport aircraft pilots. Its computer-aided design renders it compatible with current and future flying equipments. Combined with a ventilation unit with an adjustable airflow, it provides good respiratory comfort.

The mask consists of a black face mask in waterproof and fire-resistant bromobutyl rubber. There are two hardened and anti-toxin treated thermoplastic eyepieces, an internal half-mask to reduce the dead volume and an anti-mist deflector. Also provided are a sound capsule for ground use, differential microphone for in-flight communications, drinking device and an offset ventilation unit pipe.

The airflow is adjustable as required between 50 and 120 litres/min. A positive overpressure is provided of 1 hPa to the extent of an instantaneous inhalation flow of 2 litres/s (120 litres/min).

Status

Facepiece: pre-series production for the French Army.
Ventilation unit: series production for US Army helicopter pilots.

Manufacturer

Giat Industries.

VERIFIED

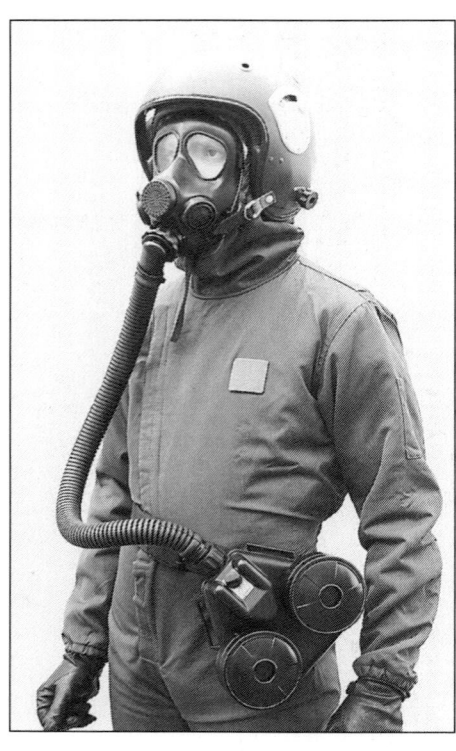

Giat NBC protection system for helicopter pilots

ISRAEL

Supergum armoured vehicle crew respirator system (ACBS)

Description

ACBS is a new system, developed on behalf of the Israeli MoD is designed to provide facial and respiratory protection to AFV crew members. The hood is made from high tensile strength 3-layer co-extruded polymer film with both high mechanical strength and high barrier resistance to toxic agents and fire. The spherical form goggle eye-pieces are designed to be compatible with vehicle-mounted optical sighting and other systems found in Israeli AFVs accommodating eyesight correction. The integral silicone nose-cap accommodates a drinking device and carries the air valves, speech amplification membrane, and a standard dynamic microphone (Israeli designation M-116). There is a flexible silicone neck collar which seals the hood assembly.

ACBS is a single-size respirator which can use either normal lung-powered filtered respiration or a fan-powered blower unit, offering an average protection factor of 10,000 for 95 per cent of the full range of wearers. The blower unit supplies air at 90 litres per minute through a double-filter pack which can be carried either on a waist belt or as a back-pack. LiSiO$_2$ batteries give 15 hours of continuous operation under normal environmental conditions.

ACBS is fully compatible with other elements of standard-range IPE and with standard AFV crew helmet and communications equipment.

Status

In production. In service with the Royal Norwegian Air Force.

Manufacturer

Supergum Ltd.

NEW ENTRY

....

Supergum Chemical Team Respirator (CETER)

Description

This new system, developed on behalf of the Israeli MoD is designed to provide facial and respiratory protection to teams engaged in arduous and dangerous operations including reconnaissance, DECON, EOD and casualty-handling activities. The hood is made from high tensile strength 3-layer co-extruded polymer film with both high mechanical strength and high barrier resistance to toxic agents and fire. The panoramic polycarbonate visor allows a wide field of vision to the user as well the ability to accommodate eyesight correction. Its proprietary scratch-resistant coating also eliminates misting. The integral silicone nose-cap accommodates a drinking device and carries the air valves, speech amplification membrane, and a standard dynamic microphone (Israeli designation M-116). There is a flexible silicone neck collar which seals the hood assembly.

CETER is a single-size respirator which can use either normal lung-powered filtered respiration or a fan-powered blower unit, offering an average protection factor of 10,000 for 95 per cent of the full range of wearers. The blower unit supplies air at 90 litres per minute through a double-filter pack which can be carried either on a waist belt or as a back-pack. LiSiO$_2$ batteries give 15 hours of continuous operation under normal environmental conditions.

The CETER is fully compatible with other elements of standard-range IPE and with standard helmet and communications equipment.

Status

In production. In service with the Israeli Armed Forces.

Manufacturer

Supergum Ltd.

NEW ENTRY

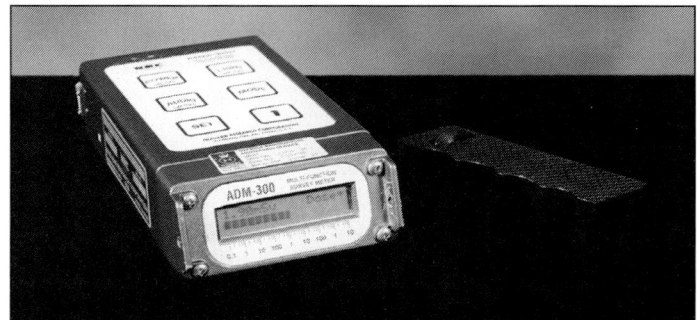

The Chemical Team respirator (CETER) (Supergum Ltd)   **2002**/0137899

NORWAY

Aircrew NBC protection aircraft integrated system Mod2B

Description

The Aircrew NBC protection aircraft integrated system Mod2B was developed to deliver a regulated 60 litres/min of filtered air to the pilot's NBC protective mask assembly. The assembly includes the (Pilot Eye Respiratory System) (PERS) face shield and the MBU-12/P standard oxygen mask and is designed to deliver ventilation and demist under all conditions of the flight envelope. The system features an automatic change of primary (Environmental Control System - ECS) and secondary (oxygen) sources while using the primary ECS source as much as possible to conserve oxygen. The system is permanently installed and constantly ready for use while being easy to operate and maintain. A portable Filter Blow Unit (FBU) is necessary during transit to and from the aircraft.

The Mod2B is employed as part of the pilot's PERS, a system which provides chemical and biological protection for a minimum of 24 hours with a protection factor of 10,000. PERS has been ejection tested.

In operation, primary pressure from the ECS from a filter mounted in a pressure container and secondary pressure from the oxygen system are combined in an NBC shuttle/regulator valve. This unit regulates the exit pressure, regardless of source, to a fixed pressure of 60 litres/min in the pilot's visor. The valve is designed to ensure delivery from the ECS as long as it is capable of delivering filtered air, after which the secondary (oxygen) source will take over from the decreasing ECS primary pressure. The pilot will not notice any changes in his air supply during the changeover stage. One-way valves are installed in both supply lines as a safety measure and a restrictor is located in the ECS supply line for use during any malfunction.

When the pilot is in transit to and from the aircraft, his air supply is provided via a hand-held FBU. During the strapping-in procedures in the cockpit, the NBC supply hose is connected to a quick disconnect on a console and oxygen is delivered automatically as soon as the FBU has been disconnected. During engine start-up and shutdown, the change of source to and from the ECS occurs automatically. After engine shutdown, the FBU can be reconnected; a built-in shutdown valve stops the oxygen supply as soon as disconnection occurs.

The Mod2B is in course of replacement by developed systems including the Mod3 and the Mods 4, 5 and 6 for special applications.

Status
In production. In service with the Royal Norwegian Air Force.

Manufacturer
NBC Aerotech A/S.

VERIFIED

UNITED KINGDOM

CBRR and AR5 aircrew respiratory systems

Description
Originally designed and developed in conjunction with the former DERA CBD (UK's Porton Down CBD authority, now part of the new DSTL organisation), the Institute of Aviation Medicine and Royal Aircraft Establishment, both at Farnborough, Hampshire, the aircrew NBC respiratory system AR5 provides full physiological protection in the air and on the ground against NBC agents in any known form. CBRR is a development of the AR5 manufactured by Cam Lock (UK) Limited. Protection is provided against agents entering the respiratory system or contacting the eyes and skin above the shoulders. It uses an under-helmet hood concept, supplied with filtered air which allows the existing range of personal equipment to be used.

The basis of the AR5 and CBRR respirators is a dual sized oronasal mask mounted on a close-fitting polycarbonate combined optical faceplate and exoskeleton. The edges of the faceplate are bonded to a flexible bromobutyl cowl worn over the head and beneath an aircrew protective helmet. Both the hood and mask are separately supplied with filtered air, or air/oxygen, by way of a chest-mounted manifold. A natural rubber neckseal bonded to the cowl effectively isolates the head from the surrounding environment. The neck seal is overlaid with

Aircrew respiratory system AR5 in use, with aviator carrying a portable ventilator

The CBRR aircrew respiratory system   0052834

a butyl rubber bellows and apron to protect the neck area whilst allowing full and free head movement. The optical surfaces offer an undistorted visual field that is almost identical to that obtained when using a conventional oronasal mask and protective helmet. An anti-abrasion coating also provides resistance to CW agent penetration. The CBRR can be configured in either the oxygen or non-oxygen variants and filtered air or oxygen is supplied separately to the hood and mask via a chest-mounted manifold assembly. For oxygen, the manifold is connected directly to the aircraft oxygen supply via a conventional pressure demand oxygen regulator. The flow of air to the hood maintains a positive pressure that prevents the ingress of contaminants and greatly reduces the thermal stress of the user in hot and humid conditions and prevents misting of the visor.

In the non-oxygen variants, in the event of a failure in the supply of blown air, filtered air may still be obtained by normal breathing demand. In the oxygen variants, a proportion of the oxygen supply may be diverted to demist the optical surfaces by the operation of the cross-over valve in the manifold.

In the event of a parachute descent into water or a helicopter ditching, an optional rip panel allows rapid removal of the complete facepiece. A quick disconnect in the mask supply can also be incorporated.

The CBRR can be fitted with a range of dynamic microphones to suit particular aircraft applications. It features a device for occluding the nasal passages to allow the valsalva manoeuvre to be performed. The mask incorporates a microphone and drinking facility. By utilising close-fitting spectacle frames, corrective lenses can be worn. Tactical and portable ventilators are also available.

A portable test set is available for the CBRR system.

The basic CBRR respirator assembly weighs 720 g and the aircraft-mounted ventilator, 780 g. The portable ventilator weighs 5.9 kg (with battery) and the tactical ventilator, 2.6 kg.

Status
In production. In service with the Royal Air Force, the Canadian Air Force and the US Marine Corps.

Manufacturer
Cam Lock (UK) Limited.

UPDATED

UNITED STATES OF AMERICA

ILC TAERS HELO™ chemical protective mask

Description
The ILC Tactical Aircrew Eye Respiratory System (TAERS™) helicopter (HELO) mask is a low-cost variant of the MBU-19/P PIHM mask (see separate entry in this section). It has the same advantages as the PIHM mask but without the specialised oxygen regulator and manifold system required for use in tactical aircraft. The system consists of a neckseal hood mask assembly with a standard MBU-12/P oxygen mask, a standard C2 or NATO thread filter canister and a filter blower for helicopter and ground operations. The TAERS HELO™ incorporates a standard microphone and drink tube assembly. The mask interfaces with the SPH-4 or other

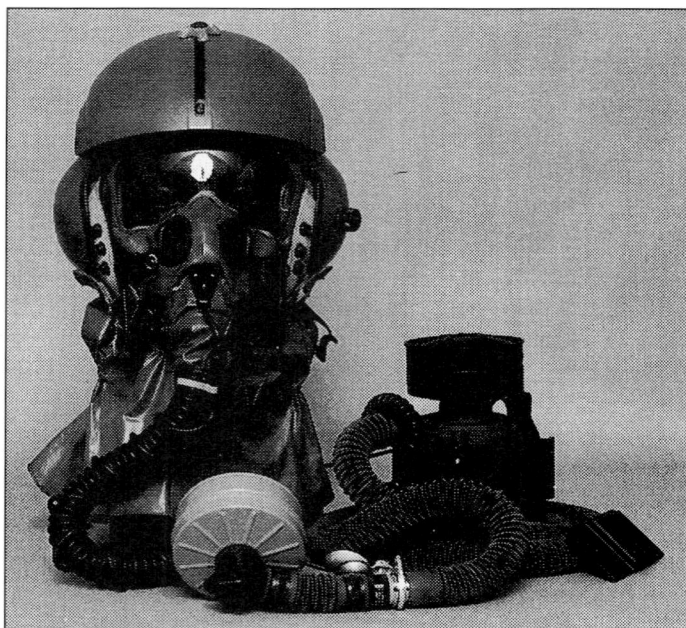

ILC TAERS HELO™ chemical protective mask

standard helicopter helmets, the HGU-55/P, or can be used in a non-helmeted configuration, depending on its intended use.

Manufacturer
ILC Dover Inc.

VERIFIED

M45 aircrew chemical biological mask system

Description
The M45 (formerly the XM45) aircrew chemical biological mask system is the replacement for the US Army's M48/M49 aircrew mask system and is specifically aimed at helicopter crews or other special operations personnel. The M45 mask system is intended to have a lower unit cost than the M48/M49 (unit cost US$350 as opposed to US$1,600 for the M48/M49 - see entry in this section).

The facepiece has a microphone port for aircraft communications, drinking tube device for liquid nutrients, close-fitting eye lenses, front and side voicemitters for face-to-face and telephone communications, low-profile canister interoperability hose assembly for both hose- and face-mounted configurations and interchangeable nose cups mounted in a silicon rubber facepiece with an in-turned peripheral seal. Injection-moulded composite materials are used to reduce weight and cost, with the components snap-fitted or ultra-sonically welded to the facepiece to assist production.

The M45 mask provides protection without the aid of forced ventilation air while maintaining compatibility with rotary-wing aircraft sighting systems and night vision devices.

M45 aircrew chemical biological mask (T J Gander)

Companies known to be involved in the M45 pre-production stage include Venture Plastic Inc (plastic composite components) and Wirtz (rubber tools). Negotiations were in progress during early 1997 to determine a contractor to produce the initial limited production run.

Status
Ready for production by 1997. Fielded 2000. Other aircrew respirator systems have also been adopted by the US Armed Forces (see this section for further details).

Development agency
Edgewood Chemical Biological Center.

UPDATED

M48/49 Series protective masks, aircraft AH-64

Description
The M48/M49 series of protective masks was specially developed for aircrew flying the AH-64 Apache helicopter and other US Army aircraft; they were originally designated the M43 and M43A1.

The mask consists of a form-fitting bromobutyl/natural rubber facepiece with replaceable prescription (if required) lenses fitted close to the eyes. The mask provides chemical and biological protection while on the ground or flying the aircraft. Type I masks are compatible with the Integrated Helmet and Display Sighting System (IHADSS) used on the AH-64 Apache. The M49 (originally the M43E1 and then the M43A1) improves on the design of the M48 (originally the M43) by providing improved chemical agent protection, a standardised battery on the filter blower, a specialised butyl outer hood for thermal protection, an oxygen adaptable system and an auxiliary motor/blower assembly for use in event of an aircraft power failure or during an emergency exit.

In all, there were four masks in the M43 series. The M48 Type I (originally the M43 Type I) has a notch in the right eyepiece to accommodate the IHADSS sighting device used by AH-64 pilots; it uses the M171/AIC microphone. The M48 Type II (originally the M43 Type II) is issued to all US Army aircrew other than Apache pilots. This model has two spherical lenses and is provided with the M133/U dynamic microphone. The M49 Type I (originally the M43A1 Type I) has the notched right eyepiece for use with the Apache IHADSS sighting system, while the M49 Type II (originally the M43A1 Type II) has spherical lenses.

The M48/M49 series of masks is available in sizes; small, medium and extra large.

Status
In service with the US Army.

Manufacturer
MSA Defense Products.

VERIFIED

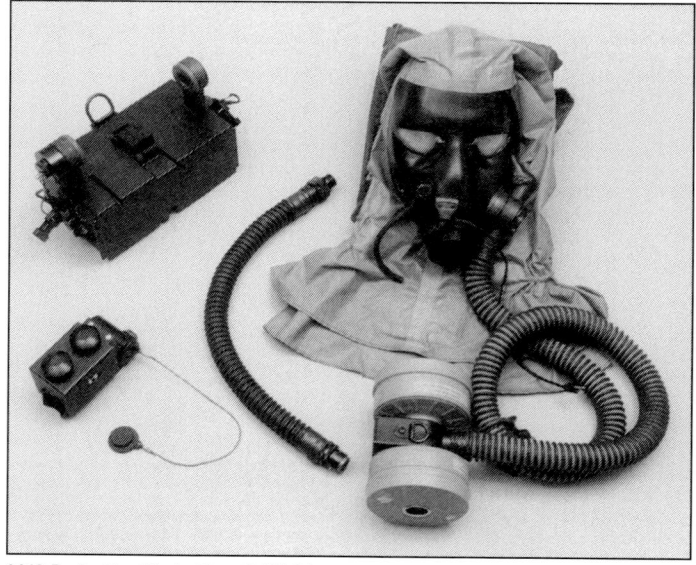

M49 Protective Mask, Aircraft AH-64

MBU-19/P Protective Integrated Hood Mask (PIHM™)

Description
The MBU-19/P Protective Integrated Hood Mask (PIHM™) is a US Air Force-qualified mask and is designed for use in transport, cargo, observation, helicopter and fighter aircraft.

MBU-19/P Protective Integrated Hood Mask (PIHM™)

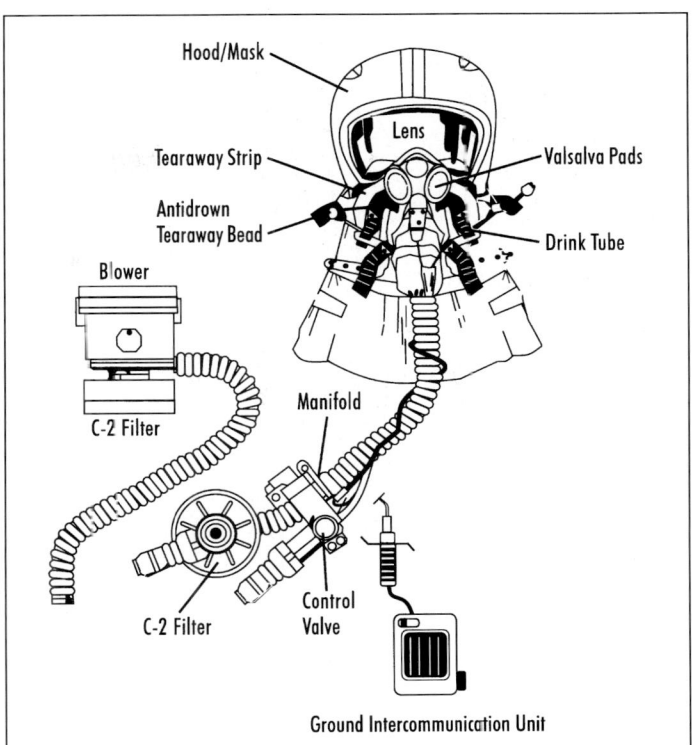

Outline diagram of MBU-19/P Protective Integrated Hood Mask (PIHM™)

The PIHM™ system consists of a neckseal hood mask assembly with the standard MBU-12/P oxygen mask, a specialised crossover valve/filter assembly and a filter blower subsystem for aircraft and ground operations (a C2 filter canister is employed). The PIHM™ incorporates a standard microphone and drink tube assembly, permitting the aircrew member to drink fluids while wearing the mask. A 'non-helmeted' version of the PIHM™ has been delivered for use in the C-9 transport aircraft. The MBU-19/P has a protection factor of 10,000.

Since October 1990, the US Air Force has awarded contracts for more than 21,000 PIHM™ units for all US Air Force Commands; 15,600 MBU-19/Ps were fielded during Operation Desert Storm alone. Eleven other countries have also purchased systems.

ILC Dover Inc and Allied Materials provide the hood assembly for the MBU-19/P. Hunter Inc provides the CQU-7/P blower subsystem and Primatec supplies the MXU-835/P intercom unit.

Status

In production for the US Air Force. Also in service in other countries. Under evaluation by 11 countries.

Manufacturer

ILC Dover Inc.

VERIFIED

PROTECTION (individual)

Filters
Body Protection
Medical countermeasures

PROTECTION (individual) - Filters

AUSTRIA

Canister Swivel Connector SMK

Description
The SMK Canister Swivel Connector is a simple adaptation which allows respirators with centrally suspended filters to swing the canister either side of the facepiece. This is to facilitate activities or postures which would be obstructed by a central canister. In addition to easing certain operations such as small arms action, it also aids left-handed personnel. The connector is a simple, fully airtight swivel elbow designed to fit any standard threaded facepiece and filter.

Status
Available. In service with the Austrian Army.

Manufacturer
J Blaschke Wehrtechnik GmbH.

VERIFIED

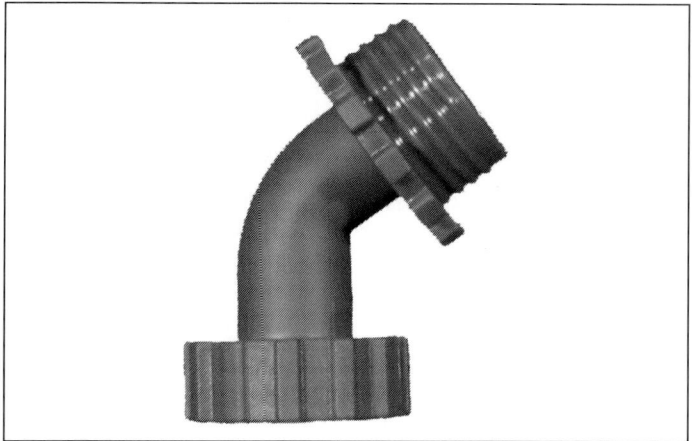

The Canister Swivel Connector SMK　　　　　　　0052133

The SMK connector in use, showing the filter swivelled to the left and right
0052132/0052134

J Blaschke NBC ventilating equipment SAB-87

Description
The J Blaschke NBC ventilating equipment SAB-87 is intended for use with heavy-duty NBC protective equipment and casualty bags and is used in conjunction with an air distribution system installed in the protective equipment involved. As a

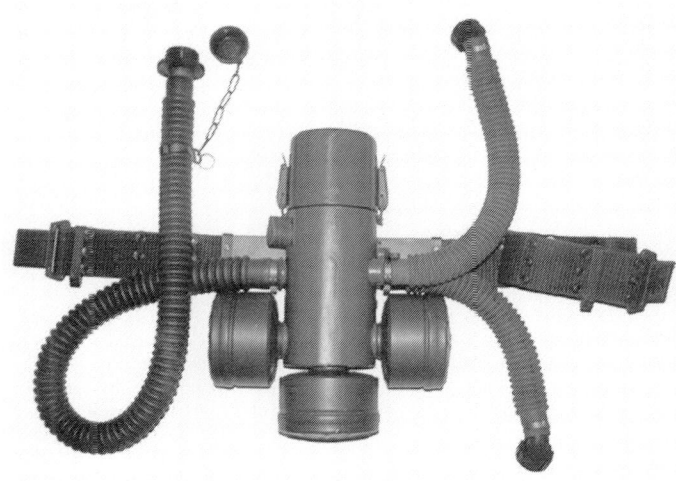

J Blaschke NBC ventilating equipment SAB-87

typical example, using this equipment, the working time for personnel wearing heavy-duty protective clothing can be extended from a maximum of 35 minutes to 3 hours.

The unit is worn attached to a field belt. The housing is light metal with three screw-filter connections; the entire unit is resistant to chemical agents and can be decontaminated. The unit provides separate fresh air feeds to a jacket and trousers (the latter can be cut off if required) with a breathing tube directed to a respirator if needed.

Two outputs are available. One is 250 litres/min for suit ventilation and breathing support. The other is for breathing support only and provides 120 litres/min. Power is provided by five 1.5 V disposable batteries or five rechargeable 1.2 V 4 Ah Ni/Cd batteries. The unit can operate for up to 20 hours on one set of batteries. The batteries are held in a sealed compartment to allow them to be changed under contaminated conditions.

Status
Available. Introduced into service with the Austrian Army.

Manufacturer
J Blaschke Wehrtechnik GmbH.

VERIFIED

CANADA

3M Canada Company filter canisters

Description
3M Canada Company is dedicated globally to the development and manufacture of standard and custom-designed filter canisters for military, police and industrial applications. The product range includes:

C2A1
The C2A1 canister is the standard NBC canister used by the US armed forces. Over 10 million have been manufactured in the Brockville facility since 1985. The design

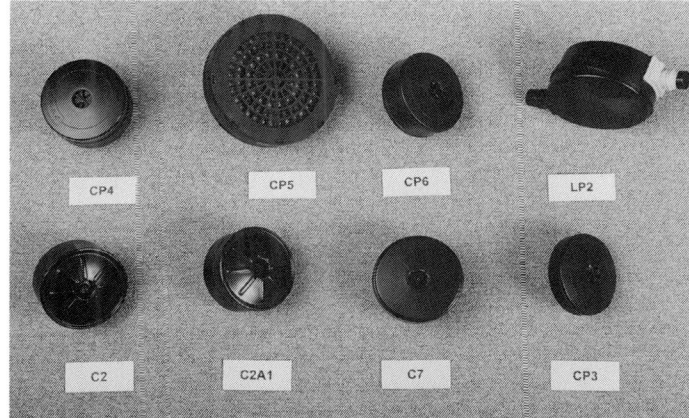

The complete range of Racal Filter Technologies (RFT) filters

Specifications						
Type	C2A1	C7	CP3	CP6/CP3N	CP4	LP2
Height	77 mm	80 mm	51 mm	70 mm	85 mm	66 mm
Diameter	106 mm	115 mm	110 mm	110 mm	110 mm	136 mm
Weight	310 g	320 g	115 g	205 g	325 g	400 g
Charcoal volume	170 cm³	250 cm³	2 layers	110 cm³	225 cm³	200 cm³
Charcoal type	ASZM-TEDA	ASC-TEDA	Cloth	ASC-TEDA	ASC-TEDA	ASC-TEDA
Thread type	NATO	NATO	NATO	NATO	NATO	None
Thread size	40 mm	40 mm	40 mm	40 mm	40 mm	
Metal/Plastic (M/P)	M	P	P	P	P	P

incorporates a high-efficiency particulate filter and a bed of impregnated charcoal. Alternative types are ASC, ASC-TEDA (ASC treated with triethylenediamine), and ASZM-TEDA which is the charcoal used in the C2A1.

C7 and C7 variants
The C7 is a new high-performance plastic canister developed for the Canadian armed forces. It has a large (113 mm diameter) plastic body and utilises a high-efficiency particulate filter element and a 250 cm³ charcoal bed. The C7 is rugged in construction with low breathing resistance and high gas life performance. Variants include the Filter 90, made for the Swedish ministry of defence and the AMF12 which is manufactured under licence from the UK (see separate entry this section).

CP3
The CP3 plastic canister is a tough, durable, lightweight, low breathing resistance canister designed to meet the needs of police, military and paramilitary users in riot control or siege situations.

CP4
This plastic canister is rugged in design and has a high-performance particulate filter and a 225 cm³ TEDA charcoal bed. The CP4 is constructed in an industrial-sized casing. The CP4 is in service with the Slovenian armed forces.

CP6/CP3N
This plastic canister is an improved alternative to the CP3 where more gas life performance is required.

LP2
This low-profile plastic canister (a replacement for the CRU80P) is a torso-mounted, lightweight unit designed specifically for aircrew use. Design features include quick release connectors and low breathing resistance at breathing rates in excess of 210 litres/min as experienced by combat aircrews under stress. The LP2 has recently been purchased by the Royal Norwegian Air Force.

FR
A new family of first responder canisters has been developed and submitted for approval to protect personnel responding to emergency situations. The canisters will provide the user with protection against both TICs and CW agents.

Status
C2A1, C7, Filter 90, AMF12, CP3/CP3N, CP4, LP2 - in production. FR canisters designed and awaiting approval.

Manufacturer
3M Canada Company.

VERIFIED

CHINA, PEOPLE'S REPUBLIC

G05 filter paper and T-02 carbon

Description
G05 filter paper is a HEPA filter medium comprising glass fibre impregnated with chemicals to deal with known CW agents. It offers flexibility, high efficiency, low air resistance, physical strength, fungal and fire resistance and it claims a minimum efficiency of 99.9999 per cent. The particle diameter is 0.3 μm and the pressure drop across the medium is less than 52 mm of water. It is suitable for use in personal or COLPRO filter canisters.

T-02 carbon
T-02 is manufactured from high quality activated carbon and offers individual and COLPRO filter manufacturers a further choice in constituents. The carbon is impregnated with copper, silver, zinc, molybdenum, vanadium, tartaric acid and TEDA. It is chrome free and ages more slowly than competing products in humid air. It has high volumetric activity.

Status
In production.

Manufacturer
Research Institute of Chemical Defence.

VERIFIED

FINLAND

Pro2000 filter canisters

Description
Scott Health and Safety specialises in the design and manufacture of high-quality powered and negative pressure filter respirators and filter canisters.

Scott's military canister range includes products that will meet most operational requirements. This includes the current UK in-service canister as issued with the Avon S10 respirator, which was originally designed to meet UK and NATO requirements. The technology incorporated has been developed to meet the needs of other armed forces throughout the world.

The canisters use high-strength thermoplastic for the canister bodies and body lid. The particulate filters (high-efficiency, water-repellent, glass fibre paper) are made integral with the canister lids which are then sealed into the canister bodies during assembly. The resulting canisters have low weight and compact overall dimensions and offer a high level of performance compared with other NBC canisters. The dimensions are such that in a side-mounted configuration the canisters present a low profile and create minimal interference with general vision and weapon sights.

Protector filter canisters fully meet the specifications of the NATO triptych AC/225 (Panel VII) D103. Performance against normally specified agents (AC, CK, CG and PS) also meets these NATO requirements. The particulate filter efficiency against solid aerosols (paraffin oil at 95 litres/min) exceeds 99.999 per cent. The canister withstands a range of climatic conditions and levels of humidity without deterioration when sealed. Canisters can either be supplied sealed in aluminised trilaminate foil bags or in thermoplastic containers to provide storage lives in excess of 10 years.

Canisters for specific applications have been delivered to several armed forces. The Pro2000 range includes a filter canister specifically developed to counter all current riot control agents including CN, CS and CR.

Status
In production.

Manufacturer
Scott Health and Safety Oy.

UPDATED

Scott Health & Safety NBC filter canister M95

Description
The Scott Health & Safety NBC filter canister M95 protects against all chemical warfare agents, radioactive particles, biological agents and many industrial gases.

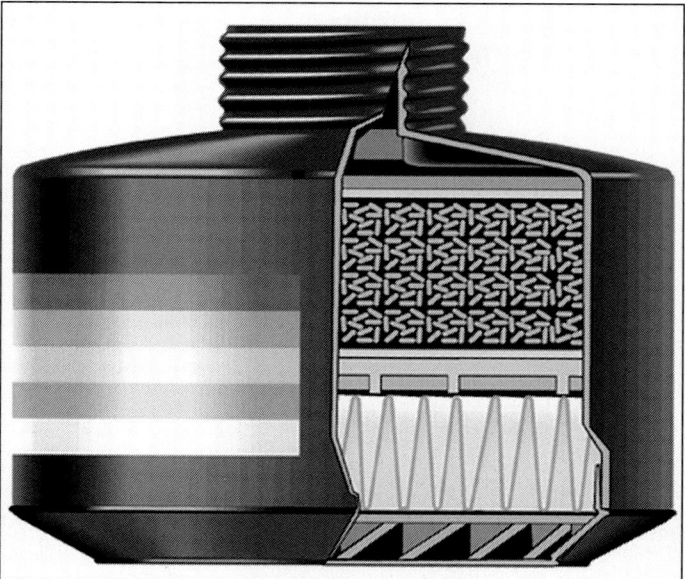

Kemira Safety NBC filter canister M95   0011435

Developed for use with the M95 NBC mask, it has a low breathing resistance, offering resistance less than 120 Pa at 30 litres/min and less than 420 Pa at 95 litres/min.

The M95 filter canister has a polyamide body with high-impact resistance down to −40°C. It weighs 250 g, is 90 mm high and the diameter is 109 mm. The screw thread is to 40 mm EN 148 standard.

Status
In production.

Manufacturer
Scott Health & Safety Oy.

UPDATED

FRANCE

Giat Industries NBC filter canisters

Description
Giat Industries manufactures various types of NBC filter canister, as follows:

Modèle FCA 1
This is an NBC filter canister capable of absorbing all known NBC agents and is provided with particulate and charcoal bed filters. It is capable of withstanding four typical chemical agent attacks. Weight is 240 g and shelf-life is 15 years. It is no longer in full production.

Modèle CF63/67
The modèle CF63/67 was the standard NBC filter canister of the French Army and uses particulate and charcoal filters. It is stated to be capable of withstanding six typical chemical agent attacks. Shelf-life is 15 years. It is no longer in production but is held in reserve.

NBC filter canister Type A1B1E1K1P3
The NBC filter canister Type A1B1E1K1P3 is intended for use against organic gases and vapours with a boiling point of up to +65°C, including sulphur dioxide, ammonia and amine derivatives. It is fitted with a standard connector complying with the French standard NF EN 148-1 and is intended for use on a full mask.

The canister is fitted with a large-capacity particle filter and uses an aluminium case with a plastic lid. The particle and activated carbon filter were selected to

Typical Giat Industries NBC filter canisters

Giat NBC filter canister Type A2B2E2K2P3 (Giat Industries)   **2002**/0121888

provide breathing assistance and minimum weight in accordance with the requirements of the French standard NF EN 141. Respiratory resistance is 95 Pa at 30 litres/min and 290 Pa at 95 litres/min. Weight of the filter canister is 370 g.

Pre-series models are undergoing approval.

NBC filter canister Type A2B2E2K2P3
Similar to A1B1E1K1P3. Respiratory resistance is 140 Pa at 30 litres/min and 445 Pa at 95 litres/min. Weight of the filter canister is 370 g.

Pre-series models are undergoing approval.

NBC filter canister Type CFF
The NBC filter canister Type CFF is intended for use against military chemical agents. It is fitted with a standard connector complying with the French standard NF EN 148-1 and is intended for use on a full mask.

The canister is fitted with a large capacity particle filter and uses an aluminium case with a plastic lid. The particle and activated carbon filter were selected to provide breathing assistance and minimum weight in accordance with the requirements of the French standard NF EN 141. Respiratory resistance at 30 litres/min is less than 110 Pa and less than 350 Pa at 95 litres/min. Weight of the filter canister is 240 g.

The Type CFF filter canister is in production and in service with the French armed forces.

NBC filter canister Type CFF4
The NBC filter canister Type CFF4 is intended for use against military chemical agents. It is fitted with a standard connector complying with the French standard NF EN 148-1 and is intended for use on a full mask.

The canister is fitted with a large capacity HEPA particle filter and uses a lacquered aluminium case with a plastic lid. The particle and activated carbon filter were selected to provide breathing assistance and minimum weight in accordance with the requirements of French standard NF EN 141. Respiratory resistance at 30 litres/min is less than 110 Pa and less than 350 Pa at 95 litres/min. Weight of the filter canister is less than 240 g.

Advanced research NBC filter canisters
Under the terms of the 'Eurofinder' Project, Giat Industries and TNO-PML, together with French and Netherlands industry, have developed an advanced range of filter canisters offering enhanced protection. These include the 'NBC three weeks' canister, the Advanced BIO/P3 canister and the Advanced A2B2E2K2P3/NBC/BIO canister. These designs overcome the additional pressure drop caused by incorporating enhanced protection (more layers of filtration) by using a miniaturised ultra-lightweight micro-ventilator (fan), regulated remotely according to the internal pressure inside the respirator facepiece.

Status
FCA1 and CF 63/67: in stock. A1B1E1K1P3 and A2B2E2K2P3: in production. CFF and CFF4 in production. CFF4 in service with the French Army. The advanced canisters remain subject of a research programme. Trials continue (November 2001).

Manufacturer
Giat Industries.

UPDATED

....

SP Défense filter canisters

Description
The SP Défense ABEK/NBC range of filter canisters is designed to prevent transmission of both TICs and NBC agents. The filter canisters comprise a paper anti-aerosol filter and an activated charcoal filter housed in an aluminium case. Fully compliant with the European Union EN 141 standard for industrial filter performance and the latest NATO defence standard for military NBC challenge, they are aimed at armed forces, internal security agencies, the emergency services and police forces. An efficiency of 99.997 per cent is claimed against aerosols.

Status
In production. In service with the French armed forces and police.

Manufacturer
SP Défense.

VERIFIED

Specifications
Model	1786354	1785622	1784003
Dimensions (mm)	107 × 72	107 × 84	107 × 94
Weight (g)	230	285	405
Thread	Rd 40 × 11/7 in	Rd 40 × 11/7 in	Rd 40 × 11/7 in
Initial pressure drop at 30l/min (Pa)	120	170	200
Protection	NBC	$A_2B_2E_2K_1P_3$/NBC	$A_2B_2E_2K_1P_3$/NBC

GERMANY

Kärcher Filtersafe 1001 NBC ventilation kit for NBC protective suits

Description
The stresses involved when wearing an impermeable NBC protection suit can be so heavy that the clothing can only be worn for short periods before the physical and physiological condition of the wearer is impaired. By using the Filtersafe 1001 ventilation system the operational wearing time can be considerably lengthened.

The Filtersafe 1001 filters incoming contaminated air and leads it to the middle of two flexible respirator hoses by means of an air diffusing system, without the respiratory resistance of a normal respirator. Additional filtered air is blown into the interior of the suit through an inlet connecting valve. The suit ventilation improves the microclimate and produces a permanent overpressure inside the suit. This effectively prevents the ingress of contaminated air due to leaks caused by damage to the suit material.

Filtered air is partly conducted to the respirator mask via a respirator connection and partly blown into the suit interior via an inlet connection valve located in the back of the Kärcher Safeguard 6004 suit (see entry under Protection (individual) - Body protection for details). Power for the system is provided by a Ni/Cd accumulator pack.

The Filtersafe 1001 therefore provides a positive pressure to assist the wearer's breathing and improves the internal microclimate inside an impermeable NBC protection suit such as the Safeguard 6004. Tests have indicated that wearers of suits equipped with the Filtersafe 1001 can increase their operational wearing times by up to 500 per cent.

The Filtersafe 1001 is modular and consists of a ventilation unit with a blower unit and housing for the Ni/Cd accumulator pack; the air supply consisting of a Y-connector for screw-on filter canisters and a Y-connector for the respirator hoses; a battery charger with a power lead and charging lead; and carrying devices including a belt and carrying pocket.

Status
In production. Under test by several countries.

Manufacturer
Alfred Kärcher GmbH & Co.

VERIFIED

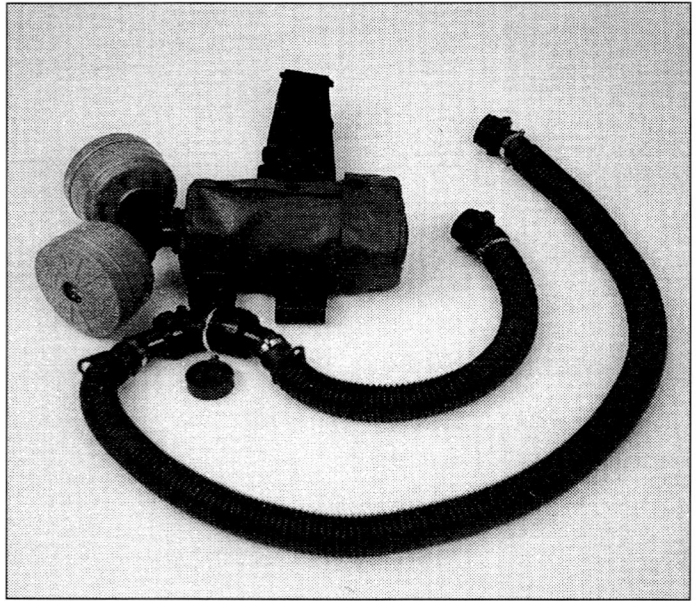

Kärcher Filtersafe 1001 NBC ventilation kit for NBC protective suits

GREECE

SUPERCOMBI NBC filter canisters

Description
The SUPERCOMBI NBC filter canisters were designed to provide protection against all known chemical warfare agents and toxic industrial substances. They are constructed using plastics to provide durability and ensure used canisters can be disposed of by incineration.

The filter system used by the SUPERCOMBI filters has two main parts. The first is a folded glass fibre cellulose paper particulate filter followed by a compressed bed of impregnated activated charcoal to absorb gases. The efficiency of the filter against 0.3 μm diameter paraffin oil is greater than 99.997 per cent.

Specification Size	Thread	Diameter (mm)	Height (mm)	Weight (gm)	Resistance to inhalation (mm H$_2$0)
I	Rd 40 × 1/7 in 114 (DIN 3182)		106	340	20
II			92	280	17
III			86	270	16

Status
In production.

Manufacturer
BIANA SA, Personal Protective Equipments.

UPDATED

HUNGARY

ABEK2P3 NBC filter canister

Description
The ABEK2P3 NBC filter canister is designed to give full individual protection to all known CW agents. The aluminium casing houses a HEPA filter and activated carbon bed to remove threat agents. The threaded connection is to standard EN 148-1.

Status
In production.

Manufacturer
Respirátor Rt.

VERIFIED

ISRAEL

NBC filter canister Type 80

Description
The NBC filter canister Type 80 was developed to provide protection against all known NBC agents in the form of vapours or aerosols. It is supplied as an integral part of all SHALON personal respiratory protection systems.

The Type 80 is a robust durable canister intended for military field use. It was designed for comfort and extended use at a minimum physiological load, having low breathing resistance and low weight. The design assures optimum performance complying with Israel Defence Forces (IDF) specifications, as well as the performance requirements of the US Army C2 canister. The canister is supplied with a standard NATO 40 mm thread. Type 80 canisters are supplied sealed with leakproof plastic caps, assuring a shelf-life exceeding 15 years.

Status
In production.

Manufacturer
SHALON-Chemical Industries Limited.

VERIFIED

NBC filter canister Type 80

SHALON SB 35/45 miniblower

Description
The SHALON SB 35/45 miniblower was designed and developed by SHALON Chemical Industries to supply air and maintain positive pressure within the SHALON protective systems for infants and children (see under PROTECTION (individual) — Masks (General issue) for details). The SB 35/45 may also be used with NBC protective masks to provide improved comfort and protection, especially for persons with hard-to-seal abnormal face contours.

The SB 35/45 is a miniature radial blower capable of drawing in ambient air, passing it through a filter canister and delivering the purified air either directly, or via a flexible rubber hose, into a personal protective system.

The SB 35/45 miniblower weighs 55 g, is 60 mm in diameter and 30 mm deep. Operating at a speed of 12,300 rpm it can deliver 45 litres of air per minute against a pressure drop of 35 mm of water. Current consumption from a dry cell battery is 0.3 to 0.35 A.

Status
In production.

Manufacturer
SHALON-Chemical Industries Limited.

VERIFIED

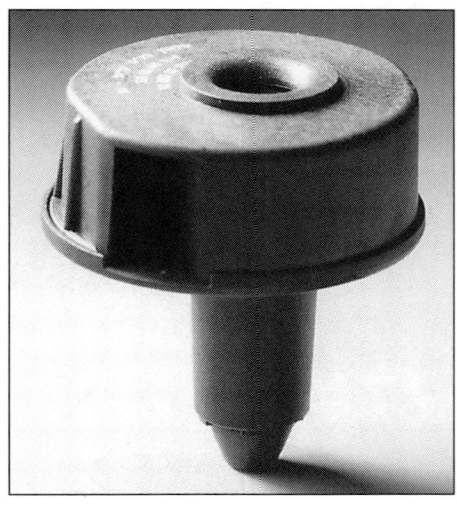

SHALON SB 35/45 miniblower

S. T. Safety Technologies NBC filter canister ST-80M

Description
The S. T. Safety Technologies NBC filter canister ST-80M is produced for its CHIPS-M1 and CCPS-M2 protective systems for children (see under PROTECTION (individual) — Body protection for details) but the ST-80M is also in service with the Israel Defence Forces (IDF) and is distributed throughout the Israeli populace.

The lightweight ST-80M has a standard NATO 40 mm thread and exceeds the most stringent IDF and US Army C2 canister performance requirements while offering extended life under extremely hostile conditions. It has a glass fibre particulate and aerosol primary filter with 99.98 per cent efficiency (DOP test) for 0.3 µm particles, and a special bedded activated and impregnated charcoal secondary gas filter. Gas life exceeds US military and IDF requirements for chloropicrin, cyanogen chloride, phosgene, hydrogen cyanide and DMMP.

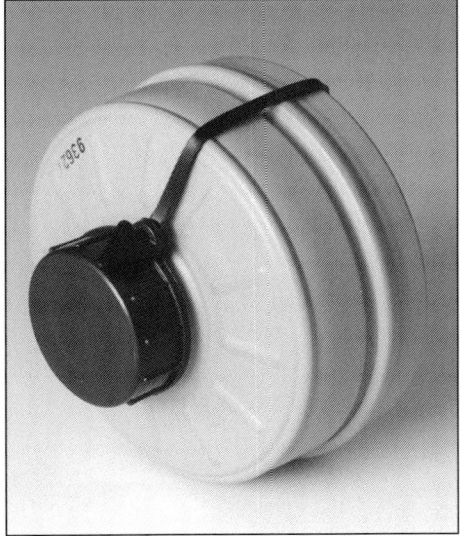

S. T. Safety Technologies NBC filter canister ST-80M

Status
In production. In service with the IDF and distributed throughout the Israeli populace.

Manufacturer
S. T. Safety Technologies Limited.

VERIFIED

ITALY

M90 NBC filter canister

Description
The M90 filter canister was developed to be used in conjunction with the M90 NBC protective mask (qv) but may be used with other masks. Its efficiency is claimed to exceed all NATO requirements (efficiency against 0.3 µm aerosols is above 99.997 per cent), having a protection factor above 10,000 and was designed to counter threat levels three to four times greater than those covered by NATO specifications. The weight of the M90 filter canister, without caps, is approximately 300 g and it contains approximately 290 cm³ of treated activated charcoal. The connector is a standard NATO 40 mm thread conforming to STANAG 4155.

Status
In service.

Manufacturer
Aero Sekur SpA.

UPDATED

NETHERLANDS

Arbin Filter NBC canisters

Description
There are three main types of NBC filter canister in the Arbin Filter range. They are the NBC20, NBC33 and NBC37. All have a diameter of 110 mm and use a NATO standard 40 mm thread. The canisters are aluminium, epoxy-coated internally and with the outside surfaces lacquered and painted army green. The carbon used has a maximum moisture content of 3 per cent and minimum efficiency is 99.997 per cent measured with DOP or paraffin oil at a concentration of approximately 100 mg/m³ and at an air flow rate of 95 litres/min. Median particle diameter is 0.3 µm.

The NBC20 pressure drop at an air flow rate of 30 litres/min is less than 160 Pa; weight is less than 255 g. The corresponding figures for the NBC33 are less than 190 Pa and 310 g, while for the NBC37 they are less than 200 Pa and 320 g.

Arbin Filter also produces a canister for protection against CS. They can supply NBC and other canisters according to customer requirements.

Status
All the above are in production. NBC filters have been supplied to Austria, Belgium, Germany, Norway, Sweden and Turkey, among others.

Manufacturer
Arbin.

VERIFIED

RUSSIAN FEDERATION AND ASSOCIATED STATES (CIS)

ShM filter canisters

Description
Operational filter canisters used with the Model ShM series of protective masks include the MO-2 or MO-4U canisters which can be fitted either directly on to the mask or on to a length of fabric-reinforced corrugated rubber hose. A GP-2 carbon monoxide filter canister can be fitted between the mask and the standard canister and training or carbon monoxide canisters can be fitted in place of the standard canister if required. The carbon monoxide GP-2 filter canister may be either rectangular or cylindrical, and weighs about 635 g.

These canisters have been produced in Egypt for the BSS and CM3 protective masks (qv).

Specifications

Model	MO-2	MO-4U
Length	195 mm	205 mm
Width	135 mm	135 mm
Depth	70 mm	68 mm
Weight (approx)	875 g	850 g

Status
Manufactured by Russian state factories prior to privatisation and under licence in Egypt by the National Organisation for Military Production (assumed) - see entry on the Egyptian CSS CM masks. Current Russian manufacturer unknown. Remains in service with Egypt, RFAS and other former Warsaw Pact forces.

Manufacturer
Unknown.

VERIFIED

SOUTH AFRICA

HAZMAT CB and riot control filter canister

Description
HAZMAT Protective Systems (Pty) Limited produces a CB (Chemical and Biological) and riot control filter canister with an international DIN screw thread attachment to fit all types of full face mask. The canister contains a double filter system, the primary filter being an HEPA media filter and the secondary consisting of impregnated activated carbon. The carbon characteristics can be altered according to the threat perception.

The filter canisters are packed in durable cardboard boxes, each holding 50 canisters and with a five-year shelf-life.

Status
Available.

Manufacturer
HAZMAT Protective Systems (Pty) Limited.

VERIFIED

HAZMAT CB and riot control filter canisters

SWEDEN

Sundström Safety 381 NBC filter canister

Description
The Sundström Safety 381 NBC filter canister uses a polyamide and aluminium construction, resistant to DS-2 decontaminating agent, mustard agent and oil-based products. Weighing approximately 280 g, the NATO standard thread filter canister is issued with protective covers on both sides. The filter contains approximately 220 cm³ of Whetlerite-type activated carbon impregnated with TEDA and has a particulate filter containing hydrophobic glass fibre paper.

Resistance time against GB is 160 minutes and 18 minutes against AC. Its performance complies with the NATO Triptych on filtration, offering protection against HCN, CNCl, PS and GB.

Sundström Safety 381 NBC filter canister (Sundström Safety AB)

The canister has a diameter of 115 mm at its widest point and is 85 mm high including the thread. The connector has a standard NATO 40 mm thread. Packed in a sealed alumina bag, the filter offers a storage life of over 20 years.

Status
In production. In use by the Norwegian armed forces.

Manufacturer
Sundström Safety AB.

UPDATED

SWITZERLAND

MICRONEL C420 Compact Air Supply Unit (CASU)

Description
The MICRONEL C420 compact air supply unit (CASU) is a lightweight high-efficiency compact air supply unit which, when connected to two standard NBC screw mount filter canisters, provides a constant filtered air flow between 90 and 120 litres/min. Using a standard lithium-sulphur dioxide battery Type BA 5800, the continuous running time is up to 20 hours, depending on the air flow demand. As an option, the C420 can also be supplied with a rechargeable battery (nickel metal hydride), giving 4 hours operational time. A connection for an external 6 to 28 V DC power source is available.

The unit is equally effective for vehicle or helicopter crew members, ground personnel or special forces. It can also be used with casualty bags. The blower unit

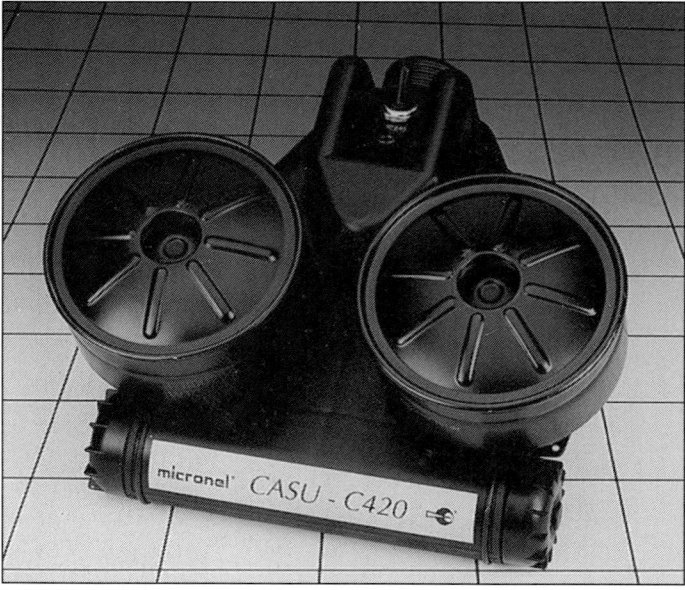

MICRONEL C420 compact air supply unit

provides a high safety factor due to the combination of high breathing comfort and the overpressure in the system.

The unit may be carried on a waist belt or shoulder strap. Weight, including the battery and two canisters, is approximately 1.4 kg.

Status
In widespread use by the UK, US and other armed forces.

Manufacturers
MICRONEL AG.

VERIFIED

MICRONEL C440 Compact Air Supply Unit (CASU)

Description
The MICRONEL C440 Compact Air Supply Unit (CASU) was designed for use with NBC casualty bags to allow for easy breathing of the occupant. The unit has threaded adapters for up to four NBC filter canisters and is powered by five D-size batteries (LR20/UM1) or via a 6 to 28 V power source. An internal automatic electronic adapter makes it possible to use all existing batteries such as alkaline, lithium and Ni/Cd. Operating time is 5 to 6 hours for Ni/Cd, 6 to 8 hours for alkaline and 15 to 20 hours for lithium batteries.

With fresh filters the flow rate is approximately 200 litres/min. The unit is driven by a high-efficiency DC blower; the on/off switch can be operated by gloved hands.

The filter canister thread is RD 40 × 1/7 (DIN 3183-T2), conforming to the NATO standard, and the canisters can have a maximum diameter of 120 mm. The unit is made using plastics that can be decontaminated by rinsing with water and there are provisions for fixing the unit to a stretcher.

Optional components that can be used with the unit include a battery level indicator, an acoustic warning system to indicate a battery low condition and a three-step switch to control the air flow (200 litres/min; 150 litres/min; 100 litres/min - all figures approximate). A connecting cable is available on special request.

The C440 is 480 mm long at its maximum point. The approximate height is 120 mm and width 100 mm. Weight fully equipped with batteries and four filter canisters is approximately 3.2 kg.

Status
Ready for production.

Manufacturer
MICRONEL AG.

VERIFIED

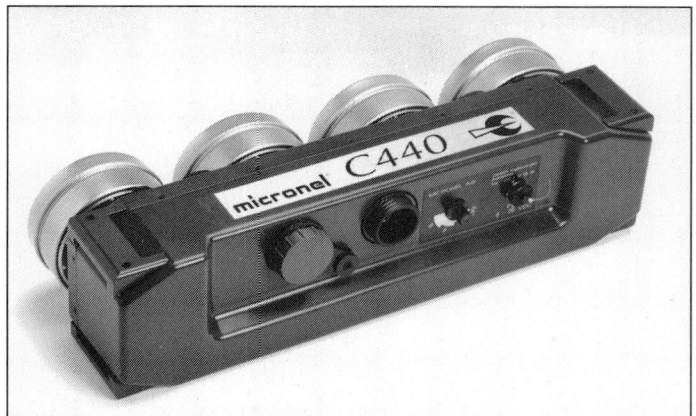

MICRONEL C440 compact air supply unit

MICRONEL C411 Compact Air Supply Unit (CASU)

Description
The MICRONEL C411 Compact Air Supply Unit (CASU) is a small unit consisting of a high-performance radial blower with an integral battery set. NBC filters are easily attached via a standard screw thread. The unit generates an overpressure within a respirator or hood to provide an improvement in the mask safety factor by eliminating leakage and makes breathing by the user easier, even when performing heavy work tasks under extreme climatic conditions. Removing the filter canister from the mask lightens the load on the wearer and makes the operation of weapons and equipment easier.

The C411 CASU can be used in combination with virtually all current respirators and hoods. The filter canister is removed from the mask and attached to the CASU (the clamping mechanism can be adapted to suit the filter involved). The mask or hood and filter are then connected by a hose and the unit is ready for operation. In order to obtain optimum performance, the inlet diaphragm should open at a pressure of less than 0.4 mbar at an air delivery of 55 litres/min.

MICRONEL C411 compact air supply unit

MICRONEL C411 compact air supply unit in use

The standard version can be equipped with six alkaline or lithium 1.5 V batteries (AA/LR6/UM3).

The C411 CASU is 210 mm long and 70 mm in diameter without the filter canister. Weight with a filter canister and batteries is approximately 650 g; without the filter canister the weight is approximately 370 g.

Status
In production.

Manufacturer
MICRONEL AG.

UPDATED

Specifications
Nominal voltage: 3 V DC
Voltage operational range: 2-3 V DC
Pressure/airflow rate at filter outlet: 360 Pa / 70 litres min⁻¹ (C2 filter)
Expected working life of blower: 500 h
Weight (filter + batteries): approx 700 g
Filters: Most canisters with a central intake orifice. An adapter is available for orifices wider than 34 mm in diameter

UNITED KINGDOM

Avon NBC filter canister AMF12

Description

The NBC filter canister AMF12 is designed to meet the latest NATO requirements while the materials used and the manufacturing techniques employed provide a product with good shock and impact resistance.

The particulate filter is a high-efficiency glass fibre/vinyon copolymer co-pleated with polypropylene netting. Its filtration efficiency against particulates and gases exceeds NATO requirements. The charcoal employed is an extruded, pelletised charcoal impregnated with metallic salts (Cu and Cr) and triethylenediamine (TEDA). The canister, which is coloured black, is made from glass-filled Noryl, combining easy decontamination with a robust construction.

The AMF12 is packed sealed in a Tyvek trilaminate bag to ML-B-131H Type 1 Class 1.

Status

In production.

Manufacturer

Avon Technical Products, Protection Group.

UPDATED

Specifications

Weight: <270 g
Diameter: 115 mm
Height: 61 mm
Breathing resistance:
   (at 30 litres/min) 95 Pa, (at 50 litres/min) 170 Pa, (at 80 litres/min) 300 Pa
Thread: STANAG 4144 NATO

Avon NBC filter canister AMF12 (Avon Technical Products)

YUGOSLAVIA, FEDERAL REPUBLIC OF

NBC filter canister M-2

Description

The NBC filter canister M-2 was developed for use with the Serbian M-2 series of protective masks but can also be used with other similar masks. The filter canister weighs 310 g and has an aerodynamic filter resistance at an air flow of 30 litres/min of 180 pa. Filter efficiency against particles with diameters from 0.3 to 0.5 μm is 99.995 per cent.

The manufacturer also produces a wide range of special purpose filter canisters for both military and industrial purposes, including custom-built units.

Status

In service with the former Yugoslav armed forces.

Manufacturer

Trayal Corporation.

Marketed by

Yugoimport-SDPR.

VERIFIED

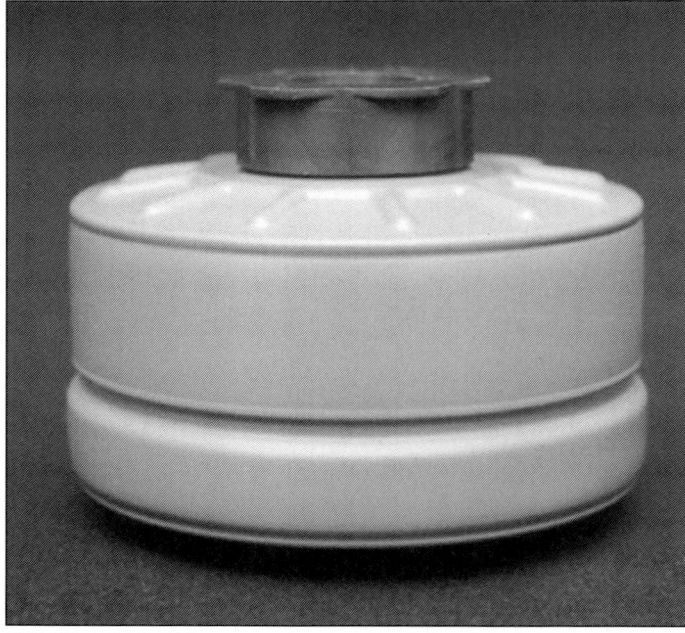

NBC filter canister M-2   0011452

PROTECTION (individual) - Body protection

AUSTRIA

EUROLITE NBC cover poncho

Description

The EUROLITE NBC cover poncho provides protection for the individual and his personal equipment. It has an integrated hood which is closed with a string fastener. It is also produced with an elasticated hood. It is delivered in sealed standard packing with a quick-opening system, partially vacuum sealed. For details on the material, see separate entry in this section.

Status

In service in Austria, Netherlands, Sweden, Switzerland and USA.

Manufacturer

Goetzloff GmbH.

Marketing agency

Goetzloff (USA).

NEW ENTRY

EUROLITE NBC cover poncho (Goetzloff)
***2002**/0137489*

EUROLITE NBC protective suit

Description

The EUROLITE is a two-piece overgarment. The jacket has an integral hood and gloves with sleeve grips. The trousers include integral double-soled overshoes. Available in four sizes, it is designed to be donned over existing clothing, including footwear. It is constructed of Rolamit NBC-Barrierfilm material developed for its NBC protective qualities and its resistance to tearing, puncture and snagging (see

EUROLITE NBC protective two-piece suit (Goetzloff GmbH)
0019554

separate entry within this section). The suits are washable and lightweight with a long shelf-life. They are vacuum-packed for low-volume storage and, once contaminated, can be disposed of by incineration with no toxic residue.

The EUROLITE suit was worn by Austrian medical personnel during Operation Desert Storm.

Status

In service in Austria, Netherlands, Sweden, Switzerland and the USA.

Manufacturer

Goetzloff GmbH.

Marketing agency

Goetzloff (USA).

UPDATED

Model CAVE-95 series NBC protection suits

Description

The Model CAVE-95 NBC protection suit series is a new development (1999) available in two variants. The CAVE-95 suit uses a standard rubber-coated tarpaulin material which remains resistant against the ingress of chemical agents for at least 8 hours. The CAVE-95-HR variant is made with the proprietary 'Schichtbarriere 2000' high-resistant tarpaulin, which raises the level of resistance against chemical agents to over 2,000 hours. The design comprises a one-piece coverall with boots and gloves attached. The integral hood has a wide panoramic face-screen to accommodate most types of facepiece and the back of the suit is contoured to enclose a self-contained breathing apparatus. The facepiece and the interior of the suit can be supplied with filtered air, either from an SCBA or from a fan-augmented filter pack, similar to the SAB-87 and produced by the same manufacturer. See under PROTECTION (individual) - Masks (general issue) for details of the blower assembly.

The Model CAVE-95-HR material has a weight of 520 g/m^2.

Status

Available.

Manufacturer

J Blaschke Wehrtechnik GmbH.

VERIFIED

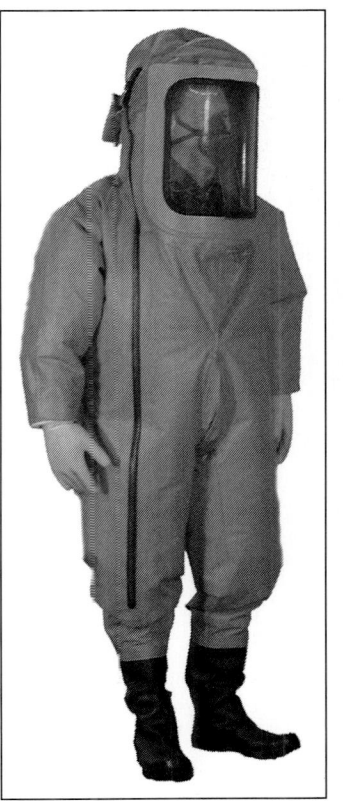 

Model CAVE-95 series NBC protection suit: front view   0055626

Model CAVE-95 series NBC protection suit: side view   0055627

Model CFR93 full protection suit

Description
The full protection suit Model CFR93 is a universal full protection suit manufactured from a novel compound of fabric, foil and rubber materials with an area weight of approximately 500 g/m².

Model CFR93 full protection suit

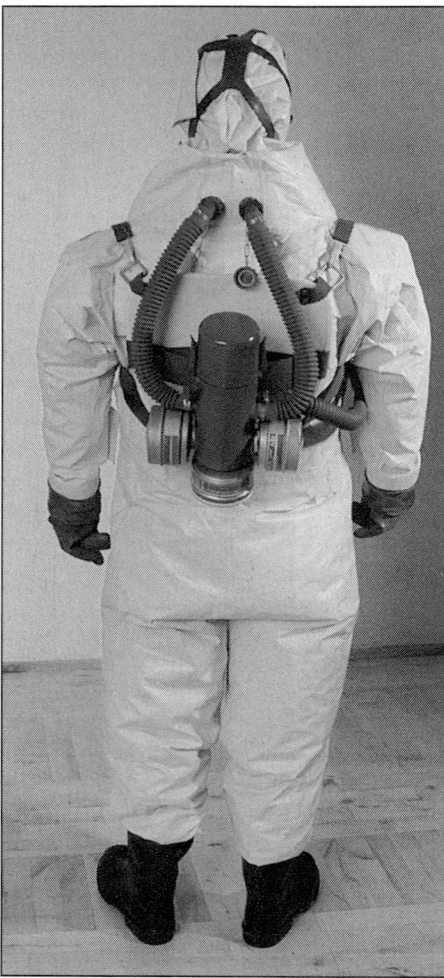

Model CFR93 full protection suit from rear showing fresh air supply unit

The suit is easy to put on and remove as the opening facility is divided into three parts, with one gas-tight zip running diagonally from the shoulders at the outer connection areas of the hood to the centre of the chest from the right and left, thus allowing the helmet to be folded downwards onto the back. A further gas-tight zip is fitted in the centre of the suit, from the centre of the seat to the centre of the chest area, ending in a central bolting device meeting the shoulder zips.

To ease heat stress during periods of heavy work a fresh air distribution system is provided, either permanently fitted to the inside of the suit or carried on a self-contained pack frame inside the suit. This system is supplied by a fresh air fan with an output of up to 240 litres/min via two inlet valves at the rear of the helmet connection. Uniform air distribution is regulated by up to 11 outlet valves, each protected against blockage. Air is supplied via a third hose from the fresh air fan on the back via the protective mask and flanged into the front of the helmet for supplies via an oxygen cylinder. This arrangement ensures that exhaled warm, moist air is passed to outside the suit.

With this ventilation system 66 per cent of the air capacity generated is passed into the suit by two air connections and the remaining 34 per cent is conducted to the mask area for respiration. If three filters are used with this type of unit the output can be approximately 240 litres/min, of which approximately 80 litres/min is for breathing and approximately 160 litres/min is for body cooling, both for a period of at least 8 hours (dependent on the energy supply).

It is suggested that the Model CFR93 could be worn with a suitable oxygen cylinder unit during NBC agent reconnaissance and detection operations. Following analysis of a situation, the oxygen cylinder unit can be exchanged for a battery-powered fresh air supply unit. This is mounted on a pack frame and provided with the necessary filters, thereby reducing weight.

Various modifications can be made to the Mode CFR93 to meet specific customer requirements.

Status
Available.

Manufacturer
J Blaschke Wehrtechnik GmbH.

VERIFIED

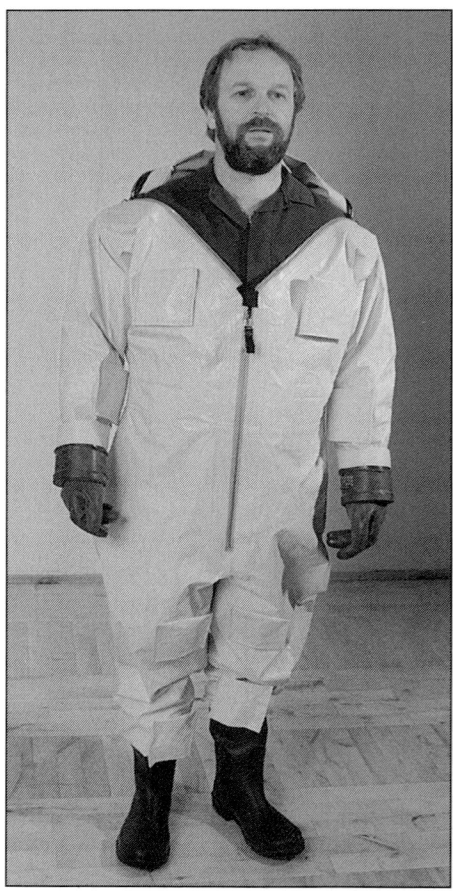

Model CFR93 full protection suit with hood removed

NBC protection suit Model ABC-92/1T and -HR

Description
The NBC protection suits Model ABC-92/1T and ABC-92/1T-HR are military one-piece suits provided with a fresh air blower for internal ventilation and reduction of heat stress. Model ABC-92/1T uses a standard rubber-coated material which remains resistant against the ingress of chemical agents for at least 6 hours. The Model ABC-92/1T-HR employs 'Schichtbarriere 2000' high-resistance material which raises the level of resistance against CW agents to over 2,000 hours. This

NBC protection suit
Model ABC-92/1T
0011430

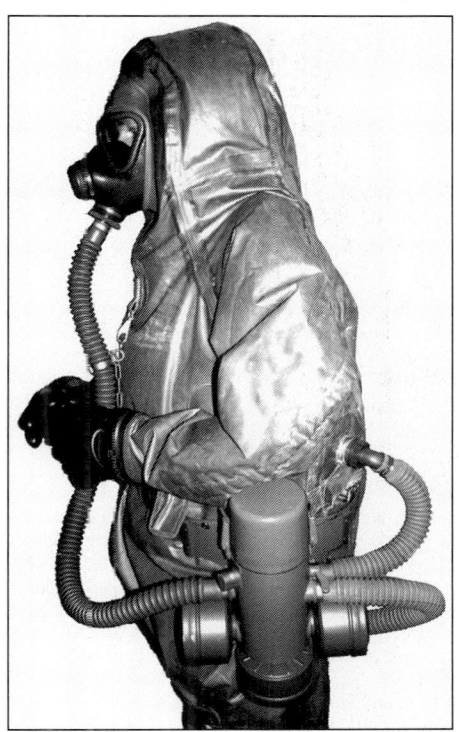

NBC protection suit
Model ABC-90
0011428

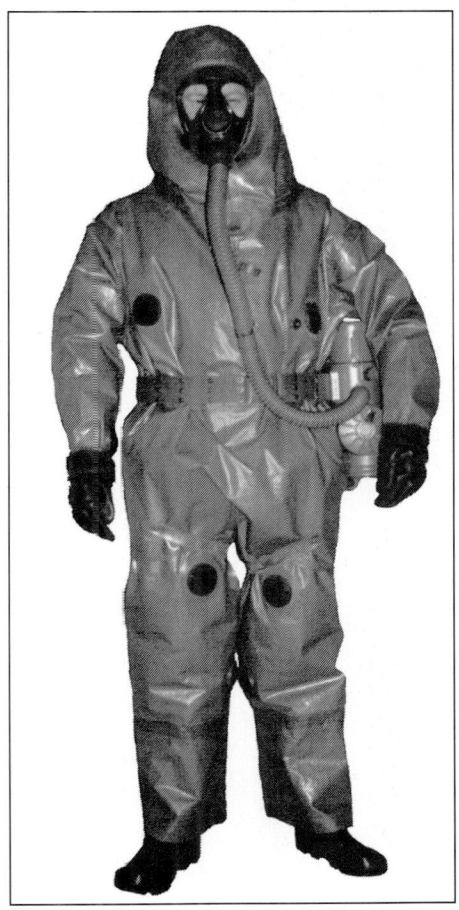

NBC protection suit
Model ABC-90
0055628

material has a unit weight of 520 g/m². Both suits are closed by a one-piece zip from the right knee to the left shoulder. Secure seals are provided between the suit, the face mask and the NBC protective boots.

The air distribution system is supplied by a fresh air blower with an output up to 250 litres/min; the blower can be carried on a belt. Uniform air distribution is regulated by up to 11 outlet valves protected against fouling.

Status
Available.

Manufacturer
J Blaschke Wehrtechnik GmbH.   *VERIFIED*

....

NBC protection suit Models ABC-90 and ABC-90-HR

Description
The NBC protection suit Models ABC-90 and ABC-90-HR are basically identical but vary in the type of material employed and the level of protection afforded. Both are military two-piece suits formed by a jacket with an integral hood and trousers; the joint between the two garments is completely gas proof. Both are provided with a fresh air blower for internal ventilation and reduction of heat stress.

The Model ABC-90 suit uses a standard rubber-coated tarpaulin material which remains resistant against the ingress of chemical agents for at least 6 hours. The Model ABC-90-HR employs 'Schichtbarriere 2000' high-resistant tarpaulin which raises the level of resistance against chemical agents to over 2,000 hours. This material has a weight of 520 g/m².

The air distribution system is supplied by a fresh air blower with an output up to 280 litres/min; the blower can be carried on a belt. Uniform air distribution is regulated by up to 11 outlet valves protected against fouling.

Status
Available.

Manufacturer
J Blaschke Wehrtechnik GmbH.

VERIFIED

BELGIUM

Seyntex NBC protective clothing

Description

Seyntex specialises in protective clothing and began developing NBC protective clothing and equipment during the mid-1970s. They produce a wide range of, the main items of which are outlined below.

NBC suits

Seyntex produces both two-piece NBC protective suits and one-piece NBC overalls. Their 'first-generation' suit was introduced into Belgian Army service from 1975 onwards. This suit is built up from two fabrics, a liquid repellent outer layer of core-spun polyester cotton and an absorbent inner layer of carbon-impregnated polyurethane foam. Total weight of the material is 400 to 450 g/m². The suit provides protection against NBC attacks for 24 hours or more and the high degree of protection, combined with the strength of the materials used, enables the suit to be used as a combat suit. A degree of heat flash protection is provided by this suit.

The suit consists of a smock and trousers, the smock having an integral hood and one breast pocket with a Velcro closure. The trousers have two patch pockets.

A newer 'second-generation' suit has been introduced with improved flameproof qualities. The outer layer is a polyester-cotton core-spun cloth or a 100 per cent cotton cloth which has been flameproofed. The inner layer consists of a flame retardant carbon-impregnated polyurethane foam laminated between a white reflectant flameproof fabric and a polyamide knit fabric. A 'Tropical' version of this suit has been developed which can be worn in hot climates for up to 8 hours without problems.

Both types of suit are issued in vacuum-packed and heat-sealed polythene bags.

Aircrew under-coverall

Specially designed for use by aircrew this under-coverall is worn over cotton underwear or a training suit and under a fireproof flight overall or NBC suit. The garment is made of a thin compressed polyurethane foam, coated with active charcoal. The foam is bonded between two layers of lightweight and flameproof jersey material. Total weight of the material is 120 g/m².

NBC gloves and boots

Seyntex NBC gloves are made from poly-isobutylene rubber 0.35 to 0.4 mm thick. NBC overboots are made from butyl rubber coated on a polyamid fabric 0.7 mm thick; three sizes are available. Both gloves and overboots provide protection against chemical agents for a minimum of 24 hours and both can be decontaminated.

Anti-gas fabric

Seyntex produces an anti-gas fabric which is coated with a liquid-repelling polyurethane laminate. The fabric can be used to manufacture light tents and

Seyntex NBC two-piece protective suit with single front smock pocket. Integral smock and hood (Seyntex)
***2002**/0137738*

Seyntex NBC one-piece front zipped protective undergarment with integral hood (Seyntex)   ***2002**/0137739*

shelters, bivouac or other tents and sleeping bags. Seyntex also manufacture an NBC casualty bag.

Status

All the above are in production.

Manufacturer

Seyntex nv.

UPDATED

BULGARIA

LZK-75 light protection overall

Description

The LZK-75 light protection overall bears a resemblance to the RFAS L-1 lightweight protective suit (qv) and is intended for the same 'special task operator' role but is issued as a one-piece overgarment. The LZK-75 consists of a hooded

Seyntex NBC two-piece protective suit. Integral smock and hood (Seyntex)
***2002**/0137736*

Seyntex NBC two-piece DPM camouflage protective suit with front smock fastening (Seyntex) ***2002**/0137737*

LZK-75 light protection overall

suit, gloves and a carrying bag with a combined weight of 2.2 kg. The material used is butyl rubber coated polyester fabric which provides a penetration resistance to liquid chemical agents for a minimum of 180 minutes. Time to don the suit is given as 2 to 3 minutes.

The LZK-75 is available in three sizes. Size I is for personnel up to 1.7 m tall, Size II for personnel between 1.7 and 1.8 m and Size III for personnel over 1.8 m tall.

Status
In production. Offered for export sales.

Marketing agency
Kintex.

VERIFIED

OKIZ Combined Arms disposable protective kit

Description
The OKIZ Combined Arms disposable protective kit is intended for emergency use only. It is intended to provide a penetration resistance to aerosols and liquid chemical agents for up to 24 hours. The kit consists of a protective coverall overgarment with a front button fly and integral hood, protective overboots and gloves, both held in place by rubber bands. Weight of the complete kit is 500 g.

The tensile resistance of the material used for the OKIZ kit is 3.5 kg/15 mm and tear resistance 1.5 kg.

The OKIZ is available in three sizes. Size I is for personnel up to 1.7 m tall, Size II for personnel between 1.7 and 1.8 m and Size III for personnel over 1.8 m tall.

Status
In production. Offered for export sales.

Marketing agency
Kintex.

VERIFIED

OKIZ Combined Arms disposable protective kit

CANADA

Acton CB-moulded glove

Description
Developed by Acton International, in collaboration with the Canadian Defense Research Establishment Suffield, the Acton CB-moulded glove uses many new design and manufacturing techniques to produce this next generation in CW protective gloves.

The Acton CB glove was created using extensive new sizing data and Computer Aided Design (CAD) techniques to optimise fit and comfort. These ambidextrous, 0.020 in (0.5 mm) thick gloves have corrugations at key points to improve dexterity. A snug wrist keeps the glove firmly on the hand. Ergonomically correct finger and hand dimensions assure good fit. Rounded fingertips further facilitate manipulation of small items (ammunition, keyboard use). Textured fingertips and palms improve grip. Available in multiple compounds based on end user needs.

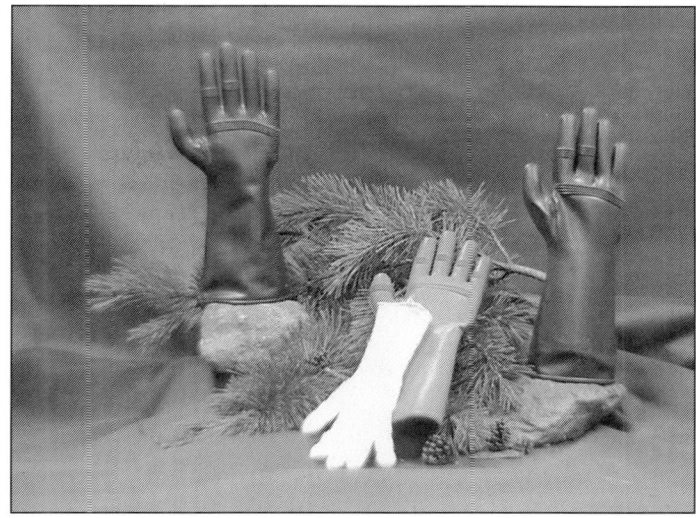

Acton CB-moulded glove (Acton International Inc)   ***2001**/0101450*

A newly designed liner using a novel material combination provides comfort and helps transfer perspiration from the skin.

Status
Available. In production.

Manufacturer
Acton International Inc.

NEW ENTRY

Acton NBC Lightweight Overboot (ALO). Overboot NBC MK 5 (UK designation)

Description
Developed by Acton International Inc, the Acton NBC Lightweight Overboot (ALO) has been adopted by a wide range of armed forces worldwide including the UK MoD as their Overboot NBC MK5. New improvements have been incorporated in the year 2000 design making these overboots even more decontaminable. The ALO provides well over 24-hour protection against chemical warfare agents and meet NATO stock numbers 8430-99-869-0394 to 0399. The ALO is available in six sizes from extra small to extra/extra large and will cover primary footwear from size 4 to 15 (Mondopoint 220 to 310).

Two outsole patterns are available. The standard and most popular outsole is designed for ground force (army) use and has an abrasion-resistant, anti-slip, self-cleaning outsole that provides good traction. The optional outsole design has a

Specifications
- 24-hours protection against CW agents
- Smooth upper butyl surface allows full decontamination
- Fully antistatic (EN 344 and DIN 4843)
- Custom features can be incorporated
- Outsole tread pattern available in 2 designs: land-based forces and naval forces
- Light and compact
- 6 sizes

Acton NBC Lightweight Overboot (ALO) with Standard Army Outsole (Acton International Inc)   ***2001**/0101451*

softer nipple pattern and is for use on smooth slippery surfaces such as wet decks and walls of buildings (for example, special forces abseiling).

Three adjustable, fully decontaminable, loop fasteners assure fast donning and doffing and allow the user to snug the overboot over the primary footwear to ensure a good fit and comfort. The smooth surface of the ALO facilitates cleaning and decontamination. In addition, the boots are fully antistatic. The close-fitting design assists in the operation of equipment and vehicles while the light weight reduces operational stress for the user.

Status
In production. In service in Australia, Canada, Japan, Kuwait, Netherlands, New Zealand, Portugal, Spain, UK and the USA.

Manufacturer
Acton International Inc.

UPDATED

Irvin NBC IPE System

Description
Irvin NBC IPE is in service with the Canadian Forces and other countries and fully meets the NATO triptych standard for protection, environmental range, endurance and interoperability. Irvin, with Defence Research Establishment Suffield (DRES) has developed this adaptable range of clothing to meet the needs of the defence, emergency service and first response community. A variety of designs is available in both normal weight and lightweight fabrics. The IPE range can be custom-designed for other mask and equipment interfaces. This IPE has been field-proven during Operation Desert Storm and other Canadian forces operations on land, air and sea.
- **General Service IPE**. The current Canadian forces permeable suit has a camouflage/coloured layer with waterproofing properties. The inner fabric has a bonded charcoal layer inside which absorbs B or C agent vapour or particles. The suit offers up to 24 hours protection following liquid chemical contamination.
- **Lightweight IPE**. This one-piece coverall is the result of extensive post-Gulf CF/DRES/IRVIN trials to produce a lightweight garment for hot weather use which trades protection time (slightly less than 24 hours) against performance decrement through heat stress. The design principle is similar to the general service IPE.

Canadian Forces experience suggests a one-piece garment as the safest solution, most popular with the wearers. The IPE is available in a full range of sizes.

Canadian IPE with the C4 mask and the Acton M5 overboot (Irvin Aerospace Canada Limited)
2001/0106868

Status
Available. In service with the Canadian Forces and other countries.

Manufacturer
Irvin Aerospace Canada Limited.

VERIFIED

Lightweight CW protective safety boot

Description
These safety boots meet CE requirements (certified to EN 345:1993 - category S5) for safety footwear. They provide more than 24 hours CW agent protection. Fully oil, fuel and flame resistant, they incorporate a steel toe, antistatic outsole and reflective strip to assure user protection. Featuring Acton's tractor anti-slip, self-cleaning outsole, these boots also have finger grips to facilitate donning and a kick-off bar above the heel to facilitate doffing. A removable 50 per cent wool liner also incorporates an antistatic sole, provides thermal protection against heat and cold and is held in place with a Velcro attachment. Available in European sizes 34 through 46 (North American Sizes 4 through12).

Status
In production. In service with the Swedish Räddnings Verket.

Manufacturer
Acton International Inc.

UPDATED

Acton lightweight CW protective safety boot (Acton International Inc)   0011454

Rubber boot 90/K (FMV specification: 11657 A)

Description
This multipurpose boot is designated by the Swedish Försvarets Materielverk (FMV) as Rubber Boot 90/K. It is an integrated, multipurpose protective boot providing more than 24 hours of CW protection. Fully POL resistant, this boot is also flame retardant and has a tall leg of 15 to 16 in (38 to 41 cm). The snow guard top has a laced closure at the back of the leg assuring a snug fit. The specially formulated antistatic non-marking outsole is ski-adaptable and will also take snow spikes. A hard rubber toe-cap makes the boot good for rugged use. A removable 'Thinsulate' liner provides thermal insulation and a removable inner sole made of a special rigid fast-drying felt completes the system. This multipurpose boot is

Acton rubber boot 90/K (Acton International Inc)   0011455

specified for cold/wet or cold/dry use in skiing, marching, decontamination and general daily use. The boot is available in 18 sizes (Mondopoint sizes 230 to 315) and the removable sock liner comes in 5 sizes which cover the same size range as the boot.

Status
In production. Procured by FMV, these boots are in service with the Swedish armed forces.

Manufacturer
Acton International Inc.

UPDATED

CHINA, PEOPLE'S REPUBLIC

M-02 NBC protective suit

Description
The M-02 permeable NBC protective suit comprises a fly-fronted, zipped, hooded smock and over trousers. The material is a two-layer combination. An outer flame-retardant cotton layer provides the wicking and weathering medium for liquid agent contamination. It also provides the colour (camouflage pattern is standard) and the wear protection for the underlying vapour absorbent layer. The latter is formed of activated charcoal impregnated cotton flannel, enhanced with oil-repellent and flame retardant properties.

The suit can be laundered and is durable enough to act as working combat clothing, with a wear life of four weeks.

The M-02 suit weighs 1.7 kg and has a life of up to 6 hours against persistent liquid agent contamination before break-through occurs. Shelf life is over five years.

Status
In service with the Chinese armed forces.

Development agency
Research Institute for Chemical Defence.

UPDATED

M-02 NBC protective suit (Research Institute for Chemical Defence)
0055072

M-66 protective suit, butyl

Description
Intended for use when carrying out decontamination procedures, the M-66 protective suit is manufactured using a butyl-based material. It is a one-piece suit with an integral hood and boots. Butyl rubber gloves are also worn. Elasticated closures are provided at the cuffs and around the integral hood, while laces are provided at the ankles and waist.

The M-66 suit weighs 2.3 kg and has a life of up to 130 minutes against 20 mg droplets of HD at +36°C.

Status
In service with the Chinese armed forces.

Development agency
Research Institute for Chemical Defence.

Manufacturer
Guilin Rubber Products Factory.

UPDATED

M-66 protective suit, butyl (Guilin Rubber Products Factory)

M-82 protective suit, permeable

Description
The material used for the M-82 protective suit, permeable, is made of vinylon-cotton blend fibre (outer layer) and cotton flannel (inner layer). One surface of the inner layer is finished with an active charcoal-acrylate mixture and the other surface is

M-82 protective suit, permeable (Huajiang Machinery Plant)

finished with an oil-repellent agent. The basic suit uses two garments, a fly-fronted jacket with an integral hood, and over-trousers. The suit is completed by butyl rubber gloves and overboots, the latter held in position by laces. The suit is supplied in a camouflaged and waterproof finish.

The suit weighs 1.3 kg and has a life between 20 and 50 hours with mustard or G-type nerve agents in vapour form and up to 6 hours against mustard and nerve agents (both G and V agents) in liquid vapour form (2.5 g/m^2), (0.1 mg droplets at +20°C, relative humidity 80 per cent).

Status
In production. In service with the Chinese armed forces.

Development agency
Research Institute for Chemical Defence.

Manufacturer
Huajiang Machinery Plant.

UPDATED

POO version
0011436

CZECH REPUBLIC AND SLOVAKIA

Air-permeable NBC protective suit BF

Description
The BF air-permeable NBC protective suits BF are designed for army and civil defence. They offer high user comfort and effective protection of the body surface against radioactive fallout, BW and CW agents in solid, aerosol or droplet form. The BF suits consist of a blouse with a hood, trousers and braces. Clients can specify additional components such as rubber under-gloves, over-boots and carrying bags for the ensemble including respirator.

The suits are constructed from material usually containing three layers. The outer camouflage cover fabric is proofed against POL and is flame-retardant and water-repellent. The absorptive layer textile is treated with powdered or spherical activated carbon supported by a firm, lightweight lining.

The suit is designed for low physiological load on the wearer, allowing comfortable long-term use. Versions include an overgarment or a special suit which replaces normal battledress. All versions offer excellent protective performance against liquid H agent (more than 24 hours).

BF series NBC protective suit: FOP-96 version
0019557

POO-LIGHT version
0011437

The BF suit series is produced in six sizes, with gloves and boots in three sizes. The basic suit without boots, gloves or mask weighs (according to the size and version) from 1.5 to 3.5 kg.

The client can specify particular requirements according to need and BF suits can be produced in versions which vary according to the different number and composition of the layers, their properties and design. All suits in the range have a long shelf life (up to 10 years) and a service life in training up to two years. Protective integrity in operational use of two to three months is typical for the BF suits. The BF series of suits have been trialled by accredited testing laboratories and the Czech Army authorities.

Status
In production. Used successfully by three UNSCOM teams during the inspection of chemical warfare agents in Iraq (POO-A and POO- LIGHT-A versions). In use by the civil defence corps in the Slovak Republic (FOP-90 version). In use by the Czech Army (FOP-96 version).

Manufacturer
B.O.I.S.-FILTRY Ltd.

VERIFIED

....

Protective anti-chemical garment OPCH-90 PO

Description
The protective anti-chemical garment OPCH-90 PO was developed for rescue and other tasks in extremely hazardous chemically contaminated environments. It is used in conjunction with a series CM-4 respirator and depends to a large extent on the use of an overpressure interior for the all-enveloping overall, with the over-pressure provided either by a small back-mounted blower unit or via a hose line.

The pressurised garment has tight trouser legs and high boots equipped with gas-proof zip fasteners. The garment is normally worn with a panoramic vision hood worn over another hood which covers the entire head apart from the aperture for the respirator facepiece. The sleeves are provided with plastic rings which connect with the butyl rubber protective gloves provided as part of the garment.

The outer material used for this garment is yellow or grey and made using a polyester coating combined with butyl rubber. All seams are covered with solutions containing natural rubber. The coatings provide protection against most aggressive and toxic chemicals and can also provide a measure of protection against extremes of heat.

Accessories available for use with the OPCH-90 PO include cotton/propylene inner gloves, PVC boots with steel toecaps and special underwear made using a cotton/propylene knit.

The garment is available in three sizes.

Status
Available.

Development agency
Originally manufactured by the Czech Research Institute of Rubber and Plastics Technology State Corporation. Current manufacturer unknown.

VERIFIED

Protective anti-chemical garment OPCH-90 PO being worn without the panoramic vision hood

DENMARK

Statens Konfektion NBC suit

Description
The state-owned Statens Konfektion produces a one-piece NBC coverall described as a suit. The suit provides full protection against chemicals in aerosol form for at least 6 hours and in liquid form for at least 24 hours. Protection is provided against radioactive dust particles for at least 6 hours. Provided the suit is not damaged it will continue to provide protection for at least 28 days. These properties remain unchanged after washing and cleaning. Storage life is 10 years. The suit can be worn as an overgarment in winter or direct on the body in summer.

The material used for the suit is a 2:1 twill polyester/cotton core-spun fabric which is flame-resistant as well as water and oil repellent. The filter material is a three-layer sandwich construction with activated carbon spheres based on the SARATOGA™ concept.

The suit has a long frontal zip fly closure and an integral hood which fits closely around an NBC mask to be fastened at the side with a Velcro strap. Similar closures are used at the cuffs and over NBC overboots. The pockets of the suit were designed to hold nerve gas tablets and chemical detection papers.

Status
Available. In service with the Danish armed forces.

Manufacturer
Statens Konfektion.

VERIFIED

EGYPT

Impermeable NBC protective suit

Description
This impermeable NBC suit is intended for heavy-duty tasks such as NBC decontamination and consists of a jacket with hood and over-trousers. The jacket has integral gloves while the over-trousers have integral overboots. A drawstring is provided around the neck and another at the lower hem of the jacket which reaches down well below the waist area for added protection.

Both garments are manufactured using a 'special cloth' coated with an isobutylene-isoprene co-polymer which can be repeatedly decontaminated.

Status
Available. In service with the Egyptian armed forces.

Manufacturer
National Organisation for Military Production.

VERIFIED

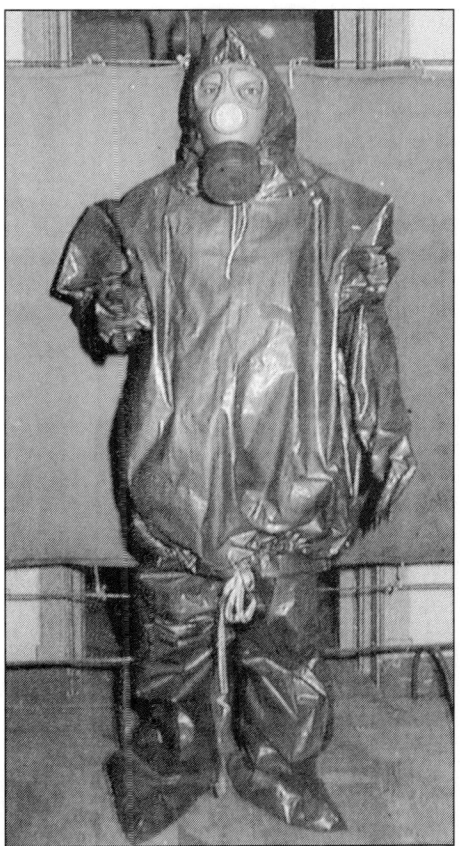

Impermeable NBC protective suit (T J Gander)

M81 NBC protective clothing

Description
The complete M81 NBC protective clothing outfit consists of four items. They are the M81-A protective suit, M81-B protective gloves, M81-C lightweight protective overboots and M81-D protective poncho.

The M81-A protective suit consists of a jacket (with integral hood) and trousers intended to be worn over a normal combat uniform. The material used is a filter fabric with a permeability rate of at least 200 litres/min^2. The material provides protection against mustard agent vapours for at least 10 hours and against mustard agent aerosols for 6 hours. The material also provides protection against a high degree of heat, fire and napalm-type weapons.

The other three garments in the M81 family are made of lightweight impermeable rubber-impregnated material and are usually issued as the Protective Kit No 1. The M81-B gloves have a thumb and index finger and are held in place by a tie-wrap tape. The M81-C lightweight overboots are held in position by laces around the ankle and top. The M81-D protective poncho is worn placed over the head and may be used as an anti-rain garment. It provides protection against chemical warfare agents for 6 hours.

Status
In production. In service with the Egyptian armed forces. Offered for export sales.

Manufacturer
National Organisation for Military Production.

VERIFIED

M81-A NBC protective suit being worn with M2-E protective mask and M81-C overboots

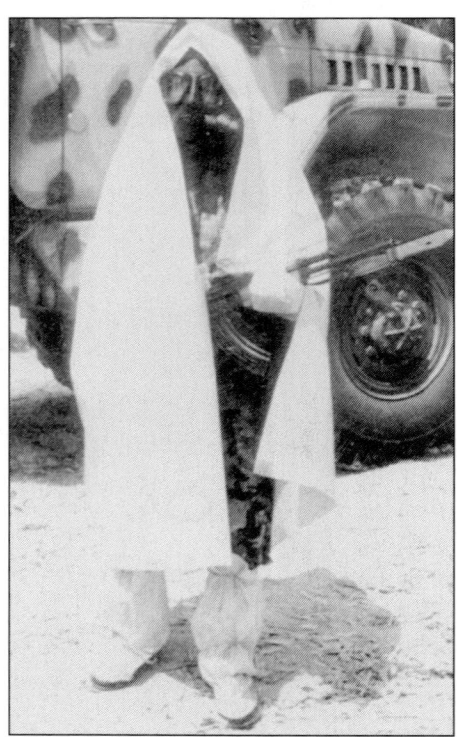

M81-D NBC protective poncho

FRANCE

Megajoule 2001-Px1 cooling system for IPE

Description
The Megajoule 2001-Px1 is a lightweight cooling undergarment working on the principle of the sublimation of solid carbon dioxide (dry ice). Tests have been performed successfully, under the auspices of DGA/CEB, by the Pompiers de Marseilles, the Pompiers de Chambery, WRC Peugeot sport and others. Versions are available offering 90 minutes duration or 3 hours. The garment is worn underneath full protective clothing and, in addition offers defogging of protection suit visors. The cooling power available is 400 and 600 W and the system has performed well even in temperatures above 100°C. and 100 per cent humidity. Suits can be connected to vehicle or aircraft systems to allow further control of the cooling temperature.

Status
Under trial. Available.

Manufacturer
Eurodéfhi.

NEW ENTRY

CO_2 reservoir for cooling suit stowed horizontally at rear of protective clothing (Eurodéfhi)
2002/0121881

Reservoir shown stowed vertically to accommodate SCBA (Eurodéfhi)
2002/0121880

NBC protective suit series S3P, T3P and NBC/F

Description
Paul Boyé, an official supplier of military protective clothing and uniforms for the French Army and other armed forces, produces a range of adaptable NBC protective suits as outlined below.

NBC protective suit Model S3P
This is a two-piece overgarment developed to meet the requirements of the French Army and is normally worn over a combat uniform together with NBC protective gloves, undergloves and socks. The two pieces comprise a jacket (with integral hood) and trousers which can be worn in a contaminated area for up to 24 hours. Both garments are made using three layers of material. The outer layer is high-density polyamide taffeta fabric (75 g/m²); the secondary layer is non-woven fabric (95 g/m²); the inner layer is 2 mm thick activated coconut charcoal impregnated polyurethane (between 220 and 300 g/m²). Weight of the complete suit is 1.7 kg and the shelf-life is at least 10 years when sealed.

The S3P suit has been produced for the French and Swiss armies. 45,000 suits were purchased by the latter in 1986; it is known as the *habit de protection* C86.

Protective coverall T3P
This one-piece garment is intended for prolonged wear by air force aircrew (T3P Pilote) and ground crew (T3P PNN) and can be worn directly over the skin together with NBC protective socks, gloves and undergloves. The material used has two layers. The outer layer is fire-resistant Kermel (Aramide)/Viscose (rayon) which is also oil and water repellent; weight is 205 g/m². The inner layer is 1.2 mm thick compressed material impregnated with activated charcoal, laminated on one side to a nylon jersey and a cotton jersey fabric on the other. The coverall has an integral hood and a front angled fly closure. A pilot's version could have a hood adaptor for a flying helmet. Weight of the suit is approximately 1.8 kg and shelf-life is up to 10 years if sealed. A repair kit is available.

The T3P Pilote and PNN are in production for the French Air Force.

Paul Boyé NBC protective suit Model S3P

Paul Boyé NBC/F Protective Coverall T3P PNN for aircrew

Paul Boyé NBC/F Protective Coverall for tank and helicopter crews

NBC/F Protective Coverall
Offering enhanced fire protection as well as NBC protection, the patented NBC/F protective coverall is designed for use by tank crews and helicopter aircrew. It can be worn next to the skin together with NBC protective socks, gloves and undergloves. The general design and material employed are the same as those for the T3P coverall but the NBC/F does not normally have an integral hood - it is available as an option.

Other protective suits produced by Paul Boyé include the one-piece NBC/F Protective Coverall Model SAMU, in production for the French Ministry of the Interior and Ministry of Health and the NM143 NBC Overgarment for the Norwegian Army.

Status
See text.

Manufacturer
Paul Boyé.

VERIFIED

Piercan NBC gloves

Description
Piercan produces a wide range of gloves for industrial and other uses and also produces NBC gloves made from black bromobutyl rubber. These five-fingered gloves provide protection against chemical warfare agents but are very flexible. The gloves are produced in a standard length of 335 mm and thickness of 0.3 mm. All other lengths and thicknesses can be produced to suit customer requirements.

Status
In service with the French Air Force.

Manufacturer
Piercan.

VERIFIED

Tactical Operations Multipurpose Protective Suit (TOMPS)

Description
This lightweight protective suit is aimed principally at first responders. It comprises a tunic with integral hood and separate trousers, supported by adjustable suspenders. In addition to its NBC protective properties, it is designed to reduce the wearer's environmental stress from heat and humidity. The outer material (KERMEL viscose™) is treated for water and POL resistance and is highly resistant to abrasion. The inner layer is a Whetlerite charcoal impregnated foam, laminated

Specifications
Size range: S, M, L, XL, XXL
Protection: 24 hours against H agent at 10g/m²
Wear tolerance (uncontaminated): 30 days
Shelf life: 10 years (sealed)
Weight: 2 kg (tunic and trousers)
8-set box dimensions: 80 × 45 × 40 cm
Boxed weight: 26 kg

to cotton jersey. The tunic's elasticated hood is designed for use with the Giat Industries ARFA mask (see under PROTECTION (individual) - Masks (general issue)).

The ensemble is available in a large range of sizes and offers good all-round survivability. NBC protective inner and outer gloves are available as well as protective socks.

Suits are also available in boxed sets for use at an incident site by teams of responders. The NBC Hazards Risk Intervention Kit contains 8 full sets of clothing, gloves and socks in a range of sizes. A user's manual is also included in each box.

Status
Available.

Manufacturer
Giat Industries.

Agency
Centech Group Inc.

VERIFIED

TLD range of lightweight NBC decontamination suits

Description
The TLD range of Paul Boyé lightweight two-piece decontamination suits was developed to meet the requirements of personnel engaged in decontamination operations and has been adopted by the French armed forces (in 1993), the Gendarmerie and Civil Defence. The suits are available in a range of colours for military, civil, tropical or DEMIL use. The TLD range is stated to cost 10 times less than butyl-based decontamination suits but provides protection for over 24 hours against a 100 g/m2 challenge of H or G agent. It is also resistant to decontamination solutions, offering 8 hours protection against most strong chemicals including 65% nitric acid.

The material used for the TLD range has a weight of approximately 120 g/m². Suits are fitted with proofed seals and fasteners. The hood provides a good seal with the standard ARFA and other designs of NBC respirator. The suits are compatible both with standard NBC gloves and overboots and with purpose-made examples from the same manufacturer. The butyl overboots have non-slip soles. Although TLD range suits are intended to be discarded after use, they may be decontaminated and reused. The weight and cut of the garment materials involved considerably improve the wearing comfort.

Other colours or a camouflage finish can be supplied.

Paul Boyé lightweight decontamination suit

Status
In production.

Manufacturer
Paul Boyé.

VERIFIED

Tropical NBC combat suit

Description
The Paul Boyé tropical NBC combat suit produces no more heat stress than a normal combat uniform. It is a two-piece suit worn with NBC protective socks, gloves and undergloves with the two pieces being a jacket (with integral hood) and trousers.

The material involved has two layers. The outer layer is 190 g/m² cotton/polyester oil and water repellent twill. The inner layer is 1 mm thick compressed polyurethane material impregnated with activated charcoal laminated to a cotton or nylon jersey fabric. The suit provides protection for up to 24 hours and is washable. Weight is approximately 1.8 kg. Shelf-life is at least 10 years if sealed. A repair kit is available.

The suit is available with an aramide fibre fire-resistant outer fabric. Suits are available in printed camouflage or plain colours.

Status
In production for the French Army and several other armed forces worldwide, including Hungary, Indonesia and Singapore. Spain and the Swiss Ministry of the Interior have selected a fire-resistant version of this suit.

Manufacturer
Paul Boyé.

VERIFIED

Paul Boyé tropical NBC combat suit

Paul Boyé tropical NBC combat suit in camouflage finish as produced for the French Army

GERMANY

Blücher SARATOGA™ NBC protective clothing

Description

SARATOGA™ is a composite filter fabric based on highly activated carbon in the form of small spherical adsorbents with a very hard shell, fixed onto textile carrier fabrics.

With a carbon density of up to 200 g/m², the adsorbent capacity of SARATOGA™ is claimed to be the highest on the market. It provides at least 24 hours protection even after 45 days of wear under battlefield conditions, including several (>10) field launderings and has a shelf-life of up to 20 years.

The technology of adhering the carbon spheres to the textile carrier fabric makes more than 85 per cent of the outer surface available to the external (toxic) atmosphere. The result is an extremely rapid and firm adsorption in the balanced pore system of the adsorbents without any tendency to desorption, even at higher temperatures. Thanks to its sophisticated construction, low thickness and high air permeability, SARATOGA™ provides the lowest possible heat insulation and water vapour resistance. A pleasant microclimate is created by the moisture buffering capacity of the activated carbon and the ventilation effect due to the flexibility of the composite fabric. The hardness of the spheres, the quality of the adhesive and the textile components guarantee the excellent mechanical durability of SARATOGA™.

SARATOGA™ NBC protective garments can be worn under all climatic conditions with the lowest possible impact on combat effectiveness. The US JSLIST (NBC protective suit for the US forces - see separate entry within this section) is made with SARATOGA™ material which was tested at 10 locations worldwide including arctic, tropical and desert environments.

A highly versatile fabric, SARATOGA™ is used in the creation of a large range of products for the personal NBC protection of both military and civilian personnel. Examples include:

- Overgarments (worn over regular clothing)
- Duty uniforms (worn over underwear)
- Hybrid garments (worn as overgarment or duty uniform)
- Undergarments (worn instead of underwear in combination with regular clothing)
- Lining systems (inner liner or intermediate garment to be integrated into an existing clothing ensemble).

Material and other accessories available from the manufacturers include:

- SARATOGA™ filter fabrics as roll goods
- Ready-made SARATOGA™ garments
- Complete sets, consisting of a SARATOGA™ garment, protective gloves and overboots in a protective carrier bag.

Overgarment - US armed forces (JSLIST). US MilSpec: PD 97-04. Outer layer: Nylon/ cotton ripstop woven (woodland and desert camouflage). Filter layer: SARATOGA™ based on polyester knit (Blücher)
2002/0102791

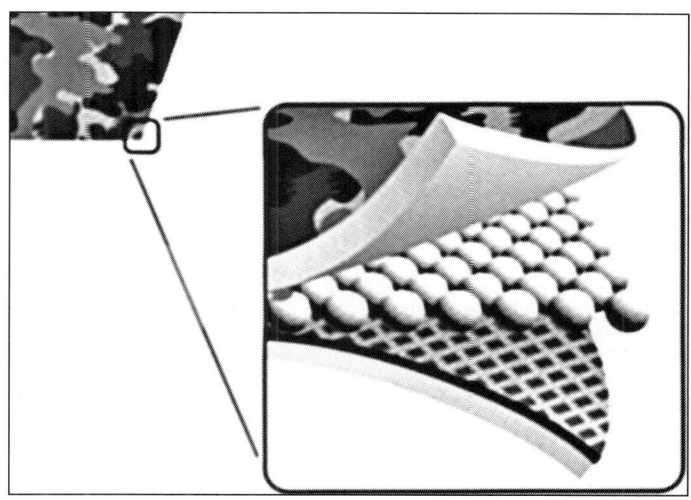

Diagram of the adsorbent layer (Blücher)   **2002**/0102787

Overgarment - US Marine Corps. US MilSpec: MIL-C-29462. Outer layer: Cotton ripstop woven (woodland and desert camouflage). Filter layer: SARATOGA™ based on polyester knit (Blücher)
2002/0102790

Chemical protective undergarment for the Finnish Army (over standard battledress). Mil Spec: 1096-E. SARATOGA™ composite fabric ('pyjama') based on cotton knit (white) (Blücher)
2002/0102795

Chemical protective suit (OPCW). US MilSpec: MIL-C-29462. Outer layer: Cotton ripstop woven (UN blue). Filter layer: SARATOGA™ based on polyester knit (Blücher)   ***2002**/0102792*

Status
In service in 18 countries. In spring 1997, SARATOGA™ became the new standard for all US forces.

Manufacturers
Blücher GmbH.
Tex-Shield Inc (USA).

UPDATED

Helsa-Werke NBC protective clothing

Description
Helmut Sandler GmbH and Co has been producing NBC protective material since 1977 and has delivered complete NBC protective outfits for the German, Norwegian, Swedish and some Middle East armed forces, including Saudi Arabia.

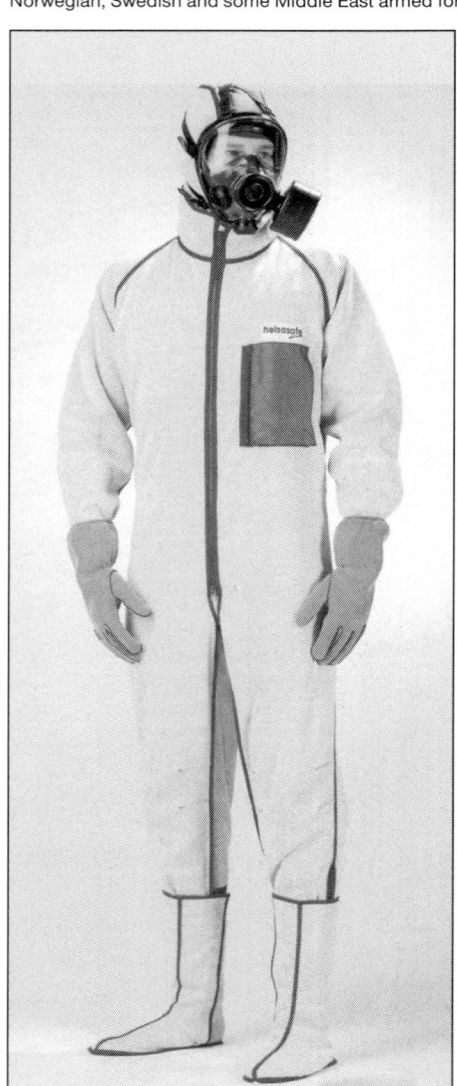

Helsa NBC protective undergarment system complete with hood, gloves and socks
2000

The NBC protective outfits include a jacket with integral hood, overtrousers, overboots and gloves, all packed in a carrying bag. The smock and overtrousers are generously cut and are provided with pockets. They are made from two layers of material, the outer layer being either a solid colour or a camouflage print. The outer layers have a water, oil and liquid agent repellent finish which prevents the ingress of moisture but allows body transpiration. The inner layer can be either a foam-based filter material, standard or compressed, or a textile carrier with a layer of finely distributed active charcoal particles or synthetic adsorbers of different kinds. This protective suit provides protection against NBC attacks for 24 hours or more while retaining a high degree of comfort.

Also produced is an NBC protective undergarment which can be worn with normal NBC protective over-clothing to provide additional protection and comfort.

The Helsasafe complete pilot programme includes NBC overalls, hoods and gloves.

Status
In production.

Manufacturer
Helsa-Werke.

VERIFIED

Kärcher encapsulation NBC suit - Safeguard 6004

Description
The Kärcher Safeguard 6004 encapsulation NBC suit is a one-piece, gas-tight full protection suit with integrated boots and protection mask. The Safeguard 6004 was designed to be worn by NBC defence personnel during decontamination operations or emergency operations in contaminated or polluted areas, as well as in the chemical industry.

Kärcher encapsulation NBC suit - Safeguard 6004

Protective gloves are linked to the Safeguard 6004 suit through a special rubber arm collar with an inner banding and an outer rubber ring which make it easy to change the gloves when necessary.

The mask is fitted with external bands to enable the wearer to adjust both the tightness and fit of the mask as required.

The suit features a diagonal gas-tight zip fastener across the front. To close the suit, the fastener is zipped from top to bottom so that the wearer can check at any time whether the fastener is closed. A broad strip ensures the protection of the zip fastener against contamination and mechanical damage. The material used for the suit is a polyester fabric coated on both sides with butyl rubber. The butyl rubber is made from pure hydrocarbon to ensure high impermeability. All seams are double-lapped and provided with extra adhesives and vulcanisation along a tight band on both the inner and outer surfaces.

Wearing time during operations is increased by the use of the Kärcher-Filtersafe 1001 ventilation kit (for details of this kit refer to the entry under Protection (individual) - Filters). There is an inlet connection valve in the back of the suit which ensures the connection to the respirator hose. Six overpressure valves ensure a permanent overpressure inside the suit to prevent the ingress of contaminated air should the suit leak as a result of damage.

The Safeguard 6004 is issued packed in a carrying bag which also contains one pair each of rubber over-gloves and cotton inner gloves, one pair of socks, a drinking hose, gas filter and a maintenance and repair kit. The latter consists of a zip fastener maintenance kit comprising two grease pencils, a piece of fabric, adhesive and maintenance and repair instructions. Each carrying bag contains instructions, written in the appropriate language, on how to use the suit and its accessories.

A suit similar to the Safeguard 6004 is used by the German Army under the name of 'Zodiac'. Unlike the Zodiac, the wearer of a Safeguard 6004 can don and remove the suit by himself without assistance. He is also able to check at any time whether the suit is closed or not, thus ensuring full protection.

Status
In production and in service.

Manufacturer
Alfred Kärcher GmbH & Co.

VERIFIED

Kärcher NBC undergarment concept Safeguard 3002-A1/Undergarment

Description
The Kärcher NBC undergarment concept Safeguard 3002-A1/Undergarment consists of a total of three fabric layers. The materials are comfortable on the skin and thus can be worn directly on the skin together with normal protective clothing to provide additional protection against chemical warfare agent vapours. Because

of the smooth and flexible structure of the material, the Safeguard 3002-A1/Undergarment offers its wearer freedom of movement to a considerable extent.

An activated carbon textile (Safeguard ACK-100) is laminated between two cotton/polyester knitted fabrics. Special polyester fibres allow body moisture to escape quickly and thus ensure that the wearer's skin remains relatively dry for a long time. The complete fabric composition (monopack) has a weight of approximately 320 g/m². For additional protection against thermal effects such as convective and radiant heat, the activated carbon textile can also be laminated between two Nomex/Viscose-FR knitted fabrics.

Status
In production.

Manufacturer
Alfred Kärcher GmbH & Co.

VERIFIED

Specifications
Thermo-physiological values:
  (water vapour resistance (R_{et})) <60.10^{-3}m² mbar/W
  (thermal insulation (R_{ct})) <25.10^{-3}m² kW
  (moisture vapour transmission rate) > 12.000 g/m²
in 24 h
Protective capability against chemical warfare agents (initial):
  (vapour - 20 mg HD/m³) >24 h
Protective capability against chemical warfare agents (after 10 washings, +40°C):
  (vapour - 20 mg HD/m³) >24 h

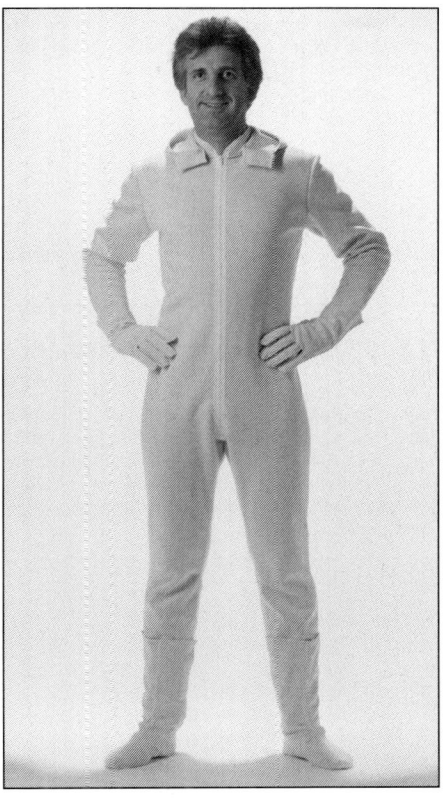

Kärcher NBC undergarment concept Safeguard 3002-A1/ Undergarment in one-piece pyjama form
0055068

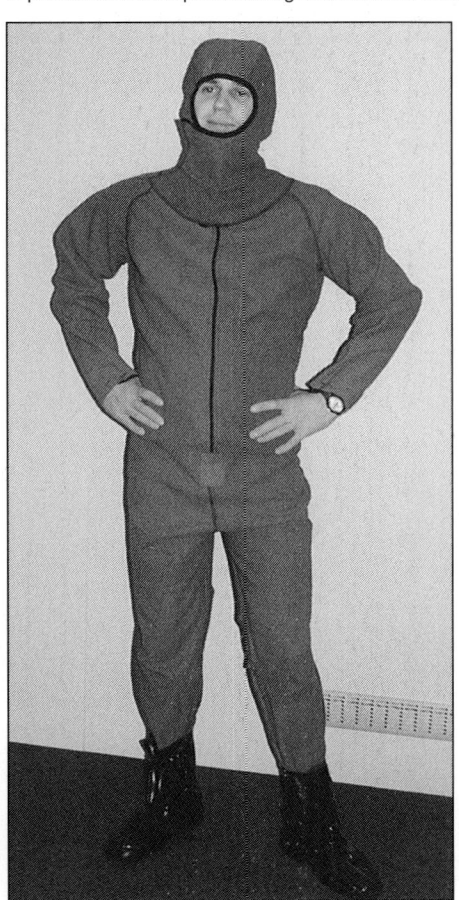

Kärcher NBC undergarment concept Safeguard 3002-A1/ Undergarment with hood

Kärcher personal protection concept Survival Kit

Description
The Kärcher Survival Kit offers an efficient breathing and body protection system for widespread use. The Survival Kit has been engineered for the protection of personnel handling dangerous goods as well as for special rescue and emergency services. The growing need of the public for personal protection was also taken into account during the design of the system, so effective protection equipment is available for families, from children to adults.

The personal protection system consists of the following matched individual components: a full-face respirator mask with air-purifying filter; a permeable protective overall; protective gloves; special overshoes; a carrying bag; and instructions for use. All parts of the Survival Kit are packed in the carrying bag which is convenient for home storage or for travel.

The microclimate inside the protective coverall is improved by the use of breathable protective fabrics which permit body heat and moisture to escape. This ensures greater comfort for considerably longer wear and working times while using the complete protective equipment.

The protective coverall consists of several layers of fabric. Each layer has a specific function to perform. The permanently flame-retardant outer layer provides

Kärcher personal protection concept Survival Kit

the wearer with short-term protection against the effects of thermal radiation and fire, as well as against the penetration of radioactive fallout particles or pathogens through the textile compound onto the skin. The outer layer is impregnated with a special oil and repellent finish to prevent the penetration of liquid chemical agents.

The inner layer (filter laminate) fulfils the function of a textile filter. In order to protect the wearer against chemical weapons in aerosol and gaseous forms, a specially developed activated-carbon component has been integrated into the filter laminate. As a result of its large effective surface area, the activated carbon traps noxious molecules and thus prevents possible damage to the wearer.

The protective overall can be laundered several times without losing its protective capacity.

Status
In production and in service.

Manufacturer
Alfred Kärcher GmbH & Co.   *VERIFIED*

....

Kärcher Safeguard™ 2002 NBC-Protective Garment

Description
The Safeguard™ 2002 NBC-Protective Garment is designed to be worn over existing combat dress or directly over undergarments. The outer shell fabric weighs 180 g/m² and is formed from a COM4 tear-resistant cotton cloth with flame retardant and oil and water repellent properties. The underlying filter laminate, at a weight of 300 g/m², is a non woven fabric with an activated carbon adsorption layer laminated onto a PES warp knitted fabric.

Specifications	
Combined weight:	480 g/m²
Heat stress data (water vapour resistance):	R_{et} = 6.86 m²Pa/W. For comparison, German army overgarments: 17.6 m²Pa/W and undergarments: 4.3 m²Pa/W (100% CO)
Durability data	
Tensile strength (ISO 50581) (weft/ warp):	
(shell fabric)	>800 N / >600 N
(filter laminate)	>400 N / >300 N
Tear resistance:	
(shell fabric)	>20 N / >30 N
(filter laminate)	>30 N / >50 N
Bursting load:	2.5 bar
Wear tolerance:	At least 30 days in battle conditions
Flame resistance	
10 s test flame (EN 532 / EN 531):	
(performance level)	A
(afterflame time)	2 s
(afterglow time)	Zero
CW agent resistance:	>24 h against a 10 g HD/m² challenge
Cleansing resistance	
Laundry (gentle wash at 40°C):	>5 cycles
Hot gas or hot steam decontamination (170°C 30-60 min):	>2 cycles
Storage:	>10 years (airtight packed)
Size range:	S, M, L, XL, XXL

Safeguard™ 2002 NBC-Protective Garment
0055065

Status
In production. Available.

Manufacturer
Alfred Kärcher GmbH & Co.

VERIFIED

....

Kärcher Safeguard™ 3002-A1 NBCF-Protective Aircrew Suit

Description
The Safeguard 3002-A1 NBCF-Protective Aircrew Suit is designed to be worn in place of a standard flying coverall and offers good NBC and fire protection

Specifications	
Combined weight:	395 g/m²
Heat stress data (water vapour resistance):	R_{et} = 6.28 m²Pa/W. For comparison, German army overgarments: 17.6 m²Pa/W and undergarments: 4.3 m²Pa/W (100% CO)
Durability data:	
Tensile strength (ISO 5081) (weft/ warp):	
(shell fabric)	>800 N / >600 N
(filter laminate)	>300 N / >200 N
Tear resistance:	
(shell fabric)	>20 N / >30 N
(filter laminate)	>14 N / >18 N
Bursting load:	2.5 bar
Wear tolerance:	At least 60 days in battle conditions
Heat resistance:	
10 s test flame (EN 532 / EN 531):	
(performance level)	A
(afterflame time)	2 s
(afterglow time)	0 s
Convective burning behaviour (EN 367 / EN 531):	
(performance level)	B2
(time to reach pain level)	5 s
(time to reach 2nd degree burn level)	8 s
Radiant burning behaviour (EN 366 / EN 531):	
(performance level)	C2
(time to reach pain level)	25 s

Safeguard 3002-A1 NBCF-Protective Aircrew Suit
0055067

Kärcher Safeguard 3002-A1NBCF-Protective Suit
0055064

properties. Its construction allows high permeability, alleviating heat stress, and good durability. The design and treatment of the material layers offers good protection against nuclear dust, CW agent vapour or liquid and BW organisms. The outer layer has a weight of 165 g/m². and is coloured according to the user's needs (for camouflage or uniform requirements). The intermediate fire protection layer is a fresco-fabric made of aramide (100% Nomex Delta C) twisted yarns. The under layer weighs 230 g/m² and is a non-woven fabric with activated carbon bonded to a layer of PES warp knitted fabric.

A full range of sizes is available and the suit design is compatible with standard flying and life-saving equipment.

Status
In production. Available.

Manufacturer
Alfred Kärcher GmbH & Co.

VERIFIED

....

Kärcher Safeguard™ 3002-A1 NBCF-Protective Garment

Description
The Kärcher Safeguard™ 3002-A1 NBCF-Protective Garment (NBC and Fire) combines the benefits of NBC protection with those of a working combat suit. It offers a high level of protection with a very low thermal load and high durability. The COM4 cotton shell fabric is tear-resistant, flame retardant, water and oil repellent. It weighs 180 g/m2. The underlying filter layer is a non-woven active carbon impregnated fabric, laminated onto a layer of PESwarp knitted fabric. The latter weights 230 g/m².

With its integrated NBC and fire retardant properties, the suit offers good protection against nuclear flash and thermal radiation as well as against CW and BW agent liquid, particles or vapour and radioactive dust.

Specifications

Combined weight:	410 g/m²
Heat stress data (water vapour resistance):	R_{et} = 6.28 m²Pa/W. For comparison, German army overgarments: 17.6 m²Pa/W and undergarments: 4.3 m²Pa/W (100% CO)
Durability data	
Tensile strength (ISO 5081) (weft/warp):	
(shell fabric)	>800 N / >600 N
(filter laminate:)	>300 N / >200 N
Tear resistance	
(shell fabric)	>20 N / >30 N
(filter laminate)	>14 N / >18 N
Bursting load:	1.5 bar
Wear tolerance	at least 60 days in battle conditions
Heat resistance	
10 s test flame (EN 532 / EN 531):	
(performance level)	A
(afterflame time)	2 s
(afterglow time)	0 s
Convective burning behaviour (EN 367 / EN 531):	
(performance level)	B2
(time to reach pain level)	5 s
(time to reach 2nd degree burn level)	8 s
Radiant burning behaviour (EN 366 / EN 531):	
(performance level)	C2
(time to reach pain level)	
(time to reach 2nd degree burn level)	25 s
CW agent resistance:	34 s
Cleansing resistance:	>24 h against a 10 g HD/m²
(laundry (gentle wash at 40°C)	>10 cycles
(hot gas or hot steam decontamination (170°C)	>5 cycles
Storage:	>15 years (airtight packed)
Sizes:	S, M, L, XL, XXL

Status
In production and in service.

Manufacturer
Alfred Kärcher GmbH & Co.

VERIFIED

....

Kärcher Safeguard HiPerm range of NBC protective clothing

Description
A collaborative venture between Kärcher and Freudenberg led in 1998, to the joint development of a new range of protective materials. Kärcher's production of technically advanced NBC protection systems and Freudenberg's experience in large-scale production of non-woven fabrics for industrial and defence roles, combine to produce an NBC protective suit material which offers high permeability

Specifications	
Combined weight: 460 g/m²	
Heat stress data (water vapour resistance):	R_{et} = 6.74 m²Pa/W. For comparison, German army overgarments: 17.6 m²Pa/W and undergarments (100% Cotton): 4.3 m²Pa/W
Durability data	
Tensile strength (ISO 50581) (weft/warp):	
(shell fabric)	>800 N / >600 N
(filter laminate)	>200 N / >25 N
Tear resistance	
(shell fabric)	>20 N / >30 N
(filter laminate)	>25 N / >25 N
Bursting load:	1.5 bar
Wear tolerance:	At least 30 days in battle conditions
Flame resistance	
10 s test flame (EN 532 / EN 531):	
(performance level)	A
(afterflame time)	2 s
(afterglow time)	Zero
CW agent resistance:	>24 h against a 10 g HD/m² challenge
Cleansing resistance	
Laundry (gentle wash at 40°C):	> 5 cycles
Hot gas or hot steam decontamination (170°C):	>2 cycles
Storage:	>10 years (airtight packed)

Kärcher combat suit using Safeguard™ and HiPerm™ components
0055066

without degradation in protective factor. The result is a series of products which significantly alleviate the heat stress which afflicts NBC-protected operators in hot climates. The Kärcher material is branded Safeguard™ and the Freudenberg element, HiPerm™.

The combined material firstly comprises an outer tear-resistant 180 g/m² COM4 cotton fabric with oil and water-repellent and fire-retardant properties. This layer is coloured to offer camouflage or other signature reduction properties. The under layer consists of a 280 g/m² non-woven fabric containing activated carbon laminated to a PES warp knitted fabric.

The new material will be incorporated into the Kärcher range of suits.

Status
In production. Available.

Manufacturer
Alfred Kärcher GmbH & Co.

VERIFIED

HUNGARY

Air-Permeable Protective Kit M89A1

Description
A complete Air-Permeable Protective Kit M89A1 consists of a protective suit, a foil overcoat, foil overboots, protective overboots and gloves and a carrying bag.

The Air-Permeable Protective Suit M89A1 protects the wearer against chemical agent vapours, small droplets and aerosols. It comprises a jacket and suspended trousers with the active protective layer consisting of a foam material with impregnated active charcoal carried on textile tissue. The outer cover is a water repellent polyester tissue, the lining being moisture-absorbing cotton.

The suit may be worn over or in place of a combat uniform. It can be worn for extended periods (up to several days) in temperatures below 25°C. The suit provides protection against aerosols for up to 6 hours; against vapours, the time is 24 hours. Weight of the M89A1 suit is 1.7 kg. Eight sizes are available.

A set of M89A2 protective underwear may be worn under the protective suit in place of a combat uniform. The material used for this underwear is silk on the outside and cotton for the inner liner. The protective layer is a self-extinguishing foam material impregnated with activated charcoal. The set comprises an anorak-type garment with a hood and trousers. Eight sizes are available to match the outer protective suit.

For further protection an M89 protective foil overcoat is available. This overcoat is closed by Velcro-type fasteners. The same foil material is used for a pair of overboots which are secured by knee and ankle straps. These are worn under a pair of rubber and plastic protective overboots. Both types of overboot are supplied in three sizes.

The M89A1 kit is completed by a pair of five-fingered rubber gloves with an interior sweat absorbent layer. The complete kit is carried in a special bag made of the same material as the protective suit. It can be carried using shoulder straps.

Air-Permeable Suit M89A1, one component of the Air-Permeable Protective Kit M89A1

Status

In service with the Hungarian armed forces.

Manufacturer

(Unknown).

VERIFIED

Impermeable Protective Suit M75

Description

The Impermeable Protective Suit M75 is intended to be used by decontamination personnel and is stated to provide total body protection against all known chemical and biological agents. The material used for the suit is fabric coated with a synthetic rubber mixture. The complete suit consists of trousers, a hooded jacket and gloves. When not in use the suit may be carried in a special bag. The suit may be worn in conjunction with the M89A2 protective underwear provided as part of the Air-Permeable Protective Kit M89A1 (see previous entry).

The M75 impermeable suit will provide a minimum penetration resistance of at least 4 hours against HD droplets. It can be decontaminated at least five times.

Status

In service with the Hungarian armed forces.

Manufacturer

(Unknown).

VERIFIED

Lightweight Impermeable Protective Suit M89

Description

The Lightweight Impermeable Protective Suit M89 is intended to fulfil two roles. The first is a means of increasing the protection provided by the Air-Permeable Suit M89A1, part of the Air-Permeable Protective Kit M89A1 (qv), while protecting the suit against the effects of rain or other external moisture. The second role is that of an individual protective suit in an emergency. As such, it can be used as a relatively low-cost NBC protective suit for civilian populations.

Status

In service with the Hungarian armed forces.

Manufacturer

(Unknown).

VERIFIED

IRAQ

NBC protective clothing

Description

The Iraqi NBC protective suit is a two-piece garment with an integral hood. It provides protection against Mustard agent aerosols for a minimum of 4 hours and 6 hours against Mustard agent vapours. Storage life is two years.

Iraqi military protective suit

The protective outfit is completed by rubber gloves 0.8 mm (±0.1 mm) thick and rubber protective overshoes. Both gloves and overshoes are supplied in three sizes. The gloves weigh 280 g/pair and the overshoes 810 g to 1.03 kg/pair.

Status

In service with the Iraqi armed forces.

Manufacturer

Iraqi state factories.

VERIFIED

IRELAND

NBC suit Mark 8Z

Description

The NBC suit Mark 8Z is the latest of a series of NBC suits produced by the Protective Clothing Company Limited of Dublin. It is in production for the Irish Permanent Defence Forces.

The Mark 8Z suit is a two-layer floating shell design. The outer layer is assembled complete with pockets, hood and all accessories as requested by a purchaser. The anti-gas layer is assembled using 'J' seams which provide the same protection as uncut anti-gas cloth. Panels are sewn together at the perimeter only. No other

NBC suit Mark 8Z

NBC suit Mark 8Z
0018696

stitching passes through both layers, thus preventing wicking of liquids to the inside of the suit. Seams are reduced to a minimum by cutting the jacket in one piece. Adjustments at each hip are made using a hem cord. Touch closure tapes are used at the ends of the sleeves, down the front, hem, hood and pockets. There is a coil zip with the double flap front. A cord adjusts the hood to most respirators, being permanently sewn at one end and with a cord lock at the other.

The trousers are assembled in the same manner as the jacket. They have a double fly, braces with a comfort bar at the back, adjusting rings at the front and an auto-adjust feature for sitting. There are touch closures at the end of each leg.

The Mark 8Z suit is intended to be worn in place of normal combat clothing. The hood folds away when not required. Weight of the suit is less than 2 kg.

The anti-gas inner layer is woven-activated reinforced carbon. It is liquid-proof, air permeable, can absorb large volumes of gas agent but is washable and can be decontaminated without loss of protection. Shelf-life is indefinite while the service life is from six months to many years, dependent on wear. Repair kits are supplied, with instructions, for the inner and outer layers.

It is possible to retrofit the anti-gas lining to aircrew coveralls and an anti-gas undergarment is available. Full- and half-cover casualty bags are also available and are in service.

Status
In production. In service with the Irish Permanent Defence Forces.

Manufacturer
The Protective Clothing Company Limited.

VERIFIED

ISRAEL

NBC disposable protective clothing

Description
Chemoplast Limited clothing is manufactured using a clear three-layer laminate of polyethylene and polyamide. Continuous heat welding is used throughout to ensure seam integrity.

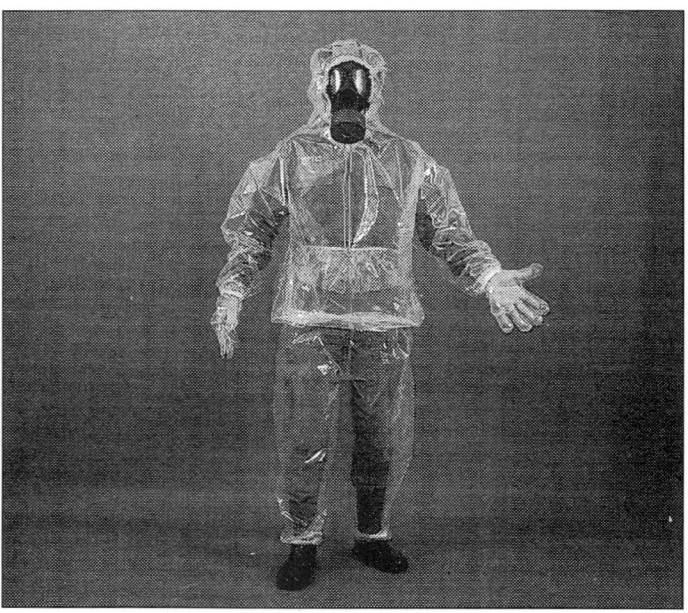

Chemoplast disposable NBC protection suit (Chemoplast Industries Limited)

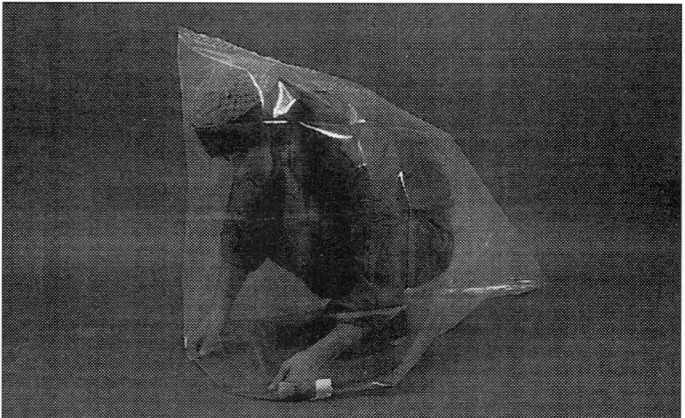

Chemoplast disposable personal NBC protection cape
(Chemoplast Industries Limited)

The disposable NBC suits consist of overtrousers sealed at the ankle and waist by elasticated bands and an overjacket with an integral contoured hood. Gloves are also available. The suit material is rigorously tested for gas permeability, mechanical strength, thickness tolerances, heat seal strength and other defects. The suits are delivered in sealed pouches.

A disposable NBC personal cape is also available, manufactured using a polyethylene/polyamide laminate. It is intended for emergency use only and acts as a coverall for the entire head and body with the user in a crouching position.

Status
In production.

Manufacturer
Chemoplast Industries Limited.

UPDATED

NBC protective modules for infants and children

Description
In order to provide NBC protection for infants and children, Supergum (1956) produce NBC protective modules using transparent high-barrier film manufactured by Plastopil. The modules intended for children, provide clean air overpressure protection for ages three to eight years. Those intended for infants can be used from the time of birth to approximately three years.

These lightweight and mobile modules allow full upper body freedom of movement and unimpeded vision. The complete module includes a protective hood, blower with a long-life battery, NBC canister, self-contained drinking system and instruction manual. The modules are stored in travel cases when not in use. Effective against all types of CW agent, they deliver full efficiency for more than 10 hours in a contaminated environment.

The materials, blower unit and battery have a shelf-life of 10 years. Supergum is certified to ISO 9001.

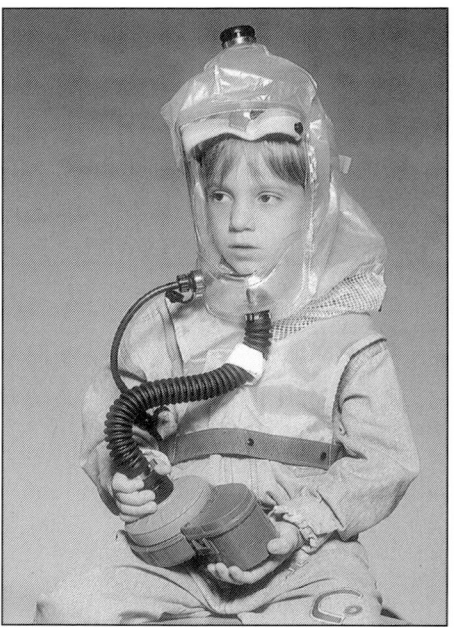

Supergum NBC protective module for children aged three to eight years

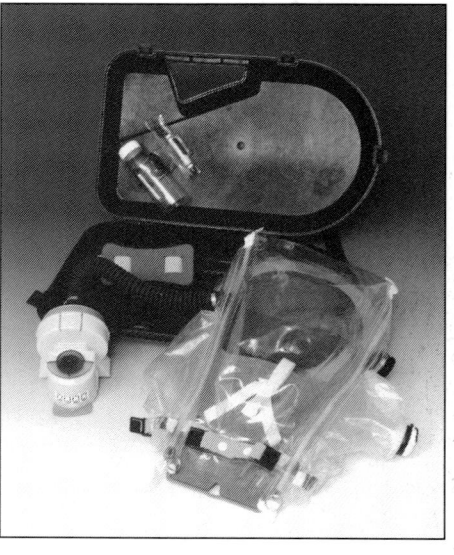

Supergum NBC protective module for babies
2001

Status

In production. In use by the Israeli Ministry of Defence.

Manufacturer

Supergum Ltd.

VERIFIED

Supergum NBC multipurpose overboot

Description

A new multipurpose overboot has been designed by Supergum to meet the requirements of the Israel Defence Forces and approved by the IDF. The overboots are made of PVC with a minimum thickness of 1.3 mm to provide optimum weight/ protection ratio for the soldier. The material provides almost unlimited shelf life in storage. The overboots are supplied in three sizes - small, medium and large. Additional sizes are possible according to customers' specific requirements. Supergum is certified to ISO 9001.

Status

In production. In service with the Israel Defence Forces.

Manufacturer

Supergum Ltd.

VERIFIED

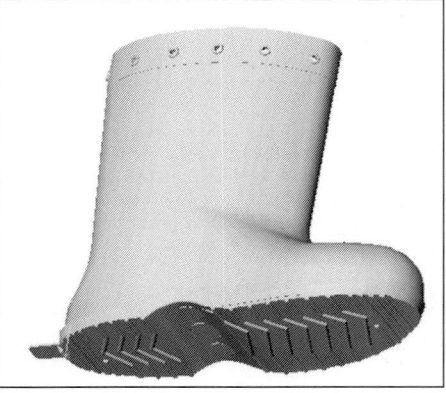

Supergum multipurpose overboot (Supergum Ltd) **2001**/0077959

Supergum NBC protective clothing

Description

Supergum NBC plastic and rubber protective clothing is intended for emergency use by both military and civilian personnel and is available in a range of sizes to suit adults and children. The clothing consists of a blouse, overtrousers, gloves and footwear covers, all intended to be worn over normal clothing or uniforms. All items

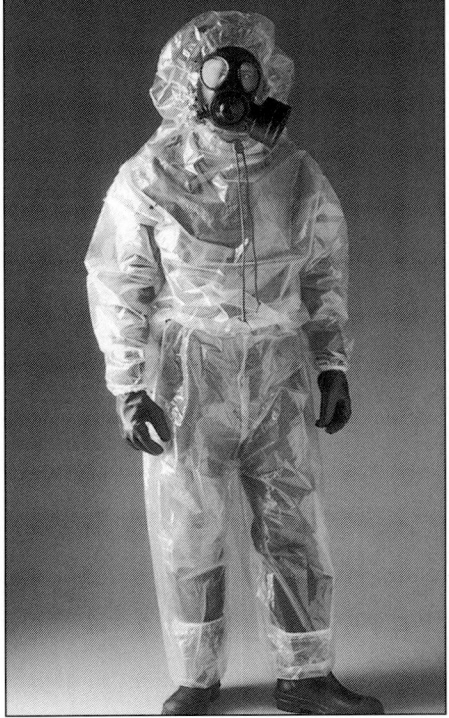

Supergum NBC impermeable NBC protective suit 0011458

are available for issue as a complete outfit contained in a sealed pouch. The material used for all the garments is a transparent five-layer film laminate with a total thickness of 0.13 mm. The outer layer is low-density polyethylene with the inner layer being high-barrier film. All seals are high-frequency welded, which ensures impermeability to NBC agents and gases for at least 6 hours.

The blouse has a cut-out contoured hood which fits around any NBC respirator or mask. A quick-fastener cord is provided for easy and secure closing of the hood. Once fitted, the plastic blouse permits freedom of movement. The protective plastic gloves are disposable while the footwear covers have thickened soles in rubber. Supergum also produces NBC rubber gloves and boots.

A protective personal cape is also available. It can be used as an emergency individual shelter with the user crouching under the cape and holding down or securing the cape periphery. Supergum is certified to ISO 9001.

Status

In production.

Manufacturer

Supergum Ltd.

VERIFIED

Supergum NBC protective rubber boots

Description

This is a new design of NBC protective rubber boot. Made from butyl rubber. The boots provide high durability and excellent protection. With a shelf life of 20 years in storage, the rubber boots are available in five sizes. They complement a new overboot design (see separate entry in this section).

Status

In production. In service with the Israel Defence Forces.

Manufacturer

Supergum Ltd.

VERIFIED

Supergum NEC protective rubber boots (Supergum Ltd) **2001**/0097681

ITALY

Aero Sekur SpA heavy-duty NBC clothing

Description

Designed for use by personnel who have to work in a decontamination role or in hazardous conditions, the Aero Sekur SpA heavy-duty NBC clothing is impermeable to any NBC agents. The main garment is a one-piece coverall with an integral hood, made of nylon cloth coated on both sides with butyl rubber. Heavy rubber boots are glued to the ends of the trouser section of the overgarment to provide a complete seal and the butyl latex gloves are sealed to the garment sleeves using rubber rings.

Being impermeable to liquids, the outfit is also impermeable to body heat and perspiration so it cannot be worn for other than very short periods. To overcome

Aero Sekur SpA heavy-duty NBC clothing (Aero Sekur SpA)

this inconvenience, the suit can be equipped with an internal ventilation and filtered air system powered by a portable turbo unit which employs two standard NBC filter canisters.

Protection against liquid agents is afforded for more than 6 hours. The external finish is a NATO infra-red reflecting green colour.

Status
Endorsed by the Italian Ministry of Defence. In service.

Manufacturer
Aero Sekur SpA.

UPDATED

KOREA, SOUTH

Samgong NBC protective clothing

Description
The Samgong NBC protective clothing ensemble comprises a 2-piece NBC suit, rubber gloves and rubber overboots. The suit smock and trousers are designed to be worn over existing battledress. The suit material consists of an outer polyester/cotton fabric and the inner (adsorptive layer) is formed from a 2.4 mm thick activated charcoal impregnated polyurethane foam. Good protection against mustard agent aerosols is claimed but would appear to offer adequate protection against nerve agents also.

The gloves are cotton-based, dip-moulded with butyl rubber.

Samgong protective clothing ensemble
0055089

The butyl rubber overboots are located by laces passing through large peripheral eyes and appear very similar in design to the UK Mark 4 overboot (see separate entry this section).

Status
Available.

Manufacturer
Samgong Industrial Company Limited.

VERIFIED

LUXEMBOURG

DuPont Tyvek® range of NBC protection materials

Description
Tyvek® is a flash spun polymer material which forms the basis of a range of products with different levels of comfort and protection:

- **Tyvek® "C"** - Offers a medium level of protection. Tyvek® "C" is an extremely light material (81 g/m²), widely used for the manufacture of contamination avoidance coveralls and ponchos. Available colours are olive green or desert sand. Garments manufactured from Tyvek® "C" occupy an extremely small volume when packed and suits weigh only 280 g. With this level of protection, users would remain safe from CW injury for several hours. Suits would normally be discarded after liquid contamination, obviating the need for lengthy and costly DECON at unit level.
- **Tyvek® "F"** - With a slightly greater area density (115 g/m²), Tyvek® "F" is designed for the manufacture of protective garments likely to be exposed to more severe contamination. This material is replacing the current heavy and expensive fabrics such as butyl rubber. Available in olive green and grey, it is used to manufacture conventional decontamination suits where its lower heat stress burden makes it a good choice. Tyvek® "F" is also used for casualty bags, casualty hoods and equipment covers. Its tensile strength makes it suitable for large area chemical resistant covers, offering liquid CW protection to logistic stocks, water supply units, living areas and key equipments.
- **Tychem® "TK"** - This material delivers an exceptionally high level of protection for sustained operations in a heavily contaminated environment, such as DEMIL and CBW EOD activities. Tychem® "TK" is a complex, high-performance multilayer polymer fabric used for the manufacture of full-coverage suits where operators have to wear stored air breathing apparatus. It is fully resistant to constant, heavy liquid contamination for up to 8 hours.

Status
Available. Widely used in many of the suits and covers shown in this section. See also under Protection (Individual) - medical countermeasures.

Manufacturer
DuPont Engineering Products SARL.

VERIFIED

NETHERLANDS

Avatech NBC protective gloves

Description
Avatech NBC protective gloves are made of high-quality butyl rubber and are manufactured using injection moulding. This system of production results in gloves that offer the optimum balance between tactile sensitivity and protection against

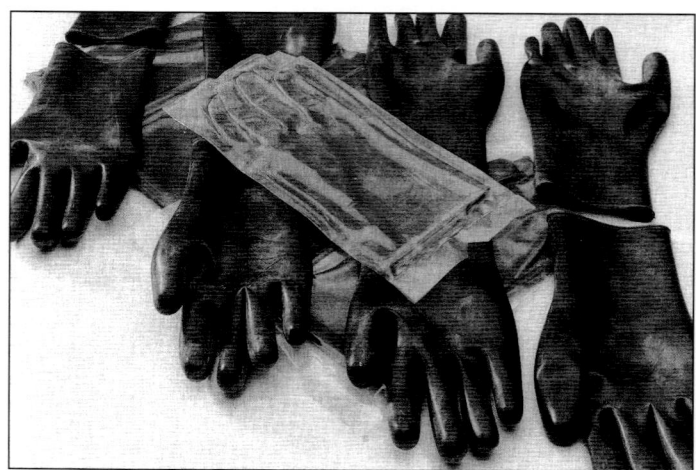

Avatech NBC protective gloves

the penetration of CW agents. The gloves have a penetration resistance of more than 8 hours against H agents (Finabel Test: 37°C) and more than 6 hours against L agent.

The gloves are available in five sizes and have an overall length of 290 or 360 mm. The tensile strength is >13 Mpa (ASTM D412) and the elongation at breaking point is >600 per cent. The gloves are normally worn with cotton inner gloves which can be supplied on request. They are also available with an oil resistant property. NATO stock numbers are NSN 8415.17.039.3515 to 3519 for the five different sizes.

Status
In production.

Manufacturer
Avatech BV.

VERIFIED

NORWAY

Aircrew full coverage NBC protective ensemble

Description
The aircrew full coverage NBC protective ensemble comprises an NBC protective undersuit, a head and neck hood and socks. Each item is made from similar material. The suit is a one-piece overall design which replaces the winter type thermal under garment worn by most aircrew. The material comprises three layers. The outer layer is a coloured cotton knit. The double inner layer carries the NBC adsorptive medium. Activated charcoal spherical adsorbers (average 0.23 mm diameter) are laminated onto a flame retardant PES tricot base.

The hood is designed to cover the head underneath the flying helmet and has a wide collar to ensure adequate vapour hazard protection to the neck area. The socks have Velcro fasteners and are designed to provide a gas-tight seal at the bottom of the under trouser and fit inside the wearer's normal size of flying boot.

The suit and socks are available in five sizes and the hood in three sizes.

Status
Available.

Manufacturer
K Stormark Konfeksjonsfabrikk A/S.

VERIFIED

Specifications
Combined weight (DIN 53854): 400 g/m²
Thickness: 1.1 mm (to DIN 53854)
Air permeability: 1.50 cm² min (DIN 53887)
Stress-strain performance: 150 N / 25% (DIN 53857 Part 1)
Stiffness: 30% (warp) and 20% (weft) at 2 mJ
Delamination: to DIN 54310
Flame resistance: to DIN 53438 Part 3 (surface) - category F2; to DIN 54336 (edge) - not resistant
CW agent breakthrough resistance: 4 mg/cm² after 6 h of challenge at 10 mg/cm² (DB-3 test)
NATO standard gas test: 500 mg min/m³ after 6 h (C_t value)

Chemical Cover Poncho CCP-99

Description
The Chemical Cover Poncho CCP-99 is made from a composition of Luflexen (high impact strength proprietory polymer) and Metallocene catalysed polyethylene, offering good protection against liquid CW agent contamination. The material is 50 µ thick. The poncho weighs less than 300 g and the vacuum-wrapped volume is 500 cm³. The facial opening of the hood can be custom-made to fit various national headgear and mask requirements.

Status
Available.

Manufacturer
K Stormark Konfeksjonsfabrikk A/S.

NEW ENTRY

RUSSIAN FEDERATION AND ASSOCIATED STATES (CIS)

Cooling-type hooded coverall

Description
The limits to the length of time rubberised protective garments can be worn can be partially offset by special cooling overgarments which are designed to be worn over heavy protective clothing. The standard garment for this purpose, which was also issued to most of the former Warsaw Pact forces, is made from cotton. It consists of a single-piece coverall with an integral hood, a five-button overlapping front opening and securing straps or ties at the neck, wrist, ankle and mid-calf. There is a belt and a pocket for the protective mask filter canister. In use the suit is soaked in water and the subsequent evaporation of the water cools the suit beneath. The efficiency of the cooling process depends on humidity and temperature, but the suit is not intended for use at temperatures below +15°C.

Status
Available from Russian state factories before privatisation. In service with the RFAS and former Warsaw Pact forces. Current manufacturer unknown.

Manufacturer
(Unknown).

VERIFIED

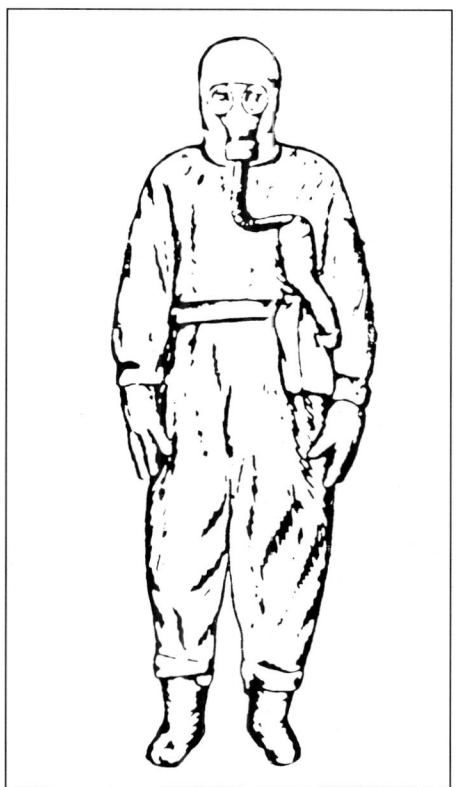

Cooling-type hooded coverall

Heavy protective suit

Description
Designed for use by decontamination personnel and personnel handling toxic agents and munitions, this suit is produced in two basic versions: a one-piece garment and a two-piece suit. As a further variation both can have integral boots. They are made of cloth, coated with a thick layer of rubber. They have integral hoods, sewn-in bibs, wrist, ankle and neck straps, thumb loops and belts. To complete the suits there are cloth liners and heavy gloves and boots. They provide excellent protection against all agents but as with other types of similar impermeable suit they are most uncomfortable to wear. Body heat build-up limits the wearing time to a maximum of 2 hours and in warmer climates to as little as 15 minutes.

Status
Available from Russian state factories before privatisation. In service with the RFAS and former Warsaw Pact forces. Current manufacturer unknown.

Manufacturer
(Unknown).

VERIFIED

Heavy rubber gloves

Description

These gloves are intended for use by decontamination teams or personnel handling toxic agents or fuels. Two types are produced, one with three fingers and one with five. Both are manufactured from heavy black synthetic or natural rubber on a fabric backing. The three-fingered glove is moulded flat and can fit either hand; a wrist strap forms a tight seal. The five-fingered gloves are issued 'handed' and lack the sealing strap.

Both types of glove are intended for use with the heavy protective suit and weigh about 340 g per pair.

Status

Available from Russian state factories before privatisation. In service with the RFAS and former Warsaw Pact forces. Current manufacturer unknown.

Manufacturer

(Unknown).

VERIFIED

L-1 lightweight protective suit, rubberised

Description

Normally worn by RFAS and former Warsaw Pact countries reconnaissance personnel, the L-1 lightweight protective suit consists of a protective jacket with fitted hood, protective overtrousers with integral overboots, two pairs of two-fingered gloves, a helmet liner and a carrying satchel. It is designed so that it can be donned rapidly. The trousers have a strap at the left and a bib joins to straps fitted with elasticised sections. The trouser legs can be fitted with additional straps just below the knee and there are overboot straps at the instep. The jacket, which can be worn over or under the trouser straps (over seems to be the more likely), has an elasticated waist and another strap provides a neck seal. The jacket cuffs and the gloves have elastic sealing bands. When worn with a face mask the L-1 lightweight suit provides complete protection against most NBC agents. The basic fabric is butyl rubber-covered cotton fabric, except the satchel, which is canvas.

The L-1 protective suit weighs 3.31 kg and is available in three sizes. Size 1 is for personnel up to 1.65 m tall; Size II for personnel up to 1.72 m tall; and Size III for individuals over 1.72 m tall. It can be used over a temperature range of from –40 to +40°C.

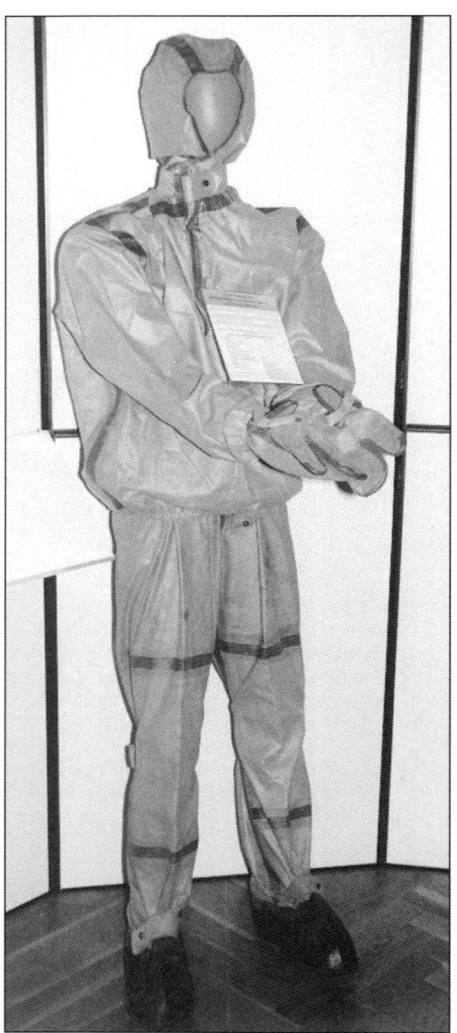

L-1 rubberised lightweight protective suit (T J Gander)

Status

In service with the RFAS and former Warsaw Pact forces.

Manufacturer

Slavyanskaya Clothes Factory.

VERIFIED

OZK individual protective kit

Description

The OZK individual protective kit is intended to protect individuals against chemical agents and radioactive dust. It can also provide a measure of protection against the thermal radiation produced by nuclear explosions, fuel-based flammable mixtures and open flames.

The basic components are a protective cape with a head cover, leggings and protective gloves. The protective cape may be worn as either an overall or as a cape which can also provide some measure of protection for personal weapons and similar equipment. The kit is completed by either rubber or rubberised gloves during warm weather or insulated mittens for cold climates.

The complete kit weighs approximately 4 kg. The kit is produced in four sizes to suit personnel under 1.66 m in height to those over 1.84 m tall. The materials used for the kit are frost-proof at temperatures down to -40°C.

Status

In service with the RFAS.

Manufacturers

(Garments) Slavyanskaya Clothes Factory.
(Gloves) Armavir Rubber Articles Factory.

VERIFIED

Protective Suit, Combined Arms

Description

The Protective Suit, Combined Arms is still held as the standard protective suit for all arms of the RFAS and most of the former Warsaw Pact forces and although there are a few variations, the suit consists of a protective coat-overall (the OP-1), overboots and gloves.

The coat-overall OP-1 is a multipurpose garment produced in five sizes which can be worn in several ways. With the sleeves inside the garment it can be worn as a cape or with the sleeves outside as a conventional overgarment. The lower portions may be wrapped around the legs and secured in place by straps. Apart from its

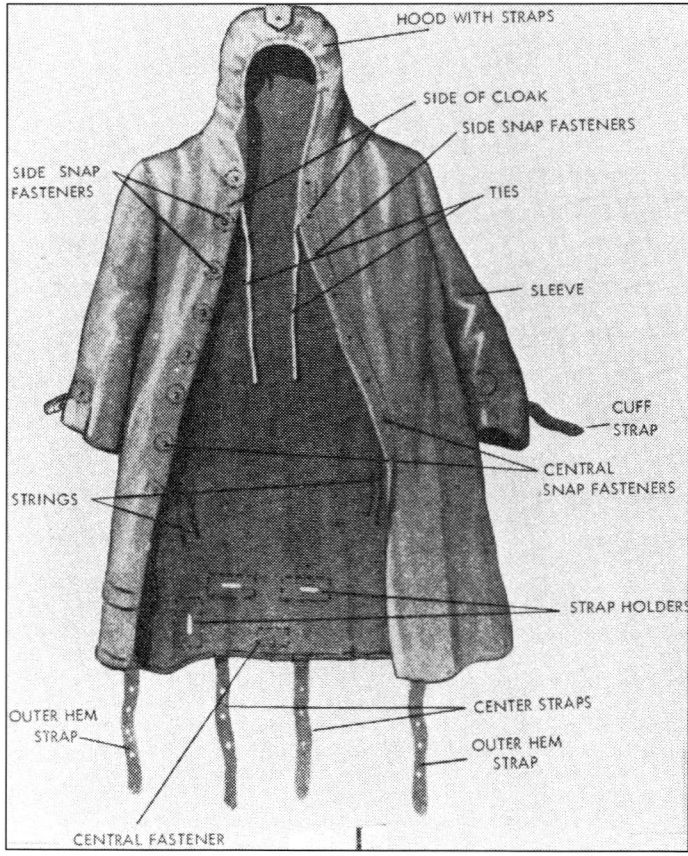

Main component parts of OP-1 protective suit

RFAS NBC reconnaissance troops wearing OP-1 protective suits and showing hoods and Model ShMhelmet masks

protective use, the OP-1 can also be used as a raincoat, groundsheet, field shelter, flotation bag and as an emergency pannier for casualties.

The overboots are usually knee-length but hip-length versions are available. The knee-length versions are secured below the knee by tie-wraps and the hip-length versions are usually secured to the wearer's uniform belt by straps.

The gloves come in two versions: a temperate climate version with five fingers and a cold weather version with two fingers and a thermal lining.

All items are manufactured from rubberised fabric and provide reasonable protection against most NBC agents. However, the suit is not completely airtight and extra protective undergarments must be worn for complete protection.

The coat-coverall OP-1 weighs approximately 1.6 kg, the overboots 1 kg and the gloves 400 g.

Status
Available from Russian state factories before privatisation. In service with the RFAS and former Warsaw Pact forces. Current manufacturer unknown.

Manufacturer
(Unknown).

VERIFIED

ZFO-58 impregnated coveralls

Description
The Model ZFO-58 impregnated coveralls are intended to protect against toxic chemical aerosols and vapours, but not against liquid agents. They can also provide partial protection against biological and radiation contamination. The coveralls are manufactured from chemically impregnated porous cotton fabric and are intended to be worn with underwear, socks and a hood liner which is also impregnated with the same chemical. The suit is sealed by elasticated cuffs and tie-wraps and gas flaps are provided. The coveralls can be worn either by themselves or (more normally) under protective clothing and they are issued in three sizes.

Status
Available from Russian state factories before privatisation. In service with the RFAS and former Warsaw Pact forces. Current manufacturer unknown.

Manufacturer
(Unknown).

VERIFIED

SPAIN

Induyco NBC combat uniform

Description
The Induyco NBC combat uniform replaced an earlier, mainly impermeable set of NBC protection overgarments. The uniform consists of a long jacket having a front opening with a zip fly and an integral hood, trousers, gloves, overboots and a carry bag.

The material used for the garments has an outer fabric which is fireproof, oil and water repellent and treated to provide a low infra-red signature. The inner fabric is formed using 1 mm thick thermo-compressed polyurethane foam impregnated with activated charcoal; the foam is carried on a polyamide protection mesh. The combined layers have an air permeability of more than 200 litres/dm²/min and are resistant to mustard agent aerosols for at least 6½ hours.

Induyco NBC combat uniform (T J Gander)

The uniform is completed by overboots with a chloroprene sole with butyl uppers. The boots are resistant to mustard agent aerosols for more than 18 hours. Gloves are made of butyl rubber with a minimum thickness of 0.35 mm.

Status
In production. In service with the Spanish armed forces.

Manufacturer
Industrias y Confecciones SA (Induyco).

VERIFIED

SWEDEN

New Pac disposable C-Cover S/89

Description
In 1985, the Swedish Defence Material Administration (FMV) invited the Swedish plastics industry to take part in a development competition to design a disposable plastic cover for chemical warfare protection following FMV requirements. In close co-operation with FMV and following troop trials, the New Pac-designed C-Cover S/89 was accepted in December 1987. By the end of 1992 nearly one million units had been produced, including some for delivery to nations involved in the conflict with Iraq.

A complete C-Cover S/89 consists of two packages, a body cover and a pair of foot covers, both tightly rolled and capable of being carried inside the face cavity of a stowed respirator or in a uniform pocket. When required, the rolled body cover

Swedish troops wearing the New Pac disposable C-Cover S/89 for emergency protection (New Pac Safety AB)

package is taken and the disposable stretch film cover is removed. The body cover is then unrolled to its full length and the entry fold is found. The thumbs are then placed into the centre fold and the user assumes a crouching position. The body cover is then placed over the head and shoulders. At this point the disposable outer bag falls to the ground and the cover's patented zig-zag folds unravel downwards to cover the entire body. To put on the entire C-Cover S/89 takes 10 seconds and can be accomplished even when the wearer is lying on the ground. There is space on the back of the body cover to accommodate a backpack and two rubber bands and a plastic belt is stored inside one of the end-sealed sleeves.

The protection process is continued by placing the respirator onto the head and then pushing the respirator through a perforated membrane to obtain a tight fit. Before moving out the foot covers are put into place. They are long enough to protect the entire length of the leg and fastening bands are used to tie them to a combat waist belt or webbing. The foot covers have soles with extra reinforcement around the toe and heel and two pairs of integrated shoe strings are used to fix the soles around the boots. The sole material can withstand at least 5 km of walking across rough ground.

The C-Cover S/89 barrier material consists of five layers of co-extruded polyethylene/polyamide blown film 0.05 mm thick which are guaranteed to be proof against chemical agents for at least 10 hours. The foot cover sole material consists of two layers of non-woven polypropylene fibre fabrics weighing 200 g/m².

The weight of a complete C-Cover S/89 is 560 g. A tightly rolled body cover is 65 mm in diameter and 150 mm long. The foot covers package measures 150 × 150 × 60 mm.

Status
In production. In service with the Swedish Air Force, Army and Marine Forces and the Swedish Rescue Administration. Exported to various nations involved in the 1990-91 Gulf War with Iraq.

Manufacturer
New Pac Safety AB.

UPDATED

New Pac lightweight disposable C-Cover Dress S/97

Description
Using the experience gained from production of the C-Cover S/89 (see entry in this section), New Pac Safety AB developed a low-cost lightweight protective suit known as the C-Cover Dress S/91. The suit provides protection against mustard agents for more than 12 hours and is disposable. In close co-operation with the Swedish Material Administration (FMV), the S/91 version has now been further optimised to meet the higher demands of marine force operations and designated S/97. A complete set of C-Cover Dress S/97 consists of a jacket, trousers and gloves. The jacket has an integral hood. On the front of the hood a reinforcement ring is sealed outside and around the circular mask opening. This prevents ruptures when the ring is stretched to fit tightly around the filter canister, air outlet and eyepieces of the protective mask. There is a patented sleeve construction that makes it easier to produce and to don. The sleeves have flexible wristbands that can be tightened around the gloves. A belt is attached to the jacket and the jacket itself has an extended length below the belt to provide sufficient overlap with the trousers.

New Pac lightweight disposable C-Cover Dress S/97
(New Pac Safety AB)
0055069

The trousers have integrated foot covers with extra fibre fabric reinforcements to be fixed around the boots by two pairs of integral shoelaces. Integral waistband and braces allow the trousers to fit any person. If required, the trousers may be designed without the integrated foot covers. Three-finger protective mittens are also available as options.

The C-proof barrier material employed with the S/97 consists of multiple layers of co-extruded ethylene co-polymers and polyamide, into a natural dull and military green blown film 0.15 mm thick. The foot cover sole material consists of two layers of non-woven polypropylene fibre fabrics with a weight of 200 g/m² and the capability to walk at least 10 km across rough ground or complete a full day's operations without penetration.

A complete set of C-Cover Dress S/97 is packed into a flip-over bag made of the same barrier material as the suit. The package measures approximately 250 × 250 × 70 mm and weighs 1 kg. Complete outfits can be delivered in cardboard boxes containing 10 sets and 240 sets can be delivered on a pallet (24 boxes/pallet). The C-Cover Dress S/97 is offered in sizes L and XL.

Status
In production. A large number has been supplied to the Swedish marine forces and smaller quantities to the Austrian, Danish, Finnish, Netherlands and Swiss armies.

Manufacturer
New Pac Safety AB.

VERIFIED

New Pac Military Service Kit

Description
The New Pac Military Service Kit KS/97 is designed to stow in a haversack and be carried by service personnel on standard webbing belts or with a shoulder strap. It allows individual protective actions prior to a CW attack and, with its DECON kit and anti-nerve agent combopen, offers some immediate post-attack protection. The kit comprises one C-Cover Dress S/97 a disposable C-Cover S/89, a decontaminant puffer bottle similar to the UK DKP-2, a combopen and a Forsheda F2 mask (see separate entries within *Jane's Nuclear, Biological and Chemical Defence 2001-2002*).

Status
In production.

Manufacturer
New Pac Safety AB.

NEW ENTRY

New Pac Poncho CD/100 and N/60

Description
The New Pac Poncho is available in two thicknesses. The 100 μ CD/100 offers protection against long-term penetration by liquid agent. The lighter 60 μ N/60 is

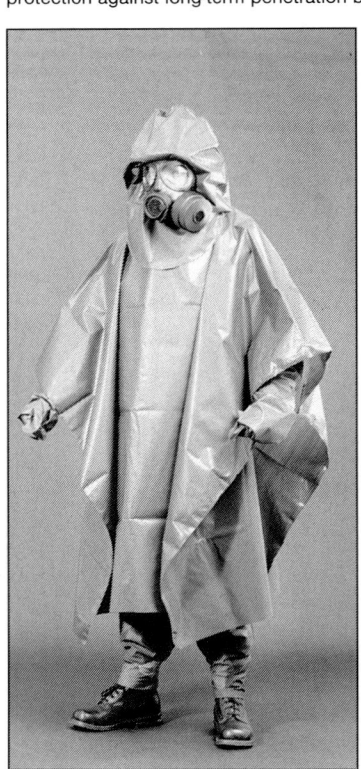

New Pac Poncho NP/60
(New Pac Safety AB)

described as 'splash proof' against liquid. Both comprise a large sheet of plastic film with a central hood which is large enough to cover the head whilst wearing protective mask and helmet. The poncho is designed to be worn over a charcoal-based combat suit as a temporary 'splash cover' to prevent gross liquid contamination (for example, with H agent).

The poncho hood has a circular elasticated hole in the front to allow the filter canister, air outlet and eyepieces of a protective mask to protrude externally. The material is High Molecular Weight, Linear Low-Density Polyethylene (HMW LLDPE), in olive drab. It can also be used as a tarpaulin and covers an area of 1.5×2.5 m (3.75 m²). The Poncho is vacuum-packed, ready for immediate use.

Status
In production. Fully approved by Norwegian Army. Exported to Hungary.

Manufacturer
New Pac Safety AB.

UPDATED

....

Personal Protective Equipment from SWEDE (Swedish Emergency Disaster Equipment)

Description
The Swedish government has strongly supported the creation of a range of response equipment available to the emergency services for use in HAZMAT intervention and recovery incidents. SWEDE forms part of this. This individual equipment is designed to offer augmented breathing capability, ease and comfort of wear and full protection for the user against a range of toxic materials.

Status
Available. In production. In service with the Swedish emergency services.

Full protective ensemble (SWEDE)
2002/0055636

Specifications
Complete ensemble comprises:
- DECON suit is sizes S-M or L-XXL
- Suspenders
- Protective hood
- Blower unit with carrying belt and flow meter
- Battery holder
- 6 × R14 batteries
- Rubber gloves
- Cotton inner gloves
- Rubber boots in sizes 36-47
- 2 × Filter 100 ABEKP3 Radiak
- Drinking facility
- High visibility vest
- Kit of spares
- Instruction manual
- Transport valise

Blower unit 47-6200 EN 164
Minimum airflow: 140 litres min⁻¹
Power source: 6 × 1.5 V standard batteries
Operating time: >6 hr
Weight 850 g excluding batteries and filters

Warning system
  2 signals = unit ready to operate
  Short interval between signals = low voltage
  Long interval between signals = low airflow

Filter 100 (72-1000 ABEKP3 radiak)
Hood (46-6010 EN 146)
Material: Butyl-coated polyester
Neck-collar material: Neoprene
Half-mask: Silicone
Seams: Double seam, taped
Tape: Butyl tape
Safety factor: 30,000
Size: One size fits all adults
Weight: 700 g

Overall
Material: Polyurethane coated polyamide
Seams: Taped
Elastic cuff and collar material: Neoprene
Pockets: 2 chest pockets, 2 leg pockets, 1 for scissors and pen
Sizes: S-M or L-XXL
General: Equipped with braces
Transport valise: Contains all components except boots
Dimensions: 60 × 33 × 33 cm
Weight: 6.5 kg

Resistance against chemicals

Chemicals		Penetration time	Repellency inclination 45°
Ammonia aqua 25%	NH_3	>3 h	98.0%
Sulphuric acid 30%	H_2SO_4	>3 h	96.5%
Hydrochloric acid 24%	HCl	3 h	99.2%
Sodium hydroxide 40%	NaOH	>3 h	99.5%
Benzene	C_6H_6	1 h	91.5%
Toluene	$C_6H_5CH_3$	0.5 h	91.5%
H agent (liquid)	$S(CH_2CH_2Cl)_2$	0.5 h	

Tests performed by FOA (Swedish national defence research establishment), IFP Research (Institute of fibre and polymer technique) and TST (Textil Skyddsteknik AB).

Manufacturers
Swedish Emergency Disaster Equipment.
Textil Skyddsteknik AB.

NEW ENTRY

SWITZERLAND

CW protective gloves

Description
A range of CW protective gloves has been developed by Lonstroff in co-operation with the Swiss Army procurement office. Use of computer-aided design methods has allowed the design thickness to be varied precisely over the glove's area to achieve the right balance between the defined level of NBC protection and the

Specifications
Construction: Proprietary Lonstroff B7129 butyl
Resistance to H agent: >15 h
Tensile strength: 10 N mm²
Elongation: 500% min
Ozone resistance: >72 h (without cracking)

finger tactility requirements of users. The butyl compound used in construction of the glove also varies to take account of the varied suppleness and stiffness requirements at various points on the glove.

The result is a tough glove which produces excellent CW protection, whilst at the same time, allowing prolonged wear without discomfort or heat stress and a low risk of tear damage.

Status
In service with the Swiss Army.

Manufacturer
LONSTROFF AG.

VERIFIED

Dätwyler CoverGlove Y15

Description
The CoverGlove is an injection moulded butyl elastomer, long-sleeved protective NBC glove designed to provide over 15 hours protection against H agents. Available in sizes 8, 9 and 10, the glove is injection moulded and robust enough to be decontamination laundered up to 10 times at 30°C. The operational temperature range is −50 to +100°C.

Status
In service with the Swiss Army.

Manufacturer
Dätwyler Limited.

VERIFIED

Specifications
IRHD hardness: DIN 53505/ISO 868: 45 +/-5
Tensile strength: DIN 53504/ISO 37: >8.5 MPa
Tear strength: DIN 53507/ISO 34: >9.0 N/mm
Elongation at break: DIN 53504: >450%

Dätwyler NBC overboots

Description
Dätwyler NBC overboots, marketed under the name LiveBoots, use a built-up vulcanised construction with the upper section being black BR-butyl-coated cotton etamine. The black sole uses a non-slip pyramid cross-section profile with a

Dätwyler Y15 NBC overboots

reinforced portion in the heel section. The overboots are held in place by laces with an after-foot stopping device. The operational temperature range is −30 to +60°C and the tear strength is to 20 N/mm (DIN 53857). Available in sizes Small (36-40), Medium (41-43) and Large (44-49).

The Dätwyler NBC overboots provide excellent resistance against mustard-type agents (see specifications). Dätwyler also produces NBC protection gloves (see separate entry in this section).

Status
In service with the Swiss Army.

Manufacturer
Dätwyler Limited.

VERIFIED

ISCO multipurpose NBC laminates

Description
Laminates are combinations of different materials brought together to offer several different capabilities within one product. For NBC applications, this allows a better balance between cost and protection factor, better matching of the outer layer material to a specified sealing or welding technology and offers enhancements such as flame retardation, transparency, stiffness, mechanical strength and low-temperature resistance.

ISCO multipurpose NBC laminates have a wide range of applications ranging from NBC hoods to casualty bags and are available in proven standard or tailor-made specifications for particular applications. The finished products are ideal for demanding environments, such as nuclear power plants and civil emergency incidents. They may also be used for the packaging and storage of NBC devices and equipment and for the covering and protection of exposed stores and vehicles.

Approved standard specifications for two types of ISCO laminate are:
- ISCOFLEX WLLW 90.20.20.90 FR: Suitable for hood applications. Used in large quantities for different types of hood by a variety of specialist producers.
- ISCOFLEX WLW 200.30.200: This material is highly transparent and offers 24-hour protection against H agent. It is suitable for visor construction in non-combatant applications and for casualty bags and similar applications.

Status
In production. Widely used in various European countries for the production of hoods and NBC casualty bags. Other tailor-made laminates are available for industrial and other specialist applications.

Manufacturer
ISCO.

VERIFIED

Specifications Type	Y15	Y06	Y06 (1995)
Upper material:	BR-Butyl / double coated cotton (etamine)	BR-Butyl / single coated cotton (etamine)	BR-Butyl / single coated cotton (etamine)
Sole:	Bonded profiled sole extended at front and sides: Natural rubber	Bonded profiled sole extended at front and sides: Natural rubber	Flat, bonded profiled sole
Resistance:	>20 hours (H)	>6 hours (H)	>6 hours (H)
Fastener:	Lace (separate panels on each side of sole and upper)	Lace (separate panels on each side of sole and upper)	Lace (single panels on each side of sole and upper)
Laundry decontamination resistance:	10 cycles at 30°C	5 cycles at 30°C	5 cycles at 30°C
Material area density:	>1100 gm/cm²	>700 gm/cm²	>700 gm/cm²
Impact penetration resistance:	>150 N/mm	>100 N/mm	>100 N/mm
Tensile strength:	DIN 53507, N/5 cm 550-600	DIN 53507, N/5 cm 300-350	DIN 53507, N/5 cm 300-350
Sole thickness:	5.0 ±0.2 mm	3.5 ±0.2 mm	3.5 ±0.2 mm

ROLAMIT barrier films

Description
The standard ROLAMIT product is a biaxially orientated, laminated film designed to offer more than 24 hours impermeability to nuclear, BW and CW agents. It is suitable for making up into a variety of NBC protective products.

By integrating the standard barrier film with cotton fabric, the physical properties are enhanced, providing greater comfort for the wearer and a lower noise factor (see other ROLAMIT entry this section).

Status
In service with the Swiss Air Force.

Manufacturer
PAVAG AG.

VERIFIED

ROLAMIT NBC overboots (disposable)

Description
Impermaplast AG produces ultra-lightweight NBC overboots for the Swiss Air Force. These overboots, made with ROLAMIT, protect the boots of pilots when walking from their shelter through a contaminated airfield to their plane. The overboots maintain their integrity even after walking more than 3.2 km over gravelled ground.

Status
In service with the Swiss Air Force.

Manufacturer
Impermaplast AG.

VERIFIED

SARATOGA (Wattwil) AG Alpine NBC protective suit

Description
SARATOGA (Wattwil) AG is a joint venture between Heberlein Textildruck AG of Wattwil, Switzerland and Blücher GmbH of Erkrath, Germany. The company produces the SARATOGA-Alpine NBC protective suit which can be worn as a

SARATOGA (Wattwil) AG Alpine NBC protective suit

SARATOGA (Wattwil) AG Alpine NBC protective suit

combat suit with all details matched to the user's needs and optimised through numerous fittings. The seal between the suit hood and the protective mask is based on a unique design, as is the special shape of the hood which offers unrestricted freedom of movement, even when the user is wearing a steel helmet.

For production SARATOGA (Wattwil) AG have established a continuously evolving quality assurance system in accordance with EN (Euronorm) 29002 designed to produce zero defects at all times. Every product and accessory used in manufacturing the SARATOGA-Alpine NBC suit undergoes strict quality control at each manufacturing process step and all components and materials are placed on a computerised identification file.

Status
In production for the Swiss armed forces (330,000 suits ordered).

Manufacturers
SARATOGA (Wattwil) AG.
Herbelein textiles AG.

VERIFIED

UNITED KINGDOM

Bondina NBC protective materials

Description
Bondina NBC materials are a range of textile fabrics containing activated carbon. These fabrics offer flexible and cost effective materials which can be produced from a myriad of combinations of components to suit almost any application. Although requirements for NBC protective have changed with time, non-woven materials have always been capable of modification to meet new needs, as is demonstrated by the latest generation of Bondina NBC protection materials that have been developed.

9300G
9300G, the basis of the current UK Mark 4 NBC suit, (see separate entry) was specified by the UK MoD to protect against chemical warfare agents in their liquid or vapour state. The material also provides protection against biological warfare agents, including bacteria and spores. 9300G is both air permeable and durable and can be adapted to be used in a wide range of NBC defence products.

Applications for 9300G include NBC protective suits, undergarments for aircrew, casualty bags and patient wraps and NBC bandages.

Hi-Perm™
Hi-Perm™ is part of a new range of Bondina NBC materials specifically developed to meet the latest challenges of the perceived threat of NBC warfare. In comparison with the current NATO specified material, Hi-Perm™ offers an improved chemical protection material that is lightweight but at the same time more comfortable.

Applications for Hi-Perm™ include NBC protective suits, undergarments for aircrew, casualty bags and patient wraps, NBC bandages, haversacks, socks, filtration and smokehood or breath protection.

A typical NBC protection suit manufactured using Bondina NBC protective materials

Freudenberg has collaborated with Kärcher to produce a new range of materials which combines the Safeguard™ and Hi-Perm™ products (see separate entry under Germany in this section).

Spherical absorber material

Bondina's new spherical adsorber material is a new development using the latest spherical carbon technology. It achieves high performance levels and is lightweight, durable and comfortable to wear. It is capable of meeting any foreseen challenge levels posed by biological and chemical agents and is also launderable. It has been demonstrated that the Bondina 'Next Generation' spherical adsorber material can be laundered up to 20 times without significantly affecting performance.

Applications include NBC protective suits, facelets, casualty bags and patient wraps, bandages and wound dressings, gloves, socks, filtration and smokehood or breath protection.

Status

All the above materials are available. The Hi-Perm™ material has recently been supplied to both the Irish and the Italian armed forces. The 9300G material remains a central part of the UK Mark 4 NBC suit programme.

Manufacturer

Freudenberg.

VERIFIED

Charcoal Cloth chemical warfare protective clothing

Description

Charcoal Cloth Limited have developed a new generation of chemical warfare protective combat uniforms. They are capable of withstanding multiple full-scale chemical attacks, while offering lower levels of physiological burden than conventional chemical warfare protective coveralls.

The Charcoal Cloth Limited system utilises a unique high-performance breathable anti-gas material which can provide extremely high levels of resistance to chemical warfare agents in vapour and liquid droplet form. This fabric can be combined with any specified outer material to meet a wide range of requirements.

Charcoal Cloth Limited protective suits are worn as combat uniforms without the requirement for any form of clothing to be worn underneath. The uniforms have thermal loading characteristics similar to standard combat fatigues. Each uniform has an extended wear life and will provide complete protection against three major chemical attacks or multiple minor attacks, unlike conventional chemical warfare clothing which would normally offer one-attack protection.

The suits have a minimum 25-year shelf-life and can be laundered after training use with no loss of performance.

Status

Available. In service with Middle East and UN forces.

Manufacturer

Charcoal Cloth (International) Limited.

VERIFIED

The Charcoal Cloth NBC protective suit

Coated textiles for NBC protection

Description

Hi-Tech Polymer Proofings Limited is a specialist producer of coated textiles. The butyl range is used for the construction of hoods, ponchos and other clothing elements designed to protect personnel against NBC agents. The material currently features in aircrew NBC mask systems for the US armed forces. Heavier weight versions are used for tank covers, decontamination systems, tents and protective clothing. The HYPALON range is a coated textile system which gives water protection and infra-red absorption characteristics in aircraft cover and radomes as well as the outer layer for ballistic protection vests.

Status

Available. In service with the US armed forces and some other nations.

Manufacturer

Hi-Tech Polymer Proofings Limited.

UPDATED

Chemical protective outfit (Hi-Tech Polymer Proofings Limited)

Complete NBC kit, civilian, heavy-duty

Description

This kit is intended for use by civilian and paramilitary authorities, including civil defence units and can be produced to any size. The kit is stored and carried in a

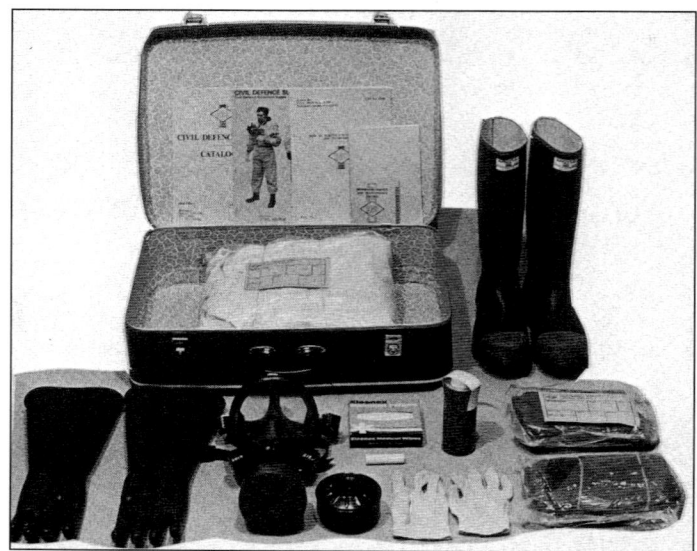

Complete individual NBC civilian kit

suitcase and consists of a heavy-duty washable and waterproof oversuit, an inner two-piece NBC protective garment, full-face respirator with two canisters, set of inner cotton and outer rubber gauntlets, mask demising pack, pair of industrial rubber boots with steel toecaps, roll of repair tape, decontamination powder and an instruction handbook.

The kit can be used as protection against either NBC or industrial hazards. Details of many of the above items can be found elsewhere in this section.

Status
In production.

Manufacturer
Civil Defence Supply.

UPDATED

....

Complete NBC kit, civilian, lightweight

Description
Intended primarily for civilian users as protection against NBC or industrial hazard, this kit can also be used by civil defence authorities. It consists of a complete lightweight kit of garments and equipment in a suitcase and can be produced in any appropriate size. It consists of a washable and waterproof lightweight oversuit, an inner NBC two-piece protective garment, full-face respirator with one filter canister, pair of industrial rubber boots with steel toecaps, set of inner cotton and outer rubber garments, mask demising pack, decontamination powder and an instruction handbook.

Details of many of the above items can be found elsewhere in this section.

Status
In production.

Manufacturer
Civil Defence Supply.

UPDATED

Lightweight civilian NBC suits in different sizes

....

Defender CB Mark 1 civilian suit

Description
The Defender CB Mark 1 civilian suit was specifically designed by Remploy to protect civilians, with particular emphasis on the family unit. When worn with a suitable respirator it will provide up to 6 hours protection in a chemical vapour environment.

Defender CB Mark 1 civilian suits

Designed to be worn over normal clothing and footwear, the Defender suit is available in five sizes. It is a one-piece coverall complete with integral hood, mitts and socks. The hood with elasticated opening is compatible with the Avon AR10/S10 respirator.

The suit is constructed from a non-woven charcoal cloth with a laminated lining. The soles of the integral socks are reinforced with butyl rubber. The suit is also equipped with adjustment straps at the hands and ankles and has a front opening zip for easy access, allowing the suit to be quickly and easily fitted when the need arises.

The suit is provided vacuum-packed in amilon and polythene and with complete wearer instructions in every pack. Storage life is eight years. Each pack measures 300 × 200 × 80 mm and weighs 1 kg.

Status
In production.

Manufacturer
Remploy Limited.

VERIFIED

....

Disposable Type 1A coverall gas tight suit and the coverall liquid tight suit

Description
Tychem TK Type 1A coverall gas tight suit
The Disposable Type 1A coverall gas tight suit is aimed at first responders, organised to deal with the aftermath of a chemical or biological warfare incident. The protective material, Tychem TK, offers protection against a wide range of toxic agents. Against the full range of chemical and biological warfare agents, Tychem TK offers an average breakthrough time of greater than 12 hours at a minimum detectable permeation of less than 0.0002 mg/cm² (DIN6 protocol). The suit is a one-piece unit designed to cover the whole body including any SCBA equipment carried by the operator. The head section is large enough to encompass most types of protective headgear and the panoramic visor allows good peripheral vision.

Tyvek F coverall liquid tight suit
Tyvek 'F' material forms the basis for this suit which is similar in design to the Type 1A but offers protection factors of 48 hours against H agents and 24 hours against GB and L.

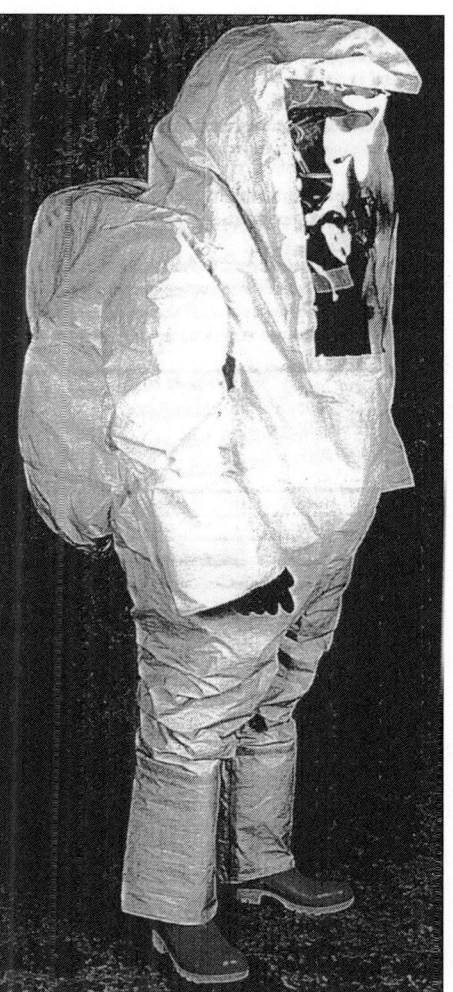

Tyvek F Disposable Liquid Tight Suit (Respirex International Limited)
0055630

Type 1A Tychem TK gas tight suit (Respirex International Limited)
***2002**/0125421*

Status
In production.

Manufacturer
Respirex International Limited.

UPDATED

Heavy-duty outer suit

Description
This is a one-piece heavyweight oversuit that is chemical resistant and flame retardant. Manufactured from PVC-coated knitted nylon, the suit has extra wear patches and double cuffs and leggings. The integral hood can be adjusted to suit a variety of respirators. It is available in large and medium sizes. Intended primarily for civilian use, it can also be used for NBC protection and by civil defence authorities, or by rescue or other such emergency squads.

A lightweight version is also produced.

Status
Available.

Manufacturer
Civil Defence Supply.

UPDATED

Combat suit protected against NBC agents in use

Lantor UK protective fabrics

Description
Lantor UK has been developing and manufacturing activated charcoal fabric for NBC protection since 1963. Original product development was for the UK MoD, based on an activated charcoal-coated non-woven fabric made into the Mk I suit. Subsequent product developments, involving non-woven fabric and activated charcoal, led to the UK MkIV suit which is still the current issue for British forces. The Lantor non-woven system has been sold to many countries as the base of their NBC protection. User requirements for higher protection levels, longer shelf life, comfort and launderability, for example, have led Lantor to develop more advanced systems based on their C-Knit technology and laminates.

Non-wovens
Lantor 9302 G is the oil and water repellent, flame retardant, permeable liner behind the MkIV suit's nylon/modacrylic military outer fabric. Activated charcoal non-wovens are used in the current undercoverall, casualty and nursing bags and bandage covers.

C-Knits
Lantor's latest developments are based on a range of activated charcoal knitted fabrics with unique capabilities, especially in laundering. Available in a range of weights and combinations with other knits and laminates, the comfortable and stretchable nature of C-Knits has found use in undergarments, overgarments and integrated suit systems. The Lantor/Remploy LR range is an example.

Impermeable materials
A small range of film/fabric laminates has been developed using Lantor's expertise in fabric selection and lamination capabilities and leading, for example, to the Remploy LR6 suit which is based on the Lantor 9306 fabric.

Status
In production.

Manufacturer
Lantor (UK) Limited.

VERIFIED

NBC clothing for aircrew

Description
The complete set of NBC clothing for aircrew which was developed at the former Chemical Defence Establishment (CDE) at Porton Down, Wiltshire, consists of an inner coverall, a hood and socks.

The NBC inner coverall permits aircrew to wear normal flying clothes without the constriction of additional outer garments. It is designed to allow the user to wear undergarments to reduce sweat pollution of the activated charcoal used in the material. The inner coverall, which can be worn under immersion suits and other

NBC hood for aircrew

NBC coverall for aircrew

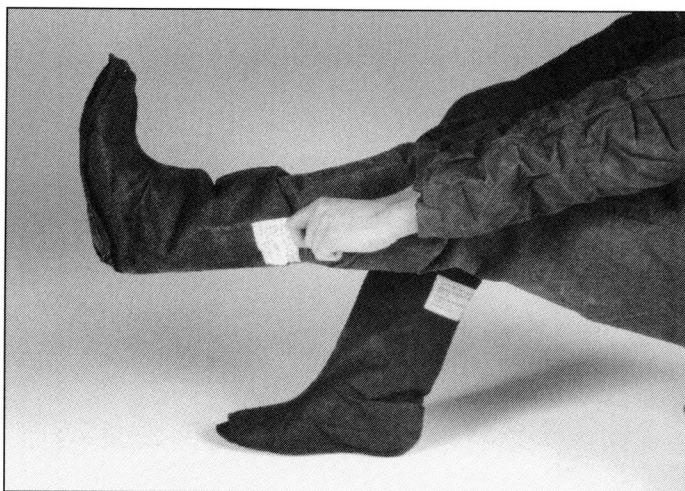

NBC socks for aircrew

garments, is a one-piece close-fitting garment made from anti-gas fabric and has a two-way front zip fastener. Pleating at the knee and elbow gives complete freedom of movement and elasticated stirrups allow a flying suit and boots to be pulled on quickly without risk of rucking. The garment can be worn around the clock and when worn with normal flying clothing, gives protection against all known chemical warfare agents. It also offers a degree of fire protection. Under chemical warfare conditions the inner coverall can protect the wearer for up to 24 hours after the outer clothing has been contaminated. The garment can be worn for up to 50 hours in non-NBC conditions. Nine sizes are available. Weight of the inner coverall is 550 g.

The NBC hood is made of the same anti-gas fabric as the inner coverall and is worn over Type G aircrew helmets and the AR5 respirator system (qv) or over leather helmets and S10 respirators. Only one size is available. Weight of the hood is 120 g.

The NBC socks also use anti-gas fabric and are worn over the user's normal woollen socks but under the NBC inner coverall, flying suit and aircrew boots. They are available in eight sizes from 253 to 308 mm toe to heel. Weight is 800 g.

All items are individually vacuum packed in envelopes with the added protection of an outer polythene bag.

Remploy Limited.and Vermilion Corporatewear are UK MoD authorised manufacturers.

Status
In production. In service with the UK armed forces.

Manufacturers
Remploy Limited.
Vermilion Corporatewear.

VERIFIED

NBC Mk 5 overboot

Description
Made of butyl rubber, the Mk 5 overboot is designed to fit over standard service footwear and meet the standards set in the NATO Triptych for individual protection.

NBC Mk 5 overboots (Silvertown UK Ltd)   0055631

There are three variants, with different tread designs. The land and maritime variants are specialised to offer grip appropriate to the environment. A new variant offers a design which works well in either. Silvertown also markets the Mk 5 overboot to other allied defence forces through Avon Technical Products.

Status
In production. In service with the UK and allied armed forces.

Manufacturer
Silvertown UK Ltd.

UPDATED

NBC poncho

Description
This garment is designed to provide protection over and above that available from existing NBC suits. Manufactured from butyl-coated fabric, there is an integral hood attached to a long poncho-type garment with large loose sleeves. It is worn over the usual NBC clothing and may double as a waterproof. There is also an internal charcoal cloth kilt/skirt. A Velcro fly front is provided and, in addition, there is a large front access zip under the poncho. If required a number of these garments can be joined together to provide a small shelter. The garments/suits may be provided in a number of colours.

Status
Available.

Manufacturers
J & S Franklin Limited.
North Safety Products Limited.

VERIFIED

NBC poncho in sand colour scheme

NBC protective gloves

Description
These protective gloves are in two parts, an inner light cotton liner and a black chloroprene rubber outer glove. The outer glove is impermeable to liquids and vapours and can provide full protection for up to 6 hours. The gloves are stored and issued in sealed plastic packs.

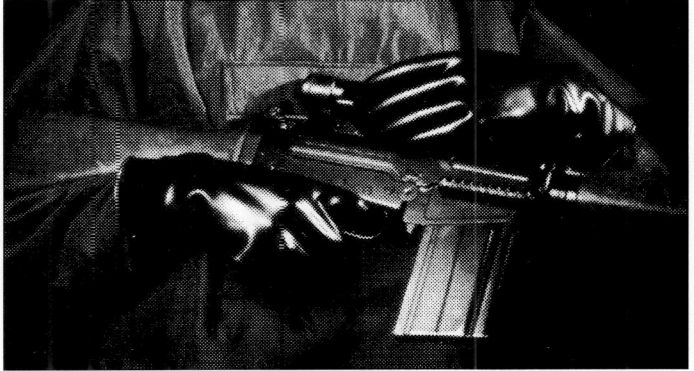

NBC protective gloves

Status
In production. In service with the UK Armed Forces.

Manufacturer
North Safety Products Limited.

VERIFIED

Protective NBC suit No 1

Description
There are four marks of the NBC suit No 1, all made from charcoal-coated, non-woven material inner lining. All are impervious to liquids but permeable to air and water vapour, making the garments more comfortable to wear than other comparable clothing. The material is also flame retardant to provide a measure of protection against nuclear flash and is also showerproof. The charcoal-coated, non-woven material was developed jointly by Dstl (CBD Porton Down), Lantor (UK) and Freudenberg Nonwovens.

The suits are pressure and vacuum packed to give extended shelf-life, which in the case of the Mark 4 is eight years. Both the Mark 2 and the Mark 3 were olive green and consisted of overtrousers and a pull over jacket. The Mark 4, now the only type available, uses a zip-up jacket in shades of disruptive camouflage. The No1 Mark 2 suit was used by the Royal Navy on vessels with citadel systems. The Mark 3 suit was the standard ground forces issue but was replaced by the Mark 4.

The Mark 4 NBC suit is a two-piece suit comprising jacket and trousers and is designed to be worn over combat clothing. The jacket has a front opening with a dual-zip fastening. It has Velcro 'touch and close' fasteners to give effective closures at the wrist and for adjustment to size at the bottom edge of the jacket. Pouch pockets are provided each side of the front opening of the garment. The hood is a close-fitting head piece, lined throughout with a plain weave cotton cloth. The hood is elasticated for close fitting with the S10 NBC respirator. The trousers have Velcro fasteners for waist adjustment and also to provide effective closures at the ankle and calf. Two tapes are stitched at the back of the trousers so that they can be used either as a belt or braces.

The Mark 4 NBC suit uses two layers of material. The outer layer is a flame-resistant modacrylic nylon twill fabric to reduce nuclear flash. The inner layer is the chief protective material constructed from a polyamide non-woven material coated with charcoal and impregnated with fire retardant chemicals. Extensive testing has shown that when the suit is worn over normal combat clothing it provides a high degree of protection against chemical agents likely to be met in the field for a minimum of 24 hours.

Status
Mark 4: in production for the UK armed forces and others.

UK MoD authorised manufacturers
(For overseas sales)
CQC plc.
Remploy Limited.
Vermilion Corporatewear.

Donning the jacket of an NBC suit No1 Mark 4

NBC suit No 1 - Desert Pattern

Marketed by:
Bondina NBC Materials.
J & S Franklin Limited.
Lantor (UK) Limited.

UPDATED

Protective overboots NBC Mark 4

Description
The service model of the NBC overboot was the Mark 3. After 1986 it was replaced by the Mark 4, manufactured from high-quality butyl rubber. The Mark 4 has a combat durability of approximately three weeks and at any one time chemical protection in excess of 24 hours is easily achieved. The materials can be readily decontaminated with standard decontaminants. Improvements introduced with the Mark 4 include an increase of 33 per cent in the thickness of the material used in the upper part of the boot. A better fitting than previously possible is assured by the choice of three sizes (sizes 3 to 5, 6 to 9 and 10 to 13) instead of the former one and an integrally moulded heel. An improved tread pattern on the sole improves traction and stability.

In 1993, the Mark 4 was replaced by the Mark 5 produced by Acton International Inc of Canada. See entry under Canada for details.

A special version of the NBC overboot has been produced for use with skis and snowshoes. Known as the Overboot NBC Ski/March No1 Mark 1, it is available in two sizes (6 to 9 and 10 to 13).

Status
In service with the UK armed forces and in Australia, France, Saudi Arabia and Turkey.

Manufacturers
Avon Technical Products.
Butyl Products Limited.

VERIFIED

Mark 4 NBC protective overboots

Remploy L.R. range of NBC protective clothing

Description
The Remploy L.R. range of NBC protective clothing utilises a series of advanced materials produced by Lantor and developed into a range of garments by Remploy. The range includes the following:

L.R.4 combat garment
The L.R.4 is a stand-alone two-piece garment capable of providing up to 24 hours protection against all known biological and chemical agents. It can be laundered up to 20 times without losing any of its chemical properties. It is produced in 12 sizes and is tailored to give maximum comfort when worn over light undergarments. The trousers are fitted with a zip and Velcro fly whilst the jacket is designed with a crotch strap to prevent any possible riding up. When vacuum packed the suit will give a 10 year shelf-life.

L.R.4 combat garment - part of the Remploy L.R. range of NBC protective clothing

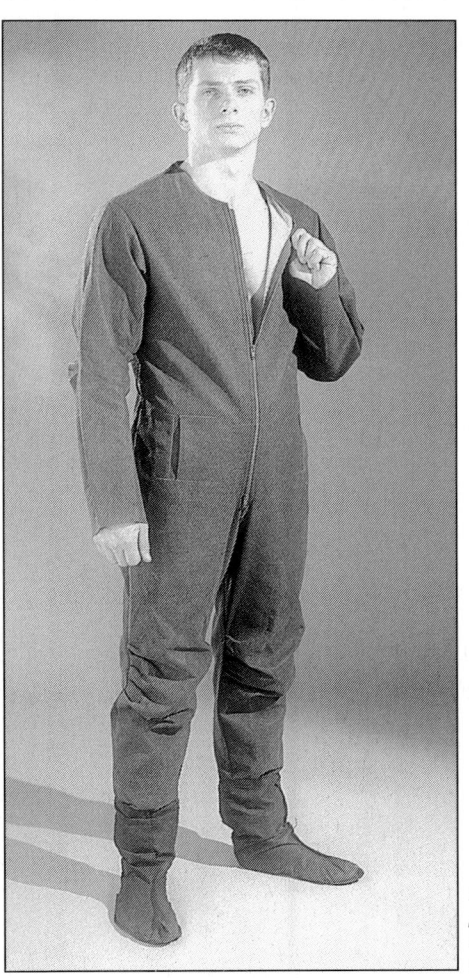

L.R.19 undergarment - part of the Remploy L.R. range of NBC protective clothing

L.R.19 undergarment
The L.R.19 undergarment is one of a series of undergarments manufactured by Remploy using Lantor advanced materials. It is basically a liner comprising a non-woven impregnate with activated charcoal and laminated with various comfort layers, giving a comfortable light garment with maximum protection. The range of undergarments is capable of being laundered up to 10 times without compromising its capability against CW and BW agents. It is designed either as part of an aircrew ensemble or as an undergarment worn in a variety of field warfare conditions.

Status
In product on.

Manufacturer
Remploy Lim ted.

VERIFIED

Remploy range - NBC protective material Tyvek® 'F'

Description
Remploy, with the use of Military Tyvek® 'F' material supplied by DuPont, presents a range of clothing giving high levels of protection against chemical agents and provides a total barrier against biological agents.

The light, strong and completely waterproof material made from a high-density polythene polymer, forms the basis for a range of inexpensive, disposable garments which are available in grey or green. The lighter Tyvek® 'C' material is used for the manufacture of chemical warfare contamination avoidance garments.

The following garments are available:
- two-piece decontamination suit (TFR3)
- one-piece decontamination suit (TFR4)
- cape
- combat rain suit (TFR2)
- casualty hood.

Status
In production.

Manufacturer
Remploy Limited.

VERIFIED

A typical NBC protective suit (TFR3) manufactured by Remploy using Tyvek® 'F' material

UNITED STATES OF AMERICA

Consumer Fuels toxic warfare chemical combination protective suit

Description
Consumer Fuels Inc has developed a toxic warfare protective work suit offering a very high degree of protection to personnel involved in Chemical Explosive Ordnance Disposal (CEOD) or the decontamination of areas affected by residual chemical warfare agents. The suit offers protection for extended periods of time and includes built-in cooling to permit work in the hottest environments, including deserts.

Development of this combination suit commenced with the US Army's proven Protective Outfit, Toxicological, Microclimate-Controlled. This all-enveloping suit, complete with helmet and environmental control backpack, was used as the 'core' suit and two major improvements were added. One was a high-technology overgarment and the other an air plenum added to the base of the backpack. The

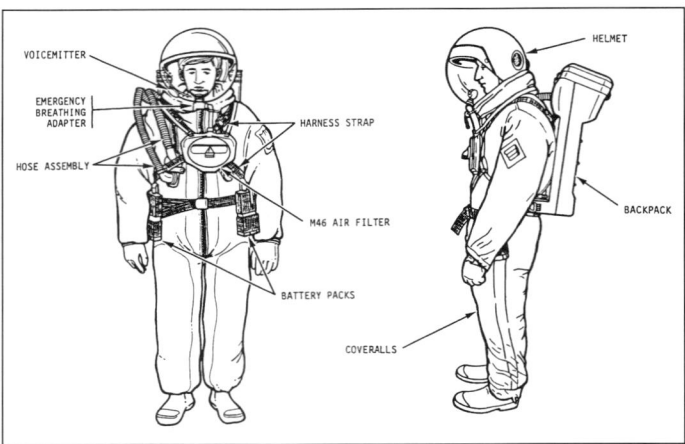

The US Army's Protective Outfit, Toxicological, Microclimate-Controlled, used as the core suit for the Consumer Fuels toxic warfare chemical combination protective suit

Consumer Fuels toxic warfare chemical combination protective suit with bubble helmet removed

The overgarment and core suit are available in various sizes and can be supplied in US Army-approved Desert Tan.

Status
In production.

Manufacturer
Consumer Fuels Inc.

VERIFIED

Consumer Fuels toxic warfare chemical combination protective suit

combination suit offers improved resistance to chemical penetration, up to five times the cooling of the original suit, the option of tethered or self-contained operation and an improved decontamination capability.

The overgarment is pressurised and sealed and is manufactured using a material known as 'Responder' produced by Kappler/Life Guard Industries of Guntersville, Alabama. This material has been used for a commercial toxic protective Level A garment and has been demonstrated as having the highest penetration resistance of any currently available material.

The air plenum engineered for the core suit allows it to be utilised either in its original battery-powered, self-contained mode as a powered air purifying suit or as a supplied air suit with a tethered air line.

When used as a supplied air suit in conjunction with the Consumer Fuels Inc Advanced High Pressure Breathing System the suit becomes one of the most highly protective items of equipment against chemical penetration produced to date. The suit can be supplied with sufficient cooling to enable its use under the hottest desert conditions. The supplied air option also provides the wearer with unlimited work time. When using the supplied air-line system the wearer can at any time revert to the self-contained mode. Full filter canister life and full battery power remain available whenever the supplied air mode is disconnected.

Apart from the core suit, features of the Consumer Fuels Inc combination suit include a charcoal-impregnated undergarment containing the air distribution system; O-ring sealed and locked gloves and heavy boots for the core suit; a large bubble helmet for the core suit, also O-ring sealed and locked; an optional helmet shade and sunshield visor for the bubble helmet; built-in hard line or optional radio communication connections; chemical penetration resistant overgarment; backpack with M41 gas particulate filter and blower; and back-up emergency chemical/biological M46 filter with helmet mouthpiece.

In addition to improving chemical protection, the overgarment holds cold air vented from core suit overpressure valves, thus cooling the outside of the core suit as well as providing shade for the core suit. This outer cooling is in addition to the direct air-flow cooling of the wearer. Overgarment overpressure valves provide a final vent to the atmosphere. The overgarment has double seam sealing on all seams and splash shields are provided at all penetrations such as the overpressure valves, overboots and so on.

Guardian chemical protective butyl gloves

These chemical protective butyl gloves are designed and manufactured in Ohio, USA to provide long-term protection against CBW agents to users for whom tactile

General Specifications			
Tensile strength:	Initial:		1,100 psi
	After ageing:		1,000 psi
	After decontamination:		800 psi
Modulus 200% elongation:	200 ±125 psi		
Ultimate elongation:	Initial:		400%
	After ageing:		350%
	After decontamination:		300%
Maximum resistance against agents (stay times):	HD		75 mins
	GB		360 mins
Resistant against other TICs:	Aldehydes		
	Ketones		
	Esters		
	Alcohols		
	Most inorganic acids		
	Most caustic compounds		
	Dioxane		
Sizes:	X-Small, Small, Medium, Large, X-Large		

Version differences			
	7 mil	**14 mil**	**25 mil**
Thickness	0.178 mm	0.356 mm	0.635 mm
Agent resistance (HD)	75 mins	240 mins	360 mins
Agent resistance (GB)	360 mins	450 mins	450 mins
Low temperature stiffness	(not reported)	180°	138°
Resistant against other TICs:	Aldehydes	Aldehydes	Aldehydes
	Ketones	Ketones	Ketones
	Esters	Esters	Esters
	Alcohols	Alcohols	Alcohols
	Most inorganic acids	Most inorganic acids	Most inorganic acids
	Most caustic compounds	Most caustic compounds	Most caustic compounds
	Dioxane	Dioxane	Dioxane
			Salts

Guardian CP Gloves - (7 mil version illustrated)   **2001**/0069685

sensitivity is important. The latter include aircrew, medical personnel, communications and sensor operators. There are three variants, of different thicknesses (and therefore different levels of resistance to CW agents), depending on the tactile requirements of the user. All three versions offer a tensile strength sufficient to make them highly tear-resistant under normal use. The designs satisfy US specification: MIL-G-43976 (revised).

Status
Available.

Manufacturer
Guardian Manufacturing Company.

VERIFIED

Joint Service Lightweight Integrated Suit Technology (JSLIST)

Description
The Joint Service Lightweight Integrated Suit Technology (JSLIST) programme was established to achieve the next generation of NBC protective clothing systems for the US Army, Navy, Marine Corps and Air Force. Concepts included adsorptive undergarments, CW agent resistant combat uniforms, overgarments, overboots and gloves. All to offer complete protection with reduced heat stress, whilst tailoring the protective level to defined mission scenarios according to current and future threat assessments.

With the aim of equipping all the armed services with a common range of solutions, JSLIST sought the best possible protective ensemble range at the lowest unit cost, thereby minimising the different numbers of suits in current service and capitalising on the resultant economies of scale and reduced life-cycle cost.

Key JSLIST programme performance criteria included: tough agent protection targets; low heat stress; launderability; durability; high degree of user acceptance; ease of maintenance; and low logistic support load.

The outcome of a six-year test and evaluation programme of fabrics and technologies from many major manufacturers was the selection of the Blücher/Tex-Shield SARATOGA™ filter fabric for the individual NBC protection of the US Armed orces (see other entry within this section).

JSLIST overgarment
The Tex-Shield JSLIST overgarment solution is a lightweight, two-piece, front-opening ensemble that can be worn as an overgarment or as a primary uniform over underwear. It has an integral hood, bellows-type pockets and high waist-length jacket. It is compatible with all current personal-issue equipment, body armour, webbing, pouches and weapons. Future compatibility is also ensured with the next generation of body armour, Generation II and future Land Warrior Systems. The overgarment provides optimum liquid, vapour and aerosol protection and is fully compatible with the current US extreme cold weather clothing system, NBC gloves and overboots. The hood design ensures compatibility with the existing US range of masks, including the M17, M40, M42 and MCU2/P series and the experimental XM45 (see under Protection (individual) - Masks (general issue)).

Status
In production. Approximately 4.5 million SARATOGA™ JSLIST chemical protective overgarments will be produced to replace the existing US DoD inventory of Battle Dress Overgarments (BDO) between 1999 and 2008.

Procurement agency
Defence Logistics Agency.

Manufacturer
Tex-Shield Inc.

UPDATED

JSLIST Pre-Planned Product Improvement (P³I)

Description
The JSLIST P³I programme follows on from the JSLIST programme. It is a preplanned product improvement scheme to enhance the performance of the JSLIST chemical protective garment and to develop chemical protective gloves, socks and undergarments. The JSLIST P³I is an army-led programme with participation by the US Air Force, Navy Marine Corps and the US Special Operations Command. The programme will seek material technologies for a lightweight suit which is to be used for missions of short duration, provide less heat stress and protect for a shorter period of time than the standard JSLIST overgarment. Material technologies will also be sought for garments with improved properties such as durability, heat stress, launderability and flame resistance, at a comparable or lower cost than the JSLIST garments. Materials and designs will be sought for chemical protective socks and for gloves suitable for general purpose use, or for duties requiring high dexterity levels.

Status
Remains under development.

Development agency
Joint Service Materiel Group.

VERIFIED

LANX fabric systems

Description
LANX fabric systems comprise a family of NBC protective products that are adsorbent, durable, air permeable and comfortable. Fire resistance is also an option. The base adsorption technology of LANX fabrics is Polymerically Encapsulated Activated Carbon (PEAC), a new and unique technology which provides extremely uniform carbon distribution and so chemical protection. PEAC maximises adsorbent performance, prevents carbon sweat-poisoning and contamination from loose carbon. At the same time, the air permeable fabric containing the encapsulated carbon is constructed to promote evaporative cooling and reduce heat stress. One of the most important advantages of LANX fabrics to the user is their comfort. They are stretchable and the degree of stretch can be specified by the user.

The product line currently includes breathable, chemically adsorbing fabrics for undergarments, overgarments and duty uniforms, gloves, boot liners and developmental shell materials. The Chemical Protective Undergarment (CPU) using LANX Type 1 fabric was developed in conjunction with the US Army Natick Research, Development and Engineering Center. The CPU is type classified by the US Military (MIL-U-44435) and used by US Army Special Operations Forces, armour crews and depot workers. LANX Type 1 fabric is also used to fabricate complimentary gloves and boot liners.

The CPUs made from LANX Type 1 fabric are the only undergarments currently approved for use by the US Military and to pass the US Joint Services Lightweight Integrated Suit Technology (JSLIST) test programme.

The LANX Fabric Systems team has developed lighter weight versions of the LANX Type 1 fabric for use in duty uniforms or as hung liners in a chemical protective overgarment system. These new fabrics are made with the same basic

NBC protective undergarments made from LANX Type 1 fabric
0019556

NBC protective overgarments made from LANX Type 1 fabric
0019555

technologies as the undergarment fabric, have similar comfort and permeability, yet weigh up to 40 per cent less. The fabrics can be combined with a variety of shell materials (including Nylon, cotton and NYCO) to provide vapour and liquid protection as well as flame resistance. LANX fabric systems can be tailored to the users' needs, are versatile and cost-effective.

Status
In service or evaluation by several countries.

Manufacturer
LANX Fabric Systems.

VERIFIED

Multipurpose Rain/Snow/CB Overboot (MULO)

Description
The Multipurpose Rain/Snow/CB Overboot (MULO) programme became part of the JSLIST programme (see separate entry within this section). It had the aim of developing a multipurpose overboot to provide protection from chemical, biological and environmental hazards. The boot incorporates two quick-release side buckles and is designed to be worn over the standard issue combat boot, desert boot, hot weather boot and Intermediate Cold/Wet Boot (ICWB).

The boot provides 60 days wear in all geographical areas without degradation of protection at temperatures equal to or greater than 0°F. It is capable of being decontaminated to an operationally safe level using standard field decontaminants. The boot is made by injection moulding an elastomer blend compounded to provide petrol, oil and lubricant and flame resistance.

Status
Development. Fielded in FY98.

Development agency
Natick Soldier Center.

UPDATED

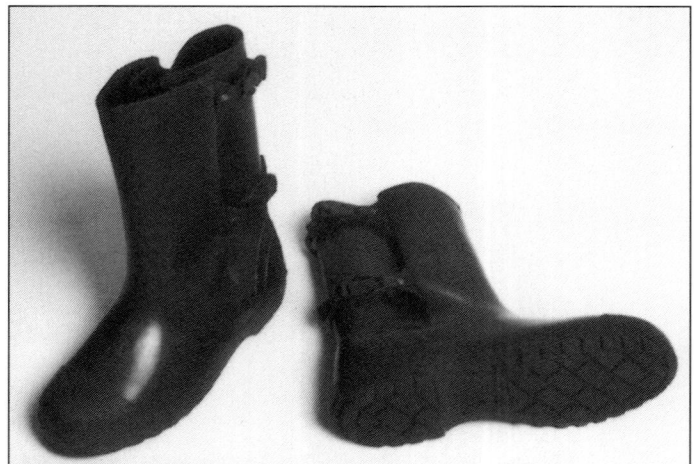

Multipurpose Rain/Snow/CB Overboot (MULO)

Responder® CSM™ protective garments

Description
Responder® CSM™ protective garments comprise a range of impermeable limited-use (disposable) one-piece protective coveralls aimed at first responders (CSM stands for Chemical Surety Materials). The material appears similar to the DuPont Tyvek® impermeable material and the suits are designed to be donned over all equipment including existing designs of mask, protective headgear or SCBA equipment.

The breakthrough time from live agent testing at Geomet laboratories is greater than 80 hours for H and G agents (breakthrough criteria: 4.0 mg/cm² for H/L and 1.25 mg/cm² for G/V agents)

The Responder® CSM™ Level A variant is a totally encapsulating vapour protective suit with a front entry and an expanded shape at the back to accommodate an SCBA (illustrated). The Level B variant is a front entry overall with an optional overhood to accommodate SCBA. The Level C variant is a coverall for wear with lung-powered respirators

Status
Available.

Manufacturer
Kappler Protective Apparel & Fabrics.

VERIFIED

Level A Responder® CSM™ protective garments
0055070

SARATOGA™ System protective clothing

Description
The SARATOGA™ System was originally developed by BLÜCHER GmbH of Erkrath, Germany. The SARATOGA™ System is a family of different materials using spherical activated carbon adsorbers providing good chemical agent protection and user comfort. SARATOGA™ System garments can be washed and reused repeatedly.

The spherical carbon adsorbers have a higher strength than any other form of activated carbon. This characteristic allows higher quantities of activated carbon per unit area than other filters while offering the best possible conditions for adsorption.

A special treatment of the inner surface of the carbon spheres with nitrogen limits water-binding properties and makes SARATOGA™ System materials practically insensitive to humidity and perspiration and allows repeated washings.

The special process of fixing the adsorbers on the carrier ensures maximum adherence with good hydrolytic stability, together with good fastness to rubbing abrasion. The adhesive covers only a small part of the adsorbers, providing a greater number of impact sites leading to improved adsorption.

Advantages of the SARATOGA™ System protection garments include a low thermal burden, repeated washings with no degradation in performance characteristics, a 15 to 20 year shelf-life, a high degree of comfort and a high degree of durability.

The SARATOGA™ System has been adapted to address a broad range of specific military applications by adjusting the fibre or texture construction. Spherical adsorber distribution can vary from 50 to 200 g/m².

Protective garments that are available using the SARATOGA™ System include the following:

US Marine Corps chemical protective overgarment
This Mil-Spec overgarment consists of a bipack construction which includes an outer shell of ripstop cotton treated with quarpel finish, a filter layer which contains

the highly activated carbon spherical absorbers and an inner liner consisting of a polyester knit. The two-piece overgarment consists of a blouse with an integrated hood and trousers with braces. It can be washed up to 10 times with no degradation in chemical agent protection. The garment is delivered in a vacuum-packed polyethylene bag and has a storage life of 15 to 20 years. Chemical protection following removal from the vacuum pack is maintained for several years.

SARATOGA™ undercoverall pyjama with integral head protection

SARATOGA™ CWU-66/P chemical protective flight suit
2002

CWU-66/P chemical protective flight coverall

This mil-spec coverall consists of a monopack construction which includes a 20/80 PBI/Nomex outer surface, a filter layer and a jersey knit inner liner containing 20/80 PBI/Nomex. The garment is a single-piece coverall which meets US Air Force chemical vapour protection requirements. The garment produces a low thermal burden, is flame resistant, provides a high degree of comfort and can be washed up to 10 times with no degradation in chemical agent protection. Shelf-life is 15 to 20 years.

SARATOGA™ pyjama chemical protective under-garment

This lightweight undergarment is worn under standard military clothing, providing protection against chemical agents – it is also flame resistant. Potential users include firefighters, special operations forces and tank crew.

Status

In production. In service with several NATO armed forces including the US Marine Corps and the US Air Force.

Manufacturer

Tex-Shield Inc.

UPDATED

SARATOGA™ US Marine Corps chemical protective overgarment

Self-contained Toxic Environment Protective Outfit (STEPO)

Description

STEPO is an extremely robust individual protected environment. It is designed for prolonged use by personnel required to work in a highly toxic ambient environment. It comprises:

- A totally encapsulating protective suit
- NIOSH-approved 1 hour open circuit self-contained PD apparatus and NIOSH-approved 4 hour closed-circuit self-contained breathing apparatus
- Personal ice cooling system
- Communications interface compatible with most military and commercial radios.

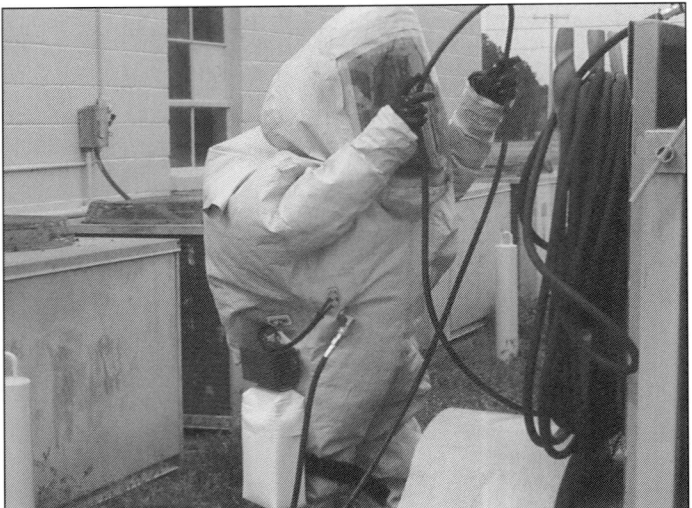

STEPO, showing external air supply and Personal Ice Cooling System (PICS)

The protective suit is constructed from a multilaminate material that provides chemical protection, flame resistance and also dissipates static electrical charges. The outside of the material is coloured light grey to reduce solar loading. The suit is available in four sizes and has a vapour-proof slide-faster closure (front entry), enlarged back to accommodate an SCBA inside the suit, an anti-fogging visor and flexible glove-cuff interface.

The STEPO system is designed to operate with NIOSH-approved open-circuit or closed-circuit breathing apparatus. The open circuit self-contained breathing apparatus can be used with either a 30 or 60 minute cylinder and can be configured as a supplied air respirator with tethered air hose. In the tethered air mode, the suit is configured with an airline pass-through assembly and supplied air shuttle valve. The latter will automatically switch from supplied air to cylinder air if the external supply is interrupted.

The closed-circuit self-contained breathing apparatus (commonly referred to as a rebreather) provides respiratory protection for a 4 hour duration. When the user exhales, the rebreather scrubs the CO_2 from the exhaled breath, adds fresh oxygen and cools the mixture prior to inhalation. This process provides 4 hours of uninterrupted breathing using only 21 ft^3 of compressed oxygen.

The Personal Ice Cooling System (PICS) is used to manage heat stress. The PICS uses ice to cool water that circulates through a liquid-cooling garment to remove metabolic heat from the users body. The ice supply is mounted on the outside of the suit so that it can be replaced for continued cooling.

Status
In production. Deliveries began in 1999 and will be completed in 2002. The STEPO system is currently used by the US Army Explosive Ordnance Disposal (EOD), Technical Escort Unit and other base applications.

Development agency
US Army Natick Research, Development and Engineering Center.

Manufacturers
Biomarine/Neutronics (CCBA Rebreather manufacturer).
Chemfab MENH (suit manufacturer).
GEOMET Technologies Inc (prime contractor and PICS manufacturer).
Interspiro (OCBA manufacturer).

UPDATED

Suit Contamination Avoidance Liquid Protection (SCALP)

Description
The MIL-S-0044384 Suit Contamination Avoidance Liquid Protection (SCALP) is a lightweight overgarment designed to extend the life of an NBC protective suit, combat uniform or work clothes. The suit consists of footwear covers, in three size choices and poncho and trousers - available in five sizes and two colours. It is optimised for comfort and lightness and is constructed of Tyvek® 'C' (100 per cent high-density polyurethane, fibre-coated to .0015 in thickness). The poncho is a pullover design with an integral hood, designed to complement NBC protection masks.

SCALP provides full protection against gross liquid contamination from VX, G agents, POL and decontamination chemicals for at least 1 hour. The complete outfit weighs 1.5 lb. Normal issue is two SCALP per individual and 10 per COLPRO shelter.

Status
Available. Type classified by the US Army in April 1990.

Marketing agency
Tradeways Limited.

VERIFIED

YUGOSLAVIA, FEDERAL REPUBLIC

NBC protective coverall M-3

Description
The NBC protective coverall M-3 is intended for the protection of an individual soldier, his personal equipment and weapon against NBC contamination in the form of droplets and aerosols. The material used for the coverall provides camouflage against detection in the visual and infra-red spectrum. It is intended that this garment is worn together with the protective mask M-2FV, protective gloves M-4 and protective overboots M-1.

The garment is available in one size only. Weight is 1.4 kg.

Status
In production. In service with the Yugoslav Army.

Manufacturer
Trayal Corporation.

Contractor
Yugoimport SDPR.

VERIFIED

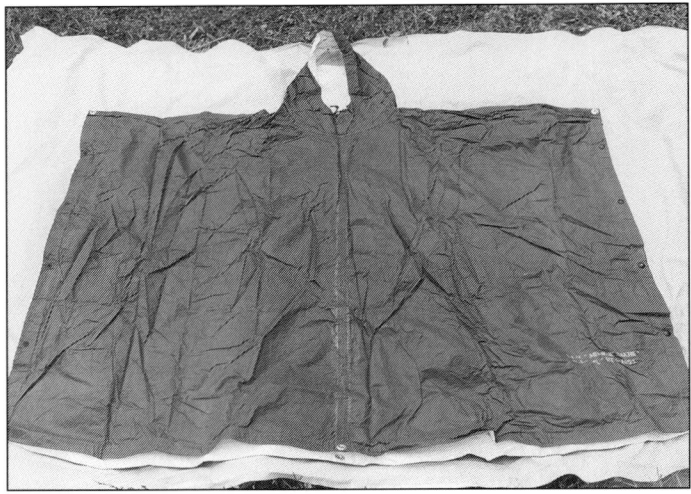

NBC protective coverall M-3   /0077676

NBC protective gloves M-4

Description
The NBC protective gloves M-4 were designed on the basis of original anthropometric data to provide an optimum balance between comfort and tactile sensing. The fingers and palm of the glove are of curved configuration and are made of a high-quality injection moulded compound of butyl and polychloroprene rubber. The gloves have a permeation resistance of up to 8 hours against mustard and nerve agents. They may be decontaminated at least five times and are flameproof. The length of the gloves is 310 mm.

Status
In service with the Yugoslav Army.

Manufacturer
Trayal Corporation.

Contractor
Yugoimport SDPR.

VERIFIED

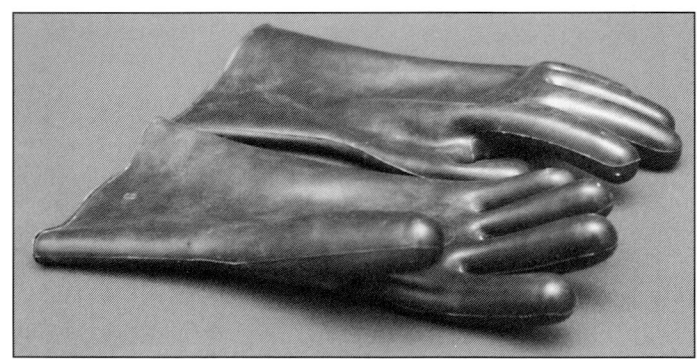

NBC protective gloves M-4   0011459

NBC protective overall M-5

Description

The NBC protective overall M-5 is a one-piece garment designed for use by NBC decontamination and reconnaissance personnel. It is manufactured from flame-resistant rubberised fabric and is provided with an integral hood and elastic sealing closures at the hood opening and at the wrist and ankle. All seams are impermeable. A front zip fastening is provided to allow for quick putting on and taking off of the overall and a drawstring can be used around the waist.

The M-5 overall is available in three sizes (small, medium and large) and weighs approximately 3 kg regardless of size.

The M-5 overall is intended to be worn in combination with the protective mask M-2, protective gloves M-4 and protective boots M-1.

Status

May no longer be in production. In service with the Yugoslav armed forces.

Manufacturer

Trayal Corporation.

Contractor

Yugoimport SDPR.

VERIFIED

NBC protective overboots M-1

Manufacturer

Trayal Corporation.

Contractor

Yugoimport SDPR.

VERIFIED

NBC protective overall M-5

NBC protective suit M-3

Description

The main type of NBC protective suit used by the Yugoslav armed forces is the M-3, a two-piece suit manufactured from polyester fabric rubberised on both sides plus an extra layer of butyl rubber. The suit consists of a blouse with an integral hood and trousers. All parts are sewn and glued together in such a way that no seams are exposed. There are extra inserts at the blouse and trouser waist, arm and leg endings and hood opening to ensure a good seal with other clothing and, in areas such as the hood opening, rubber bands are also provided.

The M-3 suit is provided in three sizes, as follows:

Small (M)	height up to 1.67 m
Medium (S)	height from 1.68 to 1.79 m
Large (V)	height over 1.8 m

Repairs can be made by glueing on rubber patches and the inserts can be replaced when damaged or frayed.

Gloves used with the M-3 suit are the protective rubber gloves M-4 (see separate entry).

Status

No longer in production. In service with the Yugoslav armed forces.

Manufacturer

Trayal Corporation.

Contractor

Yugoimport SDPR.

VERIFIED

NBC protective overboots M-1

Description

The NBC protective overboots M-1 are manufactured from flameproof rubberised fabrics. The upper parts of the overboots are thinner than the bottom and have straps sewn on them for securing them to the legs. The lower parts of the overboots have tie-wraps to secure them to the ankles and they are sewn to the uppers. The soles are ribbed for improved traction on smooth surfaces. Protection is provided against NBC agent droplets up to 360 mm.

Only one size of overboot is available and they are finished in an olive grey colour. Weight is approximately 550 g.

Status

May no longer be in production. In service with the Yugoslav armed forces.

NBC protective suit M-3 in use together with protective mask M-1

NBC protective suit OFZ

Description
The NBC protective suit OFZ has two layers. The outside layer has a camouflage finish while the inner layer has adsorption characteristics. The suit consists of a blouse with an integral hood and overtrousers. It is intended that the OFZ suit is worn over a combat uniform together with a protective mask, gloves and overboots. The suit can provide protection against military NBC agents for 6 hours.

The OFZ suit is available in three sizes (small, medium and large) and weighs approximately 2.2 kg regardless of size.

Status
In production. In service with the Yugoslav Army.

Manufacturer
Trayal Corporation.

Contractor
Yugoimport SDPR.

VERIFIED

NBC protective suit OFZ
0011460

PROTECTION (individual) - Medical countermeasures

AUSTRIA

J Blaschke NBC casualty bag KVH-96

Description

The J Blaschke NBC casualty bag KVH-96 is of the half-bag type intended for use by walking casualties or those with head wounds. The bag covers the entire upper half of the body and head and is provided with sleeves with integral gloves. A clear panel is placed in front of the face. The bag is manufactured using Rolamit A 105 film and will provide protection against the entry of chemical agents for up to 6 hours.

To provide a flow of filtered air, the KVH-96 may be used with SAB-87 (see separate entry), or SBV-93 NBC ventilating equipment worn on a belt around the waist.

Status

Available.

Manufacturer

J Blaschke Wehrtechnik GmbH.

VERIFIED

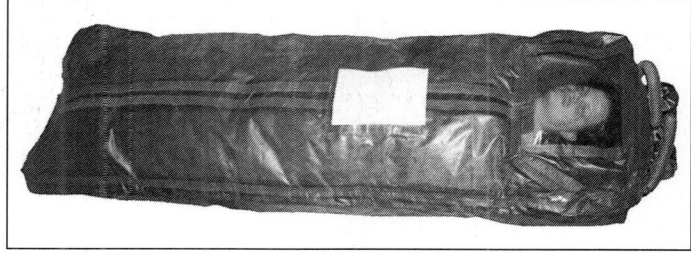

J Blaschke NBC Casualty Bag VBS-93: closed
0052124

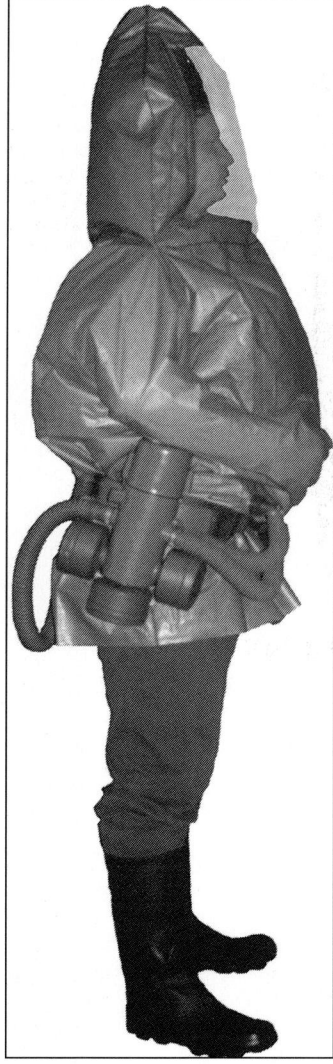

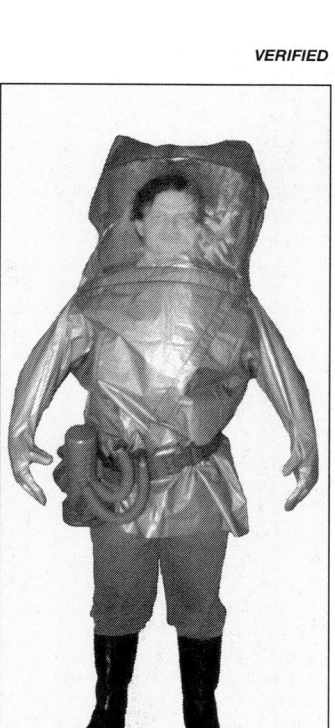

J Blaschke NBC casualty bag KVH-96 with NBC ventilating equipment SAB-87
0011429

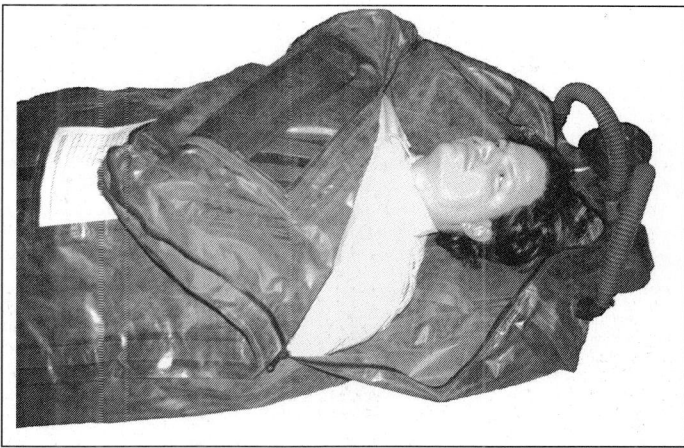

J Blaschke NBC Casualty Bag VBS-93: closed
0052126

J Blaschke NBC casualty bag KVH-96: side view
0055637

J Blaschke NBC Casualty Bag VBS-93
0052125

J Blaschke NBC casualty bag VBS-93

Description

Intended for the transport of NBC casualties from or through contaminated areas, this NBC casualty bag can be manufactured using various materials. Two standard forms are produced, one protecting a casualty for up to 4 hours, the other for

8 hours. Both types are used in conjunction with the NBC ventilating equipment SAB-87 (see separate entry) which has two filter canisters with ventilation to the casualty controlled by a two-stage switch.

If required, this casualty bag can be manufactured using the 'Schichtbarriere 2000' NBC protective material.

Status

Available.

Manufacturer

J Blaschke Wehrtechnik GmbH.

VERIFIED

EUROLITE NBC casualty bag

Description
The EUROLITE NBC casualty bag is specially designed for the transportation of casualties through contaminated areas. The casualty bag is equipped with a security harness for housing a respiration system as well as a transparent window. Important notes for further treatment may be left in an envelope attached below the transparent window. Special closure for initial medical care. For details about the material please see separate entry in Protection (individual) - Body protection.

Status
In service in Austria, Netherlands, Sweden, Switzerland and USA.

Manufacturer
Goetzloff GmbH.

Marketing agency
Goetzloff (USA).

NEW ENTRY

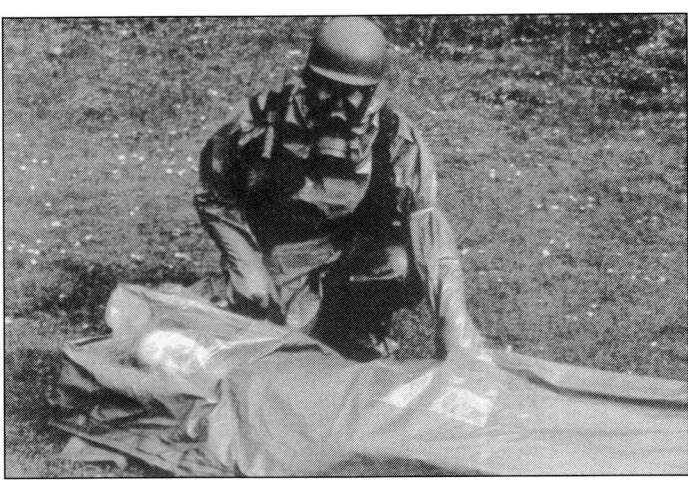

EUROLITE NBC casualty bag (Goetzloff)   **2002**/0137490

CANADA

Carleton Casualty Bag Ventilator Blower/Filter System

Description
The Carleton Casualty Bag Ventilator Blower/Filter System provides positive pressure, filtered air flow to the casualty inside the bag, offering improved protection against the intrusion of BW or CW agents and reducing the heat stress and breathing effort imposed on the casualty. The system uses field-proven components and has been selected for use by the Canadian Department of National Defense.

The primary components of the system are a lightweight motor-driven blower, filter assembly and controller/battery assembly. Total system weight, including batteries and C2 filter canisters, is less than 2.15 kg.

The blower motor and controller/battery assembly attach to the casualty bag. Ambient air is drawn through two standard C2/C2A1/C7 filter canisters, connected by a T-piece adapter to the blower and delivered to the casualty bag through a butyl

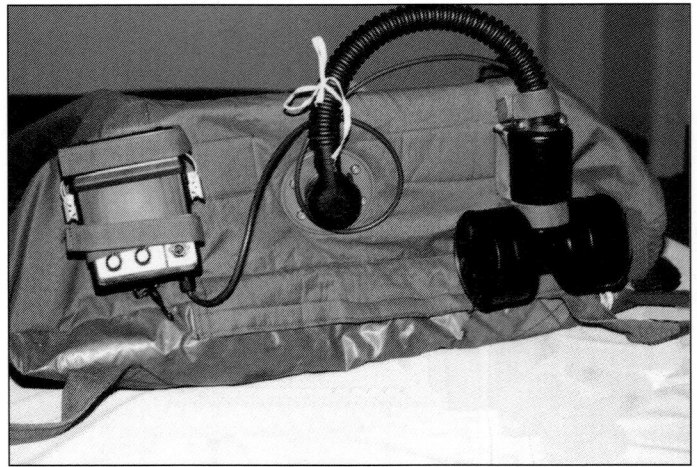

Carleton Casualty Bag Ventilator Blower/Filter System   0041302

rubber hose assembly (see entry under PROTECTION (individual) - Filters). The hose connects to the casualty bag through an adapter, which can be configured to fit a number of casualty bag designs. The ventilator system can also supply oxygen from an external source to the casualty through an optional hose assembly. All system components are oxygen compatible.

Either an external 6-28 V DC external power supply or two standard 6 V DC lithium batteries (NSN 6135-21-906-7728) contained in the controller/battery assembly powers the ventilator motor blower. Air flow to the casualty bag is controlled through a five position rotary switch on the controller/battery assembly.

The positive air pressure delivered by the system improves the protection of the casualty bag against the intrusion of toxic, BW and CW agents. The air flow provided by the ventilator blower/filter system also reduces casualty heat stress and breathing effort.

Status
Available.

Manufacturer
Carleton Life Support Technologies Limited.

VERIFIED

CZECH REPUBLIC

The protective mask for head injuries ShR-2

Description
The ShR-2 protective mask is designed to provide CW protection to personnel with head injuries. Of a similar vintage to the Russian ShM mask (see separate entry in DETECTION (individual) - Masks (general issue)), the facepiece is made from natural rubber. The head cover is large enough to fit over a head which has been treated with bandages and the side-straps allow the mask section to be gently drawn over the eyes, nose and mouth to provide an adequate seal whilst still securely retaining the mask over the injured part of the head. The carrying bag, in which the mask assembly is supplied, is made from waterproof fabric. The bag permits the storage of filter, protective mask, cleaning material and anti-demisting plates.

Status
Remains in service.

Manufacturer
VVŠ PV.

VERIFIED

Specifications
Weight (mask set): 1.75 kg
Weight (facepiece): 0.63 kg
Weight (OF-11 filter): 0.91 kg
Sizes: Universal (one size)

FRANCE

Giat NBC individual NBC survival kit

Description
This kit is intended to be issued to all personnel likely to encounter NBC warfare conditions and is carried on the person at all times, usually in a respirator case. The kit is carried in a fabric wallet which contains a book of adhesive detector papers, a

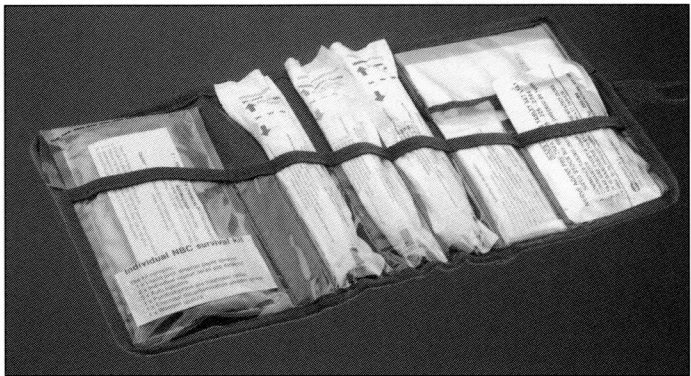

Giat Individual NBC survival kit

pack of two DET INDIV nerve agent vapour detectors, one or two mle F1 decontamination gloves filled with Fuller's earth, a box of pyridostigmin nerve gas pretreatment tablets and two or three MultiPen or ComboPen autoinjectors.

Status
Available.

Manufacturer
Giat Industries.

VERIFIED

Giat NBC individual protective kit for police and rescue services

Description
This kit is intended to equip police and rescue teams with a minimum of equipment to allow them to intervene in chemically contaminated areas. The kit, contained in a travelling bag, provides adequate protection equipment and decontamination items for an individual to carry out reconnaissance, contaminated area control and marking and decontamination missions.

The valise contains a respirator with a carrier bag, one NBC canister, one $A_2B_2E_2K_2P_3$ canister, a voice adapter for the respirator, two individual nerve agent detectors, two pairs of decontamination gloves, detector papers, a lightweight decontamination suit, gloves and overboots.

Status
In production.

Manufacturer
Giat Industries.

VERIFIED

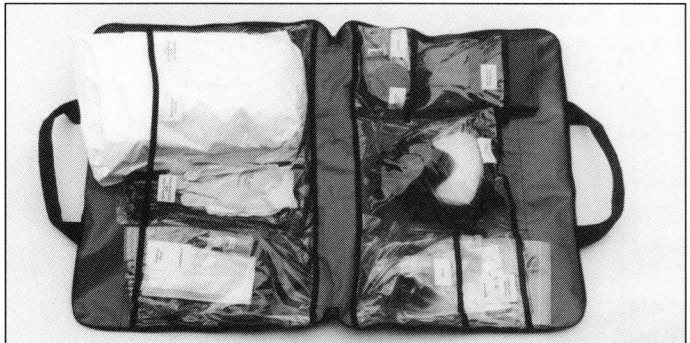

Giat NBC individual protective kit for police and rescue services

---

Giat NBC rescue and lifting kit

Description
This kit allows casualty emergency decontamination and the protection of a casualty who must be rescued by lifting from an NBC-contaminated area. The equipment carried in the kit protects the casualty's respiratory tracts during evacuation and permits the first steps of emergency decontamination.

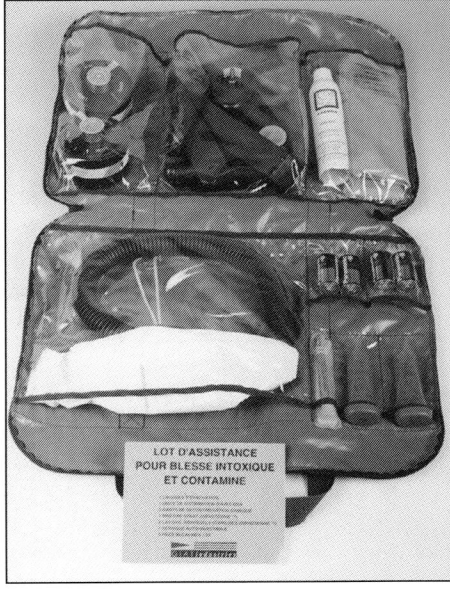

Giat NBC individual protective kit for police and rescue services

The kit contains one casualty hood, a CASU C420 NBC air filtration unit, a carrying strap, four LR20 alkaline batteries, two $A_2B_2E_2K_2P_3$ canisters, two decontamination powder gloves a mini-spray, plus two individual sterilised eyewashes for the emergency treatment of chemical burns.

Status
In production.

Manufacturer
Giat Industries.

VERIFIED

NBC casualty bag

Description
This NBC casualty bag for recumbent wounded is provided with an air filter generator which uses a bank of standard filtration canisters. The canisters, together with five batteries (15 V), provide the casualty with protection for up to 8 hours. The casualty bag is provided with a zip-fastener for access to the casualty and an interior pocket for medical care instructions.

The bag and filtration set are packed in a container measuring 780 × 420 × 420 mm.

Status
In production. In service with the French armed forces.

Manufacturer
Techniques Michel Brochier.

VERIFIED

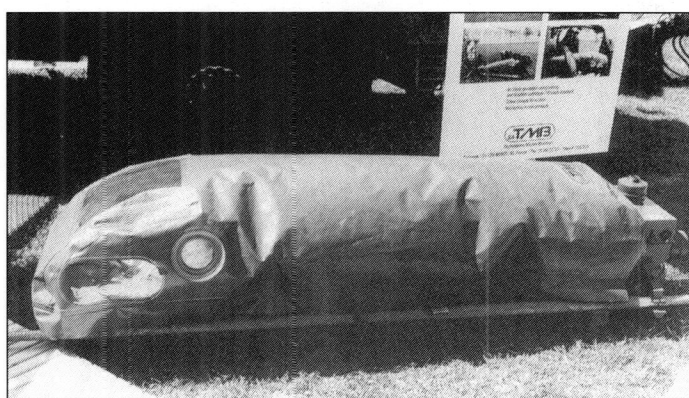

SA TMB NBC casualty bag (T J Gander)

GERMANY

Atropine Aerosol Spray (AAS)

Description
Developed as an alternative to atropine injection devices when countering poisoning by nerve agents and similar organophosphate substances, the Atropine Aerosol Spray (AAS) is a simple spray bottle device that allows atropine to be applied by medically untrained personnel or self-administration via the nose or mouth. It is claimed that each unit reduces the cost of applying identical doses by other means by as much as 98 per cent.

The AAS consists of a glass container coated with plastic foil for burst protection, a dosage valve assembly, a nebuliser, adapters for intra-nasal or oral application and cleaning materials. To use the AAS the protection cap is removed from the top of the container, one of the adapters is placed over the top of the container and one or two strokes of the dosage valve are made to prepare the device. If the nasal adapter is used it is placed into a nostril and a single stroke of the valve is made. For oral use the suitable adapter is placed in the mouth and a single stroke of the valve is made at the beginning of a deep inhalation. The breath is held for 10 seconds and then exhaled through the nose. The adapter is then cleaned, using the materials supplied and the protective cap replaced.

There are two sizes of AAS. The AAS-S contains 10 ml, sufficient for 200 applications (±3 per cent); the AAS-L contains 20 ml for 400 applications (±1.2 per cent). Each application contains 1.67 mg of atropine-free base per stroke, which is equivalent to 2 mg of atropine sulphate. In the event of nerve agent poisoning, it is suggested that three applications are made at intervals of 10 to 15 minutes.

The AAS-S weighs 50 g when full, the AAS-L 70 g; each adapter weighs 2 g. The AAS-S is 27 mm in diameter and 86 mm high, whereas the AAS-L is 34 mm and 93 mm. Storage life at less than -10°C is five years, decreasing with increases in temperature.

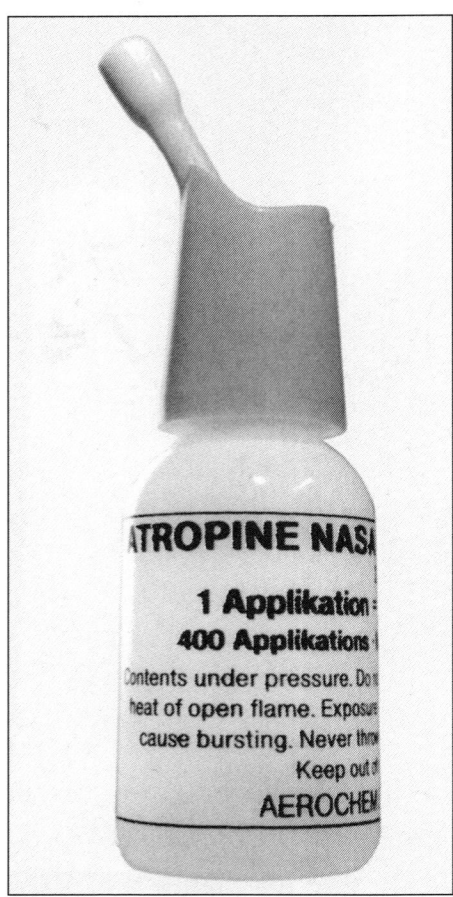

Atropine Aerosol Spray AAS-L

A method of using an AAS unit while a respirator is being worn is under development.

Status
In production.

Manufacturer
AEROCHEM.

Agency
INTER-CB, F H Schneider.

VERIFIED

Helsa-Werke casualty bags

Description
The Helsa-Werke casualty bag is intended to provide full NBC protection for casualties and is manufactured from Helsasafe chemical agent-proof material. A transparent panel is provided in front of the casualty's face and access to the interior is provided for use by medical personnel. An air blower unit supplied via

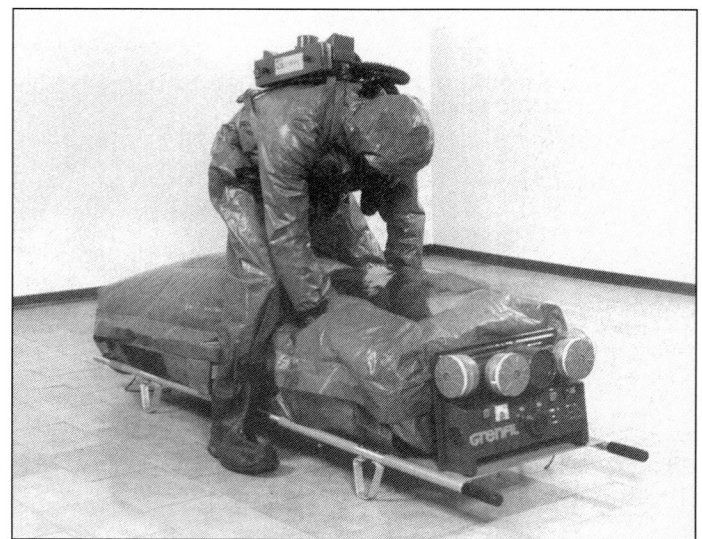

Helsa-Werke casualty bag   *2000*

Helsa-Werke casualty half-bag
2000

four standard NBC filter canisters is provided to create an internal overpressure for the bag preventing ingress of contaminated air and to assist the casualty's breathing. The casualty bag is provided with an air conditioning and pressurising device for hot and cold environments. The device can also be employed with butyl protective suits.

There is also a Helsa-Werke casualty half-bag consisting of a water and oil repellent outer layer and an adsorbing filter layer.

Status
Available.

Manufacturer
Helsa-Werke, Helmut Sandler GmbH & Co KG.

VERIFIED

Kärcher Mediclean units

Description
Described as the first of their type, Kärcher Mediclean units are intended for the pre-cleansing of wounds or areas of the body contaminated by NBC agents. The unit operates on a spray and extraction principle in which a disinfection and decontamination solution is sprayed on to an area of the body and immediately removed by suction. In this way all surface material is gently removed and sources of infection eliminated.

Both the Mediclean 1000 and the Mediclean 2000 utilise a special jet, developed for safe use on all parts of the body. Mediclean units have integral tanks for the disinfection-decontamination solution and removable tanks for the storage and disposal of used solution. Spray pressure and solution strengths can be varied according to need.

The Mediclean 2000 has an integral solution heating system and is intended for use in hospital casualty reception areas and field hospitals. The Mediclean 1000 is designed for use by paramedic units deployed at an incident.

Mediclean 2000 (Alfred Kärcher GmbH & Co)

Specifications

Unit:	Mediclean 1000	Mediclean 2000
DECON solution capacity:	10 litres	32 litres
Heating:	-	user adjustable
Delivery rate:	1.5 litres min⁻¹	1.5-3.0 litres min⁻¹ (user selectable)
System pressure:	1.0 bar	1.5-3 bar (variable)
Power consumption:	800 W	2,200 W
Hose reach:	2.5 m	4.0 m
Dimensions:	65 × 31 × 43 cm	68 × 48 × 63 cm
Weight:	11.5 kg	31 kg

Status
In production. In service with several countries.

Manufacturer
Alfred Kärcher GmbH & Co.

UPDATED

OWR NBC first aid kit

Description
The OWR NBC first aid kit is stored inside a flexible container roll which can be carried easily by soldiers in a pocket or webbing pouch. The contents are as follows:

- OWR detoxification powder in a 60 g bottle with a sprinkler top used for the treatment of contamination from mustard or nerve agents
- A package of swabs to remove droplets and smears of chemical warfare agents or decontamination powder
- A package of self-adhesive plasters for open wounds
- A wooden spatula to scrape off droplets or smears of chemical warfare agents
- Ear muffs to protect the ears following a gas alarm
- OWR decontamination soap to detoxify and remove chemical agents; after use the area involved has to be rinsed with water
- A spray bottle containing GD solution, an advanced version of DS-2. Following contamination the contents are sprayed on hands, neck and face (with the eyes firmly closed). After a few minutes the spray is rinsed off with water
- A decontamination towel impregnated with GD decontamination solution
- Up to three atropine injectors.

Status
Available.

Manufacturer
Odenwald-Werke Rittersbach GmbH (OWR).

VERIFIED

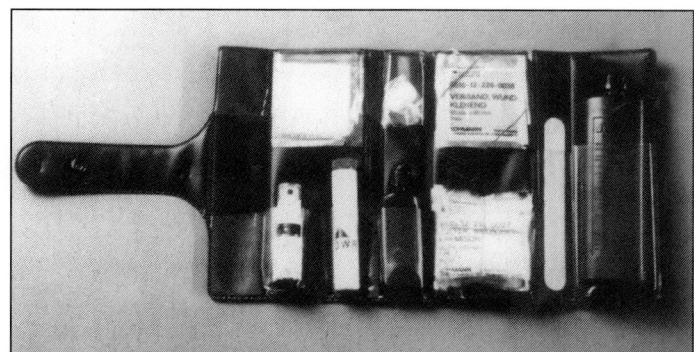

OWR NBC first aid kit showing the contents

ISRAEL

Atromat automatic atropine injector

Description
The Atromat automatic atropine injector is a lightweight and compact injector for the rapid intramuscular self-injection of atropine sulphate solution at the first indication of nerve agent poisoning. The device consists of two plastic tubes, one sliding within the other to activate its injection mechanism. The outer tube contains a yellow safety cap to prevent misfiring, while the inner tube contains a stainless steel cartridge with the sterile atropine solution and needle, as well as the firing mechanism.

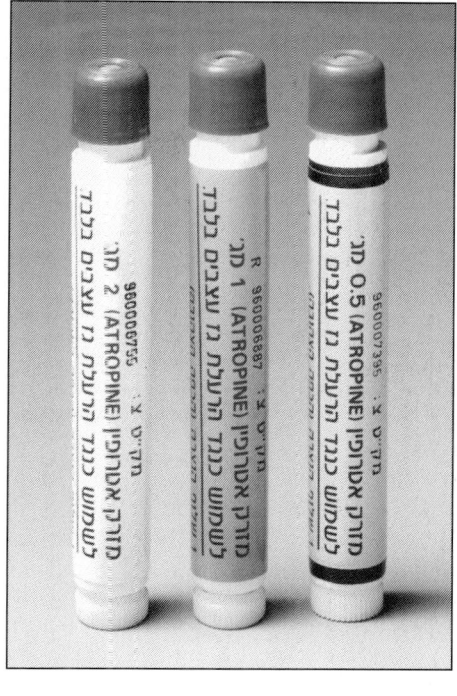

Atromat automatic atropine injectors

Each Atromat is supplied in single-pull snap-open transparent packaging. Units coloured white are intended for adults and contain 2 mg of atropine sulphate. Units coloured green are intended for children or the elderly as they contain 1 mg of atropine sulphate. Units coloured orange are for infants as they contain only 0.5 mg of atropine sulphate.

A typical Atromat weighs 20 g, is 100 mm long and has a diameter of 14 mm. The volume of atropine solution is 0.7 ml.

Status
In production.

Manufacturer
SHALON-Chemical Industries Limited.

VERIFIED

NETHERLANDS

Nerve Agent PreTreatment Tablets (NAPS)

Description
The most advanced therapy for nerve agent poisoning is the immediate (self) injection of atropine, an oxime and diazepam. This therapy is efficient against most nerve agents but the efficiency of this therapy is enhanced by taking Nerve Agent PreTreatment Tablets (NAPS) as pretreatment. The drug is available as a 30 mg tablet containing pyridostigmine and should be taken orally, under orders, three times a day. The standard packing is a plastic and aluminium foil blister pack containing 21 tablets.

Status
Procured by most NATO nations as well as various Middle East and Asian nations.

Manufacturers
Original manufacturers were Solvay Duphar BV and Roche.

UPDATED

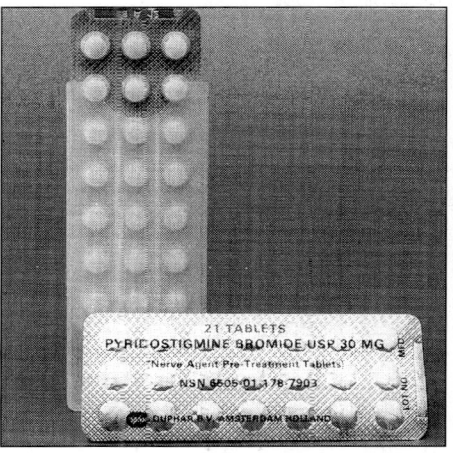

Solvay Duphar Nerve Agent PreTreatment Tablets (NAPS)

Solvay Duphar nerve agent antidotes

Description

The Solvay Duphar antidote range includes the AtroPen, ComboPen, MultiPen and Diazepam-Autoinjector. These are hypodermic devices meant for self-administration, issued to personnel likely to encounter nerve agents.

All the devices are used in the same manner. The safety cap is removed from the end and the injector is pressed against the outer thigh muscles. Further pressure releases the needle which can penetrate combat clothing and the body skin to introduce the antidote into the body. The injector is held in place for a few seconds and then discarded. Instructions for use are printed on the body of the device.

The AtroPen contains atropine only and counters the symptoms of nerve agent poisoning. It is 90 mm long and 12 mm in diameter.

The ComboPen contains both atropine and an oxime. The latter directly counters the nerve agent itself. The device is 145 mm long and 20 mm in diameter.

The Diazepam-Autoinjector is a special version of the ComboPen. Diazepam prevents possible brain damage caused by nerve agent-induced convulsions. This injector was developed under contract to the US Army.

The MultiPen is multichambered and may contain up to three medicaments which are stored separately, for example, atropine, an oxime and diazepam. The device is 170 mm long and 22 mm in diameter.

For training, an inert model without a needle or medical content is available.

Status

In service with the majority of NATO nations as well as Austria, Finland, Switzerland and various Middle East and Asian countries. Also in service with Australia, New Zealand, Saudi Arabia, Singapore and Thailand.

Manufacturer

Solvay Duphar BV.

VERIFIED

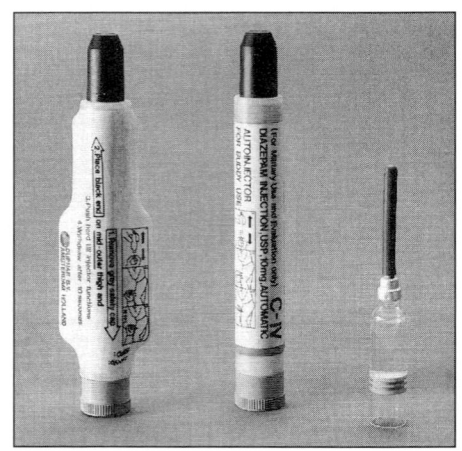

Duphar Diazepam-Autoinjectors with cartridge and needle assembly

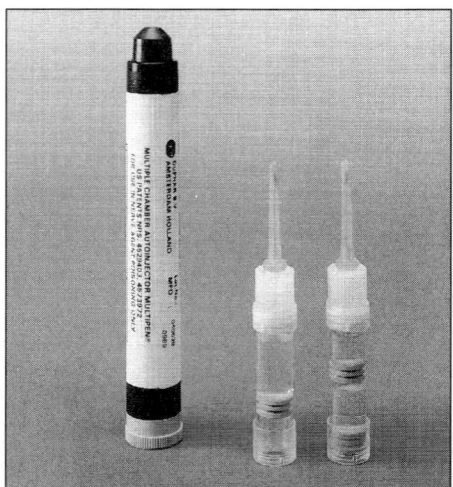

Duphar MultiPen showing multiple-chamber cartridge and needle assemblies

RUSSIAN FEDERATION AND ASSOCIATED STATES (CIS)

Model ShR head-wound protective mask

Description

There are two known versions of this mask which differ only in the positioning of the extra outlet valves. One is virtually identical to that used on the Model ShM (see entry under Protective Masks) and the other has an outlet valve on the right cheek,

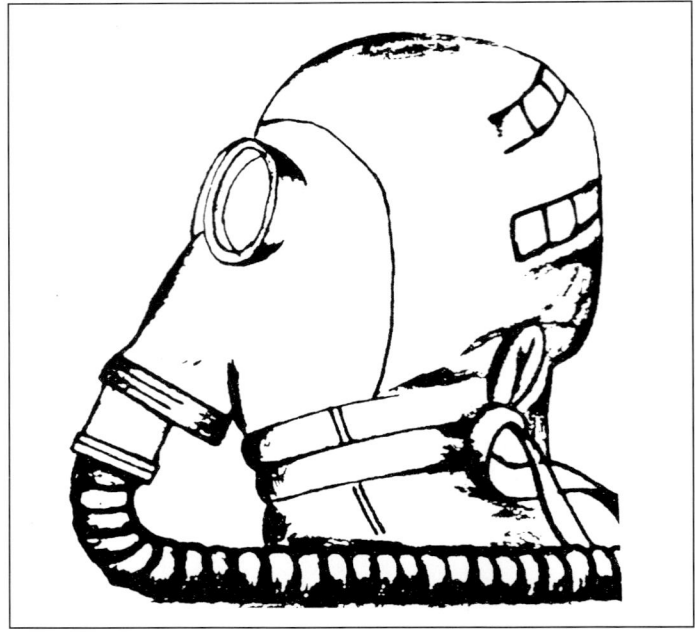

Model ShR head-wound protective mask

but both have the inlet valve at the level of the wearer's nose rather than at the normal mouth level which helps to prevent clogging of the air system by body liquids. The nose-level system also permits the easy supply of oxygen when necessary, but the usual anti-dim deflection of air across the eyepieces as used in the Model ShM is not incorporated. The same hose and filter canister as on the Model ShM is used but the hose can be fitted with a T-piece to allow two filter canisters to be fitted to ease breathing. The Model ShR covers the whole head and it is secured in place by rubber straps.

Status

Available from Russian state factories before privatisation. In service with the RFAS and former Warsaw Pact forces. Current manufacturer unknown.

Manufacturer

Unknown.

VERIFIED

SWEDEN

Astra autoinjector system

Description

The Astra two-chamber autoinjector is a flexible system for the self-administration of nerve agent antidotes. The device is reloadable and consists of a rear section

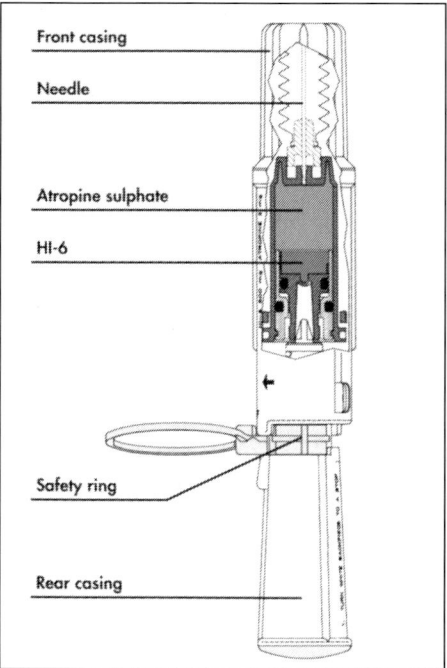

Front casing
Needle
Atropine sulphate
HI-6
Safety ring
Rear casing

Cross-section drawing of Astra two-chamber autoinjector (Astra Tech AB)

with the trigger mechanism and a front section containing the two-compartment ampoule with the antidote. The ampoule system makes possible the use of HI-6 in crystalline dry powder form in one compartment, with atropine sulphate in liquid form in the other. This arrangement enables HI-6 to maintain its stability for five years. The modular design allows for easy changeover to a new or revised antidote. Only the ampoule needs to be exchanged when replacing old stocks.

In use, a safety ring is pulled clear and the ampoule (coloured) end is placed against the thigh. The end of the trigger section is pressed and the antidote is injected into the body via a sterilised 22-gauge hypodermic needle.

A single autoinjector weighs 60 g, is 168 mm long and has a diameter of 26.5 mm. One compartment of the device contains 500 mg of HI-6 in dry powder form. The other contains 3 ml of atropine sulphate in liquid form.

Reusable training versions are available.

A version is available with a single-form antidote, with ampoule sizes of 1.4 and 2 ml.

Status
In production.

Manufacturer
Astra Tech AB.

UPDATED

UNITED KINGDOM

Casualty bags with air blower unit

Description
In order to overcome problems inherent in hot and humid climates and the resultant heat stress inflicted on casualties, a range of casualty bags has been developed. Each is fitted with a Micronel air blower which supplies a constant flow of filtered air to the wearer (See under PROTECTION (individual) - Filters). A whole-body bag is available. It can be used either on existing stretchers or, in confined spaces, on its own, using the integral carrying straps. Constructed from advanced materials, the bag gives maximum protection to walking or stretcher-bound injured personnel.

Status
In production. In service with the armed forces of Australia, Japan, Kuwait, New Zealand, Oman, Singapore, Spain and UAE.

Manufacturer
Remploy Limited.

VERIFIED

Charcoal Cloth NBC protective field battle dressing

Description
The Charcoal Cloth NBC protective field battle dressing contains a combination of modern materials and technology to provide effective immediate treatment of battlefield wounds and then protect them against the effects of chemical agents. The dressing pad contains a high-capacity liquid adsorption layer and an NBC protective strikethrough layer which will prevent the passage of chemical agents (liquid or vapour) through the dressing. The outer cover of the dressing pad is a strong green material. A long conforming bandage enables firm and even compression to be achieved around a wound to stem the flow of blood without impairing circulation. The integral strikethrough layer allows the dressing to be used for 'sucking' chest wounds without the necessity to provide a further covering.

The bandage is long enough to encircle the broadest part of the body and keep the dressing secured even during evacuation movements. The dressing is

Charcoal Cloth NBC protective field battle dressing

enclosed in transparent chemical agent-proof packaging through which an 'instructions for use' label can be easily read.

The field dressing is available in two sizes, 200 × 100 mm and 200 × 200 mm. Other sizes can be manufactured

Status
Available. In service in Ireland and the Middle East.

Manufacturer
Charcoal Cloth (International) Limited.

VERIFIED

Defence Medical Supplies NBC first aid kits

Description
Defence Medical Supplies produces a range of medical kits specifically for protection and treatment during NBC warfare. The range of kits includes the following items.

NBC field ambulance box
This was designed to be carried in a field ambulance and enables a medical team to offer effective treatment to NBC warfare casualties. The main components are a resuscitator, supply of decontamination powder, field dressings, rapid injecting syringes and prefilled syringes.

Type 5 kit
The Type 5 kit contains specific injectables, treatments and antidotes to all known chemical warfare agents. The kit is contained in a rigid and durable container which is resistant to chemical agent penetration.

Type 2 kit
This kit is intended for use by a group of three persons who have the necessary protective clothing. It contains NBC protective and treatment items suitable for self-

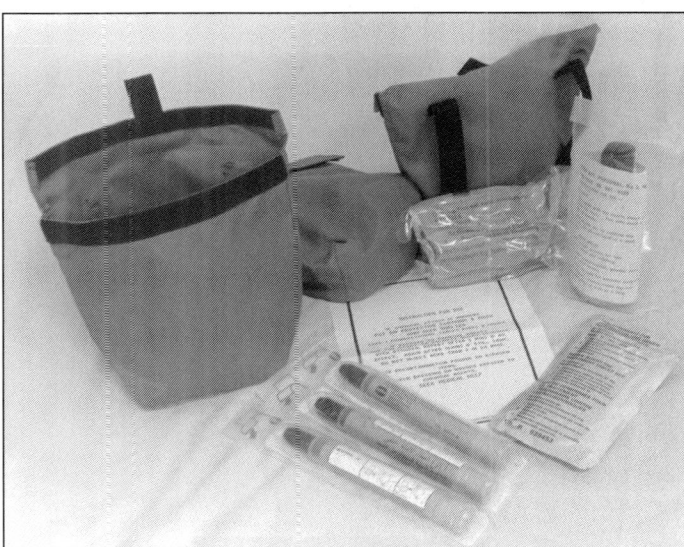

Defence Medical Supplies personal NBC medical kit

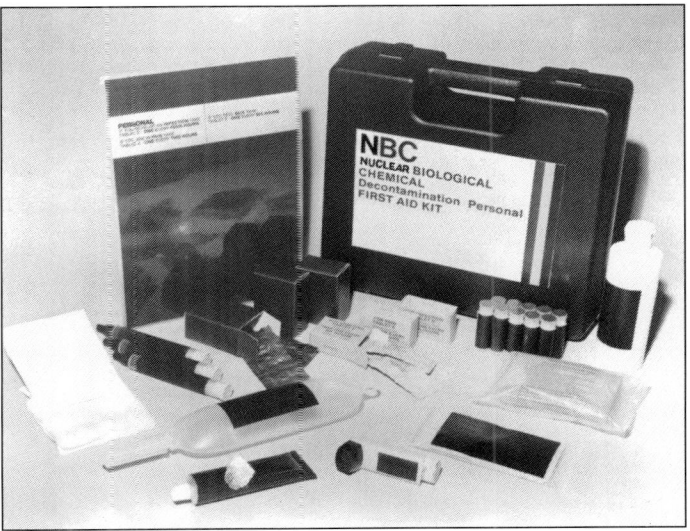

A typical Defence Medical Supplies NBC first aid kit

Defence Medical Supplies Type 5 kit

administration and also provides a comprehensive first aid/medical kit. This kit is suitable for personnel who are remotely located and require medical support to be carried with them.

Personal NBC medical kit

This kit contains all necessary items for the effective and immediate treatment of an individual exposed to a chemical warfare attack. It contains autoinjectors filled with the advanced H16 antidote for the treatment of chemical warfare agent contamination; these autoinjectors are in binary form and have a shelf-life of 10 years. Also included is a pack of tablets for pretreatment against nerve agents and an activated charcoal cloth dressing to give effective protection to wounds received during a chemical attack. Decontamination agents are also provided. The kit is contained in a waterproof pouch which can be easily mounted on a waist belt or strap.

Other kits are available to meet user requirements.

Status

Type 2 kit in service with UN forces.
Type 5 kit in service with Middle East military and special forces.
Personal NBC kit in use by various military and civil defence organisations.

Supplier

Defence Medical Supplies Limited.

VERIFIED

NBC Bandage Cover Mark 1

Description

Developed by Charcoal Cloth Limited in conjunction with the UK Ministry of Defence, the NBC Bandage Cover Mark 1 is provided as a chemical agent protective covering for first field dressings which have no integrated protection. Comprising a liquid spreading green outer layer bonded to an activated charcoal adsorbent layer, the cover measures 250 × 250 mm and is secured by a strong green 100 mm width bandage 4 m long.

The cover is compatible with all in-service first field dressings and will protect battlefield wounds from the harmful effects of chemical agents. It can also be used as a temporary repair patch for damaged NBC clothing. Packed dimensions are 125 × 50 × 50 mm.

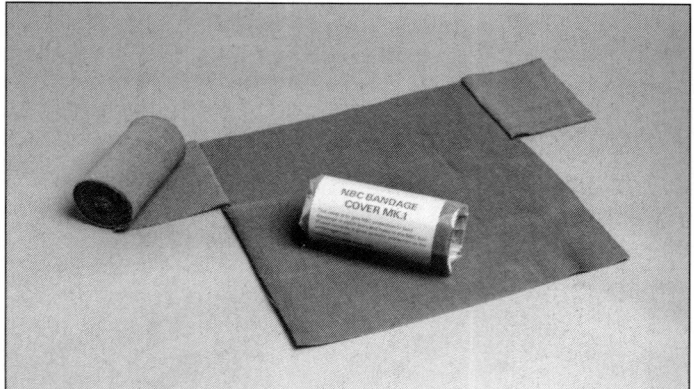

NBC Bandage Cover Mark 1

Status

Available. In service with UK armed forces.

Manufacturer

Charcoal Cloth (International) Limited.

VERIFIED

Pneupac automatic lung ventilators

Description

Artificial ventilation is an essential component of basic and advanced life support. The Pneupac range of automatic lung ventilators is designed to provide vital respiratory support. This is particularly important for NBC casualties where normal respiration has ceased. Without it there is a risk of decreasing respiratory activity leading to hypoxia and cardiac arrest. The Pneupac range of ventilators provides vital respiratory support in this situation. They are of modular construction and can provide a range of options to meet user requirements and operational logistics from field aid stations to hospitals.

On all models there is an option of a facility which monitors automatically the breathing level of the patient and changes from automatic controlled ventilation to allowing spontaneous breathing when the patient makes sufficient inspiratory effort. Return to controlled ventilation occurs automatically should the patient's breathing deteriorate. Features include an adjustable valve with an audible alarm to limit peak inflation pressure, a pressure gauge to indicate the respiratory pressure

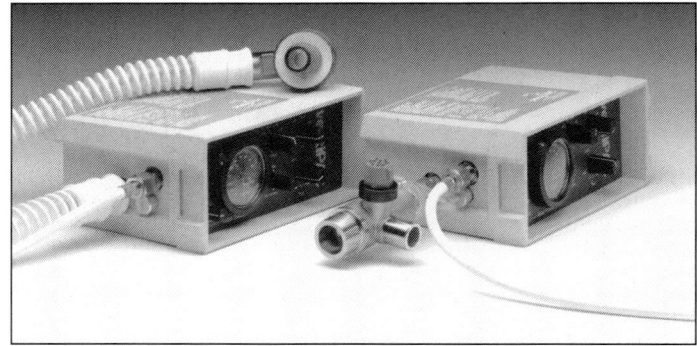

Pneupac ventiPAC automatic lung ventilator (Pneupac Limited)

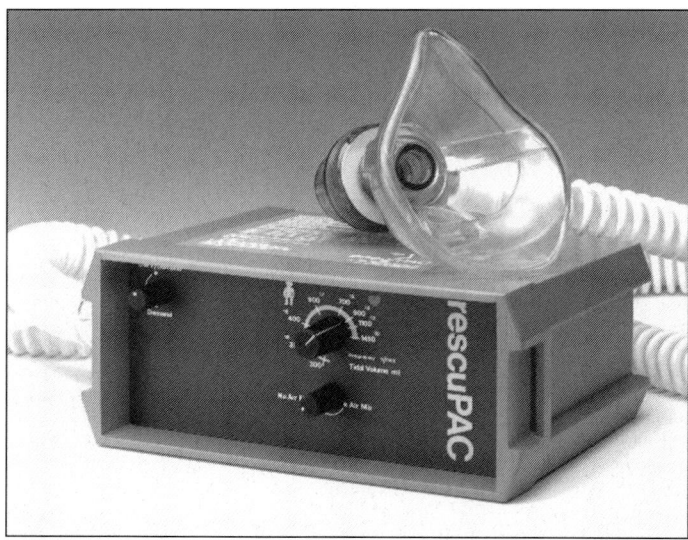

Pneupac rescuPAC automatic lung ventilator (Pneupac Limited)

Pneupac rescuPAC Model R20D showing control panel (Pneupac Limited)
***2002**/0137543*

and an air-entrainment system which reduces driving gas consumption by two-thirds. The basic units, which differ mainly in the ventilator controls, are as follows:

- rescuePAC. This is the simplest model having a single ventilatory parameter control knob operating in a manner similar to the Pneupac responsePAC ventilator (see separate entry). The single control knob controls the frequency of breathing and tidal volume.
- paraPAC. This model has two ventilatory parameter control knobs which vary the tidal volume and frequency.
- transPAC. On this model the two ventilatory parameter control knobs vary the frequency and minute volume.
- ventiPAC. This model has three ventilatory control knobs to control the inspiratory time, the expiratory time and flow. This ventilator offers a range of performance which can provide ventilation characteristics comparable with advanced hospital ventilators.

The internal control elements are housed in vibration isolating mountings within a shock-resistant rigid casing capable of withstanding abusive use. The average weight of the units is 2-3.5 kg and the dimensions are 92 × 220 × 162 mm.

All units can be operated using Pneupac airPAC compressors or the oxyPAC range of high-pressure oxygen concentrators.

Status
In production.

Manufacturer
Pneupac Limited.

UPDATED

Pneupac compPAC portable ventilator

Description
The compPAC (Model COM200) is a portable self-contained ventilator designed in conjunction with international military and emergency organisations for use by first responders in a wide range of military or civil roles. It may be powered from compressed air or oxygen sources or electrically powered using an internal battery or an external 28 V DC source. Its design allows very economical use of oxygen in situations where supply is limited.

The ability to provide early basic and advanced life support to casualties in a contaminated zone is now recognised as being of critical importance to survival in severe cases of chemical injury. The support of breathing through intermittent positive pressure ventilation is an integral part of this and can be achieved using the compPAC which is a flow generator with variable minute volume and frequency adjustment. It is capable of ventilating patients injured by CW agents who have altered airway resistance and compliance. The ComPAC features the additional assurance of an integrated electronic pressure monitoring and alarm system to help detect possible adverse changes in the patient's ventilation. It includes visual and audible alarms for high pressure, continuous pressure, low pressure/disconnect, low gas supply and low battery.

The ventilator may be deployed in the field, used in field or air ambulances for transport ventilation or used in field hospitals for medium term ventilation. It may be used in conjunction with field anaesthesia system. Its construction and built-in filtration capability enable it to be used in a stand-alone role in contaminated NBC environments.

The compPAC is housed in an easily carried chemical hardened case. The case contains a long-life battery driving a small compressor which powers a gas flow controlled ventilator module capable of ventilating patients in severe respiratory

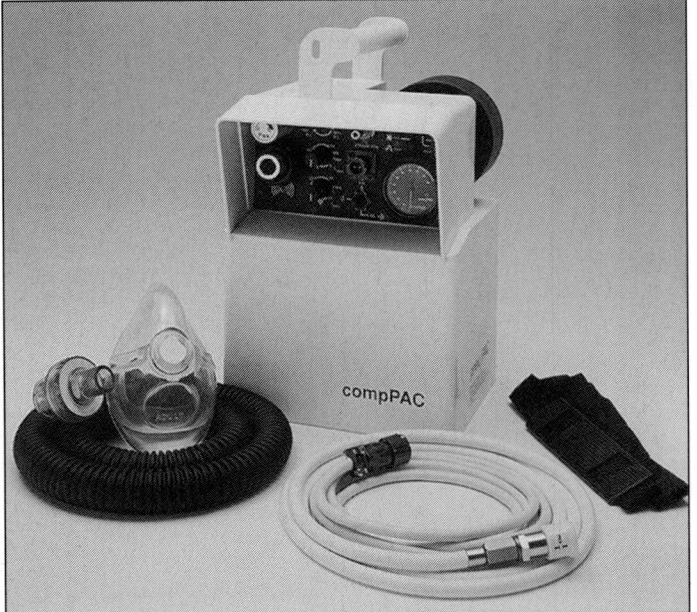

pneuPAC compPAC portable ventilator (SIMS pneuPAC)

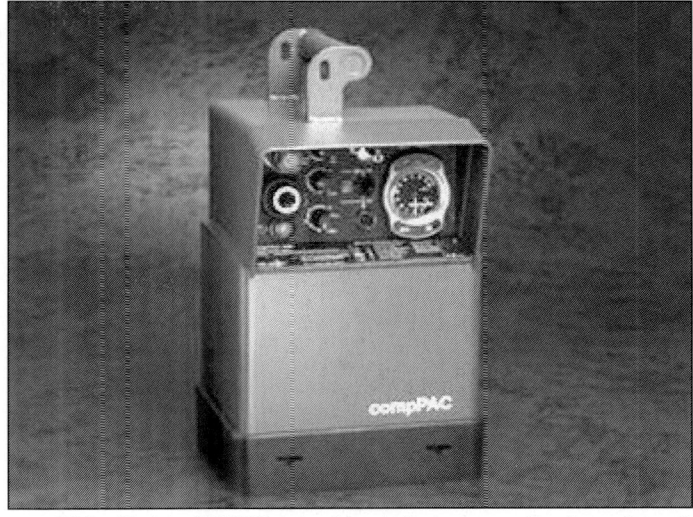

compPAC Model COM200 showing battery compartment underneath (SIMS pneuPAC)   **2002**/0137544

difficulty. Ambient air passes into the system through a standard NATO NBC filter canister. One third of the volume delivered to the patient is compressed to drive the ventilator before expansion in an entrainment mixing device which draws in the other two-thirds at breathing system pressure. This arrangement reduces the power requirements of the compressor by two thirds while the volume delivered to the patient s relatively unaffected by changes in the airway resistance. If oxygen is available from a bottled source or a concentrator it may be added to the entrained air to produce an FiO_2 of 0.27 to 0.72 using a special mixing valve. The ventilator may also be powered by compressed air or oxygen at 300 to 400 kPa when an FiO_2 of 0.45 or 1.0 will be produced.

The unit can either be driven by a battery (Ni-Cad, which can be trickle charged whilst in the unit when running on the power supply or lithium ion battery), fixed power or compressed gas. With the battery, the compPAC weighs 8 kg. It is 350 mm high, 210 mm wide and 210 mm deep.

Status
In production and service with UK and other armed forces and in civil use for HAZMAT emergencies. Extensive clinical and operational trials have been completed by the French armed forces and the equipment is scheduled for future purchases. It has also been scaled by the UK armed forces and will be incorporated in new field ambulances and base hospitals.

Manufacturer
Pneupac Limited.

UPDATED

Pneupac ground role resuscitation system

Description
The Pneupac ground role resuscitation system was developed in consultation with military medical advisers from several governments. It was designed specifically for war zone field hospital and major incident applications, where the supply of liquid or compressed oxygen is limited or non-existent and there is a need for equipment

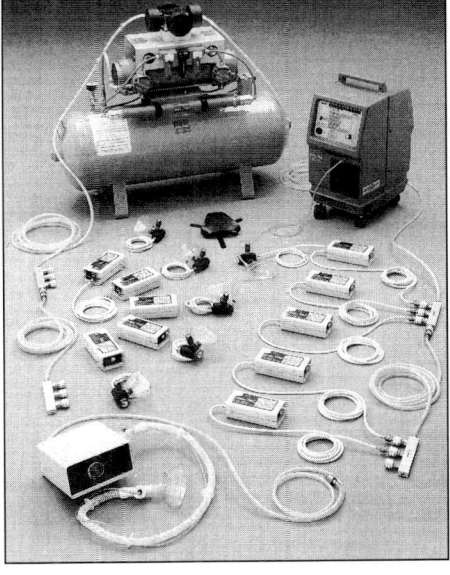

A typical Pneupac ground role resuscitation system

capable of ventilating mass casualties who may have been subjected to contaminated atmospheres.

The basic feature of the system is a specially designed compressor unit which provides filtered air of the correct quality both for ventilating casualties and for driving gas-driven ventilators. The system configuration can be adjusted to suit user requirements, but a typical system comprises a compressor unit, ten Model 2R Pneupac ventilator/resuscitators, a ventiPAC ventilator, five oxygen concentrators and all the necessary hoses, circuits and connectors.

The compressor unit draws air from the atmosphere via four NBC filters and compresses it by means of an oil-free compressor to approximately 5 bar before charging a 125 litre reservoir. The air is then filtered and dried before being distributed by manifolds to the point of use.

Casualties will normally be ventilated by Pneupac ventilators (see separate entry). Oxygen enrichment can be achieved using the concentrators by means of a special fitting at the patient valve connector. Humidification is achieved by use of heat and moisture exchangers at the same point. The use of the optional Pneupac entrainment valve on each ventilator/resuscitator would enable up to 25 patients to be ventilated off each compressor unit.

Status
In production. In service with the UK armed forces.

Manufacturer
Pneupac Limited.

UPDATED

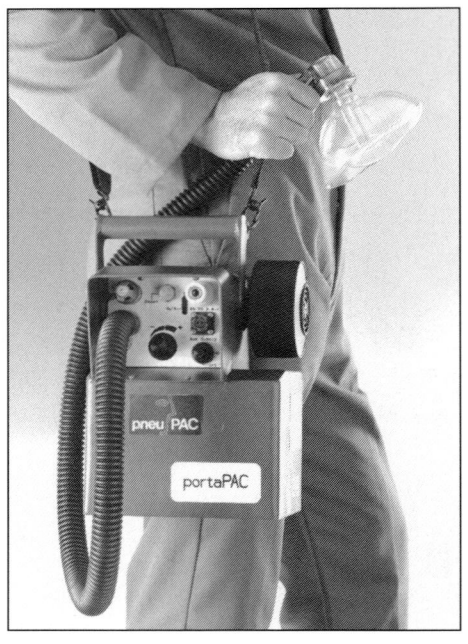

The Pneupac portaPAC resuscitation system in its portable mode

Pneupac portaPAC portable resuscitation system

Description
The portaPAC is a self-contained gas-powered automatic ventilator/resuscitator for casualties in need of lung ventilation. It can be carried in the field by military personnel for 'buddy-buddy' lung ventilation in the front line, used in field and air ambulances for transport ventilation and in field hospitals for medium-term ventilation. The units can be used as fixed installations in vehicles as well as portable units.

The portaPAC is housed in an easily carried, chemically hardened metal case. The case contains a long-life battery to drive a small compressor which provides the nutrient gas, through a fixed or adjustable 'oscillator', to the casualty. Ambient air passes into the system through a standard NBC filter canister. About one-third of the volume is compressed to drive the ventilator before expansion in an entrainment mixing device which entrains the other two-thirds at breathing system pressure. This arrangement reduces the power requirements for the compressor by two-thirds and the volume delivered to the casualty is relatively unaffected by changes in airway pressure.

The integral Clansman lithium battery provides power for more than 8 hours' ventilation. Ni/Cd batteries are also available providing ventilation for more than 2 hours. There is a socket to accept an external 24 to 28 V DC supply from a vehicle electrical supply to drive the compressor directly. Power consumption is less than 50 W.

An auxiliary gas connection is provided to allow the portaPAC to be connected to a 400 kPa gas supply so that it can be operated independently of its internal battery or external supply. In this way if oxygen cylinders or liquid oxygen are available, 45 per cent oxygen can be supplied to the casualty and the internal battery can be preserved. This facility also allows connection to compressor systems such as the Pneupac Ground Role Resuscitation System (see entry in this section). Because of the simplicity of the controls, models can be used in the front line as a 'buddy-buddy' resuscitator by service personnel, for transport and base hospital ventilation by paramedics and medical personnel.

A variable frequency and tidal volume setting enables the unit to be used by paramedics and doctors during casualty transport and in hospitals.

The portaPAC weighs 8.7 kg. It is 330 mm high, 220 mm wide and 170 mm deep. There are two basic models. One has pre-set controls providing a tidal

volume of 1,000 ml, a frequency of 12 bar/min and a flow rate of 40 litres/min. The other model has a single control providing frequencies from 9.5 to 16 bar/min, tidal volumes from 1,200 to 300 ml and a flow rate of 40 litres/min.

Variants of the portaPAC can be specified which include a patient pressure monitor and a resettable meter to provide an indication of battery life and other alarm functions.

Status
In production. In service with the UK and US armed forces and supplied to a US arms inspection organisation dismantling chemical warfare weapons in the RFAS as part of the Co-operative Threat Reduction programme (CTR).

Manufacturer
Pneupac Limited.

UPDATED

Pneupac responder ventilator

Description
The Pneupac range of rescue/emergency artificial ventilation equipment was produced originally for medical or industrial use where normal breathing is impaired or stopped after an accident or sudden collapse. The equipment is therefore of particular importance in the management of casualties from CW and BW agents where respiration is at risk, particularly from nerve or lung-damaging agents.

In its basic form, the Pneupac responder ventilator provides artificial ventilation (or intermittent positive pressure ventilation) through a mask placed over the patient's nose and mouth. Either 100 per cent oxygen, or an air/oxygen mixture can be supplied at a predetermined or variable rate. A pressure relief valve automatically limits the pressure of the delivered gas.

In addition to ventilation by mask, Pneupac ventilators may also be used with more advanced airway management techniques such as endotracheal intubation or the laryngeal mask airway.

The responder is powered by gas and requires no battery or other electrical supply. It operates immediately when connected to a source of gas under pressure which may be a cylinder of oxygen or air, piped gas or a compressor. The gas

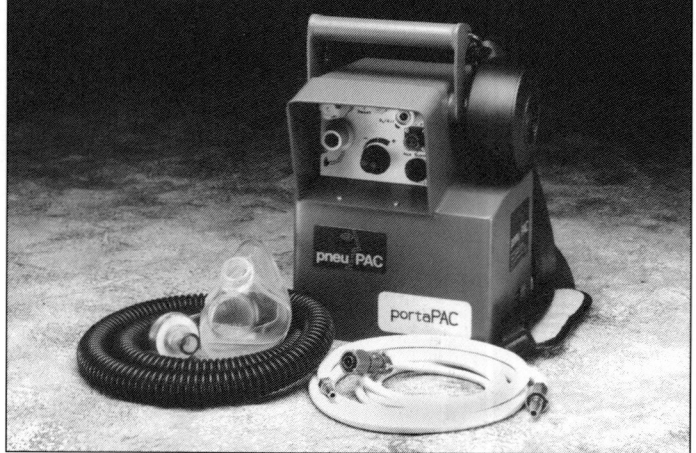

The Pneupac portaPAC resuscitation system

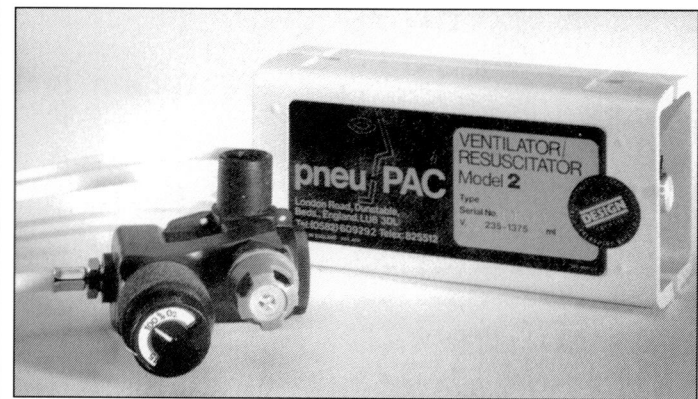

Pneupac Responder control module with air entrainment valve on left (Pneupac Limited)

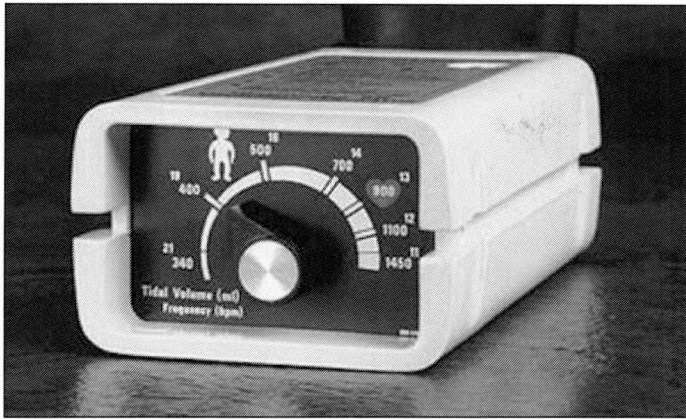

Close up of the Pneupac Responder Model 2R control unit   ***2002**/0137742*

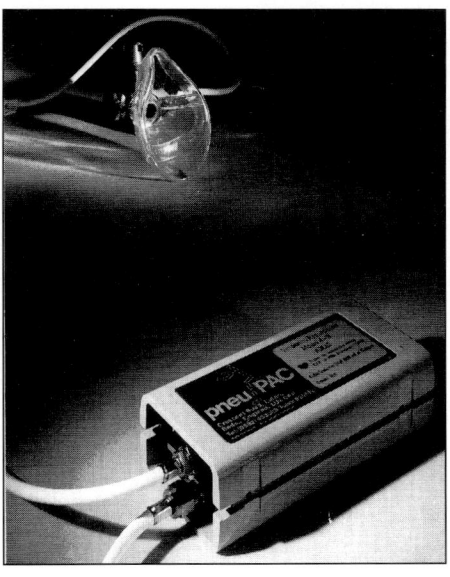

Pneupac Responder control module (foreground) with casualty valve and mask (Pneupac Limited)

Specifications
Performance specifications at different settings, Model 2R, 4R and 4R(R).

Factor	Min	Max	CPR
Tidal volume	340 ml (+10%/−20%)	1,450 ml (±10%)	900 ml (±11%)
Frequency (breaths/min)	21 (+20%/−10%)	11 (±10%)	13 (±10%)
Inspiratory time	0.5 s	2.2 s	1.35 s
Expiratory time	2.35 s	3.3 s	3.2 s
Inspiratory/Expiratory ratio	1/5	1/1.5	1/2.4
Minute volume	7.1 l	16 l	11.7 l

Status
In production. In widespread use with many NATO armed forces, including the UK (Army and Royal Air Force) and US Air Force and in countries in North Africa, the Middle East and Asia.

Manufacturer
Pneupac Limited.

UPDATED

UNITED STATES OF AMERICA

Gentex Casualty Care System (CCS)

Description
The Gentex CCS has all the features required to allow the safe battlefield primary care of casualties in a toxic environment. Based on a standard rigid stretcher (the Raven™ litter for example), the system features:

- A flexible NBC agent proof canopy, supported by a stay system and accessed via a full-length zipper.
- Large soft windows to allow medical personnel to observe the patient.
- 3 sets of glove ports. One fitted with long-sleeved butyl gloves allowing access by unprotected personnel.
- Colour-coded patient restraint system.
- Medical interface to allow the transfer of sterile fluids or critical gasses to the patient via standard medical supply equipment.
- Filtered air supply to the canopy interior or directly to the patient's mask.

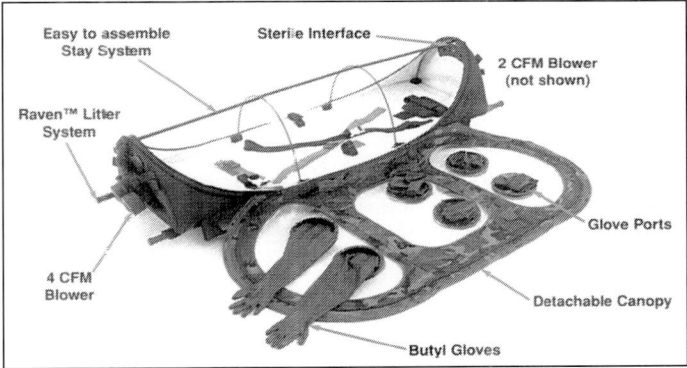

Annotated diagram of the Gentex CCS (Gentex Corporation)   ***2002**/0137852*

The Gentex CCS in use (Gentex Corporation)   ***2002**/0137853*

drives the responder by using pneumatic logic components within the module to generate the required breathing pattern. The control module weighs 1.3 kg and measures 90 × 50 × 180 mm. The pneumatic logic components are contained inside a foam-core plastic case which is strong enough to withstand being driven over by a vehicle. It can be safely dropped from a height of 1 m and immersed in water. The control module can operate continually for many hours, needs no routine maintenance and normally requires a factory service check after every three years. In storage, it only requires checking every five years.

The gas from the responder control module passes through a flexible, reinforced, non-kink hose to the patient valve. The valve is manufactured from reinforced plastics. Located on the valve is a transparent mask covering the casualty's nose and mouth. In the inspiration phase ('gas on') gas is delivered to the casualty and in the expiration phase ('gas off') the valve allows the casualty to exhale into the atmosphere. The valve, with a switching force of 500 g, is not easily jammed, is simple to clean and cannot be misassembled. An audible relief valve is fitted to prevent excess pressure in the casualty's airway and warn of obstruction.

The responder control modules have a single protected control knob which can provide a continuous variation of the ventilation parameters (tidal volume and frequency of respiration) to suit the casualty. It is suitable for a range of age groups from 5 year old children to adults. It has a click top setting for adult cardiopulmonary resuscitation (CPR).

The patient valve on the mask can be supplied with the option of air entrainment, that is, mixing and augmenting the supplied oxygen with air. Most resuscitation operations, especially following exposure to nerve agents, commence with the supply of 100 per cent oxygen. After satisfactory breathing control is established, the entrainment valve can be switched to draw in atmospheric air so that the casualty can receive the more suitable level of 45 per cent oxygen for long-term ventilation. This facility can extend the life of a supply cylinder by a factor of three and can be used once the casualty has been removed from the toxic environment.

The Responder can be supplied in portable self-contained resuscitation sets with their own cylinder. Two typical sets are the Compact Instant Action Set which includes as standard a Pneupac aluminium cylinder which gives, on CPR setting, up to 17 minutes resuscitation or 50 minutes with air entrainment. The larger Comprehensive Instant Action Set is designed for use with a Pneupac aluminium 'D' cylinder which gives, on CPR setting, up to 25 minutes resuscitation, or 75 minutes with air entrainment.

The ventilator may be mounted in ambulances or static installations. A wide range of accessories is available. A test set is available to check the calibration of the tidal volume, frequency and relief valve settings. Other accessories include oxygen therapy sets, aspirator systems, patients' airway pressure monitoring systems, different relief and other valves, carrying pouches, mounting brackets and knapsacks. Portable and other compressors are available.

The CCS can utilise standard blower systems such as the Micronel C420 (see under PROTECTION (individual) - Filters). There are ports at either end allowing a 4 CFM or 2 CFM arrangement. Negative pressure can alternatively be applied at 4 CFM for the removal of aerosol contamination evaporating from the patient's clothing inside the canopy.

Status
Available and in production.

Manufacturer
Gentex Corporation.

NEW ENTRY

Meridian auto-injectors

Description
Meridian auto-injectors are designed to allow military personnel quickly and easily to self-inject a fixed dose of required life-saving pharmaceuticals.

The auto-injectors are small, rugged, lightweight and field proven, spring operated, pressure activated delivery systems for military pharmaceuticals. They are activated by removing a safety cap and pressing the front end of the injector against the outer thigh with moderate pressure. Upon activation, the spring is released, moving the needle rapidly into the intramuscular site and delivering the drug. The complete self-injection sequence can be completed in about 2 seconds.

The fielded product range features nerve agent antidotes including anti-convulsants (atropine, 2-PAM, obidoxime, TMB-4 and diazepam), analgesics, and morphine. Meridian can provide performance testing and stability information on all its auto-injector systems. Product registration approvals with civilian agencies have been granted as required in some countries.

The various types of Meridian auto-injectors are as follows:
- ComboPen® - for atropine, oximes and diazepam.
- AtroPen® - for atropine, oximes and morphine (10 mg).
- Mark 1 Kit - the Mark 1 Nerve Agent Antidote Kit (NAAK) is the standard item for the US Department of Defense. The kit comprises two separate injectors, one containing atropine sulphate equivalency (AtroPen®) and the other containing pralidoxime chloride (ComboPen®).
- Morphine - morphine (10 mg morphine sulphate in 0.7 ml).

Dosages for paediatric, low-weight individuals and other special formulations have been produced by Meridian for some nations and are currently fielded. AtroPen and ComboPen are registered trademarks in the USA of Meridian Medical Technologies Inc.

Other devices are available for 'Homeland Defense'. These include:
- Pralidoxime ComboPen® - (2-PAM Cl). Delivers 600 mg pralidoxime chloride in 2 ml.
- Diazepam (CANA) - Delivers 10 mg diazepam in 2 ml.

Trainers
To train personnel in the use of auto-injectors, Meridian has developed training devices for military use. These include the AtroPen® Trainer, the ComboPen® Trainer and the Mark 1 Trainer. Each mimics the exact size, shape and appearance of the auto-injector system it simulates, operates in the same manner and can be recocked for repeated use. The training devices contain no drug or needle.

Status
Meridian auto-injectors are in production and service in the US and many other national armed forces and civilian agencies.

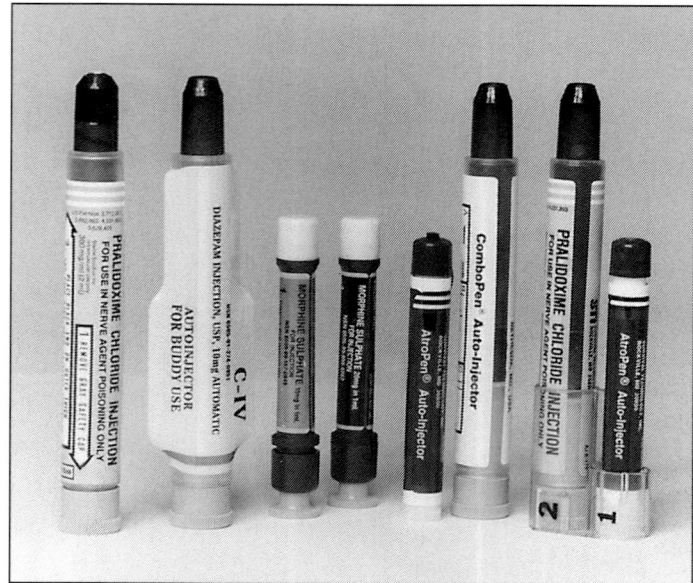

The complete range of Meridian auto-injectors
(Meridian Medical Technologies Inc)   *2001*/0063374

Manufacturer
Meridian Medical Technologies Inc.

Agency
Meridian Medical Technologies Limited (UK).

UPDATED

Multi Shield TSP barrier cream

Description
Multi Shield TSP barrier cream was developed by Interpro Inc and the US Army Medical Research Institute of Chemical Defense for use in the Persian Gulf region during Operation Desert Shield. The cream has been extensively and successfully tested under both laboratory and high-temperature field conditions by the US Army Medical Research Institute of Chemical Defense.

Multi Shield TSP is an easily applied, non-greasy skin protectant cream which can be applied in anticipation of chemical attack. The cream will provide individual skin protection by inhibiting the penetration of Distilled Mustard (HD) agent. The cream will also enhance the effectiveness of field decontamination systems.

When applied, Multi Shield TSP can be worn for extended periods and will not adversely affect face mask seals. In the event of a chemical attack involving HD, a 0.15 mm layer of Multi Shield TSP will mean that decontamination can be delayed for 60 minutes, after which the cream can be removed using soap and water or an alcohol wash. The cream will not cause skin or eye irritation, although it is recommended that any cream contacting the eyes should be rinsed away as soon as possible.

Multi Shield TSP is supplied in 4 oz/118 ml or 16 oz/454 ml plastic dispenser bottles and may also be supplied in 1 oz/28 ml bottles. The bottles are marked with directions for use.

Status
In production.

Manufacturer
Interpro.

VERIFIED

YUGOSLAVIA, FEDERAL REPUBLIC

POLYJET VAZ multipurpose auto-injector

Description
The POLYJET VAZ multipurpose auto-injector is a compact lightweight device providing long-term storage and immediate intramuscular self-administration of various first aid solutions in an emergency. The contents of the internal ampoule can be varied and may be in separate liquid and powder form to be mixed automatically immediately before use, or in single solution form. The device can be reused if required.

The POLYJET VAZ is approximately 150 mm long and 19 mm in diameter; weight is 20 g without the filling. The ejected volume of solution from the double-chamber system is approximately 3 ml. To use the device the cylindrical body is twisted to allow the internal liquid(s) to pass into a chamber for mixing. The yellow end of the device is placed against the thigh and a downward push applied. An internal spring then releases the needle and the contents are injected. It is claimed that the special shape of the needle reduces administration trauma.

The POLYJET VAZ 1 is an essentially similar device that cannot be reused.

Status
May no longer be in production.

Supplier
ELKOM TSN.

VERIFIED

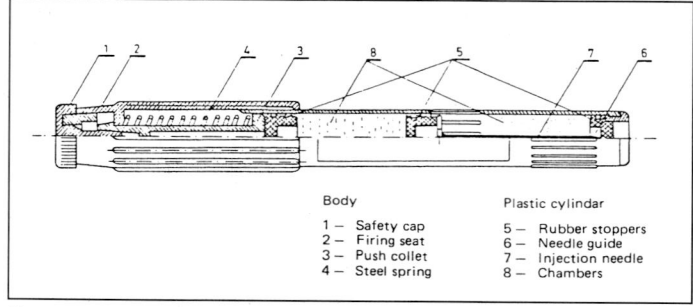

Cross-section of POLYJET VAZ multipurpose auto-injector

PROTECTION (collective)

PROTECTION (collective)

AUSTRIA

NBC Tarpaulin Material Schicht-Barriere 2000

Description
NBC Tarpaulin Material Schicht-Barriere 2000 was developed to counter the limitations of existing tarpaulin materials, used for field hospital tents or other shelters, which have severe NBC protection limitations. Schicht-Barriere 2000 has a protective foil embedded in one side of a strong tarpaulin fabric which is also provided with a rubberised finish. The resultant material has a minimum area weight of from 520 to 1,500 g/m², according to customer requirements and a resistance against chemical agents of at least 2,000 hours.

NBC Tarpaulin Material Schicht-Barriere 2000 can be supplied in rolls 1.5 m wide or may be used for the construction of ready-made finished products such as tents and other shelters, complete with frames. It may also be used for the manufacture of NBC casualty bags (see under Protection (individual) – Medical countermeasures). Camouflage finishes are available.

Status
In production. In service with the Austrian Army.

Manufacturer
J Blaschke Wehrtechnik GmbH.

VERIFIED

Multipurpose tent manufactured using NBC Tarpaulin Material Schicht-Barriere 2000

BELGIUM

NBC air filtration unit Air-Unit-90

Description
The Air-Unit-90 system was developed to provide individual crew members of armoured and unarmoured military vehicles with purified air without the need for vehicle positive pressure. The air delivery module is connected upstream of the individual respirator filter, reducing the respiratory load and the agent challenge. When required to dismount from the vehicle, the user simply removes the air

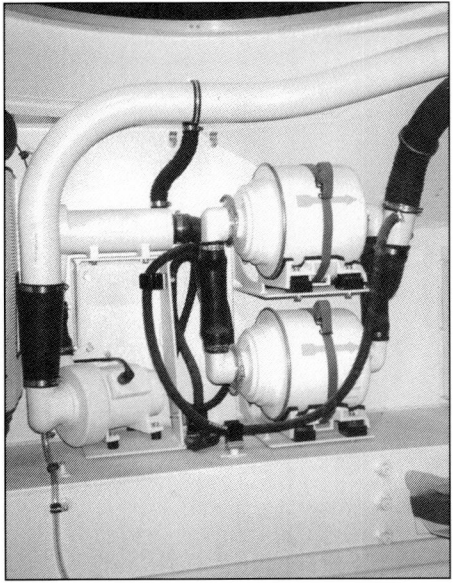

NBC air filtration unit Air-Unit-90 installed in APC
(Engicom Systems NV)

delivery hose, allowing the respirator canister to resume the filtration role. The unit is able to supply approximately 1,500 litres/min of air to up to 15 crew members equipped with gas masks. A smaller Air-Unit-45 has a capacity of 600 litres/min to supply up to six crew members. In both cases excess filtered air is blown into the vehicle interior to dilute any contaminated air.

The Air-Unit-90 has three modules. The first, the air generating module, has a collection system for condensed water, a motor with ventilator and a cyclone prefilter. Next, the filter module comprises two cylindrical filters, each containing 10 litres of impregnated charcoal to remove gaseous and particle contaminants. Quick-release connections allow users to replace filter units easily. This module also includes the shock-mounted frame. The third stage, the air distribution module, has: two compensation valves to allow for varying numbers of users; individual air supply hoses; and individual flow regulators to maintain masks under positive pressure.

Status
In service in Belgium and Turkey.

Manufacturer
Engicom Systems NV.

UPDATED

BULGARIA

FVA-100/50 filter ventilation unit

Description
The FVA-100/50 filter ventilation unit is intended for use in both static shelters and field constructions. It consists of an electrically driven VAP-1 blower unit, FP100/50 filter unit, ventilator safety vent, UVR-2 air flow meter and various other components.

The unit can provide filtered air at a rate of between 100 and 150 m³/h with a constant air flow resistance at 100 litres/min of 650 Pa. The VAP-1 blower unit, which has a manual start facility, weighs 41 kg while the FP-100/50 filter unit weighs 67 kg. The total weight of the complete unit, when stored in four wooden cases, is 261 kg.

Status
Available.

Marketing agency
Kintex.

VERIFIED

Main components of the 100 to 150 m³/h FVA-100/50 filter ventilation unit

CANADA

Carleton Ventilated Respirator System (VRS)

Description
The Carleton Ventilated Respirator System (VRS) is a vehicle-mounted system connected to the vehicle's own power source. When in operation a ventilator (blower) assembly draws filtered air through a standard C2 filter canister and

directs it to a C4, AC4 or equivalent mask. VRS can take any canister that has a NATO standard thread. The rate of air flow can be regulated to an individual's preference by adjusting a variable-speed control knob on the control box assembly, mounted close to the user's station on the vehicle.

The pressurised or positive air flow delivered by the ventilator (blower) significantly eases the burden of breathing through a C2 canister. When connected to a gas mask the ventilator increases the protection factor of the mask, prevents misting of the mask's eyepieces and provides a cooling effect to the wearer's face. In the event of a power or blower failure, the user can continue to breathe normally through the system.

For vehicle entry or egress the VRS includes a quick-disconnect within the reinforced butyl rubber connecting hose assembly. The blower remains with the vehicle but the user retains the mask and C2 canister assembly.

Status
In production.

Manufacturer
Carleton Life Support Technologies.

UPDATED

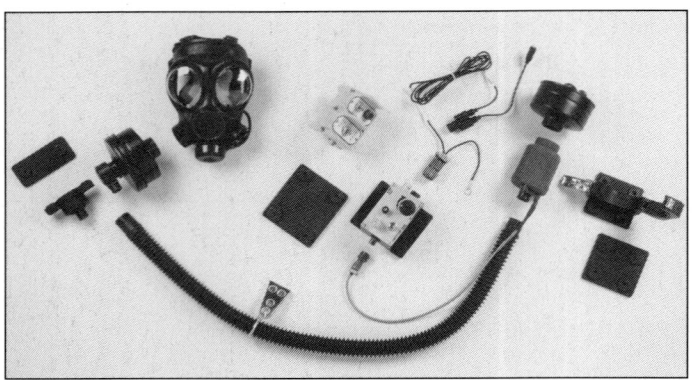

Carleton Ventilated Respirator System (VRS) components

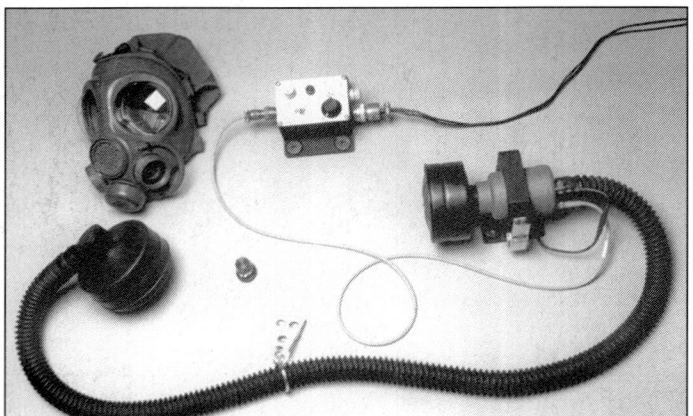

VRS - Components   *2000*/0084495

Carleton Ventilated Respirator System (VRS) in use

CHINA, PEOPLE'S REPUBLIC

M-07 gas-particulate filter and ventilation unit for field shelters

Description
The M-07 gas-particulate filter and ventilation unit for field shelters comprises a blower unit which draws external (contaminated) air through a HEPA pre-filter and then through an NBC charcoal filter. The blower unit is normally powered electrically to provide COLPRO filtered air at overpressure, but can be powered by hand. Airflow is flap controlled manually according to the required manometer pressure. The connection pipes are quick to fit, easing filter change operations. The unit equips command posts, communication centres, hospitals, rest and relief stations at divisional and regimental level.

Status
In service with the Chinese armed forces.

Development agency
Research Institute for Chemical Defence.

Manufacturer
Shanxi Xinhua Chemical Factory.

UPDATED

Specifications
Nominal flow (direct clean air): 300 m³/h
Nominal flow (filtered): 200 m³/h
Filter life (CNCL challenge - 1.0 mg/litre): >60 min
Filter life (DMMP challenge - 4.0 mg/litre): >100 min
Aerosol penetration: <0.001%
Airflow resistance at 200 m³/h: <0.95 kPa
Electric power: <0.3 kW
Hand power (at 150 m³/h): <0.12 kW
Weight: £100 kg

M-07 gas particulate filter and ventilation unit for field shelters (Shanxi Xinhua Chemical Factory)
0055074

M-73/1000 gas filter

Description
The M-73/1000 gas filter has an air flow capacity of 1,000 m³/h and an air flow resistance of less than 343 Pa at full capacity. Life when operating at full capacity against 2 mg/litre of cyanogen chloride vapour is over 2 hours at +20°C.

Weight of the M-73/1000 is 240 kg. Dimensions are 1,000 × 700 × 670 mm.

Status
In service with the Chinese armed forces.

M-73/1000 gas filter (Shanxi Xinhua Chemical Factory)

Development agency
Research Institute for Chemical Defence.

Manufacturer
Shanxi Xinhua Chemical Factory.

UPDATED

M-75/2000 particulate prefilter

Description
The M-75/2000 particulate prefilter has an air flow capacity of 2,000 m³/h and an air flow resistance at full capacity of less than 290 Pa. Using artificial dust for testing at full capacity the filtration efficiency is more than 95 per cent.
   The M-75/2000 weighs 35 kg and measures 650 × 650 × 570 mm.

Status
In service with the Chinese armed forces.

Development agency
Research Institute for Chemical Defence.

Manufacturer
Shanxi Xinhua Chemical Factory.

UPDATED

M75/2000 particulate prefilter (Shanxi Xinhua Chemical Factory)

M-77/500 gas-particulate filter

Description
The M-77/500 gas-particulate filter has an air flow capacity of 500 m³/h and an air flow resistance at full capacity of less than 686 Pa. The filtration efficiency is stated to be not less than 99.9999 per cent against aerosols when operating at full air flow capacity. Life when operating at full capacity against 2 mg/litre cyanogen chloride vapour is over 2 hours at 20°C.
   The M-77/500 weighs 160 kg and measures 920 × 604 × 604 mm.

M-77/500 gas-particulate filter (Shanxi Xinhua Chemical Factory)

Status
In service with the Chinese armed forces.

Development agency
Research Institute for Chemical Defence.

Manufacturer
Shanxi Xinhua Chemical Factory.

UPDATED

FINLAND

NanoBio® air purification system

Description
Genano has developed a novel system to overcome the danger caused by microscopic particles in working spaces. The technology, (described as 'MFI® technology' and trademarked as NanoBio® and NanoCap®), works on particles that are often identified as the size-range most dangerous to human health - those between 1 to 100 nanometers (1 × 10⁻⁹ nm) which include viruses, bacteria and smoke or fume particles. The MFI® technology method is based on the use of ion jets and electrodynamic forces, acting on the airflow inside a duct, to separate out the ultra-fine particles without using a filtering medium.
   The length of duct required depends on the volumetric airflow and the likely particle loading. Three specific effects are applied to the airflow:
• A strong negative ion impact which forces the particles to flow towards a collection area.
• The particles receive a negative charge from the negatively-charged ions (this only applies to those particles that can be electrically charged).
• A positive electric field in front of the collecting area attracts particles towards the collecting surfaces.
The collected concentrations of particles are removed by flushing. The residue, containing material that is environmentally harmful or dangerous to human health, is separately disinfected by neutralising agents.
   The MFI® method is fully automatic and offers reduced energy consumption advantages compared with HEPA, cyclone or micronic filtration. The air flows cleanly through the duct, unimpeded by in-line filters (although an EU9 or similar filter can be used if extra security is desired). The energy consumption of a typical unit is 20 to 160 W. Other advantages include simple maintenance and surveillance. The system is not affected by moisture like TEDA filters and avoids the problem of microbial growth which plagues conventional systems and renders them susceptible to infection by *Legionellae*. The cleaning process is fully automatic and systems can be scaled to meet the volumetric requirement. The three standard duct diameters currently available are 300, 400 and 500 mm.

Status
Available for defense uses such as BW protection for headquarters, vehicles, field hospitals and shelters against nuclear fallout.

Manufacturer
Genano Limited.

NEW ENTRY

TEMET special shelter equipment

Description

The product range of TEMET covers all the special equipment required in civilian shelters and military hardened facilities to make them withstand blast and shock effects and to provide protection against toxic gases and similar threats.

Finland has a network of 30,000 shelters providing protection for 2/3 of the country's 5 million inhabitants. The aim is to provide civil defence shelters for people at their homes as well as their places of work. The Finnish construction requirements for both military and civilian shelters and their equipment are among the most stringent in the world. Second World War experience and subsequent research formed the basis for current shelter standards and TEMET has been involved in this development in close co-operation with the Finnish authorities and the Government Technical Research Centre.

Each piece of equipment and system in the TEMET range fully complies with rigid performance requirements, with tests continually being performed by the Finnish Army, government testing laboratories and several international test institutes. TEMET has its own quality assurance system based on the stringent standards set by the Finnish Civil Defence authorities and certified in accordance with ISO 9000 quality system standards.

The comprehensive TEMET Sheltering System ensures that each shelter is provided with the ideal shelter components to provide maximum protection against destructive effects. This ensures that adequate protection is provided for all ventilation openings, wall penetrations, passageways and life support systems.

TEMET's product range covers all types of shelter doors and hatches such as blast-resistant doors, airtight doors and combinations of the two. In addition to standard specification doors, TEMET can also manufacture custom-built doors to specific requirements.

To protect ventilation openings and wall penetrations, TEMET manufactures blast valves and airtight closing valves in various protection classes. Their design concept is based upon the principle that the valve will continue to function after being repeatedly subjected to maximum blast loading. The TEMET blast valve product line incorporates blast valves for three blast load ranges. Low pressure

TEMET concrete arch blast-resistant doors provide optimum structural protection for large shelter passageways   *2001*/0079812

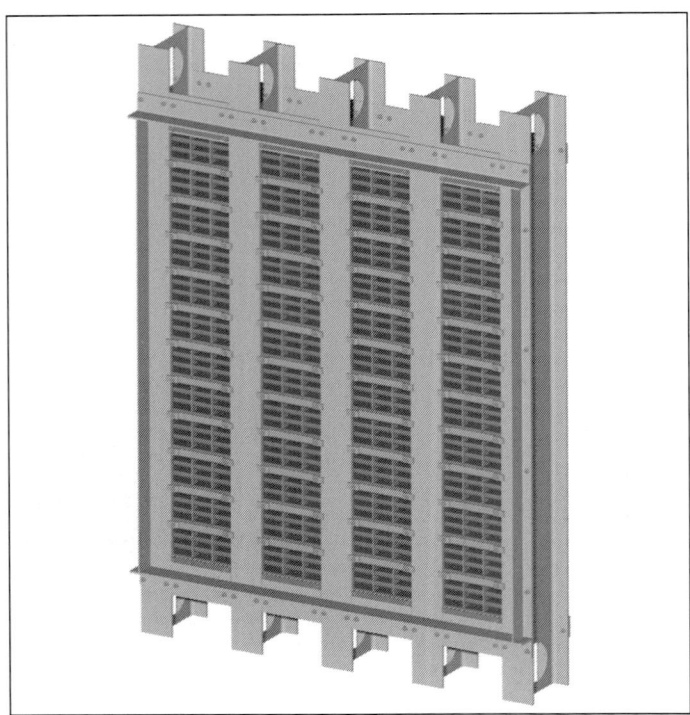

TEMET blast valve walls are designed for blast protection of large ventilation openings   *2001*/0079810

blast valves operate up to 15 bar. Intermediate load range blast valves operate up to 60 bar and high pressure valves up to 125 bar explosive pressure.

To meet emergency ventilation situations, TEMET produces a full range of NBC filtration systems including prefilters, and NBC filters which include a particle filter and a gas filter. With the latter, the particle filter is a microfilter able to arrest solid particles and aerosols; the gas filter absorbs toxic gases and vapours. The filter casing is a cylindrical steel container designed to provide hermetic sealing of the filtration units and to protect against the effects of pressure and ground shocks.

TEMET also provides ground shock and vibration isolator systems designed to protect mechanical and electrical equipment installed in shelters. The isolator systems are based on elastic springs that absorb repeated shock loadings without loss of efficiency.

TEMET was involved in the development of the ECAS fixed-site CW agent detection and alarm system which is based on the Environics M90 CW agent detector (see under Detection (sensor systems) - Chemical). The ECAS system is type approved for use in shelters by the Finnish Ministry of the Interior and is now required to be installed in the large underground S6-Class shelters.

Status
The above-mentioned equipments are in production and in service within Finland.

Manufacturer
TEMET Oy.

VERIFIED

TEMET NBC filters provide clean air for all types of COLPRO   *2001*/0079811

TEMET NBC air filtration unit for small capacity COLPRO   *2001*/0079809

Temet stand-alone inflatable NBC shelters

Description
In addition to hardened shelters, Temet Oy has produced a series of modular temporary NBC shelters in tented form. Manufactured from 500 gm/m² PVC-coated polyester fabric (other strengths available according to customer needs), the systems are easy and quick to erect and provide good protection against known CBW agents.

Biological and chemical protection system LSS-80
This system is designed for erection inside a building to provide temporary clean air refuge facilities for up to 6 occupants. It comprises a protective tent with overhead rope supports and an NBC filter/blower unit capable of being erected and made operational in 15 minutes. Access is via a zipped airlock chamber. The nominal internal overpressure of 25 Pa is maintained by the blower unit which can be operated manually via a crank handle if electrical power is unavailable. The wheeled packing container makes it easy to transport.

Stand-alone inflatable NBC shelter INT-1
The INT-1 stand-alone inflatable shelter is a man-portable rapid deployment modular COLPRO facility providing efficient protection against CBW agents and radioactive fallout dust. It is a free-standing structure requiring neither support poles, pegs nor a continuous inflation supply for the supporting pneumatic frames. The material is waterproof, fire-retardant and resistant to abrasion and UV decay. An IR-reflective finish is available. Positive pressure is maintained by the Temet VIL-150P NBC filtration unit.

A proprietary spring-controlled access door arrangement (Lip-Door™) has been developed to balance loss of overpressure against ease of access.

Modules can be fitted together at the sides and ends allowing flexible structures to be developed to satisfy the operational need.

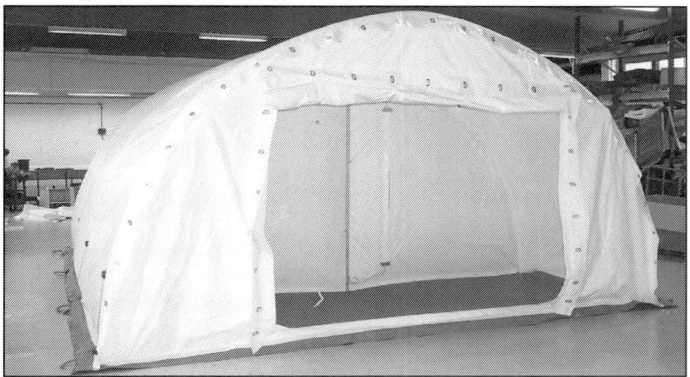

Temet INT-1 COLPRO shelter showing the air lock and, at rear, the slotted inner access door (Lip-Door™) (Temet Oy)   **2002**/0122651

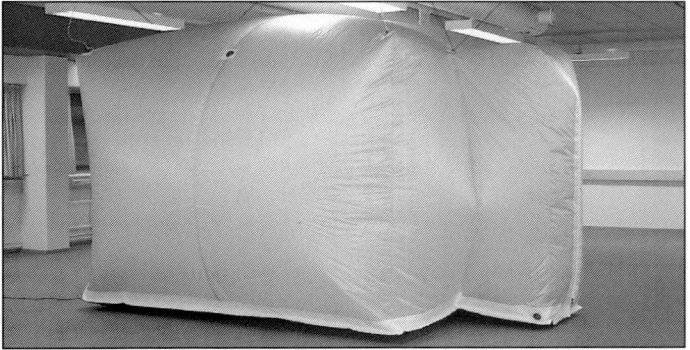

Temet LSS-80 internal COLPRO system erected inside a building (Temet Oy)   **2002**/0122652

Temet INT-1 air-framed modular COLPRO shelter (Temet Oy)   **2002**/0122650

Specifications
Packed container size: 58 × 79 × 98 cm
Weight: 70 kg
Erected size: 250 × 371 × 210 cm
Airlock chamber: 75 × 75 cm
Total inside floor area: 7.4 m²
Blower unit output: 65 m³/h

Specifications

Module erected dimensions:	500 × 600 × 250 cm
Weight:	150 kg
Deployment time:	5 - 10 min
Material:	PVC-coated 1100 dtex polyester at 650 g/m²
Filtration unit capacity:	150 m³/h (sufficient for 1.5 to 2.5 total air changes per hour)
Sustainable overpressure:	35-50 Pa
HEPA filter efficiency:	99.9995% for 0.3μ particles
Carbon filter efficiency:	
Chloropicrin (CCl_3NO_2) 13.7 g/m³ @ 2,000 ppm	>150 m
Cyanogen Chloride (CICN) 5.2 g/m³ @ 2,000 ppm	>100 m
Hydrogen Cyanide (HCN) 2.0 g/m³ 1,800 ppm	>140 m
Radioactive Methyl Iodide (CH_3I^{131}) at 8.5 MBq	99.995%

Status
Available. In production and in service with the Finnish Armed Forces.

Manufacturer
TEMET Oy.

NEW ENTRY

FRANCE

Giat NBC 150 m³/h collective protection unit

Description
The Giat NBC 150 m³/h collective protection unit is intended for installation in a confined area such as a sealed room. It supplies a flow of filtered air at 150 m³/h at a pressure of 200 mPa, sufficient to provide air for 30 to 50 personnel within the sealed area.

The unit comprises the filter unit within a casing secured to a wall or on the ground and a control box. The sheet metal casing, which can be locked, encloses the filter and its valves, a ventilator and the 100 mm diameter connection hose. It is suggested that the air inlet to the unit is located next to an interior corridor or an adjoining room rather than in direct contact with the external atmosphere. The lockable control box contains the unit controls as well as instructions, a radio and battery, an overpressure gauge and a roll of adhesive tape to seal the premises. The key to the control box is normally located behind a glass cover which has to be broken for access.

The filter unit casing measures 1,000 × 400 × 400 mm and the control unit 360 × 300 × 160 mm. Weight complete with filter is 65 kg. Power is provided by a 220 V 50 Hz mains supply with a power consumption of 200 W.

Status
In production.

Manufacturer
Giat Industries

VERIFIED

Giat NBC 150 m³/h collective protection unit

Giat NBC 1,500 m³/h naval collective protection unit

Description

Naval NBC collective protection units are designed to protect personnel and equipment within a 'citadel' pressurised area against contamination from nuclear and chemical weapons. The citadel is overpressured with filtered air and the crew can work without recourse to protective clothing.

The Giat Industries 1,500 m³/h collective protection unit consists of a low-magnetic stainless steel housing containing a washable prefilter, 16.2 kW electric heater, two HEPA filters, a 4 kW high-pressure motor blower and six composite (HEPA and vapour) filters. The unit may be mounted on an internal or external bulkhead and is 1.872 m high, 2.107 m wide and 1.35 m deep.

Status

In production.

Manufacturer

Giat Industries.

VERIFIED

section. The filters may be placed within protective cases on the exterior of shelters or vehicles.

Status

In production.

Manufacturer

Giat Industries.

VERIFIED

Specifications				
Nominal throughput	12 m³/h	90 m³/h	170 m³/h	600 m³/h
Max throughput	15 m³/h	100 m³/h	180 m³/h	600 m³/h
Weight	1.4 kg	18 kg	18 kg	100 kg
Height	103.6 mm	120 mm	370 mm	917 mm
Outside diameter	160 mm	N/A	325 mm	500 mm

Giat NBC 1,500 m³/h naval collective protection unit

Giat NBC collective protection composite filters

Description

Giat Industries has developed a range of NBC collective protection composite filters for use in NBC protection systems for vehicles, aircraft and mobile military shelters.

One of the Giat NBC filters is in the form of a prismatic box and has a nominal throughput of 90 m³/h. This filter measures 566 × 254 × 126 mm and weighs 18 kg.

Four of the filters in the range are cylindrical with air flowing along the axis of the cylinder. Air is first treated in an anti-aerosol section and then in an anti-toxin

Giat NBC 60 m³/h collective protection composite filter

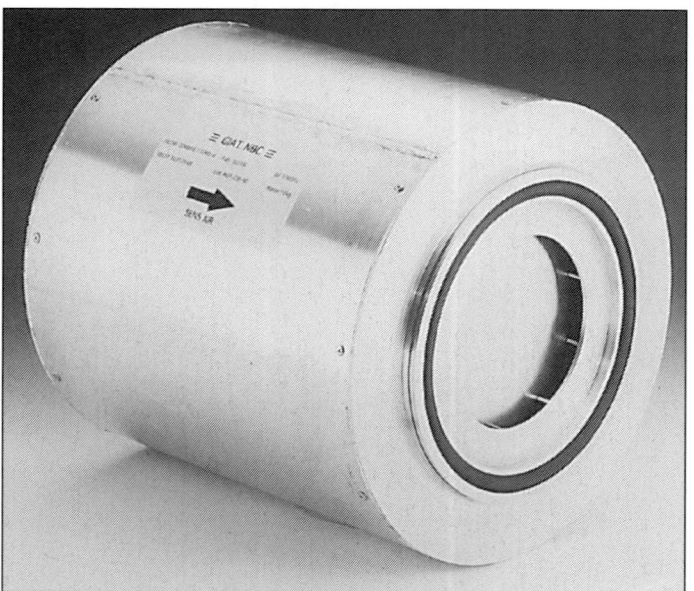

Giat NBC 170 m³/h collective protection composite filter

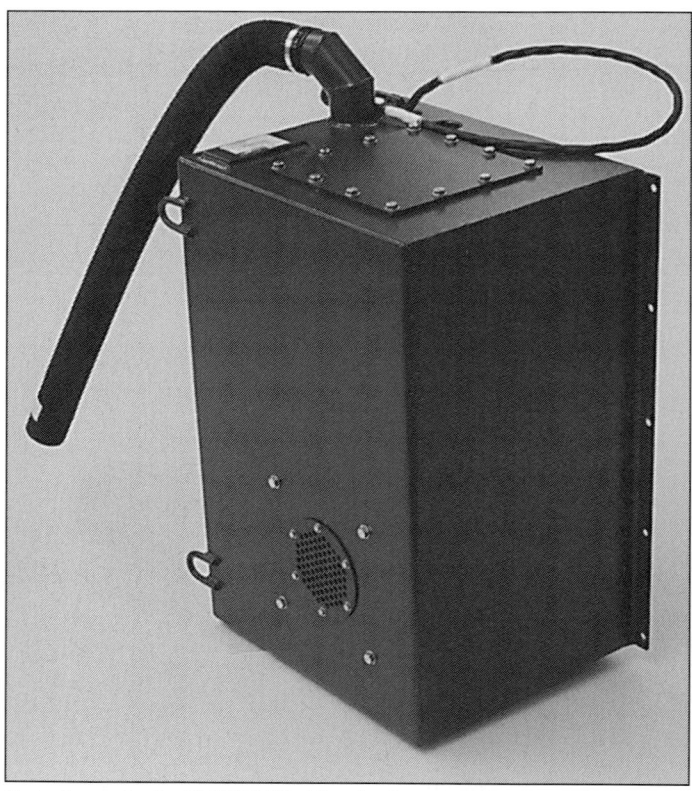

Protective casing for Giat NBC 60 m³/h collective protection composite filter

Giat NBC collective protection for armoured vehicles

Description

Giat Industries has installed NBC collective protection systems on all generations of armoured vehicles, including those produced by Giat Industries. These systems fall into two main categories, collective crew space protection systems and collective protection systems with individual protection. Both of these are combined in the so-called hybrid collective protection systems.

The hybrid collective protection systems and those with individual distribution, form the new-generation of NBC systems which provide protection in all operational situations, in particular when personnel are entering and leaving armoured personnel carriers, or in battle tanks when the main gun is being fired. Such systems can be retrofitted to existing vehicles without necessarily requiring improvements to existing vehicle seals.

The concept of 'microcooling' combined with an individual distribution system can offer an ideal compromise between the ergonomic needs of an individual and the power supplies and space available in a vehicle.

Giat Industries can design and manufacture collective protection systems on armoured personnel carriers incorporating overpressure and individual distribution, with or without air conditioning. One collective protection system employs a canister filter rated at 170 m³/h. The system can distribute air individually to 11 personnel, with or without air conditioning and is available in two versions. One system has a central collective NBC protection unit with air conditioning supplying a volume of 260 dm³; it weighs 180 kg. A second system uses a central collective NBC protection unit supplying 200 dm³; weight is 30 kg.

These new-generation systems, with their modular component design, can be adapted to suit any layout or operational configuration.

A Giat NBC collective protection system was selected for installation in the Swedish CV 90 IFV. An essentially similar system is installed in Sisu XA-186 APCs delivered to Norway for UN deployment in Bosnia.

Status

In production.

Manufacturer

Giat Industries.

VERIFIED

Giat NBC collective protection system for armoured personnel carrier

Giat NBC collective protection system for Leclerc

Description

The NBC collective protection system developed by Giat Industries for the Leclerc main battle tank has four main components inside a cast aluminium casing. They are a ventilation or NBC collective protection system; a hybrid ventilation, air

NBC collective protection system produced by Giat Industries for the Leclerc main battle tank

conditioning and/or heating system; nuclear detection; and a computer control system. The flow rate is 180 m³/h.

Air flow and interior temperature are automatically controlled by a computer which takes into account the temperature of air entering the blower to adjust the air flow. The computer constantly monitors the overpressure inside the vehicle and provides an alarm to the vehicle commander in the case of loss of pressure. The computer also monitors the hybrid ventilation temperature and controls the air conditioning and heater.

Status

In production. In service with the French Army and the armed forces of the United Arab Emirates (UAE).

Manufacturer

Giat Industries.

VERIFIED

Giat NBC EVATOX emergency evacuation kit

Description

The Giat NBC EVATOX emergency evacuation kit is intended for use during the evacuation of the public from an area affected by a toxic atmosphere. The kit contains a set of protective respiratory hoods for 1,000 people, including adults, children and infants.

The full kit has two main parts. One is a standard ISO 6.096m/20 ft container. The second part of the kit is the container's contents, 68 cardboard boxes containing a total of 1,124 hoods.

The container provides long-term storage and a means of transferring the kit to where it is required. It can also be used as a base for providing public warnings. The container has a dehumidifying ventilation system for long-term storage, a lifting eye and a mast. The mast carries a two-tone siren, rotating beacon, siren issuing a warning code audible over a range of 300 m and a public address system. The container also has two electronic display panels to depict warning messages and a

One of the ventilated protective hoods supplied in the Giat NBC EVATOX emergency evacuation kit

A Giat NBC EVATOX emergency evacuation kit container

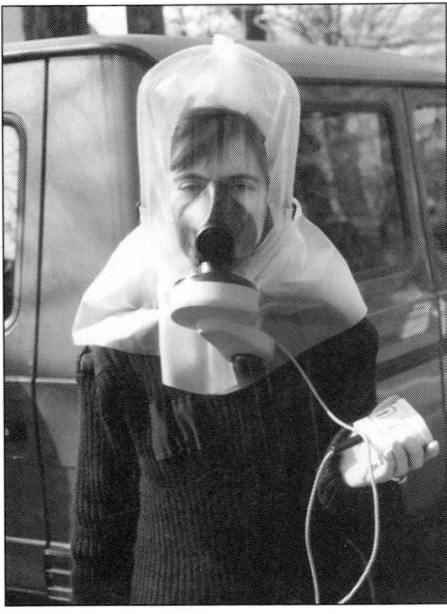

One of the child's protective hoods supplied in the Giat NBC EVATOX emergency evacuation kit

spotlight for internal and external use. There are also power supply connections for 12 V DC and 220 V AC.

The container's contents are 60 Standard cardboard boxes and eight Special boxes. Each of the Standard boxes contains 12 adult protective systems, two ventilated adult protection systems, two child's protective systems, one infant protection system and a set of user instructions with illustrated message boards for requesting additional equipment (to be displayed at windows). All the items are placed inside vacuum-packed bags.

The eight Special boxes contain additional equipment. Three of the boxes each contain 15 ventilated adult protection systems. Three further boxes each contain 15 ventilated child's protection systems. The two final boxes each contain seven infant's protection systems.

Status
In production. In service with some French fire brigades.

Manufacturer
Giat Industries.

VERIFIED

Giat NBC filtering and pressurisation unit for soft skin structures

Description
Giat Industries produce a 340 m³/h NBC filtering and pressurisation unit intended for use with soft skin structures, mobile shelters and NBC liners within existing buildings. The unit is a composite material container provided with handles and is easily transportable, weighing 30 kg without the filter and 66 kg with the filter installed.

In use, outside air is drawn in through a cleanable static prefilter. The air can then be heated, if required, and delivered by a fan into a NATO standard 200 cfm NBC

Giat Industries NBC filtering and pressurisation unit for soft skin structures

composite filter. A flexible duct connects the unit to the area to be supplied, either directly or via an air conditioning unit.

The pressure flow rate of the unit is 340 m³/h at 500 Pa. The unit measures 630 × 550 × 630 mm and requires a 220 V single-phase electrical supply. The optional heater has a power consumption of less than 3,000 W.

Status
In production.

Manufacturer
Giat Industries.

VERIFIED

Giat NBC intervention kits

Description
Giat NBC have developed a series of NBC collective intervention kits providing operational units, such as fire service, civil defence, emergency services and so on, with equipment suitable to deal with the threat or aftermath of NBC incidents. The scope of the kits is sufficient to provide long-term support in the event of major incidents as well as being able to deal with limited emergencies.

The kits are contained in prepacked portable cases held in standardised, pallet-stackable containers for long-term storage and transport. Each container is designed to meet a specific function. The inherent modularity of the system enables a choice of kits to be made in accordance with the mission to be handled.

The kits are categorised in two major groups. The Elementary intervention kits consist of a few cases containing sufficient equipment to enable a limited number of personnel to intervene and deal with small scale chemical and nuclear incidents. The second category, Functional intervention kits, are more comprehensive and are intended to deal with seven functions following a major incident. These are: individual protection; chemical and nuclear detection; sampling; the indication of affected area by signs; population alert; the evacuation of casualties; and decontamination of personnel and equipment. In addition, further kits are available for training purposes.

Status
In production.

Manufacturer
Giat Industries.

VERIFIED

A sample of Giat NBC intervention kits

Giat NBC ventilation/filtration system with integrated air conditioning

Description
The Giat NBC ventilation/filtration system with integrated air conditioning is intended to protect the crew of an armoured vehicle from NBC contamination and ensure comfortable conditions for the crew, by distributing filtered and conditioned air together with breathing assistance when masks are worn. Values of air flow and cooling can be set as required. The system is electrically operated from the vehicle power supply and is self-contained, with its own operating and control systems.

The system consists of three subassemblies: the complete box unit, a distribution circuit and a control panel.

The box unit can have optional ballistic protection and contains the NBC filtration system consisting of an external air intake, cyclone prefilter, HEPA/carbon NBC filter, fan and the filtered air outlets. The box also contains the air conditioning system with a cooling unit, heat exchanger, light alloy casing for the support structure and a power unit.

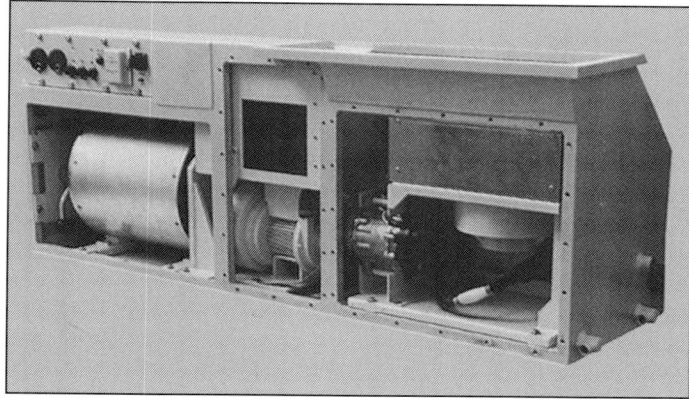

Giat NBC ventilation/filtration system with integrated air conditioning

The air distribution circuit comprises insulated pipework, air distribution boxes and the connecting hoses between the distribution boxes and gas mask filter canisters.

The control unit carries the controls to start and monitor the system.

The NBC flow rate is adjustable from 120 to 160 m³/h with a minimum flow to each distribution box of 9 m³/h. Cooling power of the system is approximately 2,800 W. Up to 10 passengers and crew can be protected by the system.

Weight of the complete system is 140 kg (less the filter) and dimensions are 1,500 × 550 × 400 mm. Power consumption is 2,800 W at 24 V DC.

Status
Preproduction.

Manufacturer
Giat Industries.

VERIFIED

GVFS Portable filtration-pressurisation unit (300 m³/h)

Description
The GVFS is a compact, lightweight, portable unit comprising a prefilter, air heater, fan and NBC composite filter. It is fully weatherproof and designed to supply filtered air through a flexible hose to tented fixed or mobile temporary NBC

Specifications
Nominal airflow: 300 m³/h
Dimensions: 630 × 600 × 512 mm
Weight (with NBC filter): 75 kg
Power supply: 220 V AC, 50 or 60 Hz
Fan power consumption: 450 W
Air-heated power consumption: 830, 1,660 or 2,490 W (3 positions)
Outlet pressure: 500 Pa
Efficiency against aerosols: 99.997%
Gas resistance: complies with STANAG 4447

GVFS Portable filtration-pressurisation unit (300 m³/h)   0011471

COLPRO facilities in the field. It is compatible with other equipment, causing no interference with medical, radar or communications facilities.

The GVFS meets the military specifications of GAM-EG-13 and satisfies STANAG 4447 for performance.

Status
In service with and in production for the French armed forces.

Manufacturer
SP Défense.

UPDATED

NBC modular collective protection shelters

Description
The NBC modular collective protection shelters protect personnel and equipment against conventional and nuclear blasts and NBC hazards. Several types are available: command posts, signal centres, casualties clearing posts, personnel survival shelters. These shelters are especially suitable for the protection of Air Force bases. The units can shelter between 50 and 150 people. Each NBC modular shelter is self-supporting, with its own power supply, drinking water and food (stocked for between two days and three weeks). Protection against the NBC hazard is sustainable for the same period.

The structure consists of 2.5 or 3.2 m diameter concrete prefabricated modules arranged in a line and separated by neoprene joints. The shelter is either buried or semi-buried under a berm. Its resists nuclear blast to a peak static pressure of 350 kPa (that is 1,400 kPa for the reflected wave). Anti-blast doors and anti-blast valves protect the personnel entrances, the air inlets and outlets.

The inside of an NBC modular shelter is divided into three areas. The TFA, where personnel can work or rest without the need for IPE, is entered via a CCA where personnel entering the sheltered off contaminated IPE and are monitored for contamination. This area is joined to a Technical Area which contains mechanical and electrical equipment.

Pressurised with NBC filtered air, heated or conditioned, the TFA contains berths, tables, chairs, toilets, showers, a kitchen, drinking and waste water tanks and the food storage facilities. The modular design allows the precise internal arrangements to be tailored to the individual requirements of the base according to the operational need.

The Technical Area shelters two diesel generators, fuel tanks, an air conditioning unit (if required) and the NBC filters and blowers. One generator remains

Installing an AP 60 modular shelter (Bonna Sabla)   0011482

Artist's impression of a CF-CPS-100 modular 100-man shelter (Bonna Sabla)   0011473

constantly on load when the shelter is operational. The other remains on standby, starting automatically if the other stops. The NBC filters and blowers (1,200 m³/h of nominal airflow per filter) provide the TFA with NBC filtered and pressurised air.

The CCA is equipped with stowage for contaminated IPE and can process about 40 people per hour. Personnel detailed for essential exposed activities, exit via the CCA airlock wearing full IPE.

NBC modular shelters have been subjected to full-scale tests in the USA for resistance to both conventional and nuclear blasts. Nuclear blasts were simulated in the Direct Course and Minor Scale experiments (simulating 1 kT and 8 kT nuclear blasts respectively). Dispersion of toxic agent simulant on an air base during realistic exercises confirmed the effectiveness of the shelter design against NBC hazards.

Status
In service with the French, Canadian and US Air Forces.

Manufacturers
Bonna Sabla.
SP Défense.

UPDATED

View from inside the tunnel (Bonna Sabla)　　　　　**2002**/0116918

Inside the accommodation section showing side-mounted sleeping berths (Bonna Sabla)　　　　　**2002**/0116919

SP Défense NBC composite filters for collective protection

Description
The SP Défense range of composite NBC COLPRO filters are designed to remove all known CW and BW agents, radioactive fallout, particulates and gases. They comprise a paper HEPA filter to remove particles and aerosols supported by a downstream activated charcoal anti-vapour filter housed in a single metal frame. This filter series is aimed at COLPRO systems for vehicles, tented and fixed shelters or ship's NBC citadels or CPS. It is reported as 99.997 per cent effective against aerosols.

The FMM7 model meets STANAG 4447.

Specifications

Model	Nominal airflow	Initial pressure drop (Pa)	Dimensions (L × W × H)	Weight (kg)	Designed for
FMM3	30 m³/h	600	256.4 (dia) × 156	5.5	Vehicles, mobile shelters
FMM4	90 m³/h	1,200	571 × 254 × 126	22	Vehicles, mobile shelters
FMM5	180 m³/h	1,650	616 × 395 × 157	37	Vehicles, mobile shelters
FMM2	255 m³/h	1,000	460 × 460 × 336	34	Ships, mobile shelters
FMM7	300 m³/h	1,600	544.5 (dia) × 270	29	Ships, tents, mobile shelters
FMM1	1,200 m³/h	960	790 × 790 × 1,600	340	Fixed shelters

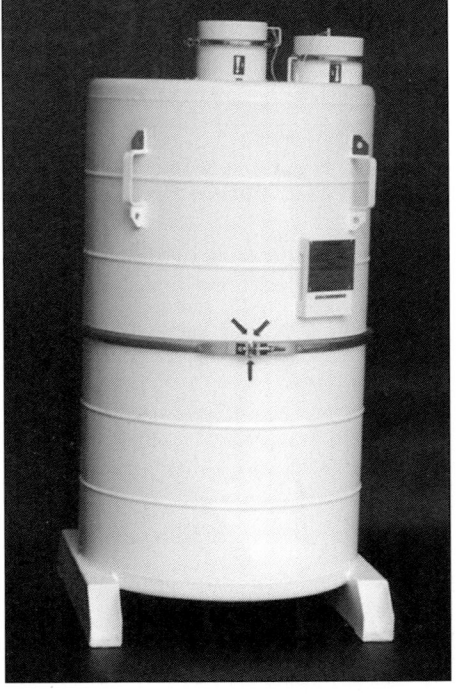

FMM1 1,200 m³/h NBC filter - for shelters
0011470

FMM7 300 m³/h NBC filter　　　　　0011481

Status
In service with the French armed forces and others.

Manufacturer
SP Défense.

UPDATED

SP Défense NBC filtration-pressurisation units for ships

Description
The SP Défense filtration-pressurisation design for ships delivers NBC filtered pressurised air to surface ship NBC citadels or other protected areas.

A typical unit comprises a metal casing (usually stainless steel) containing a media prefilter, air heater, high efficiency HEPA filter, motor fan and NBC filters. Other components can be provided according to user requirements, such as anti-blast valves, a shower to wash the prefilter and bypass valves to bypass the NBC filters in peacetime. Bolted to an external bulkhead, the unit offers easy access to allow removal and replacement of any component from the outside, without risk of contamination inside the citadel. The nominal airflow of a unit varies from 600 to 3,000 m³/h according to the tonnage of the ship. These units meet military national and NATO specifications for blast, shock, vibrations, noise or low magnet signature, according to requirements. Examples are shown in the table.

Specifications (examples)

Unit type	4129-0101	4510-0001	4313-0500
Ship	Frigate *Jean Bart*	Landing Barge Transport *Sirocco*	Aircraft carrier *Charles de Gaulle*
Airflow	900 m³/h	1,200 m³/h	3,000 m³/h
Dimensions	1,900 × 1,300 × 1,900 mm	1,900 × 1,300 × 1,900 mm	2,850 × 1,900 × 1,900 mm
Weight	1,615 kg	1,450 kg	2,660 kg
Power consumption	11.5 W	11.5 W	37.2 W
Power supply	3-phase 440 V 60 Hz	3-phase 440 V 60 Hz	3-phase 440 V 60 Hz

Status
In service with and in production for the French armed forces and other navies.

Manufacturer
SP Défense.

VERIFIED

900 m³/h filtration-pressurisation unit for a corvette   0011474

SP Défense NBC filtration-pressurisation units for vehicles and mobile shelters

Description
SP Défense offers a range of filtration-pressurisation units designed to be tailored to suit particular vehicles or mobile shelters. The parameters of two currently deployed types are illustrated in the table.

The alloy casings are designed to fit the space parameters defined by users. Housed within the casing are cyclonic prefilters for dust and sand, one or more NBC filters depending on the required output, a motor and fan system. Other components can be incorporated, such as a media prefilter or a 4-way valve to allow the filters to be bypassed for non-NBC operations.

Nominal airflow for these systems ranges between 30 and 180 m³/h according to the system design target. All systems offer 99.997 per cent efficiency against

60 m³/h filtration-pressurisation unit for the Panhard VBL   0011472

Specifications (examples)

Unit type	4318-0001	4127-0500
Vehicle	VBL (Panhard)	VAB II (RVI*)
Airflow	60 m³/h	180 m³/h
Dimensions	940 × 485 × 320	720 × 720 × 320
Weight	37 kg	75 kg
Power consumption	200 W	800 W
Power supply	28 V DC	28 V DC

*Renault Véhicules Industriels

aerosols and take full account of current specifications for shock, vibration, noise, EM compatibility and EMP protection.

Status
In service with and in production for the French armed forces.

Manufacturer
SP Défense.

VERIFIED

TMB collective NBC protection tent

Description
The TMB collective NBC protection tent covers an area of 45 m² and uses a double-layer construction. The outer layer acts as a sun and weather cover and is made of cotton. It also creates an insulating air gap between the two layers. The inner layer is impermeable to toxic chemical agents for up to 24 hours and is maintained at a positive overpressure by a ventilation and filtration unit. A cooling unit is used to maintain an even internal temperature of 20°C, although this can be varied.

Status
In production. In service with the French armed forces.

Manufacturer
Techniques Michel Brochier.

VERIFIED

TMB collective NBC protection tent

GERMANY

Kärcher NBC protective foil

Description
Kärcher NBC protective foil is designed for the temporary large-scale protection of exposed equipment against liquid contamination by CBW agents. It is delivered in the form of a tough multilayer flexible film. The material can also be used as a pathway protection barrier to prevent contaminated personnel from treading

Specifications	
Protection capability:	up to 24 h against HD
Thickness:	approx 25 μm
Weight:	approx 30 g m⁻²
Tensile strength (machine direction):	approx 180 Mpa
Tensile strength (transverse direction):	approx 55 Mpa
Temperature resistance:	93°C (max continuous use temperature)
Water vapour transmission rate:	180 g m⁻² in 24 h

contamination into clean surfaces. Examples include aircraft cockpits, routes between covered facilities and vehicle interiors. For this purpose, the foil can be delivered in 1 metre-wide rolls, 100 metres long.

The material is also easy to shape and weld into other forms. For example, it can constitute temporary overboots and carrier bags for contaminated equipment or samples.

Status
In production. In service with OPCW.

Manufacturer
Alfred Kärcher GmbH & Co.

VERIFIED

···

Kärcher SPS 2000 long-term conservation system

Description
The Kärcher SPS 2000 long-term conservation system is a tent which can be erected easily and which has been designed for the local conservation of large-scale material. The tent is completely self-sufficient and the tent scaffold and outer awning have been designed to withstand harsh environmental conditions. An inner tent prevents humid air from entering the conservation compartment while air in the inner tent is circulated by a dehumidifier; the optimum long-term results are achieved with air of approximately 40 per cent humidity. Energy is supplied by solar panels mounted on the outer tent and is stored in accumulators via a control circuit.

An oil protection sheet collects oil or other liquids from leaks in the stored material. Granular mats and track boards are unrolled on the oil protection sheet. After placing material inside the inner tent it is hermetically sealed using a sliding seal.

The outer tent is closed and protected by overlapping seams on the outer awning. The awnings are secured with lashing ropes and tent pegs, with joining washers between the tent poles and the awning. Wind bracing is provided for the front scaffold segments.

Kärcher SPS 2000 long-term conservation system

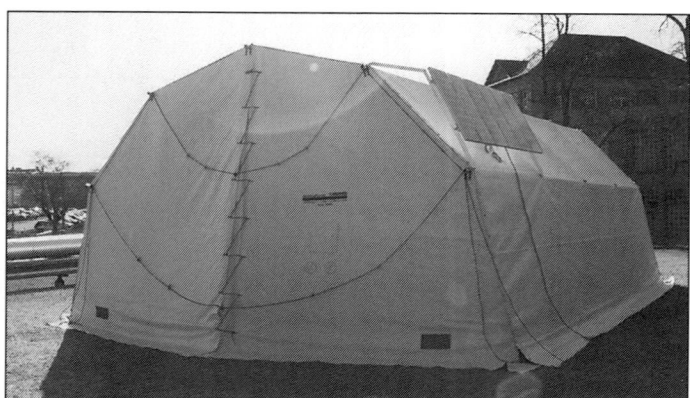
Kärcher SPS 2000 long-term conservation system

Status
In production. In service with Saudi Arabia and under test or in service with other countries.

Manufacturer
Alfred Kärcher GmbH & Co.

VERIFIED

IRAQ

NBC shelter filter

Description
This filter is intended to remove chemical, biological and radioactive toxic agents from air used inside a collective shelter. The filter has an air flow rate of 100 m³/min and a maximum air resistance of 650 Pa. It will remain operational in an atmosphere free of toxic agents for two years, out of which a maximum of 500 hours will be spent operating. Storage life under controlled conditions is five years.

The filter is tested with HCN at 220 g/min and phosgene at 450 g/min.

Weight is a maximum of 37 kg. The filter body is 410 mm high (±5 mm) and 460 mm in diameter (±5 mm). Inlet and outlet diameters are 105 mm (±5 mm).

Status
In service with the Iraqi armed forces.

Manufacturer
State factories.

VERIFIED

Iraqi NBC shelter filter

ISRAEL

Kinetics NBC conditioning systems

Description
Kinetics Limited manufactures a range of NBC, air conditioning and microclimate cooling systems for Israeli tanks, aircraft and armoured vehicles.

LSS
The Life Support System (LSS) fitted in the M60 tanks offers a wide choice of operational options. The crew can select between cabin cooling, microclimate cooling or personal cooling using cooling vests. NBC protection can also be delivered as mask-free overpressure conditioning or directly to individual masks from fixed points in the vehicle.

Overpressure/collective protection NBC system
A generic overpressure/collective protection NBC system has been developed for tanks and AFVs which uses widely available, common-range components to deliver a high standard of collective protection.

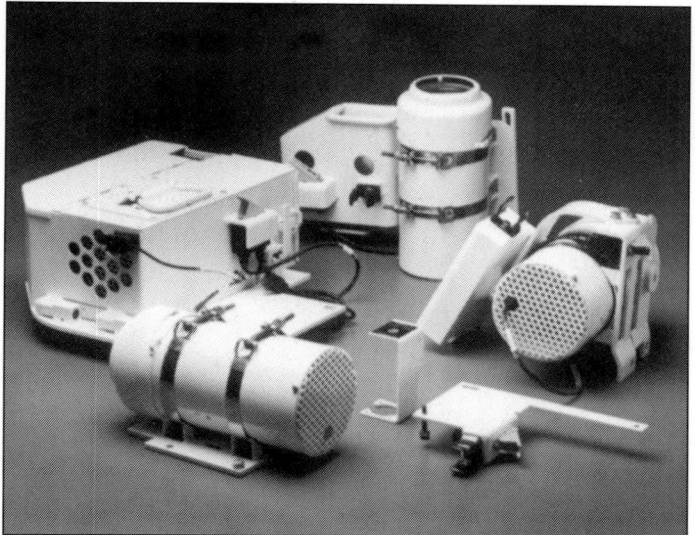

Kinetics NBC personal protection system for a tank   0050793

Specifications
Rated flow: 2.83-5.66 m³/min
Power consumption: 63 A max at 28 V DC
Voltage: 18-32 V DC (MIL-STD-1275)
NBC filtration: HEPA and activated charcoal
Dust separation efficiency: 97% at SAE coarse dust

Status
Available.

Manufacturer
KINETICS Limited.

VERIFIED

FA 100/50 T NBC filtration and ventilation system for vehicles and containers

Description
The FA 100/50 T is a COLPRO filtration and ventilation system for vehicles and containers, providing full NBC protection. The system creates overpressure in the compartment to prevent penetration by contaminated air and relieves the occupants from the need to wear respirators and IPE. Heating or air conditioning can be connected downstream of this system. The purified air exhausts through an overpressure valve, which regulates the positive pressure in the compartment. The system is designed for permanent installation on the side of the vehicle, and operates with vehicle stationary or in motion. The system comprises:
- Blower
- Prefilter
- NBC filter
- Overpressure regulation valve
- Positive pressure gauge (on request)
- 220 V/110 V power supply interface (on request)

Specifications
Number of people: 4-8 persons
Overpressure inside the vehicle in filtration mode (adjustable): 12-50 mm water column
Air flow rate:
   50 m³/h (filtration)
   100 m³/h (ventilation)
Voltage: 27.5 V DC
Operating voltage range: 18-32 V DC
Weight:
   27 kg (air supply unit)
   4 kg (overpressure valve)
Operating temperature range: +60 to −20°C
Operating altitude range: −457 to +3,048 m

NBC Filter
The NBC filter is a sealed unit attached to the side of the vehicle, disconnected from the ventilation system. In the event of an NBC alert, the seals are broken, the lids removed and the blower airflow diverted through the NBC filter. Changeover is simply achieved without the use of specialist tools within 1 minute.

Specifications
NBC filter model: HF 50 T
DMMP absorption capacity: >775,000 (mg/min/m³)
CK absorption capacity: >170,000 (mg/min/m³)
Aerosol filtration efficiency: 99.995%

Status
Available. Certified by the Standards Institution of Israel, the Israeli Institute for Biological Research and by the Israeli Defence Forces.

Manufacturer
Beth-El Zikhron Yaaqov Industries Ltd.

NEW ENTRY

FA series NBC filtration systems for tents

Description
The FA range of COLPRO filter systems are portable high-capacity units designed to provide filtered air for tents (*ad hoc* or temporary COLPRO) against BW and CW contaminants. The systems are connected by quick connectors and a flexible hose to a sleeve in the tent sidewall. The filtered air creates overpressure inside the tent and relieves the occupants from the need to wear respirators and IPE. The system comprises:
- Blower unit
- Filter units (comprising prefilter, HEPA and activated carbon vapour adsorption matrix)
- Differential pressure gauge

Type	FA 90 N	FA 150 N	FA 150 NC	FA 300 N	FA 300 NM
Air flow rate (single filter)	90 m³/h	150 m³/h	150 m³/h	300 m³/h	150 m³/h
Pressure drop	450 Pa	420 Pa	750 Pa	600 Pa	750 Pa
Weight of active carbon	11.5 kg	17.0 kg	11.5 kg	25.0 kg	11.5 kg
Separation efficiency Particle size of 0.3 µ	>99.995%	>99.997%	>99.995%	>99.997%	>99.995%
Adsorption capacity (DMMP)	>2.0 × 10⁶ mg m m⁻³	>1.5 × 10⁶ mg m m⁻³	>1.2 × 10⁶ mg m m⁻³	>1.2 × 10⁶ mg m m⁻³	>1.2 × 10⁶ mg m m⁻³
Motor (voltage/current)	230 V 1.5 A	230 V 1.8 A	230 V 1.7 A	230 V 2.0 A	230 V 1.5 A

Status
Available. Certified by the Standards Institution of Israel, the Israeli Institute for Biological Research and by the Israeli Defence Forces.

Manufacturer
Beth-El Zikhron Yaaqov Industries Ltd.

NEW ENTRY

Rainbow 36 family protection system

Description
According to planning law, every Israeli household has to have a room designed for collective protection for all the family. The Rainbow 36 family protection system is an innovative new system which permits full protection for a six-member family in their standard protected room without the need to wear respirators or IPE. It is cost effective and reliable, giving several weeks of safe protection against all known

Specifications
NBC filtration unit
Airflow rate: 36 m³/h
Pressure drop @ 36 m³/h: 24 mm WC
Weight of activated carbon: 3.8 kg
Total weight: 7.3 kg
Separation efficiency against particles of 0.3 µ: 99.995%

Air inlet valve
Airflow rate: 36 m³/h
Pressure drop @ 36 m³/h: 5 mm WC
Side-on pressure resistance: 300 kPa

Ventilation unit
Airflow rate: 36 m³/h
Overpressure drop @ 36 m³/h: 52 mm W
Connection with electrical supply: 230 V AC
Current: 0.2 A
Noise level: 68 dB(A)

NBC warfare agents. The system operates from a standard AC power source, with an integrated battery back-up system in case of power failure, delivering clean air at a maximum output of 36 m³/h. The battery will run the system for up to 10 hours, recharging as soon as the system is reconnected to a live AC power source. For additional back up there is a hand-operated blower.

The system draws air from outside through a filter mounted on the inside wall which adsorbs any agent onto activated charcoal. Wall-mounted inlet and exhaust valves offer blast protection and a pinwheel airflow indicator shows the occupants when the correct overpressure inside the protected room has been reached.

The system is supplied in two hand carrying cases which are convenient to store inside the home. It can be assembled in one minute without the need of tools. Domestic users would mount the system on receipt of a warning from the Army or local authorities in times of emergency. No routine maintenance is necessary although the battery should be checked every six months and recharged if necessary. Battery condition is continuously monitored even when switched off

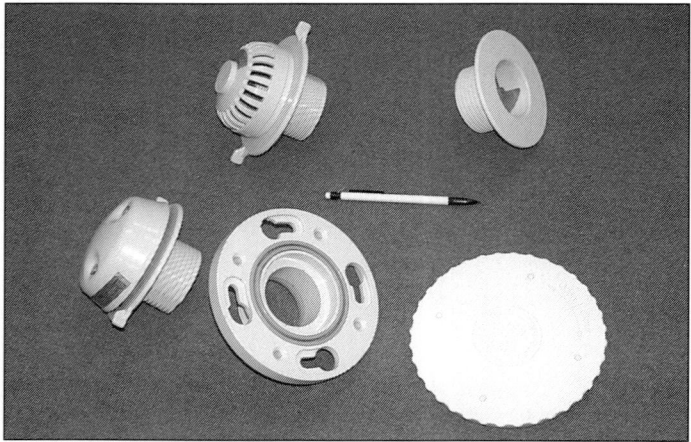

To supply filtered air to the protected room, an external wall is pierced with supply and exhaust trunks. The air inlet valve (bottom left) fits over the external end of the air supply trunking. The bayonet connector flange (bottom centre) is fitted on the inner end to receive the filter unit (see above). The blast overpressure valve (top left) fits on the external end of the exhaust trunking. An airflow indicator valve (top right) fits on its interior end, showing the occupants when the correct overpressure has been reached (John Eldridge)   *2002*/0137262

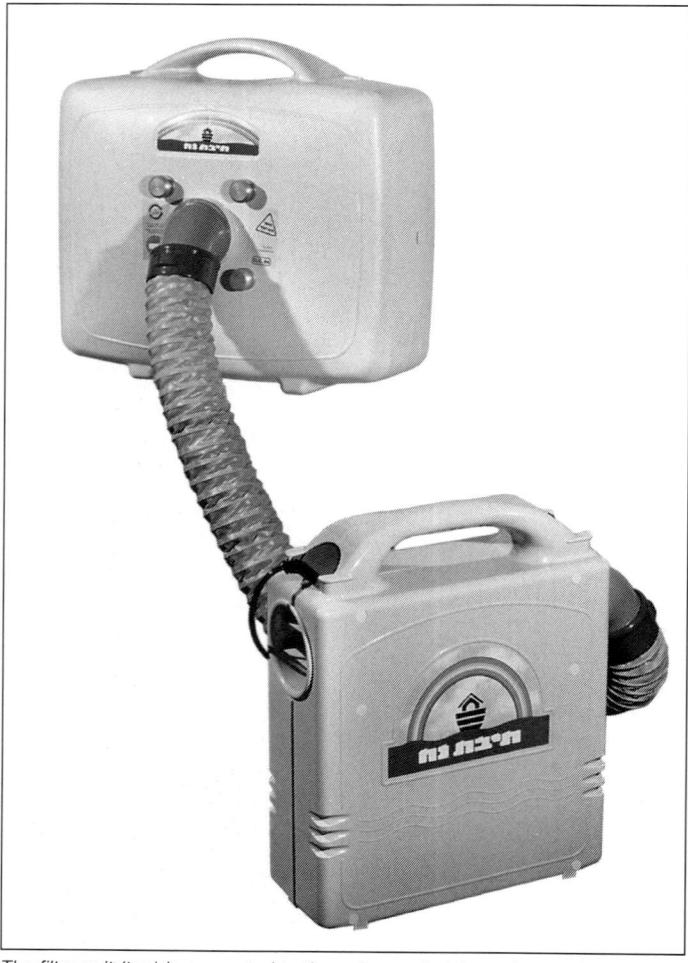

The filter unit (top) is connected to the wall-mounted flange by a quick (bayonet) connection. The hose connects the filter unit to the portable ventilator unit (below) whose motor draws the filtered air into the compartment (Beth-El Zikhron Yaaqov Industries Ltd)   *2002*/0137263

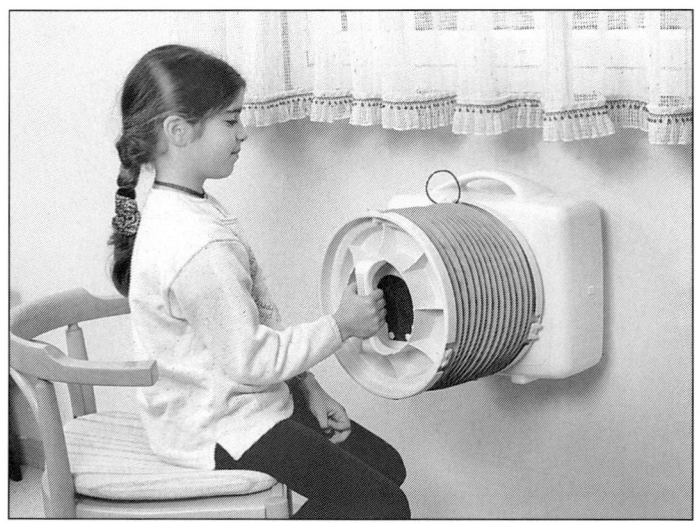

Following a total electrical failure (the Rainbow 36 has a 10-hour duration battery), air can be drawn through the filter unit using the hand-powered bellows unit shown here (Beth-El Zikhron Yaaqov Industries Ltd)   *2002*/0137272

and an audible alarm sounds when the battery is discharged below a pre-set trigger level.

Status
Available. Certified by the Standards Institution of Israel and by the Israeli Civil Defense Command.

Manufacturer
Beth-El Zikhron Yaaqov Industries Ltd.

NEW ENTRY

SafeAir series of NBC COLPRO filtration systems

Description
SafeAir FAH range of filtration systems
SafeAir is a range of COLPRO filtration systems designed to allow sustained operations by a variety of fixed facilities, such as command shelters, whilst under NBC attack. Filtered air capacities range from 50 to 600 m³/h and the systems also deliver normal, unfiltered air at the outputs shown below. Filtered air is delivered at a slight overpressure (between 18 and 27 Pa depending on the arrangements) to the protected area, creating a safe and comfortable environment without the need for respirators or IPE. The systems are electrically operated from the mains power supply, with manual back up in the event of power failure. Systems with additional battery back up are in the final stage of development.

Model HF series filter units
HF series units are modular, allowing them to be connected and linked together in banks to deliver high capacity filtration to major facilities. They are manufactured in heavy-duty coated steel and contain a HEPA section upstream of an activated carbon section. The units have castor wheels for mobility and shock absorbent feet. All units have side carrying handles and the heavier units have lifting lugs.

A modular high-capacity COLPRO filtration installation showing Model L 1800 Multi-Purpose Blowers (bottom) connected to four banks of filter units comprising the Models HF 600 C and HF 300 C (Beth-El Zikhron Yaaqov Industries Ltd)   *2002*/0137269

*Another view of a high-capacity COLPRO filtration plant
(Beth-El Zikhron Yaaqov Industries Ltd)* ***2002**/0137268*

Specifications

Model	FAH 100/50	FAH 180/90	FAH 480/180	FAH 800/300	FAH 1600/600
Fresh air output	100 m³/h	180 m³/h	480 m³/h	800 m³/h	1600 m³/h
Filtered air output	50 m³/h	90 m³/h	180 m³/h	300 m³/h	600 m³/h
Supplied compartment occupancy guide	8 people	15 people	30 people	50 people	100 people

Filters		
Model	HF 600 C	HF 900 C
Output	600 m³/h	900 m³/h

Multipurpose blowers			
Model	L 1600/600	L 1800	L 3000
Power consumption	1.15 kW	1.50 kW	2.35 kW
Designed pressure	1,700 Pa	2,400 Pa	2,400 Pa

Blast and overpressure valves
Specially designed valves are fitted to the FAH series and to larger filter plants to reduce the effects of blast damage to the filtration components and to regulate the overpressure inside the facility.

Status
Available. Tested and certified by the Standards Institution of Israel, the Civil Defence Command and the Israeli Institute for Biological Research. Field tests have been carried out in realistic conditions and approvals for use have been issued by several laboratories worldwide. Widespread installation throughout Israel, supplying facilities in the civilian, medical, public, educational and government sectors. Exported and in service with several armed forces.

Manufacturer
Beth-El Zikhron Yaaqov Industries Ltd.

NEW ENTRY

SHALON collective NBC filtration systems

Description
The collective NBC filtration systems produced by SHALON-Chemical Industries Limited comply with Israeli civil defence specifications but are suitable for many

Interior of a shelter employing a SHALON FA-75 collective NBC filtration system

military requirements. Each system comprises a washable prefilter employing synthetic foam, blast valve, NBC gas-particulate filter, fan unit, flowmeter and an overpressure valve. The fan unit has an electric motor but can be operated manually by one person.

SHALON produces four basic systems:

(a) The FA-75 is suitable for shelters housing up to 12 people. The NBC protection mode air flow is 75 m³/h; the normal ventilation flow rate is twice that figure

(b) The FA-150 is suitable for shelters housing up to 25 people. The NBC protection mode air flow is 150 m³/h; the normal ventilation flow rate is twice that figure

(c) The FA-5 is suitable for shelters housing up to 50 people. The NBC protection mode air flow is 300 m³/h; the normal ventilation flow rate is twice that figure

(d) The FA-10 is suitable for shelters housing up to 100 people. The NBC protection mode air flow is 600 m³/h; the normal ventilation flow rate is twice that figure.

(e) The FA-15 is suitable for shelters housing up to 150 people. The NBC protection mode air flow is 900 m³/h; the normal ventilation flow rate is twice that figure.

Status
In production.

Manufacturer
SHALON-Chemical Industries Limited.

UPDATED

ITALY

Aero Sekur SpA NBC filter units

Description
Aero Sekur SpA produces a range of NBC filter units to suit a variety of purposes, ranging from the protection of building interiors to mobile installations on vehicles and warships.

All the Aero Sekur SpA NBC filter units follow the same basic operating pattern. There is a blastproof valve that may also be used as an outlet valve. The first filter stage is a prefilter to remove dust and other coarse particles. The pre-filtering process is achieved by the use of either a dynamic (cyclone) filter or a static unit

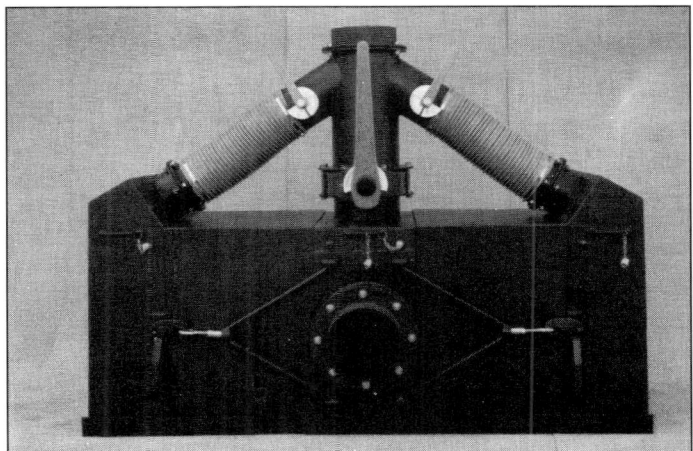

Aero Sekur SpA SP-1200 NBC filter unit for large structures

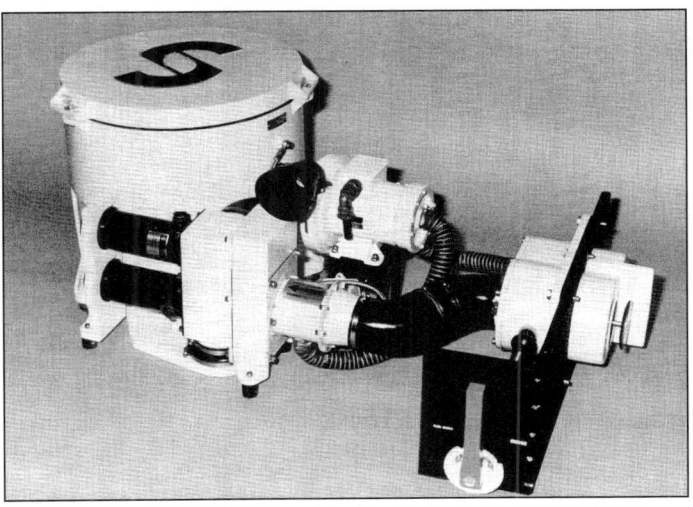

Aero Sekur SpA SP-180 NBC filter unit as fitted to the Ariete main battle tank

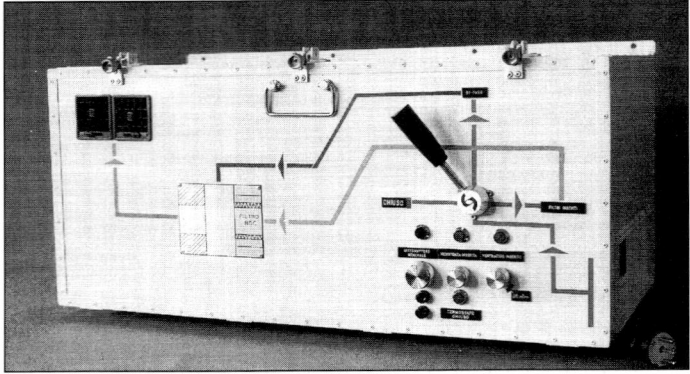

Aero Sekur SpA SP-35 NBC unit for use in mobile shelters or small armoured vehicles

using washable synthetic foam. Both are up to 95 per cent effective. For naval use, a special prefilter to retain sodium chloride particles is available. Downstream, an aerosol filter unit retains particles larger than 0.3 μm. The efficiency of this stage is ≥ 99.995 per cent. The pleated filters consist of a special glass fibre paper resistant to heat and chemical agent corrosion. The final stage in the filtration process is an activated charcoal layer to remove remaining CW agents and radioactive particles. All these stages are contained in stainless steel or galvanised steel casings.

Air intake and delivery pipelines, intercept valves, air flow dampers and electric fans are all available from Aero Sekur SpA as compatible accessories.

Status
In production. Aero Sekur SpA COLPRO systems are in service in several of the principal Italian army vehicles (see *Jane's Armour and Artillery* for vehicle details). These include (Original Irvin Part Nos):
- Ariete MBT: 43956020 (main), 43956011 (driver)
- Centauro Armoured car: 43950200
- Dardo IFV: 43956200R04
- Puma APC: 4 × 4 vehicle: 43950110R03
- Puma APC: 6 × 6 vehicle: 43956210R01
- Shelters: 43956138R00
- VBC Troop carrier - wheeled: 43956136

Manufacturer
Aero Sekur SpA.

UPDATED

JAPAN

N-KRAFTON protective shield materials

Description
N-KRAFTON is a proprietary technology incorporated into a variety of products to offer significantly increased protection against γ-radiation and neutrons. It is being used in a wide variety of applications in the defence and nuclear industries. This technology can be incorporated into fixed and temporary shielding barriers or into surface coatings where it reduces the penetration of γ-rays and neutrons. The lead/acrylic transparent material is suitable for the shielding of personnel operating vehicles in support of clean-up operations following a nuclear accident, for example. Similarly, the non-transparent rigid shielding material can be used as cladding on vehicles and as semi-permanent shielding for personnel. N-KRAFTON technology can also be incorporated into personnel protection shields for those operating in dangerous, high-radiation environments. In the defence arena, N-KRAFTON technology clearly has a major role to play in the nuclear protection of both armoured and soft skinned vehicles as well as in aircraft and ships. There are several types of product in which the N-KRAFTON technology is incorporated. These include:

N-5 Series
A neutron radiation barrier constructed of a rigid lightweight composite material.

C-1 Series
A neutron radiation barrier constructed of a rigid, transparent, lightweight, composite material. Used with the K-XA series (see illustration).

K-XA (lead acrylic) Series
A primary and secondary γ-radiation barrier constructed of a rigid, transparent, lightweight, composite material.

XP-3 Series
A neutron and γ-radiation barrier constructed of a medium-density composite material. XP-3 comes in the form of a flexible thermo-setting mastic material suitable for providing shielding plugs where electricity or data wiring and pipe-runs penetrate shielding walls in, for example, a nuclear reactor facility.

XP-5 Series (high lead/iron content)
A neutron and γ-radiation barrier constructed of a rigid high-density composite material. XP-5 (and XP-3) is widely used in the Japanese nuclear industry as a lining material for nuclear waste containers.

Coating material
A surface coating material offering protection against radon gas. It dries at room temperature and offers a hard enamelled finish.

Status
In production. In widespread service.

Manufacturer
Nippon Tokuso Co Ltd.

UPDATED

KOREA, SOUTH

KC 100 NBC filter

Description
In the KC 100 NBC filter, contaminated air first meets a HEPA filter with an effective area of 19 m² made from pleated and spaced media. It then passes through an NBC gas filter charged with 150 litres of activated charcoal in a 140 mm deep bed. The unit is a vertical drum shape resting on flanged feet, secured to supporting structure by four bolts. KC 100 filters can be used singly or in banks to deliver high

KC 100 NBC filter – industrial installation
0050790

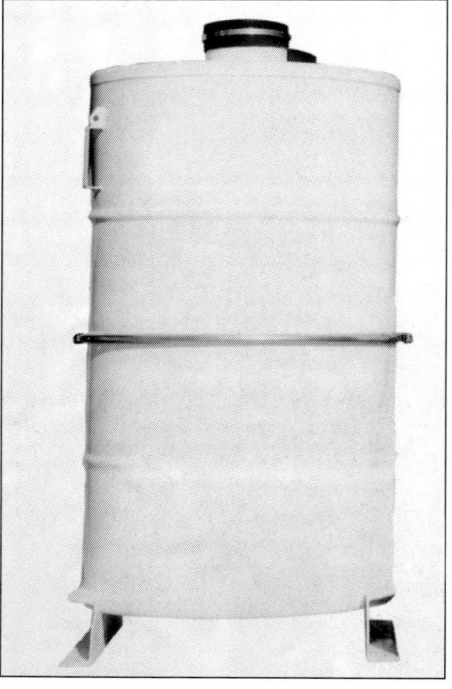

KC 100 NBC filter
0050792

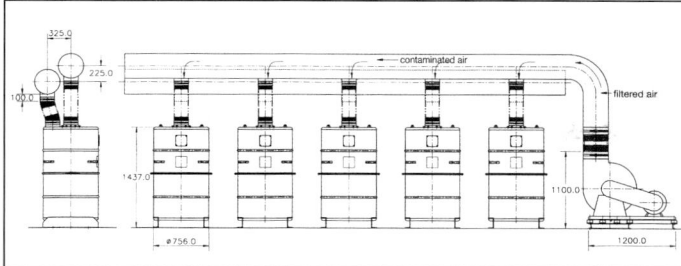

KC 100 NBC filter bank   0050791

Specifications
Rated Flow: 18 m³/min
Pressure gradient across filter: 500 (±50) Pa
Voltage: 18-32 V DC (MIL-STD-1275)
Efficiency: 99.995% (0.3 mm aerosol)
Dimensions: 790 × 760 × 1600 mm
Weight: 350 kg

Initial capacity against selected simulants:

DMMP	CK (CNcl₂)		PS (chloropicrin)
	CK (CNcl₂)		PS (chloropicrin)
Challenge concentration (mg/m³)	4,000	5,000	5,000
Breakthrough concentration (mg/m³)	8	5	0.04
Filter life	≥40	≥50	≥125
Cₜ value (mg min/m³)	≥60,000	≥250,000	≥625,000

volumes of filtered air in military or civil defence facilities, hospitals or command centres.

Status
In service with Korean defence, civil defence and civilian authorities.

Manufacturer
Samgong Industrial Company Limited.

VERIFIED

NETHERLANDS

AstroCarb NBC filters

Description
AstroCarb NBC filters combine a HEPA filter for the filtration of NBC agents and an adsorber for the filtration of toxic gases and vapours into a complete NBC filter. The two main components are enclosed in one filter cell made of anodised aluminium cell sides. To seal the filter against a sealing frame it is provided with a one-piece polyurethane foam gasket on the intake side.

The AstroCarb NBC 200 HC collective protection filter unit has a nominal capacity of 200 m³/h and an initial (clean) resistance of 1,750 Pa. The unit measures 460 × 460 × 329.5 mm and weighs approximately 35 kg.

AstroCarb NBC filter in NBC collective protection unit (AAF-International BV)

AstroCarb NBC 90 HC collective protection filter unit (AAF-International BV)

The AstroCarb NBC 90 HC collective protection filter unit has a nominal capacity of 90 m³/h and an initial (clean) resistance of 410 Pa. The unit measures 620 × 302 × 170 mm and weighs approximately 17 kg.

Status
In production.

Manufacturer
AAF-International BV.

UPDATED

Stork Bronswerk NBC filter station

Description
This Stork Bronswerk NBC filter station was developed and produced to STANAG 4447 (and other appropriate STANAGs such as 4192). In standard form the capacity is 900 m³/h but other capacities are available on request (see illustrations). The ship's system will normally consist of one or more identical NBC filter units which, at the highest state of NBC readiness, supply filtered decontaminated air to the ships citadel while maintaining an overpressure of ±500 Pa. For normal cruising and during training exercises the NBC filter units can be bypassed.

The NBC filter station is equipped with built-in prefilters and preheaters. Optional blast protection valves can be incorporated. The casing is made of 316 stainless steel while an additional multilayer coating protects the weather-exposed front area. The filter station has a low magnetic signature and is shockproof up to 50 g (depending on the version). The station can be delivered with standard indicators and pilot lights or with an electronic control and monitoring system, including local and remote operation. Filter systems are available for protected communications facilities and mobile hospitals.

The production facilities comply fully with ISO 9001.

Status
In production. Applications include the Spanish 'Caraminas' minehunters, the Netherlands and Spanish 'Amsterdam' class replenishment vessels and the Netherlands LCF frigate programme.

A free-standing Air Filtration Unit NBC 600 is in production for the UK MoD. With a planned capacity of 600 m³/h, the unit can be fitted inside or outside the ship's citadel.

Manufacturer
Stork Bronswerk BV.

VERIFIED

RUSSIAN FEDERATION AND ASSOCIATED STATES (CIS)

FPT-100B and FPT-200B NBC filter units

Description
These two NBC filter units are used to filter air contaminated by chemical warfare agents and radiation before passing the filtered air into the interiors of tanks and other armoured combat vehicles.

FPT-100B NBC filter unit
0019378

FPU-200 standard filter
0019380

FP-300 standard filter
0019381

FPT-200B NBC filter unit
0019379

The FPT-100B can deliver 100 m³/h of filtered air. It weighs 9.5 kg, is 332 mm high with a diameter of 305 mm.

The FPT-200B can deliver 200 m³/h of filtered air. It weighs 15 kg and is 355 mm high with a diameter of 334 mm.

Status
In production. Offered for export sales.

Manufacturer
Unknown.

Marketing agency
Rosoboronexport.

VERIFIED

FPU-200 and FP-300 NBC filter canisters

Description
FPU filter canisters are designed to provide air filtration for conditioning systems in important fixed COLPRO-equipped facilities and in ship-installed citadels or other CPS systems. They can be installed singly or connected together (from 1 to 3 canisters). The FPU-200 when connected in this way can offer 600 m³/h and the FP-300: 900 m³/h volumetric flow.

Status
In production and available for export.

Specifications			
Type	FPU-200 (fixed installations)	FPU-200 (ship citadels)	FP-300
Volume flow of air	100 m³/h	200 m³/h	300 m³/h
Max initial airflow resistance	55 mm H₂O	120 mm H₂O	85 mm H₂O
Height	650 mm	1,190 mm	530 mm
Diameter	455 mm	455 mm	580 mm
Working temperature limits	(figure not supplied)	(figure not supplied)	50°C
Weight (less container)	30 kg	60 kg	66 kg

Manufacturer
Unknown.

Marketing agency
Rosoboronexport.

VERIFIED

FVUA-100A NBC filtration unit

Description
The FVUA-100A NBC filtration unit removes radiation particles, chemical agent aerosols and dust from the atmosphere before passing the filtered air into vehicle pressurised bodies and crew cabs. The complete unit consists of a housing and frame, dust prefilter assembly, filter assembly, 12/24 V DC motor and an installation kit.

The unit can provide a flow of filtered air at a flow rate of 100 m³/h with a filter efficiency of 99 per cent. The filters have to be changed every 1,000 hours. Weight of the complete unit is 50 kg.

Status
In production. Offered for export sales. Unable to trace current manufacturer or supplier of service and parts.

FVUA-100A NBC filtration unit   0019377

Manufacturer
Unknown.

Marketing Agency
Rosoboronexport.

FVU vehicle collective protection system

Description
Following the introduction of the PAZ collective protection system (see entry in this section) the development of a second generation of more sophisticated systems began during the early 1960s. By 1967 a system known as the FVU (*Fil'troventilyatsionaya Ustanovka*) was introduced to the T-64 tank. It is also employed on the late production BMP-1 infantry combat vehicles and BMD airborne combat vehicles, MT-LB armoured transporters, BRDM-2 reconnaissance vehicles, 2S1 122 mm self-propelled guns and 2S3 152 mm self-propelled howitzers. Also equipped are SA-4 Ganef, SA-6 Gainful, SA-8 Gecko, SA-9 Gaskin and SA-13 Gopher air defence missile launchers. Some of these vehicle types are also provided with the PAZ system.

If the FVU system as fitted to the BMP-1 is taken as typical, the vehicle must be hermetically sealed by manually closing all hatches (the BMP-1 does not have the PAZ). Once the vehicle is closed, a centrifugal blower/dust separator is turned on drawing air into the FVU system through an air intake directly behind the turret. Contaminated outside air is drawn through ducts in a race around the turret ring and radioactive particulate matter is removed by impeller action. The air can then be fed through an NBC filter and, in winter, into a heater chamber. Once decontaminated, the air is passed through 11 vents located around the crew compartment to provide purified air and a slight overpressure to prevent the infiltration of contaminated outside air.

Extra protection against gamma radiation for lightly armoured vehicles such as the BMP-1 is provided by internal liners (or blankets) known as *svintsov'iye kovriki*. These liners consist of lead pellets embedded in a plastic blanket about 19 mm thick. Later versions of these liners are believed to have neutron-trapping properties.

The latest form of this type of system produced within Russia is known as the PKUZ-1.

Status
Manufactured by Russian state factories prior to privatisation. Current manufacturer unknown. Available and in service with the RFAS.

Manufacturer
Unknown.

PAZ vehicle collective protection system

Description
Development of what was to emerge as the PAZ vehicle collective protection system (PAZ - *Protivoatomnoy Zaschita*) was undertaken by the Morozov tank design bureau as part of the M-45 project to modernise the T-54A main battle tank, a project which resulted in the T-55.

The heart of the PAZ system is an RBZ-1M radiation detector located in the turret compartment and designed to detect the initial pulse of gamma or neutron radiation emitted by a nuclear explosion. If the detector senses radiation it actuates a series of explosive squib mechanisms to close the engine louvres, the gunner's sight aperture and the engine firewall vent fan aperture. Normally these apertures are kept open by detent pins but when the squibs are detonated the detent pins are blown out of place. This allows the apertures to close and hermetically seals the tank protecting the crew from the subsequent blast wave overpressure. The RBZ-1M also shuts off the engine to warn the crew of an impending blast and activates the turret blower/dust separator. This centrifugal separator has a set of fan blade/impellers that spin the air at 7,000 rpm to remove particulate matter. The system also overpressurises the interior.

The PAZ was introduced to the B-45/T-55 tank in 1957-58 and was later incorporated into the T-62 tank and the IMR combat engineer vehicle.

Status
Manufactured by Russian state factories prior to privatisation. Current manufacturer unknown. In service with the RFAS and other nations to whom RFAS military aid was extended.

Manufacturer
Unknown.

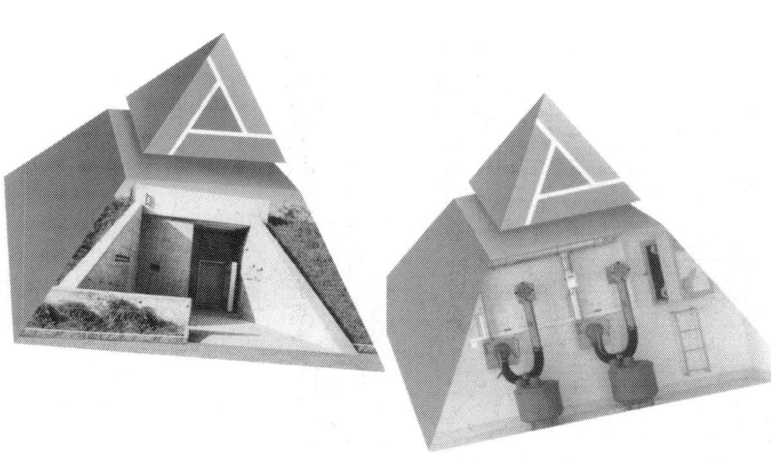

SWEDEN

DALOC SHELTEC shelter equipment

Description
DALOC SHELTEC AB specialises in the development and manufacture of components and systems designed to protect installations against weapon effects and achieve specified levels of protection.

Blast valves and blast doors, with auxiliary equipment, are produced to match almost any level of protection against mechanical weapon effects. Product performance has been verified by military agencies and full-scale tests. Door sizes range from personnel doors to doors large enough for large vehicles. Doors and valves are available in protection classes from 0.1 to 10 MPa.

Blast valves are produced for fresh air intake and exhaust and also for high-temperature diesel exhaust fumes. Modular air purification units are supplied as complete systems together with accessories and are available with capacities from 75 to 4,800 m³/h.

EMP shielding components and systems are designed and produced to meet specific requirements.

Status
The full range is in service with many armed forces.

Manufacturer
Daloc Sheltec AB.

UPDATED

JP SHELTEC NBC protective door installed in an underground shelter (Daloc Sheltec AB)

Air purification plant for a large shelter (Daloc Sheltec AB)

Filtrator NBC Filters – Type CV-90

Description
In addition to manufacturing the Type CV-90 filter series, Filtrator AB manufacture and supply a variety of NBC filtration and airconditioning equipment to the Swedish armed forces, including systems for the minor war vessels and for armoured vehicles. A range of filter packs is available.

Type CV-90 series NBC filters　　　0050892

Type YSB NBC system fitted in Swedish MCMVs　　　0050891

Specifications

Filter Type:	CV-90/50	CV-90/150
Air flow:	30-50 m³/h	130-170 m³/h
Pressure gradient:	690-1290 Pa	670-880 Pa
Aerosol penetration (DOP test):	0.003%	0.003%
Weight:	7.70 kg	16.30 kg

RBS-23

Total air flow:	600 m³/h. Filtered air supplies for 3 computers and the operators.
NBC capacity:	30-50 m³/h at designated overpressure
NBC Filter:	CV-90/50
Temperature control:	Automatic
Heating capacity:	6 kW
Cooling capacity:	6.5 kW

YSB

Total air flow:	340 m³/h
NBC Filtration:	Two CV-90/150
Overpressure:	200-600 Pa
Power requirement:	400 V AC 50 Hz
Air preheater:	1,200 W
Weight:	118 kg

RBS 23

The RBS 23 system is an integrated NBC and air conditioning system designed for the BAMSE mobile missile system (see *Jane's Land-Based Air Defence* for further details).

YSB

The YSB system is installed in the four Styrsö class MCMVs (see *Jane's Fighting Ships* for further details). It is constructed of non-magnetic components.

Status

In production.

Manufacturer

Filtrator AB.

VERIFIED

TrellTent® chemically hardened hospital tent system

Description

TrellTent® is a versatile, rapid deployable inflatable tent and shelter system. It can be used in all climates for treating injured people or for DECON purposes. It is lightweight and low volume when packed, making it easy to transport. It takes only 4 to 5 minutes to erect. The hospital is generally operational within one hour after erection in a new area.

Individual arches inflated with a small high-pressure blower, a foot pump or compressed air, support the system. Each arch is a closed unit with a diffusion-proof, removable inner tube. The semicircular arches make the unit very stable and resistant to snow and wind. Snow slides off, and there are no pockets where water can collect. Floor, ceiling, walls, windows and doors are all integrally welded to form a one-piece unit and the only loose parts in the tent are the divisible spacers between the arches. An erected tent is easy to move since the floor and the tent are welded together.

The TrellTent® can be made of different material and in different colours depending on customer specification. The most common choice is a strong, flame-retardant, synthetic fabric coated on both sides with PVC. The surface can be matt, gloss, monochrome or camouflaged. Standard colours are green and sand for military use and yellow for civil use. The ceiling always has a light green colour to meet medical requirements. Sewn and welded joints give a tight seal and every opening has sturdy zippers. The doors can be secured rolled up and large windows of transparent plastic allow for plenty of light. Alternatively the windows can be fitted with mosquito nets. The PVC material is easy to keep clean and the floor is made of a strong, PVC-coated fabric. The TrellTent® can be fitted with an extra non-slip removable inner floor to offer tougher wear and better hygiene.

The standard units are produced in one, two or three sections and can be joined door-to-door to amalgamate them into bigger units. Linking sections can be used to join several units together or to connect a unit to a vehicle or another tent. There are four different sizes. TrellTent® 3/2 and 3/4 weigh approximately 180 to 200 kg and when packed can be carried by four men. There are also special lightweight versions in a different material and a TrellTent® 1/2 weighs only 90 kg.

A range of specialist accessories is available, including sun nets, interlinings, 'light sluices' (blackout sections for use during military night operations). The latter also offer extra storage space and weather shelter. There are also 'transitional' sections to compensate for obstacles and irregular terrain.

The chemically hardened version of TrellTent® is available for use under threat or attack with CW agents. It provides the occupants with intact COLPRO, offering over 24 hours protection following a liquid attack. Tents may be connected together without loss of protection and filtered air can be supplied from external NBC filtration, environmental control and conditioning units at a slight overpressure.

The CW-proof tent material is a flame-retardant, self-extinguishing synthetic material coated on both sides with PVC. The same material is used for four inflatable self-supporting ribs and the high-frequency welded integrated floor. Each tent has two or four zip-fly entrances, with window lighting as required. Tents can be joined with other similar tents by using inflatable gaskets and securing with the use of eyelets, grommets and ropes. A single CW-proofed tent weighs 228 kg, is 7.25 m long and 5.2 m wide; maximum height is 2.6 m. The floor area is 38 m².

The filter unit for the tent consists of a fan unit, compressor, particle filter and an activated charcoal type ASC 12-30 TDKA filter. The unit is contained in a portable steel housing. A radial fan unit is employed to pass clean air through either an air conditioning unit (ACUTE 75) or through a VACAN 100 heater controlled from within the tent.

Status

The TrellTent® was developed in collaboration with the Swedish Army and has been deployed with the Swedish Defence Forces since the 1970s. Also in service with NATO and other countries. Available.

Manufacturer

Trelleborg Industri AB.

NEW ENTRY

SWITZERLAND

Andair AG shelter components

Description

Andair AG produces a wide range of COLPRO shelter components, including ventilation equipment. The company claims the creation of enough blast-proof and COLPRO shelter capacity over the years to protect 7,000,000 people. Switzerland's national NBC protection scheme has several thousand COLPRO

Ventilation unit showing filter bank and motor/hand-crank operated valve (Andair AG) **2002**/0122983

Shelter entrance, showing blast doors (Andair AG) **2002**/0122982

facilities for command, warning and medical installations. These cover the armed forces and the civil defence organisation.

Andair facilities include explosion protection up to 80 bar reflection pressure, gas-tight shut-off valves (150 - 1,250 mm diameter), filters (from 20 to 1,500 m3/h) air handling units, airconditioning units and power packs (diesel generator sets).

All equipment produced by Andair AG is produced according to Swiss Federal Office of Civil Defence directives and is type-tested and approved by the Swiss Defence Procurement Agency.

Gas filter Type GF-600 (Andair AG)
***2002**/0122984*

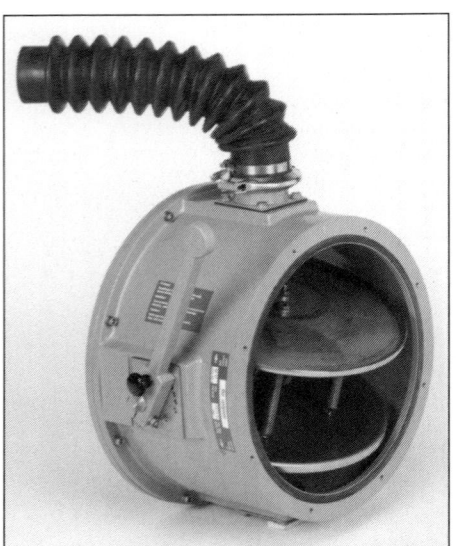

Gastight shut-off valve with connection for air barrier Type GAK-D (Andair AG)
***2002**/0122981*

Explosion protection valve Type ESV 4 (Andair AG)   ***2002**/0122980*

Status
In production. Adopted by over 30 countries worldwide.

Manufacturer
Andair AG.

UPDATED

....

BERICO shelter components

Description
All Swiss civil defence constructions have to utilise components that have been authorised by the Bundesamtes für Zivilschutz (BZS), the Swiss civil defence authority. BERICO AG of Niederglatt holds such an authorisation and produces a range of armoured doors, gates, covers and sliding walls for shelters and compression doors for use inside structures. It also produces ventilation plant for shelters of varying sizes.

The armoured doors and similar components are 200 mm thick to withstand one atmosphere excess pressure. Stronger components, to withstand up to three atmospheres excess pressure, are 250 or 350 mm thick. All components have standard height and width dimensions, for example a typical armoured door (Model PT1) is 1.85 m high and 0.8 m wide. Also available are sealable armoured windows.

Three basic models of NBC ventilation equipment are produced. They are the VA 40, VA 75 and the VA 150, with the numbers referring to the filter air flow capacity in m/h. The VA 40 is intended for shelters holding between 8 and 13 people, the VA 75 for shelters holding 14 to 25 people and the VA 150 for shelters holding between 26 and 50 people. All three models can be either hand operated or powered by a three-phase electrical supply.

Status
In production.

Manufacturer
BERICO AG.

VERIFIED

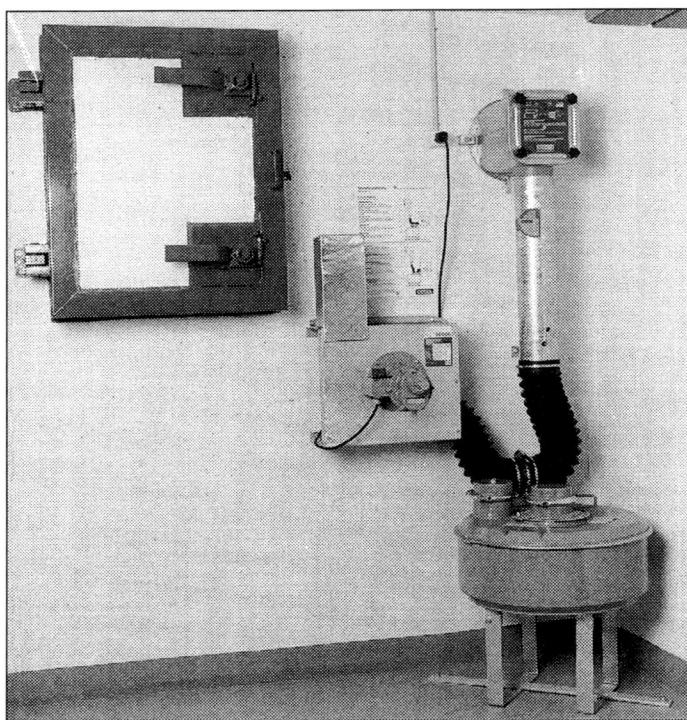

BERICO AG shelter components including a filtered ventilation system for NBC shelter

UNITED KINGDOM

AEA Technology NBC shelter system (NBCSS)

Description
The NBC Shelter System (NBCSS) designed by AEA Technology is a development of the Modular Containment System used to contain toxic materials in portable structures. The NBCSS will prevent the ingress of toxic materials, nerve agents, biologically active materials and bacteria and radioactive materials. It is adaptable, lightweight, robust and easy to build.

AEA NBC shelter system in use

The NBCSS consists of GRP panels which can be assembled by two people into almost any configuration required. The external surfaces are then covered with an easily strippable coating. Should this become contaminated, a further coat is applied before both are removed with the contamination trapped between the layers.

The clean working area is made up of a series of GRP panels bolted together through flanges and then sealed. The panels are usually 2.4 or 1.8 m high and 0.9 m wide. A mobile shower tunnel entrance is available for decontamination operations if required.

Filtration and ventilation systems can be tailored to meet user requirements.

Status
Available.

Manufacturer
AEA Technology plc.

VERIFIED

Aircontrol Technologies collective NBC protection for armoured vehicles

Description
Aircontrol Technologies are specialists in the design and manufacture of environmental control systems for all types of armoured vehicle. An example is the GKN Warrior infantry fighting vehicle. Aircontrol Technologies systems are fitted in all variants to provide collective NBC protection and crew temperature control. To mitigate the limitations of available space and power, major system components are supplied as separate line-replaceable units, with the compressor mechanically driven off the engine and interconnecting pipework forming part of the vehicle installation. The system provides 170 m³/h of filtered air for collective NBC protection as well as 5.5 kW of cooling and 5 kW of heating for crew temperature control. A specialised variant of this combined NBC and air conditioning system

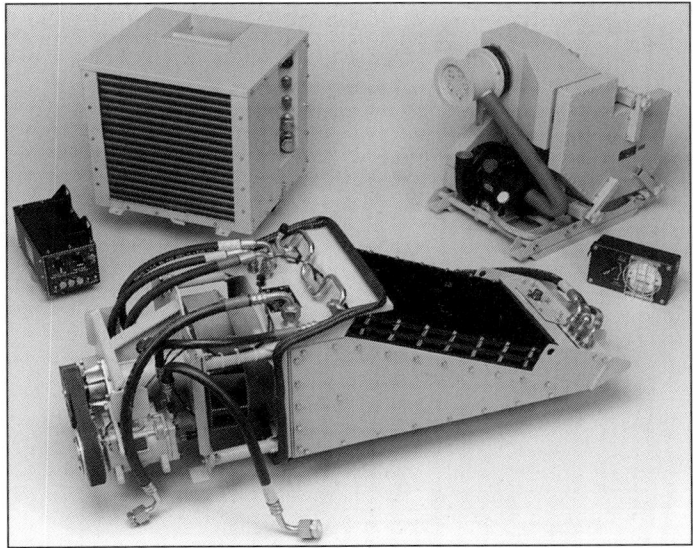

Aircontrol Technologies combined NBC/air conditioning system for the Warrior Desert Fighting Vehicle (Aircontrol Technologies Limited)   0011476

Typical Aircontrol Technologies NBC collective protection unit for armoured vehicles (Aircontrol Technologies Limited)   0011477

design is supplied for the Warrior Desert Fighting Vehicle to cater for the high-ambient conditions of desert environments.

Status
Warrior systems in service with the British Army.

Manufacturer
Aircontrol Technologies Limited.

UPDATED

Aircontrol Technologies Environmental Life Support Systems (ELSS) for fighting vehicles

Description
An Environmental Life Support System (ELSS) integrates the functions of heating, cooling, air filtration and oxygen replenishment to provide operationally acceptable conditions for vehicle crew members over a wide range of external conditions. To maintain a high level of crew operational efficiency, it becomes necessary to consider effective methods of heating or cooling the crew compartment air supply.

Under combat conditions with vehicles closed down and the crew in NBC clothing, vehicle temperatures can reach unacceptably high levels even in a European climate. In cold conditions, unless heated, the NBC system will deliver air at sub-zero temperatures. Aircontrol Technologies Limited systems provide full environmental control to maintain crew efficiency. Air from the NBC filtration system is cooled or heated before being delivered to the vehicle interior.

An Aircontrol Technologies (ELSS) for the British Army's AS90 155 mm self-propelled gun is supplied complete with armoured enclosures and mounted on the rear of the turret. Air from the collective NBC filtration system is heated or cooled before discharge through louvres at each crew position. The NBC filtered air provides spot cooling, heating, fume removal and vehicle overpressure. The NBC system is rated at 340 m³/h; 4.5 kW of cooling and 4 kW of heating are provided.

Status
In production. In service with the British Army for the AS90. Other systems are in production with the British armed forces.

Manufacturer
Aircontrol Technologies Limited.

UPDATED

Aircontrol Technologies Environmental Life Support System (ELSS) unit fitted to AS90 155 mm self-propelled gun (Aircontrol Technologies Limited)   **2000**

Aircontrol Technologies NBC filtration for container bodies

Description

Aircontrol Technologies produces a range of NBC filtration kits for use on military container bodies, shelters and soft-skinned vehicles. The externally mounted filtration pack will supply 170 m³/h clean air into a container where a preset pressure relief valve will maintain an overpressure to overcome leaks in the container. Control and monitoring of the system is carried out from within the container using a combined control and pressure indicating unit. The NBC filtration pack contains a fan and two stages of filtration. The first filtration stage is a disposable pre-filter to remove airborne dust and sand as it enters the pack. The second stage is a composite filter having two elements; the first is a HEPA element to remove fine particles and biological agents; the second is a deep bed activated-carbon anti-vapour element which absorbs chemical agents. Pack dimensions are 865 × 510 × 454 mm and the weight is 87 kg. A thermostatically-controlled ambient air heater is mounted in the filtration pack 'clean' ductwork. It is switched on automatically when air entering the shelter from the NBC filtration pack falls to 0°C.

Status

In production. In service with the British armed forces.

Manufacturer

Aircontrol Technologies Limited.

UPDATED

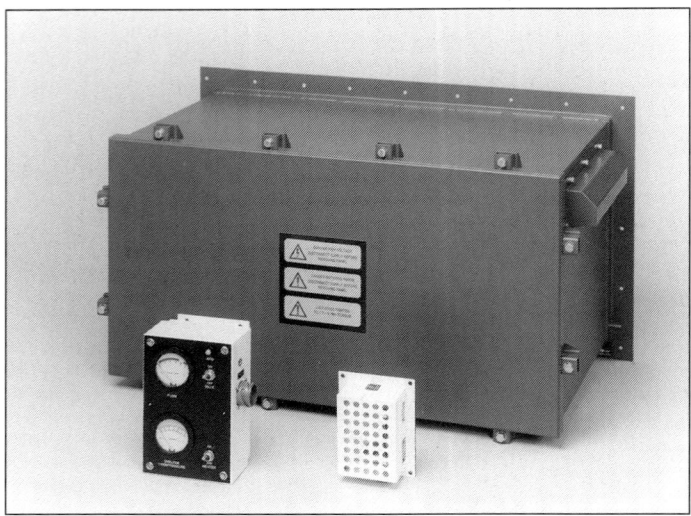

Complete NBC filtration system for container bodies
(Aircontrol Technologies Limited)   0011475

Aircontrol Technologies/PALL Aerospace collective NBC filtration

Description

Air Control Technologies Ltd and PALL Aerospace have teamed together to develop a NBC filtration package that incorporates regenerative filtration rather than non-reusable activated carbon elements. This is in response to recent evaluations which have shown the benefits to be gained from this type of filtration package. Current COLPRO systems provide a high level of protection but the filters are not regenerative and must be discarded and replaced following a chemical attack.

The regenerative NBC system uses Pressure Swing Adsorption (PSA) filtration technology with air cycle air temperature control. The PSA regenerative filter assemblies developed by PALL use filter beds filled with sorbent materials which adsorb gases under pressure and desorb them when the pressure is removed. A typical system has two filter beds, one on stream which is fed with contaminated air under pressure adsorbing the gases, while the other off stream bed is regenerating. The desorbed gases are purged from the system with unpressurised clean filtered air. The required primary high pressure air and heating or cooling of the filtered air is provided by a purpose-designed AirControl Technologies air cycle system.

To address the practical aspects of installation in a typical fighting vehicle, a PSA regenerative life support system has been developed by Aircontrol Technologies Ltd and PALL Aerospace.

Status

Currently under trial.

Manufacturers

Aircontrol Technologies Limited.
PALL Aerospace.

NEW ENTRY

Aircontrol Technologies portable NBC filtration units

Description

The Aircontrol Technologies portable NBC air filtration units deliver filtered air, providing collective overpressure protection to personnel within chemical agent resistant shelters. The filtration units are suitable for the pressurisation of unhardened shelters to a pressure of 1.25 mb when erected outside or within existing buildings for use as rest and recuperation areas, field hospitals and so forth. The filtration units are electrically driven and can be supplied for use with any specified standard electrical supply and for a range of air flows.

The units are fitted with two filters. A washable prefilter removes coarse particles and has a dust holding capacity in excess of 1 kg based on No 2 test dust. The second filter is a composite filter having two elements: the first is a high-efficiency particulate element to remove fine particles and biological agents; the second is a deep bed activated-carbon anti-vapour element to remove gases.

The filtration units have a robust aluminium housing containing the filters and the fan/motor assembly, mounted through anti-vibration shock mountings to a carrying/support frame. The units are designed to provide ease of maintenance. They are supplied with an electrical flying lead up to 25 m long and a 100 mm diameter flexible air supply ducting up to 5 m long, complete with a quick-connect coupling at one end for assembly to the filter housing.

Filtration units are available with air flows of 212 and 340 m³/h.

Also available is a portable Air Conditioning Unit (ACU) which has a nominal capacity of 7 kW cooling and 4 kW heating, for use with the portable NBC filtration unit. The ACU is designed to function in either a 'passive' or 'active' mode. In the 'passive' mode it can operate with up to two filtration units to provide toxic-free cooled or heated air to a toxic-free area. For situations where there is no imminent NBC threat, the operation of a lever changes the ACU into the 'active' mode, utilising its own fan to run independent of the NBC filtration unit providing a nominal 750 m³/h of unfiltered cooled or heated air. The ACU is designed to operate in ambient temperatures of up to 55°C.

Status

In production. In service with the British Army and Royal Air Force.

Manufacturer

Aircontrol Technologies Limited.

UPDATED

Aircontrol Technologies portable air conditioning unit
(Aircontrol Technologies Limited)

Aircontrol Technologies portable NBC air filtration unit in use
(Aircontrol Technologies Limited)

Chemical Agent Resistant Material (CARM)

Description

CARM is used primarily to cover supplies and other material to protect them from liquid contamination produced by chemical agent droplets falling through the atmosphere. It is a multipurpose material consisting of two layers of low-density polyethylene incorporating a polyester grid reinforcement. The membrane is produced by laying the grid on to a preformed sheet and extruding another sheet directly on top. Uses for CARM, other than chemical agent protection for supplies, include covers and liners for unhardened collective shelters, emergency tents and camouflage coverings.

CARM is infra-red reflectant and thermally stable between −40 and +75°C. Although strong, CARM is light and easy to handle even when wet. A full range of fixing accessories is available and eyelets can be sealed on to the surface during manufacture.

Several grades and types of CARM are available. One is Tactical CARM which is infra-red reflective and NATO Green on both sides; a further type of Tactical CARM is NATO Green on one side and Urban Grey on the other. Logistic CARM is coloured NATO Green while Desert CARM is sand-coloured. All these types of CARM are 0.25 mm thick and weigh 0.27 kg/m². By contrast Lightweight CARM (Desert Green) is 0.15 mm thick and weighs 0.16 kg/m².

Status

In service with the British Army.

Manufacturer

Monarflex Limited.

VERIFIED

A typical example of the liner for an NBC collective shelter

Deployable Shelter System

Description

The Deployable Shelter System (DSS) is being developed by INSYS Limited for the Joint Force Air Component Headquarters (JFACHQ) which will operate at the operational/tactical level of command and communications in support of deployed operations.

The shelter system comprises a network of COLPRO shelter facilities connected together by protected tunnels. Its modular construction allows for a highly adaptable layout, according to the operational requirements of the force commander. Each shelter facility is supplied with filtered conditioned air from transportable AFU/ACU units.

Status

In development. The £1 million contract has been awarded by the UK MoD.

Manufacturer

INSYS Limited.

UPDATED

Entrance to the liner for an NBC collective shelter

enter the protected areas. Special decontamination areas are also set up to allowing personnel and patients to enter field hospitals, aid posts and command posts after or during chemical attack.

Status

In service with the UK MoD.

Manufacturer

J & S Franklin Limited.

VERIFIED

Deployable Shelter System (DSS) (INSYS Limited)   ***2001**/0059321*

Liners for NBC collective shelter

Description

The current UK Ministry of Defence general purpose shelter system is provided with a fully sealed liner to provide NBC protection. The shelter system is used for command posts, observation posts, regimental aid posts and modular field hospitals. It comprises four module units: the Shelter, General Purpose 7.3 × 5.5 m (24 × 18 ft) - GP 240; Shelter, General Purpose 3.66 × 3.66 m (12 × 12 ft) - GP 120; Passageway, Straight; and Passageway, four-way Connector. NBC liners are produced for all four units and all can be connected in any configuration depending on requirements.

The liner, sometimes known as a 'Porton Liner', is suspended from the same shelter frame that supports the polycotton canvas outer and is kept inflated by a portable NBC filtration unit. This unit inflates the tent to slightly over standard atmospheric pressure to prevent ingress of chemical or biological agents. Pressure is controlled by vents in the liner.

The shelter liner is made from chemical-resistant butyl-coated nylon, with a heavy-duty reinforced base. The liner is fully sealed and all seams are sewn and taped to be airtight. All doors are fully zipped.

The Liner, Passageway, Straight is constructed with an integral airlock with a sealable door at each end. This unit is placed at each entrance for personnel to

MDH Defence maritime NBC filtration systems

Description

MDH Defence has developed a series of maritime NBC filtration systems to supply an agent-free environment for ship citadels and sanctuaries. Maritime systems are

MDH Defence maritime NBC filtration systems equip the Swedish 'Göteborg' class corvette   0062052

NATO naval filter 300 m³/h   **2000**/0081485

MDH Defence ship NBC filter station
2000/0081484

installed in the Swedish 'Göteborg' class corvette and the Australian 'Collins' class submarine.

Status
Available.

Manufacturer
MDH Defence.

UPDATED

MDH Defence NBC COLPRO for armoured vehicles, ships and shelters

Description
MDH Defence COLPRO systems are successfully operating in mobile and fixed installations with defence forces in the UK, US and several non-NATO countries. The product range includes:
- Vehicle NBC systems
- Vehicle crew cooling units
- Ship NBC systems
- Unhardened Collective Protection Systems (UCPS)
- NBC filter systems

MDH Defence NBC pack for the UK Challenger 2   0050889

MDH Defence range of filter from 85 to 375 m³/h   0050890

- Support equipment, including inertial separators, pressure relief valves and engine air filters

The Vickers Challenger 2 main UK battle tank uses a 340 m³/h NBC system incorporating a NATO standard 170 m³/h composite filter. The system is designed to remove both 99 per cent of BW and CW agents. The Challenger 2 system also incorporates an inertial separator, first-stage dust separator and pressure relief valve. It has a control box fitted with a membrane keyboard giving a range of functions and system information.

Status
In production. In service worldwide.

Manufacturer
MDH Defence.

UPDATED

MDH Defence NBC systems (CV90)

Description
MDH Defence has developed a 170 m³/h vehicle NBC system for the prototype Swedish Combat Vehicle 90 (CV90). First-stage filtration comprises a cyclone particulate filter and a downstream ultra-high efficiency (UHE) filter. The UHE stage

MDH Defence dual-mode bypass NBC filtration system   0050888

system includes the latest NBC filter technology, is portable by two people and chemically hardened for ease of decontamination. The filters can be changed by one person without the use of tools.

When used in chemical and biological warfare conditions, this UCPS allows personnel to discard their protective clothing and respirators for rest periods within a contamination-free area. In-line heating and cooling units complement the UCPS.

Status
Available.

Manufacturer
MDH Defence.

UPDATED

Swedish CV90 carries the MDH Defence dual-mode bypass NBC filtration system
***2000**/0081487*

is designed to remove dust, smoke, BW agents and nuclear particulate material (from fallout). The second-stage NBC vapour filter uses activated charcoal in compliance with current NATO agreements on CW filtration. The system can be operated in full NBC filtration mode or the bypass mode which allows non-NBC filtered cabin purging and conditioned fresh air supply to the crew.

Status
Available.

Manufacturer
MDH Defence.   *UPDATED*

MDH Defence Unhardened Collective Protection Systems (UCPS)

Description
MDH Defence designed and developed this UCPS for the UK Ministry of Defence. It comprises two 375 m³/h NBC filter units serving a 25-person rest and relief or medical facility. Incorporating lessons learned during Operation Desert Storm, this

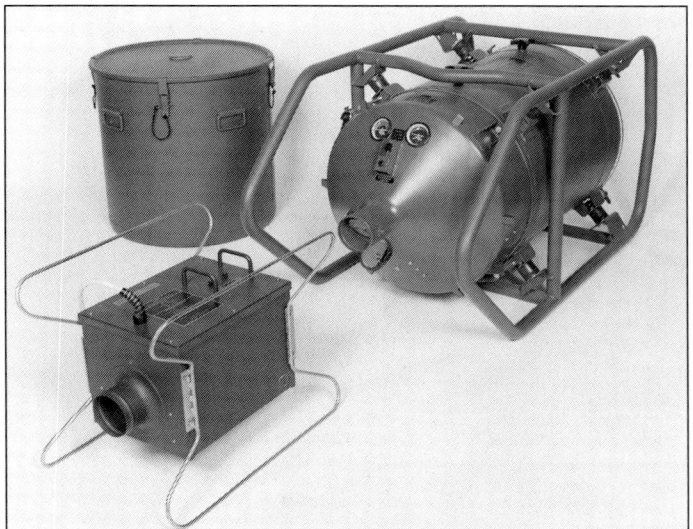
Transportable NBC filter unit, heater and filter container   0050893

MDH Defence 25-person Unhardened Collective Protection System (UCPS)

Microgenix BW Air Purification Systems

Description
The Microgenix Air Purification System is designed to remove all pathogens in air circulation systems. A series of trials at Dstl (CBD Porton Down), proved that this new approach to BW COLPRO filtration showed a 100 per cent elimination rate for simulated pathogens in 7 out of the 12 tests so far conducted. The system uses a two-stage process to eliminate bacteria, viruses and prions from the air. The air passes first through a filtration matrix coated with a proprietary material called 'Bio-green 3000' which presents microbe-scale 'spikes' to the incoming flow. The spikes are designed to pierce the cell membranes of bacteria or the capsids of viruses, thereby destroying them. The air is then swirled through an ultraviolet-lit chamber which disrupts the DNA of any remaining organisms. Although VOC and CW agent removal requires separate airstream processing, this application appears to offer significantly lower running costs (30 per cent claimed) for clean, conditioned air compared with current HEPA/cyclone air conditioning systems. Additionally, BW filter-changes are no longer required, eliminating the risk from the germs they retain. This technology appears to offer a promising counter to the increase in illnesses caused generally in the modern conditioned environment. Candidate applications therefore include field hospital operating rooms, electronics compartments, and BW COLPRO filtration systems in fixed structure, vehicles, ships and aircraft.

The air purification units are scalable according to the output capacity required.

Status
Available.

Manufacturer
Microgenix Limited.

UPDATED

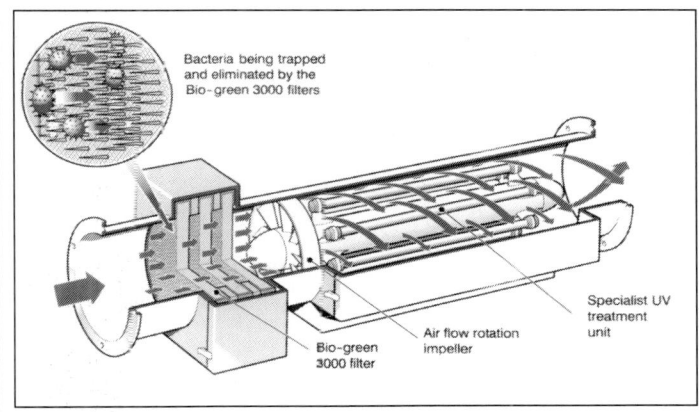
Diagrammatic view of the Microgenix air purification unit   ***2001**/0059318*

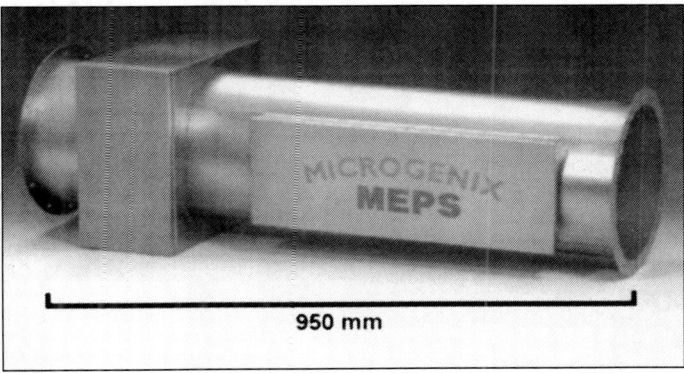
Microgenix air purification unit   ***2001**/0059320*

SFP Services on-site filter testing equipment

Description

This on-site filter testing equipment is used to test the efficiency of air filters and filter installations used in a wide variety of applications. The test equipment operates by introducing a controlled amount of sodium chloride (salt) into the air flow in front of the filter concerned. It then measures the amount of sodium chloride left in the air after passing through the filter, thus determining the filter's efficiency. A portable version of this test equipment has been developed to cover shipborne systems.

The testing system employs two items of test equipment, a portable sodium chloride detector and a thermal generator.

The portable sodium chloride detector is a compact portable flame photometer with its own built-in sampling fan. It features an LCD panel meter and automatic sampling of air flow rate through a feedback system to the fan motor which is positioned downstream from the flame. The sampled air flows through a closed system so that, if necessary, it may be returned to a contaminated duct. In this application the sampling fan is required only to move the air around the closed loop, allowing systems at low pressure to be tested.

The Light Sensitive Detector is a fast acting photoconductive diode. The linear response to sodium chloride concentrations is utilised for display on the digital panel meter. Where flame emission is non-linear (at high concentrations above 2 mg/m³) it is necessary to refer to calibration graphs. An outlet socket is available as standard for a chart recorder.

The portable sodium chloride detector is powered by a 110/220 V 50/60 Hz power supply but can operate from an optional rechargeable lead acid battery pack. The unit uses a 1 litre hydrogen bottle with a consumption of 500 ml/min. The weight of the detector is 17 kg which increases to 22 kg when the battery pack is used. Dimensions are 480 × 200 × 480 mm.

The thermal generator is available in two models, standard and mini. Both produce an aerosol by burning a 'stick' of sodium chloride in an oxy-propane flame at a constant controlled rate. The sodium chloride evaporates to recondense in the form of a dry sub-micron aerosol.

The standard thermal generator weighs 6.4 kg and measures 400 × 180 × 270 mm. It requires oxygen and propane gas supplies and can be powered by eight 1.5 V AA batteries. The aerosol range is variable up to 6 g/m.

The mini model of thermal generator weighs 3.2 kg and measures 245 × 120 × 150 mm. It also requires oxygen and propane gas supplies and is powered by eight 1.5 V AA batteries. The aerosol range is variable up to 0.5 g/m.

Status
Available.

Manufacturer
SFP Services Limited.

VERIFIED

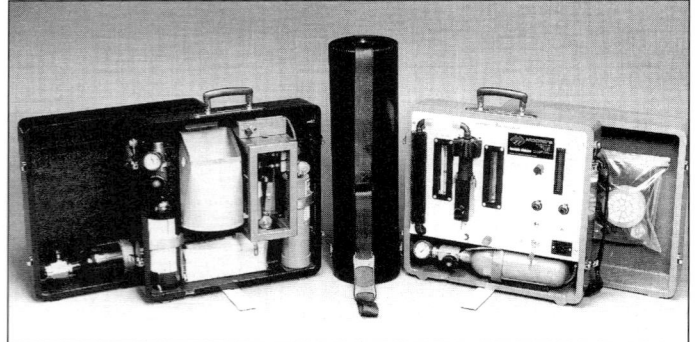

SFP Services on-site filter testing equipment with (left) thermal generator in its carrying case complete with associated hoses, and (right) the portable sodium chloride detector

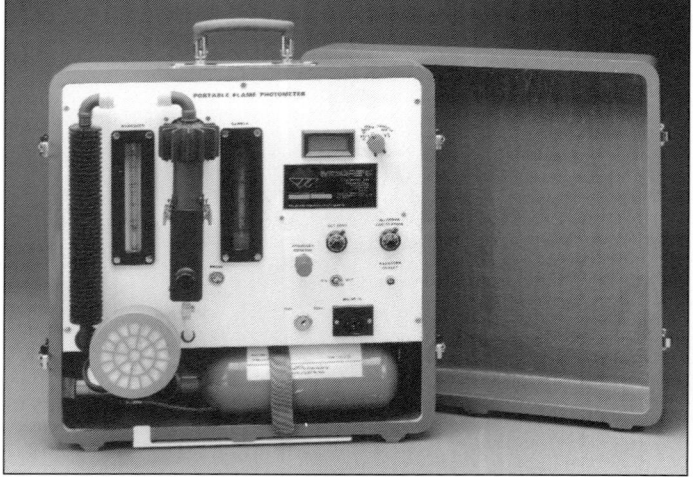

Portable sodium chloride detector used with the on-site filter testing equipment

UCPS

Description

UCPS was developed by a consortium of British companies headed by Monarflex Limited. The system was designed to meet NATO standards of collective protection against liquid and vapour hazards produced by chemical agent attacks in desert and other environments, where the usual solid structure accommodation for liner-type collective shelters is not available. A single system can allow 30 occupants to feed and 24 to sleep. Two systems can support a company sized unit in sub-units of 24 personnel each, over a period of 24 hours. A complete UCPS can be erected by eight people in 30 to 35 minutes.

The UCPS is contained within a metal-framed 12 × 4 m tent constructed from Chemical Agent Resistant Material (CARM — see separate entry). It is contained in a valise and weighs a total of 150 kg. It can be erected by a team of six people in 15 minutes.

An airtight CARM inner fits inside the outer CARM tent with sufficient space left inside the tent to form a liquid hazard area. The inner is fitted with two 1 × 0.75 m airlocks and is normally maintained at an overpressure to ensure that any leakage will be outwards. The upper part of the inner is made of translucent CARM to admit daylight. There are two small screened-off cells for use as toilets — two chemical closets are supplied as part of the system. The inner is supplied in a valise and weighs 105 kg. When erected, the UCPS is kept in shape by suspended sandbags acting as weights; there is also an inner system of tensioning cords connected to neoprene eyelets at each end.

Two Howden Aircontrol Limited Air Filtration Units (AFUs) provide a supply of toxic-free air to the shelter system. Each is coupled to the toxic-free area by flexible air supply hoses. One AFU can supply the system when filters are being changed or if one AFU fails. Each AFU is supplied with 20 m of power cable and 2.5 m of flexible ducting. Maximum air flow from each AFU is 290 m³/h. Air conditioning units can be supplied.

Trailmaster Trailers Limited designed the special 1.5 tonne single-axle trailer that carries a complete UCPS; it can be towed by any standard 4 tonne truck. The trailer is fitted with Land Rover wheels and tyres and a standard NATO towing eye. Dedicated stowage space is provided for each UCPS component and the whole cargo bed is covered by a CARM tilt cover.

Two separate 12.5 kVA generating sets are mounted on a common frame carried transversely across the trailer above the axle. The generators are powered by 18 hp petrol engines.

Each system is provided with a power distribution system. Fluorescent lighting units for the shelter interior can be provided.

Status
Ready for production. Has been tested by the British Army.

Major contractor
Monarflex Limited.

VERIFIED

Unhardened Collective Protection System (UCPS) fully assembled

Erected inner CARM liner for Unhardened Collective Protection System (UCPS) inside tent frame

UNITED STATES OF AMERICA

200 CFM NBC filter

Description
The M48A1 delivers 340 m³/h (200 CFM) and comprises an integrated filter housing with an annular HEPA particulate filter surrounded by a chromium free ASZM-TEDA activated carbon filter. The 200 CFM filter is suitable for fixed installations ashore or at sea, where up to three filters can be stacked, delivering 1,020 m³/h of filtered air on an installation footprint of less than 1 m².

Status
In production.

Manufacturer
Parmatic Filter Corporation.

UPDATED

Specifications
Rated flow: 340 m³/h
Particulate media: Glass micro fibre (US Mil-Spec MIL-F-51079)
Efficiency (HEPA): >99.97% at 0.3 m
Life (DMMP at 5.0 mg/litre): >50 mins
Life (CK at 4.0 mg/litre): >50 mins
Activated carbon type: ASZM-TEDA
Activated carbon weight: 12.84 kg
Airflow resistance (HEPA): 498 Pa
Airflow resistance (gas): 1,119 Pa
Airflow resistance (overall): 1,617 Pa
Outside diameter: 544.3 mm
Filter length: 250.9 mm
Inside diameter: 307.0 mm
Unit weight: 21.8 kg

200 CFM NBC Filter (Parmatic Filter Corporation)   *2002*/0050896

Advanced Deployable Collective Protection Equipment (ADCPE)

Description
The Advanced Deployable Collective Protection Equipment (ADCPE) system was developed to illustrate that regenerable protection technology can be applied whilst simultaneously providing heating and cooling to various shelters and tents. The system shown here during field demonstration trials is designed for field tactical use. The primary mission in the implemented application was providing clean, conditioned, breathable air to a 19.5 × 6 m chemically-hardened tent. The integrated system incorporates heating/cooling, dust and dirt elimination and filtration for both vapour and aerosol CBW contaminants. Air is continuously circulated from the tent through the ADCPE at the rate of 5,040 m³/h while the temperature is maintained at a comfortable level. The system also develops sufficient overpressure to ensure a net outflow from the unit. The compound heat/cooling system allows operation even at very low temperature. Innovative

Advanced Deployable Collective Protection Equipment (ADCPE)   0050763

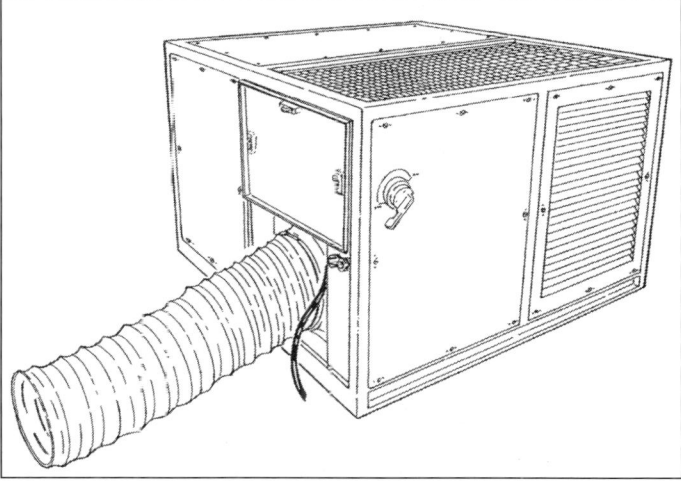

Advanced Deployable Collective Protection Equipment (ADCPE) - filter unit 0050762

Specifications

Cooling capacity:	189,560 kJ
Heating capacity:	94,780 kJ
Operating temperature range:	−29 to 50°C
Recirculation air flow:	5,040 m³/h
Make-up air flow:	1000 m³/h
Tent over-pressure:	up to 0.8 in H₂0
Power source:	208 V 3-Phase, 200 A or 60 kW generator
Power consumption:	17 kW (average) for heating/cooling. 25 kW (peak) for NBC filtration.
Dimensions:	1.8 × 1.8 × 1.27 m
Weight:	1,270 kg

---

techniques provide excellent energy utilisation. Two-stage particulate filtration, (standard particulate/HEPA) is used to counter aerosol-borne threats. Gaseous agents are removed by an advanced temperature-swing adsorption subsystem. The table below provides additional details.

The unit is skid-mounted and, being microprocessor controlled, can run unattended. The processor is also designed to assist diagnostic testing and maintenance. ADCPE satisfies US MIL-STD 810 or equivalent.

Status
In production.

Manufacturer
Guild Associates Inc.

VERIFIED

Advanced Integrated Collective Protection System (AICPS)

Description
The Advanced Integrated Collective Protection System (AICPS) is an advanced NBC air filtration system integrated with environmental control and auxiliary exportable power. It is designed for installation on tactical vans and shelters and is manufactured by Lockheed Martin Librascope.

The AICPS is a single modular unit, replacing the separate diesel generators, heaters/air conditioners and NBC collective protection and filtration systems currently used for NBC protection. The AICPS provides power to support mission equipment as well as providing a heated or cooled overpressure environment free of NBC or other contaminants.

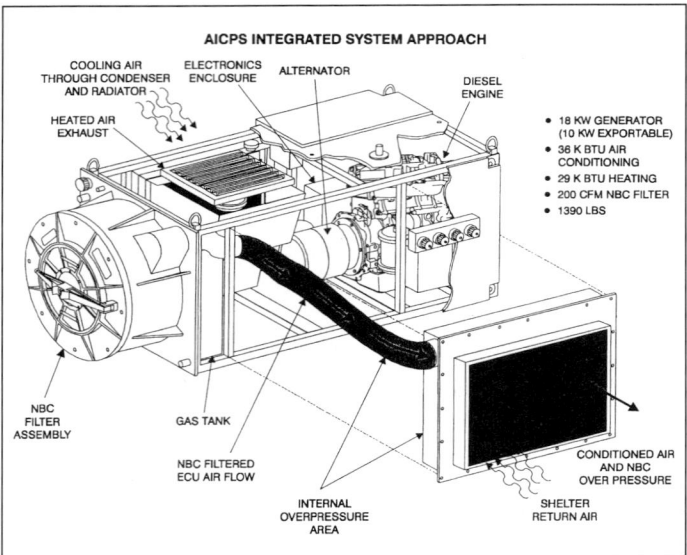

Outline drawing of the main features of the Advanced Integrated Collective Protection System (AICPS) 0011478

AICPS mounted on S788 shelter on HMMWV 0011479

Specifications		
Model	**XM32**	**XM31**
Air flow	200 cfm	400 cfm
Cooling/heating	36,000 BTU	60,000 BTU
Available power	10 kW	10 kW
Volume	1.67 m³	2.77 m³
Weight	630.5 kg	905 kg

The AICPS is available in two sizes corresponding to the three most commonly used shelters. These are the Standard Integrated Command Post Shelter (SICPS), S-280 shelters and ISO-Expandable shelters and vans. The AICPS is designed as a single modular unit to be mounted on the front of a shelter, thereby eliminating the previous requirements for external cables, external air ducts and towing equipment.

There are five main subsystems within the AICPS. They are an NBC survivable enclosure, engine/alternator (involving a lightweight turbocharged diesel engine), NBC HEPA deep bed carbon gas filter system, Environmental Control Unit (ECU) and a system control unit.

AICPS can interface with various chemical alarm systems and external protective entrances. Installation is rapid and simple and there is a high degree of commonality between the three main types of AICPS.

Status
Advanced development.

Manufacturer
Lockheed Martin (prime).

Manufacturer
Hunter Protective Systems (200 cfm gas filter).

VERIFIED

Biochemical Filter Blower Unit (BFBU)

Description
The Biochemical Filter Blower Unit (BFBU) is a collective protection device designed to provide fresh air and overpressure to the occupants of the crew compartment of a military vehicle.

The BFBU typically has three operational modes: bypass, low-flow NBC and high-flow NBC. In the bypass mode, designed for use in non-threat conditions, the BFBU delivers 100 to 140 cfm (170 to 238 m³/h) of make-up air. The air is filtered for dust using a two-stage filtration approach. The NBC filter modes can be set at up to 140 cfm (238 m³/h) and 210 cfm (357 m³/h) for the low- and high-flow modes respectively. Both NBC modes use two standard M48 or M48A1 gas/particulate filters with precleaners. The BFBU can also be delivered as a 2-mode system, without bypass.

The BFBU's control electronics, monitor the overpressure in the crew compartment and will switch between high and low NBC modes to maintain any required overpressure at the minimum current draw. The control system also provides full built-in test and diagnostic output and provides switching from bypass to NBC modes either manually or automatically on receipt of an alarm signal. The control electronics monitor the pressure drop through the filters and deliver an analogue signal to drive an external display. This allows the occupants of the interior to monitor filter status from the inside.

The system has been qualified through MIL-STD-810 environmental testing and has met US Army requirements during simulant testing. The electronics have additionally been through combined environment/multi-axis vibration to prove ruggedness of the packing used. The system is designed for EMP survivability and an acoustic attenuator limits the noise transmitted to the crew compartment.

The BFBU system is installed in the Ground-Based Common Sensor – Heavy (GBCS-H), the M4 Command and Control Vehicle (C2V) and the Armored Medical Transport Vehicle (AMTV). All these vehicles are variants of the United Defense LP Fighting Vehicle Systems Carrier (for details see *Jane's Military Vehicles and Logistics 1998-99*). The BFBU can be fitted to other platforms and can be used on shelters.

Status
Preproduction.

Manufacturer
General Dynamics Armament and Technical Products.

UPDATED

A sealed Biochemical Filter Blower Unit (BFBU)

A Biochemical Filter Blower Unit (BFBU) with the side covers and front manifold removed

Chemical Biological Protective Shelter System (CBPSS)

Description

The Chemical Biological Protective Shelter System (CBPSS) is a collective protection system developed for use as a mobile front line Battalion Aid Station (BAS) or Division Clearing Station (DCS).

The CBPSS consists of a 27.87 m² (300 ft²) tent fully integrated with the M1113 expanded capacity version of the HMMWV M998 4 × 4 vehicle which carries a Lightweight Multipurpose Shelter (LMS), Environmental Control Unit (ECU) and medical equipment. The HMMWV also tows a 10 kW tactical quiet generator on a High-Mobility Trailer. The tent includes both litter and ambulatory airlocks for ease of ingress and egress under operational conditions. The system includes heating and cooling which is integrated with the filtration systems contained in the dedicated vehicle's ECU. The ECU is also designed to deliver sufficient output to heat or cool the system in extreme environmental conditions. The tent has an insulated liner to assist the ECU to operate under these extreme conditions.

The tent can be erected in less than three minutes and the entire system can be made operationally ready by four people in less than 20 minutes. This is aided by the permanent connections between the LMS and the rapidly erectable airbeam supported shelter. The tent is fabricated from a high-performance fluoropolymer/aramid laminate that provides a high degree of liquid and vapour protection and is readily decontaminated. Side-to-side modularity connectors allow for the joining of individual CBPSS units.

Status

Initial production in progress (May 2001).

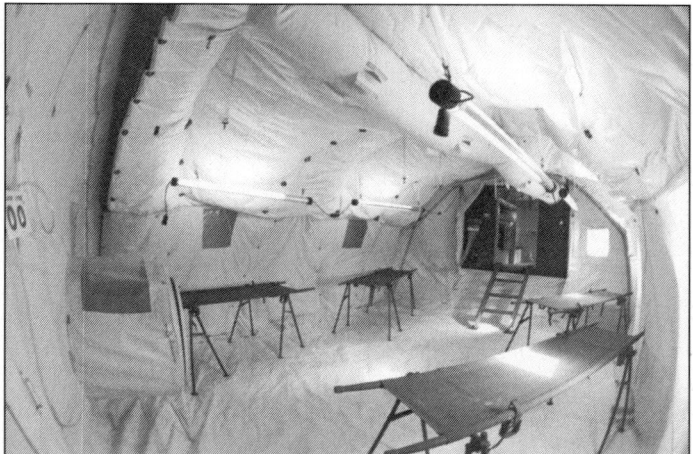

Chemical Biological Protective Shelter System (CBPSS) interior

Chemical Biological Protective Shelter System (CBPSS) 0011480

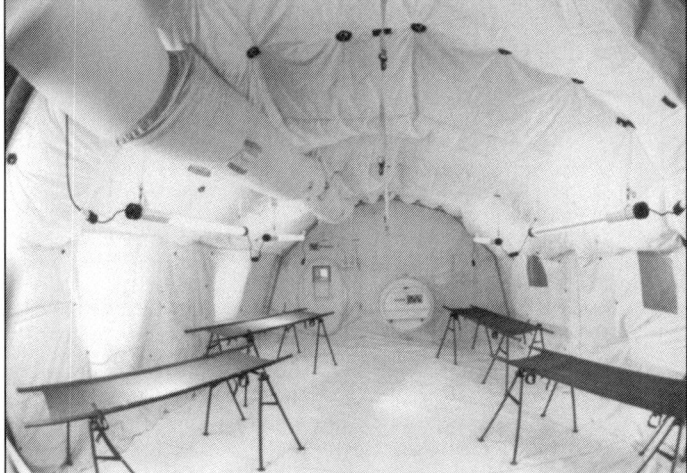

Chemical Biological Protective Shelter System (CBPSS) interior

Manufacturer

Engineered Air Systems Inc (prime).

Development agency

US Army SBCCOM.

VERIFIED

Chemically Hardened Air Management Plant (CHAMP)

Description

The Chemically Hardened Air Management Plant (CHAMP) is a large capacity supply unit which has been developed to deliver a range of services to mobile hospital units. It integrates with the US Army's Chemically Hardened Air-Transportable Hospital (CHATH). The unit is designed for minimum weight and footprint, reducing these parameters by 60 to 80 per cent from previous designs. Two units can be loaded on a single 463L pallet.

The unit delivers air conditioning, heating, unit overpressure, NBC filtration, main and standby electrical services to the CHATH. Standby power is designed to maintain supply within as little as 10 seconds following a mains failure incident. The unit can convert a range of external power supplies or generate its own from generators capable of running on a wide variety of fuels.

Status

In service with the US Air Force.

Manufacturer

Engineered Air Systems Inc.

VERIFIED

Specifications

Make up air filters:	Type FF2790 1B prefilter and NBC filter to MIL-F-51527
Make up air flow:	up to 1,344 m³/h
Heating capacity:	208,516 kJ
Cooling capacity:	189,560 kJ
Recirculation air filter:	Type FF2790 2D
Rated recirculating airflow:	13,440 m³/h
Flexible duct size:	Two 0.76 m (supply and return)
Operating temperature range:	–31 to 52°C
Onboard fuel tank:	83.27 litres (6 h operation)
Fuel requirements:	Including JP4, JP5, JP8, DFA, DF-1, DF-2
Integral diesel generator capacity:	60 kW
Refrigerant:	R-22
Maximum electrical input:	Cooling: 200 A
	Heating: 227 A
Electrical power utilisation range:	208-240 V, 50/60 Hz, 3-phase, 4-wire
Dimensions:	1 × 2.7 × 2.4 m
Weight:	2,270 kg

The CHAMP NEC air management plant 0050764

ECS-600AL NBC protection system

Description

The ECS-600AL is a cost-effective, lightweight filter-blower system designed to deliver up to 600 CFM (1,020 m³/h) of clean, filtered air into portable or fixed COLPRO shelters. The ECS-600AL is of modular construction and uses a variable-speed motor controller. This allows the processing capacity to be rated between 250 and 600 CFM (45 and 1,020 m³/h) in four manually-controlled steps. The unit has a unique (patent pending) feature that allows each filter to be replaced whilst the unit is operating. Filtration is delivered through six standard M48A1 filters (see separate entry within this section). The unit can be moved from location to location with the detachable wheels and handling system. Construction is of steel or stainless steel (EPS-600SS) and the exterior is treated with an epoxy-based CARC resin coating. The power requirement is to 20 V, 60 Hz single phase. Optionally, other power supplies can be utilised.

Status
In production.

Manufacturer
Parmatic Filter Corporation.

VERIFIED

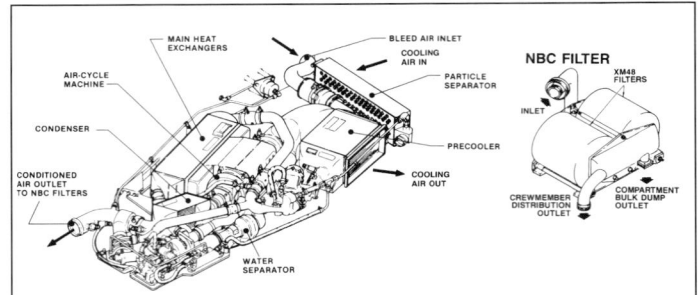

AlliedSignal Environmental Control/Nuclear, Biological and Chemical (EC/NBC) protection system components for M1A1 Abrams MBT

ECS-600 NBC protection system

Description
The ECS-600 NBC protection system is an NBC protection unit designed for forward deployed shelter systems or small building protection. The use of six modular filters in parallel allows filter replacement without system shutdown or contamination (patent pending for this feature). Designed to operate at 1,020 m³/h (600 cfm) maximum, this unit has blower electronic speed control, allowing the airflow to be reduced to 25 per cent of maximum. Additionally, the airflow can be automatically controlled to maintain a desired over pressure.

The ECS-600 NBC protection system utilises the proven M48A1 filter element produced by Parmatic and in use with the US Army Abrams Main Battle Tank. Designed to run on a 220 V, 50/60 Hz single phase electric current, this unit has a 3 hp totally enclosed fan-cooled motor and commercial blower for ease of maintenance. The unit has a skid base for fork lift access.

Options available include aluminium construction where weight is a consideration: stainless steel construction for high humidity environments, optional maximum air flows through modular design, moisture separation on air intake, heavy dust filtration and an umbilical cord for remote pressure change monitoring. Parmatic Part No 690398.

ECS-600/1200
The ECS-600/1200 is a modular NBC blower system based on the 200 CFM filter set (see separate entry within this section). It is easily transported by four people, allowing rapid deployment. The filters can quickly be changed without the use of hand tools. The system can readily be field-reconfigured to deliver airflows between 400 CFM and 1,200 CFM. It is designed for smaller COLPRO applications.

Status
In production.

Manufacturer
Parmatic Filter Corporation.

VERIFIED

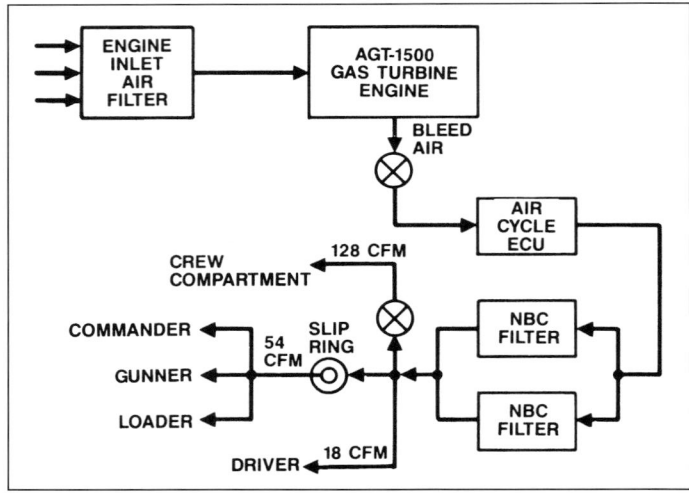

Schematic layout of AlliedSignal EC/NBC protection system for M1A1 Abrams MBT

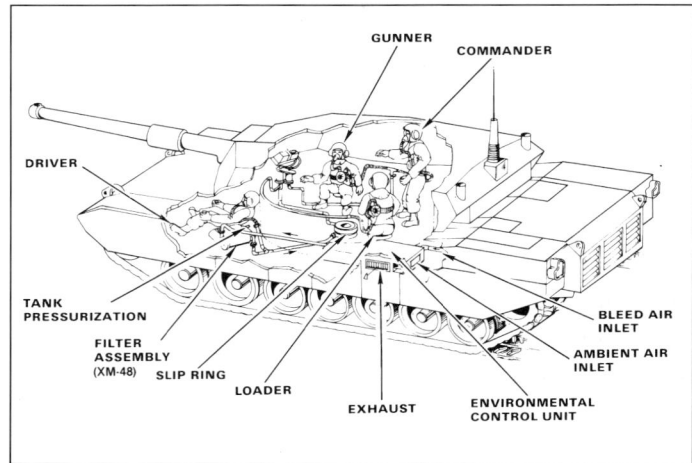

Layout of AlliedSignal EC/NBC protection system in M1A1 Abrams MBT

Environmental Control/Nuclear, Biological and Chemical (EC/NBC) protection system for the US Army's M1A1 Abrams main battle tank. AlliedSignal produced the system under contract to General Dynamics Land Systems Division, the prime contractor for the M1A1.

The M1A1 Abrams was the first of the US Army's armoured vehicles to be equipped with an EC/NBC system. In 1977, the US Congress passed Public Law 95-79 requiring the US Army to begin equipping its main battle tanks, mechanised infantry vehicles, armoured personnel carriers and similar combat vehicles with such a system to protect personnel assigned to them.

AlliedSignal began full-scale development engineering of the EC/NBC system in 1982 and successfully completed field testing in mid-1984. As a result of these tests, which were performed from Fort Greeley, Alaska, to the Panama Canal Zone, AlliedSignal was awarded an initial production contract beginning production deliveries in early 1985.

The AlliedSignal EC/NBC system provides the M1A1's crew of four with conditioned air for breathing as well as personal heating and cooling as required, while they are wearing their protective suits and masks. It also provides positive air pressure within the tank hull to prevent the entry of any NBC contaminants. The system uses an aviation-type air cycle refrigeration unit in which bleed air from the tank's gas turbine engine provides the energy required for powering the air-cycle machine. The system has five major components: a precooler, high-pressure water extractor, air-bearing cooling turbine, primary heat exchanger and an NBC filter assembly plus valves and controls. The bulk of the system is packed into a 152 mm high sponson toolbox below the turret. The filter canisters, which remove contaminants from the air by physical and chemical means, are furnished by the US Government.

ECS-600 NBC protection system (Parmatic Filter Corporation)   0050898

Environmental Control/NBC (EC/NBC) protection system for M1A1

Description
Honeywell (formerly AlliedSignal Aerospace Systems and Equipment, a unit of AlliedSignal Aerospace based at Torrance, California), produced the

Status
In service with the US Army.

Manufacturer
Honeywell.

UPDATED

....

FR-65/FR-100 NBC protection system

Description
The FR-65 and FR-100 first-responder air filtration units can be used either as a rapid response mobile NBC system, or as a fixed installation. The system is easy to transport and install and can operate either as a suction system, to evacuate contaminated air, or to deliver pressurised and filtered air inside a COLPRO unit. The maximum airflow is 170 m³/h, drawing 6.7 A from a dedicated domestic supply. With the optional 65 CFM (100 m³/h) version, a lower-rated motor (560 W) allows it to operate from a domestic supply with no load shedding. The ECS-100 utilises the proven M48A1 filter element in use with the US Army (see separate entry within this section). The use of a COTS blower and motor eases both maintenance and re-supply. Power can be drawn from existing single-phase domestic utility outlets at 110 V, 15 A, 50/60 Hz. The unit has a footprint of 457 × 457 mm and is 991 mm high. Optional pre-filters and mobility kits are available.

Status
In production.

Manufacturer
Parmatic Filter Corporation.

UPDATED

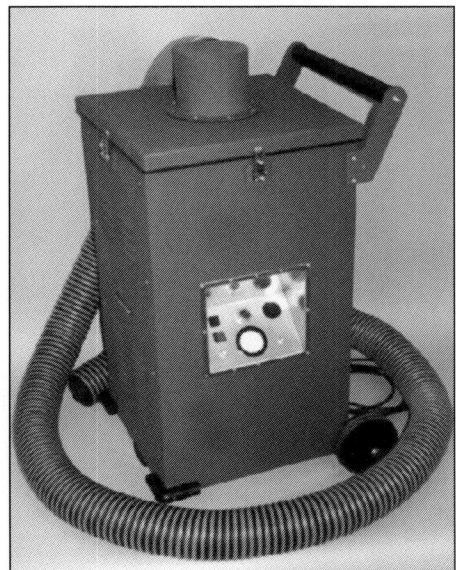

ECS-100 NBC protection system (Parmatic Filter Corporation)
2001/0059317

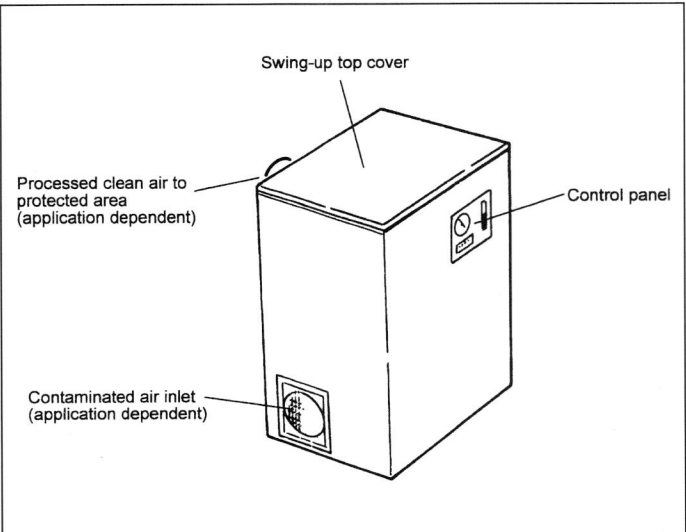

ECS-100 NBC protection system (Parmatic Filter Corporation) 0050899

Hunter Protective Systems NBC filtration and distribution systems for vehicles

Description
Hunter Protective Systems manufactures the two Individual Distribution (ID) systems in wide use by the US military today: the 12 cfm M8A3 and the 20 cfm M13A1. ID systems deliver individual protection against NBC agents when, because of the vehicle design, positive pressure cannot be maintained. Also called micro- or ventilated facepiece-collective protection systems, the ID approach is employed either as the primary NBC system in vehicles that cannot maintain overpressure or as back-up NBC protection for overpressure NBC systems.

Originally designed for the M60 tank, these systems are currently installed in virtually all classes of combat and tactical vehicles, among them the MIA2 Abrams, M2A2 Bradley, M109A6 Paladin, M113s and variants, Multiple Launch Rocket System (MLRS), Piranha, HMMWV variants and the Heavy Equipment Transport System (HETS).

In ID systems, small filter units supply 3 to 4.5 cfm (5.1 to 7.7 m³/h) of purified air through NBC-resistant hoses to each crew member's protective mask. ID systems have the lowest volume, weight and power consumption of any collective protection system. An ID system relieves the crew of the effort required to breathe unassisted through a mask as well as reducing the discomfort of wearing a mask by increasing sweat evaporation on the face.

The M8A3 and M13A1 NBC filtration systems employ both a HEPA filter that is 99.97 per cent efficient at 0.3 μm, as well as an activated carbon filter using the newly developed ASZM-TEDA carbon. Other types of carbon may be employed if required.

M8A3 filter system. The M8A3 is the smallest NBC system offered by Hunter Protective Systems. Ideally suited for three persons, the M8A3 provides 3 to 4.5 cfm of purified air for up to four crew members, with a maximum airflow of 12 cfm. All major filtration components are contained in the single M2A2 housing, which is connected by NBC-resistant air supply hoses to protective masks.

M13A1 filter system. The 20 cfm M13A1 filter system uses the same housing and blower assembly as the M8A3, but it employs two gas filters separate from the main filter unit. The M13A1 supports up to five people but its output is optimised for four.

Multiple filter units are used in APCs and other vehicles with more than five people. For example, the M88A2 Hercules recovery vehicle uses two M8A3 filter units, one for its crew of three and one for the crew of the recovered vehicle.

M3 Heaters, standard on the M13A1 and optional on the M8A3, raise the temperature of the air supplied to the mask to comfortable levels, necessary during cold weather operation.

Standard hoses on the M8A3 and M13A1 systems are manufactured of NBC-resistant rubber reinforced with a metal coil and surrounded by a woven nylon covering for protection against wear. These hoses are available in diameters from 22 to 38 mm (0.875 to 1.5 in) and in lengths of 15.2 to 548.6 cm (6 in to 18 ft). Hunter Protective Systems also offers insulated and heated air supply hoses. The insulated hose minimises heat loss within the hose, whilst the heated hose eliminates the need for a separate M3 heater.

IDL Defence manufactures all the components of the M8A3 and M13A1 filter systems: the main filter units, filters, heaters, electrical components and mounting hardware. Though the M8A3 and M13A1 systems are available in standard configurations, Hunter Protective Systems recommends customisation according to client needs.

Status
In service with the US Army.

Manufacturer
Hunter Protective Systems.

VERIFIED

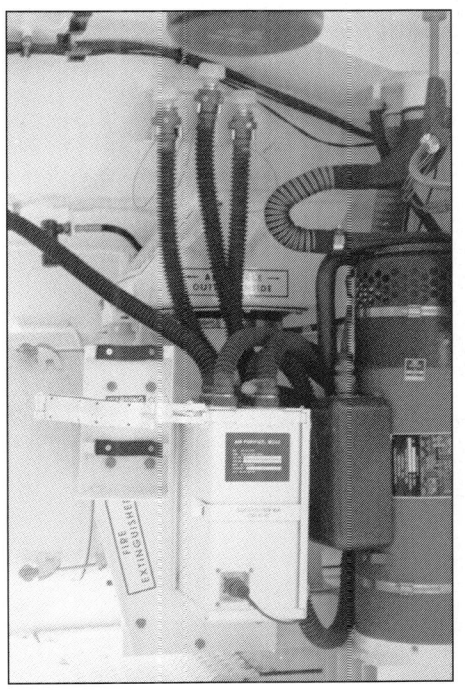

The M8A3 air purification unit shown fitted to the Hercules recovery vehicle (BFBU)
0019558

M20A1 Simplified Collective Protection System (SCPS)

Description

The M20A1 Simplified Collective Protection System (SCPS) room liner system inflates inside existing rooms or shelters for use as chemically protected rest/relief and command/control shelters. The system consists of room liners; protective entrance; accessory kit with motor blower, hoses, repair material and a drop light; recirculation filter blower; filter canister; and Environmental Control Unit (ECU) interfaces. The M20A1 is resistant to liquid and vapour threats.

The liners are 3.04 m high and 4.87 m in diameter and are extendable in 90° increments. The shelter can be deployed in 20 to 30 minutes by a team of two; this includes the inflation time.

Apart from the added liquid agent protection the M20A1 SCPS differs from the original M20 in that the protective entrance purges more efficiently, allowing entry and exit in 1.5 minutes rather than 3 minutes. A recirculation filter blower is provided to scrub the shelter air continuously, electromagnetic interference protection is improved and the ECU interfaces are provided to allow the air conditioning/heating of shelter air.

Status
In production.

Marketing agency
Tradeways Limited.

UPDATED

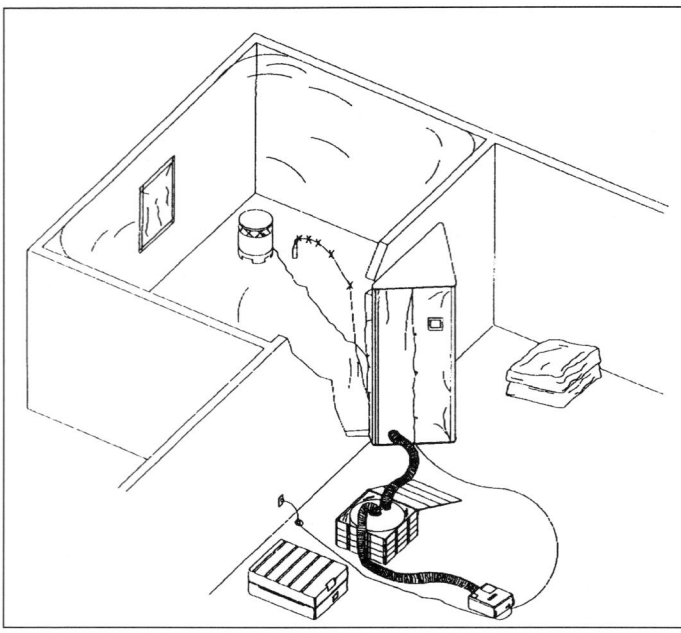

Outline drawing of M20A1 simplified collective protection room liner system

M28 Collective Protection Equipment (CPE)

Description
The M28 Deployable Medical Collective Protection Equipment (CPE, also known as DEPMEDS) was developed as a field-deployed inflatable collective protection system for use inside the Tent Extendable Modular PERsonnel (TEMPER). The

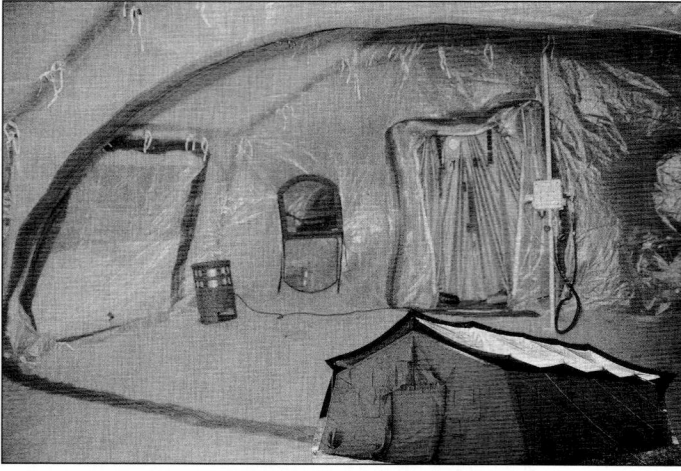

Interior of M28 Collective Protection Equipment (CPE) with inset showing an erected TEMPER tent inside which the system is installed

liquid- and vapour-resistant polyethylene liner system attaches to the TEMPER frame and allows for a contamination-free shelter area. All existing TEMPER accessories can be deployed inside the shelter without affecting normal operations, whether in a contaminated environment or under normal conditions.

The M28 CPE maintains the expandable nature of the base system and provides for interface to other shelter systems. Rest/relief, command and control and medical operations from battalion aid stations to hospitals in excess of 500 beds having all levels of care are possible in contaminated environments.

The system is equipped with an air filtration system, a Tunnel Entrance for Litter Patients (TELP), shelter sequencing adaptors, protective entrance walkways and tent liners. The M28 CPE shelter system interfaces with the TCPS (see entry in this section) and standard environmental control systems.

Status
In production.

Manufacturer
General Dynamics Armament and Technical Products.

UPDATED

M48A1 NBC filter

Description
The M48A1 delivers 170 m³/h and comprises an integrated filter housing with a HEPA particulate filter and chromium free ASZM-TEDA activated carbon filter. The M48A1 is lightweight (approximately 14 kg) and easy to handle. It is widely installed in US armoured vehicle NBC packs including those fitted to the Abrams MBT. The M48A1 is also used extensively in NBC protection systems for buildings and ships.

TF-100 CFM
The TF-100 CFM is designed as a training version of the M48A1. It has exactly the same footprint and incorporates dual-stage particulate filters. The primary stage serves as a pre-filter whilst the second stage has an HEPA rating. It has the same flow rate and differential pressure characteristics as M48A1 and is clearly marked 'for training use only'.

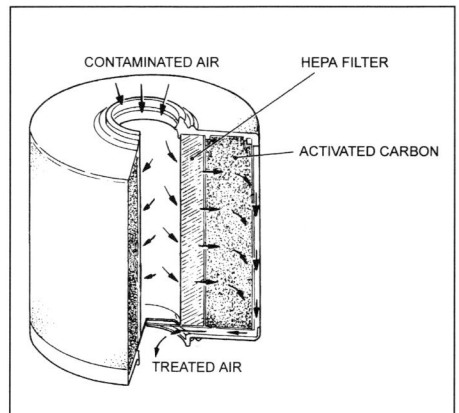

M48A1 NBC filter - cut away view (Parmatic Filter Corporation)
0050895

M48A1 NBC filter (Parmatic Filter Corporation)
0050894

Specifications

(M48A1)
Rated flow: 170 m³/h
Particulate media: Glass micro fibre (US Mil-Spec MIL-F-51079)
Efficiency (HEPA): >99.97% at 0.3 m
Life (DMMP at 5.0 mg/l): >82 min
Life (CK at 4.0 mg/l): >55 min
Activated carbon type: ASZM-TEDA
Activated carbon weight: 6.9 kg
Airflow resistance: 1,743 Pa
Outside diameter: 301.8 mm
Filter length: 320.8 mm
Overall length: 367.5 mm
Inlet outside diameter: 105.4 mm
Outlet outside diameter: 122.4 mm
Unit weight: 12.9 kg

Status

In production.

Manufacturer

Parmatic Filter Corporation.

VERIFIED

Modular chemically hardened tent

Description

This chemically hardened tent is a multipurpose frame-supported collective protection tent with an initial application as a command post tent. It features four interchangeable removable wall panels and a modularisation connector for unlimited flexibility in the configuration of tents, command posts and so forth. Included in the selection of removable wall panels are an entry airlock and a vehicle connector vestibule. An individual tent covers 11.24 m² (121 sq ft). The material used to configure the tent skin is a high-performance fluoropolymer/aramid laminate that provides chemical protection and is readily decontaminated. The tent is supported by 200 cfm of filtered air.

Status

32 in service with the US Army (1998).

Manufacturer

Chemfab Corporation.

Development agency

Natick Soldier Center.

VERIFIED

Modular chemically hardened tent configured as a command post

Parmatic shipboard Collective Protection System (CPS) and Chemical, Biological and Radiological (CBR) filter systems

Description

The Parmatic shipboard CPS is a complete installation designed to provide warships with sustained capability to operate under an NBC threat or within an NBC environment. It comprises filter intakes with blast dampers, HEPA filter banks, AC

Navy CBR-02 marine COLPRO System (Parmatic Filter Corporation) 0050897

fans and monitoring devices. Other essential components include cleansing (DECON) stations for entry to the CPS, airlocks, overpressure valves, pressure sensors, alarm devices and exhaust fans. The Navy CBR filter assembly (shown) comprises four major components: the HEPA filter, the CW agent vapour filter, after-end assembly and the cover assembly. The latter two units bolt into the filter casing from opposite ends and the whole forms a shock and vibration-resistant housing for the filtration plant. This unit is located upstream of the high-pressure supply fan for the ships heating, ventilation and air-conditioning (HVAC) system. The NBC filters are Parmatic's 200 CFM (340 m³/h) gas particulate filter sets (see separate entry within this section). Typically, three gas particulate filter sets are mounted within each NBC housing tube. This commits various flow combinations in multiples of 600 CFM (1, 020 m³/h). The HEPA filters remove 99.97 per cent of 0.3 μ contaminants. The vapour filters contain chromium-free activated carbon which adsorbs toxic vapour and are mounted downstream of the HEPA filters.

Status

In production. Approved for use by the Department of Defense and installed on close support vessels such as the DDG-51 and LPD-1

Manufacturer

Parmatic Filter Corporation.

VERIFIED

Pressure Swing Adsorption (PSA) Advanced COLPRO

Description

NBC protection for concentrations of essential personnel at headquarters, intelligence and armoured units is made harder by the need for facilities to be mobile and responsive. Working with the ERDEC engineering team at Aberdeen Proving Ground as well as major combat equipment commands, Guild Associates evolved highly advanced adsorption and catalytic oxidation technology.

The most advanced technology for application to combat vehicles is Pressure Swing Adsorption (PSA). PSA technology removes CBW vapours and aerosolised agents by manipulating beds of adsorbent material through a cycle in which filter beds are alternately charged and purged. The heat exchangers and air cycle components ensure equal quantities of contamination-free air to the crews and purging air to remove agent concentrations from the charged bed. A variety of adsorbent materials have been evaluated in several combinations to obtain optimum performance. For mobile systems, such as armoured vehicles, the system needs to operate within strict weight and footprint criteria at minimum power consumption.

Status

In development.

Manufacturer

Guild Associates Inc.

VERIFIED

Transportable Collective Protection System (TCPS)

Description

The Transportable Collective Protection System (TCPS) was developed as an expedient field-deployable, inflatable collective protection system for emergency base operations. The system provides a covered area for personal decontamination. It uses proven processing procedures in the contamination control area and rest/relief or light work tasks in a shirt-sleeve environment within a toxic-free area. The system is equipped with emergency lighting, low-pressure warning units, heater/air conditioner, chemical defence hardening kits and personal hygiene equipment.

TCPS interfaces with the M28 and ES/C shelters and can also function as a stand-alone system.

Status
Ready for production.

Development agency
US Air Force HSC/YA.

VERIFIED *Transportable Collective Protection System (TCPS)*

DECONTAMINATION

DECONTAMINATION

AUSTRIA

J Blaschke expendable foil decontamination collecting troughs

Description

During training exercises or NBC decontamination operations in times of peace or conflict, there exists a situation where any form of decontamination agent employed has to be collected following the operation. For this purpose, J Blaschke developed a low-transport volume and weight decontamination agent collecting trough system to collect such residues.

The collecting trough consists of a frame which can be assembled at any suitable site using handy-sized components. The trough size can vary from 2 × 3 m up to 20 × 30 m or more, the lengths of foil involved being connected with a special adhesive tape as required. The standard design used when decontaminating wheeled vehicles employs a foil width of 6 m, resulting in a trough width of 5 m.

Austrian Air Force J35 Drakken undergoing decontamination in an expendable foil decontamination collecting trough   0052127

Austrian Army M60A3 MBT undergoing decontamination in an expendable foil decontamination collecting trough during a training exercise

The components involved in a typical J Blaschke expendable foil decontamination collecting trough set

The self-priming diaphragm pump used with the expendable foil decontamination collecting trough

For tracklaying vehicles the foil width is 8 m, resulting in a trough width of 7 m. The trough length depends on the length of the vehicle(s) involved. When aircraft are involved extra foil lengths may be connected together. The foils are supplied folded in widths of 1 or 1.5 m — an unrolling device is provided with each set. All the equipment in each trough set is resistant to NBC warfare agents.

The foil involved can be reused or disposed of without hazards. For special applications, various foil thicknesses can be supplied.

For drawing off decontamination liquids from the trough a diaphragm self-priming pump is provided. Parts of the pump which come into contact with any aggressive liquids are highly resistant to any chemical reactions.

Status

In production. In service with the Austrian armed forces.

Manufacturer

J Blaschke Wehrtechnik GmbH.

VERIFIED

BRAZIL

ENGESA EE-25 (4 × 4) NBC decontamination truck

Description

The ENGESA EE-25 (4 × 4) NBC decontamination truck was designed and produced specifically for the NBC decontamination of troops, roads and vehicles, and uses the 4 × 4 chassis of the ENGESA EE-25 2,500 kg truck (vehicle details from *Jane's Military Vehicles and Logistics*).

The load-carrying area of the EE-25 truck has a 3,000-litre capacity water tank integral with the platform, 200-litre fuel tank for the heater, decontaminant solution tank with a 160-litre capacity, heater with fuel consumption of 12 litres/h, decontamination solution dispenser with a normal mix of 1 litre of decontaminant to every 20 litres of water (graduated from zero to 30 litres/h), water pump with a flow of 48 litres/min, 16 showers, two 100 mm diameter hoses, road or terrain decontamination set and canvas covers for the shower areas.

ENGESA EE-25 (4 × 4) NBC decontamination truck

ENGESA EE-25 (4 × 4) NBC decontamination truck

Status
In service with a North African country, believed to be Libya.

Manufacturer
ENGESA Engenheiros Espacializados SA.

VERIFIED

BULGARIA

DK-5 vehicle decontamination kit

Description
The DK-5 vehicle decontamination kit is intended for the decontamination of a carrier vehicle and utilises pressure taken from a point on the vehicle's exhaust system.

The complete kit weighs 14 kg and is issued in a metal box containing four kits. Each kit consists of a number of spare parts and a tool kit to assemble and maintain the kit. A complete kit contains a 40 litre rubber tank, plastic container for the decontaminating agent, wash nozzle, scrubbing and kit cleaning brushes, length of hose and various spare parts. Operating pressure of the system is 80 Pa (±10 Pa).

To decontaminate a ZIL-130 truck, it will require between 50 and 60 litres of decontaminating agent; the amount required to decontaminate a smaller vehicle such as a GAZ-66 truck or BRDM-1 armoured vehicle will be between 30 and 40 litres. The amount of decontaminating solution required to treat 1 m² is between 1 and 1.5 litres.

Status
In production.

Marketing agency
Kintex.

VERIFIED

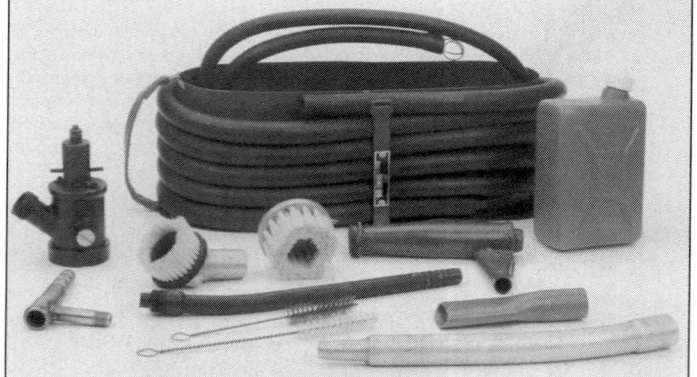

DK-5 vehicle decontamination kit

DKV-M decontamination system

Description
The DKV-M decontamination system is intended for decontaminating vehicles and weapons and is the latest version of the RFAS decontamination system, portable, Model DKV (see entry under RFAS for available details). In its Bulgarian-produced

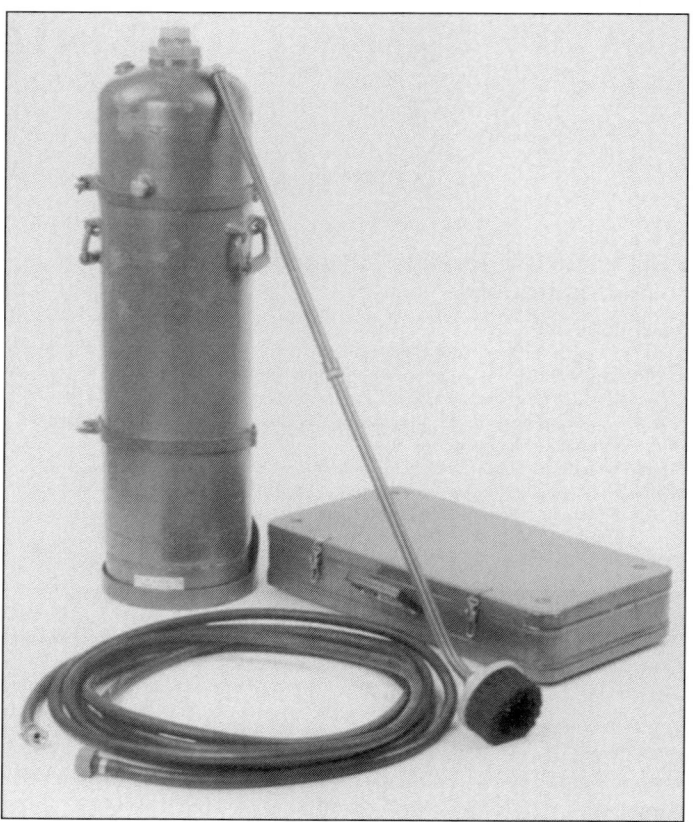

A complete PAD decontamination kit, one of the components of the DKV-M decontamination system

version, the DKV-M is based on a specially equipped ZIL-131 6 × 6 or similar truck which carries either 48 or 96 PAD decontamination kits.

Each PAD kit consists of a tubular pressure container holding 22 litres of decontamination agent solution ready for dispensing via an 8 m hose and brush system; the kit is completed by a metal box containing spare parts and tools. The decontaminating agent involved may be either calcium-hypochlorate solution or a multi-agent solution. The PAD operates at a pressure of 6.5 Pa and can treat up to 100 m² with a multi-agent decontaminating solution. After use the PAD is returned to the DKV-M vehicle for refilling.

Status
In production.

Marketing agency
Kintex.

VERIFIED

KBSO vehicle-mounted decontamination kits

Description
There are two sizes of KBSO vehicle-mounted decontamination kit, large and small. The large KBSO kit consists of a pressurised aerosol spray canister, a toolkit in a metal box and a plastic container for 5 litres of decontaminating agent. The small kit lacks the plastic container.

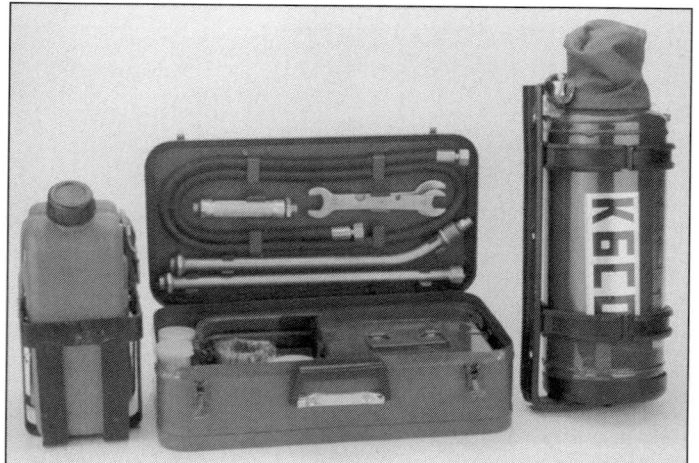

A large KBSO vehicle-mounted decontamination kit

Although vehicle-mounted, the KBSO kits are intended primarily for decontaminating weapons and equipment. The small kit can decontaminate an area of up to 20 m² and the large kit up to 40 m². The time required to prepare the equipment for use by one operator is up to 5 minutes.

A large KBSO kit weighs 25 kg and a small kit 19 kg.

Status
In production.

Marketing agency
Kintex.

VERIFIED

KID-6 and KID-12 individual decontamination kits

Description
The KID-6 and KID-12 individual decontamination kits differ only in the number of individual units they contain; the KID-6 kit contains six individual units and the KID-12 has 12. In both cases the individual kit units are identical, consisting of a decontamination agent dispenser weighing 220 g and a package of a further decontamination agent (bentonite) weighing 90 g. Intended for the decontamination of personal weapons and clothing, the dispenser can be used to decontaminate a rifle in about 1 minute; the time taken to decontaminate personal clothing is between 10 and 15 minutes.

Both kits are contained in metal boxes, a complete KID-6 kit weighing 4.5 kg and the larger KID-12, 7.5 kg.

Status
In production.

Marketing agency
Kintex.

VERIFIED

KID-6 individual decontamination kit

CANADA

Canadian Aqueous System for Chemical-biological Agent Decontamination (CASCAD)™

Description
Owing to the drawbacks of current decontamination compounds, which include surface damage to equipment, DRDC/RDDC, (Defence R&D Canada - Suffield) at Medicine Hat, Alberta, sought to develop an effective replacement that would successfully deal with contamination by all known CBW agents. It had to be stable when mixed as a compound and cause minimal damage to equipment or treated surfaces.

The Canadian Aqueous System for Chemical-biological Agent Decontamination (CASCAD)™ formulation contains active and other ingredients derived from families of readily-available industrial chemicals delivered as concentrates. These concentrates, when mixed with fresh or sea water, generate a foam which effectively decontaminates surfaces or material contaminated with G, H, L or V agents, BW agents and organophosphorous compounds and which has been shown to effectively remove radioactive dusts. One concentrate contains the active ingredients and, optionally, stabilisers, whilst the other concentrate comprises a co-solvent to increase the potential to retain agent in solution. It also contains a

surfactant which forms foam and ensures adequate wetting of the target surface. All the ingredients are commercially available.

Status
In production.

Developer
DRDC/RDDC Suffield.

Manufacturer (under licence)
NBC Team Limited.

UPDATED

NBC-DEWDECON-2L decontamination device

Description
The DEW 2-litre DS2 applicator meets the requirements of STANAG 2253. It disseminates DS2 decontaminating agent in a controlled spray to remove chemical warfare agents from contaminated surfaces. The device is a smaller version of the in-service DEW 3-litre unit (see separate entry) and shares many interchangeable parts. The DEW 2-litre device uses nitrogen cartridges as the primary method of pressurisation, with an attached hand pump as back-up. The device comes complete with a mounting bracket, spare parts, tools and spare nitrogen cylinders. It is reusable and can be filled, pressurised and operated while wearing full NBC protective clothing. Instructions are provided in English, French and Arabic.

The device, when stowed in its mounting bracket, measures 150 mm wide, 160 mm deep and 440 mm high. Dry weight is 4 kg.

DEW has teamed for North America with Cristanini of Italy and the DEWDECON-2L has been adapted to use the Cristanini BX24 decontaminant.

Status
In production and in service in the Middle East.

Manufacturer
DEW Engineering and Development Limited.

UPDATED

NBC-DEWDECON-2L decontamination device (DEW Engineering and Development Limited)

NBC-DEWDECON-3L decontamination device

Description
This portable DEW 3-litre decontamination device was designed to meet the requirements of STANAG 2253. It disseminates DS2 decontaminant in a controlled spray to remove chemical warfare agents from the surface of military equipment and can be filled charged and operated while wearing full NBC protective clothing. The device is pressurised by hand or air compressor and disseminates DS2 in a fan spray pattern from 1 to 3 m. It is deployed on wheeled and tracked vehicles, on aircraft ground support equipment and on exterior bulkheads on ships. The NBC-DEWDECON-3L is supplied with a mounting bracket, pressure gauge, safety relief valve and operator instructions in English, French and Arabic.

The device is reusable, made of high-quality materials and designed for a long service life. It is corrosion-resistant to DS2 and adaptable to other decontaminants. All required maintenance can be carried out by the operator using the spare parts and tools provided with each unit. The integral hand pump is interchangeable with the NBC-DEWDECON-20L decontamination device (see entry in this section).

The device, when stowed in its mounting bracket, measures 150 mm wide, 160 mm deep and 635 mm high. Dry weight is 5.4 kg.

NBC-DEWDECON-3L decontamination device (DEW Engineering and Development Limited)

DEW has teamed for North America with Cristanini of Italy and the DEWDECON-3L has been adapted to use the Cristanini BX24 decontaminant.

Status
In production. In service with Australia, Canada, Saudi Arabia and other countries.

Manufacturer
DEW Engineering and Development Limited.

UPDATED

NBC-DEWDECON-20L decontamination device

Description
The DEW 20-litre decontamination device disseminates C8-C type decontaminant when used with a standard 5-gallon (22.7 litre) plastic jerrican. It can be filled, pressurised and operated while wearing full NBC protective clothing and an operator with a charged device can cover an M113 APC within 8 minutes. The device is filled by the NBC-DEWDECON-M emulsion mixer and the C8-C decontaminant is effective for at least 72 hours.

The device is reusable, made of high-quality materials and designed for a long service life. It includes an integral pump, compressor fill valve, pressure gauge, safety relief valve, quick-disconnect hose, two quick-disconnect wand extensions, three-piece pole with a scraper, scrubbing brushes, jerrican contents identification ring, basic tools and operator/maintenance instructions in English, French and Arabic. The tank has a decontaminant capacity of 18.5 litres. It is pressurised by an external air source or by the integral hand pump. The integral hand pump is interchangeable with the NBC-DEWDECON-3L decontamination device (see entry in this section). All required maintenance can be carried out by the operator using the spare parts and tool kit provided. An optional DS2 conversion kit is available to disseminate DS2.

The NBC-DEWDECON-20L is supplied stowed in a rugged fabric bag that fits into any available space on a vehicle. The kit weighs 10 kg and measures 600 mm long, 200 mm wide and 150 mm high.

DEW has teamed for North America with Cristanini of Italy and the DEWDECON-20L has been adapted to use the Cristanini BX24 decontaminant.

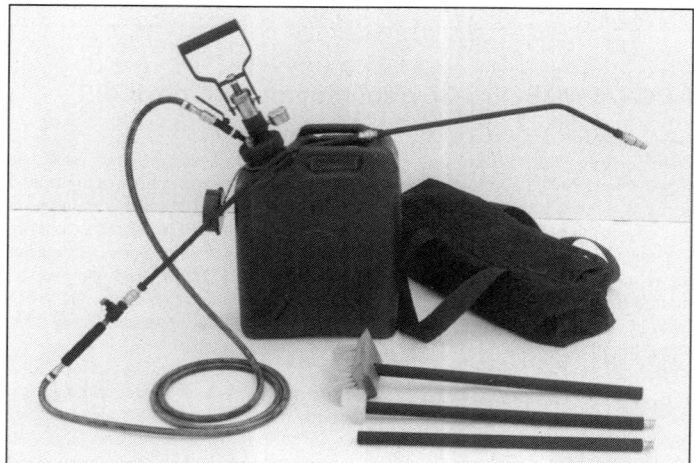

NBC-DEWDECON-20L decontamination device (DEW Engineering and Development Limited)

Status
In production. In service with Australia, Canada, Saudi Arabia and others.

Manufacturer
DEW Engineering and Development Limited.

UPDATED

NBC-DEWDECON-M decontaminant mixer/ applicator

Description
The DEWDECON-M Mixer and C8-C decontaminant were developed to provide a non-corrosive, stable and effective decontaminating system for both the operational and thorough decontamination of ships, aircraft, vehicles and equipment.

The mixer can be set up in 10 minutes by two people and will produce a continuous, online calcium hypochlorite-based emulsion at a rate of up to 2,200 litres/h. The mixer can be used as a direct applicator or to fill the DEWDECON-20L device (see entry in this section) for remote decontamination. The mixer also has a built-in rinse capability and a top-mounted accessory box to store all hoses, wands, spare parts and tools. Both diesel- and petrol-powered units are available.

Overall dimensions are 130 × 97 × 89 cm. Dry weight is 357 kg.

The C8-C emulsion produced by the mixer quickly and effectively neutralises chemical agents such as thickened GD and HD and VX. The emulsion is very stable and is effective for 24 to 72 hours, depending on ambient temperature.

A toluene-based perchloroethylene solvent replacement is available.

Status
In production. Tested and approved for service in Canada and other countries.

Manufacturer
DEW Engineering and Development Limited.

UPDATED

NBC-DEWDECON-M decontaminant mixer/applicator (DEW Engineering and Development Limited)

Reactive Skin Decontaminant Lotion (RSDL)

Description
Reactive Skin Decontamination Lotion (RSDL) is a liquid, broad spectrum, chemical warfare agent decontaminant for personnel. The lotion encapsulates, neutralises and destroys the known CW agents including vesicants (H and L*) and nerve agents (G and V). Even when it has reacted to destroy an agent on the skin, the RSDL residue is non-toxic* and can be safely left on the skin to be washed off at

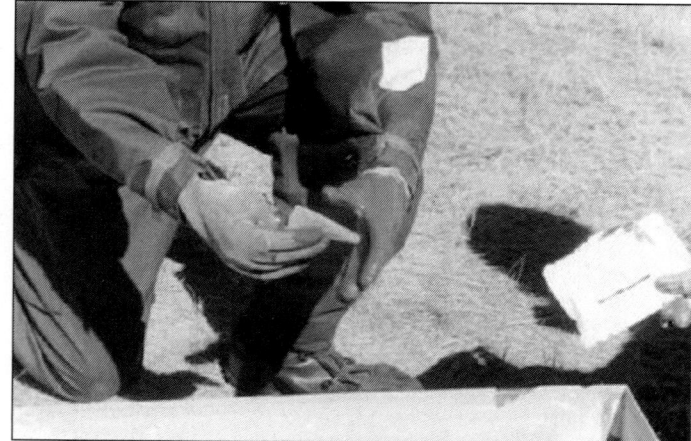

The RSDL foam applicator in use (O'Dell Engineering Limited)   0019369

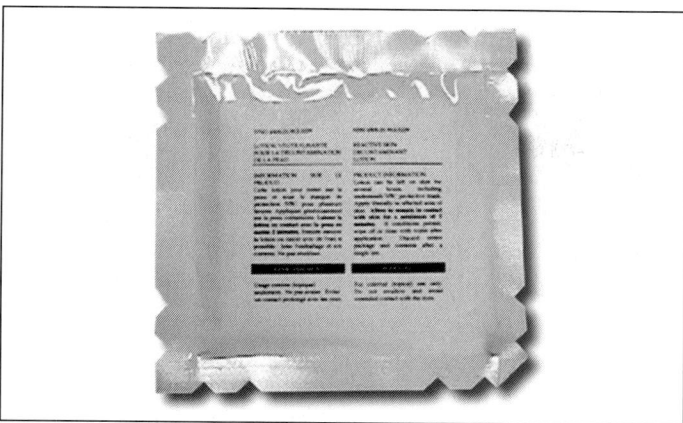

Individual RSDL tear-open pouch (O'Dell Engineering Limited) 0019382

a later opportunity. Canadian Forces testing has confirmed that RSDL is compatible with all skin types and can also be used safely on the eyes. RSDL is provided in two forms. The individual use pouch (shown) has a unique foam applicator for personal decontamination. Bulk containers are available for larger-scale casualty or equipment decontamination.

RSDL has been tested by the Canadian Forces and others where it has proven to be compatible with all types of personal weapons. As a liquid it is easier to handle in all climatic conditions than decontaminant powders such as Fuller's Earth (which merely absorb, not destroy agent). It will not bind or clog mechanical systems or obscure sight lenses.

Use of RSDL involves simply tearing open the pouch, removing the foam applicator and liberally applying RSDL to the suspected contamination site. The foam pad is used to scrub the skin surface to ensure full coverage and thorough mixing with even thickened agents. Reacted RSDL is non-toxic* and both the lotion and its reacted residue can be safely washed off with water.

A low cost Training Lotion is available in pouches and bottles for troop training and familiarisation with RSDL. Training Lotion allows full, accurate and realistic training in decontamination drills for individuals and medical personnel handling casualties. Other applications of RSDL are being developed.

(*Lewisite contains arsenic. This is reduced to a less toxic compound but not destroyed.)

Status
Available. In service with the Armed Forces of Australia, Canada, Ireland and Netherlands and with the OPCW. RSDL is under evaluation for procurement by several additional countries.

Manufacturer
O'Dell Engineering Limited.

UPDATED

Skin decontaminant lotion

Description
This skin decontaminant lotion is formulated to deactivate chemical warfare agents such as mustard (H), nerve agents and lewisite (L) on contact. It can also be used to decontaminate undamaged skin and durable personal items such as weapons, protective gloves and hand tools.

The lotion is supplied packed under a blanket of inert gas inside a sealed barrier material pouch. Each single-use pouch, measuring 155 × 155 × 50 mm, contains a towelette impregnated with 45 ml of lotion. Pouches are supplied in four-pouch sets. They can be opened with gloved hands.

In use, the impregnated towelette is applied liberally to the affected area and then wiped off after application. The lotion can be left on the skin for several hours, including underneath a mask if necessary. The lotion was designed for external use only and should not be taken internally or placed in contact with the eyes. The operating temperature range is from −10 to +50°C.

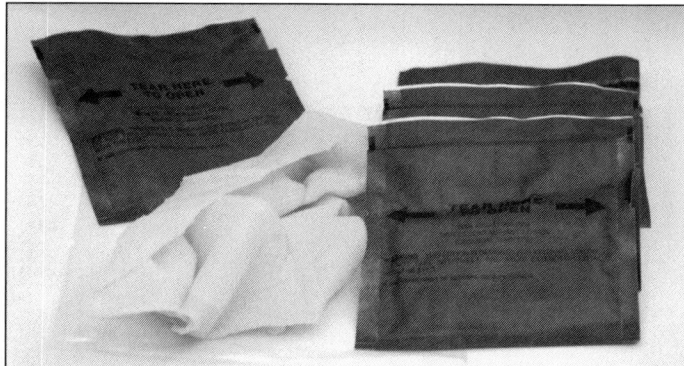

Skin decontaminant lotion pouches and towelette (Anachemia Canada Inc)

Status
In production.

Manufacturer
Anachemia Canada Inc.

UPDATED

Zenon Advanced Double Pass Reverse Osmosis Water Purification Unit (ADROWPU)

Description
The Zenon Advanced Double Pass Reverse Osmosis Water Purification Unit (ADROWPU) was designed for field use and can be used to treat water contaminated by NBC agents as well as fresh, brackish and seawater. The ADROWPU employs a double pass reverse osmosis process, this is a pressure driven membrane separation process that separates dissolved solutes and suspended substances from water. The double-pass process is used for NBC contaminant removal and seawater treatment.

The ADROWPU is a fully integrated self-contained system with its own 40 kW diesel power generator and a semi-automatic control system digital controls and instrumentation. Automatic self-cleaning and pretreatment features are built-in and a pillow tank for storing processed water is provided with each system. Installation time is 20 minutes.

The system is arranged in a self-contained palletised enclosure which enables it to be transported via all modes of military transport, including NATO standard palletised loading systems. The ADROWPU can also fit inside a standard ISO container.

The ADROWPU weighs 6,400 kg and measures 5.5 × 2.1 × 1.7 m. It can be operated at temperatures ranging from −40 to +40°C.

Status
In production. In service with the Canadian and Taiwanese armies and the UN. Has been deployed operationally to Cambodia, Qatar, Rwanda, Saudi Arabia, former Yugoslavia and Haiti.

Manufacturer
Zenon Environmental Systems Inc.

VERIFIED

Specifications
Typical outputs (litres/day)

Input water	Fresh	Brackish	Sea
Basic unit:			
Without NBC	84,700	81,520	52,390
With NBC	59,250	58,950	52,390
Expanded unit:			
Without NBC	112,650	108,450	69,680
With NBC	73,830	78,380	69,680

Zenon Advanced Double Pass Reverse Osmosis Water Purification Unit (ADROWPU) 0018697

CHINA, PEOPLE'S REPUBLIC

M01 decontaminating system

Description
The M01 system comprises a tanker vehicle, carrying decontamination solution and a truck-mounted contamination delivery vehicle. The latter uses an aerospace gas turbine to deliver the hot contaminant to the target vehicle in a similar way to the Polish WUS-3 system (see separate entry this section). The gas turbine is mounted on a platform at the rear of the vehicle, allowing it to be trained and elevated under

M01 decontaminating system (Research Institute for Chemical Defence) 0050761

the control of the decontamination operator. The operating position is enclosed and supplied with filtered air. The endurance of the system is claimed as over 4 hours, drawing from the accompanying tanker. The institute also claims a 3 to 5 minute decontamination time for vehicles of MBT size or below.

The operating environment is −20 to +40°C.

Status
In service with the Chinese armed forces.

Development agency
Research Institute for Chemical Defence.

UPDATED

M04 personnel shower vehicle

Description
The M04 personnel shower vehicle is based on the chassis of the EQ141 (6 × 6) 7,500 kg truck (see *Jane's Military Vehicles and Logistics* for details). The personnel decontamination equipment is housed in a metal cab, supplied with filtered air. A petrol generator drives a water heater containing a diving pump and a variable delivery pump. The water or decontaminant is delivered to an eight-point shower unit with rotating nozzles inside a deployed shower tent. The system can process approximately 120 people per hour and the air heater can maintain the tent temperature to +15°C. The operational range is −10 to +40°C.

Status
In service with the Chinese armed forces.

Development agency
Research Institute for Chemical Defence.

UPDATED

M04 personnel shower vehicle (Research Institute for Chemical Defence) 0050760

M-73-1 decontamination vehicle

Description
The M-73-1 decontamination vehicle is based on the chassis of the CA-30 (6 × 6) 2,500 kg truck. The vehicle carries a 2.5m³ tank for DECON agent at the rear. Decontaminant is delivered through nozzles on deployable spray bars. The rate of dispensing is controlled by two operators seated at the rear of the tank and over the spray bars when the vehicle is operating. For travelling purposes the two operators are seated in an enlarged crew cab behind the driver and vehicle commander.

The M-73-1 has a gross weight of 9,300 kg. Vehicle length is 6.86 m, width 2.4 m and height 2.42 m.

M-73-1 decontamination vehicle (Research Institute for Chemical Defence)

Status
In service with the Chinese armed forces.

Development agency
Research Institute for Chemical Defence.

UPDATED

M-82 personnel shower vehicle

Description
The M-82 personnel shower vehicle is based on the chassis of the CA-30 (6 × 6) 2,500 kg truck with an enclosed box body. When in use for personnel decontamination, the box body slides to the rear where it is supported on folding legs; this provides more internal space for preparatory disrobing and putting on fresh clothing after showering. Extra compartments enclosed by canvas covers are then erected on either side of the rear body. Access to the shower section is via ladders through one of three rear entrances and egress is through side exits towards the front of the canvas extensions. The M-82 carries a water pump and heater with water supplied to the system through a flexible hose normally carried on the front of the box body.

Status
In service with the Chinese armed forces.

Development agency
Research Institute for Chemical Defence.

UPDATED

M-82 personnel shower vehicle (Research Institute for Chemical Defence)

M-82 personnel shower vehicle (Research Institute for Chemical Defence)

M-84 decontaminating apparatus, tank

Description

This portable decontamination apparatus is a rechargeable cylinder containing 1.5 litres of a decontamination agent known as T-191. It is carried on a variety of vehicles for local first aid decontamination.

Weight of the M-84 is 3.2 kg. It is 380 mm high and 108 mm in diameter.

Status

In service with the Chinese armed forces.

Development agency

Research Institute for Chemical Defence.

UPDATED

Decontaminating apparatus, tank, M-84 (Research Institute for Chemical Defence)

CZECH REPUBLIC AND SLOVAKIA

ACHR-90 NBC decontamination vehicle

Description

The ACHR-90 NBC decontamination vehicle is based on the chassis of the TATRA T 815 26WR25, 26255 (6 × 6) truck (see *Jane's Military Vehicles and Logistics 1998-99* for technical details). Although developed primarily for military purposes, the ACHR-90 has been adapted as the APZ-94 for civil defence and similar applications.

The vehicle cab is covered by a layer of material intended to protect the crew against nuclear radiation. On the rear of the vehicle is a 6,000 litre capacity three-chambered stainless steel tank for water (one chamber) and decontaminants (two chambers) plus decontamination lances, piping, system accessories and nozzle bars for area spraying. Also carried are components for a personnel shower system, hoses and stowage for decontaminants and other solutions. The ACHR-90 also carries a power generator, pumps and mixing devices.

ACHR-90 NBC decontamination vehicle

APZ-94 DECON vehicle 0018698

APZ-94 DECON vehicle in operation 0018699

Using two Cristianini SANIJET C921 high-pressure systems, the ACHR-90 is capable of delivering decontamination agent solutions and/or water at a maximum flow of 50 litres/min at a maximum pressure of 0.4 MPa. It can generate high-pressure warm and cold water or steam and can provide 12 V electrical power or 220 V at 2,000 W.

The flux pump carried on the ACHR-90 may be used to pump hazardous agents from filled containers or an incident area up to a maximum output of 140 litres/min. The stainless steel, hydraulically driven META pump is fixed to the truck body and delivers pumping capacity for the distribution main, allowing inter-chamber transfer, fluid draw from external sources, system drainage and supply to external containers. The pumping rate is 1,000 litres/min. Included in the equipment suite carried on the ACHR-90 are two 2,000 litre flexible rubber tanks, together with high-pressure units and other components allowing the vehicle to be used as an independent decontamination station. The vehicle may also be used to spray terrain and/or roads over a width of 12 m while on the move and may also provide warm water for personnel showers. The vehicle can be used to fight fires using either water or special fire extinguishing solutions from a distance up to 240 m.

The vehicle is provided with a 100 kN self-recovery winch.

Overall dimensions of the ACHR-90 are length 8.4 m, width 2.5 m and height 3.32 m. Total weight is of the order of 23.8 tonnes; kerb weight is 17 tonnes. The operating range is approximately 1,000 km.

Status

ACHR-90 is ready for serial production. APZ-94, (derivative designed for emergency services use) is available.

Manufacturer

VOP 025 Nový Jicin, sp.

VERIFIED

Decontamination apparatus, truck-mounted, Model TZ-74

Description

Based on the chassis of the TATRA T 48 PPR 15 (6 × 6) truck, this equipment is a progressive development of the RFAS Model TMS-65 and TMS-65M. It follows the same general lines, in that it employs a small gas turbine to spray decontaminating agents over vehicles and equipment. As with the TMS-65, the gas turbine is mounted on the rear of the vehicle with the operator's cabin on the right-hand side. The rest of the vehicle rear is taken up with tanks for the decontaminating liquids and fuel for the gas turbine. The gas turbine, which is designated Type M 701 c-500, can be traversed through 120°, elevated 30° and depressed 20°. It can dispense liquids at the rate of 900 litres/h. A fuel tank for the gas turbine holds 2,000 litres. The main decontaminant tank has a capacity of 5,000 litres. Fully loaded, the Model TZ-74 weighs 22,000 kg and 14,000 kg empty. Length is 9.5 m, width 2.5 m and height 2.95 m. It has an operational crew of two and, as well as being used for equipment decontamination, can be used to create smoke screens.

Model TZ-74 truck-mounted decontamination apparatus (Michael Jerchel)

A training system known as the TTZ-74-B was developed for use with this system. Images of equipment to be 'decontaminated' are projected on to a screen placed in front of the jet exhaust. A sound system simulates the noise of the equipment's operation and other vehicle sounds and an electronic system records the operator's reactions to the ranges of the equipment being 'decontaminated'. Engine faults can be fed into the system and the training system may be used under cover or in the field.

Status
In service with the Czech, Slovak and German armies.

Manufacturer
Truck chassis: TATRA.

VERIFIED

DP2 decontamination apparatus

Description
The DP2 decontamination apparatus has a capacity of 1 litre and is designed to spray liquid decontaminants including hypochlorite suspensions, OS 3 decontaminating solutions and their variants. One filling is sufficient to decontaminate a vehicle, a unit's personal weapons or to cleanse a vehicle crew. The decontamination solution is applied from the trigger-operated spray head and the reservoir is charged with nitrogen from 8 g pressurised capsules. The illustrations show that the DP2 is delivered in a steel boxed set which includes the reservoir, spray-head, vehicle-mounting bracket (with screws), operating manual and 8-spare nitrogen capsules. A separate box carries three 1 litre cans of decontaminant solution. Spares for all parts are readily available.

The reservoir can also be pressurised by using a compressor (at 8 bar) with a safety valve. An adapter is supplied.

DP2 reservoir and spray-head, assembled in vehicle-mounting bracket (EST + as)
2001/0077985

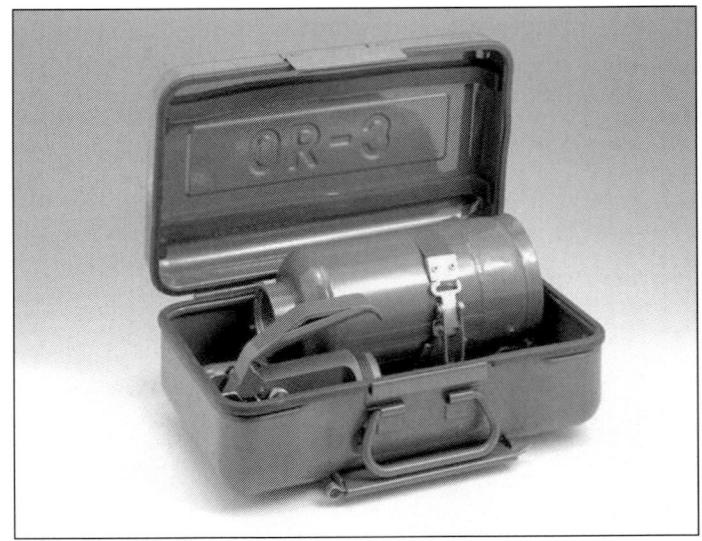

DP2 DECON apparatus stowed in carrying box (EST + as)　　*2001*/0077986

Specifications

Operating temperature:	30 to 50°C
Maximum air or nitrogen pressure:	9 bar
Time to exhaust full reservoir:	50 to 90 s, depending on type of nozzle
Weight:	Approx 50 g (not including decontaminant cans)
Carrying case dimensions:	320 × 120 × 190 mm

Status
Available.

Manufacturer
EST + a.s.

VERIFIED

LINKA 85 NBC decontamination system

Description
The LINKA 85 NBC decontamination system has been in service with the Czech and Slovak armed forces for some time. It consists of a spray frame capable of being set up under field conditions to spray decontaminating agents or water over vehicles passing through the frame. The spray bars are supplied via a pumping and agent mixing unit carried on a trailer or a system such as the TATRA ST-T815 or ACHR-90 NBC decontamination vehicles (see preceding entries).

Status
Originally manufactured at various state factories under communism. Current manufacturer unknown. In service with the Czech and Slovak armed forces.

Manufacturer
(Unknown).

VERIFIED

152 mm self-propelled gun-howitzer DANA passing through a LINKA 85 decontamination system spraying frame (T J Gander)

Model OS-3 decontamination kit

The OS-3 decontamination kit series is designed for hand-operated decontaminate of tanks and other armoured or wheeled vehicles. The complete kit is delivered in a boxed set which includes the spray gun, a 1 litre pre-packed container and a suction extension to allow decontaminant to be drawn from any suitable container. A separate box holds three containers. The spray unit is electrically powered and the power lead allows vehicle DC power points to be used. The suction extension is provided with a filter and connector to allow decontaminant to be drawn from an open container. The following modifications are available:

- OS-3/24 V. Works from 24 V DC vehicle supply
- OS-3/12 V. Works from 12 V DC vehicle supply
- OS-3 M. This stainless steel version offers user selectable vehicle power supply (12 or 24 V).

Versions differ only motor type. All other components are identical and interchangeable.

The OS-3 series is robustly designed to allow all current types of decontaminants to be used, including DS-2, OR-3 and hypochlorite solutions.

OS-3 is also suitable for cooling and cleansing personnel wearing impermeable IPE.

OS-3 DECON sets were used extensively during UNSCOM operations in Iraq where the units were invariably found reliable and effective especially in high-ambient operating temperatures.

Status
Available.

Manufacturer
EST + a.s.

VERIFIED

Specifications

Preparation time (including donning IPE):	5 mins
Time to deliver 1 litre of DECON solution (in viscosity range from 7.2×10^{-3} Pa - ethanol for example to 3.5×10^{-1} - oil for example):	Approx 90 s
Max period between working cycles (10 cans of decontaminant):	20 mins
Operating temperature range (equipment):	−30 to +50°C
Working operating temperature:	−25 to +50°C
Operation:	Single operator
Power supply:	Internal lamp socket
Power supply ranges:	
OS-3/24V:	21.6-30 V. 4.3 A
OS-3/12V:	10.8-15 V. 8.5 A
OS-3 M:	10-30 V. 4-8 A
Cable length:	15 m
Minimum guaranteed suction head:	3.5 m
Recommended spray distance from target surface:	10 to 50 cm

An OS-3 decontamination kit in use

Loading an OS-3 decontamination kit from a prepacked canister

PZ 9S and PZ 18S foam generators

Description

The PZ 9S and PZ 18S foam generators are designed to deliver and apply cleansing, DECON or disinfecting foam on large areas. Both types can be used for cleansing and disinfecting production equipment, especially in sensitive cases where high-pressure devices are inappropriate. The PZ series is designed to minimise the consumption of water and DECON agents. The design ensures that the aqueous foam created is stable, homogeneous and durable. Both generators are easy to set up, use and control and are robustly constructed for reliability and high performance. The waste residue is small and, after application, the foam can be removed for destruction with the vacuum pump (see illustration).

Technical documentation includes guidance on the various foams and additives that are suitable for use with the equipment, together with the operating conditions for use. EST supplies complementary DECON equipment, including vacuum cleaners, compressors and transport.

Specifications

Equipment type:	PZ 9S	PZ 18S
Maximum vessel capacity:	9 litres	18 litres
Air inlet pressure:	8 bar	8 bar
Working pressure:	4 bar	4 bar
Air consumption at 3 bar:	Approx 9 Nm³/h	Approx 9 Nm³/h
Range of working temperature:	2 - 50°C	2 - 50°C
Filling neck dimensions:	80 × 100 mm	80 × 100 mm
Weight:	5.2 kg	8.9 kg

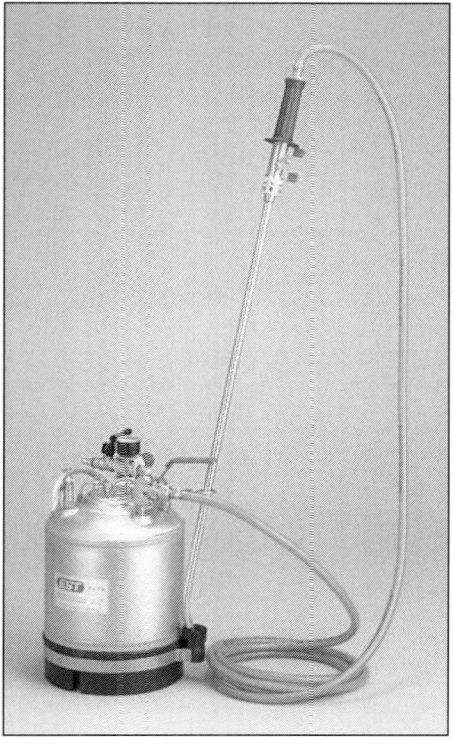

PZ 9S foam reservoir and lance (EST + as)
***2001**/0077981*

*PZ 18S trolley-mounted
reservoir and lance
(EST + as)*
2001/0077982

Status
Available.

Manufacturer
EST + a.s.

VERIFIED

EGYPT

Aboukir decontamination trailer

Description
This trailer is a purpose-designed unit offering a comprehensive suite of facilities for the decontamination of equipment and personnel. The equipment is arranged around the trailer on several levels and is protected from contamination and the weather by proofed fabric covers supported on steel frames. Equipment includes a compressor, pumping equipment, decontamination lances, spray nozzles and stocks of contaminant fluid. Hose-reels are stowed beneath the main platform fore and aft of the wheels. Mounted on a single-axled, wheeled chassis, the trailer can be towed by most military and other heavy-duty vehicles.

The manufacturer claims similar performance to its multipurpose decontamination vehicle (see separate entry within this section).

Status
In production. In service with the Egyptian armed forces.

Manufacturer
Aboukir Engineering Industries Co.

VERIFIED

Aboukir multi-purpose decontamination vehicle

Description
The equipment for this vehicle is mounted on a flat platform protected by proofed fabric covers supported on steel frames. The decontamination equipment suite comprises a compressor and pumping equipment, decontamination lances, hose reels, gantries, spray nozzles and drums of contaminant fluid. There is also a water tank and a foam canon can also be mounted for large scale DECON. An enclosed cabinet allows decontamination of clothing or small equipment, retaining the residue for later disposal.

Motive power is provided by the vehicle which is a locally manufactured version of the Magirus Deutz Mercur (see *Jane's Military Vehicles and Logistics* for further details of the vehicle).

The manufacturer claims that personnel can be decontaminated at the rate of 540 people per hour and the unit can deal with between five and seven tanks or armoured vehicles and between 10 and 15 medium-sized wheeled vehicles per hour. The vehicle also carries extendable spray bars, allowing it to be used for terrain decontamination. In the latter role, the vehicle can decontaminate a 3 m wide strip of terrain with the vehicle underway at speeds of approximately 6 km/h.

Aboukir multi-purpose decontamination vehicle

Status
In production. In service with the Egyptian armed forces.

Manufacturer
Aboukir Engineering Industries Co.

VERIFIED

FRANCE

ACMAT UMTH 1000 vehicle-mounted decontamination system

Description
The UMTH 1000 is a variant of the ACMAT VLRA series of vehicles and is purpose-designed to offer a mobile, autonomous DECON capability in the field. Clearly, it can also be tasked to provide DECON assistance, mobile area disinfection and firefighting resources following industrial or domestic terrorist incidents. System mobility is provided by an ACMAT VLRA (6 × 6) TPK 6.40 CSD truck chassis which carries the equipment and a crew of three.

The system is air-transportable by Transall C-160 and C-130 transport aircraft with little preparation required. There are four main components:
- An equipment platform mounted on the rear of the vehicle. The side and back panels can be folded down to provide DECON working platforms for the crew.
- A hydraulic conversion unit (the Hydraulic Transformation Mobile Unit UMTH 1000).
- Motor-driven pump.

ACMAT UMTH 1000 vehicle-mounted decontamination system (Pierre Touzin)

ACMAT UMTH 1000 DECON system in use (ACMAT)   **2001**/0109619

ACMAT UMTH 1000 annotated diagram (ACMAT) **2001**/0109620

Specifications
Crew: 3
Weight: 11,800 kg
Length overall: 6.91 m
Width: 2.31 m
Height overall: 2.6 m
Max speed: 90 km/h
Range: 1,600 km

- 3,000-litre water tank.
- Demountable high-pressure hot water and steam generator.
- Lockers for various accessories.

The carrier vehicle is fitted with NBC detection and carries protective clothing for the crew and contaminated area markers. A bracket crane is provided for unloading the hydraulic conversion unit and there is stowage space for a 3.5 m lightweight ladder.

See *Jane's Military Vehicles and Logistics* for full details of the ACMAT VLRA (6 × 6) TPK 6.40.

Status
In service with the French Army and several other countries.

Manufacturer
Vehicle: ACMAT (Ateliers de Construction Mécanique de l'Atlantique).

VERIFIED

EURODECONT decontamination system for sensitive material

Description
This container-based system is designed for the rapid DECON of small or sensitive pieces of equipment, such as personal weapons, hand-held electronic equipment, masks, helmets and so on. Giat Industries is the French partner in this project (see entry under Germany, within this section).

Status
Under development for the French and German armed forces. French prototype handed over in October 1999.

Manufacturers
Alfred Kärcher GmbH &Co.
Giat Industries.

VERIFIED

Giat Thorough Decontamination System (SDA)

Description
The Giat Thorough Decontamination System (SDA - *Système de Décontamination Approfondie*) is a mobile facility, based on a Renault Véhicules Industriel TRM 200.13 chassis (see *Jane's Military Vehicles and Logistics* for vehicle details). This supports two large water tanks and a decontaminant tank. The decontamination medium drawn from the tanks is heated and pressurised before supply to the manually-operated decontamination lance. The operator, situated in a howdah on the end of a gantry, can move above and around the target equipment or vehicle to complete the decontamination operation. The gantry can be steered from the chassis or from controls on the howdah itself. The chassis allows full off-road operation and the system claims to clean a small tank in less than 15 minutes. Used

Giat thorough decontamination system 0050765

Giat thorough decontamination system 0050756

Specifications
Capacity (water): 2,400 litres in 2 tanks
Capacity (decontaminant): 500 litres in 2 tanks
Water delivery: 700 litres/min at 70 bar
Water temperature: 70°C
Steam delivery: 210°C at 20 bar
All-up weight: 13 t
Readiness: within 15 min

in conjunction with TCC (see separate entry), it offers mobile units a comprehensive contamination control solution.

SDA can be used for CW, BW or radioactive decontamination and the tanks can be replenished from any water source.

Status
In service with the French Army and several other countries.

Manufacturer
Giat.

VERIFIED

Temporary Camouflage Coating (TCC)

Description
TCC is a surface coating designed to offer equipment operators a quicker and more assured method of removing contamination. Prior to operations in an NBC risk environment, the coating is easily applied over the surface of any types of equipment such as tanks, aircraft or vehicles. Little surface preparation is required and the coating can be applied using brushes, spray guns or aerosol cans within an ambient temperature range of −5 to +40°C.

The coating itself offers other benefits such as rapid change of camouflage scheme to take account of changed location or climatic conditions. It can also incorporate other defensive measures according to need, including IR, UV, radar and visual signature reduction. Intumescence can be added for nuclear flash protection. Biocides and other more sophisticated and pathogen-specific BW defensive measures can be incorporated, as well as signature reduction against 'smart' weapons and laser designators.

The resultant coating is tough and durable, long-lasting and highly resistant to POL contamination. Surfaces are operationally available 15 minutes after treatment. Treatec Mirage aircraft have been flying for over two years with this

TCC-protected 'contaminated' vehicle being sprayed with emulsion (pH³ 10) prior to cleansing   0050757

TCC and 'contamination' being washed off with water   0050758

Cleaned vehicle showing underlying permanent coating, revealing the thoroughness of the decontamination effort   0050759

coating and tanks and armoured cars have been successfully treated with TCC for service with SFOR.

TCC is designed to absorb NBC contamination without transferring it to the surface below. Removal simply requires the application of a mild alkaline gel or domestic detergent (pH 10 or greater) to loosen the surface. The residue is then removed with water. Any of the physical decontamination equipment shown in this section is suitable for the removal process. The residue is non-toxic and can be disposed of through normal drainage. With a TCC coating of a different colour to the permanent coating underneath, decontamination teams can quickly see how thorough their efforts have been.

Status
In service with the French armed forces and others serving on UN assignments.

Manufacturer
International Celomer sfpv.

VERIFIED

GERMANY

EURODECONT decontamination system for sensitive material

Description
This decontamination system for sensitive material is under development for the Federal German Armed Forces and the French Army as the subject of a joint venture. The EURODECONT consortium comprises Giat Industries (France) and Alfred Kärcher (Germany). It is the first such mobile system to have been designed specifically for decontaminating the complete range of sensitive materials and equipment including optical, electrical and electronic devices, protective masks, hand guns and helmets. The unit also carries a track-mounted Decont Shuttle for the internal DECON of confined areas in which sensitive equipment is fitted, such as cockpits or vehicle and tank cabs/compartments (see separate entry within this section).

The vacuum DECON facility, mounted on the left side of the container is of particular interest. Accessed through a hatch, the chamber inside is fitted with a heating element comprising a conductive jacket and an IR radiator. The heat inside can be precisely adjusted from ambient to 200°C. With the hatch closed, the contaminated air is removed from the heated chamber by a vacuum system and passed through a HEPA filter before being released to atmosphere. This element of the EURODECONT system weighs 300 kg and its pump can deliver 10^{-2} mbar of vacuum.

EURODECONT system for sensitive material shown mounted at the rear of its trailer   **2001**/0098355

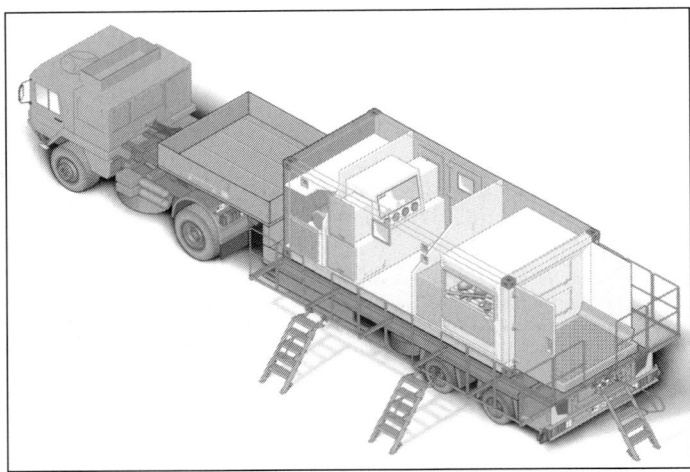

EURODECONT system - diagram of container compartments   **2001**/0098350

EURODECONT system - left side view of container   **2001**/0098351

The operator at the glove box needs no respiratory protection to decontaminate equipment in the tray (the outer door remains closed during this period)
2001/0098352

Specifications
Dimensions: Standard ISO container: 6.058 × 2.438 × 2.438 m
Weight: 8,000 kg
Performance data: Operational temperature range: −32 to +49°C
Capacity: Approx 54 items/h (including masks, helmets, personal weapons, optical or electronic equipment). Simultaneous DECON facility for interiors including cockpits, tanks compartments, vehicle cabs and helicopters.

The EURODECONT system's various facilities fit inside a standard 6 m ISO container and it has its own integral COLPRO unit (NBC filtration plant and air-conditioning) for the protection of operators. The system is designed to operate autonomously for sustained periods. The EURODECONT container locates onto the truck where it is securely fixed by a twist-lock system.

Status
Under development for the German and French armed forces. German prototype handed over to the German Army in May 1999.

Manufacturers
(EURODECONT Consortium):
Alfred Kärcher GmbH.
Giat Industries.

VERIFIED

Kärcher AEDA1 decontamination equipment

Description
The Kärcher AEDA1 system combines four main components as one set and is used for internal decontamination within field hospitals, command centres, radio installations, vehicles and containerised systems. The four components are: an aerosol spraying dispenser; hot air generator; remote-control unit; and a surface cleaning system.

To utilise the system the aerosol sprayer first sprays the interior concerned with a disinfection solution, generating 0 to 100 μm droplets; it is possible to control the spray quantity and air quantity separately. The fine spray fog will remain in the air for a long period of time to decontaminate any bacteria in the air. Bacteria on ground surfaces will also be decontaminated by the droplets once they land.

After the reaction time, the hot air generator is used to heat up the interior being decontaminated. This will help to remove any remaining decontamination solutions which might be dangerous. The hot air can be adjusted to temperatures up to 150°C; maximum air flow rate is approximately 3,000 m³/h. The hot air generator and the internal temperature are controlled and monitored by the remote-control unit.

Any remaining liquid decontamination solution on the surface or ground can be removed with the surface cleaning system. This module incorporates a spraying pump and a vacuum turbine. It is possible to spray out liquid disinfection solutions from an integral water tank and simultaneously suck them back into an integral waste water tank. The output capacity is 1 litre/min at a pressure of 1 bar. Differing surface nozzle sizes allow a safe application and surface treatment.

Status
In production. In service with the German armed forces and Crisis Reaction Forces.

Manufacturer
Alfred Kärcher GmbH & Co.

VERIFIED

Kärcher DADS Direct Application Decontamination System

Description
The Kärcher DADS direct application decontamination system was designed for the whole range of decontamination tasks, including deradiation, detoxification and disinfection of terrain and large items of military hardware such as armoured vehicles and aircraft.

The system is driven by an air-cooled diesel engine which drives a high-pressure (75 bar) water pump, a solvent pump for liquid agents and two hydraulic pumps. The diesel engine may be started by hand or by an electric starter. Once started, an automatically regulated hydraulic system guarantees the correct mixing of the various decontamination chemicals. The heavy-duty high-pressure pump can draw water up to a height of 5 m. All important parts of the system are constructed using corrosion-resistant materials.

With the DADS system, optimum detoxification is accomplished using the Munster emulsion, once surface dirt has been removed using cold water at high-pressure. The Munster emulsion, developed by the German Research and Science Centre in Munster, is a composition of a special calcium hypochlorite known as C8,

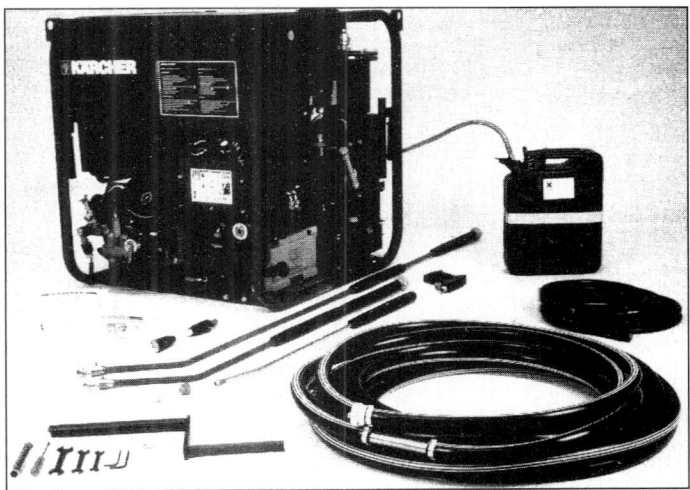

Kärcher DADS Direct Application Decontamination System

Truck-mounted Kärcher DADS Direct Application Decontamination System in use spraying Munster emulsion onto Leopard 1 MBT

Specifications
Weight: 320 kg
Length: 0.96 m
Width: 0.85 m
Engine: 4-stroke diesel developing 13 hp at 3,000 rpm
Electrical system: 12 V

tetrachloroethylene (which may be added to act as a solvent), a phase transfer catalyst and water. When used with the DADS system the emulsion is mixed automatically under the control of a single switch. The emulsion is then applied under pressure using a special spray lance. For tanks and large vehicles with chemical-resistant paintwork, approximately 150 litres of emulsion are required. For vehicles with non-resistant paintwork 300 litres are needed. The emulsion is applied in two coats. The DADS system is also designed for mixing and application of other kinds of decontamination emulsions (TDE 202 and so on).

Optimum deradiation (10 mrad/h) is guaranteed by a hot foam treatment. For this, two deradiation powders are mixed within the system and best results are achieved at a water temperature of 80°C. Hot water is provided by the Kärcher MultiPurpose Decontamination System (MPDS — see separate entry in this section).

Disinfection of biological and bacteriological warfare agents is carried out with the help of aqueous C8 solution.

When decontaminating terrain the decontaminating liquids may be sprayed using nozzles mounted on vehicles. The basic DADS module can mix up to 254 kg of C8 with water (providing 1,300 litres of aqueous solution). To increase this quantity, additional pump systems can be deployed.

The DADS system may be carried on a truck with the components mounted in a tubular steel frame and can be dismounted for use. Accessories include 20 litre cans for fuel and chemicals, a 25 m application hose, hand spray guns, a 1.04 m long spray lance, various suction hoses, emulsion application lances with lengths of 1.2 and 2 m, other spray lances, nozzles and spare parts.

Status
In production. In service with Australia, Austria, Egypt, France, Germany, Taiwan, Thailand, NATO headquarters and some nations in Asia. The system has been tested by the US Army.

Manufacturer
Alfred Kärcher GmbH & Co.

VERIFIED

Kärcher DECOCONTAIN 1500 decontamination system

Description
The Kärcher DECOCONTAIN 1500 decontamination system is designed for deployment at NBC defence units and forms the technical basis for the establishment of a decontamination site. The DECOCONTAIN 1500 is a compact system integrated into a container. Using the DECOCONTAIN 1500 practically all essential decontamination operations can be performed following an NBC attack, the main application being the decontamination of material.

The independent diesel-powered Kärcher MPS 3200, DADS and MPDS modules are used for the pre-, main- and post-treatment of material decontamination with optimum results. The decontamination procedure used is dependent on the type of decontamination. Radioactive decontamination uses a hot foam procedure while disinfection uses various solutions. The detoxification of all known chemical warfare agents involves the Kärcher detoxification emulsion TDE 202.

In addition, the decontamination of personnel, their clothing and equipment can be carried out. Personnel decontamination is achieved using a two step, pulsating

Kärcher DECOCONTAIN 1500 decontamination system

shower procedure. Detoxification and disinfection of clothing and equipment is accomplished using hot steam. In addition, terrain decontamination can be carried out using an aqueous detoxification solution.

By using an integral 1,500 litre water tank the DECOCONTAIN 1500 can be used to perform relatively independent decontamination tasks.

The following are typical operating capacities for a DECOCONTAIN 1500 system:

Material decontamination — six to eight tanks or 12 to 16 vehicles/h
Personnel decontamination in a non-contaminated area — approximately 120 persons/h
Decontamination of clothing — 15 to 20 sets/h
Terrain decontamination — up to 10,000 m²/h
A DECOCONTAIN 1500 is 5 m long, 2.2 m wide and 2.2 m high; the weight is 4,900 kg. The operating temperature range is from −20 to +50°C.

Status
Available. In service with the Hungarian armed forces.

Manufacturer
Alfred Kärcher GmbH & Co.

VERIFIED

Kärcher DECOCONTAIN 3000 decontamination system

Description
The Kärcher DECOCONTAIN 3000 decontamination system was designed for deployment at battalion level or higher and forms the technical basis for the establishment of a decontamination site. The DECOCONTAIN 3000 is a new generation compact and high-performance system integrated into a standard 6 m ISO container. Using the DECOCONTAIN 3000 practically all essential decontamination operations can be performed simultaneously following an NBC attack, the main applications being the decontamination of material, personnel, clothing and equipment.

The independent diesel-powered Kärcher MPS 3200, DADS and MPDS modules are used for the pre-, main- and post-treatment of material decontamination with optimum results. The decontamination procedure used is dependent on the type of decontamination. Radioactive decontamination uses a hot foam procedure while disinfection uses various solutions. The detoxification of all known chemical warfare agents involves the Kärcher detoxification emulsion TDE 202.

Personnel decontamination is accomplished using a two-step, pulsating shower procedure in special decontamination tents. Detoxification and disinfection of clothing and equipment is accomplished using hot steam inside a collapsible tent. In addition, terrain decontamination can be carried out using an aqueous detoxification solution.

By using an integral 3,000 litre water tank the DECOCONTAIN 3000 can be used to perform relatively independent decontamination tasks.

Kärcher DECOCONTAIN 3000 decontamination system

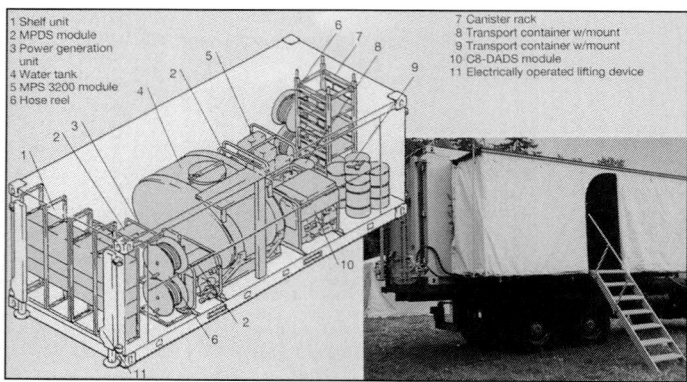

Diagrammatic outline of the Kärcher DECOCONTAIN 3000 decontamination system

Specification

Material decontamination: 6 to 8 tanks or 12 to 16 vehicles/h
Personnel decontamination: approximately 120 persons/h
Decontamination of clothing: 20 to 30 sets/h
Terrain decontamination: up to 10,000 m²/h
Length: 6.058 m
Width: 2.438 m
Height: 2.438 m
Weight: 11,500 kg
Operating temperature: −20 to +50°C

Status

In production. In service with NATO and Saudi Arabian forces and many civil defence organisations.

Manufacturer

Alfred Kärcher GmbH & Co.

VERIFIED

Kärcher DECOCONTAIN 3000 ELV decontamination system

Description

The Kärcher DECOCONTAIN 3000 ELV is an enhanced version of the DECOCONTAIN 3000 (see separate entry within this section). It is designed as a battalion-level facility, offering a wide range of DECON utilities. Its compact layout is based on a standard 6 m ISO container. The generous power output, decontaminant supply and delivery systems allow most DECON operations to be conducted simultaneously, including equipment DECON and personnel cleansing.

The central power generator guarantees the electrical power supply of all decontamination modules. The arrangement of the system depends on the DECON task detailed:

- Radioactive decontamination - hot foam
- Biological - appropriate disinfectant solutions
- Neutralisation of CW agents - Kärcher detoxification emulsion TDE 202.

For personnel cleansing, the system is linked to a container-based shower section (not shown). Terrain DECON is also effective with this system.

The integral 3000 litre capacity water tank allows the DECOCONTAIN 3000 ELV considerable endurance in detached DECON operations.

Status

In production. In service with the United Arab Emirates armed forces.

Manufacturer

Alfred Kärcher GmbH & Co.

VERIFIED

Specifications

Material decontamination: 12 to 16 tanks or 24 to 32 vehicles/h
Personnel cleansing: Up to 120 persons/h
Decontamination of clothing: 20 to 30 sets/h
Terrain decontamination: up to 5,000 m²/h
Length: 6.058 m
Width: 2.438 m
Height: 2.438 m
Weight: 11,500 kg
Operating temperature: −20 to +50°C

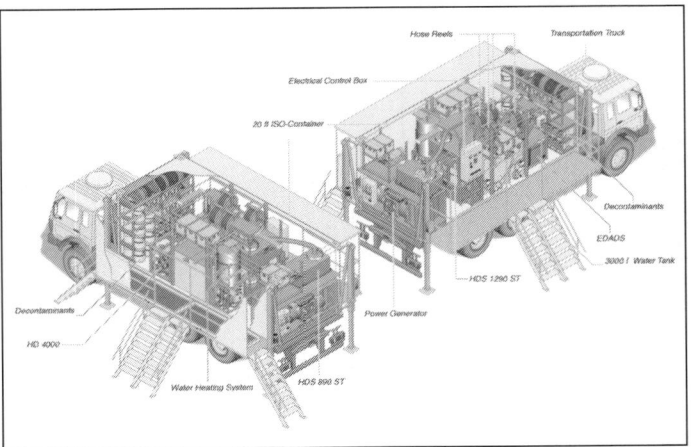

Kärcher *DECOCONTAIN 3000 ELV decontamination system*   **2001**/0098354

Kärcher DECOCONTAIN decontamination system

Description

With the Kärcher DECOCONTAIN decontamination system all the necessary equipment to meet most military decontamination requirements is integrated in a standard 6.096 m/20 ft ISO container — a typical DECOCONTAIN weighs 12 tonnes. Tasks that can be carried out using the DECOCONTAIN system include complete deradiation; disinfection and detoxification of material such as trucks, armoured vehicles and aircraft; complete decontamination of clothing and equipment; full decontamination of personnel; terrain decontamination and water recycling. The system can also be used as a firefighting and water supply system, for field showers, general maintenance and cleaning work using the integrated high-pressure steam jet cleaners. In addition, the system can be used for de-waxing and de-icing material and steaming explosive from various forms of munitions.

By placing the various equipment and components inside a standard container, the DECOCONTAIN system can be readily moved by standard handling systems and transport methods to wherever it is required. The container is fitted with interchangeable corner fittings and is normally located on two pairs of skids for locking the container onto a carrier vehicle and for loading by crane or helicopter. Various container options are available including an NBC filter ventilation system with airlocks and equipment designed to produce an internal overpressure. Features such as EMP protection, heat insulation and fire protection are integral.

The DECOCONTAIN is divided into the following sections: supply; equipment; water treatment; water supply; air-conditioner and NBC air supply; shower and an ancillaries section. A gas turbine is integrated into the container to drive a three-phase generator and provide hot exhaust gases for clothing decontamination.

A typical DECOCONTAIN carries an MPS 3200 module, a DADS module and an MPDS module (see entries in this section for details). Water treatment is carried out by a WATERCLEAN 1000 unit with a capacity of approximately 1,000 litres/h (see entry under PROTECTION (individual) — Medical countermeasures for details). Clothing decontamination is carried out in a separate tent using the exhaust from the gas turbine used to provide power for electrical generation. Two further tents are carried for use during personnel decontamination. All three tents are identical, have double walls and are supported on inflatable tubular frames. Two people can erect one tent in about 5 minutes.

The following are typical operating capacities for a DECOCONTAIN system:
Decontaminating tanks or heavy trucks — six to eight tanks or 12 to 16 vehicles/h
Personnel decontamination in a non-contaminated area — approximately 120 persons/h
Decontamination of clothing and equipment — up to 60 sets/h
Water recycling — approximately 1,000 litres/h.
A DECOCONTAIN is 6.698 m long, 2.418 m wide and 2.585 m high when transported.

Status

In production. In service with the Portuguese Air Force.

Manufacturer

Alfred Kärcher GmbH & Co.

VERIFIED

Kärcher *DECOCONTAIN decontamination system being prepared for transport on a 6 × 6 truck*

Kärcher *DECOCONTAIN decontamination system in use*

Kärcher Decojet decontamination system

Description

The Kärcher Decojet is a mobile, self-contained decontamination system integrated in a single light metal frame carried on a Mercedes-Benz, Peugeot P4 4 × 4 or similar light vehicle. It is used as a quick reaction first aid decontamination system close to the front line or as a company level self-decontamination measure. It can also be used in conjunction with the Kärcher DECOCONTAIN system at locations such as airbases for decontaminating vehicles, aircraft and their weapon systems.

The Decojet system can be used to decontaminate personnel using a pulsating shower apparatus which is part of the system; it provides a prewash for material decontamination, post treatment for material decontamination with hot steam (140°C), the main decontamination treatment for material using an injector for prepared decontamination solutions (such as calcium hypochlorite and water), for deradiation and disinfection with specific agents, decontamination of clothing and equipment in an evaporating container, for terrain decontamination and for daily high-pressure and steam cleaning purposes.

The complete Decojet system consists of a Kärcher MultiPurpose Decontamination System (MPDS — see entry in this section) module with a cold and hot water circuit using a two-spray lance operating principle, a 435 litre water tank and run-back pipe preheating the water in the tank, an additional 200 litre water tank with an injector system for mixing and applying decontamination solutions, hose reels for prewash and post treatment, two-stage personnel shower with an injection system for adding chemicals to the water jets, boiling vessel for clothing and equipment decontamination, decontamination chemicals such as calcium hypochlorite for material decontamination, RM21 chemicals for personnel decontamination and RM54 for deradiation and material decontamination. The system is completed by various spray lances, tools and other accessories.

All operations are controlled from a central operating panel at the rear of the Decojet unit.

Specifications

Weight: 1,055 kg
Length: 1.653 m
Width: 1.41 m
Height: 1.24 m
Water tank capacity: (main) 435 litres
(additional) 200 litres

Status

In production. In service with France, various countries in the Middle East and Asia.

Manufacturer

Alfred Kärcher GmbH & Co.

VERIFIED

Kärcher Decojet trailer ADT-1000

Description

The Kärcher Decojet trailer ADT-1000 is a new mobile decontamination system, mounted on a single-axle trailer. The system is designed for the simultaneous decontamination of vehicles, material, personnel and equipment. Its integrated nozzle system allows terrain decontamination as well. Mounted on the trailer itself, there are three modules - a MultiPurpose Decontamination System (MPDS), a Motor Pump System (MPS) and a decontamination emulsion generator (DADS). A water tank, holding 1,000 litres of water, a shower system and all the necessary chemicals and accessories for the NBC decontamination task are located on the

Specifications

2-axle trailer with detachable platform
Weight: 3,000 kg
Length: 5,000 mm
Width: 2,300 mm
Height: 2,700 mm
Volume: 1.67 m³

Platform dimensions
Length: 1,600 mm
Width: 1,900 mm
Height: 1,200 mm

Kärcher Decojet ready for transport on Peugeot P4 (4 × 4) light vehicle

Kärcher Decojet being used to decontaminate trucks

Kärcher Decojet providing showers for personnel decontamination

Kärcher Decojet-trailer decontamination system

Decojet-trailer decontamination system in operation cleansing heavy trucks

trailer and its platform. This allows fully autonomous operation from the outset as soon as the system is set up on site. The detachable platform mounted on the rear of the trailer holds a second DADS and MPDS, offering further flexibility in operation. The platform can be detached and redeployed by vehicle to operate independently at a separate DECON site. While still attached to the trailer, it doubles the output performance. The system can also be used for firefighting or general cleaning and maintenance, with a flow rate of up to 11,000 litres/h of water.

Status
In production.

Manufacturer
Alfred Kärcher GmbH & Co.

VERIFIED

Kärcher Showerjet 15 field shower unit

Kärcher decontamination accessories

In addition to the decontamination units and systems covered in the previous entries, Alfred Kärcher GmbH & Co also produces the following associated units which can be used as part of various decontamination processes.

MPS 3200 motor pump set
The MPS 3200 motor pump set can be used as part of any decontamination system but is normally used for the prewash stage to remove dirt prior to the application of decontamination chemicals. It allows the operation of two spray lances at one time with a water flow rate of 1,600 litres/h for each lance. If required, one of the lances can be replaced by a shower unit. The MPS 3200 can also be used for firefighting, in which case the flow rate can be increased to 11,000 litres/h by a high-capacity injector. The high-pressure pump is sea and waste water-resistant and can raise water up to a maximum suction height of 5 m. All components are integrated in a light metal frame and all functions are controlled automatically.

The MPS 3200 weighs 280 kg and dimensions are 0.96 × 0.7 × 1 m.

Field shower unit
This is a two-stage shower unit for personnel decontamination. It can be supplied with water by the Kärcher MPDS MultiPurpose Decontamination System jet or the HDS 1200 EK high-pressure steam cleaner unit. The five shower nozzles operate to two sides, so one side may be used to deliver water with a decontaminant chemical such as RM21 (which does not irritate the skin) while the other side can be used for rinsing with clean water. Shower water is prepared by an injector where heated water and cold water are blended to produce comfortably warm shower water. A heat-compensation reservoir in the shower base regulates the temperature which is set by a thermostat. Pulsating nozzles are used.

The weight of the shower unit is approximately 48 kg and it is 2 m high.

Showerjet 15
The Showerjet 15 can decontaminate up to 15 people simultaneously and can be operated with the Kärcher MPDS, HDS 1500 D or HDS 1200 hot water, high-pressure modules. The shower contains an injector kit with a temperature safety valve to ensure a safe and agreeable shower temperature. An integrated chemical injector permits the addition of personal decontamination solvents (for example RM21) and decontamination agent consumption is up to 2 litres/min. The water flow rate is 3,400 litres/h at an operating pressure of 5 bar.

Collapsible water tank
The Kärcher collapsible water tank has a maximum capacity of 2,500 litres and can be used to supply fresh or drinking water to a decontamination system or unit. It weighs 26 kg when empty and when full measures 2.3 × 1.5 × 0.7 m.

Status
All the above units are in production. The MPS 3200 is in service with NATO headquarters, Australia, Austria, Portugal and countries in North Africa and the Middle East. The shower unit and collapsible water tank are in service in several countries. The Showerjet 15 is in service with the British and New Zealand armies and with some other countries.

Manufacturer
Alfred Kärcher GmbH & Co.

VERIFIED

Kärcher decontamination and cleaning agents

Description
Kärcher decontamination and cleaning agents were developed and tested in accordance with NATO standards and guarantee the necessary safety involved in decontamination, disinfection and detoxification following the use of NBC weapons. Included in the range are the following agents:

RM21
A very mild and dermatologically tested but effective cleaning agent for all kinds of routine decontamination.

Kärcher MPS 3200 motor pump set

Packs of Kärcher decontamination and cleaning agents   ***2001**/0098391*

Selection of decontaminant: Kärcher decontamination and cleaning agents

	Material	Clothing	Equipment	Personnel	Interior	Terrain
Radioactive DECON	RM 54		RM 54	RM 21		
BW DECON	Di 60 + RM 54 *or* RM 35	RM 35	RM 35	RM 21	RM 35	Di 60 + RM 54 *or* RM 35
CW DECON	TDE 202	RM 21	TDE 202 or RM 31	RM 21	RM 21	Di 60 + RM 54

RM31

A liquid alkaline agent used for removing persistent contamination.

RM35

A non-abrasive, extremely effective disinfecting agent for sanitation.

RM54

A foam cleaner for cleaning sensitive weapon systems and for radioactive decontamination. It is non-abrasive and completely biodegradable.

Di60

An effective decontamination agent for the detoxification and disinfection of terrain.

Detoxification emulsion TDE 202

TDE 202 is a patented universal decontamination emulsion. It is highly effective against all known chemical warfare agents and warfare agents with thickener, even if those agents have penetrated paintwork. TDE 202 is tetrachloroethylene-free and can also be used in winter. Emulsion TDE 202 is stable for a minimum of 48 hours and provides improved environmental compatibility in comparison with conventional decontamination emulsions.

Training detoxification agent

Training detoxification agent is employed during field exercises as a simulant for detoxification emulsions (TDE 202 German emulsion).

Status

All the above agents are in production and are in service worldwide.

Manufacturer

Alfred Kärcher GmbH & Co.

VERIFIED

Kärcher decontamination trailer

Description

The Kärcher decontamination trailer is a custom made system to meet customer requirements for a mobile decontamination system. The system consists of two subsystems; the trailer with the decontamination modules fixed on to the chassis and a removable platform, also equipped with decontamination modules. Both the trailer and the removable platform are equipped with all the necessary accessories, hoses and chemicals to commence decontamination operations directly on arrival at a site.

The trailer-mounted subsystem is designed for the decontamination of vehicles, material, personnel and protective equipment. For a simultaneous treatment of vehicles three different modules are mounted; a cold water high-pressure unit for the prewash, a decontamination emulsion generator for the main treatment and a steam generator for the post-treatment with hot steam. For personnel decontamination there is a shower unit and decontamination tent carried on the trailer. Warm water is generated by a second steam generator. A 1,000 litre water tank ensures an immediate start-up and independent operations on site.

The platform subsystem is primarily designed for the decontamination of vehicles and material. If required, this subsystem can be lifted from the trailer and mounted on a truck platform. Two modules are provided, a decontamination emulsion generator for the main treatment and a steam generator providing cold water for the pretreatment stage and steam for the post-treatment.

Status

Available. In service within Europe.

Manufacturer

Alfred Kärcher GmbH & Co.

VERIFIED

Specifications

Weight: approx 5,000 kg
Length: (incl towbar) 6.5 m
Width: 2.3 m
Height: 2.6 m

Kärcher Decont Jet 21

Description

The Kärcher Decont Jet 21 decontamination system for large vehicles is a sophisticated solution to the problem of quickly reducing the contamination hazard from a squadron of vehicles in the field. It is part of a range of equipment developed by Kärcher which offers the simultaneous decontamination of personnel, field equipment and small or sensitive material, following the contamination of a unit after an NBC attack. See several other entries in this section.

The Decont Jet 21 mobile transporter deploys a bridge gantry, large enough to allow the largest type of field vehicle to pass beneath. The gantry, supported and deployed hydraulically from vehicle power, delivers decontaminant at the required temperature and pressure through downward and sideways-pointing nozzles. The transporter vehicle has ground support feet to add stability and the decontamination process is controlled by an operator housed in a COLPRO cabin at the rear of the vehicle. By use of levers, the operator can change the target area for decontamination, moving the gantry over a static vehicle, or allowing vehicles to drive underneath. Also, individual nozzles are electrically rotated to ensure full coverage and 'smart decontamination' is achieved by laser-based vehicle contour recognition. The operator can direct either decontaminant fluid at high pressure or hot gas at the target vehicle.

Kärcher Decont Jet 21 decontamination system for large vehicles   0055098

Kärcher Decont Jet 21 decontamination system for large vehicles   0055099

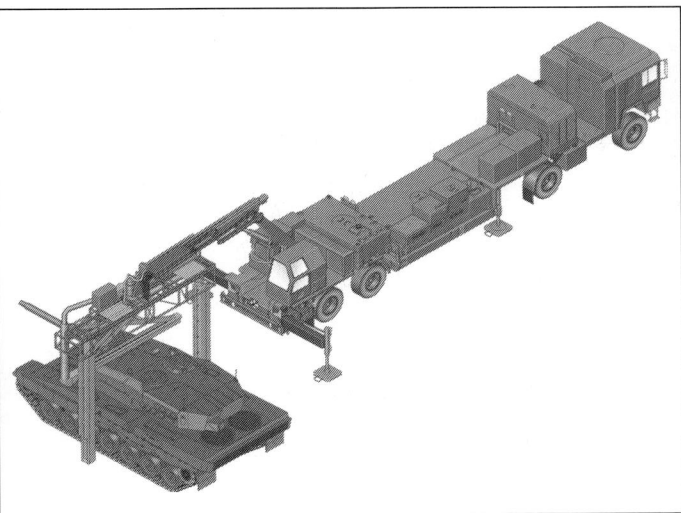

Kärcher Decont Jet 21 decontamination system for large vehicles 0055097

Specifications
Tank capacity (water): 6,000 litre
Tank capacity (decontaminant): 300 litre
Gas generation: 2,000 litre fuel tank
Dimensions: 12,700 mm (L), 2,500 mm (W), 3,500 mm (H)
Weight (unladen): 16,000 kg
Weight (laden): 27,000 kg

With a crew of two, driver and operator, the system is capable of thoroughly decontaminating up to six tanks or 10 wheeled vehicles per hour. Decont Jet 21 is air portable and can be ready to operate 15 minutes after arrival.

Status
Available. In service with the German armed forces as part of the HEP 90 programme (Main Decontamination Site of the Nineties). This programme includes a comprehensive range of equipment for simultaneous unit-level personnel and equipment decontamination. See separate entries in this section, including the DSSM system for sensitive material and the DaimlerChrysler Aerospace AG Mobile NBC decontamination semi-trailer.

Manufacturer
Alfred Kärcher GmbH & Co.

VERIFIED

Kärcher Decont Jet 21 ST fixed DECON facility for air and naval bases

Description
The Kärcher Decont Jet 21 ST decontamination system uses a similar technique to the Decont Jet 21. In this case, the facility is designed around a collection pit covered with a strengthened steel grid over which vehicles drive. Two pipes along the length of the pit are fitted with nozzles at intervals which spray DECON fluid onto the underside of the target vehicle. The gantry has evolved from the successful mobile Decont Jet 21 design, with several performance improvements. Spray nozzles on the gantry direct DECON fluid from supply tanks at the vehicle via

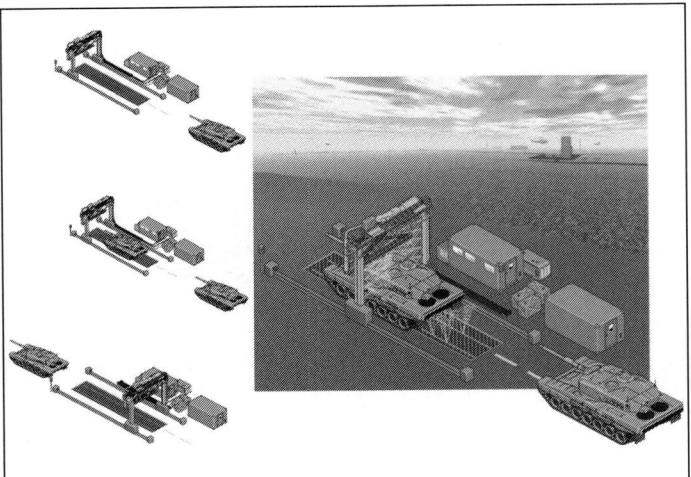

Kärcher Decont Jet 21 ST fixed site DECON system (design concept) **2001**/0098364

Specifications
Dimensions: 21 (L) × 6.5 (W) × 6 (H) m
Average throughput: 6 tank-sized tracked vehicles or 10 wheeled vehicles per hour

a series of spray nozzles. The contaminated residue is collected in the pit and removed for safe disposal.

Status
Available.

Manufacturer
Alfred Kärcher GmbH & Co.

VERIFIED

Kärcher Decont Shuttle

Description
The Decont Shuttle is a compact and effective DECON facility aimed at the decontamination of sensitive equipment inside enclosed spaces. Applications include the decontamination of helicopter and fixed-wing aircraft cockpits, C⁴I systems and tank crew compartments. This unique system applies decontaminant under relatively high temperature and pressure. Simultaneously, it removes the contaminated residue through a spray-extraction system sited alongside the applicator, serviced by a vacuum-pump. Mounted on a tracked chassis with its own prime-mover, the Decont Shuttle is powered by a small diesel engine and is operated independently by one man. The 'eco-friendly' decontamination removal device minimises the spread of contamination inside the target or to neighbouring areas.

The tank, located on top of the chassis, is divided internally, containing the decontaminant solution in one compartment. Whilst operating, the second compartment receives the contaminated residue. The Kärcher steam extraction system, aerosol generator or vacuum cleaner (see separate entries within this section) can also be used with the Decont Shuttle via an external connection.

Specifications
Dimensions: 1,600 × 800 × 1,500 mm
Transport mass: Approx 300 kg (500 kg with DECON solution)
Method: Combined applicator and spray extractor
Connection: Aerosol generator, steam extraction unit and vacuum cleaner
Max speed: 6 km/h
Operational temperature range: −32 to +49°C

Kärcher Decont Shuttle mobile DECON system for decontamination of interior/sensitive equipment and facilities **2001**/0098382

Kärcher Decont Shuttle mobile DECON system is optimised for one-man operation
2001/0098363

Status

Under field evaluation by the German Armed Forces.

Manufacturer

Alfred Kärcher GmbH & Co.

VERIFIED

Kärcher Decont tent

Description

The Kärcher decont tent is a tent system designed for decontamination tasks. The tent with its inflatable tubular frame can be pitched in 5 minutes using a manual air pump supplied with the tent. Once set up the tent can be used for weather protection when decontaminating personnel, in which case a field shower system is set up inside the tent, or for the decontamination of clothing and equipment. In the latter case saturated steam (at temperatures up to 210°C) or hot gas is fed into the tent using either the Kärcher MultiPurpose Decontamination System (MPDS - see separate entry in this section) or a similar module.

The tent tubular frame is reinforced over the floor area and has a safety overflow valve. The tarpaulin used in the tent construction is resistant to chemical warfare agents and can be decontaminated. It has a watertight inner lining and, a canvas groundsheet doubles as a cover for the folded tent when it is being transported or stored. The tent seams are sewn and treated with a high-temperature resistant adhesive. Two hose connections are provided at opposite corners for waste water disposal. Once erected the tent is fastened securely by four ground loops and guy lines.

Status

In production. In service with several countries.

Manufacturer

Alfred Kärcher GmbH & Co.

UPDATED

Kärcher HDS 1200 EK high-pressure steam jet cleaner unit

Description

The Kärcher HDS 1200 EK high-pressure steam jet cleaner unit can be used for a variety of cleaning, maintenance and decontamination tasks. The unit can produce cold or hot water, steam or dry steam and is powered by a 4 kW electric motor.

On the HDS 1200 EK, the electric motor is switched on as the spray gun trigger is squeezed and switched off as it is released. This drive system ensures that the system pump operates only when it is needed, to save energy and extend service life. The system uses a high-pressure (up to 50 bar) pump producing enough suction to raise water up to 5 m from streams and rivers. The pump feed rate is between 240 and 1,200 litres/h and decontamination or other chemicals can be introduced via the pump at a rate of up to 60 litres/h. The heater burner fuel consumption is controlled via a closed-circuit system: diesel fuel is injected when the burner is on and returned to the fuel tank when the burner is turned off. A continuous duty ignition system is employed. An integral electronic water-softening unit is provided to prevent scale forming on the heater coil, this is adjustable to suit local water conditions.

There are two independent safety systems. A thermostat and low-water cutout prevents thermal overload and a pressure switch and safety valve prevent overpressure.

All components are mounted on skids with fold-down wheels. The entire unit is enclosed in a corrosion-resistant casing.

Specification

Components: Decont tent unit including hand air pump, 1 bag with 8 tent pegs
Environmental
Outer tent: −30 to +80°C
Inner tent: −30 to +140°C
Dimensions (erected): 2.15 × 2.36 × 2.40 m
Useable space: 2.00 × 2.00 × 2.15 m
Weight: 43 kg

Specifications

Weight: (HDS 1200 EK) 280 kg
Length: 1.45 m
Width: 0.75 m
Height: (without exhaust pipe) 1.11 m
Pressure range: 15 to 50 bar

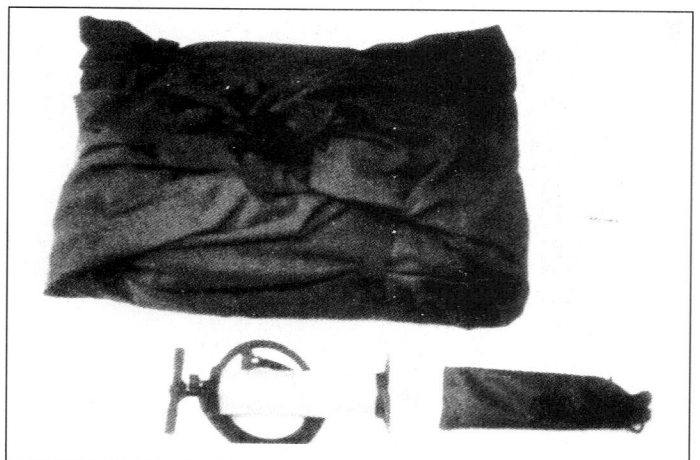

Kärcher decont tent ready for transport

Kärcher HDS 1200 EK high-pressure steam jet cleaner unit

As well as being used for general decontamination and cleaning tasks the unit can be adapted for de-icing aircraft and missile systems, for cleaning fuel tanks and steaming explosives from munitions. The unit can also be used to supply water for field showers.

Status

In production. In service with the German Armed Forces and 40 armed forces worldwide.

Manufacturer

Alfred Kärcher GmbH & Co.

VERIFIED

Kärcher hot air generator FB 20

Description

The Kärcher FB 20 hot air generator was designed for the decontamination of clothing and other equipment placed inside a decontamination chamber or tent. The system is a compact and handy unit. It can be transported easily, as the entire control unit, fan and burner is totally covered with a shell and integrated into a tubular frame system. The system can be employed under all climatic conditions.

The hot air generator has a heating capacity of 15 kW/51,000 BTU and an air flow rate of 830 m³/h; maximum output temperature is +90°C. The fuel consumption of the burner is 2.2 litres/h using diesel fuel for the heater. Electrical power requirements are 230 V 50 Hz.

Status

In production. In service with Canada, Denmark, Germany, Norway, Sweden, USA and other NATO forces.

Manufacturer

Alfred Kärcher GmbH & Co.

VERIFIED

Specifications
Weight: 78 kg
Length: 1.355 m
Width: 0.46 m
Height: 0.605 m

Kärcher FB 20 hot air generator

Kärcher hot air generator FB 60 E

Description

The Kärcher hot air generator FB 60 E was designed for the hot air decontamination of indoor areas as well as NBC protective clothing and other equipment. It can also be used for the ventilation of NBC decontamination stations.

Decontamination of interiors is carried out using the FB 60 E and an interior decontamination extension set with special NBC filters. Vehicles, aircraft and vessels can be decontaminated on a purely physical basis by heat application with a remote thermostat controlling the required temperature. Clothing and equipment can be effectively decontaminated using the hot gas/hot steam process. For example, NBC protective garments may be hung in the Kärcher Decont Tent (see entry in this section) treated with hot air and a small quantity of hot steam.

The FB 60 E can also be used as a hot air generator to heat cabins, cockpits, tents or shelters and during Summer it can be used as a fan to cool these installations.

All the main components of the FB 60 E are installed in a torsionally stiff tubular aluminium frame with pick-up points for forklifts or air transport. It can be easily moved by one person using the integral handles and wheels.

The main fan and fuel pump are driven by a central 220 V electric motor. Fresh air is drawn in via a radial fan and is then heated by a burner and heat exchanger. Optimised combustion is achieved by the specially shaped burner sphere, which is made of high temperature-resistant material, as is the heat exchanger; both are gastight welded to ensure that no combustion gases are mixed with the heated fresh air.

Specifications
Weight: 200 kg
Length: 1.72 m
Width: 0.74 m
Height: 0.91 m
Nominal burner heating capacity: 60 kW
Nominal radiator air flow: 3,000 m³/h
Fuel consumption: 7.5 litres/h

Kärcher hot air generator FB 60 E

Kärcher hot air generator FB 60 E with accessories

Due to its two integrated fuel heaters the FB 60 E is able to work in temperatures down to −30°C. The generator can operate using diesel, diesel-petrol and kerosene fuels.

Status
In production. In service with the Australian, German and US Armed Forces.

Manufacturer
Alfred Kärcher GmbH & Co.

VERIFIED

Kärcher M600 Decontaminant Mixer System

Description
The Kärcher M600 Decontaminant Mixer System allows both powdered decontaminant and water to be metered, mixed together in the correct proportions and delivered to a high-pressure decontamination system, online. The tubular frame has a wide-mouthed filling hopper for the DECON powder (holds 15 kg), a feed screw, relay station, safety devices and an operator's control panel on the front. The illustrations show the M600 in operation, connected to a high-pressure DECON system. The hopper, electrical supply, water supply and output to the DECON system via a 25 m hose can also be seen. A spray lance is included with this equipment, allowing the user to start or stop the application process at will. The user can allow the system to be rinsed automatically from time to time, using a control device on the lance, to prevent clogging by the powder. The powder/water ratio can be adjusted on the control panel. The designed water flow rate is 10 litres/min and powder is added at a recommended rate of 2 kg/min.

M600 Decontaminant Mixer System in operation

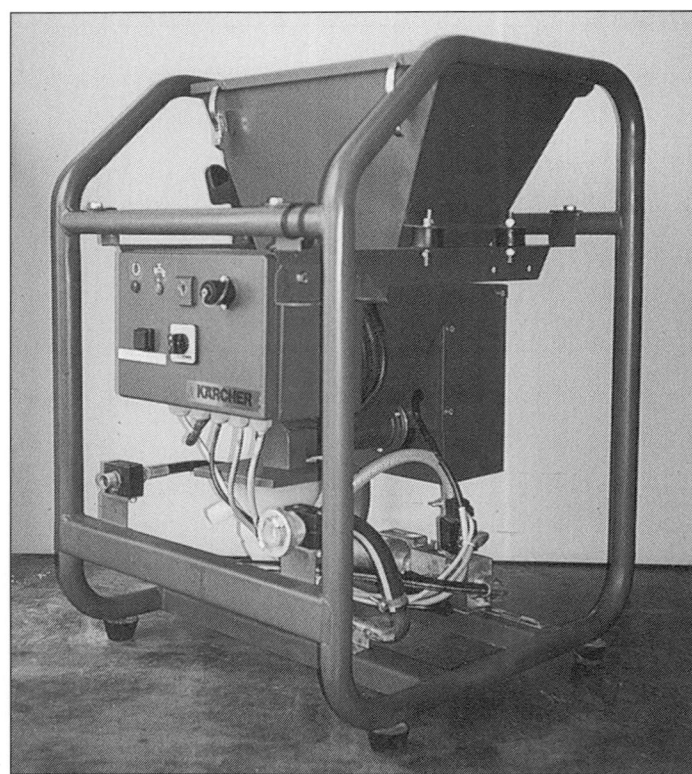

M600 Decontaminant Mixer System

Specifications
Weight: 40 kg
Length: 560 mm
Width: 600 mm
Height: 840 mm
Power supply: 12 or 24 V

Status
In production. In service within NATO.

Manufacturer
Alfred Kärcher GmbH & Co.

VERIFIED

Kärcher mobile field laundry CFL 60

Description
The Kärcher mobile field laundry CFL 60 was developed in close co-operation with the German armed forces with regard to the special requirements of washing and disinfecting all kinds of laundry, including clothing, coming from all arms of the services; mobile hospital and emergency service requirements have also been taken into consideration. All essential laundering needs can be met including disinfection, washing, drying, ironing, folding and packing. The CFL 60 can be used to decontaminate biologically contaminated clothing once appropriate safety precautions have been taken.

The strict separation between dirty and clean areas makes this equipment suitable for the treatment of items suspected of contamination with BW agent.

The CFL 60 was successfully tested by German armed forces operating with the UN in Somalia; 38 systems were subsequently ordered by them and are in service.

The CFL 60 is installed in a 6.096 m/20 ft ISO container which can be secured to a truck or trailer with a twist-lock system. The system is air-transportable in C-130 or C-160 Transall transport aircraft. It may be handled using an electrical device with four corner supports and may be utilised when placed directly onto the ground. Once installed, the system is self-contained apart from the provision of water supplies for the top-up reserve. Power is supplied by an integral 125 kVA electrical generator. When used in conjunction with the Kärcher WATERCLEAN 1000 water

Kärcher mobile field laundry CFL 60 with washing section open

Loading laundry into the 'dirty wash' end of a CFL 60 mobile field laundry

treatment system (see entry this section), the mobile laundry system can be utilised independent of available water supply quality.

Once emplaced, the container may be opened up to form attached tents and covered working areas. The container carries laundry items such as folding work tables, mangles, ironing boards and handling containers.

The CFL 60 utilises the latest hospital and household laundering and cleaning technologies. It ensures a strict separation of contaminated or dirty sections and clean sections inside the container. The required washing process is initiated by the selection of the appropriate program on an electronic operation panel and all operations are then carried out automatically. Water consumption is reduced by means of a water recycling plant, an integral component of the CFL 60 and an important contribution to the system's independent operation and environmental safeguards. The dosage of liquid washing agents and disinfectants is regulated automatically. Effective sound insulation as well as a sound-insulated generator maintain noise at a low level. All machines, devices and components are located for ease and rapidity of maintenance.

The CFL 60 system can be used under extreme climatic conditions by the provision of air conditioning in tropical climates and integrated auxiliary heating in Arctic zones.

Status

In production. In service with the German and Norwegian armed forces in Bosnia (IFOR) and elsewhere.

Manufacturer

Alfred Kärcher GmbH & Co.

VERIFIED

Specifications
Weight: approx 10,000 kg
Washing capacity: up to 60 kg dry laundry/h
Drying rate: up to 60 kg/h
Generator: 124 kVA

Kärcher containerised CFL 60 mobile field laundry mounted on lorry chassis

The washing end of a Kärcher mobile field laundry CFL 60 with protective covers in place

Kärcher MultiPurpose Decontamination System (MPDS)

Description

The Kärcher MultiPurpose Decontamination System (MPDS) evolved from the Kärcher HDS 1200 BK high-pressure steam cleaning system (see separate entry this section). It is a self-contained high-pressure (60 bar) system that can be operated using cold water, hot water (80°C), hot steam (140°C) or dry steam (210°C).

The diesel engine and burner draw fuel from a NATO jerrican. A pre-heater eases starting at low ambient temperatures. The system uses a double-walled high-performance burner with the flame being electronically controlled via a photocell. There is a high-pressure water pump with a maximum suction height of 5 m, allowing water to be drawn from a nearby water source such as a stream or river. The pump is also seawater-resistant. Liquid DECON chemicals are drawn from a reservoir via the high-pressure pump up to a rate of 60 litres/h. A single main switch allows pre-selected operations to initiate and all user functions are centralised on an electrical control panel located at the front of the lightweight metal carrying frame. The engine is started by and electric starter motor, powered by a 12 V battery carried on the frame, but can also be started by hand.

DECON operations use hot steam (140°C) through a steam spray nozzle. DECON fluid is added via the high-pressure pump and, with the introduction of an injector accessory kit, more aggressive chemicals such as calcium hypochlorite can be used. Processes such as decontaminating personal equipment, structural interiors, NBC protective clothing and respirators, are carried out using dry steam at 210°C.

For radiological DECON, a special foam nozzle and a rotating brush can be used. Using chemicals such as RM54 foam cleaner and a calcium inhibitor it is possible to carry out an effective foam treatment.

The MPDS can also be used to supply field showers and for many maintenance and general cleaning tasks. It can also be used to steam explosive from munitions and for de-icing aircraft and missile systems. A sandblasting set is available for removing corrosion from equipment.

Status

In production. In service with Australia, Austria, Canada, Portugal, Sweden, UK, USA, NATO Headquarters and countries in North Africa and the Middle East. MPDS equips the British Army's NBC Defence Regiment (formed 1994). The UK MPDS is to receive a mid-life improvement programme from NBC Team Limited of Canada.

Manufacturer

Alfred Kärcher GmbH & Co.

UPDATED

Specifications
Weight: 220 kg
Length: 1.25 m
Width: 0.575 m
Height: 0.85 m
Engine: 4-stroke diesel developing 5.5 hp at 3,000 rpm
Operating temperature: −30 to +60°C

MPDS in use. Two units can be seen mounted on the back of a UK Drops trailer and supplying the operator in the foreground who is decontaminating a Fuchs reconnaissance vehicle (see entry in DETECTION (reconnaissance systems) (Crown copyright 2000)   **2001**/0055104

Kärcher portable lightweight decontamination system DS 10

Description

The Kärcher portable lightweight decontamination system DS 10, also known as the Decosprayer DS 10, has a filling capacity of 10 litres. It was designed for the NBC decontamination of vehicles, aircraft, protective suits and other equipment.

The DS 10 is equipped with an integrated mixing device to generate different solutions or emulsions for NBC decontamination. These solutions or emulsions can include hypochlorites, various solvents, emulsifiers, salts and water. The DS 10 consists of a pressure tank, a mixing device and an air pump to pressurise the

Specifications

Weight: 9.5 kg
Height: 704 mm
Diameter: 210 mm
Capacity:
  (total) 15 litres
  (filling) 10 litres
Max operating pressure: 6 bar
Max operating temperature: +60°C
Application area: 50 m²

system. All items are made from a special chemical-resistant stainless steel. The gauge, hand spray gun, application device and all sealing materials are also resistant to the most commonly used decontamination agents.

The mixing device is directly driven and operated by the hand pump so that liquids and powders can be thoroughly mixed or emulsified. Mixtures can be dispensed with the hand-held trigger gun after the container has been pressurised. The container can be easily refilled once emptied. A pressure relief valve will open if the maximum operating pressure (6 bar) is exceeded.

The DS 10 is manually operated and can treat an area of at least 50 m² with one filling. Once filled and pressurised it is possible to work for more than 5 minutes without interruption. The operating pressure is sufficient to empty the container, leaving no residual material.

The DS 10 is supplied with an application hose, trigger gun, spray lance and nozzle. Carrying straps are available. Other accessories include a vehicle mounting kit, a trolley and special nozzles.

Status
In production. In service with Austria, Belgium, Germany, Norway, Sweden and some countries in the Middle East.

Manufacturer
Alfred Kärcher GmbH & Co.

VERIFIED

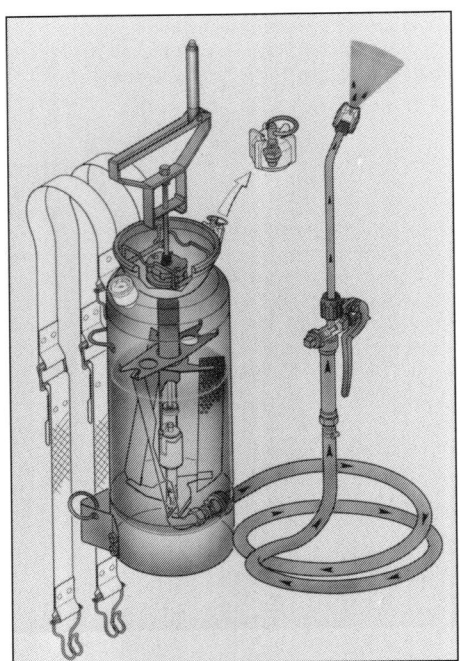

Cutaway of Kärcher portable lightweight decontamination system DS 10

Kärcher portable lightweight decontamination system DS 10 in use

Kärcher Rapid Intervention Lightweight Decontamination System RILDS

Description
The Kärcher RILDS system is designed to offer, in one transportable framed unit, all the equipment and materials required to set up and run a quick-response DECON utility for personnel and equipment. Fork lift pockets are fitted for easy handling and it can be mounted on most heavy-duty types of trucks or trailer. It is especially designed for deployment at battalion level or higher and forms the technical basis for establishing a full decontamination site. Features include:

- Decont module MPDS for carrying out the decont task
- Hot steam decontamination with up to 140°C/280°F
- 2,500 litre/660 g collapsible water tank for independent operation
- Decontamination tent, inflatable, for the personnel field shower
- Field shower for decontamination of personnel
- Engine start capability down to –30°C/22°F.

Status
In production. In service with the United Arab Emirates Armed Forces.

Manufacturer
Alfred Kärcher GmbH & Co.

VERIFIED

Kärcher SCS 1800 DE decontamination system

Description
The Kärcher SCS 1800 DE is a high-performance decontamination module developed for an increased output volume of high-pressure water with decontamination agents, steam and superheated steam. The system is suitable for the generation of a hot foam treatment for nuclear decontamination. Chemical and biological decontamination is accomplished, by generating and applying decontamination solutions, using integrated chemical mixing injectors. Personnel decontamination can be supported by supplying shower units supplied with warm water and chemicals.

The self-contained system consists of a diesel engine, a high-pressure pump, two different injector systems, a blower, heat exchanger and safety devices. All

Kärcher SCS 1800 DE decontamination system   0103641

Kärcher SCS 1800 DE decontamination system in use to cleanse an aircraft
0103856

Specifications
Weight: 410 kg
Length: 1.1 m
Width: 0.85 m
Height: 1.035 m
Engine: diesel
Electrical system: 24 V

these components are integrated in a solid frame system which is totally covered. The module is powered by a four-stroke diesel engine. The water output rate and pressure are variable, with the output rate being from 300 to 1,800 litres/h and the pressure range from 20 to 110 bar.

The engine is started electrically. There is an automatic reduction of engine speed to the minimum rpm when the trigger gun applicator is switched off. This keeps noise levels low and prolongs the service life of the engine and water pump.

Water is heated via an upright standing coil, with anti-condensation heat transfer, in the heat exchanger unit. A continuous ignition system and automatic burner shutdown in the event of a low heating oil level combine to provide system safety. An integrated electronic water softener unit is provided to prevent scaling in the heating coil and is adjustable according to local water conditions.

A chemical intake injector allows the addition of liquid detergents to the water flow. A second injector, made of a special chemical-resistant stainless steel, is suitable for the addition of aggressive decontamination chemicals. An integrated high-volume water output injector allows the thorough cleaning of tracked vehicles and a quick water tank refill. This injector system can also be used for firefighting, using an output rate of 5,000 litres/h.

The system generates pressurised cold water, hot water, hot steam and dry steam. Maximum steam output temperature is 140°C while the dry steam temperature in the system is 200°C. Two spray lances can be used with the system.

The SCS 1800 DE can be set up and operated by one person using an instrument panel. Integrated retractable levers and a wheel set make the unit mobile and manoeuvrable. The unit can be air-transported as an internal or external load.

Status
In production. In service within Europe.

Manufacturer
Alfred Kärcher GmbH & Co.

VERIFIED

Kärcher Team decontamination kit

Description
The Kärcher Team decontamination kit is specially designed for decontamination activities by inspection teams and is in service with OPCW. Compactly arranged, the constituents can be transported in two large 40 kg boxes. The kit includes the DS-10 portable lightweight decontamination system (see separate entry this section) and sufficient decontaminant chemicals for three operations against N, B or CW agents.

Specifications
Length: 800 mm
Width: 600 mm
Height: 400 mm
Weight: 40 kg (each box)

Team decontamination kit ***2000**/0103264*

Status
In production. In service with OPCW, Netherlands armed forces and others.

Manufacturer
Alfred Kärcher GmbH & Co.

VERIFIED

Kärcher USC AB-DEKO Mobile Decontamination System

Description
The Kärcher USC AB-DEKO is a universal mobile DECON system aimed at the domestic preparedness market. The specially equipped ISO container, mounted on a 4 × 4 truck chassis, is internally arranged to allow the decontamination and cleansing of contaminated personnel, following an incident involving a toxic hazard. Simultaneously, it can decontaminate, wash and dry between four and eight sets of contaminated CW protective clothing per hour. It can also decontaminate equipment or vehicles at the rate of between four and eight vehicles per hour.

Status
In production. In service with the German Fire Brigade services.

Manufacturer
Alfred Kärcher GmbH & Co.

VERIFIED

Specification
Length: 5.50 m
Width: 2.45 m
Height: 2.30 m
Max payload: 8,000 kg
Average personnel DECON rate: 4-8/h
Average rate of processing decontaminated clothing: 4-8/h
Material DECON rate: 4-8 vehicles/h

Kärcher USC AB-DEKO mobile decontamination system (rear view) ***2001**/0055106*

Kärcher USC AB-DEKO mobile decontamination system shown being positioned by supporting vehicle ***2001**/0098398*

Kärcher vacuum Decontamination System for Sensitive Material (DSSM)

Description

The Kärcher vacuum Decontamination System for Sensitive Material (DSSM) ensures the effective detoxification of sensitive material in case of contamination with chemical warfare agents. It is especially designed for the decontamination of electronical, opto-electronical equipment and detection equipment.

A heating element (jacket and IR heating) is integrated into the vacuum chamber which can be continuously adjusted between the environmental temperature at approx 200°C. The contaminated air sucked out of the vacuum chamber is transported via an NBC protection filter to the outlet of the vacuum pump. The vacuum chamber can also be integrated into containerised decon systems.

Status

In field test evaluation with the German Armed Forces.

Manufacturer

Alfred Kärcher GmbH & Co.

UPDATED

Specifications
Dimensions: 1,000 (L) × 800 (W) × 800 (H) m
Weight: approx 300 kg
Vacuum: 10^2 mbar
Heating system: up to 200°C
Performance data: Operational temperature range (−32 to +49°C)

Kärcher vacuum decontamination system for sensitive material   **2001**/0098396

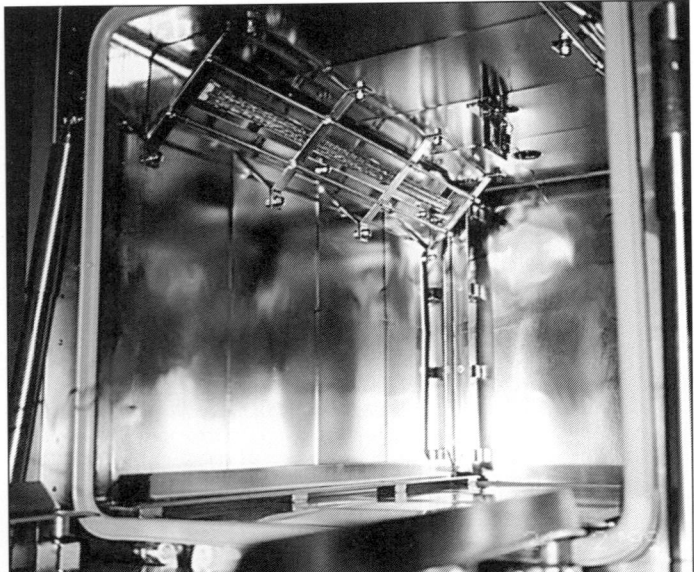

Kärcher vacuum decontamination system for sensitive material   **2001**/0098390

Kärcher Vehicles for material and personnel decontamination

Description

During 1989, Alfred Kärcher GmbH & Co supplied the Austrian armed forces with new decontamination systems mounted on two trucks.

One system is mounted on a 5-tonne truck and is intended for material and terrain decontamination. Carried on the truck are the following equipments: MPDS 3200 for high-pressure prewashing with one or two spray lances which can also be

Kärcher decontamination vehicle for Austria

Rear of Kärcher decontamination vehicle

used for fire-fighting; DADS (qv) for the main decontamination treatment using different types of emulsion; MPDS (qv) for post treatment with high-pressure water and steam as well as for deradiation with hot foam treatment; a 2,000-litre water tank; decontamination chemicals; four hose reels with preconnected hoses; an electrical generator and accessories. The system was designed for the effective decontamination of material (such as trucks, tanks, weapons and so on) and terrain and can carry out all three decontamination steps simultaneously. Owing to the modular design and the hydraulic lift platform used on the truck, the modules and hose reels can be rapidly unloaded for use individually.

The second system is intended for the decontamination of personnel and their personal equipment. It is also carried on a 5-tonne truck and consists of the following equipments: one MPDS (qv) for operating a field shower with adjustable water temperatures; one MPDS (qv) for decontaminating protective clothing and personal equipment; two shower tents with the integrated Kärcher shower system; one hot air generator for tent heating; one special vacuum cleaner for dry deradiation; five foldable containers for the steam treatment of personal equipment; decontamination chemicals; a 2,000-litre water tank and accessories. Owing to the modular design and the hydraulic lift platform used on the truck, the equipments can be rapidly unloaded for use individually.

Status

In service with the Austrian Armed Forces.

Manufacturer

Alfred Kärcher GmbH & Co.

VERIFIED

Kärcher Waterclean water treatment systems

Description

Kärcher produces a series of Waterclean water treatment systems for all qualities of drinking water. The series includes the following systems:

Waterclean WTC 100

The Waterclean WTC 100 is a table model equipped with a microfiltration membrane and an activated carbon filter. It operates using the pressure (0.7 to 2.9 bar) of a drinking water supply and its usage is limited to pretreated drinking water. It has a total capacity of 5,000 litres at a maximum of 240 litres/h.

The WTC 100 has a diameter of 110 mm and is 260 mm high; weight is 0.8 kg.

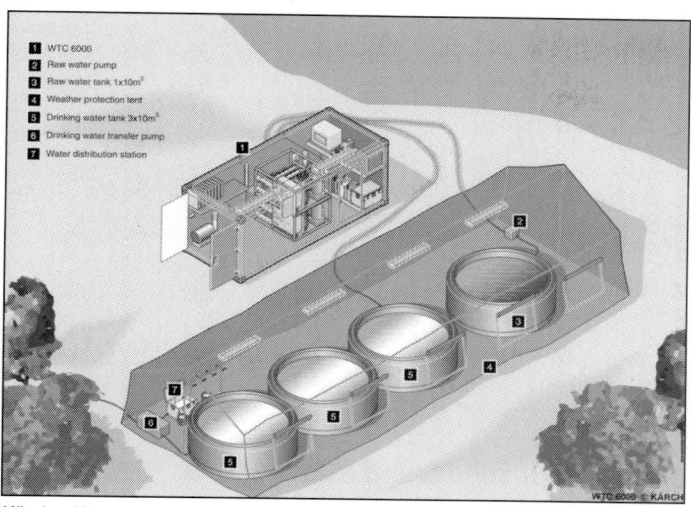

Kärcher Waterclean 6000 water purification system 2001/0098389

Kärcher Waterclean 1600 2001/0098387

Waterclean WTC 500-RO

The Waterclean WTC 500-RO uses multimedia or microfilter systems for pre-treatment. Pre-treated raw water is pumped through a safety fine filter in a Reverse Osmosis (RO) membrane where it is desalinated at a maximum pressure of 69 bar.

Two types of membrane are available, brackish water and sea water. The RO permeate is resterilised with an ultra-violet light. Finally the permeate is chlorinated, minerals are replaced and the pH neutralised.

The WTC 500-RO has a maximum capacity of 500 litres/h, dependent on the salt content and temperature; input can be up to 2,000 litres/h. An external 400/200 V, 50/60 Hz, 5.8 kW external power supply is required. Dimensions are 1,185 × 785 × 585 mm and weighs 197 kg.

Waterclean WTC 500-BC

The Waterclean WTC 500-BC has pretreatment multimedia and microfilter systems. Pretreated water is combined with ozone for a brief period to oxidate and disinfect before flowing through a tubular reactor with a defined reaction time. The ozone-treated water passes over an ultra-violet lamp for further disinfecting by activated rest ozone. Finally, the water is treated in an activated carbon filter where

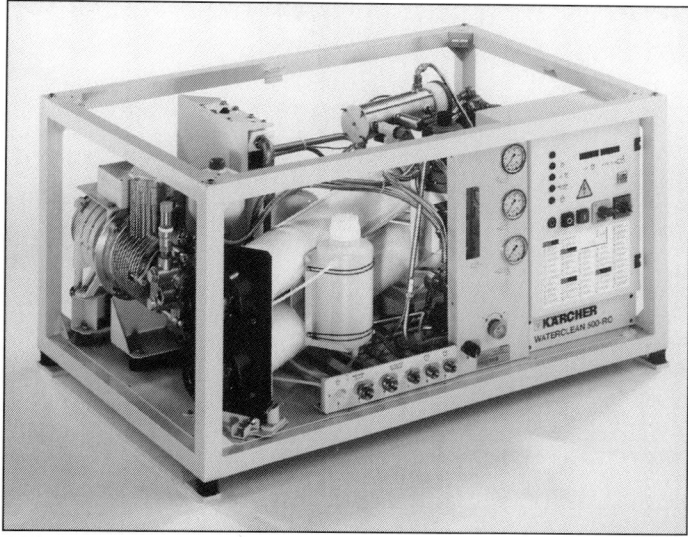

A Kärcher Waterclean WTC 500-RO

A Kärcher Waterclean WTC 1000 water treatment system in operation during a disaster relief operation in the USA

any remaining pollution and products of oxidation are separated off. Post chlorination of the clean water prevents germ formation in a storage tank.

The WTC 500-BC has a maximum capacity of 500 litres/h. An external 230 V, 50/60 Hz, 580 W external power supply is required. Dimensions are 1,185 × 785 × 585 mm and weighs 180 kg.

Waterclean 1000

With the Waterclean 1000, untreated water is pumped over a water preheater in a tubular reactor by an underwater pump. Chemicals for oxidation (to correct the pH value), for adsorption and precipitation are added at time defined intervals. Flakes are then separated off into a sedimentation container. The clean water is then further treated by a fine filter and an optional activated carbon filter. Treated water can then be used as drinking water or, if necessary, can be desalinated by an integrated RO process without further pretreatment or with a reduced use of chemicals.

The Waterclean 1000 can operate as an autonomous equipment with an integral generator to supply power and auxiliary heating. Maximum output can be up to 1,000 litres/h depending on the salt content and mode of operation. Dimensions are 2.405 × 1.755 × 1.495 m and weighs 1,445 kg.

Waterclean 1600

The Waterclean 1600 is a fully automated system designed to remove contamination from water sources such as large surface-water areas, ponds, rivers or the sea. It is therefore equally effective against the type of water contamination likely to be encountered either in the conventional battlefield environment or following a natural emergency (a flood for example) or contamination by NBC agents. It includes a 0.5 mm pre-filter, single- and double-pass RO and a post chlorination facility. The output is approximately 1,600 litres/h depending on the supplied water quality. Output quality complies with current WHO standards. The system is air portable or can be shipped on a single- or double-axis trailer. Its noise signature is very low - 53 dbA at 10 m from the unit. Dimensions are 3,200 × 1,600 × 1,100 m and weight is 1,550 kg.

Waterclean WTC-Containerised systems

The modular construction of both the WTC 500 and 1600 series enables them to be installed inside standard 7 m ISO mobile containers. Units can thus be combined to deliver treatment flow rates of up to 5,000 litres/h, operating autonomously with their own power plants and air conditioning systems.

Waterclean 6000 system

The Waterclean 6000 is a fully automated, high-capacity system designed to remove NBC agents and deliver potable water to a large detached unit. Daily output is up to 120,000 litres - sufficient for continuous supply to a deployed unit of above battalion size, or to 12,000 personnel in an emergency environment. The modular system comprises a 0.5 mm pre-filter, single- and double-pass RO and post-chlorination facilities. The container can be delivered with air conditioning and integral generator according to customer needs. Dimensions are 6,058 × 2,438 × 2,438 m and weight is 10,000 kg.

Status

In production.

Manufacturer

Alfred Kärcher GmbH & Co.

VERIFIED

Mobile NBC decontamination semi-trailer

Description

In January 1992, Deutsche Aerospace AG (Munich), now EADS, was awarded a DM2 million contract by the German Federal Procurement Agency for the manufacture of a mobile decontamination facility as a pilot project for NBC

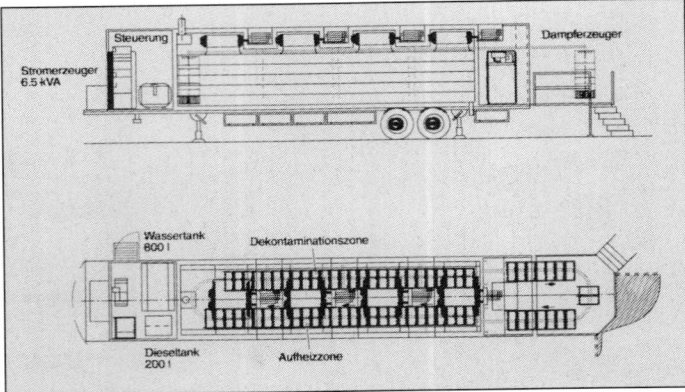

Outline drawings of mobile NBC decontamination semi-trailer

decontamination. The facility is a component of the HEP 90 ('Main Decontamination Site for the 90s') project conducted by the armed forces. The project was carried out by the Command and Information Systems Division of the Defence and Civil Applications Group at Dornier GmbH in Friedrichshafen. Techniques used were based on research carried out by the *Wehrwissenschaftliche Dienststelle der Bundeswehr* (Federal armed forces Research Centre for NBC Protection) in Munster.

The facility is based on a 13 m long semi-trailer and used for the NBC decontamination of equipment and clothing. Using the system, contaminated equipment and protective clothing are hung on a 'racetrack' conveyor system to pass through an airlock into a heated area where fans are used to blow away contaminated particles. The heating stage lasts 30 minutes, after which the contaminated equipment undergoes the next stage where it is subjected to superheated steam at 170°C. For the full decontamination process, carried out after the mechanical removal of dust particles, a foam treatment and rinsing are carried out using spray pipes. Up to 63 protective suits can be treated every hour.

Water for long-term decontamination operations is supplied to the semi-trailer from a water tanker vehicle via hoses. The semi-trailer has its own integral 800 litre water tank. Contaminated water is collected in an underfloor vat and passed into separate tanks. The semi-trailer has a 200-litre diesel fuel tank for the 6.5 kVA generator used to supply electrical services to the system. The latter are located in a control compartment at the front of the semi-trailer.

Status
Entered service with the German armed forces in April 1993.

Contractor
Dornier GmbH.

UPDATED

NBC decontamination truck

Description
The standard Bundeswehr NBC decontamination truck is a MAN (6 × 6) 7,000 kg truck adapted for the role by the addition of special mounting points and equipment stowage on the load area. The truck carries decontaminating equipment for terrain and equipment decontamination in the field. It can also be used as a firefighting vehicle dispensing water and/or foam.

The decontamination equipment carried weighs approximately 5,400 kg. The centre of the load area is occupied by two insulated 1,500 litre tanks. All the other equipment carried is located around the tanks. This includes a water heater with a variable heating capacity of 40.5 to 128 kW and a flow capacity of 600 to 3,600 litres/h. Also carried is a dry mixer for decontamination chemicals with a maximum flow capacity of 200 litres/min. Other equipment includes four area application spray bars with a flow capacity up to 465 litres/min and a spray range of 8 m; their

MAN 7,000 kg (6 × 6) truck converted for NBC decontamination by Minimax GmbH (Michael Jerchel)

coverage capability is 1.16 litres/m². The vehicle carries a range of other decontamination equipment and chemicals, including a pump and hose to draw water from local sources.

Fully loaded the truck weighs 18,700 kg. It is 8.62 m long, 2.5 m wide and 3.48 m high. Top speed is 90 km/h.

Status
There are 210 units in service with the German Army. Starting in early 1992 the trucks and decontamination sets underwent a general overhaul to extend their service life.

Manufacturer
Minimax GmbH (decontamination equipment).

VERIFIED

OWR CLEAN 6000 G and DECO-CLEAN 7000 G decontamination units

Description
The OWR CLEAN 6000 G and DECO-CLEAN 7000 G decontamination units can be used both as high-pressure cleaners and steam jet devices. The DECO-CLEAN 7000 G has a decontamination module with a special pump that can be connected to the cleansing system to cleanse all known chemical warfare agents. Both units are sled-mounted and have small wheels for easy movement.

A total of 300 systems were ordered by the US Army for the cleaning and decontamination of vehicles and equipment prior to their return from Saudi Arabia following the 1991 operations in the Gulf region. The units were used under severe conditions for extended periods without respite under OWR technical guidance.

Status
In production. Accepted as US Standard.

Manufacturer
Odenwald-Werke Rittersbach GmbH (OWR).

VERIFIED

OWR DECO-CLEAN 7000 G decontamination unit in operation

OWR COBRA Personal Self Decontamination Unit

Description
The COBRA Personal Self Decontamination Unit contains sufficient GD-5 to allow an individual to quickly decontaminate himself, following a liquid CW attack. The bottle is pressurised using the hand pump on top of the bottle and the user directs the nozzle to apply the GD-5 to the affected areas. It can be used in any environment and stows in a zipped insulated bag looped onto any type of service webbing waist belt.

GD-5 is a proprietary decontaminant that has passed trials protocols set by CEB (France), Dugway Proving Ground (Utah, USA) and FOA (Umeå, Sweden), against all known CBW agents. GD-5 is claimed to be as effective as DS-2 whilst also being

---

Specifications
Size of bottle: 25 × 7.3 cm
Size of unit with bag: 26 × 12 cm
Weight empty: 350 g
Weight with GD-5 solution: 850 g
Internal volume: 0.5 litres
Length of hose: 80 cm

---

non-corrosive, non-conductive and environmentally safe. It has a minimum 10-years shelf life.

Status
In production.

Manufacturer
Odenwald-Werke Rittersbach GmbH (OWR).

NEW ENTRY

OWR Containerised MultiPurpose Decontamination System MPD12 and PD12

Description
The OWR Containerised MultiPurpose Decontamination System MPD12 is used for the NBC decontamination of personnel, equipment, weapons, vehicles, aircraft and helicopters and terrain. It was developed considering the proposals, requirements and concepts of NATO and FINABEL, the practical experience of the NBC defence unit of the German Army and self-protection school in Sonthofen, the findings of the NBC defence research centre of the German Army in Munster and using inputs from various associates and others involved in NBC decontamination.

The MDP12 has 12 shower positions for personnel and is carried as a container unit on a vehicle; container handling devices are included in the system. The system can be transported by helicopter. The unit used with the system is the PD12 which is primarily used for the decontamination of personnel and their personal clothing and equipment. However, the MPD12 system does incorporate a steam jet cleaner.

In use, the PD12 unit incorporates a pre-decontamination station provided with two 10 litres/min shower units equipped for the addition of decontaminants. The steam jet cleaner could be used on equipment at this stage. Removed clothing is provided with an identification number for decontamination at another station as personnel remove underclothing in an entrance station. At this stage adjustable time interval audio and visual signals indicate the intervals at which personnel should proceed into the shower area. In the shower area 12 shower units, each distributing 5 litres/min (approximately 3,600 litres/h), can be supplied with continuously adjustable amounts of decontaminants. After showering, personnel can then put on fresh underclothing in a further exit station before their decontaminated or fresh uniforms are returned along with any decontaminated equipment.

The PD12 has an onboard 1,400 litre aluminium water tank with automatic maximum and minimum feed levels for the associated feed pump. This is backed up by three 1,000 litre flexible tanks. If required, a water pump can supply up to 30,000 litres/h from a water source up to 15 m below the PD12 unit.

Status
In production.

Manufacturer
Odenwald-Werke Rittersbach GmbH (OWR).

VERIFIED

OWR PD12 with entrance and exit section

OWR DECOFOG III decontamination system and GD solution

Description
DECOFOG III involves a portable decontamination system which emits NBC decontamination agents in the form of fine aerosol clouds which resemble a form of natural fog. The system is normally used with a decontamination agent known as

OWR DECOFOG I'I ir operation (OWR)

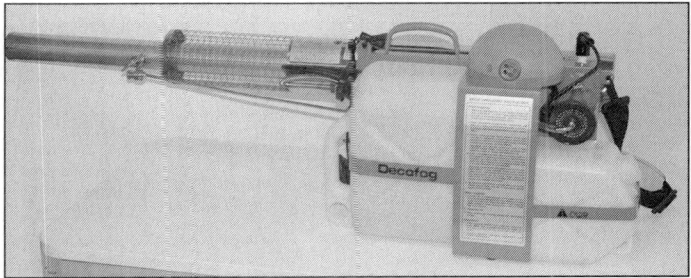
OWR DECOFOG DECON agent applicator (John Eldridge) **2002**/0137901

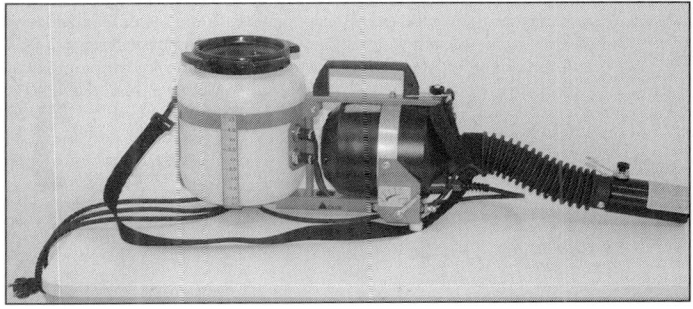
OWR DECOFOG smaller capacity DECON applicator (John Eldridge) **2002**/0137902

GD-5 which is applied to the surfaces to be decontaminated in a thin coat formed from droplets only 4 µm in diameter. This thin coat can reach areas of equipments which cannot normally be accessed by other cleansing systems. The thin coating which needs to be applied means that less agent is required to cover any particular surface area, a typical case being quoted as one tenth the normal amount. GD-5 is quoted as having the same decontamination efficiency as DS2 but it is not corrosive.

The DECOFOG III dispenser uses a 24 hp pulse jet engine which produces an agent flow rate of between 10 and 25 litres/h according to the size of nozzle fitted. A 0.3 litre combustion chamber is used to form the decontaminating fog. The GD-5 solution tank on the applicator is detachable and holds 5 litres; as the tank is transparent the content status can be easily monitored. The DECOFOG III uses quick-start electronic ignition powered by four 1.5 V dry batteries and when running has an average fuel consumption of between 1.5 and 1.9 litres/h; the fuel tank holds 2 litres.

The DECOFOG III applicator is 1.06 m long, 290 mm wide and 330 mm high.

Status
In production.

Manufacturer
Odenwald-Werke Rittersbach GmbH (OWR).

UPDATED

OWR DEDAS 100 decontamination unit

Description
DEDAS stands for Decontamination Emulsion Direct Application System. It is intended to be used as a supplement to existing decontamination systems or as a central module for future systems. It can also be used for the direct and continuous production and application of NBC decontamination solutions or emulsions.

DEDAS can produce 100 litres/min of decontamination emulsion (6,000 litres/h) or 200 litres/min of diluted solutions (12,000 litres/h) and operates at a pressure of

DEDAS 100 decontamination unit

3.5 bar. Liquid and solid decontaminants are mixed automatically and constantly. All elements are corrosion-resistant and construction is rugged. The unit weighs approximately 200 kg and may be operated either ground-mounted or carried on a vehicle. Power may be provided by any electrical or internal combustion unit providing between 3 and 4.5 kW. Controls include a malfunction indicator, a flow and level controller and an automatic switch-off sequence.

Various accessories are available, including low-temperature operation components.

Status
In production.

Manufacturer
Odenwald-Werke Rittersbach GmbH (OWR).

VERIFIED

OWR DEKON Trailer 6000

Description
The OWR DEKON Trailer 6000 is intended for the decontamination of personnel, equipment and terrain but it can also be used as water or foam fire extinguishing equipment, as mobile disinfecting equipment or as a mobile shower station. It is carried on a single-axle trailer which can be towed by any vehicle with a 3,000 kg capacity and upwards.

On the move the trailer is enclosed by a framed cover which can be removed to act either as a dressing station or a weather cover for a shower installation. The trailer has its own 1,000 litre water tank and carries several further 1,000 litre foldable water tanks. There is an integral heater that can heat 3,600 litres of water by 28°C every hour for use by the shower units or for other purposes. The canvas cover can enclose a shower unit with eight shower heads which is sufficient to process approximately 100 persons/h. Power to pump the water is provided by an engine on the trailer which also provides power for a generator to supply the circulation heater and possible lighting units. The trailer also has facilities for mixing decontamination and other agents with the water and, by using the mixing

Complete OWR DEKON Trailer 6000 with shower unit in use

injector and a projection tube, it can be used to produce foam for decontamination or firefighting. The trailer can also carry a work bench, fire extinguishers, entrenching tools, spare parts and tools, first aid kits and extra supplies of decontaminating and other agents.

Status
In production.

Manufacturer
Odenwald-Werke Rittersbach GmbH (OWR).

VERIFIED

HUNGARY

Decontamination trailer type FMU

Description
The Type FMU decontamination trailer is based on a single-axle trailer and contains all the equipment carried on the decontamination vehicle type FMG-85 (see entry in this section) except that the trailer does not have any facility for mixing decontamination solutions on board — a separate tank is required. The centrifugal pump is driven by a petrol engine carried on the trailer and all the equipment is carried inside a box body.

Status
Available. In service with the Hungarian armed forces.

Manufacturer
Budapesti Vegyipari Gépgyár Rt.

VERIFIED

Decontamination vehicle type FMG-85

Description
The decontamination vehicle type FMG-85 is based on the chassis of a 3,000 kg truck. It can be used for chemical and nuclear decontamination and may also be used as a water tanker, for firefighting and for the supply of heated water on a large scale.

The chassis carries a special body with a 2,000-litre water tank, mixing equipment for the preparation of decontamination solutions, centrifugal pump driven by the vehicle's engine and controlled from the driver's position, heating boiler, pipelines and valves, plus storage and other cabinets.

The filling time for the tank is 5 minutes using the centrifugal pump; suction height is 5 m. The fuel tank for the boiler heater contains 40 litres of diesel oil which is sufficient for a minimum of 2 hours. The approximate time taken to heat 2,000 litres of water to +60°C is 60 to 80 minutes.

Status
Available. In service with the Hungarian armed forces.

Agency
Budapesti Vegyipari Gépgyár Rt.

VERIFIED

Field shower unit FF

Description
The field shower unit FF is intended to provide decontamination or other showers for between 250 and 280 personnel every hour. The unit is based around four shower frames, each with six shower heads and one 40 mm diameter threaded outlet; each frame is supported on two steel bipods. Also supplied are eight rubber mats to assist drainage; each mat measures 7,000 × 120 × 3 mm. A collapsible rubber water tank holds 2,000 litres and four M63 tents are provided with the kit; two to protect the showers and one each for dressing and undressing. Also included in the kit are 800 bags to contain personal belongings during the showering. A further 500 numbered tags are provided to identify personal clothing and equipment.

Up to 48 personnel can use the shower unit at one time and the showering time for each group is estimated at 10 minutes. The recommended amount of water required for each individual is estimated at 15 litres.

Status
Available. In service with the Hungarian armed forces.

Agency
Budapesti Vegyipari Gépgyár Rt.

VERIFIED

Individual decontamination kits

Description
The Hungarian NBC protection industry produces a number of NBC decontamination kits. The Individual NBC Survival Packet, which contains examples of personal decontamination equipment, is mentioned in the PROTECTION (individual) - Medical countermeasures section.

Individual decontamination packet FVCS-78
The contents of this packet are intended for the decontamination of an individual soldier's small arms and other personal equipment. The kit consists mainly of a plastic bag containing calcium hypochlorite powder, into which water is poured before use.

Decontamination kit for weapons FVCS-M
Normally carried as part of a soldier's individual equipment, this kit is supplied in a rubber-coated polyethylene bag inside a sealed linen bag. It contains 120 g of calcium hypochlorite, 5 g of an emulsifier and 6 g of activated aluminium foil. The contents are added to 0.5 to 0.7 litres of water, which will then warm up to 60 to 70°C within 10 minutes. The solution is applied to weapons using a piece of soft cloth.

Individual decontamination spray FVS
This consists of a plastic bottle containing a decontamination solution known as UDL, plus a spray head. The spray can be used for the chemical decontamination of equipment, including sensitive optical and electronic equipment.

Status
Available. In service with the Hungarian armed forces.

Agency
Budapesti Vegyipari Gépgyár Rt.

VERIFIED

Truck-mounted universal decontamination equipment FMG-90

Description
The truck-mounted universal decontamination equipment FMG-90 is used for chemical and radiological decontamination of weapons, vehicles, terrain and other military material. It can be used to transport, mix and heat decontamination solutions, wash down surfaces, pump water from sources and conduct firefighting operations.

The FMG-90 is carried on a truck chassis. The special body design contains a pump unit driven by a diesel engine and containing a low-pressure centrifugal pump and a high-pressure pulsating pump. Also on the body is a 5,000 litre tank unit with a heater and boiler, pipelines with controlling valves and a control panel. There is also a decontamination agent mixing unit with a hand winch for loading materials. Other accessories carried include vehicle washing appliances and other fittings.

The decontamination system can operate at either high or low pressures. When the low-pressure system is in use, up to 40 working stations at distances up to 90 m can be supplied. Output is up to 13 litres/s. Under high-pressure operation, the number of workstations is reduced to eight, up to 300 m distant. The high-pressure operating output is up to 220 litres/min. The time required to heat water to +70°C is between 60 and 90 minutes, depending on the initial temperature. Source water can be raised to 5 m by the diesel-driven pump system.

Status
In service with the Hungarian armed forces.

Agency
Budapesti Vegyipari Gépgyár Rt.

VERIFIED

Truck-mounted Universal Decontamination Equipment FMG-90

Vehicle decontamination kits and materials

Description
Vehicle decontamination kit MK-67P
The Vehicle Decontamination Kit MK-67P is intended for the full or partial decontamination of armoured vehicles. The kit can also be used for heating water or other liquids, cleaning equipment or for personnel washing and so on.

The kit comprises a container with lid, stove, foot pump, petrol burner and hoses. Also included is a rubber collapsible container with a capacity of 8 litres, funnel, cloths and some spare parts.

The container is made from steel plate and can contain the entire kit. When empty it is used to prepare the various decontamination solutions; the internal capacity is 35 litres. There are two carrying handles and the lid can form part of the stove. The stove itself is a four-part assembly used to heat the various solutions and accommodates the kit's petrol burner; if this is not available wood or coal can be used. The foot pump is used to distribute the prepared solution and can deliver 3 litres/min at 40 strokes/min. Hoses supplied with the kit are a 2 m suction hose and two 7 m delivery hoses connected to brushes.

The kit is arranged so that it can be secured to and carried on the exterior of a vehicle. Dimensions are 380 × 800 × 220 mm and the kit weighs 35 kg.

Decontamination kit MK-67
This kit is similar to the MK-67P but intended for the decontamination of vehicles and other equipments. The kit is packed inside a metal container which can be used as a dish for the preparation of the decontamination solutions. The kit contains a foot pump to direct the prepared solution through two brushes.

Small decontamination kit MK-67CS
This is a variant of the MK-67 kit intended for the decontamination of small vehicles such as Jeeps. The kit is similar to the MK-67 but uses a smaller container and only one brush.

Chemical decontamination package VM-80
This is a plastic bottle containing calcium hypochlorite powder intended for use with the portable decontamination kits mentioned above.

Radiological decontamination package SM-80
This is a plastic bottle containing detergent in powder form intended for use with the portable decontamination kits mentioned above.

Universal Decontamination Agent (UDA)
Universal Decontamination Agent (UDA) is used for the decontamination of equipment and terrain contaminated with liquid blister or nerve agents and may be used with the kits mentioned above and other decontamination equipments. UDA is an aqueous calcium hypochlorite slurry containing 7 to 10 per cent calcium hypochlorite, 0.1 per cent detergent and 1 per cent thickener.

Status
Available. In service with the Hungarian armed forces.

Manufacturer
Budapesti Vegyipari Gépgyár.

VERIFIED

ISRAEL

DP-2 decontamination powder

Description
DP-2 decontamination powder is intended for decontamination of the skin and personal equipment following contact with liquid NBC agents. DP-2 is a homogeneous finely ground powder with a large surface area to make it a highly

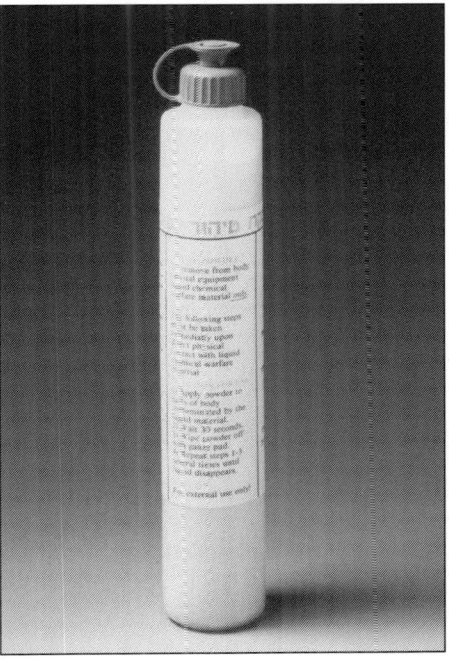

DP-2 decontamination powder

effective absorbent. Tests have demonstrated that 1 g of DP-2 powder absorbs at least 15 ml of 1 per cent quinine bisulphate solution. Bulk density of the powder is 0.8 to 0.9 g/ml.

DP-2 is issued in plastic dispensers containing 130 g of powder. The powder can be applied to affected areas direct from the dispenser and then left in position for 30 seconds. The powder is then wiped off with a gauze pad and the process can be repeated as required.

Status
In production.

Manufacturer
SHALON Chemical Industries Limited.

VERIFIED

ITALY

Cristanini BX24 SPECIAL decontamination product

Description
BX24 is a CW decontaminant which is claimed to be extremely effective against the known threat agents and to be superior to other commonly available decontaminants. Its success is a function of its ability chemically to convert contaminant agent into neutral compounds and to suspend these materials into solution, allowing them to be removed easily through the normal manual decontamination process.

BX 24 is a powder and is delivered in 50 kg drums or in special rechargeable cartridges designed to be used with the SANIJETGUN decontamination lance (see entry in this section). As a finely-ground powder it is easy to prepare for use, requiring simply to be mixed with water in the proportions shown. It disperses and mixes easily in water without forming lumps and requires no additives, catalysts or specialist tools for preparation. The residue remaining after decontamination is free of chlorinated solvent, hydrocarbons or other environmentally unfriendly compounds and is therefore non-toxic. It is designed to be removed easily through rinsing and dispersed through normal drainage.

BX24 is highly efficient against known CW agents, neutralising the materials quickly, without release of hazardous vapour or generation of excessive heat. It is non-corrosive to metal or other surfaces protected by paint coatings and is easy to handle, store and distribute. BX24's effectiveness against BW agents is as a result of the release of free active chlorine (8.7 per cent) with its strong bactericidic action. This decontaminant remains chemically stable in storage for at least five years, even in extremes of temperature (stable range is -20 to +60°C).

Successful laboratory tests have been undertaken to illustrate the compound's effectiveness against HD, VX and specific simulants for the organic phosphates coated on surfaces such as aluminium, glass, natural or butyl rubbers, polymer fabrics, alkyd, single and multicomponent paints. The substrate has remained unaffected by the compound and, unlike calcium hypochlorite, BX24 and residues are easily removed with just water even after prolonged exposure. The graph illustrates the results of the laboratory tests.

Status
In production. Supplied to many NATO and non-NATO countries.

Manufacturer
Cristanini SpA.

VERIFIED

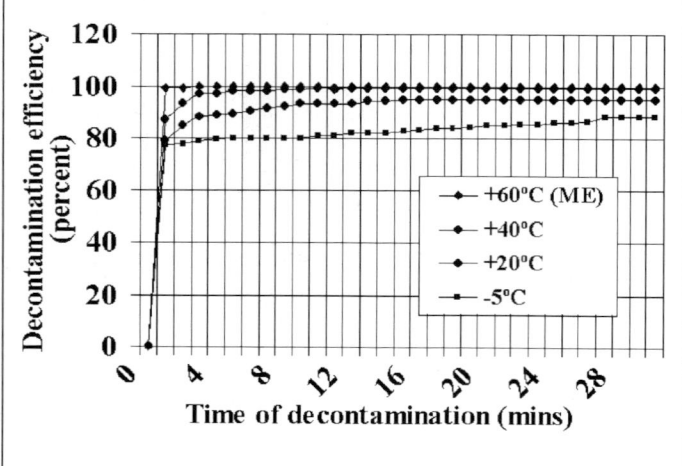

Decontamination efficiency of BX24 against VX (20:1 ratio) at four different ambient temperatures
0019368

Cristanini decontamination and shower tent

Description
The Cristanini decontamination and shower tent is available in two sizes (6 × 9 m and 6 × 12 m) and does not require special training for installation. When provided with a waste water sump under the shower floor, it can be connected by means of a constant level water pump to a waste water tanker or to a water recycling unit.

Construction is PVC fabric over a stainless steel tubular frame. Internal partitions are optional. Capacity is from 60 to 250 personnel/h, depending on the capacity of the water supply unit.

The tent can be supplied with an air heating system to either heat the interior or dry personnel by blown air.

Status
In production. In service in Italy and Spain.

Manufacturer
Cristanini SpA.

UPDATED

Internal view of the Cristanini decontamination and shower tent (Cristanini SpA)
0018702

Cristanini SANIJET 3000/3 containerised decontamination system

Description
The Cristanini SANIJET 3000/3 containerised decontamination system makes use of a single 6.096 m/20 ft ISO standard container which encloses a fully self-contained unit for the decontamination of personnel, vehicles and other equipment and clothing.

The container used for the SANIJET 3000/3 system is RFI- and EMP-resistant and uses an NBC filtration system to produce an internal overpressure. Also provided is an air conditioning unit with a capacity of 1,000 m³/h and capable of working in temperature conditions from −32 to +52°C. An integral water tank has a total capacity of 2,000 litres and is supplied by a pump and filter system that can collect water from various sources. Electrical power for the complete system is provided by a 39 kVA generator.

The container is divided into a number of compartments with a central compartment acting as a control room. Other compartments are used for showers, garment decontamination and access/exit. One entire end of the container is occupied by an engine compartment housing detergent tanks, compressor, main engine, electrical generator and water boiler. The engine room has its own ventilation and access hatches.

For personnel decontamination, entry to the container is via an end hatch. Once inside, personnel remove their clothing with the NBC mask being the last item to be removed; the masks are then discarded through a hatch to outside the container. The interior of the entrance hall where disrobing takes place is kept at an overpressure of 25 mbar. Personnel then enter a shower section where there is a differential pressurisation of 30 mbar. The shower unit has a flow of 18 litres/min with a selection of hot water from 20 to 40°C. It is possible to add detergents to the water flow. After the initial shower a further one is provided for rinsing. Once showering is completed, personnel pass to the main control compartment for drying and dressing. They then leave through a pressurised exit compartment which has two airtight doors.

For decontaminating vehicles and equipment, an internal compartment houses spray lances and other accessories. These include two cold water lances, each operated at a pressure of 110 bar and with a capacity of 18 litres/min, one of which may be replaced by a hot water lance with a similar flow rate and operating pressure. Also provided is a foam application lance operating at a pressure of 35 bar and with a flow rate of 6 litres/min and a vapour lance operating at 16 bar and with a flow rate of 10 litres/min. The compartment housing them also contains hose reels that can be unwound for external use.

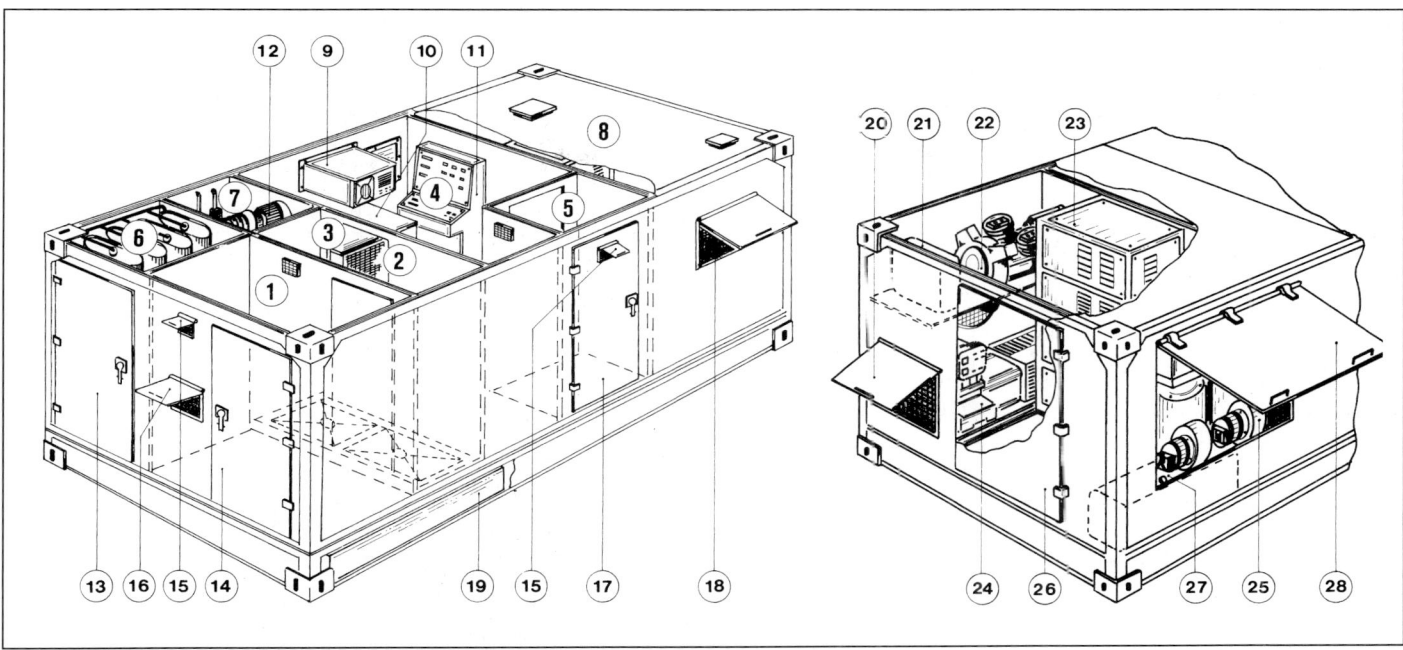

Cristanini SANIJET 3000/3 containerised decontamination system **(1)** entrance compartment **(2)** decontamination shower **(3)** rinsing shower **(4)** main control compartment **(5)** pressurised exit compartment **(6)** garment decontamination compartment **(7)** lances and accessories **(8)** engine compartment **(9)** NBC pressurisation filter unit **(10)** air conditioning unit **(11)** internal control panel **(12)** hose reels **(13)** entrance hatch to garment decontamination compartment **(14)** entrance hatch **(15)** differential pressure valves **(16)** hatch for mask disposal **(17)** exit hatch **(18)** air vents for engine compartment **(19)** water cistern **(20)** engine ventilation hatch **(21)** detergent tank **(22)** compressor unit **(23)** modular high pressure pump unit **(24)** generator unit **(25)** boiler unit **(26)** entrance hatch to engine compartment **(27)** fuel tank **(28)** side hatch (Cristanini SpA)

Garments can be decontaminated in a separate compartment which is equipped with racks, clothes hangers, bags and a sump for waste water collection. Decontaminating vapour is fed into the chamber at a pressure of 16 bar, with a flow rate up to 10 litres/min and at a temperature between 120 and 140°C.

Status
In production. In service in the USA.

Manufacturer
Cristanini SpA.

VERIFIED

Cristanini Sanijet C.921 ASP/3 MIL DECON utility (Cristanini SpA)   **2001**/0077723

Cristanini Sanijet C.921 ASP/3 MIL

Description
This transportable unit is a high-capacity field DECON resource. Its modular construction allows units to be used together, deployed at a static site or mounted on vehicles. Five configurations are described.

Configuration A
In this configuration, three units are co-located to deliver enough hot water, steam and BX24 decontaminant to deal simultaneously with three operations:
- DECON of up to 15 tank-sized vehicles per hour. Hot water at 90 Bar, 90°C, 1,020 litre/h
- Continuous facility for DECON of light equipment such as personal weapons, masks and clothing. Steam at 180°C, 436 kg/h
- Field DECON shower unit capable of cleansing 180 people per hour (assuming a 2 minute shower stay per person). Hot water at 3 Bar, 38°C, 3,000 litre/h and cold water as required. The portable DECON unit has six shower points and 8 litre/min is delivered to each point.

Configuration B
This is designed for the mass cleansing and DECON of contaminated personnel. Six field DECON shower units (see above) are supplied from three C.921 units. The hot water output is 3,000 litre/h at 90 Bar, 90°C. Mixed with the 5,940 litre/h cold water supply and decontaminant, the facility allows 540 personnel to be processed per hour with a 2 minute stay-time, delivering 8 litre/min to each shower head. The combined output is 9,000 litre/h of mixed water at 38°C.

Configuration C
This arrangement is designed for the mass decontamination of light equipment and clothing. The maximum output is 1,310 kg/h of steam at 180°C.

Configuration D
Used for the mass decontamination of vehicles, this arrangement again uses three of the C.921 units together, supplying 3,060 litre/h of hot water and decontaminant at 90°C. The facility claims a DECON performance of 45-50 tank-sized vehicle per hour.

Configuration E
Used for ground DECON, this mobile configuration uses three C.921 units mounted on a standard 4 × 4 truck chassis, together with tanks and pumps for the supply of water and decontaminant. At a speed of 8.4 km/h, the vehicle-mounted unit claims a decontamination capacity of 21,000 m²/h over a path of width 2.5 m.

Unit specification		
Parameter		**Value**
Thermal capacity		210.000 kCal/h
Steam output at 20 Bar	at 210°C	1,105 kg/h
	at 180°C	1,310 kg/h
	at 150°C	1,620 kg/h
Hot water production	at 90 Bar/90°C	3,060 litre/h
	at 3 Bar/38°C	9,000 litre/h
Fuel consumption (diesel)	(boiler and motor)	30 litre/h
Power output		2.5 kW
at 220 V 50 Hz		
Starter motor power requirement	(12 V DC battery)	45 Ah
(manual start facility also fitted)		
DC power output		200 VA
Fuel capacity		60 litre
Pressure range		20-90 Bar
Approx all-up weight (including		1,400 kg
accessories and skid)		
Footprint		240 × 210 × 130 cm

(Data based on ambient atmospheric conditions and supplied water temperature of 20°C)

Configuration F

Designed primarily for fire suppression, this arrangement is similar to configuration E but without the requirement for decontaminant. The unit delivers 9,000 litre/h of fire-fighting water at 3 Bar allowing a reach of 15 m per hose.

Status

In production.

Manufacturer

Cristanini SpA.

UPDATED

Cristanini SANIJET C921 decontamination system

Description

The Cristanini SANIJET range of decontamination systems are available in several versions. The two types of SANIJET C921 are self-contained units. The SANIJET C921 D 18/50 has an electrical system operating on 50 Hz while the D 18/60 operates on 60 Hz. The SANIJETONE is a silenced version.

The system is powered by an air-cooled diesel engine which drives a water pump, enabling the system to operate by delivering cold or hot water (up to 95°C), steam (120°C) or dry steam (190°C). The pump has a capacity to raise water up to 5 m from streams and rivers via a suction hose. The engine uses preheating devices to assist starting under cold weather conditions. Once in the unit, the water can be heated or superheated by a coil burner assembly with a 2 kVA alternator feeding the burner and other auxiliary electrical equipment. The engine also provides a 24 V DC supply for recharging a dry battery which can be used to start the engine and a 24 V DC socket used to supply other auxiliary equipment, such as lighting, or for connecting into another 24 V DC supply in case of battery failure.

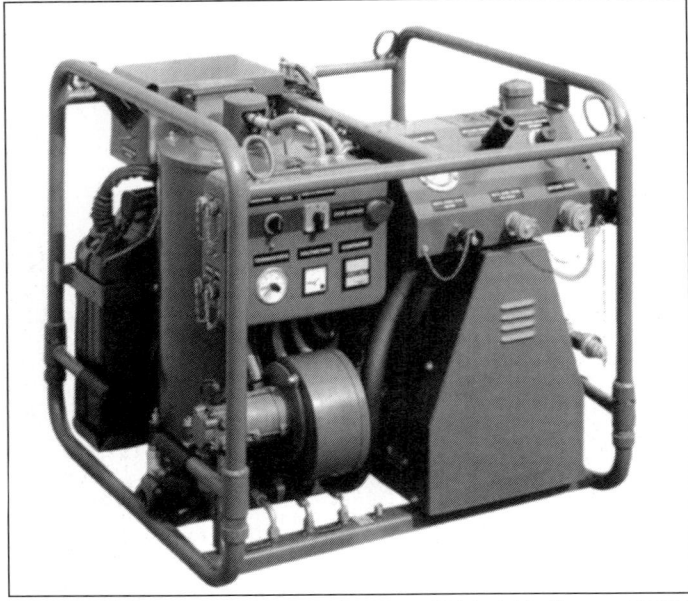

Cristanini SANIJET C921 decontamination system (Cristanini SpA) 0018700

Cristanini SANIJETONE decontamination system (silenced version) (Cristanini SpA) 0018701

Specifications

		SANIJET C921 versions	SANIJETONE
Weight:		230 kg	450 kg (incl batteries)
Length:		0.85 m	0.86 m
Width:		0.85 m	0.94 m
Height:		0.8 m	1.1 m
Engine:		Air-cooled diesel developing 7.5 kW	
Capacity:	(hot water)	90 bar, 840 litres/h at 95°C	110 bar, 800 litres/h at 90°C
	(steam)	20 bar, 550 kg/h at 120°C	20 bar, 490 kg/h at 150°C
	(superheated steam)	20 bar, 300 kg/h at 190°C	20 bar, 295 kg/h at 210°C

Using the special SANIJETGUN decontamination lance, with its decontaminant cartridge system, the SANIJET C921 can spray decontaminant, including BX24, without wasting time during preparation. The system can be coupled to a large range of spray lance types and deliver a variety of different decontaminants. It has provision to vary the strength of the aqueous decontaminant as required. All operations are controlled from a waterproof control panel which also houses the pressure and temperature adjustment controls. Safety devices against overpressure and over-temperature are provided.

Using only one lance and without changing the nozzle, the SANIJET C921 system provides a primary wash for material decontamination, application of decontaminant solution without previous preparation and final rinse.

The SANIJET C921 system can be used with various accessories such as a water softener, a foam emulsion production unit, various types of field shower unit and devices intended to steam explosive from various types of munitions.

The SANIJET C921 system forms part of several decontamination systems. It forms part of the suite carried by the Czech ACHR-90 NBC decontamination vehicle and ACMAT UMTH 1000 (see this section).

All Cristanini systems conform to AQAP 110.

Status

In production. In service with Belgian, French, Italian, South Korean, Spanish and US Armed Forces, and others.

Manufacturer

Cristanini SpA.

UPDATED

Cristanini SANIJETGUN

Description

The SANIJETGUN is a flexible single-operator solution for vehicle and large equipment DECON. The unit is designed to carry out the three key DECON activities: prewashing with high-pressure cold water, mixing and delivery of water-based decontaminant emulsion and rinsing with hot water. For emulsion delivery, the SANIJETGUN is designed to draw up the decontaminant powder and mix it in

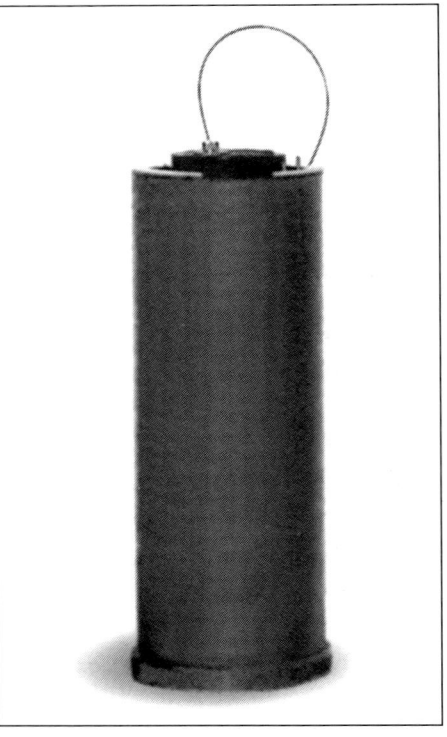

Rechargeable cartridge for the SANIJETGUN (Cristanini SpA) 0018704

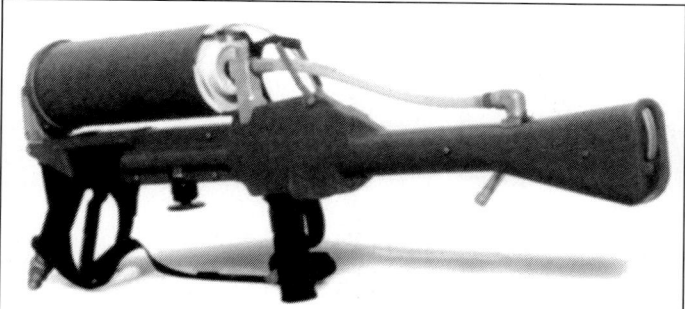

The SANIJETGUN decontamination lance (Cristanini SpA)   0018703

The SANIJETGUN in use (emulsion mode) (Cristanini SpA)   0018705

exactly the right proportions with water before delivering it in a low-pressure spray form. The BX24 decontaminant powder (see entry this section) is supplied in rechargeable cartridges which are mounted on the top of the unit. A simple push-button switch provides high pressure (washing) or low pressure (emulsion) operating modes. The SANIJETGUN is delivered with a shoulder strap and a set of maintenance tools, stowed inside the front handle. The controls can easily be operated whilst wearing thick NBC protective gloves.

Using the SANIJETGUN with the SANIJET C921 and the BX24 decontamination product (see separate entries this section), a single operator can conduct decontamination independently of others.

Status
In production.

Manufacturer
Cristanini SpA.

VERIFIED

....

Cristanini small decontamination set

Description
The Cristanini small decontamination set is available for autonomous operation. It works like a hand-held fire extinguisher and is similar in character to the Tirrena SDS T155 (see separate entry this section). The unit is made from AISI-304 standard stainless steel and is charged by a nitrogen pressurising bottle.

Status
In production.

Manufacturer
Cristanini SpA.

UPDATED

Specifications
Weight
(full) 4.3 kg
(empty) 3.2 kg
Height: 340 mm
Diameter: 130 mm
Working pressure: 15 kg/cm²
Testing pressure: 50 kg/cm²
Contents
(max) 2.0 litres
(working) 1.5 litres

Cristanini trailer C90-120/2 MIL decontamination system

Description
The Cristanini trailer C90-120/2 MIL decontamination system is based on a single-axle trailer towed by any vehicle with a 3,000 kg capacity and is designed for the decontamination of vehicles, equipment, personnel and terrain. When used with the Cristanini decontamination and shower tent (see separate entry), this trailer system allows up to 250 persons to be showered every hour.

The two-lance system is powered by an air-cooled diesel engine and supplies 6,000 litres/h at low shower pressure, 1,920 litres/h at 90 bar of hot water and from 800 to 1,200 kg/h of wet and dry steam.

The trailer's canvas cover, mounted on a special frame, does not need to be removed to operate the system and provides weather protection for operating personnel. A 3.5 kVA generator recharges the system batteries, operates the burner and also supplies all external electrical accessories.

The system is supplied with a 1,500-litre water tank and can be used for mobile disinfection or fire fighting, using appropriate equipment. The C90-120/2 MIL can be removed from the trailer base and mounted on skids. The system is air portable.

Status
In production. Tested and accepted by the Italian Army. In service with NATO armed forces.

Manufacturer
Cristanini SpA.

UPDATED

Specifications
Weight:		3,100 kg
Length:		4.55 m
Width:		2.2 m
Height:		2.7 m
Engine:		Air-cooled diesel developing 24 kWp
Capacity:	(hot water)	90 bar - 32 litres/min
	(wet steam)	20 bar - 20 kg/min
	(dry steam)	20 bar - 13.3 kg/min

Cristanini trailer C90-120/2 MIL decontamination system (Cristanini SpA)
2002/0137898

....

Tirrena Small Decontamination Set SDS T155

Description
Most of the chemical solutions used to decontaminate vehicles and equipment after exposure to chemical warfare agents are corrosive to varying degrees. In practice this means that decontaminants have to be mixed or otherwise prepared immediately before use or their dispenser/containers will become corroded to the point of uselessness. This has several disadvantages, not only in a timescale; for example there will be certain areas or sets of circumstances in which the water essential to produce the working solutions will be either unavailable or scarce.

For these circumstances the Small Decontamination Set SDS T155 produced by Tirrena SpA was developed. It is a small fire extinguisher-type decontaminant dispenser, designed to be carried on a vehicle for the immediate first aid decontamination of the vehicle or other equipment. The SDS T155 dispenser can be carried fully loaded for long periods, as it is made entirely of AISA-304L stainless steel which is proof against chemical corrosion. The equipment can be loaded at a remote point and carried in its special bracket until required. DS2 decontaminant is propelled from the container by compressed nitrogen which is supplied with the equipment in a special loading container which weighs 300 g and has a diameter of 40 mm. The 95 cm³ of nitrogen stored in the loading container pressurises the contents of the dispenser to a level of 20 kg/cm² and the loading container is itself reloadable. The dispenser is tested to a pressure of 55 kg/cm². The dispenser cylinder has a maximum capacity of 1.85 litres but the normal working capacity is

1.5 litres. This is sufficient to decontaminate an area of 6 to 8 m² at a spray range of 2 m.

Status
Over 35,000 units produced for the Italian Army.

Manufacturer
Tirrena SpA.

VERIFIED

Specifications
Weight
  (full) 4.3 kg
  (empty) 2.8 kg
Height: 310 mm
Diameter: 125 mm
Working pressure: 20 kg/cm²
Testing pressure: 55 kg/cm²
Contents
  (max) 1.85 litres
  (working) 1.5 litres

Loading container
Weight: 0.3 kg
Diameter: 40 mm
Contents: 0.95 litres
Loading pressure: 150 kg/cm²
Testing pressure: 400 kg/cm²

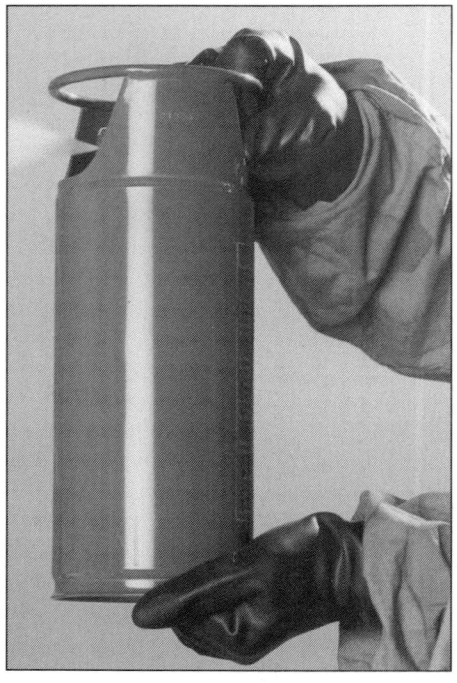

Tirrena small decontamination set SDS T155

JAPAN

Japanese NBC Systems

Description
Both the vehicles shown here offer decontamination facilities. The SU 60 vehicle is part of the Japanese Army's NBC reconnaissance capability (see under

Decontamination vehicle based on 6 × 6 truck chassis

Chemical protection (defence) vehicle based on chassis of Type SU 60 APC (K Nogi)

DETECTION (reconnaissance systems)). The decontamination vehicle is a mobile tanker capable of decontaminating terrain through the fixed nozzles at the front and sides of the chassis.

Status
Both vehicles are in service with the Japanese Army. Manufacturers are unverified although likely to be Mitsubishi Heavy Industries.

Manufacturer
(Unknown).

VERIFIED

KOREA, SOUTH

T4-86 lightweight decontaminator

Description
The T4-86 lightweight decontaminator is a transportable, diesel-powered system designed to deliver high-pressure water or steam at high- or low-temperature to a man-operated DECON Lance. The output can also be delivered, at the correct temperature, to a shower system for personnel DECON. The prime-mover has electric ignition (12 V battery) or can be started by hand. Handles can be extended from underneath the unit, allowing it to be carried by four people.

Status
In service with Korean defence, civil defence and civilian authorities.

Manufacturer
Pao-Chang Company.

VERIFIED

The T4-86 is a transportable, diesel-powered DECON system (Pao-Chang Company)
                           2001/0059323

NETHERLANDS

ACD Deko Basin and Deko Pool

Description

ACD Salvage Techniek make a range of containment facilities for the small-scale fluid-based decontamination of equipment and personnel, suitable for use by first responders. Useable areas of between 200 and 450 cm square are available.

The Deko Basin 250 and 450 use a PVC dam whose collar can be inflated by a pump or a BA cylinder. The Deko Basin 200 and the Deko Pool dams are formed from quick-connect frame elements.

Specifications				
Model	Deko Basin 200	Deko Basin 250	Deko Basin 450	Deko Pool
Length	200 cm	250 cm	250 cm	270 cm
Width	200 cm	250 cm	450 cm	270 cm
Dam height	35 cm	30 cm	30 cm	35 cm
Capacity	1,000 l	800 l	1,500 l	1,700 l
Weight (empty)	30 kg	15 kg	24 kg	3.7 kg
Material	Galvanised steel frame. 850 gm/m² PVC	Inflatable PVC frame with agent-proof insert	Inflatable PVC frame with agent-proof insert	6 mm Vikuprop frame. 0.2 mm PE liner

Deko Basin 200 (ACD Salvage Techniek BV) 0053991

Deko Basin 250 (ACD Salvage Techniek BV) 0053992

Deko Pool (ACD Salvage Techniek BV) 0053993

Status
Available.

Manufacturer
ACD Salvage Techniek BV.

UPDATED

ACD Deko Circle

Description

The ACD Deko Circle is a lightweight, hand-held hoop. Water from a pump or the main supply, feeds eight spray nozzles on the inner circumference of the hoop. The operator moves the hoop over the contaminated person to ensure thorough decontamination of his impermeable protective suit. Used in conjunction with one of the decontamination pools, the Deko Circle is an effective tool in contamination control.

Status
Available.

Manufacturer
ACD Salvage Techniek BV.

VERIFIED

Specifications
Length: 245 cm
Hoop diameter: 120 cm
Height: 1,140 mm
Nozzles: 8 × 0.9 mm
Capacity: 30 litres/min at 4 bar
Weight: 3.8 kg
Material: stainless steel

ACD Hazmat shower

Description

The ACD Hazmat shower is a lightweight, portable shower unit, designed to facilitate the cleansing and decontamination of first responders and decontamination teams wearing impermeable protective clothing. Erectable in approximately 15 seconds, the unit is simple and effective. A flat plate forms the base on which the contaminated responder stands and the vertical pipe and shower head are joined to the main water supply through a standard European hermaphrodite connection. A shut-off valve half-way up the vertical pipe can isolate the shower. This allows the heavier contaminated lower parts of the responder's clothing to be dealt with first with less risk of the spread of contaminated spray. A further refinement is a tubular PE curtain around the shower to contain the spray. A mixer unit is an optional extra feature, allowing decontaminant to be drawn from an external container and mixed with water from the main supply.

ACD Hazmat Shower (folded) (ACD Salvage Techniek BV)
0050780

ACD Hazmat Shower in use (ACD Salvage Techniek BV)   0050779

Specifications
Length: 390 mm
Width: 205 mm
Height: 1,140 mm
Capacity: 50 l/min at 7 bar
Weight: 23 kg
Material: stainless steel

Status
Available.

Manufacturer
ACD Salvage Techniek BV.

VERIFIED

ACD mass decontamination system

Description
The ACD mass decontamination system, incorporated into an Airshelter IV tent, offers decontamination for up to 50 persons per hour. It is ready for use in less than 10 minutes. Integral compartments within the tent provide 2 shower cabins both with 4 highly effective nozzles. Water consumption is only 12 litres per person. There is ample room for placing a casualty rescue stretcher in the cabin. The system is easily transportable and can be made to customer demand.

Status
Available.

Manufacturer
ACD Salvage Techniek BV.

NEW ENTRY

Specifications

Component	Description	L × W × H (cm)	Weight (Kg)
Airshelter IV S	Tent inflated by BA cylinder or blower in 5 min by 1 person	140 × 100 × 70	125
Integrated shower (2)	Check compartment for both male and female	100 × 95 × 20	30
Hydrophor Pump 710	Instantaneous water supply for nozzles, shower and brush	62 × 55 × 32	34
Hot Box 200	Comfortable heated water, automatic temperature controlled	62.5 × 69.5 × 96.5	80
Inducer 201	Accurate mixing percentage decontaminant, detoxificant or disinfectant regardless of flow or pressure	14 × 20 × 47	1.8
Waste Pump 720	Automatic drainage of DECON waste water into reservoir	41 × 36 × 35	9
Fluidbag 500	Foldable reservoir for pure water or waster water, 500 litres	16 × 86 × 85	3

ACD quick-reaction on-site decontamination facilities

Description
This range of equipment comprises two types of facility. The Airshelter III is designed to offer high-speed processing for contaminated personnel following an NBC incident. The Airshelter Deko Unit comprises two separate, connected inflatable shelters which allow doffing of contaminated clothing and personnel cleansing, in a two-stage process, with a minimal risk of cross-contamination.

Airshelter III
This PE shelter is supported on four inflatable arches. Inside the shelter is a CBW agent-proof liner which prevents ingress of liquid contamination. There are 12 personal shower decontamination points inside the shelter allowing a decontamination throughput of over 100 personnel per hour.

Specifications
Dimensions (erected): 600 × 400 cm
Dimensions (packed): 0.3 m³
Weight: 65 kg
Frame arch minimum air pressure: 0.3 bar
Material: Polyamide coated outer skin. Rubber arches. Impermeable liner

Airshelter Deko Unit
The Deko Unit comprises two separate inflatable tents which are connected by an airtight collar. The larger unit is equipped for the safe doffing of contaminated clothing from personnel involved in the management of a contamination incident. The smaller unit contains six shower stations. Both units have integral floors which can be connected to waste fluid removal pipes. The combined unit can be erected in 4 minutes, using an integral air cylinder.

Specifications
Footprint (undressing area): 500 × 450 cm
Footprint (shower area): 350 × 350 cm
Nozzles: 6 × 0.9 mm
Capacity: 50 to 250 litres/min at between 1 and 4 bar
Material: Polyamide coated outer skin. Rubber arches. Impermeable liner

Status
The Airshelter III system is in use with the Royal Dutch Army and the US Army in Saudi Arabia. Airshelter Deko Units are deployed with the Belgium, Chinese and German civil defences.

Manufacturer
ACD Salvage Techniek BV.

UPDATED

Airshelter Deko Unit (ACD Salvage Techniek BV)   0053994

NORWAY

LDS NBC-SANATOR® III lightweight decontamination system

Description
The LDS NBC-SANATOR® III is a man-portable, self-contained system operating independently of all external power sources and under all environmental conditions. It provides up to 24 litres/min of superheated water (150°C) through one or two spray wands for equipment decontamination. The system can also provide up to 80 litres/min of warm water to 12 shower heads at a desired temperature for personnel decontamination.

Powered by an 8.5 hp two-stroke air-cooled engine, the system supplies a continuous water flow. It can draw from any water source, with a suction height of up to 3 m and is operable at temperatures as low as −40°C. A heat exchanger, burning petrol or alternative fuels, heats the water to the desired temperature. Fuel for both the engine and the heater is drawn from jerricans. Weighing only 165 kg, the unit was designed with automatic controls and safety devices. It can be

LDS NBC-SANATOR® III lightweight decontamination equipment in use with a Royal Norwegian Air Force F-16A fighter (Karl H Høie & Co A/S)   ***2001***

Specifications
Basic unit
Weight: 165 kg
Length: 1.02 m
Width: 590 mm
Height: 860 mm

Standard accessory case
Weight: 65 kg
Length: 1.06 m
Width: 520 mm
Height: 390 mm

6,000 litre water tank
Weight: 36.5 kg
Height: 1.22 m
Max diameter: 2.96 m

12,000 litre water tank
Weight: 60.5 kg
Height: 1.62 m
Max diameter: 3.79 m

---

operated by one person, with only periodic checking and refuelling. System features include a temperature selector control switch (45°/70°/150°C) with automatic heater ignition, plus instrumentation including a water thermometer, a temperature adjustment control, fuel and water manometers and an hour meter. Safety devices include thermostats, pressure sensors, a photocell and a pressure relief valve. An auto-injector for concentrated chemical decontaminants is optional.

Apart from NBC decontamination the LDS NBC-SANATOR® III can be used for hot or cold drinking water supplies, water storage, water purification and supplying hot water to field hospitals. The system is also used by numerous humanitarian aid agencies and construction contractors working in remote areas.

A standard accessory case for the LDS NBC-SANATOR® III includes two high-pressure spray wands, 12 shower points, a 10 m suction hose with filter, two 20 m high-pressure hoses and a high-volume injector for chemical decontaminants. Associated with the system are LDS NBC-SANATOR® water tanks made from abrasion-resistant, high-quality PU material approved for drinking water. The basic 6,000 litre tank weighs 36.5 kg empty, including the top cover, a groundsheet/carrier bag, couplings and a repair kit. A 12,000 litre model weighs 60.5 kg.

The LDS NBC-SANATOR® III is licence produced in the USA as the M17 LDS lightweight decontamination system, Sanator (see separate entry).

Status
In production. In service with the Australian, Finnish, Norwegian, Saudi Arabian, Spanish, Swedish, UK and US (licence production) armed forces. More than 10,000 units are in use all over the world.

Manufacturer
Karl H Høie & Co A/S.

UPDATED

POLAND

Decontamination apparatus, Model UDU

Description
The decontamination apparatus, Model UDU is carried to its place of use by a truck and is unloaded to be placed on the ground. It is operated by two people and used to remove radioactive particles from clothing, tents, tarpaulins and similar equipment. The equipment has a small internal combustion engine which powers a system of beaters and a vacuum system. The contaminated clothing is fed into the equipment where it is beaten and the particles are then removed by suction to a special container for disposal. The Model UDU can process up to 120 pieces of clothing/h. The weight of the equipment is 270 kg.

Status
Previously manufactured by Polish state factories. Current manufacturer unknown. In service with the Polish armed forces.

Manufacturer
Unknown.

VERIFIED

Decontamination apparatus, truck-mounted, model IRS

Description
The Model IRS is used to decontaminate large areas of terrain, large pieces of equipment, vehicles and buildings. It can also be used as a water carrier, to provide heated water for showers or firefighting. The equipment is mounted on a Star 66 (6 × 6) 2,500 kg truck and consists of a 2,500 litre tank, a pump driven by the vehicle engine, hand pump, heater, various hoses, pipes, nozzles and shower equipment. The tank is fitted with internal baffles and s internally corrosion proofed. The inlet manhole is screened and the system can be used to discharge solid or liquid decontaminants. Liquid solutions are mixed by the internal cycling of the pump which is driven by a power take-off and when discharging the pump, can deliver 600 litres/min at a working pressure of 4 kg/cm². The tank contents are heated by a fuel-fired heater fed from the vehicle fuel tank by compressed air from the vehicle's airbrakes. The heater can warm 2,000 litres/h of solution to a temperature of +70°C, which is then maintained. The various accessories are stowed along the sides of the tank in cases, and drums of decontaminant are stowed over the cases. The equipment consists of 14 10 mm hoses, each with a spray pipe, nozzle and nozzle brush. For cleansing large areas or buildings (or firefighting) three 20 m long hoses, 25 mm in diameter, can be fitted. The shower unit has eight heads. For terrain cleansing, fan-shaped nozzles can be fitted to the front and rear of the vehicle.

Total weight of the Model IRS is 9,560 kg.

Status
Previously manufactured by Polish state factories. Current manufacturer unknown. In service with the Polish armed forces.

Manufacturer
Unknown.

VERIFIED

IRS decontamination vehicle based on Star 66 (6 × 6) 2,500 kg truck chassis

---

WUS-3 truck-mounted vehicle decontamination apparatus

Description
The WUS-3 truck-mounted vehicle decontamination apparatus was developed along the same lines as the Czech TZ-74 and the RFAS TMS-65/TMS-65M in that it uses a jet engine to spray decontaminants over vehicles that pass through the efflux. The system can also be used to decontaminate roads, hard surfaces and runways, for firefighting, and for the generation of large smoke screens.

The equipment is based on the chassis of the Star 660 (6 × 6) truck and mounts an SO-3 jet turbine engine at the rear behind the three-baffle decontaminant tank. The SO-3 jet engine can be traversed 90° right and left, elevated 15° and depressed 25°. The system is controlled from a transparent housing mounted on the right-hand side of the vehicle cab roof.

Status
In service with the Polish armed forces.

WUS-3 truck-mounted vehicle decontamination apparatus

Manufacturer
Wojskowe Zaklady Lotnicze NR 2.

Marketing agency
Cenzin Limited.

VERIFIED

ROMANIA

ADE 84 mobile DECON facility and shelter

Description
The ADE 84 mobile DECON facility and shelter is designed for the washing and decontamination of hand-held or other small equipment and clothing. It is also described as capable of on-site 'reproofing' of cotton clothing (presumably fireproofing). The combination acts also as a field power source for electricity and steam.

The prime mover is a 6 × 6 DAC 665-T type chassis, powered by a 215 HP diesel engine, type D2156 HMN8 (see *Jane's Military Vehicles and Logistics* for further details). It tows a twin-axle covered trailer, type 2RPF-7, and a single-axle type GTE-38/400 trailer. The prime mover contains the washing and drying units, treatment machines and centrifuge. The steam generation plant, together with supplies of water and decontaminant, are carried by the twin-axle trailer. The single-axle trailer carries the 38 kVA generator set.

The system carries sufficient supplies for up to 30 hours of continuous operation.

ADE 84 mobile DECON system   0050785

ADE 84 prime mover - internal view: washing machine   0050788

ADE 84 system - filtration unit and centrifuge   0050789

Status
In production. In service with the Romanian armed forces.

Manufacturer
AEROSTAR SA.

VERIFIED

ADTT-1 truck-mounted jet DECON system

Description
The ADTT-1 jet decontamination truck-mounted installation follows the same general lines as the Czech TZ-74 (see separate entry this section) and the now obsolete RFAS TMS-65/TMS-65M (see earlier editions of *Jane's NBC Protection Equipment*) in that it employs a jet engine exhaust to decontaminate vehicles and other equipment. In the case of the ADTT-1 the carrier vehicle is the DAC 15.215 DFAEG (6 × 6) 5,000 kg truck (also known as the DAC 665T — see *Jane's Military Vehicles and Logistics* for details).

The turntable-mounted jet engine used with the ADTT-1 is an RD 45 FA (M05) turbine operating at between 8,000 and 11,000 rpm. With a 17 litre/min oil pump, the engine delivers a gas jet up to 9 m long into which decontaminating agents can be added from tanks carried on the vehicle. Water output is 25 m³/h at a maximum suction head of 4 m. A type AE-3000 AC 3-phase generator is fitted and the crew compartment is supplied with 3.5 m/h of filtered air. The engine exhaust can be controlled and directed by an operator seated in a cab to the right of the engine and connected to the vehicle driver by an intercom system.

The ADTT-1 can also be used for snow clearing from runways and other surfaces, for aircraft de-icing and may be employed to fight fires.

A civilian fire extinguishing variant carried on a ROMAN 19215 truck chassis is known as the TURBOJET R37 and is intended to fight large area or oil well fires.

Status
In production. Offered for export sales.

Manufacturer
AEROSTAR SA.

VERIFIED

ADTT-1 jet DECON system in operation   0050784

RUSSIAN FEDERATION AND ASSOCIATED STATES (CIS)

Artillery decontamination kit, Model ADK

Description
The Model ADK kit is issued to the crews of artillery pieces and large calibre mortars for decontaminating their weapons and is effective against both blister and nerve agents. The kit is contained in a metal case which holds four 1 litre cans of decontamination solution, two smaller plastic containers, two application brushes with handle extensions, two metal scrapers, about 500 g of cotton wool, a 150 mm long roll of sealing tape and four cork gaskets. Instructions are inside the case lid and four rubber blocks keep the cans in place. Two of the four cans are embossed with the number '1' and have red lids: they are for the decontamination of blister agents and V-type nerve agents. The two black-lidded cans are embossed with the number '2' and are for use against G-type nerve agents. The scrapers are used to remove mud and dirt from the equipment as the solutions, normally two, are being mixed.

The first solution is made by emptying one 80 g packet of DT-6 decontaminant powder into a can with a red lid, which contains 1 litre of dichloroethane, and shaking for approximately 5 to 10 minutes. All possible surfaces are wiped with the cotton wool and then the mixture is applied with the brushes supplied. The cans with the black lids contain the ready-mixed No 2 aqueous solution of 2 per cent sodium hydroxide, 5 per cent monoethanolamine and 20 per cent ammonia. Even if there are no G-type nerve agents present, the solution No 2 has to be used to remove the corrosive solution No 1 from unpainted metal surfaces. When any contaminating agent is unidentified the solutions are used in numerical order.

After use the kit can be refilled with solutions and powder by an ADM-48 decontamination vehicle (qv) and the spare gaskets and tape supplied with the kit are used to reseal it ready for reuse. The scrapers can also be used to loosen the solution container caps.

Status
Available from Russian state factories before privatisation. In service with the RFAS and former Warsaw Pact forces. Current manufacturer unknown.

Manufacturer
(Unknown).

VERIFIED

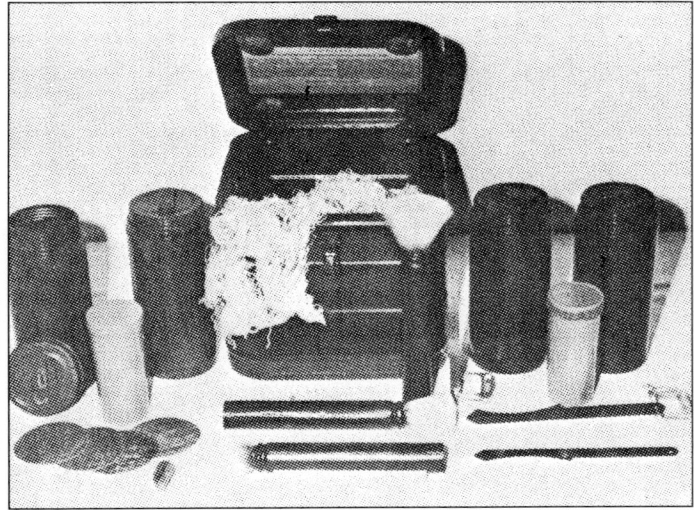

ADK artillery decontamination kit

BKSO decontamination kit

Description
The BKSO decontamination kit is carried on trucks and armoured personnel carriers for the vehicle crews to carry out initial decontamination procedures following a chemical attack. The kit is carried on each vehicle inside a metal case and contains the equipment, solutions and components required for the connection of the kit to the vehicle's exhaust and compressed air brake systems. By mixing decontaminating agents in an aqueous solution the resultant aerosol is sprayed over the vehicle exterior using the vehicle exhaust/ compressed air pressure. Consumption of decontamination solution for an aqueous solution is from 0.6 to 1.5 litres/min. For solid decontaminants the rate is 0.2 to 0.8 litres/min.

The kit weighs 26 kg when packed in a metal case measuring 650 × 530 × 190 mm, or 16 kg when in a canvas bag. The operating temperature is from −40 to +50°C.

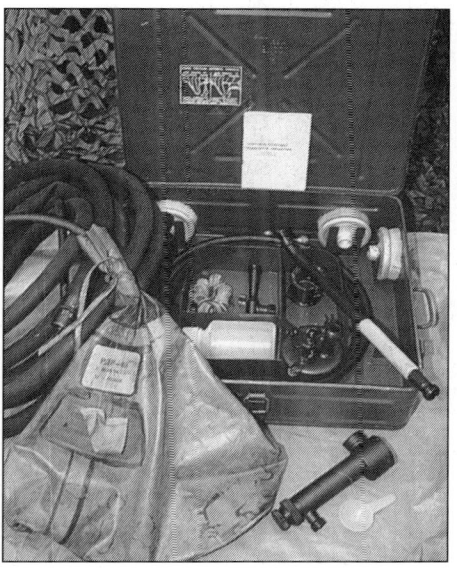

BKSO decontamination kit

Status
Available.

Marketing agency
Rosoboronexport.

VERIFIED

Decontamination apparatus, backpack, Model RDP-4V

Description
This hand-operated backpack equipment s used to decontaminate vehicles, equipment, weapons, small buildings and small areas of ground. The backpack is held in place by a pair of straps at the waist and shoulders. It consists mainly of a tank with a large filling aperture, filter screen, clamp-on pressure cap, piston-type air pump operated by the left hand, outlet tube and hose and a spray pipe and nozzle to which a brush may be attached. A wrench and a small canister containing tools are fitted to the bottom of the tank. The equipment can be used to spray decontamination solution No 1 or 2, depending on the nature of the contamination. The pump handle is normally clipped in the vertical position but in use it is horizontal and operated by the left hand: a pumping rate of 25 to 30 strokes/min gives a discharge flow of about 0.7 litres/min. When full, the Model RDP-4V weighs approximately 17 kg; empty weight is 8.5 kg. t is 355 mm high, 290 mm long and 190 mm wide.

Status
Available from Russian state factories before privatisation. In service with the RFAS and former Warsaw Pact forces. Current manufacturer unknown.

Manufacturer
(Unknown).

VERIFIED

Model RDP-4V backpack decontamination apparatus

Decontamination apparatus, truck-mounted, Model ARS-14

Description

This equipment is a development of the earlier Model ARS-12U. The main change is the use of a ZIL-131 (6 × 6) 3,500 kg truck chassis, but the full extra tank capacity this provides (2,700 litres, weighing 2,500 kg when filled with a normal decontaminant load) cannot be used because extra drums of decontaminant are carried on specially fitted racks. However, the extra drums enable the equipment to be used over a longer period without reloading. Some changes were made to the piping system in that the previously fixed outlet pipes are carried separately and fitted to the equipment by hoses. The wide-spreading DN-3 nozzle is fitted at the vehicle front as well as the rear, and the eight hoses are wound onto four drums instead of the former eight; these drums are at the left rear. Other changes were made to the general 'plumbing', but an innovation is an extra rubber hose which can be fitted to the vehicle exhaust in Winter to thaw out any frozen parts of the equipment. Filling pistols can also be fitted to the system for loading or reloading decontamination kits.

A loaded ARS-14 weighs 10,185 kg. It is 6.856 m long, 2.47 m wide and 2.48 m high.

Status

No longer in production. In service with the RFAS, Czech Republic and Slovakia.

Manufacturer

Unknown.

VERIFIED

Model ARS-14 truck-mounted decontamination apparatus

Decontamination apparatus, truck-mounted, Model ARS-14K

Description

The decontamination apparatus, truck-mounted, Model ARS-14K follows similar overall lines to the earlier Model ARS-14 but is carried on the chassis of a KamAZ-4310 cross-country truck. It may be used to decontaminate weapons, equipment and ground vehicles, carrying all the equipment, consumables and other items for this role.

Operated by a crew of three, the ARS-14K carries one 2,700 litre tank, one 1,040 litre tank, two mechanical pumps, a hand pump, pipelines, an intake for free-flowing chemicals, ten 20 litre canisters and some other items of equipment. The overall system can provide between five and eight decontamination stations at one time. Between six and eight large items of equipment can be decontaminated every hour. Using the pipelines and dispensers provided, a strip of terrain 5 m wide can be decontaminated over a length of 1,400 m.

Setting up or removing the equipment from operation takes between 8 and 15 minutes. The operating temperature range is from −40 to +50°C.

Status

No longer manufactured. May remain in service with some former Warsaw Pact armed forces.

Model ARS-14K truck-mounted decontamination apparatus

Manufacturer

Kraneks Machinery Company Limited.

Marketing Agency

Rosoboronexport.

VERIFIED

Decontamination apparatus, truck-mounted, Models ARS-12D, ARS-12U and ARS-12M

Description

Apart from use as decontamination equipment, the Models ARS-12D, ARS-12U and ARS-12M can be used as water carriers, for firefighting and to provide cold showers for various purposes. The three models are basically similar but the Model ARS-12D is mounted on a ZIL-151 (6 × 6) 2,500 kg truck chassis and the Model ARS-12U on a ZIL-157 (6 × 6) 2,500 kg truck chassis. The Model ARS-12M appears to have been produced for the Czech and Slovak armed forces only, as it is based on the chassis of the Praga V3S (6 × 6) 3,000 kg truck.

Each consists of a 2,500 litre tank divided by two baffles with a large manhole for access and filling. Also provided is a depth gauge, self-priming pump driven from the vehicle engine, hand pump, piping system, hoses, nozzles and other spares and accessories. The main pump drive shaft turns at 1,400 to 1,600 rpm and can turn the pump to deliver 300 to 400 litres/min. The hand pump, operated at 45 strokes/min, can deliver 4.5 to 5.5 litres/min. Decontaminants are mixed in the tank as water is poured in and a thorough mixture is made by internally recycling the solution through the pump, otherwise the vehicle motion and the filling mixing is all that is necessary. Many different decontaminants may be used but the standard solutions are numbers 1 and 2. There are several administration methods. To clear roads or terrain a wide-spreading nozzle (the DN-3) can be fitted directly to the main discharge pipe and the vehicle is then driven over the affected road or terrain. A full tank can then clear a strip 500 m long and 5 m wide. For decontaminating vehicles and other equipment, up to eight 18 m long hoses can be fitted, each with spray pipes, nozzles and nozzle brushes. Up to four vehicles can be cleansed at one time and there is sufficient solution in a full tank to cleanse 12 MBTs, 13 APCs, 15 trucks or 45 artillery pieces. To decontaminate buildings (or for firefighting) four 25 mm diameter hoses can be used. Racks on top of the tank hold drums of decontaminant and several RDP-4V backpack equipments can be carried for remote use. The Model ARS-12D can be used to emit smoke screens by discharging chlorosulphonic acid.

Status

No longer in production. Available from Russian state factories before privatisation. In service with the RFAS, Czech Republic and Slovakia. Current engineering support source unknown.

Manufacturer

(Unknown).

VERIFIED

Czech Army ARS-12M mounted on V3S (6 × 6) 3,000 kg truck chassis

Decontamination apparatus, truck-mounted, Models ARS-15 and ARS-15M

Description

The decontamination apparatus, truck-mounted, Model ARS-15 entered production in 1983 and is basically the same as the ARS-14 (see separate entry in this section) but based on the chassis of the Ural-375 (6 × 6) 4,000 kg truck. Some improvements have been introduced to facilitate operations in humid environments and it can operate at temperatures of down to −15°C.

The ARS-15 has a crew of three and takes about 15 minutes to bring into operation. Once ready for use it can decontaminate and disinfect weapons, vehicles and other equipment and can also decontaminate areas of terrain. The equipment carried on the vehicle can also heat water and solutions for various purposes, including mobile bath units and may be employed to fight fires. It is also possible for the ARS-15 to provide heated air.

Part of the decontamination equipment carried by the decontamination apparatus, truck-mounted, Model ARS-15

The ARS-15 is provided with a TsN-245 mechanical pump constructed of titanium. Also provided is a 2,300 litre tank and a heater which can heat the tank contents to 70°C in 1 hour. The tank may be used for the temporary storage and transport of various liquids. A heated air blower unit is also of titanium construction.

The ARS-15 can decontaminate up to 12 vehicles every hour. One ARS-15 can decontaminate between 23 and 150 vehicles (depending on vehicle size) for each tank filling. The system can provide warm water to allow a maximum of 24 individuals to bathe every hour.

The Model ARS-15M is the latest version of this equipment, based on the chassis of the Ural-4320 (6 × 6) 4,500 kg truck. There are a few design differences from the earlier model.

The operating temperature range for both models is from −40 to +50°C.

Status
In service with the RFAS and other armed forces.

Manufacturer
Ocher Engineering Plant.

Marketing agency
Rosoboronexport.

VERIFIED

Decontamination apparatus, truck-mounted, Models DDA-53, DDA-53A, DDA-53B and DDA-66

Description
These equipments are used by decontamination units and medical units for sterilisation, disinfecting and disinfestation. With chemical units the Model DDA-53 series is used for steam-decontamination of chemically and biologically contaminated clothing and small items of equipment, the apparatus can also provide hot water for showers.

Each equipment consists of a vehicle borne system with two steam chambers, vertical boiler (the RI-3), fuel oil tank, water pump, formaldehyde tank, 12-head shower unit and all the associated hoses and fittings. The system does not have its own water tank, so water has to be provided by another tanker vehicle or a stand tank if no natural source is available. The water boiler, which produces steam or hot water, contains 250 litres and is normally fired by fuel oil, although wood may be used with a loss in heating efficiency. The fuel oil tank holds 55 litres which is enough for 8 to 10 hours operation. Pipes connect the boiler output to steam chambers, each of which can contain 25 to 30 Summer uniforms, 20 Winter uniforms or 12 sheepskin jackets. Decontaminants can be added to the steam if necessary. The system can thus decontaminate up to 80 uniforms/h in Summer and up to 48 in Winter. When used for showers, the system is normally used in conjunction with tents which are carried on other cargo trucks; enough water can be provided for up to 100 showers/h in Summer and 70 to 72 in Winter.

DDA-53A truck-mounted, decontamination apparatus based on GAZ-63 (4 × 4) 1,500 kg truck chassis

There were four models of the DDA-53 (sometimes known as the ADA), some of the oldest of which have now been withdrawn. The Model DDA-53 was mounted on the GAZ-51 (4 × 2) truck chassis, the Model DDA-53A on the GAZ-63 (4 × 4) chassis and the Model DDA-53B on a ZIL-130 (4 × 2) 4,000 kg truck chassis. The DDA-53B differs also in having the boiler and steam chambers enclosed in a metal body. The Model DDA-66 is mounted on a GAZ-66 (4 × 4) 2,000 kg truck chassis.

Status
Available from Russian state factories before privatisation. In service with the RFAS and former Warsaw Pact forces. Current manufacturer unknown.

Manufacturer
(Unknown).

VERIFIED

Decontamination kit, Model PKhS

Description
The Model PKhS kit is a personnel and clothing decontamination set for areas which are too large to be covered effectively by an IPP kit. It is contained in a plywood case and consists of three 0.5 litre bottles, two large and two small packets of decontaminant agent, two packets containing 10 gauze pads each, a mixing dish and a wooden stirrer. Two of the bottles are sealed with white wax and the other has a red wax seal. The powdered contents of one large and one small decontaminant agent packet are emptied into the dish and mixed with the contents of one of the white wax sealed bottles and the resultant solution can be freshly used as an anti-blister agent decontaminant. The contents of the red wax sealed bottle can be used direct from the bottle as a nerve agent decontaminant. Whichever solution is used, it is normally applied using the gauze pads.

The carrying case, which is supplied with a carrying sling, is 305 mm wide, 305 mm high and 90 mm deep.

It is possible that this kit is in the process of being replaced by a modernised version.

Status
Available from Russian state factories before privatisation. In service with the RFAS and former Warsaw Pact forces. Current manufacturer unknown.

Manufacturer
(Unknown).

VERIFIED

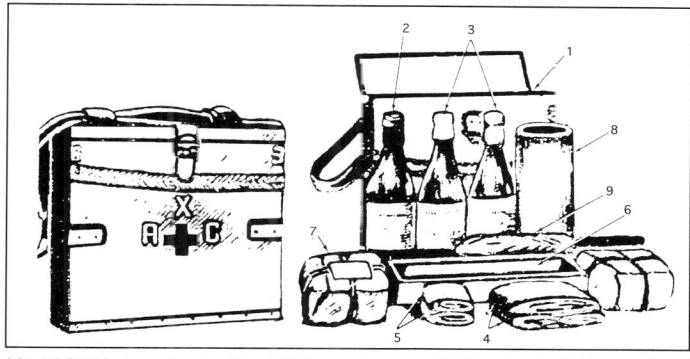

Model PKhS decontamination kit (1) carrying case (2,3) solvents (4,5) powdered decontaminants (6) mixing dish (7) gauze pads (8) container for powdered decontaminants (9) wooden stirrer

Decontamination packet, model DPS

Description
The Decontamination packet, model DPS is used for the personal decontamination of clothing once a contaminated area has been left. The packet is made from clear plastic and is issued sealed. Inside are an instruction sheet and a fabric dusting bag which is used to dust the agent-absorbing brown powder over all outside clothing and headgear. Once applied, the dust is rubbed into the clothing with either the bag or rubber gloves. The dust is then shaken or brushed from the clothing before the protective headgear is removed.

It is possible that this kit is in the process of being replaced by a modernised version.

Status
Available from Russian state factories before privatisation. In service with the RFAS and former Warsaw Pact forces. Current manufacturer unknown.

Manufacturer
(Unknown).

VERIFIED

Decontamination station, Models AGV-3M and AGW-3M

Description

The decontamination station Model AGV-3M is used for steam decontamination of chemically or biologically contaminated clothing and small pieces of equipment. It consists of four special vehicles: one Model AGV-3M truck-mounted decontamination steam and hot air generator; two Model AGV-3M truck-mounted decontamination steam chamber equipments and one cargo truck carrying a drying tent, shower tent, collapsible water tank and accessories.

The steam and hot air generating unit is mounted on a ZIL-157 or ZIL-151 (6 × 6) truck chassis. It has a van-type body which contains an oil-fired heater, 500 litre boiler with a superheater, petrol engine for driving the fuel oil and water pumps, single-stage turbo-blower and a heat exchanger. The steam generated in the boiler is superheated to between 160 and 200°C and then passed to the steam chamber vehicles. The turbo-blower supplies 350 m³/min through the heat exchanger at a pressure of 0.5 kg/cm², which is then supplied to the drying tent via a large diameter hose. The boiler also supplies hot water for the crew's shower unit. The steam chamber units are mounted on either a ZIL-130, ZIL-150 or ZIL-164 (4 × 2) truck chassis. Each unit has three 2 m³ pressure chambers and all the pipes and other fittings. Clothing is hung inside the steam chambers which are then pressurised by the steam from the steam generator vehicle. Extra decontaminants can be added to the steam if necessary. After a period in the steam chambers the clothing is taken to the drying tent. The cargo truck used with the station is usually a ZIL-130 or a ZIL-164. The station can process between 50 and 150 uniforms/h, depending on the type of contamination.

The decontamination station Model AGV-3M replaced the earlier Model AGV-2. In East German service the Model AGV-3M was known as the Model AGW-3M.

Status

Available from Russian state factories before privatisation. In service with the RFAS and former Warsaw Pact forces. The earlier Model AGV-2 may remain in service with some civil defence organisations. Current manufacturer unknown.

Manufacturer

Rosoboronexport.

VERIFIED

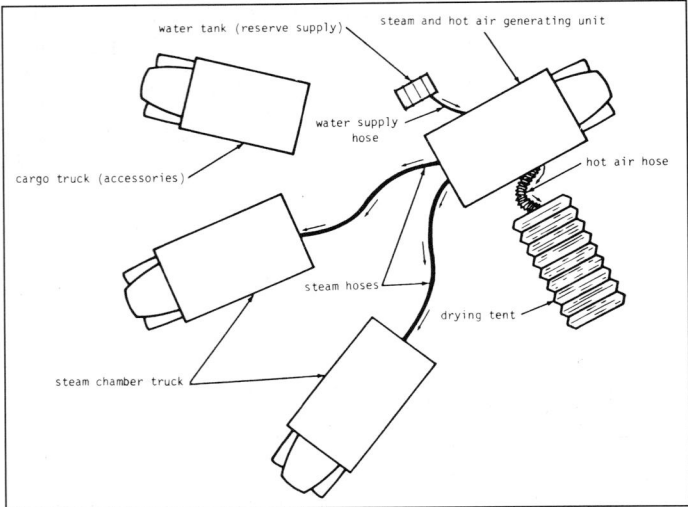

Layout of AGV-3M and AGW-3M decontamination system

Decontamination system, portable, Model DKV

Description

This portable system is used to decontaminate vehicles and is composed of 78 cylindrical tanks carried on a specially equipped truck and trailer. Each cylindrical tank is preloaded with decontamination solution No 1 or 2 and each tank can be fitted with either one or two spray pipes. The tanks are delivered to the users where they can be pressurised by either the user vehicle's airbrake system or by a separate compressor. The spray pipes are usually fitted with a brush and the contents of two cylinders are usually sufficient to decontaminate a truck; an MBT takes three cylinders. Each cylinder can take up to 30 litres. Several cylinders can be fitted to a single compressed air source and the cylinders can be collected and refilled.

The latest version of the DKV is the DKV-M - see entry under Bulgaria for available details.

Status

Available from Russian state factories before privatisation. In service with the RFAS and former Warsaw Pact forces. Current manufacturer unknown.

Manufacturer

(Unknown).

VERIFIED

Individual decontamination kit, IPP

Description

The IPP kit has largely replaced the earlier IPP-3 kit and contains decontaminants capable of dealing with nerve agents as well as blister and biological agents. It is normally carried in the protective mask carrier and is contained in a plastic case. The main items inside the case are a glass phial and a plastic phial which contains a glass ampoule. Gauze pads are also included and another small compartment contains four anti-smoke gauze-wrapped ampoules. The glass phial contains anti-nerve gas decontaminant and the plastic phial decontaminant for use against blister gases. The plastic container holds alcohol and the glass ampoule chloramine-B powder; crushing the ampoule mixes the two. The resultant mixture can be used once the plastic phial has been punctured with the metal point in the case lid. The gauze pads are used to apply the solution to the affected areas. Enough mixture is supplied to cover 5 m². In use the nerve gas decontaminant would be spread first from the glass phial, followed by the mixture from the plastic phial. The four anti-smoke ampoules can be removed quickly from the kit case by fitted drawstrings. They are crushed and placed inside the wearer's protective mask and the resultant inhalations can overcome the effects of most irritant smokes. The ampoules contain a mixture of chloroform, ethanol, ethyl ether and ammonia water.

Status

Available from Russian state factories before privatisation. In service with the RFAS and former Warsaw Pact forces. Also produced in Iraq. Current manufacturer unknown.

Manufacturer

(Unknown).

VERIFIED

Individual decontamination kit IPP produced in Iraq

Machine gun/mortar decontamination kit, Model PM-DK

Description

The kit Model PM-DK follows much the same lines as the larger artillery kit model ADK and the solutions involved and their use are the same. The difference lies in the smaller sizes of the items concerned. Only two solution cans are contained in

Model PM-DK machine gun/mortar decontamination kit

the metal case, one of the No 1 and one of the No 2. There is only one application brush which fits directly onto the solution can in use and a cleaning wire is supplied to clean the hollow handle of the brush. Each solution can contains 250 ml. Like the larger artillery kit, the Model PM-DK can be refilled and reused and spare gaskets are supplied to reseal the solution cans. There are two types of carrying case, both with carrying slings. One has straight sides with rounded corners and the other has curved, body-contoured sides.

Status
Available from Russian state factories before privatisation. In service with the RFAS and former Warsaw Pact forces. Current manufacturer unknown.

Manufacturer
(Unknown).

VERIFIED

Personal weapons decontamination kit, Model IDP

Description
The Model IDP kit is contained in a drab olive metal case and, although primarily issued for decontaminating personal weapons, there is enough decontaminant in the kit for use on small crew-served weapons. The kit consists of five cotton swabs and two ampoules. One of the ampoules is marked with a red tip and is intended for use with blister agents. Its position in the case is marked by a '1' embossed in the case wall. The other ampoule (No 2) has a black tip and is intended for use against nerve agents. The weapon is first wiped with a swab to clear as much of the agent as possible. The appropriate ampoule is then opened and the decontamination solution applied, after which the weapon is wiped dry and lightly oiled. Each ampoule contains 82 ml of solution. The case is 130 mm long, 80 mm wide and 40 mm deep and the entire kit weighs 305 g. Printed instructions are glued to the side of the case.

Status
Available from Russian state factories before privatisation. In service with the RFAS and former Warsaw Pact forces. Current manufacturer unknown.

Manufacturer
(Unknown).

VERIFIED

Protector-N truck-mounted decontamination apparatus

Description
The Protector-N (also known as the Protektor) truck-mounted decontamination apparatus, described as a calorific decontamination vehicle, is essentially an updated version of the TMS-65 and TMS-65M (see separate entry in this section) and operates along the same general lines. The main changes are the mounting of the decontaminating equipment on a KrAZ-260 (6 × 6) 9,000 kg truck chassis and the relocation of the decontamination solution tank along the left-hand side of the vehicle. It appears that this tank has a larger capacity than that of the TMS-65. It is also possible that a more powerful jet engine is used to dispense the decontamination solution. As well as the usual role of decontaminating vehicles, aircraft and similar equipment, the Protector-N can also be employed to decontaminate hard-surfaced roads and terrain.

The Protector-N has a crew of two and can be made ready for use or removed from operation in 10 to 12 minutes. When operating, it can decontaminate between 10 and 40 trucks or other vehicles every hour or between 5 and 15 aircraft in the same time scale.

The operating temperature range is from −40 to +50°C.

Protector-N truck-mounted decontamination apparatus

Status
Available.

Marketing agency
Rosoboronexport.

VERIFIED

SLOVENIA

NBC decontamination vehicle

Description
Few details are known regarding an NBC decontamination vehicle built on to the chassis of a TAM 150 T11 BV (6 × 6) 3,000/5,000 kg truck. It carries a tank for decontamination solution over the rear axles with what appears to be an independently powered pump unit at the rear and hose and reel units just behind the hardtopped cab. The pump, hose and reel units are contained inside housings with upward-opening flaps. There appears to be no provision on the vehicle for other than the usual crew of two, including the driver.

For full details of the TAM 150 T11 BV (6 × 6) 3,000/5,000 kg truck refer to *Jane's Military Vehicles and Logistics 1998-99* page 429.

Status
In service with the Yugoslav armed forces.

Manufacturer
Tovarna Avtomobilov Motorjev (vehicle only).

VERIFIED

SWEDEN

Frenatus CARGO decontamination unit

Description
The Frenatus CARGO decontamination unit is designed to offer first response units a safe facility in which contaminated walking and stretchered personnel can be decontaminated before being passed to medical teams for triage and treatment.

The DECON unit is based on a towed, double-axle trailer which contains the frames and material for erecting two tented shower and treatment areas. It also carries an air heater unit, power generator and water heater. The unit is capable of processing 60 to 120 walking personnel and 15 to 30 stretcher cases per hour and can be made operational within 10 minutes.

A strict separation is maintained between the clean and contaminated areas and the personal decontamination teams wear the Textil Skyddsteknik personal protection equipment (see separate entry under PROTECTION (individual) - Body protection). The unit is designed to be stationed at the upwind end of the cordoned area, acting as a controlled exit point for potentially contaminated response teams.

The CARGO trailer requires connection to a water supply and delivers hot water to the shower and decontamination brush points at up to +35°C. With ambient temperatures at −20°C, the interior of the treatment area can be maintained at +20°C throughout. The unit can supply water at 70 to 90°C for decontaminating vehicles and equipment at 40 to 60 litres/min through high-pressure nozzles.

Status
In production.

Manufacturer
Frenatus International AB.

VERIFIED

Hot air unit VA-8

Description
The Hot air unit VA-8 is used for the decontamination of uniforms and other equipment stowed in an adjacent decontamination tent or chamber. For instance, the unit can be used to heat air to a nominal 110 to 130°C so that 150 standard uniforms can be decontaminated within a 2 hour period.

The hot air unit is mounted in a sound dampened enclosure complete with a sound suppression device and a heat distribution system. The superheated air is generated in what is described as a pulse jet motor, so that the exhaust air is mixed with fresh replacement air and then blown into the decontamination chamber.

The hot air unit VA-8 has a heat capacity of 75 kW and can produce a maximum temperature of 220°C at the maximum air flow of 800 to 1,000 m³/h. The fuel used is standard automotive petrol and the fuel consumption is 9 litres/h.

Hot air unit VA-8

Specifications
Heating unit
Weight: 140 kg
Length: 2.06 m
Width: 603 mm
Height: 660 mm

Sound suppression and distribution system
Weight: 50 kg
Length: 1.39 m
Width: 460 mm
Height: 460 mm

Status
In production.

Manufacturer
Ventilatorverken.

VERIFIED

SEDAB container decontamination system

Description
The SEDAB container decontamination system is built upon a 6 m steel container with special cargo profile for lorry transports. The container is built in two compartments: a machinery compartment and storage space. Between the two compartments there is a stainless 3000 litre water tank that functions as a partition wall.

The machinery is handled from inside the container. During transport, all required equipment is loaded into the container. There is a fire-proof door at one of the long sides, which is the entrance to the engine room. The container is equipped with a heater in order to keep the temperature of the container from going down to below zero degrees. There is a fan inside to enable air circulation and to prevent the engine room from getting too warm. The system is self-supporting in electricity, hot water, air and heat. The entire system is propelled with diesel.

It requires four people to set up the system. The entire system is operated from the engine room by an engine man. With full capacity and self-supporting on electricity, heat, hot and cold water the system is built for decontamination of up to 600 people per hour, but can easily be extended.

The container is constructed to meet the requirements of the Swedish Rescue Service Authority to have a water temperature of 37°C and to be operational within 15 minutes of arrival. Components in the system can also be used for other tasks in rescue operations, for example heated and lighted reception camp for wounded, command centre, heated shelter at scenes of accidents and storage.

Status
In production.

Manufacturer
SEDAB, Safety Equipment Development AB.

VERIFIED

SEDAB hot air decontamination system

Description
The SEDAB hot air decontamination system is based on a standard 6 m container. It is designed for the decontamination of portable equipment which can withstand temperatures up to 110°C. The system is ready to operate within five minutes after arrival. The container has the capacity to decontaminate 110 complete protective suits with tools and instruments in one decontamination session.

The hot air decontamination system can be used for biological decontamination, for hazardous waste from hospital, infected clothing and material in refugee and catastrophe camps. The system is constructed to meet the requirements of the Swedish Rescue Service Agency.

Status
In production.

Manufacturer
SEDAB, Safety Equipment Development AB.

UPDATED

SEDAB hot air decontamination system   0050787

SEDAB trailer decontamination system

Description
The SEDAB trailer decontamination system is based upon two 1,500 kg trailers or two tracked vehicles.

Trailer 1 is the decontamination trailer with all the equipment to fulfil a complete decontamination of staff and civilians who have been contaminated. Trailer 1 can solve tasks without trailer 2, but the two trailers should be grouped together at the same place in order to carry out the decontamination procedure.

SEDAB Trailer - equipment stowage   0050774

SEDAB Trailer Decontamination System   0050775

Trailer 2 is constructed like trailer 1, but with an extra 700 litre water tank, trailerheater, battery and charger. During transport all the required equipment is loaded into the trailer and car. With full capacity and self-supporting on electricity, heat, hot and cold water it has a capacity for decontamination of up to 80 people per hour.

Status
In production.

Manufacturer
SEDAB, Safety Equipment Development AB.

VERIFIED

SWITZERLAND

INTER-CB decontamination apparatus E-85

Description
The INTER-CB decontamination apparatus E-85 is a portable and rugged hand pump device for the spraying of NBC decontamination solutions. Using a plastic body, it has an internal volume of 1.5 litres which is sufficient to decontaminate an area of about 10 m². About 20 operations of the hand pump handle are sufficient to empty the apparatus. It is claimed that no maintenance of the apparatus is necessary. Training decontamination solutions can be used with this device.

Birchmeier & Cie AG also produces a 10 litre stainless steel decontamination unit operated by a hand pump.

Status
In service with the Swiss Army since 1985.

Manufacturer
INTER-CB.

VERIFIED

INTER-CB
decontamination
apparatus E-85

UNITED KINGDOM

Aireshower inflatable personal DECON units

Description
The Aireshower inflatable DECON unit has been designed in partnership with the UK Ambulance Service Association to create a stand-alone facility which can be erected swiftly at an incident location, providing a comprehensive facility for the decontamination of first responders. It is supplied with its own hot water supply, and is equally appropriate for use at the scene of chemical incidents or at hospital casualty departments.

The Aireshower is constructed of 1300 GM/M Hypalon (synthetic rubber) on polyester fabric (1100DTEX). The walls and roof are made of lightweight reinforced vinyl and the floor from 1000 GM/M2 PVC on polyester (940 DTEX). It can be inflated using a standard SCBA bottle or a foot pump. Internal pressure is 0.75 bar.

Aireshower D/W/D
The Aireshower D/W/D variant is a larger version (see below).

	Aireshower	Aireshower D/W/D
Internal measurements:	2.4 × 2.1 × 2.3 m	3.7 × 4.3 × 2.3 m (1.2 m undressing area, 2.4 m decontamination area, 1.2 m redressing area)
Weight:	70 kg	84 kg
Storage size:	1 × 0.5 × 0.5 m	

Status
Available.

Manufacturer
Airshelta Limited.

NEW ENTRY

Decontaminant chemical agent

Description
Decontaminant chemical agent is available in kit form and provides chemical reagents for decontaminating vehicles and other unit equipment contaminated with persistent chemical warfare agents, thus reducing the hazards of contact or subsequent vapour circulation. The reagents are dissolved in water and sprayed using a stirrup pump. The kit is also an efficient disinfectant for biological agents.

Status
In production. In service with the UK armed forces.

Manufacturer
Richmond Packaging (UK) Limited.

UPDATED

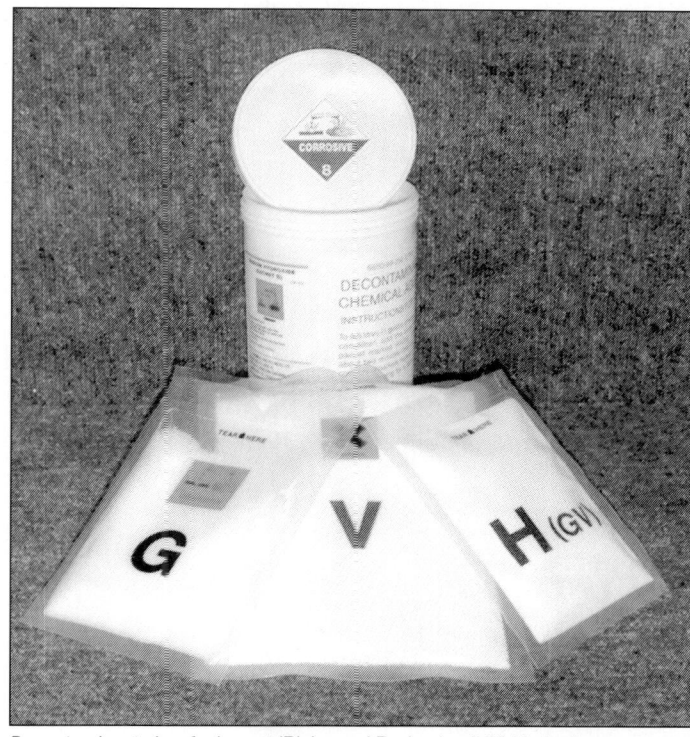

Decontaminant chemical agent (Richmond Packaging (UK) Limited) 2001/0098653

Decontamination Kit, Personal No 1 Mark 1

Description
The Decontamination Kit, Personal No 1 Mark 1, is normally issued to all field personnel and is carried as part of their standard personal equipment. It consists of a sealed clear plastic bag containing a set of instructions and four pads charged with Fuller's earth. In use, the plastic bag is torn open and one pad is taken out and placed over one glove. The pad is then banged over suspected contaminated areas of skin or clothing and equipment to dispense the Fuller's earth, which is then rubbed over using the pad to spread the powder.

Status
In production. In service with the UK armed forces.

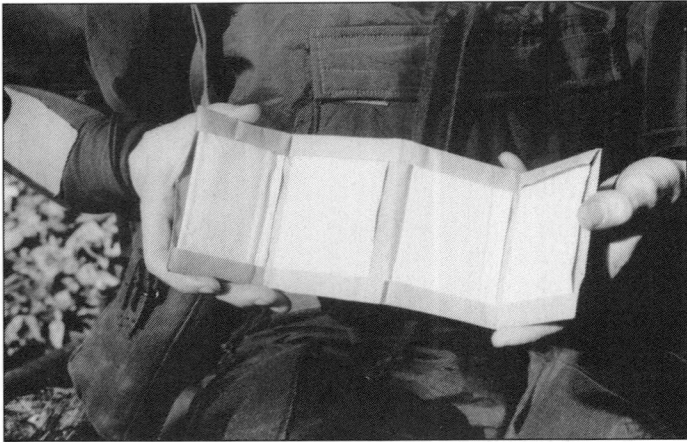

Decontamination Kit, Personal No 1 Mark 1 (Richmond Packaging (UK) Limited)

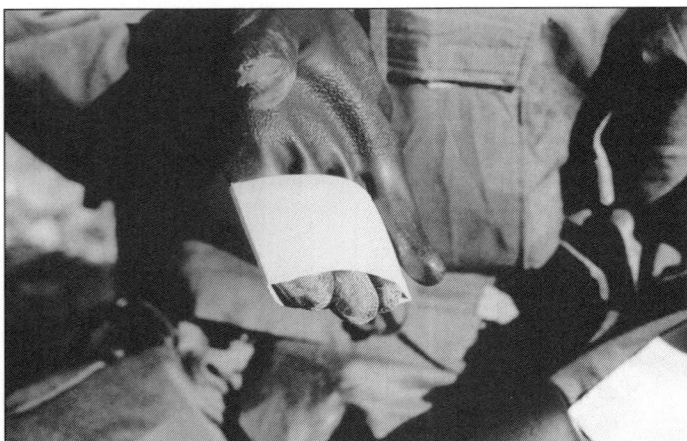

Decontamination Kit, Personal No 1 Mark 1 as worn over glove for personal decontamination (Richmond Packaging (UK) Limited)

UK MoD authorised manufacturers
(For overseas sales)
Remploy Limited.
Richmond Packaging (UK) Limited.
BCB International.

Other supplier
J & S Franklin Limited.

UPDATED

Decontamination Kit, Personal No 2 Mark 1

Description
This personal decontamination kit consists of a flat polythene dispenser containing 113 g of Fuller's earth. It is used for decontaminating the user's boots and personal

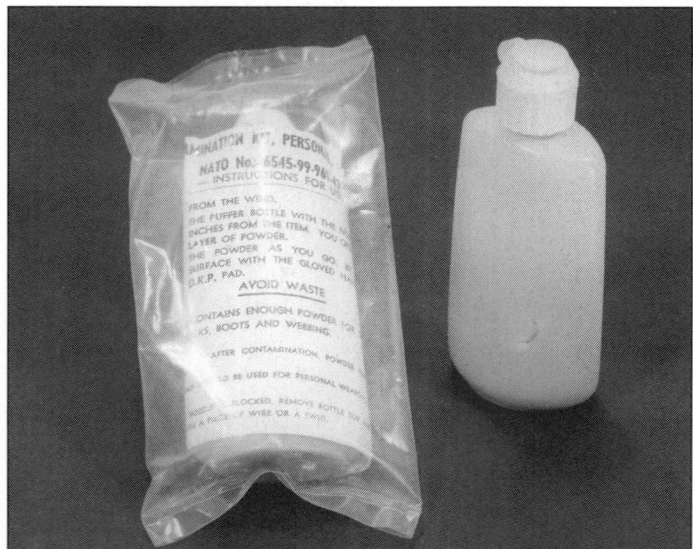

Decontamination Kit, Personal No 2 Mark 1

equipment and would normally be used in conjunction with the Decontamination Kit Personal No 1 Mark 1.

Status
Production complete. In service with the UK armed forces.

Suppliers
Richmond Packaging (UK) Limited.
J & S Franklin Limited.

VERIFIED

Hughes emergency response and decontamination units

Description
This range of products is aimed at both defence and civil emergency planning users. The decontamination facilities include electrical generation, oil fired heaters, thermostatically controlled shower cubicles, casualty handling measures suitable for stretchered and walking wounded personnel and personnel monitoring. These facilities can be delivered in standard trailer bodies, containers or de-mountable units built to any size or specification.

The facilities are capable of handling large numbers of contaminated personnel at incident scenes or, placed closer to main medical facilities, as pre-surgery decontamination rooms.

Status
Available.

Manufacturer
Hughes Safety Showers.

VERIFIED

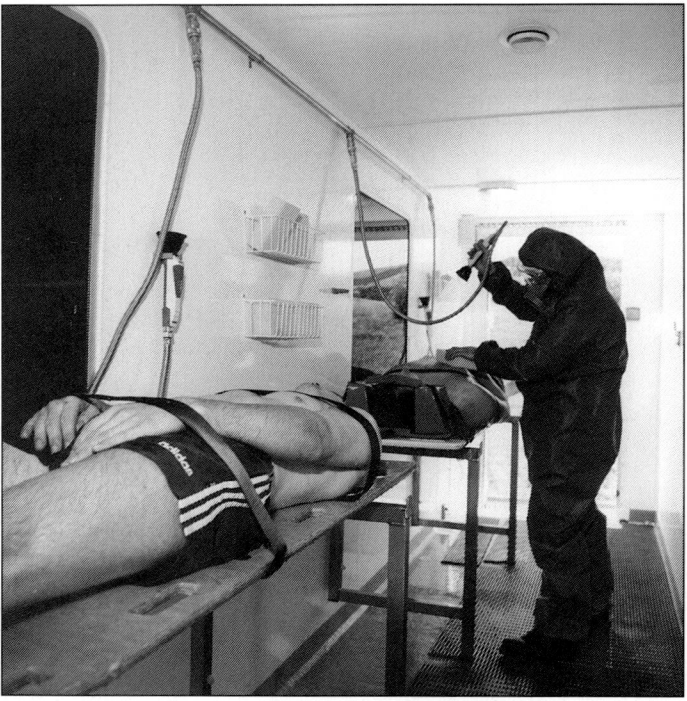

Casualty monitoring in a Hughes emergency response and decontamination unit
0018708

Plychem DECAS W decontamination unit

Description
The DECAS W casualty decontamination shower unit is a development of the original DECAS system (see the 1998-99 edition of *Jane's NBC Defence Systems*). Following consultation with UK medical and emergency service operators, the unit can now accommodate a stretcher and medical team. With a footprint of 1.9 × 2.8 m, the DECAS W packs down to fit into its large holdall sized transport bag.

The unit can be erected in under 3 minutes. It is based on a robust inflatable frame which can be inflated using a BA cylinder or external airline. Floor, roof and side screens contain the spread of contaminated fluids and the floor well has a waste outlet point, compatible with standard UK and overseas emergency service connections. The water supply, also through a standard connector, is directed to two coiled supply pipes inside the unit with drench hose attachments. An in-line water heating system can also be supplied. A ballast skirt and lugs allow the unit to

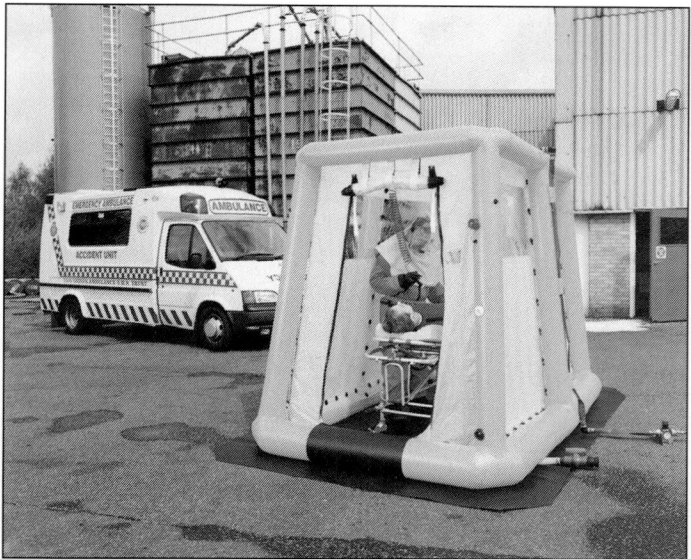

Plysu DECAS W decontamination unit   0050782

be stabilised in high winds. The inflatable structure, the contaminant containment booth and the water delivery system are connected together using secure quick-release fasteners. A double-size version, the Plychem DECASX2, is also available.

Status
Available.

Manufacturer
Plysu Protection Systems Limited.

VERIFIED

....

Plychem DPI decontamination units

Description
DPI
The Plychem DPI inflatable decontamination shower unit is designed for use by on-scene contamination incident response units. It packs into a 20 kg holdall and is easily erected by personnel, with a minimum of training. Made from durable polymer material, it allows external personnel to carry out decontamination without hazard to themselves. The brush units are connected to a pressurised supply of decontaminant fluid by flexible coiled piping, operated from outside through rubber glove sleeve units built into the sides. Showered decontaminant maintains a continuous flow over personnel inside the structure.

The 40 litre/min flow rate is maintained through a series of nozzles and can be supplied from the domestic main. Inflation is achieved from BA bottles or compressed airline supply. An all-round ballast skirt ensures stability of the structure in high wind conditions, through the weight of the drain fluid on the rubberised floor.

Plychem DPI inflatable decontamination shower unit (Plysu PLC)   **2000**/0050783

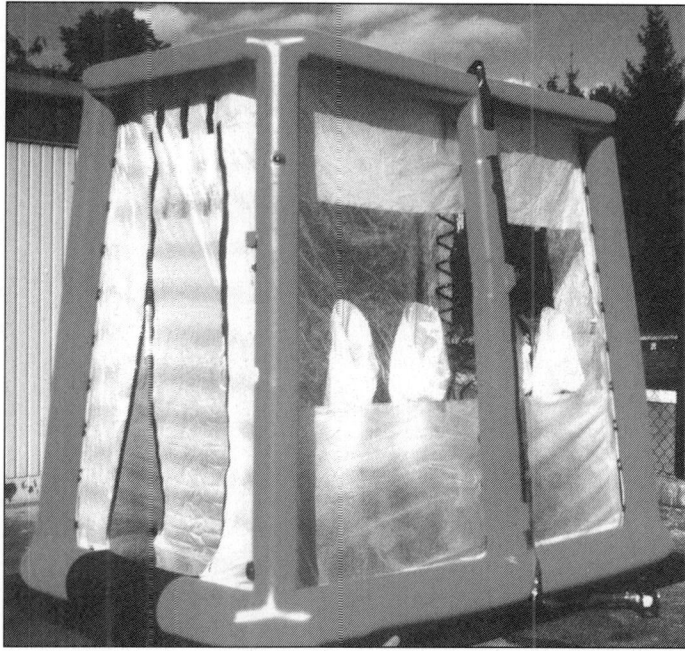

Plychem DPIX2 variant (Plysu PLC)   **2001**/0097442

DPIX2
The DPIX2 is a larger alternative to the DPI, capable of handling several personnel simultaneously through increased provision of water-jets and hand-held water-fed brushes. A water supply of 2 bar pressure (minimum) is required and detergent injection systems can be used in-line with the supply to improve the quality of decontamination. Interior valves give the choice of using either all water sprays or half the system to conserve water, depending on operational requirements. The low-flow, high-pressure water-jets create a mist that achieves effective decontamination with the minimum of water usage (80 l/min) whilst, at the same time, creating the minimum amount of contaminated residue requiring disposal. The DPIX2 weighs 42 kg and packs into a canvas bag from which it can be deployed within three minutes to create a full-size stable shower cubicle. Inflated size is 2.9 × 1.9 × 2.4 m.

Status
Available.

Manufacturer
Plysu PLC.

VERIFIED

....

Portaflex 300 decontamination shower unit

Description
The Portaflex 300 decontamination shower unit uses its rugged carrying case as a non-slip raised base platform, eliminating the necessity for a heavy-duty stainless steel tubular base. Instead the basic design is a lay-flat hose assembly.

Portaflex 300 decontamination shower unit in use
0018707

Portaflex 300 packed into carrying case   0018706

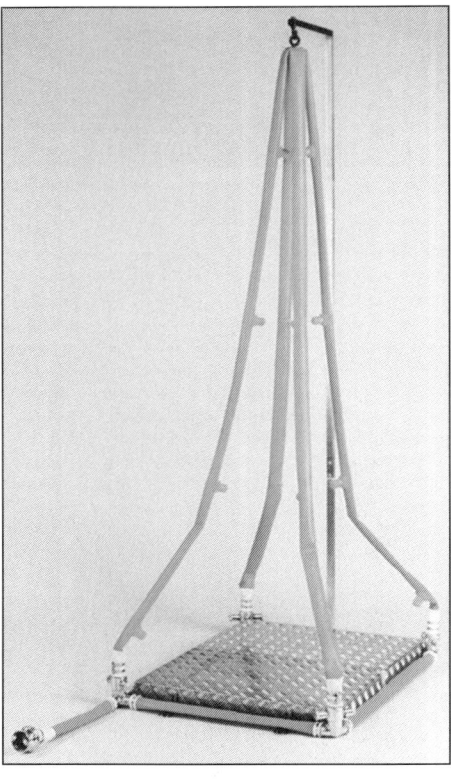

Portaflex 300 erected for use

With this shower unit, 38 mm/1.5 in lay-flat Duraline hoses form four shower legs, each of which is fitted with four spray nozzles. The base assembly, which fits around the foot platform, incorporates four 38 mm/1.5 in male instantaneous couplings which mate with female couplings on the legs. When under pressure all the hose legs and the base frame assembly become rigid, thus forming a stable frame for the decontamination of personnel. The Portaflex 300 may be used in conjunction with the Portaflex CUPOLA decontamination shelter (see entry in this section).

The Portaflex 300 shower unit can be assembled in less than 30 seconds. It weighs 23 kg and packed dimensions are 750 × 480 × 170 mm.

Status
In production.

Manufacturer
Hughes Safety Showers.

VERIFIED

Portaflex CUPOLA decontamination shelter

Description
The Portaflex CUPOLA decontamination shelter is deployed in conjunction with the Portaflex 300 decontamination shower unit (see entry in this section). It was designed for occasions where the containment of contaminated water or decontamination agents is necessary following a decontamination operation. The Portaflex CUPOLA is constructed so that personnel undergoing decontamination walk through the shelter from a dirty area to a clean area beyond, while the water (or decontaminating agent) which has been fed from the clean area is eventually directed through to the dirty area.

A Portaflex CUPOLA decontamination shelter in use   0018709

Status
In production.

Manufacturer
Hughes Safety Showers.

VERIFIED

Stella-Meta Portable Water Purification Unit WPU7(NBC)

Description
The WPU7(NBC) is a portable, skid-mounted, membrane water treatment unit which produces clean drinking water from any fresh water source contaminated with NBC agents at an output of 40 litres/hour. It is simple to operate and has been field-tested. It is powered by a single cylinder diesel engine driving a triple cylinder high pressure pump. The WPU7(NBC) together with the pre-filter, diesel engine, two reverse osmosis elements, a single carbon column and hoses are mounted within a single tubular frame and will fit inside a small utility or trailer.

Status
In service with the British Army and other armies worldwide.

Manufacturer
Stella-Meta Filtration Systems.

NEW ENTRY

Stella-Meta Water Purification Units (NBC)

Description
The Stella-Meta Water Purification Unit (NBC) Fresh - WPU(NBC)6F was developed in conjunction with the UK MoD to treat NBC contaminated drinking water sources. The treatment process includes precoat filtration, reverse osmosis, activated carbon treatment and post-disinfection, capable of removing ionic, organic and bacterial contaminants. The layout diagram provided, shows each stage of the process together with the storage and distribution equipment, using flexible storage tanks supplied with the unit.

The WPU(NBC)6F has been extensively field-tested and is in service with the British Army. All the pumps, hoses, storage tanks and accessories, together with ready-use consumables, are mounted within a self-contained space frame on a standard twin-axle trailer.

Stella-Meta WPU(NBC)S in operation   **2002**/0102995

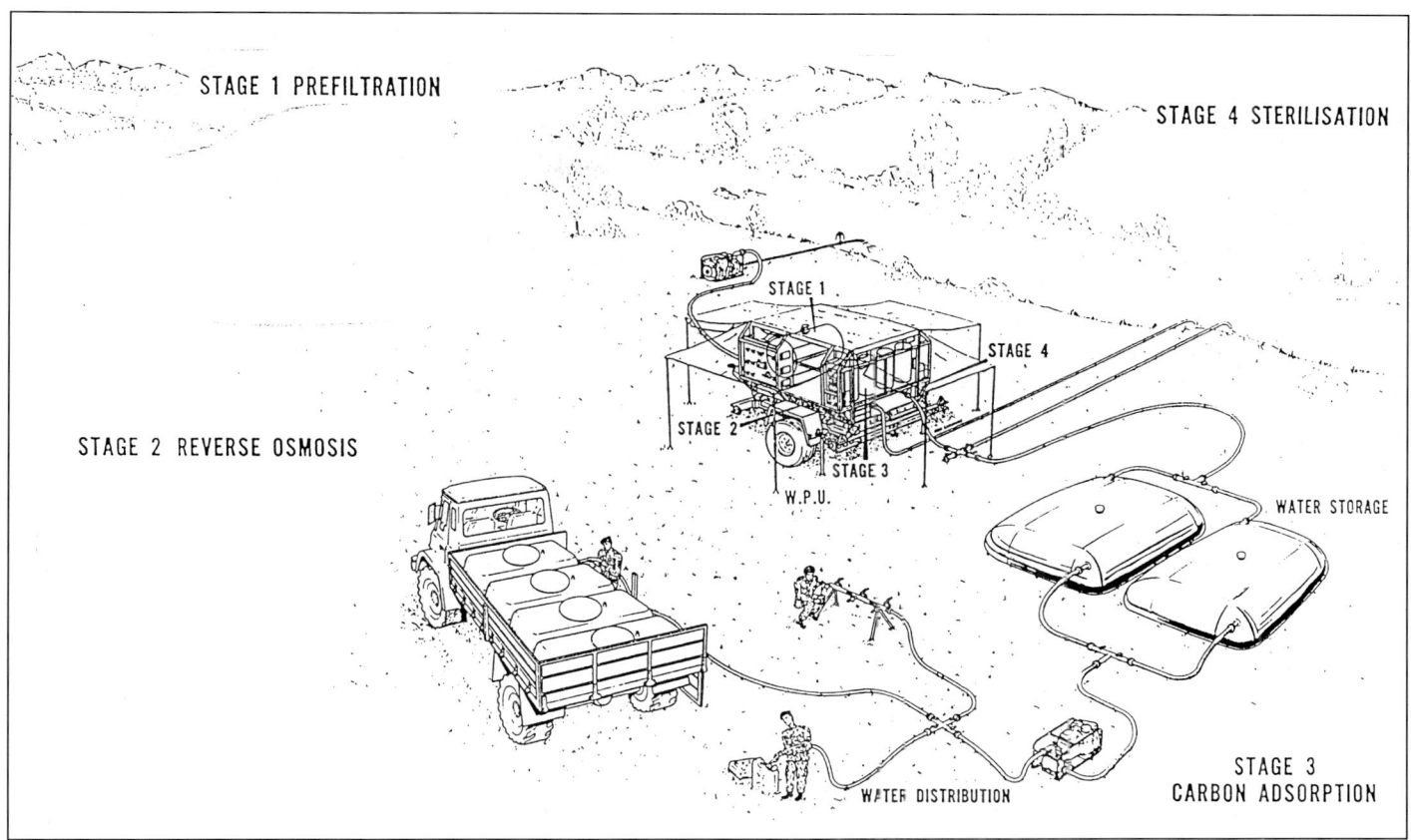

STAGE 1 PREFILTRATION

STAGE 4 STERILISATION

STAGE 2 REVERSE OSMOSIS

STAGE 1

STAGE 4

STAGE 2

STAGE 3

W.P.U.

WATER STORAGE

STAGE 3
CARBON ADSORPTION

WATER DISTRIBUTION

Artist's impression of a typical water point using the WPU(NBC)6F

Specifications
Flow Rate: 2.2 m³/h
Weight: 2,300 kg
Length: 2.4 m
Width: 2.1 m
Height: 1.7 m

Operation of the unit in full NBC protective clothing is possible as a result of the ergonomic design of the controls and valve positioning. The WPU can treat 2.2 m³/h in the full NBC mode. NBC treatment can be bypassed, allowing treatment of agent-free local fresh water supply, at the higher rate of 8.2 m³/h.

A seawater version, the WPU(NBC)S, allows drinking water to be produced from an NBC contaminated saline water source.

Status
In service with the British Army and other armies worldwide.

Manufacturer
Stella-Meta Filtration Systems.

UPDATED

UNITED STATES OF AMERICA

Decontaminating apparatus, portable, DS2, ABC-M11

Description
The decontaminating apparatus, portable, DS2, ABC-M11 is carried on a rack on nearly all US-produced military vehicles and is intended for the decontamination of vehicles or crew-served weapons. Normally the equipment is issued with the nitrogen pressure cartridge already filled and in place but the container has to be filled with DS2 decontaminating agent before the equipment is ready for use. When used, the handle safety pin seal wire is removed and the handle is lifted to puncture the pressure can. A thumb lever then controls the spray, which has an optimum range of about 2 m and the equipment will cover approximately 42 m².

The standard M11 has a filled volume of 1.26 litres. Also available are the M11 Stretch with a filled volume of 1.5 litres and the M11 Super Stretch holding 2.66 litres. All three models can be used over a temperature range from −31.7 to +49°C.

Also available is the Model M11 A/G dry sorbent dispenser. This equipment dispenses the same free-flowing resin-based AMBERGARD™ XE-555 powder as that contained in the M291 skin decontaminating kit (qv).

The range of Decontaminating Apparatus, Portable, DS2, ABC-M11

Specifications

Model	M11	M11 Stretch	M11 Super Stretch
Container volume	1.42 litres	1 67 litres	2.66 litres
Filled volume	1.26 litres	1 5 litres	2.5 litres
Weight filled	3.42 kg	3 77 kg	6.34 kg
Weight empty	2.19 kg	2 31 kg	3.07 kg
Diameter	139.7 mm	139.7 mm	139.7 mm
Height	361.9 mm	361.9 mm	525 mm
Duration of charge	30 s	35 s	60 s
Coverage	25 m²	150 m²	248 m²

Status
In service with the US Army, Israel armed forces and some other nations.

Marketing agency
Tradeways Limited.

UPDATED

Decontamination kit, individual equipment: M295

Description

The decontamination kit, individual equipment: M295 employs the same adsorption technology as the M291 SDK (see entry in this section) and consists of a kit containing four individual 'wipedown' mitts, each enclosed in a soft protective packet designed to fit comfortably in a battle dress overgarment pocket. Each mitt has adsorbent resin contained within a non-woven polyester material and a polyethylene film backing. In use, resin from the mitt is allowed to flow freely through the non-woven polyester pad material. Decontamination is accomplished by adsorption of contamination by both the non-woven pad and the resin.

The M295 mitt can be used to decontaminate protective hoods, masks, gloves and footwear as well as personal weapons, helmet and webbing. The intended basis of issue is one fibreboard container holding 20 kits for each squad or section. The container may be stored inside or outside tactical vehicles.

Each kit weighs approximately 227 g and measures approximately 220 × 140 × 50 mm. The M295 kit can be used over a temperature range of from −32 to +71°C.

Status

Entering service with the US armed forces. First issues made during Fiscal Year 1994.

Development agency

Edgewood Chemical Biological Center (ECBC).

Marketing agency

Tradeways Limited.

UPDATED

HAZ/MAT shower unit and collection pool

Description

The HAZ/MAT shower unit and collection pool is part of a flexible range of decontamination facilities available to first responders. The pool unit is large enough to deal with a contaminated casualty or it can support two separate shower units for uninjured personnel. The pool and the shower unit covers are made from heavy-duty agent-proof material to reduce the spread of contamination residue when in use. The supply piping is made from easy-connect heavy-duty PVC and is designed to be compatible with standard US emergency service supplies. System operating handles are designed for ease of use when wearing IPE gloves.

Status
Available.

Manufacturer

HAZ/MAT DQE Inc.

UPDATED

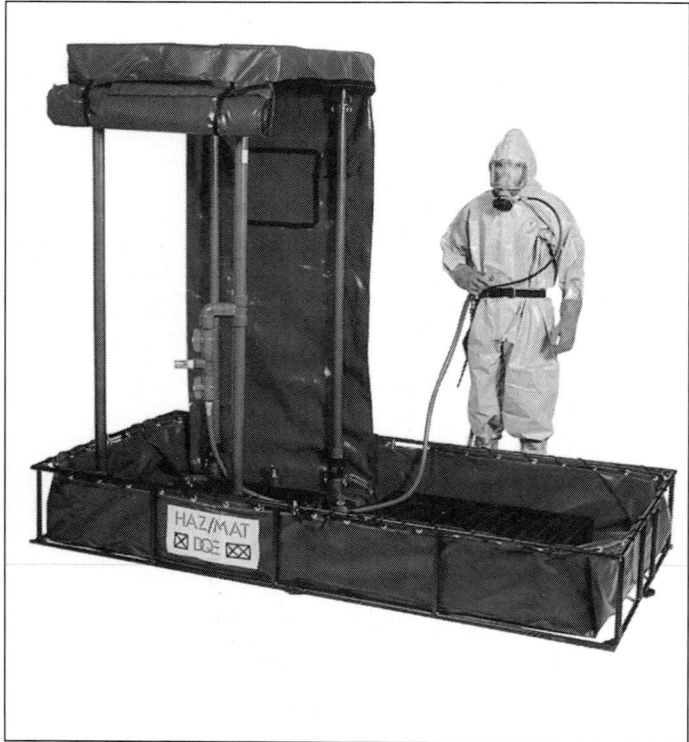

HAZ/MAT units shown packed (HAZ/MAT DQE Inc)   *2002*/0050776

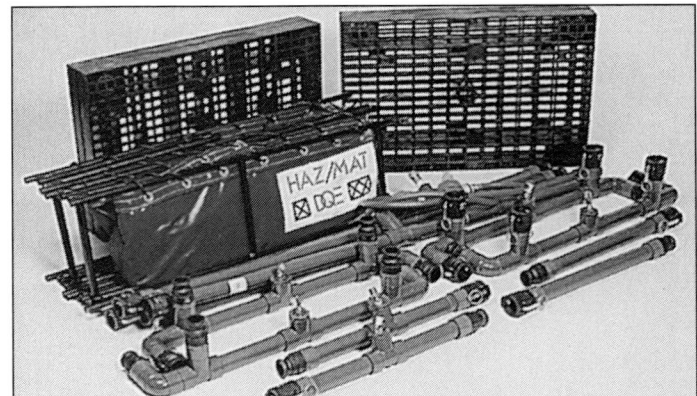

HAZ/MAT single shower unit and collection pool erected for use (HAZ/MAT DQE Inc)   *2002*/0050777

Lightweight Water Purification System (LWPS)

Description

The Lightweight Water Purification System is currently under development in the USA to deliver 0.3 m3/h of potable water in the field to support deployed operations by US special forces. It can treat salt or fresh water, contaminated with any type of NBC agent. Based on a series of portable modules, the system includes pre-filtration and reverse osmosis as the key treatment processes.

Status

Joint product development by Centech Group and Stella-Meta (UK). See also Stella-Meta Water Purification Unit (NBC) in this section. Joint bidder for 320 unit US DoD tender as part of an Engineering Materiel Development programme.

Manufacturers

Centech Group Inc.
Stella-Meta.

VERIFIED

M13 Portable Decontamination Apparatus (DAP)

Description

The M13 Portable Decontamination Apparatus (DAP) is a vehicle-mounted, man-portable, manually operated unit. It is used for the decontamination of wheeled and tracked vehicles, towed artillery and crew-served weapons described as 'larger than 60 calibre' (0.60 in/15 mm). Development began in 1979 with first deliveries commencing during 1985.

The apparatus consists of a prefilled disposable container holding 14 litres of DS2 decontaminating agent (70 per cent diethylenetriamine, 28 per cent ethylene glycol monomethyl ether, 2 per cent sodium hydroxide), an accessory container holder, a manual in-line pump, one or two wand sections and a disposable synthetic filament polypropylene brush. An accessory container is provided for the storage of all components. The apparatus may be used to dispense decontaminating agents other than DS2.

Status

In production. In service with the US Army and some other nations.

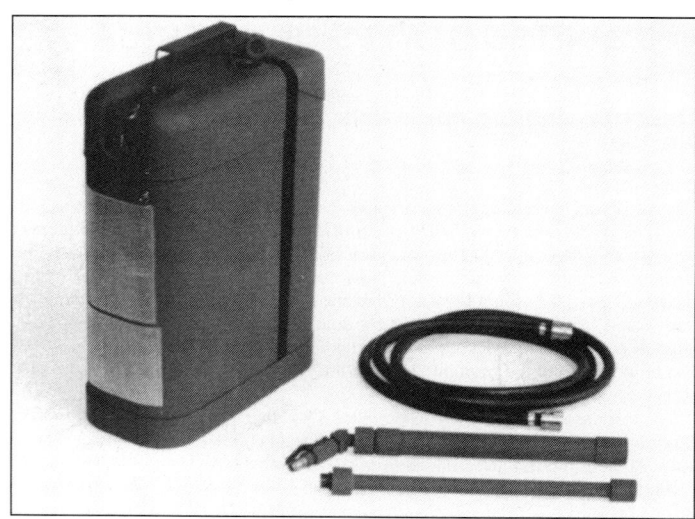

M13 portable decontamination apparatus (DAP)

Specifications

Volume: 14 litres
Weight filled: 24.5 kg
Weight empty: 10.9 kg
Coverage per filling: 112 m²
Length: 355.6 mm
Width: 168.4 mm
Height: 476.3 mm
Operating temperature: −31.7 to +49°C

Manufacturer

Dalden Corporation.

Marketing agency

Tradeways Limited.

VERIFIED

M21 decontamination pumper   0050786

Manufacturer

The Centech Group Inc.

Marketing agency

Tradeways Limited.

VERIFIED

M17 lightweight decontamination system, Sanator

Description

The M17 lightweight decontamination system, Sanator, is produced in the USA under licence from Karl H Høie & Co A/S of Oslo, Norway. For full details of this equipment refer to LDS NBC-SANATOR® III entry under Norway in this section.

Status

Type classified and in service with the US Army, Air Force and Marine Corps. Orders to date have exceeded 4,000 units.

Manufacturer

Engineered Air Systems Inc.

VERIFIED

M22 decontaminating apparatus

Description

The M22 decontaminating apparatus high pressure washer is under engineering development in the same way as the M21 Decontaminating Pumper (see entry in this section). It delivers a higher performance than the M21 and is designed as the initial part of a 5-stage vehicle DECON process being evaluated by SBDCOM (the US Army Soldier and Biological-Chemical Defense COMmand at Edgewood, Maryland). The M22 comprises a diesel engine prime-mover which powers a water heater and provides electrical and pumping power. High-volume, low-pressure and low-volume, high pressure pumps deliver water to the operators through two hose-fed decontamination lances. The operators' task is to loosen and remove the gross contamination from a vehicle, using a variety of lance tools (powered brushes, jets, sprays) before decontaminant fluid is applied as a second stage, using the M21, upwind of the first operation. The M22 is delivered in two large transit cases.

Status

In development.

Manufacturer

The Centech Group Inc.

Marketing agency

Tradeways Limited.

VERIFIED

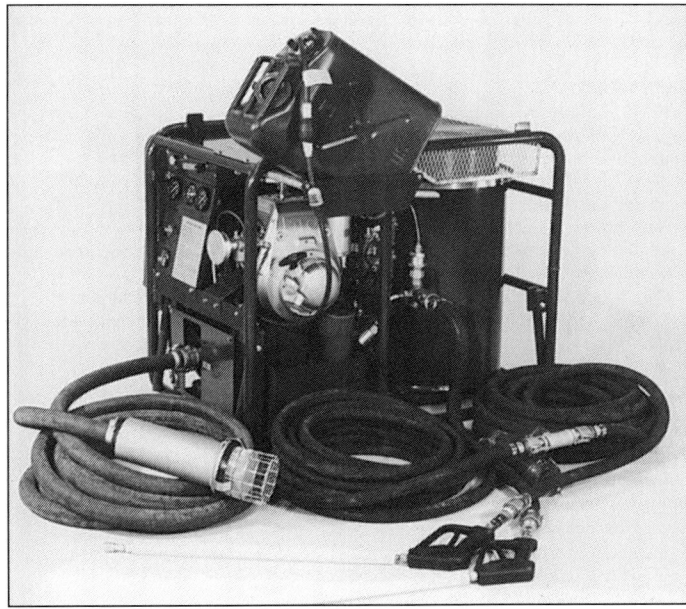

M17 lightweight decontamination system, Sanator

M21 decontamination pumper

Description

The M21 decontamination pumper has been designed to deliver high-volume, high-pressure decontaminant for the swift decontamination of vehicles and other large equipment.

The unit is based on a 4.5 kW 6-cylinder, air-cooled, fuel-injected diesel engine, consuming 1.9 litres of fuel/h. The engine drives a chemical pump with an output of 15.4 litres/min up to 30 bar. Amongst a variety of cleansing apparatus, the equipment set includes two 24 V DC motor driven 350 rpm power scrub brush assemblies.

The pumper is delivered in three modules. The diesel engine is fitted into a man-portable tubular carrying frame. A transport box carries hoses and hose reels. A second box contains the hose-end assemblies.

Status

In development.

Specifications

Engine: 14 kW air-cooled diesel
Fuel consumption: 1.9 litres/h
High-volume/low-pressure pump: Flowmax 5 MP. 9,240 litres/h at up to 2.7 bar
Low-volume/high-pressure pump: CAT positive displacement. 19 litres/h at 34-205 bar
Heater: Beckett Burner - 443,000 kJ
Dimensions: 72.5 × 120 × 100 cm
Weight: 334 kg (465 kg with all accessories)

M22 decontaminating apparatus   0050781

M291 Skin Decontamination Kit (SDK)

Description

The M291 Skin Decontamination Kit (SDK) is used to decontaminate the skin following contamination by liquid chemical agents. It can be used in an emergency to decontaminate the outside of protective masks, butyl rubber gloves, the hood and an individual weapon. It contains a non-toxic compound which will not irritate the skin; it can also be used around wounds and the face. It can also be used for training purposes. This kit replaces the M258A1 kit in US service.

The kit consists of a wallet-type pouch containing six individual packets, each containing a non-woven fibre fill laminated pad impregnated with the decontamination compound that reacts with chemical agents to absorb and neutralise them in a single step, non-toxic/non-irritating application. The free-flowing resin-based powder used is AMBERGARD™ XE-555. Instructions for use are marked on the pouch and packets.

The M291 kit weighs 45 g and measures 112 × 112 × 36 mm.

Status

In service with the US armed forces.

Marketing agency

Tradeways Limited.

VERIFIED

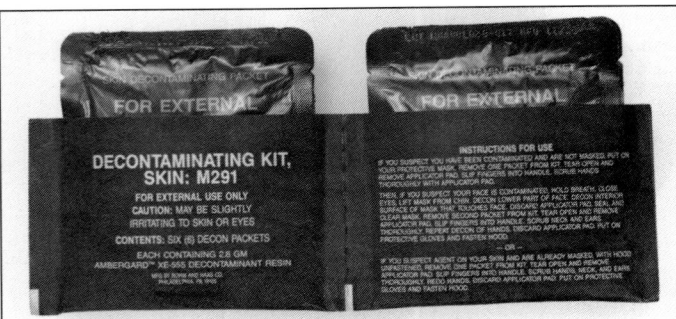

M291 Skin Decontamination Kit (SDK)

Mass Casualty Decontamination System

Description

The Mass Casualty Decontamination System is a trailer-mounted facility, based on a standard ISO container. It is designed to offer decontamination for personnel following a CBW incident It can deliver decontaminant and water washing facilities suitable to process a claimed throughput of 800 walking or stretchered people per hour.

Inside the container, there is a heating system with an output of 8 million kJ, an automated decontamination mixer and supply system and external booms and pumping nozzles. The interior of the container allows operators to control the decontamination process in a COLPRO environment, supplied with filtered air. There are also access ramps for stretcher cases and NBC medical treatment facilities.

The Mass Casualty Decontamination System can be set up and operational within 5 minutes of arrival at a site.

Status

Available.

Manufacturer

Modec Inc.

Marketing agency

Tradeways Limited.

UPDATED

Multi Shield TSP barrier cream

Description

Multi Shield TSP barrier cream was developed by Interpro Inc and the US Army Medical Research Institute of Chemical Defense for use in the Persian Gulf region during Operation Desert Shield. The cream has been extensively and successfully tested under both laboratory and high-temperature field conditions by the US Army Medical Research Institute of Chemical Defense.

Multi Shield TSP is an easily applied, non-greasy skin protectant cream which can be applied in anticipation of chemical attack. The cream will provide individual skin protection by inhibiting the penetration of Distilled Mustard (HD) agent. The cream will also enhance the effectiveness of field decontamination systems.

When applied, Multi Shield TSP can be worn for extended periods and will not adversely affect face mask seals. In the event of a chemical attack involving HD, a 0.15 mm layer of Multi Shield TSP will mean that decontamination can be delayed for 60 minutes, after which the cream can be removed using soap and water or an alcohol wash. The cream will not cause skin or eye irritation, although it is recommended that any cream contacting the eyes should be rinsed away as soon as possible.

Multi Shield TSP is supplied in 118 ml/4 oz or 454 ml/16 oz plastic dispenser bottles and may also be supplied in 28 ml/1 oz bottles. The bottles are marked with directions for use.

Status

In production.

Manufacturer

Interpro.

VERIFIED

'Nor E' series of mobile decontamination units

Description

Nor E-60T™ and Nor E-120T™

The Nor E-60T™ is a four-stall quick erect mass decontamination shower suite with the capability to decontaminate 60 walking casualties per hour of operation, or with the litter (stretcher) decontamination suite, 45 walking and 12 litter casualties per hour of operation. The system is modular, allowing expansion or customisation to meet agency requirements for increased capacity. It can either be used on its own or as part of other modules, adapted to fit existing shelters. The outer cover is proofed against CW and BW agents, easily decontaminated, flame and UV radiation resistant. The cover stretches over a quick erect frame system. It is suitable for all types of weather and meets NATO requirements for wind and snow loading. It is designed to be highly mobile, self-contained and easily deployed. Two teams of two can deploy and set up the unit, ready for operation, in 15 minutes. The system utilises a temperature-/pressure-controlled module so that all showers can be individually turned on and off to conserve water and offer a warm to hot shower without the fear of scalding users.

A larger version, the Nor E-120T™, is an eight-stall quick erect mass decontamination shower suite with the capability to decontaminate 120 walking casualties per hour of operation, or with the litter decontamination suite, 90 walking and 24 litter casualties per hour of operation. Three teams of two can deploy and set up the unit, ready for operation, in 30 minutes.

Nor E-100SCT™

The Nor E-100SCT™ is a semi-autonomous, trailer-borne DECON unit with the capability to decontaminate either 75 walking casualties, or 24 litter casualties per hour of operation.

Features (all units):

- Self-contained system collects and stores contaminated water from the showers
- Uses only 7.5 litre/min/person or 30 litre/4 min DECON cycle to decontaminate completely
- Clearly defined decontamination flow with no possibility of recontamination
- Private curtains in change area mean male and female can be co-located
- Casualty roller system for litter patients is designed to stop the spread of contamination without causing further injury or discomfort
- The complete shower facility is decontaminable and reusable

Status

Available.

Manufacturer

Nor E First Response Inc.

Specifications

Standard features for Nor E-60T™ and Nor E-120T™

Quick erect modular outer frame set and covers in NBC tan: 4.9 × 7.3 m

Quick erect shower suite

    Nor E-60T™: 4 stalls, 4 shower heads

    Nor E-120T™: 8 stalls, 8 shower heads

Secondary containment berm, raised flooring and extraction pump

Quick erect litter shower suite: Soap/solution induction system with separate rinse, secondary containment berm, raised flooring and extraction pump

Module A: water pump/heater (portable with turf wheels): 110 V, 443 MJ on demand, diesel fired, colour-coded connection hoses, camlock quick connections to Module B suction hose (5 cm × 4.6 m)

Module B: temperature/pressure control module (portable with turf wheels): Self-contained pump for auxiliary water source, colour-coded connection hoses, camlock quick connections to Module A

Collapsible NBC pillow tank for sealed fresh water: 11,365 litre

Collapsible NBC onion bladder for grey/contaminated water: 11,365 litre

Water hoses with camlock connections, colour coded: 7.6 m

Two emergency lighting kits

    Nor E-60T™: 2 fluorescent 40 W HH bulbs

    Nor E-120™: 8 stands with 2 fluorescent 500 W bulbs

NEW ENTRY

Tactical Water Purification System (TWPS)

The Tactical Water Purification System is a trailer or truck-based facility currently under development in the USA to deliver 6 m³/h of potable water in the field. Described as the primary means for US forces water purification, it can treat salt or fresh water, contaminated with any type of NBC agent. This mobile, integrated system includes pre-filtration and reverse osmosis as the key treatment processes.

Status

Joint product development by Centech Group and Stella-Meta (UK). See also Stella-Meta Water Purification Unit (NBC) and the LWPS in this section. Joint bidder for 400 units US DoD tender as part of an Engineering Materiel Development programme.

Manufacturers

Centech Group Inc.
Stella-Meta.

VERIFIED

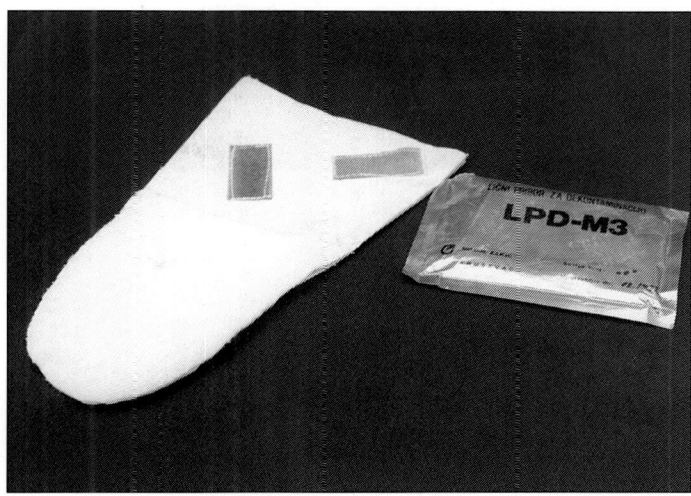

Personal decontamination kit LPD-M3   0018712

YUGOSLAVIA, FEDERAL REPUBLIC

Personal decontamination kit LPD-M3

Description

The personal decontamination kit LPD-M3 is based around a textile glove consisting of two pads containing Fuller's earth for skin decontamination and a soft fabric pad for cleaning decontaminated surfaces. When issued the kit is wrapped in a sealed pouch. The LPD-M3 is intended for use on uncovered skin surfaces, clothing, personal weapons and equipment.

Status

In service with the Yugoslav Army.

Contractor

Trayal Corporation.

UPDATED

DEMILITARISATION

DEMILITARISATION

GERMANY

NIGAS - non-destructive testing system for the identification of the contents of intact CW and explosive-filled munitions

Description

NIGAS, the Neutron Induced GAmma Spectrometer for the detection of explosives or CW agent inside intact munitions such as shells or bombs, thereby avoiding the dangers associated with dismantlement. This non-destructive testing equipment is transportable and is suitable for both ammunition depots and operational sites. The NIGAS system relies on a High Performance Germanium detector (HPGe) which picks up the γ emissions from the contents of the munition, irradiated by a neutron generator. The pulsable neutron generator uses the deuterium-deuterium reaction as a source. This method ensures there is no residual radiation risk when the unit is switched off. The γ characteristics differ depending on the element and, depending on their relative concentrations, allows precise identification of the explosive or CW agent contained in the shell or bomb.

The munition can be precisely positioned in front of the detector using a remotely programmable adapter and support device. The entire system is managed by computer, including the presentation of real-time data to the analyst.

Technical specifications
- Intrusive, non-destructive method
- Usable in ammunition destruction facilities
- Transportable. It can be used in ammunition depots and under field conditions
- Measurement at different types of shells, bombs, cans and barrels
- Identification of the characteristic elements for CW agents (H, N, F, P, S, Cl, As) and of other elements (Na, Mg, K, Ca, Ti, Zn, Br, Sr, Sn, I, Ba, Pb) which identify uncommon contents or special construction materials
- Automatic spectrum analysis and substance identification.

Neutron source:
Neutron generator: E_n=2.5 MeV
Neutron flux: 1×10^6 to 1.5×10^6 neutrons/s
Pulse length: 20 μs
Pulse frequency: 10 kHz

Physical data:
Power consumption: 1.7 kW
Dimensions (measuring system): 130 × 96 × 135 cm
Weight: 270 kg

Software

The integrated NIGAS control and analysis software gives fast access to all relevant parameters for system operation. The complete spectrum acquisition and substance identification cycle is repeated every 10 seconds. Analytical data is automatically interpreted, and results are displayed online during the measuring process.

Operation
- A pulsable neutron generator, based on Deuterium-Deuterium reaction, which does not emit radiation when switched off, is used as the neutron source

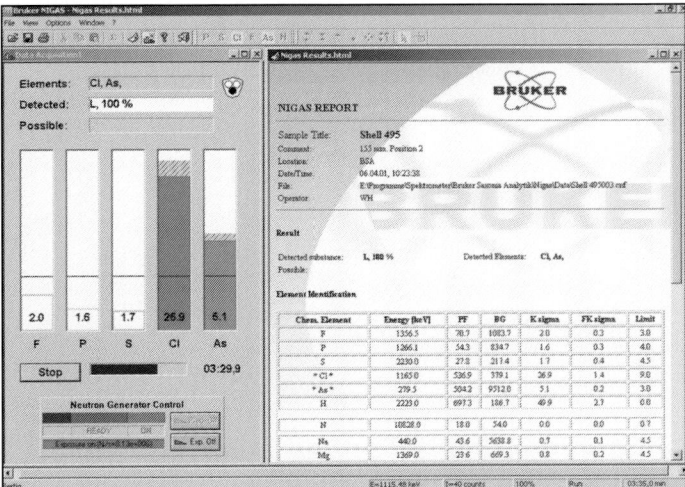

NIGAS software screen-snap showing munition contents containing high proportion of chlorine and arsenic and identified as Lewisite (Bruker Daltonics®)
***2002*/0137485**

- Reaction of neutrons with the contents of the shell results in substance type and concentration specific γ emissions, enabling precise identification
- In contrast to explosives, CW agents contain the key elements chlorine, arsenic, sulphur, phosphorus and fluorine, allowing them to be distinguished with confidence. The full range of agents, from the earliest (World War 1) to the latest are identifiable.

Status
Available.

Manufacturer
Bruker Daltonics®.

UPDATED

PLASMOX® plasma incineration process

Description

The CW destruction plant at Munster, known as Munster II, uses the PLASMOX® system. This is a pyrolysis process which differs from other thermal processes, such as induction or electric arc furnaces, by dealing with the material in two stages. The first stage uses a DC electric high-performance plasma torch, generating a temperature of 20,000°C. Together with an oxygen torch, the conditions in the plasma chamber ensure complete pyrolysis of the products, converting them either to gaseous components or into a vitrified slag which remains in the chamber. Therefore, combustible waste is destroyed, metal components melted and inert slag generated inside the single reaction chamber.

The plasma hearth has a capacity of 1 m³, which allows the processing of up to 1 tonne of material per hour, depending on the density of the smelt. The time taken for each process is several hours, allowing it to match staff shift changes in-between operations. Solid materials smelted in the plasma stage are retained as vitrified slag and dry evaporation residues until the reaction is over. Gaseous components, pretreated by the plasma torch, are transported into the post-combustion stage for controlled and complete oxidation.

The Munster II CW munitions destruction plant   0050933

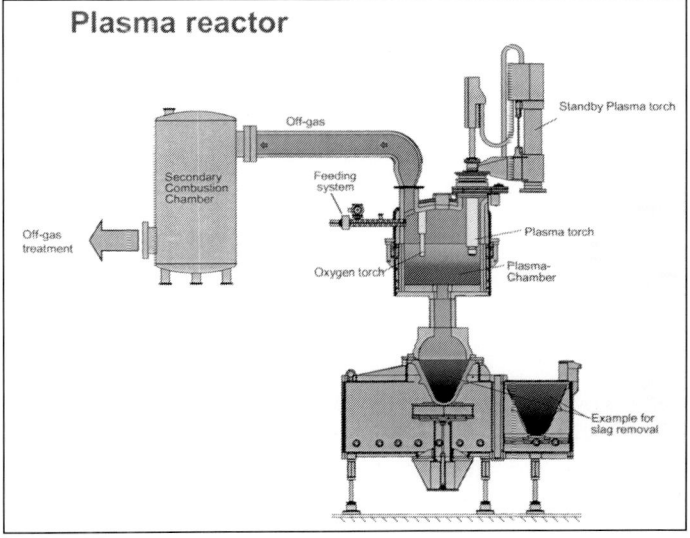

PLASMOX® plasma incineration process   0050934

One of the constituents which can cause problems is arsenic. In the process, arsenic compounds migrate to the gaseous residue where they are removed during the multistage scrubbing process downstream of the secondary combustion chamber. A range of effective arsenic residue treatments can be used. At Munster, they are disposed of as industrial waste. Equally, the arsenic can be chemically recovered or the residues recycled back into the plasma reactor where they are bound up into the non-leaching vitreous slag.

Status
In operation at Munster.

Manufacturer
TECHNIP GERMANY GmBH (formerly Mannesmann Demag).

VERIFIED

UNITED KINGDOM

On-site NBC DEMIL containment facility

Description
DSTL Porton Down is the development agency for a new design of facility for dealing with ACWs. It is particularly suitable for deactivation of CW munitions inadvertently unearthed during excavation or site clearance activities.

There are three components to the facility. Firstly, the munition itself is contained within a ballistic containment shield designed to arrest and de-energise shrapnel and solid particles should they be explosively released during the process. Secondly, a large CW agent proof, tented shelter is erected over the site of the target munition. It is designed to contain agent vapour released during the dismantling process. The sides of both the ballistic containment shield and the tented shelter are flapped like bellows, allowing them to expand under explosive pressure without compromising their containment capability. Finally, the tented shelter is purged of agent vapour by drawing the potentially contaminated air through several NBC air conditioning units (see the Aircontrol portable NBC filtration units entry within this section).

The facility is easily transportable and offers a useful means of dealing with the unexpected discovery of isolated CW munitions, especially when close to centres of population or where their removal is physically difficult or condition suspect.

Status
On trial.

Development agency
DSTL Porton Down.

Manufacturer
Aircontrol Technologies Limited.

UPDATED

DEMIL containment facilities prepared for use (Aircontrol Technologies Limited)
0050935

NBC DEMIL containment facility (end flap open) (Aircontrol Technologies Limited)
0050936

SILVER II™ Chemical Weapon Demilitarisation

Description
SILVER II™ was developed in 1987 by AEA Technology as a means of destroying organic wastes generated by the nuclear industry. The name derives from the silver ions (Ag++) which are formed during the destruction process and which attacks the CW agent or organic waste. The current system is designed to destroy both CW agent and energetic material from chemical munitions turning them to carbon dioxide, water and salts. At the heart of the SILVER II™ process is the low-temperature chemical oxidation of organic molecules by the aqueous Ag[II]ion – progressively converting it in a series of steps irreversibly to CO_2, water and residual salts. The resulting Ag[I] is then regenerated back to Ag[II] again at the anode of a commercial electrochemical cell to give a 'catalytic' Mediated Electrochemical Oxidation Process. The organic species do not have to be water-soluble to be successfully treated by SILVER II™, as has been shown by the successful treatment of solvents, oils, ion-exchange resins, tissues and so on.

This process is currently being demonstrated and evaluated as part of the US Army's Assembled Chemical Weapons Assessment (ACWA) Programme as a safe, effective and environmentally benign low temperature, atmospheric pressure alternative to incineration for the complete destruction of stockpiled CW agent and energetic materials. The integrated plant is founded on proven systems for the disassembly of all types of munitions followed by destruction of the organic fillings in SILVER II™ units. Residual low-volume solid and liquid waste streams from

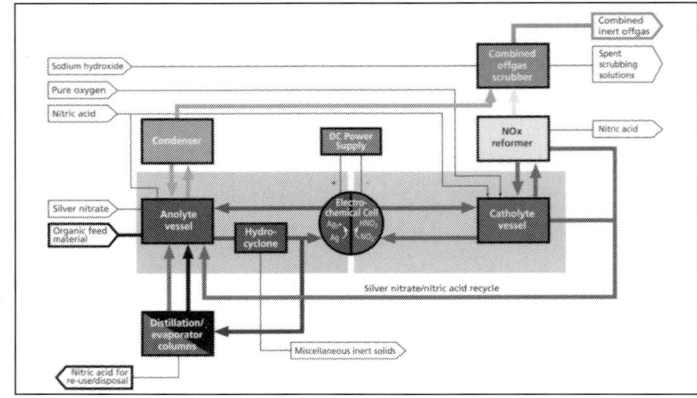

SILVER II™ process (Accentus plc) 0050929

General view of the 12 kW SILVER II™ plant (Accentus plc) **2002**/0102235

SILVER II™ – anolyte vessel (approx 1 m diameter)
(Accentus plc)
0050928

SILVER II™ are readily handled, with much of the material (Silver, water and nitric acid) being recycled for reuse. Off-gas production is low volume and lends itself to hold-up and analysis prior to release to atmosphere.

During 2000, two plants were shipped to the USA, assembled and commissioned by Accentus/CH2M Hill staff prior to operation by Army staff for over 3,500 hours with no lost-time accidents. One of the plants (2 kW) processed over 46 kg of agent, while the other (12 kW) processed over 0.5 tonne of organic material - including 272 kg of energetics. The latter incorporated the explosive equivalent of 69 M55 rockets. The process demonstrated the completeness of destruction of agent materials to >99.9999%, and rocket propellant to >99.999% with high electrochemical efficiencies. By comparison between the two plants, throughput was demonstrated to be proportional to the number of cells. As a result, costs of full-scale plant have been determined to be comparable with the existing baseline incineration technology - thus showing that there is no significant cost penalty in adopting this alternative approach. An Engineering Design Study, carried out in 2001, involved long-term operation of the 12 kW SILVER II™ plant to provide the additional information required for the design and construction of a full-scale weapons destruction facility based on the SILVER II™ process. Over 1 tonne of organic material was destroyed during the 2000/2002 programme.

The SILVER II™ process is also applicable to the demilitarisation of recovered munitions either as mobile or fixed facilities by matching with a suitable technology for extracting the fillings from items that are corroded, damaged or in a generally poor condition. Accentus has patented electrochemical munition-accessing technology, which combines with SILVER II™ to offer an integrated process package.

Status
Accentus is a wholly-owned subsidiary of AEA Technology plc and has teamed with CH2M Hill to develop SILVER II™ as a viable solution to the US Army's requirement to destroy stockpiled CW agent-filled munitions.

Manufacturers
Accentus plc.
CH2M Hill.

UPDATED

UNITED STATES OF AMERICA

Actodemil™ technology

Description
Actodemil™ is a re-agent based on naturally occurring coal-derived humic acid. It is optimised to achieve a series of useful reactions in the decomposition of highly toxic or hazardous chemicals. Humic acid is a water-soluble colloidal medium which, being a reducing agent, promotes reductive hydrolysis. Additionally, it has a strong affinity for organic molecules and metal ions. Therefore, it is effective in absorbing reaction products. The active material is a proprietary 'a-HAX' re-agent. The medium is suitable as the prime re-agent in a DEMIL process vessel operating at atmospheric pressure and a temperature of 160 to 180° fanlight. Following completion of the reaction (between two and four hours), the residue is neutralised and can be safely used in applications such as fertilisers or other means of safe disposal. Actodemil™ complies with EPA Toxicity Characteristic Leaching Procedure (TCLP) universal treatment standards. It is not mutagenic, as determined by the Ames assay test and is not phytotoxic to plants. In addition to its capability to decompose nerve and blister CW agents as well as BW agents like E Coli, it can also reduce energetic compounds such as military explosives and missile propellants.

Status
In production. Funding achieved for trials involving the recycling of energetics at Hawthorne army depot and for development of a mobile unit for multiple site demonstrations.

Manufacturer
ARCTECH Inc.

VERIFIED

Advanced Non-Destructive Evaluation (ANDE™)

Description
ANDE™ is a non-intrusive detection and identification system optimised for DEMIL operations. It can determine the presence of CW agents, precursors, hazardous chemicals, binary agents and biological growth media within a variety of containers, weapons, and vessels. Identification can be achieved in less than 1 minute without opening the container. ANDE™ uses sound waves swept through a range of frequencies to measure several physical parameters of the material in the target container. At present, there are seven parameters but this is set to increase. The parameters include sound speed, attenuation and density. The system compares the measured parameters with previously measured substances to find a match. The ANDE™ database can identify more than 11,000 TIMs and all current CW agents at discrete levels of purity. ANDE™ can use a contact sensor attached to the target container or can be used in the standoff mode using a sensor placed at a set distance from it.

The system comprises the sensors and an electronics module. The latter contains the digital programmable sensor interface board which generates and analyses acoustic and/or non-acoustic signals to determine the contents of sealed containers. It is fitted with a full-size 81-key keyboard and a large format TFT screen. The container is ruggedised and easily maintainable. There are two sealed sensor ports (acoustic and/or non-acoustic) and a full-size PCMCIA II/III port.

Status
In development. ANDE™ is being developed by the Weapons of Mass Destruction Countermeasures Coalition (WMDCC), a partnership of DTRA SBCCOM, JSOC LANL and SAIC.

Manufacturer
SAIC.

VERIFIED

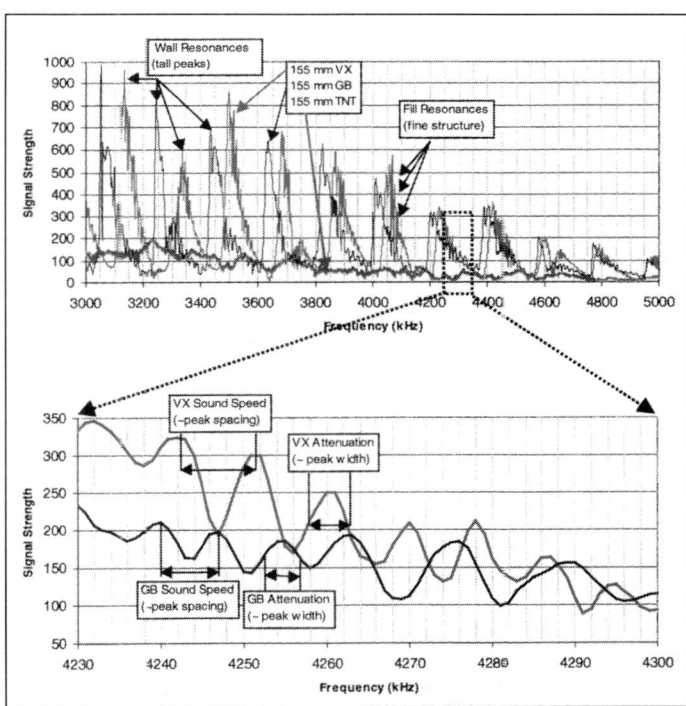

Signal strengths of a variety of ANDE™ parameters (SAIC) **2001**/0059319

General Atomics Cryofracture and Incineration Plant

Description
General Atomics' McAlester Cryofracture Facility (MCAAP MCF) uses the cryofracture principle in preparing CW munitions for destruction. It has already safely dismantled over 4,000 explosive non-CW munitions, ranging from boxed 105 mm cartridges, 155 mm projectiles, 20 mm projectiles and 10.7 mm mortars.

The munitions are mechanically manoeuvred onto a purpose-designed conveyor system as they enter the system, thereby obviating the need to place people at risk. The conveyor passes the munitions through 'cryobaths' of liquid nitrogen where the extremely low temperature renders the munition casing and contents hard and brittle. From here, the munitions are passed by robot conveyor to a cryofracture press where they are crushed into small pieces.

Downstream neutralisation of the energetic (explosive) components, the dunnage (transport and packing materials), metal components and CW agents is achieved by incineration. A rotary kiln takes the crushed components from -300°C to +1,700°C along the 21 m length of the unit, creating a solid and gaseous residue. The solid residue is discharged to a scrap conveyor where it is monitored prior to disposal. The gaseous fraction is passed to a cyclone and thence to an afterburner where secondary combustion takes place at 1,000°C.

The pollution abatement system is the final stage of exhaust gas treatment where the gases are scrubbed, filtered and quenched before final safe release to atmosphere.

A cryofracture plant, optimised for conventional munitions is being built at McAlester AAP to process conventional landmines. It will be operational in 2001.

Status
Ready for commissioning.

Manufacturer
General Atomics.

VERIFIED

Liquid nitrogen cryobaths   0050930

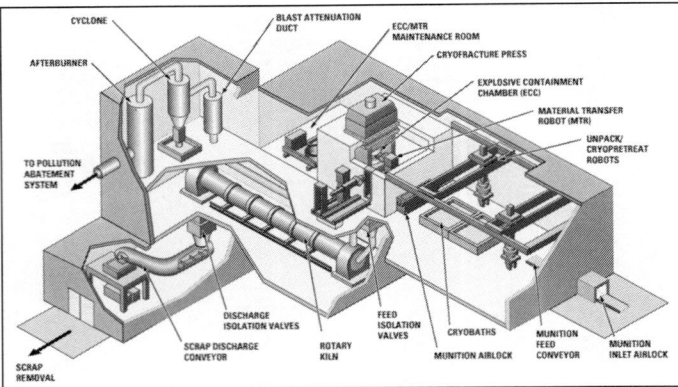

Layout of the cryofracture and incineration plant   0050931

The rotary kiln for incineration of the cryofracture products   0050932

Solvated Electron System

Description
The Solvated Electron System (SES) has been developed by Teledyne-Commodore in the USA. It is a chemical neutralisation process, generally conducted at ambient temperature. The active ingredients are metallic sodium (Na) and anhydrous liquid Ammonia (NH_3). The proportion of sodium is 3 per cent by weight. When dissolved in ammonia, the sodium turns the liquid into a deep blue homogenous mixture with a large supply of free electrons. The shade of blue is a good gauge of the need for sodium replenishment during the destruction process. CW agent, introduced to the liquid, is attacked by the free electrons which convert it into a less harmful residue. The process is complex and varies with each type of agent. The residue from nerve agent conversion for example, requires extra oxidation. However, the reaction is fast and the process scaleable. A pilot plant developed by Teledyne-Commodore is designed to detoxify 57 kg of agent (enough to fill 12 M55 rockets) per hour. It also has the great advantage of being a closed system, eliminating the need for further treatment of gases or solid waste. The principle disadvantage is that liquid ammonia is volatile, toxic and flammable.

The chemical process has been understood since 1865 but only brought to bear on waste disposal in the last 10 years. The sodium is used up in the process but all the ammonia is recycled.

Status
In development.

Manufacturer
Teledyne-Commodore.

VERIFIED

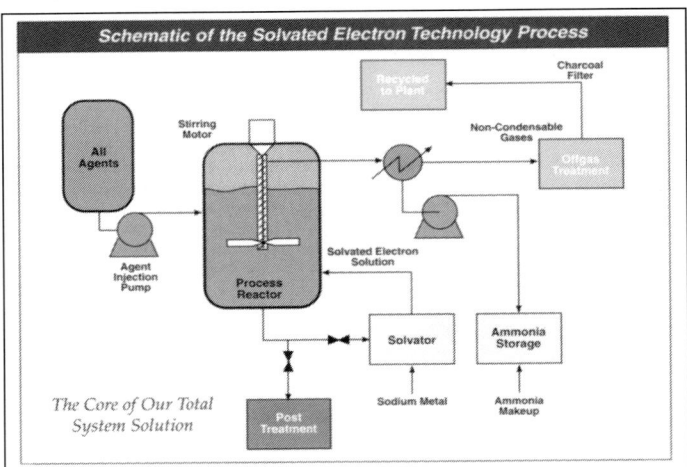

Solvated Electron System (SES) – Process diagram   0050924

Solvated Electron System (SES) – Pilot plant   0050925

Supercritical Water Oxidation (SCWO)

Description
Supercritical water oxidation (SCWO) takes place at moderate temperatures but requires high pressure. It capitalises on the property of water to dissolve hydrocarbons when in the supercritical state (above 374°C and 218 bar). The principle is already being used commercially for the treatment of toxic organic waste. CW agents dissolved in the supercritical water can similarly be oxidised by air, oxygen or hydrogen peroxide to water, CO_2 and harmless metallic salts.

In trials conducted at Aberdeen Proving Ground, Maryland USA in 1994, conversion rates of 99.999 per cent were achieved. Here the destruction environment was 250 bar and 400°C. The exothermic reaction raised the

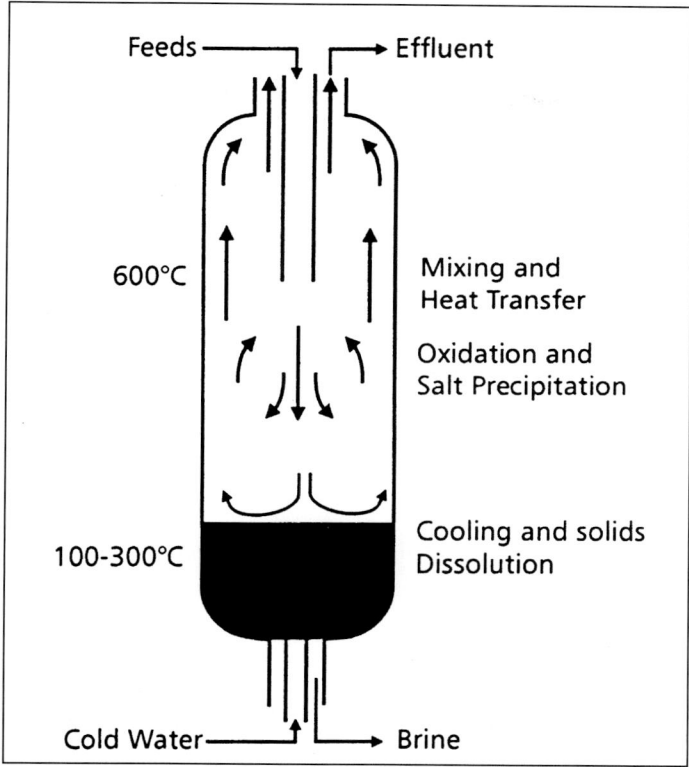

Modar patented SCWO reactor vessel   0050923

Feeds → **Effluent**

600°C — **Mixing and Heat Transfer**

Oxidation and Salt Precipitation

100-300°C — **Cooling and solids Dissolution**

Cold Water → **Brine**

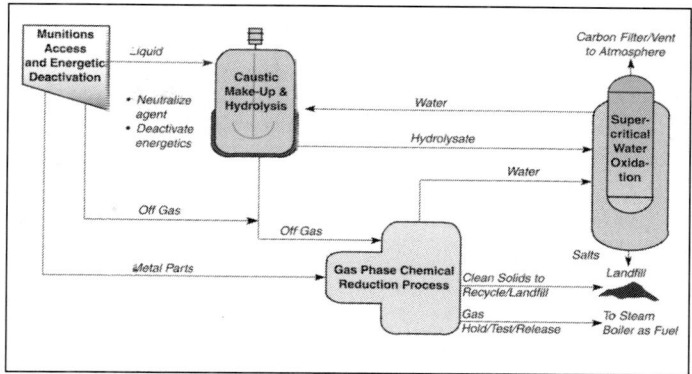

LMIDS involves SCWO but incorporates other technologies in the DEMIL process
0050922

temperature further, to 600 to 650°C. To be effective, dilution of the supercritical water by CW agent should remain in the range 2 to 25 per cent with a residence time of about 5 minutes. This time can be reduced by raising the temperature to 500°C. Even further increases in temperature (to 600°C for example) can reduce the residence time to as little as one minute.

SCWO has the advantage that it is a single phase, closed process, avoiding the need to further treat toxic gaseous or solid products. It is also quick. However, the environment is extreme and explosion is possible if firm control is not maintained on the temperature. Also a leak in the reactor would release highly toxic products under high pressure. The SCWO environment is also highly corrosive, forcing designers to use expensive materials to construct the pressure vessels. Additionally, no organic salts can form a sludge with the potential to block valves and pipes. An SCWO reactor, patented by Modar, avoids this problem by cooling the bottom of the vessel to below supercritical conditions, forcing inorganic salts back into solution again.

SCWO development has been undertaken by a number of organisations, including General Atomics and Lockheed Martin. The latter offers the Lockheed Martin Integrated Demilitarisation System (LMIDS) which recognises that no single technology can provide the complete solution and that all processes in the demilitarisation chain have to be considered in order to deliver a turn-key solution. These range from transport from site through personnel monitoring to final safe dispersal of residue into the atmosphere, ground or water waste system.

Status
In development.

Manufacturer
Lockheed Martin.

VERIFIED

TRAINING and SIMULATION

TRAINING and SIMULATION

BULGARIA

40 mm chemical warning rocket

Description
This chemical warning rocket can be operated either manually by using friction cap initiation or, when installed on an armoured vehicle, by an electrical primer cap. When fired from its disposable cylindrical projector, the rocket produces a hissing sound and discharges a yellow light. The rocket can reach an altitude of approximately 400 m.

When issued, each rocket projector has a diameter of 40 mm and weighs 450 g. The rockets are issued in sealed polyethylene packs, each containing three rockets.

Status
Available.

Marketing agency
Kintex.

VERIFIED

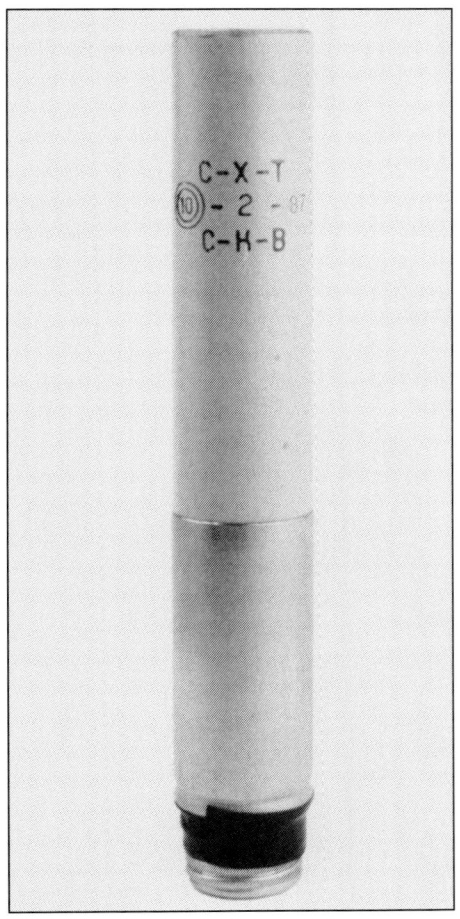

Bulgarian 40 mm chemical warning rocket

CANADA

Hands Fireworks chemical simulators

Description
The Canadian Armed Forces have evaluated two types of chemical simulator that simulate a chemical warfare attack by projecting clouds of chemical powder or liquids without using any form of mortar. These are in two basic forms. One is the Groundburst Chemical Simulator and the other the Airburst Chemical Simulator.

Both types can be fired from the ground or from the top of a vehicle. Each consists of a canister which acts as the projector. If required, the canister may be held steady by a simple spiked holder into which it is clipped once the spike has been driven into the ground. The groundburst version projects a cloud of chemical powder or liquid 6 to 10 m into the air where the wind will carry the cloud. The airburst version projects a cloud of liquid to a height of approximately 60 m.

SC-300 groundburst chemical simulator with one ground spike holder

Specifications		
Type	Groundburst	Airburst
Weight	350 g	400 g
Weight of filling	225 g	165 g
Height of canister	143 mm	143 mm
Diameter of canister	73 mm	73 mm
Height of burst	6-10 m	60 m

Both types of simulator are packed in plastic foam boxes containing 18 canisters and spike holders. The boxes measure 570 × 390 × 180 mm.

Status
Evaluated by the Canadian Forces.

Manufacturer
Hands Fireworks Inc.

VERIFIED

M28 and M29 simulant sets

Description
The M28 and M29 chemical agent simulant sets are designed for classroom training, illustrating to students the effects of the agents or the detector papers sets. The M28 deals with G agents and the M29 with blood agents such as AC and CK. They both comprise:
- A corrugated box containing the simulators
- A booklet of M8 test papers sealed in a plastic bag (M28)
- Two sets of 8 pouches containing G or blood agent simulant and inert samples
- Comprehensive instructions.

Status
In production. In service with the US armed forces.

Manufacturer
Anachemia Canada Inc.

VERIFIED

M256 Simulator

Description
The M256 Chemical Agent Simulator Detector Kit is designed for classroom use and allows trainees to observe the reactions of the M256A1 Chemical Agent Detector (see under Detection (sensors systems) - Chemical) to different conditions of the presence or absence of AC, CK, H, L or G agent.
The kit comprises:
- A corrugated box containing the simulators.
- 36 pouches containing the simulant materials.
- Instructions.

Status
In production. In service with the US armed forces.

Manufacturer
Anachemia Canada Inc.

VERIFIED

FRANCE

Lacroix ASATOX chemical attack warning device

Description
The Lacroix ASATOX chemical attack warning device is a hand-held pyrotechnic launcher intended to produce audible and visual warning signals to alert military units in the field of the approach of a chemical agent attack.

In use, the rocket container is held so that one hand can turn the main body to arm the launching system and the other hand can pull the firing grip at the base of the unit. This ignites a propellant charge to launch an assembly to a height of approximately 200 m. At that height a parachute is deployed to suspend visual signals involving a yellow-red-yellow flare pattern and an audible signal warning module. The visual signals remain ignited for 20 seconds while the audible signal lasts 10 seconds.

The complete unit weighs 690 g. Length is 347 mm and the cartridge diameter is 47.5 mm.

Status
In production. In service with the French Army and some other armies.

Manufacturer
Étienne LACROIX Défense SA.

UPDATED

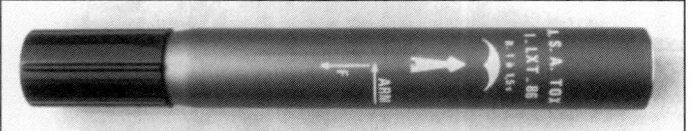

ASATOX chemical attack warning device (Étienne LACROIX Défense)

Lacroix SIMULTITOX chemical attack simulator

Description
The Lacroix SIMULTITOX chemical attack simulator consists of a six-barrelled 154 mm mortar that fires smoke or chemical training agent grenades to simulate a

A Lacroix SIMULTITOX chemical attack simulator grenade (Étienne LACROIX Défense)

Lacroix SIMULTITOX chemical attack simulator (Étienne LACROIX Défense)

chemical warfare attack. The mortar may be set up on the ground or on a light vehicle trailer. It can be used to fire a single salvo of six or four grenades, or three salvos each of two grenades. The polystyrene grenades can be fired to a range of between 125 and 240 m and can cover an area of 1 to 5 ha, depending on wind conditions and the various forms of training agent used. The launcher weighs 45 kg and takes 3 minutes to prepare for use. Each grenade weighs 300 g (empty) and has a diameter of 154 mm. The filling is a training agent, 1.5 litres of water with a washable dye in solution or any other training agent. This equipment was designed exclusively for training.

Status
In production. In service with the French Army and some other armed forces.

Manufacturer
Étienne LACROIX Défense SA.

UPDATED

NBC training suits

Description
These suits are designed for training purposes only and provide no NBC protection whatsoever. They are intended to replace normal NBC clothing during training and exercises and reduce training costs. They resemble full NBC clothing, thus allowing NBC alarm drills to be practised fully. For the army, the complete suit comprises a jacket and trousers that can be worn over normal battledress, with the jacket having an integral hood. The Air Force training suit is a coverall with an integral hood, worn directly against the skin or over undergarments. The material used for both suits is conventional cloth through which air can permeate freely. The suit is available in several sizes and is reusable.

Status
In production.

Manufacturer
Paul Boyé.

VERIFIED

Paul Boyé NBC training suit for army personnel

GERMANY

Kärcher Training Emulsion

Description
The Kärcher Training Emulsion is designed to accurately simulate the physical properties of most kinds of emulsion used in the DECON process, including the German TD 202 emulsion. It is a non-toxic and low-cost solution to the training of DECON personnel.

Status
Available.

Marketing agency
Alfred Kärcher GmbH & Co.

VERIFIED

NORWAY

Radiac RAS 100 simulated fallout radiation training set

Description

The Radiac RAS 100 simulated fallout radiation training set was developed in a joint programme between the manufacturers AME, the Norwegian National Defence Research Institute (FFI) and the Norwegian Army Materiel Command (HFK). The Radiac RAS 100 provides realistic training for NBC personnel by using the same equipment working exactly as it would during or after a nuclear attack, or in the aftermath of a nuclear power incident.

The complete set consists of a control unit/transmitter in a watertight case, receivers in the shape of simulated sensor units, 'hot spot' transmitters, helmets with receiving antennas and a directional transmitting antenna. A typical training set could consist of one control unit/transmitter with a directional transmitting antenna, three to five 'hot spot' transmitters, five to eight sensors and the same number of helmets with receiving antennas.

The training system simulates a real situation in an operational area after a nuclear explosion. The control unit/transmitter and the directional antenna represent the ground zero of an explosion. Before operating, the following parameters must be set: decay rate, simulation speed, time after explosion and time for the accumulation of fallout. The simulation system may be operated in real time but this can be increased for training purposes in nine steps.

The 'hot spot' transmitters simulate increased radioactivity and are placed at points where fallout concentrations are likely to occur. Signals from the directional and 'hot spot' transmitters are detected by the antennas carried on the top of the trainees' helmets and are sent to the simulated sensor instruments by cable. The system can be adapted to operate with various radiac meters. Instruments used with the Radiac RAS 100 include the NM 70 a Norwegian radiac detector and used by Norwegian civil defence organisations which are issued with the PDRM 82 portable dose rate meter. The programming of the simulation equipment allows breaks for instruction, briefings and so on.

The Radiac RAS 100 training system can operate off a 24 V DC supply or 110 and 220 V AC mains supplies. The control unit/transmitter operates on a frequency of 29.5 MHz. Weight is 3.7 kg (excluding the antenna) and it measures 300 × 115 × 230 mm.

Status

In service with the Norwegian armed forces and civil defence organisations.

Manufacturer

AME as.

UPDATED

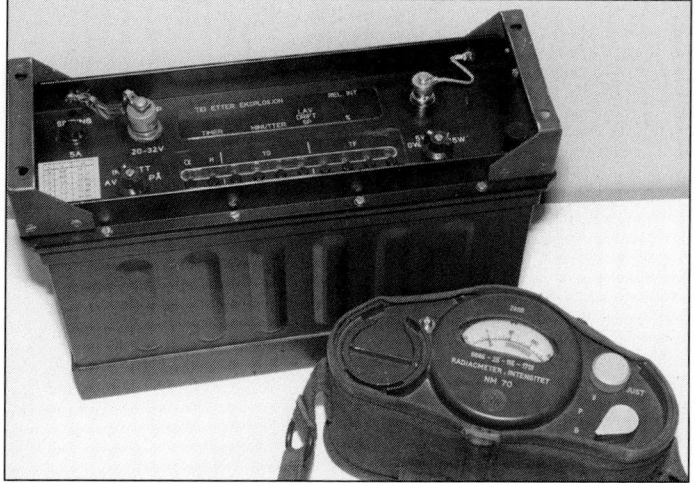

Radiac RAS 100 simulated fallout radiation training set showing the control unit/transmitter and a simulated radiac meter (the NM 70) sensor unit (T J Gander)

RUSSIAN FEDERATION AND ASSOCIATED STATES (CIS)

Model SKhM-M protective equipment workbench

Description

The Model SKhM-M protective equipment workbench, is a compartmented workbench specially equipped for the maintenance of protective clothing, protective masks and filter canisters and breathing gear. It contains special and other tools, instruments, some spare parts and various materials for the servicing, testing and repairing of all the above.

Status

Available from Russian state factories before privatisation. In service with the RFAS and former Warsaw Pact forces. Current manufacturer unknown.

Manufacturer

Unknown.

Agency

Rosoboronexport.

VERIFIED

PPKhM-1M mobile chemical workshop

Description

Based on the chassis of the GAZ-66 (4 × 4) 2,000 kg truck, the PPKhM-1M mobile chemical workshop is used for field repair, maintenance, calibration and servicing of chemical and radiological equipments. The workshop is contained within a specially equipped box body, additional tentage and equipment are carried in a towed IAPZ-738 single-axle trailer. The box body has four workspaces although more can be provided within the carried tents. Wherever possible, electrical power is taken from an external source, but an AB-2T230 portable generator can supply power for lighting and the various items of test and repair equipment. A petrol heater and an air compressor are also provided.

Status

Available from Russian state factories before privatisation. In service with the RFAS and former Warsaw Pact forces. Current manufacturer unknown.

Manufacturer

Unknown.

Agency

Rosoboronexport.

VERIFIED

Trainer for chemical reconnaissance troops

Description

Developed at the Saratovkij Higher School of Military Engineering for Chemical Defence, this trainer is used to train NBC reconnaissance troops in the use of their NBC detection equipments. The equipments involved are the GSA-12 chemical detector, DP-3B radiation detector and the ASP biological agent detector.

The trainer consists of an instructor station and up to four trainee stations. The latter are usually in one or more RKhM reconnaissance vehicles, while the instructor station could be either in a command post or another vehicle. In both cases communications between the two stations are carried out using the standard R-123 radio set. Throughout a training exercise the R-123 maintains its normal communications function.

The usual ratio of instructor to trainees is one to four, equalling one reconnaissance platoon. The instructor station has a control panel, command formulator, modulator and a timer. The system works on the principle that commands are transmitted in a number of pulses imposed on a sine wave. Following the formulation of a command the station automatically transmits the appropriate signal and then automatically switches to the receive mode; transmission time is from 1 to 1.5 seconds. Commands may be transmitted in either a manual or automatic mode. In the latter mode command updates will be transmitted every 30 to 40 seconds.

At the trainee station the R-123 radio is set to receive and decode the signals and key the correct equipment to respond. The station includes a wave form amplifier, remote-control signal decoder, simulator for the DP-3B and simulated control panels and devices that are attached to the GSA-12 and ASP sets. Received signals are passed to the wave form amplifier for conversion to a 2 kHz carrier frequency which is then boosted and converted to a pulsed form suitable for the decoder. The decoder counts and evaluates the number of pulses sending the appropriate signal to the simulated equipment concerned for actuation of the display needles or register. The circuitry for the receiver is located where the body of the DP-3B would be normally located in the vehicle involved.

Power for the system is taken from normal vehicle supplies and stabilised to a 5 V level for use.

Status

In service with RFAS chemical warfare units.

Developing agency

Saratovkij Higher School of Military Engineering for Chemical Defence.

VERIFIED

SWEDEN

Nammo LIAB chemical attack simulator

Description
Nammo LIAB's chemical attack simulator gives reality to training exercises, such as reporting the presence of nerve gas, using detection equipment, decontamination and using individual protection equipment.

The Nammo LIAB chemical attack simulator consists of a chemical exercise grenade, a launcher and a liquid chemical agent simulant. The grenade is fired from the launcher and, at a predetermined burst point of approximately 60 m, a cloud of simulant particles and droplets is formed to cover an area of approximately 800 m².

In use, the liquid chemical agent simulant (of which two separate types are available, one simulating VX, the other GD), which causes colour changes on chemical agent detection papers, is poured into a plastic container which contains 0.8 litres and forms the body of the practice grenade. A cover is then pushed into the filled body and an overflow pipe device is actuated to ensure the correct volume of contents is enclosed. A delay fuze is then placed into a housing in the lid and the practice grenade is then loaded into the launcher. The launcher, which is secured to the ground by a spade-type frame, is a simple steel pipe from which the grenade is fired electrically using a black powder charge. The firing charge also initiates the grenade fuze which subsequently detonates, scattering the contents of the grenade.

Status
In production. In service with the Swedish armed forces.

Manufacturer
Nammo LIAB AB.

VERIFIED

Loading a Nammo LIAB chemical attack simulator (Nammo LIAB AB)

UNITED KINGDOM

AMS NBCD Ship Trainer

Description
The AMS NBCD Ship Trainer is installed at the Phoenix NBCD School at Portsmouth, UK to a specification by the MoD (Navy). The simulator element allows complex exercises to be conducted from a mock-up Ship Control Centre (SCC)

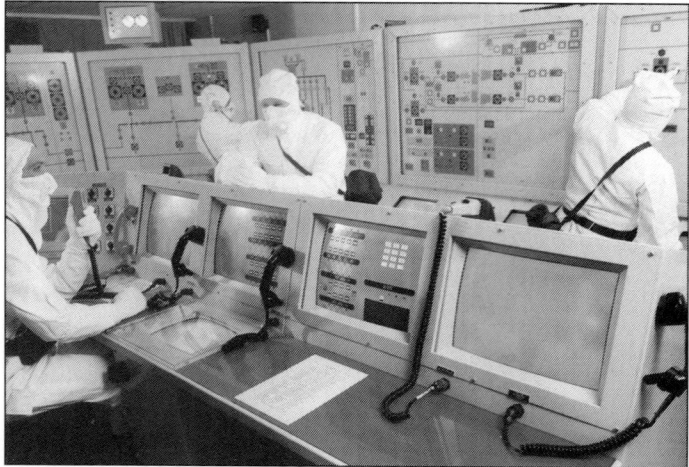

AMS NBCD Ship NBCD Trainer (AMS)   **2002**/0050771

which, using large format screen and emulation software, can recreate the SCC layout of the Type 22 or Type 23 DD/FF.

The model accurately simulates ship systems and allows NBC contamination control and fallout transit exercises to be practised, as well as damaged system management.

The trainer suite also incorporates Computer-Based Training (CBT) courseware, delivered on networked PCs, covering topics such as intact and damaged stability, contamination control and ship knowledge.

Status
In service with the Royal Navy.

Manufacturer
AMS.

UPDATED

AMS NBC Protection Training Unit (PTU)

Description
The AMS NBC Protection Training Unit (PTU) provides realistic training to NBC protection, cleansing, monitoring and decontamination teams on board naval vessels. The teams are trained to carry out their tasks by following set procedures as necessary in a contaminated environment following an NBC attack.

The PTU contains realistic training areas comprising the following: a large waterproofed area representing a surface warship upper deck, laid out with various equipment such as bollards, lockers, instrument pinnacles, a gun, hatches, fire hydrants and so on; a ship citadel airlock; NBC cleansing stations representative of the ship type involved; a chemical training balcony; various outstations; and facilities for the instructor to monitor all student activities using communications and closed-circuit television systems.

Status
In service with the Royal Navy.

Manufacturer
AMS Systems.

UPDATED

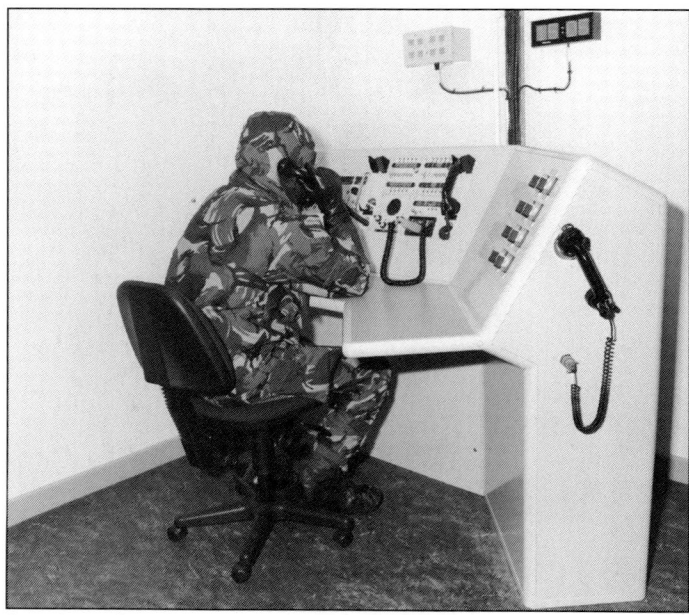

Operator at the trainee console of the AMS NBC Protection Training Unit (PTU)
2002

Argon Electronics CAMSIM001 Chemical Agent Monitor (CAM) training system

Description
Produced by Argon Electronics, the CAMSIM001 is a comprehensive training system for the Graseby Dynamics Chemical Agent Monitor (CAM is a trademark of Graseby Dynamics Limited). The simulator not only enables a user to be trained in the detection of blister and nerve agent but it also assists in the evaluation of a trainee in the correct use of the CAM.

Utilising electronic simulation technology, both wide area and point contact agent contamination simulation is possible. A simulation confidence check device is included with the system. The need for a radioactive source or environmentally harmful simulation chemicals is avoided.

CAMSIM001 realistically simulates 'contaminated' personnel, vehicles, munitions, stores and ground areas **2000**/0085454

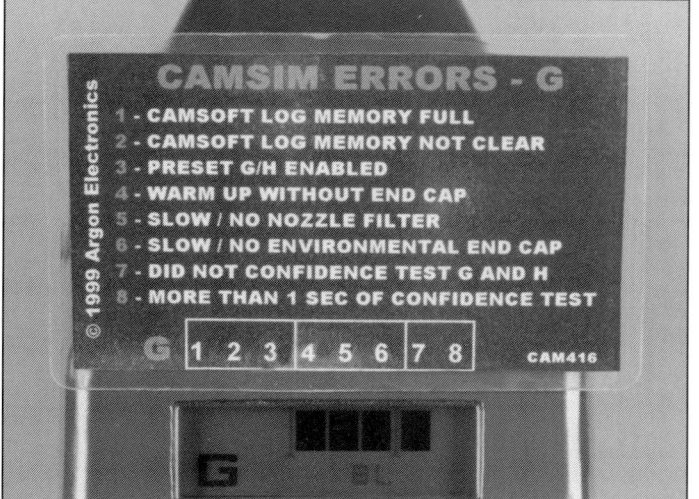

Trainee errors are logged by CAMSIM001. The error card shown here can be printed in any language **2000**/0085455

In order to ensure effective training, CAMSIM001 is able to detect and record the following user errors (see illustration):
- Failure to use the confidence check (an optional regular check error is also available).
- Incorrect mode change between agents.
- Probe contamination.
- Instrument overload.
- Incorrect instrument shut down.
- Attempt to change mode change during a 'Wait' period.

Errors are displayed on the simulator in BAR format upon the insertion of a special instructor's key. Only an instructor can read or erase errors once they have occurred.

CAMSIM001 includes the simulation of partial and full decontamination of chemical agents. Decontamination training is electronic and completely under the control of the instructor.

Comprehensive data logging permits each operator error to be recorded on the basis of the time occurrence and duration of the error. In addition, missed agent errors are also detected. This records any instance where the operator encountered either simulation nerve or blister agent but had the detector either turned off, or the incorrect mode was selected at that time.

If the simulator is switched to blister and an area contaminated with nerve agent is entered, the time, duration, detection mode and level of simulation agent encountered will be recorded. The time at which the simulator is switched on or off and selection of the agent mode is also recorded.

An interface with a personal computer permits thorough analysis of a training session together with permanent training records produced on a printer. Changes in operator ability can be readily determined by comparison with records from previous training sessions.

Standard and optional error reporting can be customised to suit specific organisational operating procedures. Radio control of both the CAMSIM001 and simulation sources is also available. A Global Positioning System (GPS) based option is available to enable large areas of vapour contamination to be simulated. Simulated nerve and blister agents can be varied in strength and location throughout a training area as required by the instructor. Options can be upgraded at a later date if required.

CAMSIM is a trademark of Argon Electronics.

Status

In production. Established in service at the Chemical Training School, Fort Leonard Wood and other US Army sites. Also in use at the British and Norwegian NBC Schools. In 1999, Argon Electronics was contracted to supply 1,280 units to the US DoD. They will enter service with the US Air Force, Army, Marine Corps, National Guard and reserve units.

Manufacturer

Argon Electronics.

VERIFIED

Argon Electronics quartz fibre dosimeter simulator

Description

The Argon Electronics quartz fibre dosimeter was designed to enable personnel to be trained in reading and interpreting quartz fibre dosimeters. The simulated instrument is a specially adapted dosimeter with a scale calibrated to suit customer requirements; the standard unit is produced with a range of 0 to 50 mSv.

The simulator responds to the standard Argon Electronics GS series of simulated gamma sources. The electronics for the dosimeter are enclosed in a small pager-type case worn on the trainee's waist. An umbilical cable connects the electronics to the simulated dosimeter.

The dosimeter is read by holding it to the light, as with a real instrument. The initial reading can be preset to any value so that a specific reading can be recorded on both entry and exit to a simulated incident.

Status

Advanced development.

Manufacturer

Argon Electronics.

VERIFIED

Defence NBC Centre (DNBCC) training courses

Description

The Defence NBC Centre offers courses in all aspects of land-based NBC defence. At the strategic end of the scale, training, symposia and study periods are offered to senior officers from the armed forces, police, health and emergency services, other government agencies and non-governmental organisations from UK and other nations authorised under the terms of the UK's ethical foreign policy.

At the operator level, courses are available for the use and unit management of current UK equipment for: detection, protection and contamination control; for the management and operation of battlefield medical facilities in NBC conditions; and for the preparation of instructors to carry out NBC defence training at all levels. All instructors are experienced NBC practitioners who normally join the staff from operational service units.

The Defence NBC Centre is also the headquarters unit for the UK's Joint NBC Regiment and is the producer of UK NBC defence doctrine and publications for land and air operations.

With its expertise and experience in defence analysis, in the development of safe and effective operational procedures and in training, the Centre is able to offer consultancy services in all aspects of NBC defence preparations and operation.

Students learn maintenance procedures on the S10 respirator (see under Protection (individual) - Masks (general issue) **2000**/0088181

Command team training: students using the Bruhn BRACIS system (see under Detection C³I systems)   ***2000**/0088185*

Status

Central UK authority for land and air NBC operational training and the development of procedures. Headquarters of UK Joint NBC Regiment.

Manufacturer

DNBCC.

VERIFIED

D-Sim Depleted uranium residue and DECON simulator

Description

The D-Sim Decontamination simulator is designed to train monitors in all aspects of their task, especially in procedures where use of live radioactive or CW agents would be risky or is prohibited by law. By using a powder or liquid-based simulant and corresponding optical detector, a wide range of agents can be simulated. Designed originally for training in civilian nuclear installations, the principles of good monitoring practice can equally be taught using this system to any nuclear or chemical monitor.

Exercises are conducted by the distribution of the simulant onto the ground, equipment or building and vehicle surfaces. The monitoring teams use D-Sim to carry out correct monitoring procedures, assessed by supervising staff. α, β and α+β simulations are available. The β selection is also used for CW simulation.

Status

Available.

Manufacturer

Argon Electronics.

UPDATED

Specifications

Detection distance (α): 10 mm (proximity sensor cuts off signal at this distance)
Detection distance (β): Up to 200 mm (depends on concentration of simulant. Decays with increasing distance)
α + β: Combines both signals within 10 mm of simulant
Probe viewing angle: 30°
Response time: Designed to mimic real instruments (time constant 1.5 s)
Size: Small briefcase
Weight: 5.5 kg
Battery life: 10 h continuous (size AA). 30 h with alkaline cells
Environmental range (instrument): –20 to +55°C
Environmental range (simulant powder): -20 to +55°C
Environmental range (simulant liquid): 0 to +55°C
Environmental range (humidity): =95% (non-condensing)

Generator Irritant CS Hand-Held L6A1

Description

The Generator Irritant CS Hand-Held L6A1 produces a highly effective concentration of irritant smoke to familiarise troops with NBC warfare conditions by simulating the effects of chemical vapours. The generator's applications include testing the effectiveness of respirators against CS smoke; demonstrating the effects of CS smoke; providing a widespread exposure of troops to a CS environment; and aiding the peacetime simulation of NBC warfare.

Generator Irritant CS Hand-Held L6A1 (PW Defence Limited)

With a burn time in excess of 150 seconds the compact generator can be used as a single unit or in series. The generator may be hand held or deployed on the ground. A cool burning process ensures that it will not set fire to ranges. The generator is designed to withstand exceptional environmental exposure and to perform under extreme conditions.

The generator is constructed using two pressed pellets of CS producing composition encased in a steel tube. The firing mechanism is contained at the base, consisting of a pull-twist striker integral with the handle. For safety, the striker is always positioned in a safety gate to avoid accidental initiation.

Each generator weighs 245 g. Diameter is 35 mm and length 247 mm. The NATO stock number for the L6A1 is 1365-99-547-8853. PW Defence reference is 58030.

Status

In production. In service with the UK armed forces and with other armed forces worldwide.

Manufacturer

PW Defence Limited.

VERIFIED

NBC Operational Simulator (NBC OpSim)

Description

The NBC OpSim is a system designed for NBC C³I training, using field data relayed by radio from simulated detectors deployed around a geographical area. It is suitable for training response teams at an airfield or logistic support area, for example. Large area and a small area systems are available and the simulator software, developed in C++, runs in a windows environment. It includes:

- a Situation Display Suite (SDS) which provides a real-time overview of the exercise on a map background. It displays the location of the field instruments, the location of NBC strikes and the hazard level contours as the hazard develops
- a Control Centre Radio Controller (CCRC), usually located at exercise headquarters, into which the exercise scenario is fed, together with the meteorological data
- a Test and Diagnostic Suite (TDS) which allows the user to test the CCRC and the field instruments
- a Communications Suite, providing an SDS - TDS and a TDS - CCRC interface
- an Exercise Database which maintains the data for the exercise.

The user interface is designed to emulate the current operational UK C³I system: BRACIS(I) (see under DETECTION (C³I systems)). Strike data is transmitted via the CCRC to the field instruments which calculate and display the appropriate hazard reading for the instrument's own time and position. The position of the field instrument is derived from an internal GPS receiver and is accurate to between 10 to 100 m. The trainee takes the reading and performs his normal reporting drills (NATO ATP45 NBC NUC or CHEM message reports). For the large area system, the coverage is 15 km radius. This can be extended by using an extra radio controller as a relay station. The frequency used for the large area CCRC is 80 MHz and the number of instruments supported will depend on the characteristics of the exercise area. The Royal Air Force requires the instruments in its system to pass back their positions and reading to the control centre. The instruments are therefore fitted additionally with internal transmitters and stub aerials. This facility allows the SDS to display the extra data on demand and it will also have high positional accuracy 5 to 10 m by the use of differential GPS, requiring very accurate surveyed positioning of the master GPS receiver. Area coverage will be 5 km. Incoming commands are received by the 80 MHz data receiver and outgoing data is sent from the instrument by a 142 MHz transmitter.

A number of small area systems may be procured to form a pool of equipment which can be drawn by organisations for training on demand. Both the large and small area systems will offer a debrief mode which will allow exercises to be played back and analysed.

Training instruments are designed to mimic their operational versions as far as possible in size, weight, display features, power consumption, controls and shape. External changes include LEDs to indicate data link faults (short flashes) GPS failure to lock (long flashes) or both faults (continuous). The field training instruments currently available are the UK radiac instruments PDRM (NIS 501), PDRM 82M and MD3 (see under DETECTION (sensor systems) - Nuclear) and the chemical detector, CAM (see under DETECTION (sensor systems) - Chemical).

Status
Production contract placed for five systems for the Royal Air Force.

Manufacturer
AEA Technology (technical demonstrator).

VERIFIED

PW Defence irritant smoke pellets, respirator testing

Description
These small irritant pellets, supplied in tins of 50, take the form of a compressed CS composition. They are designed as a testing or training agent to facilitate the correct fitting of respirators and to demonstrate the effects of CS smoke. The pellets may be used individually or in quantity depending on the density or duration of emission required. They can also be used as an aid in the simulation of NBC warfare tactics.

On ignition by a fusée match or portfire, dense irritant smoke is released immediately to last up to 25 seconds.

Each pellet weighs 1 g, has a diameter of 12.7 mm and is 5.6 mm thick. A tin of 50 pellets weighs 65 g, is 80 mm high and has a diameter of 40 mm.

Status
In production. In service with the UK armed forces and police and with armies and police forces in Europe, the Middle East and Asia.

Manufacturer
PW Defence Limited.

UPDATED

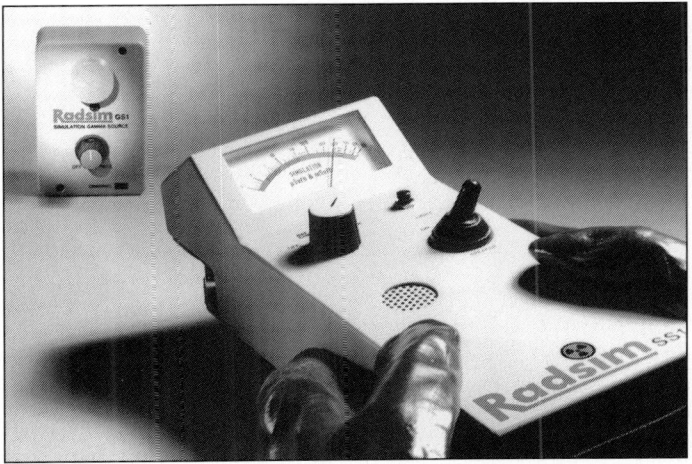

PW Defence irritant smoke pellets, respirator testing and fusée matches (PW Defence Limited)

Radsim DS1 α contamination simulator

Description
The Radsim DS1 α contamination simulator utilises discreet non-hazardous, non-radioactive simulated a-radiation sources capable of penetrating most substances without degradation in measured strength. The sources can be hidden under clothing or inside footwear for realistic detection by a simulated contamination meter. The DS1 simulates the widely used Mini-Instrument Mini-Monitor meter but other models of instrument can be simulated if required.

To simulate a person entering a contaminated area, the simulation sources are concealed underneath clothing by removable straps or inside footwear. Employing a hand-held probe, the detection system is similar to a thin window Geiger type connected to a meter via a 1.5 m coiled cable. A four-position rotary switch is located on the front panel with the first position being off, the second for battery test, the third for general operation and the fourth to switch off the speaker for covert use.

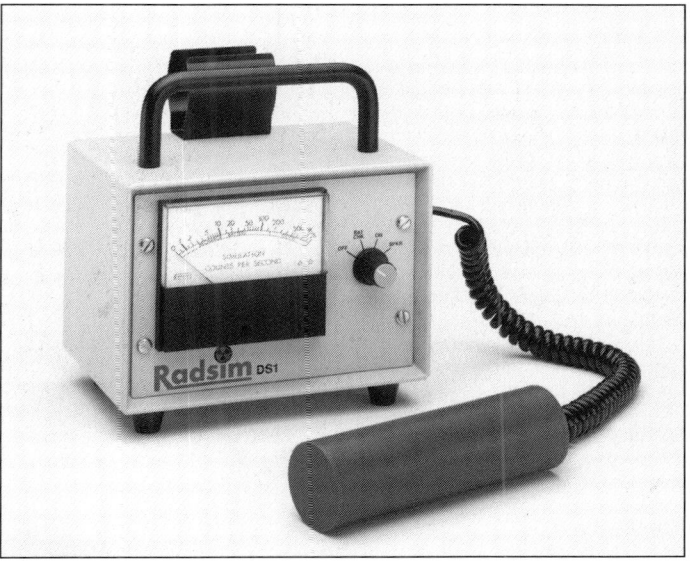

The Radsim DS1 alpha contamination simulator meter

In use, a trainee has to hold the probe close to the contaminated person without contaminating the probe by making contact. If the probe is brought too close to the surface being tested a proximity detector will sound an alarm warning. The DS1 also has a simulation background radiation generator activated by operating a rotary control located on the rear of the DS1 and variable from zero to approximately 20 counts/s. This reading is added to that from the simulation source.

Two simulation strengths are used, the first a lower power source for casualties, the second at an increased level for rescue personnel. The power source has a low count range of approximately 40 mm. The higher source count range is approximately 80 mm which overcomes the airspace between a person's body and the outer surfaces of protective clothing. The sources are supplied in two basic sizes, 75 × 30 mm and 75 × 150 mm but other sizes can be produced as required.

The DS1 is powered by rechargeable batteries which provide power for up to 4 hours of continuous use. The DS1 is supplied with three high-power and three low-power simulation sources, a battery charger unit, carry/storage case and instructions.

Status
In production and in widespread service.

Manufacturer
Argon Electronics.

VERIFIED

Radsim SS1 survey meter simulator

Description
The Radsim SS1 survey meter simulator was developed to fulfil the need to train personnel in the use of radiation survey meters. Typical user groups include military and civil defence, fire and rescue services and the nuclear industry.

The SS1 enables personnel to be trained in the location of gamma radiation sources and the determination of safety demarcation areas. The SS1 simulates the widely used Alnor RD-8 survey meter, but the techniques involved can be adapted to many other models of survey meter.

The Radsim SS1 survey meter simulator with a Radsim GS1 simulation gamma source in the background

The SS1 is supplied with a Radsim GS1 simulation gamma source. The GS1 has an infinitely variable power output and can operate continuously up to 40 hours at full power using two readily available disposable batteries. The splashproof housing of the GS1 enables it to be used in adverse conditions. Utilising ultrasound technology, the GS1 can be placed within EMI/RFI sensitive locations. The GS1 is available with three different channel allocations enabling the signal from more than one source to be overlapped in order to increase the area of simulated contamination. Further variations include a time delay to simulate a worsening situation and remote-control simulation sources for particularly aggressive environments.

The SS1 has three controls. A rotary control has positions for OFF, BAT CHK (battery check), mSv/h and μSv/h. A push-button to illuminate the meter scale and an on/off switch are also included. Rechargeable batteries provide 4 hours of continuous use.

The SS1 is housed in a robust corrosion-protected metal case and is supplied complete with one GS1 simulation gamma source, battery charger unit, neck strap, storage/carry case and instructions.

Status
In widespread service with the UK Ministry of Defence, Royal Saudi Air Force and numerous fire services.

Manufacturer
Argon Electronics.

VERIFIED

....

Simulator Chemical Agent Multi-Barrelled Launcher (SCAMBL)

Description
The SCAMBL launcher is a multi-barrelled assembly designed to project a volley of airbursting simulated chemical agent over a large open area. The role of the launcher therefore is to provide armed forces and civil service units with a means of realistic training and readiness against the threat of chemical warfare. This robust all-metal rectangular frame, housing six barrels, set at different angles, provides a wider dissipation of chemical agents. The set evaluation angles are three barrels at 40° and three barrels at 30°. The set azimuth angles are two barrels on axis, two 10° to left and two 10° to right. The ammunition used is a SCAMBL munition. The bottles are placed down the launcher barrels with the fuse wires attached to spring contact pins at the outer base of each barrel. The launcher contacts are configured in series to fire a full salvo of six units. All SCAMBL munitions are then electrically fired using a SHRIKE exploder or equivalent. The launcher is an all weather, two-man portable item with provision for carrying handles and stakes. The barrels can be stowed horizontally to enable easier carriage.

Specifications
Overall length: 217 mm
Overall diameter: 90 mm
Weight without liquid: 190 g
Liquid capacity: 1 litre
NEC: 32 g
Airburst height: 40 m
Range: 120-150 m
Min voltage: 4.5 V

PW Defence SCAMBL CW training launcher, with munitions (PW Defence Limited)
0019372

SCAMBL munition
The SCAMBL munition is designed realistically to simulate the effects of CW sprays and liquid-filled airburst shells and bombs. This is achieved by the projection and fracturing of a special bottle filled with Chemical Training Liquid, which spreads droplets over a wide area. The size of the droplets will provide a visible identification on NBC detector paper. Each container is a 1 litre, high-density polyethylene plastic bottle with an ejection and a burster charge built into the cap as an assembly. The ejection charge propels the bottle and starts a 4.5 second pyrotechnic delay. The delay then activates the burster charge. Bottles and cap assemblies are supplied with special threads for easy and leak-free assembly by personnel wearing heavy, environmental gloves and during night and day exercises. The ejection charge is initiated electrically using a Shrike exploder or equivalent. When used with a PW Defence SCAMBL launcher, the resulting salvo of six airbursting bottles will cover an area of between 10,000 and 30,000 m². The system is suitable for use in battlefield exercises involving land forces with vehicles and equipment. SCAMBL munitions are supplied with a suitable chemical training liquid.

Status
Ready for production. Contract awarded by UK MoD, December 1998.

Manufacturer
PW Defence Limited.

UPDATED

Specifications
Length: 900 mm
Width: 720 mm
Stowed height: 280 mm
Deployed height: 740 mm
Weight: 62 kg
Picket points: 4
Airburst height: 40 m
Range: 120-150 m
Area coverage: 10,000-30,000 m² (subject to weather conditions)

....

Simulator Projectile Airburst Liquid Chemical L1A2 (SPAL)

Description
The Simulator Projectile Airburst Liquid Chemical Agent L1A2 (SPAL) was designed to simulate chemical warfare spray and liquid-filled airburst weapons, by projecting a container filled with Chemical Agent Training Mixture (CATM) into the air and bursting it at a height not less than 12 m. The bursting of the container spreads the CATM in the form of droplets over an area approximately 15 m wide and 50 m long and downwind from the point of burst. The droplets are large enough to produce a visible colour on chemical agent detector papers.

The SPAL consists of a cardboard mortar-type barrel, a polythene bottle filled with CATM and an electrically initiated actuator for projecting and bursting the bottle.

Status
In production. In service with the UK armed forces.

Manufacturer
Brocks Explosives Limited.

VERIFIED

....

SMF 3 Surface Monitoring Fluorimeter

Description
The SMF 3 Surface Monitoring Fluorimeter represents a new approach to training personnel for monitoring CW agent contamination on the clothing or skin of workers handling agrochemicals, pharmaceuticals or toxic industrial materials, allowing real-time quantification.

The SMF 3 Surface Monitoring Fluorimeter is a whole body dosimeter which is used to study the spread and penetration characteristics of any hazardous material where skin contamination and absorption require quantitative measurements. A fluorescent tracer is added to the hazardous material being studied and the worker carries out a set task over a prescribed period. The tracer and the hazardous material move in unison and quantitative measurement of the tracer on the clothing or skin allows the amount of active material to be determined. The SMF 3 Surface monitoring Fluorimeter measures the entire body surface area in one step. The worker to be monitored sits inside a dodecahedron of UV lights which cause the tracer to fluoresce. A video camera, equipped with an appropriate filter, then records the fluorescence. Image processing software quantifies the fluorescence of the tracer on the worker, thereby measuring the impact of the hazardous material. This highly sensitive system measures to levels of the order of micrograms per square centimetre, either over whole body or at specific locations, such as the hands. The complete measurement takes less that 30 minutes. The system has been validated against biological monitoring studies on pesticides with excellent correlation and has been used for both aqueous and organic formulations. The SMF 3 Surface Monitoring Fluorimeter has applications in areas of health and safety monitoring, the development of techniques for handling hazardous materials, studies on protective clothing and in the decontamination of personnel.

Status
In production.

Manufacturer
Safe Training Systems Limited.

VERIFIED

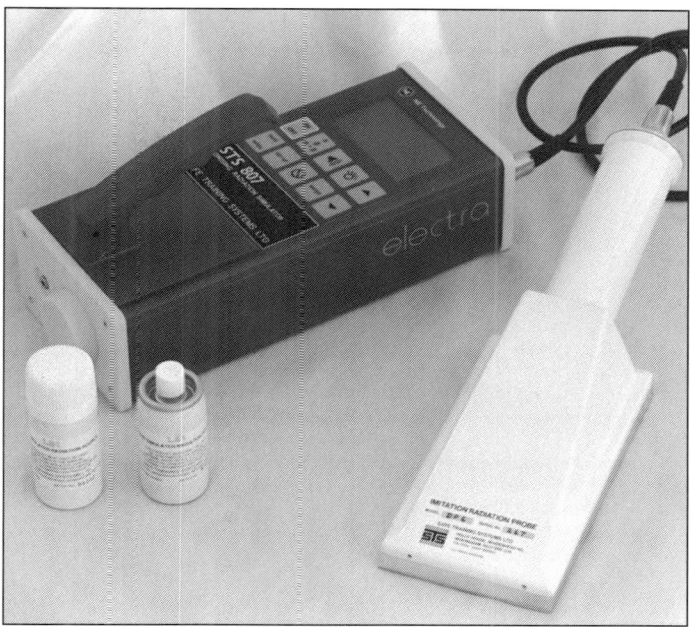

STS 807 ionising radiation simulator

STS Chemical Agent Monitor Simulator

Description
Safe Training Systems Limited produce a Chemical Agent Monitor Simulator (CAM — CAM is a Graseby Dynamics Limited trademark) which is identical to the real instrument in both appearance, handling and operation. The simulator may be used with powder, liquid and gas simulants provided by STS and may be operated in either the G or H mode. Dynamic bar displays are provided in exactly the same manner as with a real CAM.

Status
In production.

Manufacturer
Safe Training Systems Limited.

VERIFIED

STS simulators utilise the same housing as the real instrument and all are operated in exactly the same manner as real instruments.

The simulants used with STS simulators are available in powder form or as liquid in spray cans. Any contact with the liquid simulant will result in some of the simulant being transferred to the new surface in exactly the same manner as real chemical or radioactive contamination. The simulants are based on low-toxicity materials that are non-inflammable and environmentally safe. They may be removed by the usual chemical or radiation decontamination techniques.

As well as manufacturing products based on commonly used detection systems, STS also have the capability to design and develop systems to suit clients' specific needs.

Status
In production in several forms.

Manufacturer
Safe Training Systems Limited.

VERIFIED

STS plumes simulators for airborne release of NBC materials

Description
Safe Training Systems Limited has developed a plumes system for training personnel in the event of an airborne release of radioactive materials or chemical agents. The plumes system uses the Global Positioning System (GPS) and sophisticated software to produce real meter readings on simulated ratemeters during outdoor exercises to monitor simulated plumes passing over an area. A plume measuring some 14 × 3 km can be created and monitored from a vehicle travelling up to 80 km/h.

Status
In production.

Manufacturer
Safe Training Systems Limited.

VERIFIED

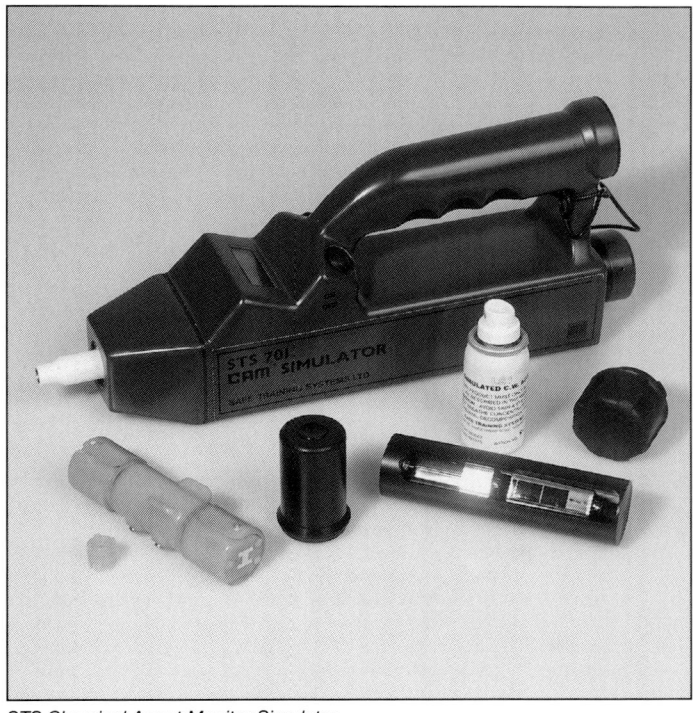

STS Chemical Agent Monitor Simulator

STS chemical and radiation monitoring training systems

Description
Safe Training Systems Limited produce a range of chemical and radiation monitoring simulators which make use of powder and liquid simulants capable of being detected by a specific gas sensor. The simulant is applied to the surfaces required to appear contaminated and a gas sensor, installed in an imitation detector probe for the system in use, senses the gas released from the simulant and triggers and actuates the chemical detector or radiac system accordingly.

The STS simulators enable trainee operators to be trained in probe manipulation, for instance to maintain the correct probe distance from a possible contaminated surface. Decontamination and other exercises may also be carried out in a very realistic manner. The simulators may be real instruments connected to the imitation probe via an electronic interface mounted within the instrument. Some

STS radiation field survey and dosimeter simulators

Description
Safe Training Systems Limited produces a range of gamma ray monitor and dosimeter simulators which allow training to take place without exposing the trainer or trainee to radiation.

The simulators operate by using a radio frequency source housed in a small box or pipe to simulate a radiation source. This emits a signal that the radio frequency detector in the simulated survey meter or dosimeter detects as 'radiation' and provides a reading on the instrument display.

The frequency at which the system operates was carefully chosen so that the system appears to reproduce the characteristics of real radiation, in particular obeying the inverse square law and also the shielding properties of materials.

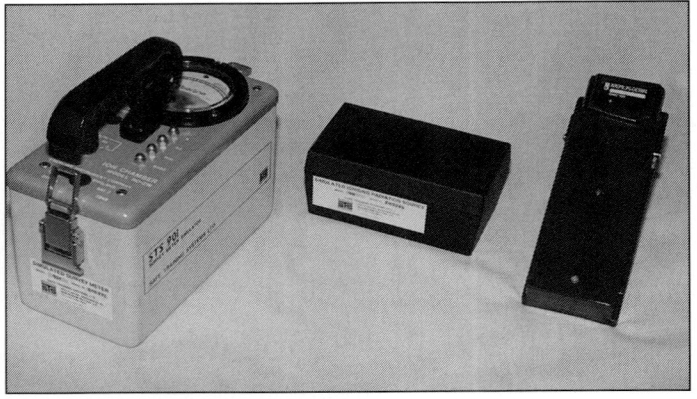

STS radiation field survey and dosimeter simulators   0050770

Several hand-held survey meters have been manufactured, including the Eberline RO 2, Ludlum 19 and NET PDRM. Dosimeters by Merlin Gerin, Rados and Stephens have been simulated.

Safe Training Systems has the capability to design simulators using the above technology but based on other manufacturers products.

Status
In production.

Manufacturer
Safe Training Systems Limited.

VERIFIED

UNITED STATES OF AMERICA

Live chemical agent test facilities

Description
Calspan SRL Corporation operates a large complex of live agent test facilities at a 700 acre site 35 miles south of Buffalo, New York. The two key facilities are the chemical surety materiel test chamber and the chemical agent laboratory.

Chemical Surety Materiel (CSM) Test Chamber
The CSM chamber is a 270 m³ facility housing a preparation laboratory with two fume hoods for storage and testing of CSM. Two additional fume hoods located inside the CSM chamber are used to facilitate chamber tests and minimise the requirement for operators to wear IPE. An adjacent structure houses a second CSM laboratory with a further two fume hoods and an analytical support laboratory. There is also a 590 m³ environmental chamber and an 8 m³ munitions chamber, originally used for the development of non-lethal chemical ordnance.

The main CSM chamber measures 6 × 12 × 4 m/20 × 40 × 12.5 ft. Constructed of 6.5 mm/¼ in welded steel plate (11 mm/7/16 in ceiling thickness), the chamber has all corners rounded and smoothed for ease of decontamination. A 500 gallon underground tank stores decontaminant for use during trials and there is a 1,000 gallon holding tank for residue. The chamber is large enough to conduct full-scale trials on armoured vehicles and aircraft for example.

Chamber access ports are designed into the structure to address the needs of any type of large-scale project and the secure volume of the chamber can be increased vertically by 7 m/24 ft through a 1 m/3 ft diameter portable flange in the ceiling. There are a variety of additional ports. Many of the glazed areas are portable for different materials to be used and long clear sight lines are featured to permit large-scale optical trials. Two important functions are the ability to discharge contaminated air from large desorption cells into the fume hoods so that unprotected access to the chamber is possible during long desorption tests, and the use of short sample lines from these cells to analytical equipment in the laboratory.

The agent preparation laboratory is built with reinforced concrete block walls and is equipped with all necessary safety equipment. It is designed with two toxic fume hoods, with a hood-ventilated, refrigerated storage vault for CSMs. In addition to its support for the chamber, this laboratory is used for small-scale chemical agent experiments. The filter trains for each fume hood and for the chamber exhaust consist of bag-in, bag-out equipment manufactured by Flanders. The secure air filtration train includes, in order, a pre-filter, HEPA filter, activated-carbon filter, sampling plenum, second carbon and HEPA filters, drawn by large fans. The two fume hood filter trains can be differentially connected for additional effort or as failure back-up. The entire facility satisfies national, international and US DoD safety standards.

Access to the facility is only via the control room entrance air-lock and cleansing station where full monitoring, personnel decontamination and medical facilities are provided.

Chemical Agent Laboratory
This is an approved facility for smaller scale CSM work, offering two recently designed agent handling fume hoods.

For detection and monitoring, the facilities include six MiniCAMS, a Varian 6000/Integrator gas chromatographic system with autosampler and, a 3DQ Discovery mass selective detector connected to a Hewlett-Packard 5890+ gas chromatograph and autosampler. Specialised solid sorbent tube and Depot Area Air Monitoring System (DAAMS) tube analysis (see under DETECTION (sensor systems) – Nuclear and Chemical).

Status
Available.

Manufacturer
Calspan SRL Corporation.

VERIFIED

Post Engagement Ground Effects Model - PEGEM

Description
PEGEM (the Post Engagement Ground Effects Model) is a powerful software simulation tool for assessment of Chemical, high-explosive and biological agent ground effects. PEGEM uses realistic digital terrain information and current meteorological data/forecasts to provide automated ground hazard assessment and generation of NATO-formatted early warning messages.

The illustration shows the effects of a CW weapon. In this example, the user can also see the population density in the area, represented by the circles. The footprint from this CW attack is shown as a series of colour-coded bands depicting the LD_n estimates for casualty predictions, where 'n' represents the likely percentage mortality in the affected population. The US government defines toxicity standards which are mapped to a colour code. Lethality is interpreted by the key in the lower left corner of this screen. As an example, LD_{95} signifies that mortality will be 95 per cent in the second band out (the darkest).

PEGEM is used by over 100 organisations and has participated in numerous military exercises since 1996. Sponsored by the Ballistic Missile Defense Organisation (BMDO), PEGEM is available at no cost to qualifying government agencies contractors.

Status
In current use by US armed forces and a wide variety of other agencies concerned with response to a CBW, nuclear or HE attack.

Manufacturer
MEVATEC Corporation.

VERIFIED

Simulated Area Weapons Effects (SAWE) - NBC I and II

Description
The Project Manager for Training Devices, US Army Materiel Command, is developing a series of devices and simulants that will simulate the effects of NBC weapons to provide a realistic NBC environment in which to train.

The Simulated Area Weapons Effects - Nuclear, Biological, Chemical I (SAWE-NBC I) includes the Casualty Assessment System (CAS), the Persistent Chemical Agent Simulant (PCAS) and the Chemical Agent Decontamination Simulant (CADS). These will provide real-time casualty assessment during force-on-force engagement simulation exercises. The CAS consists of an electronic package located inside the standard protective mask. The PCAS and CADS are simulants which replicate various chemical and biological agents on the battlefield and provide the opportunity to exercise proper chemical and biological avoidance, detection and decontamination procedures.

SAWE-NBC II will consist of biological and nuclear training device systems that will provide the means to assess casualties in real time during force-on-force engagement exercises in a simulated biological and/or nuclear warfare environment. The systems will activate casualty assessment devices worn by participants in exercises and will be MILES interoperable.

Associated with SAWE-NBC I and II is SAWE-RF (RF - radio frequency). This is a projected training system which will simulate accurately and in real time the effects of persistent NBC agent contamination, as well as the effects of indirect fire and land mines, in force-on-force training exercises. The system will make use of Global Positioning System (GPS) technology to provide accurate position location.

All three versions of SAWE will be fielded at the US Army's three Combat Training Centers (CTCs): at Fort Irwin, California; Fort Chaffee, Arkansas; and Hohenfels, Germany.

Status
In development.

VERIFIED

YUGOSLAVIA, FEDERAL REPUBLIC

KODS M2, M3 and M4 chemical agent training smoke pots

Description

The KODS M2, M3 and M4 chemical agent smoke pots are used as training aids during training and tactical exercises when a simulated chemical agent attack is required.

The KODS M2 and M3 resemble cylindrical hand grenades and may be thrown or placed on the ground for use. To prepare the smoke pots, a screw top is removed from one end of the cylindrical body to expose a cord handle. This is then pulled and the smoke pot is thrown or placed on the ground. The cord operates a friction fuze to initiate smoke production and the simulated chemical agent cloud issues through three small holes close to the fuze. The simulated agent used with the M2 is a substance known as HAF which is not dangerous but provides an unmistakable odour and a degree of tear action. The M3 uses CS and may be ignited electrically. A cloud of simulated agent will be produced for 8 to 10 minutes.

The M4 smoke pot is a much larger cylindrical device 240 mm high and 180 mm in diameter. It weighs 5.2 kg and contains 4 kg of CS to produce a

cloud of simulated agent for up to 18 minutes. Electrical or friction ignition may be used.

Status

In service with the Yugoslav armed forces.

Contractor

Yugoimport SDPR.

VERIFIED

Specifications
M2 and M3
Weight: 700 g
Weight of smoke production material: 450 g
Diameter: 80 mm
Height: 130 mm
Smoke emission time: 8-10 min
Shelf-life: 3 years

CONTRACTORS

CONTRACTORS

DETECTION – NUCLEAR

Austrian Research Centre Seibersdorf
Department of Radiation Protection, A-2444 Seibersdorf, Austria
Tel: (+43 2254) 780 25 00
Fax: (+43 2254) 740 60
e-mail: Ilse.Kraus@arcs.ac.at

BEFIC
18 rue de Villeneuve, Silic 551, F-94643 Rungis Cedex, France
Tel: (+33 1) 46 87 25 16
Telex: 263 384 F

BNFL plc, Instruments
Pelham House, Calder Bridge, Cumbria CA20 1DB, UK
Tel: (+44 1946) 78 50 67
Fax: (+44 1946) 78 50 19
e-mail: jmk3@bnfl.com
Web: http://www.bnfl.instruments.co.uk

B.O.I.S. Engineering Praha sro
Holeckova 103/31, CZ-150 95 Praha 5, Czech Republic
Tel: (+420 2) 57 32 44 02/04
Fax: (+420 2) 57 32 44 05
e-mail: info@bois-praha.cz
Web: http://www.bois-praha.cz

Bruker Biospin AG
Industriestrasse 26, CH-8117, Fällenden, Switzerland
Tel: (+411 825) 91 11
Fax: (+411 825) 96 96
e-mail: sales@bruker-biospin.ch
Web: http://www.bruker.ch
Note: Bruker Biospin AG is closely linked to Bruker Daltonics®

Bruker Daltonics®
Bruker Daltonik, Fahrenheitstrasse 4, D-28359 Bremen, Germany
Tel: (+49 421) 220 52 31
Fax: (+49 421) 220 51 03
e-mail: sales@bdal.de
Note: Bruker Daltonik (Germany), together with Bruker Saxonia Analytik GmbH and Bruker Daltonics Inc (USA), is part of Bruker Daltonics®.

CANBERRA EURISYS SA
ZA de l'Observatoire, 4 avenue des Frênes, Montigny le Bretonnoux, F-78067 St Quentin Yvelines, Cedex France
Tel: (+33 1) 39 48 57 70
Fax: (+33 1) 39 48 57 80
e-mail: info@eurisysmesures.com
Web: http://www.eurisysmesures.com
Canberra Eurisys SA is part of the French COGEMA group

GAMMA Technical Corporation
Petzvál J. u. 56, H-1119, Budapest, Hungary
Tel: (+36 1) 205 57 89
Fax: (+36 1) 205 57 78
e-mail: gamma@gammatech.hu
Web: http://www.gammatech.hu

Graetz Strahlungsmeßtechnik GmbH
Westiger Straße 172, D-58762, Altena, Germany
Tel: (+49 23 52) 70 07 16
Fax: (+49 23 52) 70 07 10
e-mail: stma-graetz@t-online.de
Web: http://www.graetz.com

Harwell Instruments Ltd
Building 528.10 Unit 1, Harwell International Business Centre, Didcot, Oxfordshire OX11 0RA, UK
Tel: (+44 1235) 83 83 00
Fax: (+44 1225) 83 83 63
e-mail: harwellsales@harwellinst.com
Web: http://www.harwellinst.com
Harwell Instruments Ltd is a subsidiary of Canberra Industries Inc

Inovision Radiation Measurements
6045 Cochran Road, Cleveland, Ohio 44139-3303, USA
Tel: (+1 216) 248 93 00
Fax: (+1 216) 349 23 07
Web: http://www.inovision.com

Jasmin Simtec Limited
Sellers Wood Drive, Bulwell, Nottingham NG6 8UX, UK
Tel: (+44 115) 916 51 65
Fax: (+44 115) 927 86 14
e-mail: sales@jasmin.plc.uk
Web: http://www.jasmin.plc.uk

Jianan Instrument Factory
PO Box 2509, Chongqing, Sichuan, People's Republic of China

Kintex
PO Box 209, 66 James Boucher Street, BG-1407 Sofia, Bulgaria
Tel: (+359) 266 23 11
Fax: (+359) 29 63 11 23

MGP Instruments
Route d'Eyguières, BP1, F-13113 Lamanon, France
Tel: (+33 0) 490 59 59 59
Fax: (+33 0) 490 59 55 18
e-mail: vely@mgpi.com
Web: http://www.mgpi.com

MoD Institute of Military Technology
HM HTI, Szilágyi Erzsébet fasor 20, Pf. 26, H-1525, Budapest, Hungary
Tel: (+36 1) 394 17 33
Fax: (+36 1) 394 30 14

Nardeux SA
ZI 'La Vallee du Parc', F-37602 Loches Cedex, France
Tel: (+33 2) 47 59 32 32
Telex: 750 808 f
Fax: (+33 2) 47 59 04 54

OMNIPOL as
Nekázanka 11, CZ-112 21 Praha 1, Czech Republic
Tel: (+42 02) 401 11 76
Fax: (+42 02) 401 21 99
e-mail: omni22@omnipol.cz
Web: http://www.omnipol.cz

Pacific Northwest National Laboratory
PO Box 999, 902 Battelle Boulevard, Richland, Washington 99352, USA
Tel: (+1 509) 375 37 76
e-mail: pnl.media.relations@pnl.gov
Web: http://www.pnl.gov/news

PerkinElmer (formerly EC&G ORTEC)
100 Midland Road, Oak Ridge, Tennessee 37830, USA
Tel: (+1 423) 483 03 96
Fax: (+1 423) 483 03 96
e-mail: info_ortec@perkinelmer.com
Web: http://www.ortec-online.com/products.htm

Philips GmbH
Unternehmensbereich Systeme und Sondertechnik, PO Box 44 87 40, Hans-Bredow-Strasse 20, D-2800 Bremen 44, Germany
Tel: (+49 421) 428 71
Fax: (+49 421) 20 46 60

RADOS Technology Oy
PO Box 506, FIN-20101 Turku, Finland
Tel: (+358 2) 468 46 00
Fax: (+358 2) 468 46 01
e-mail: info@rados.fi
Web: http://www.rados.com

RADOS Technology GmbH
Ruhrstrasse 49, PO Box 501245, D-22761 Hamburg, Germany
Tel: (+49 40) 85 19 30
Fax: (+49 40) 85 19 32 56
e-mail: info@rados.de
Web: http://www.rados.com
This Hamburg-based RADOS company specialises in detection for contamination and waste operations (see also RADOS Technology Oy, Finland which specialises in personal dosimetry and detection for area operations).

Research Institute for Chemical Defence
West Building, PO Box 925, PC 100083, Beijing, People's Republic of China
Tel: (+83 10) 62 03 37 05; 69 76 02 57
Fax: (+86 10) 62 03 37 05
e-mail: ricdcc@public3.bta.net.cn

Rheinmetall Landsysteme GmbH
Falckensteiner Straße 2, D-24159 Kiel, Germany
Tel: (+49 431) 39 99 22 92
Fax: (+49 431) 39 99 32 78
e-mail: rls-info@rheinmetall-ls.com
Web: http://www.rheinmetall-ls.de
Note: The takeover of Henschel Wehrtechnik GmbH by Rheinmetall Landsysteme GmbH was announced in October 2000.

ROMTEHNICA
5C Timisoara Blvd, R-77311, Bucharest, Romania
Tel: (+40 1 413) 01 55
Fax: (+40 1 410) 14 67; (+40 1 413) 03 17
e-mail: rth.marketing@mb.roknet.ro

Rosoboronexport
21 Gogolevsky Boulevard, 119365 Moscow, Russian Federation
Tel: (+7 095) 202 66 03
Fax: (+7 095) 202 45 94
Web: http://www.rusarm.ru

SAIC (UK) Limited
26 Craven Court, Stanhope Road, Camberley, Surrey GU15 3BW, UK
Tel: (+44 1276) 67 55 11
Fax: (+44 1276) 67 62 62

Scintrex Limited
222 Snidercroft Road, Concord, Ontario, Canada L4K 1B5
Tel: (+1 905) 669 22 80
Fax: (+1 905) 669 64 03
e-mail: scintrex@idsdetection.com
Web: http://www.idsdetection.com/products_html/pd_sesi.html
(Now part of Intelligent Detection Systems - IDS)

SE International Inc
PO Box 39, 436 Farm Road, Summertown, Tennessee 38483, USA
Tel: (+1 931) 964 35 61
Fax: (+1 931) 964 35 64
e-mail: seintl@seintl.com
Web: http://www.seintl.com

Siemens Environmental Systems Limited
Sopers Lane, Poole, Dorset BH17 7ER, UK
Tel: (+44 1202) 78 27 40
Fax: (+44 1202) 78 20 56
e-mail: epd@poole.siemens.co.uk
Web: http://www.poole.siemens.co.uk

Signal Instrument Making Plant JSC
127 Lenin Street, Obninsk 249020, Kaluga Region, Russia
Tel: (+7 095) 546 39 63

System Planning Corporation
1500 Wilson Boulevard, Arlington, Virginia 22209-2454, USA
Tel: (+1 703) 351 82 00
Fax: (+1 703) 351 85 67
e-mail: sales@sysplan.com
Web: http://www.sysplan.com

Tradeways Limited
184 Duke of Gloucester Street,
Annapolis, Maryland 21401, USA
Tel: (+1 410) 295 08 13
Fax: (+1 410) 295 08 21
e-mail: jhtradeway@msn.com

Yugoimport-SDPR pc
PO Box 89, Bulevar umetnosti 2, YU-11070 Belgrade, Yugoslavia
Tel: (+381 11) 311 27 43
Fax: (+381 11) 324 87 91
e-mail: jugoimport@sezampro.yu
Web: http://www.yugoimport-co.yu

DETECTION – BIOLOGICAL

Alexeter Technologies LLC
Suite #6, 830 Seton Court, Wheeling, Illinois 60090, USA
Tel: (+1 877) 591 55 71
Fax: (+1 847) 419 16 48
e-mail: dolson@mesosystems.com
Web: http://www.mesosystems.com

CLR Photonics Inc
655 Aspen Ridge Drive, Lafayette, Colorado 80026, USA
Tel: (+1 303) 604 20 00
Fax: (+1 303) 604 25 00
e-mail: jimro@ctilidar.com
Web: http://www.ctilidar.com

DRDC/RDDC Suffield
Defence R&D Canada/R&D pour la Défense Canada
Box 4000, Station Main, Medicine Hat, Alberta, Canada T1A 8K6
Tel: (+1 403) 544 46 72
Fax: (+1 403) 544 33 88
e-mail: Camille.Boulet@dres.dnd.ca
Web: http://www.dres.dnd.ca

DSTL CBS Porton Down
Chemical and Biological Sciences, Porton Down, Salisbury, Wiltshire SP4 0JQ, UK
Tel: (+44 1980) 61 31 21
Fax: (+44 1980) 61 00 44
e-mail: central-enquiries@dstl.gov.uk
Web: http://www.dstl.gov.uk

Edgewood Chemical Biological Center (ECBC)
Aberdeen Proving Ground, Maryland, USA
Tel: (+1 410) 436 43 37
Fax: (+1 410) 436 65 29
e-mail: cet@sbccom.apgea.army.mil
Web: http://www.sbccom.apgea.army.mil/RDA/ecbc

EDS Defence
EDS Defence Limited, 1-3 Bartley Wood Business Park, Bartley Way, Hook, Hampshire RG27 9XA, UK
Tel: (+44 1256) 74 20 20
Fax: (+44 1256) 74 23 49
Web: http://www.eds.co.uk

Engineering Computer Optecnomics (ECO) Inc
1354 Cape St Claire Road, Annapolis, Maryland 21401-5216, USA
Tel: (+1 410) 757 32 45
Fax: (+1 410) 757 86 14
e-mail: ecoinc@worldnet.att.net

Environmental Technologies Group Inc (ETG)
2202 Lakeside Boulevard, Edgewood, Maryland 21040, USA
Tel: (+1 410) 510 91 00
Fax: (+1 410) 510 94 54
e-mail: marketing@smiths-etg.com
Web: http://www.envtech.com
Note: ETG is a wholly-owned subsidiary of Smiths Group (UK). Other significant NBC-related companies in the group are Graseby Dynamics Limited (UK) and Barringer (USA) (see separate entries).

Fibertek Inc
510 Herndon Parkway, Herndon, Virginia 20170, USA
Tel: (+1 703) 471 76 71
Fax: (+1 703) 471 58 06
e-mail: bsuliga@fibertek.com
Web: http://www.fibertek.com

General Dynamics Canada
1020 68th Avenue NE, Calgary, Alberta, Canada T2E 8P2
Tel: (+1 403) 295 54 14
Fax: (+1 403) 295 67 90
e-mail: busdev.calgary@gdcanada.com
Web: http://www.computingdevices.com/land/4warn/index.htm
Note: Formerly called Computing Devices Canada

Graseby Dynamics Limited
459 Park Avenue, Bushey, Watford, Hertfordshire WD2 2BW, UK
Tel: (+44 1923) 22 85 66
Fax: (+44 1923) 22 13 61
e-mail: sara.whitel@grasebydynamics.com
Web: http://www.grasebydynamics.com

INSYS Limited
Reddings Wood, Ampthill, Bedford MK45 2HD, UK
Tel: (+44 1525) 84 10 00
Fax: (+44 1525) 40 58 61
e-mail: marketing@insys-ltd,co.uk
Web: http://www.insys-ltd.co.uk

Integrated Photomatrix Limited
Pacecombe Way, Poundbury, Dorchester, Dorset DT1 3SY, UK
Tel: (+44 1305) 26 36 73
Fax: (+44 1305) 26 36 79
e-mail: ipl@ipl-uk.com

Israel Institute for Biological Research
PO Box 19, 24 Reuven Lehrer Street, Ness-Ziona IL-74100, Israel
Tel: (+972 8) 938 16 56
Fax: (+972 8) 940 14 04
e-mail: mambi@iibr.gov.il
Web: http://www.iibr.gov.il

Joint Program Office for Biological Defense (JPOBD)
Suite 1200, 5201 Leesburg Pike, Falls Church, Virginia 22041-3203, USA
Tel: (+1 703) 681 96 00
Fax: (+1 703) 681 34 54
e-mail: trudeln@jpobd.osd.mil
Web: http://www.jpobd.net

Lockheed Martin
Lockheed Martin Naval Electronics & Surveillance Systems - Undersea Systems (Lockheed Martin NE & SS-Manassas), 9500 Godwin Drive, Manassas, Virginia 20110, USA
Tel: (+1 703) 367 19 64
Fax: (+1 703) 367 54 41
e-mail: richard.read@lmco.com
Web: http://www.lockheedmartin.com/manassas

MesoSystems Technology Inc
1021 N Kellogg St, Kennewick, Washington 99336, USA
Tel: (+1 509) 737 83 83
Fax: (+1 509) 737 84 84
e-mail: dolson@mesosystems.com
Web: http://www.mesosystems.com

New Horizons Diagnostics Corporation
Suite B, 9110 Red Branch Road, Columbia, Maryland 21045, USA
Tel: (+1 410) 992 93 57
Fax: (+1 410) 992 03 28
e-mail: NHDiag@aol.com
Web: http://www.NHDiag.com

Rheinmetall Landsysteme GmbH
Falckensteiner Straße 2, D-24159 Kiel, Germany
Tel: (+49 431) 39 99 22 92
Fax: (+49 431) 39 99 32 78
e-mail: rls-info@rheinmetall-ls.com
Web: http://www.rheinmetall-ls.de
Note: The takeover of Henschel Wehrtechnik GmbH by Rheinmetall Landsysteme GmbH was announced in October 2000.

Tetracore Inc
Suite C, 11 Firstfield Road, Gaithersburg, Maryland 20878, USA
Tel: (+1 301) 258 75 53
Fax: (+1 301) 258 97 40
e-mail: wnelson@tetracore.com
Web: http://www.tetracore.com

TNO-PML
Prinz Mauritz Laboratory, PO Box 6050, NL-2600 JA, Delft, Netherlands
Tel: (+31 15) 269 69 69
Fax: (+31 15) 261 24 03
e-mail: infodesk@tno.nl
Web: http://www.tno.nl

TSI Inc
PO Box 64394, St Paul, Minnesota 55164, USA
Tel: (+1 651) 490 28 07
Fax: (+1 651) 490 38 80
e-mail: tsiinfo@tsi.com
Web: http://www.tsi.com

DETECTION - CHEMICAL

Air Techniques
Division of Hamilton Associates Inc, 11403 Cronridge Drive, Owings Mills, Maryland 21117-2247, USA
Tel: (+1 410) 363 96 96
Fax: (+1 410) 363 96 95
e-mail: airtech1@erols.com
Web: http://www.atitest.com

Åkers Krutbruk Protection AB
PO Box 84, 60 Åkers Styckebruk, SE-640 Sweden.
Tel: (+46 159) 366 00
Fax: (+46 159) 307 28
e-mail: info@akerskrutbruk.se
Web: http://www.akerskrutbruk.se

Anachemia Canada Inc
255 Rue Norman, Lachine, Montreal, Quebec, Canada H8R 1A3
Tel: (+1 514) 489 57 11
Fax: (+1 514) 363 52 81
e-mail: msds@anachemia.com
Web: http://www.anachemia.com

BAE Systems
Integrated Defense Solutions, 6500 Tracor Lane, Austin, Texas 78725, USA
Tel: (+1 512) 929 47 53
Fax: (+1 512) 929 23 81
e-mail: gary.morris2@baesystems.com
Web: http://www.baesystems.com

Barringer Technologies Inc
30 Technology Drive, Warren, New Jersey 07059, USA
Tel: (+1 908) 222 91 00
Fax: (+1 908) 222 15 57
e-mail: info@barringer.com
Web: http://www.barringer.com

J Blaschke Wehrtechnik GmbH
Wienerbergstrasse 42-44, A-1120 Vienna, Austria
Tel: (+43 1) 810 09 09
Fax: (+43 1) 810 09 09 33
e-mail: office@blaschke.com

Bruker Biospin AG
Industriestrasse 26, CH-8117, Fällenden, Switzerland
Tel: (+411 825) 91 11
Fax: (+411 825) 96 96
e-mail: sales@bruker-biospin.ch
Web: http://www.bruker.ch
Note: Bruker Biospin is closely linked to Bruker
   Daltonics®.

Bruker Daltonics®
Bruker Saxonia Analytik GmbH
Permoserstrasse 15, D-04318 Leipzig, Germany
Tel: (+49 341) 24 31 30
Fax: (+49 341) 24 34 04
e-mail: sales@bsax.de
Web: http://www.bsax.de

Bruker Daltonik GmbH
Fahrenheitstrasse 4, D-28359 Bremen, Germany
Tel: (+49 421) 220 52 00
Fax: (+49 421) 220 51 03
e-mail: sales@bdal.de
Web: http://www.bruker-daltonik.de
Note: Bruker Saxonia Analytik GmbH, together with
   Bruker Daltonik (Germany) and Bruker Daltonics
   Inc (USA), is part of Bruker Daltonics®.

CDS Analytical Inc
465 Limestone Road, PO Box 277, Oxford,
   Pennsylvania 19363, USA
Tel: (+1 610) 932 36 36
Fax: (+1 610) 932 41 58
Web: http://www.cdsanalytical.com

Dräger
Sicherheitstechnik GmbH, Revalstrasse 1, D-23560
   Lübeck, Germany
Tel: (+49 451) 88 20
Fax: (+49 451) 882 20 80
e-mail: info@draeger.com
Web: http://www.draeger.com

Environics Oy
Graanintie 5, FIN-50190 Mikkeli, Finland
Tel: (+358 201) 43 04 30
Fax: (+358 201) 43 04 40
e-mail: sales@environics.fi
Web: http://www.environics.fi

USA Office:
Environics USA Inc, PO Box 290699, Port Orange,
   Florida 32129-0699, USA
Tel: (+1 386) 304 52 52
Fax: (+1 386) 304 52 51
e-mail: EUSAsales@aol.com
Web: http://www.environics.fi

Environmental Technologies Group Inc (ETG)
2202 Lakeside Boulevard, Edgewood, Maryland
   21040, USA
Tel: (+1 410) 510 91 00
Fax: (+1 410) 510 94 54
e-mail: marketing@smiths-etg.com
Web: http://www.envtech.com
Note: ETG is a wholly-owned subsidiary of Smiths
   Group (UK). Other significant NBC-related
   companies in the group are Graseby Dynamics
   Limited (UK) and Barringer (USA) (see separate
   entries).

ESG
7530, Suite 2, South Madison, Willowbrook, Illinois
   60521, USA
Tel: (+1 630) 323 01 30
Fax: (+1 630) 323 01 91
e-mail: info@esgsafety.com/
Web: http://www.esgsafety.com/start.htm
Note: ESG - Environmental Safety Group

FLIR Systems Inc
16 Esquire Road, Billerica, Massachusetts 01862-
   2598, USA
Tel: (+1 978) 901 82 25
Fax: (+1 978) 901 88 87
e-mail: info@flir.com
Web: http://www.flir.com

General Dynamics Armament and Technical Products
2000 Brunswick Lane, Deland, Florida 32724, USA
Tel: (+1 904) 736 17 00
Fax: (+1 904) 736 22 50
e-mail: gmilbouer@intellitec.com
Web: http://www.gdatp.com   or   http://
   www.intellitec.com/product/chemicaldefense/
Note: General Dynamics Armament and Technical
   Products formed as a new company on 14 June
   2002 and includes Advanced Technical Products
   Inc, Intellitec Division.

Giat Industries
13 route de la Minière, F-78034 Versailles Cedex,
   France
Tel: (+33 1) 30 97 37 37
Fax: (+33 1) 30 97 39 00
e-mail: sales@giat-industries.fr
Web: http://www.giat-industries.fr

Graseby Dynamics Limited
459 Park Avenue, Bushey, Watford, Hertfordshire
   WD2 2BW, UK
Tel: (+44 1923) 22 85 66
Fax: (+44 1923) 22 13 61
e-mail: sara.white@grasebydynamics.com
Web: http://www.grasebydynamics.com

Great Wall Instrument Factory
PO Box 377, Beijing, People's Republic of China

Innova AirTech Instruments
Energivej 30, DK-2750 Ballerup, Denmark
Tel: (+45 44) 20 01 00
Fax: (+45 44) 20 01 01
e-mail: msteenberg@innova.dk
Web: http://www.innova.dk

Israel Institute for Biological Research (IIBR)
PO Box 19 Ness-Ziona, IL-74100 Israel
Tel: (+972 8) 938 16 56
Fax: (+972 8) 940 14 04
e-mail: mambi@iibr.gov.il
Web: http://www.iibr.gov.il

Jasmin Simtec Limited
Sellers Wood Drive, Bulwell, Nottingham NG6 8UX,
   UK
Tel: (+44 115) 916 51 65
Fax: (+44 115) 927 86 14
e-mail: sales@jasmin.plc.uk
Web: http://www.jasmin.plc.uk

Lockheed Martin
Lockheed Martin Naval Electronics & Surveillance
   Systems - Undersea Systems (Lockheed Martin NE
   & SS-Manassas), 9500 Godwin Drive Manassas,
   Virginia 20110, USA
Tel: (+1 703) 367 19 64
Fax: (+1 703) 367 54 41
e-mail: richard.read@lmco.com
Web: http://www.lockheedmartin.com/manassas

Louis Schleiffer AG
CH-8714 Feldbach Zürich, Switzerland
Tel: (+41 1) 552 44 22 12
Fax: (+41 1) 552 44 26

MGP Instruments
Route d'Eyguières, BP1,
F-13113 Lamanon, France
Tel: (+33 0) 490 59 59 59
Fax: (+33 0) 490 59 55 18
e-mail: vely@mgpi.com
Web: http://www.mgpi.com

Microsensor Systems
62 Corporate Court, Bowling Green, Kentucky 42103,
   USA
Tel: (+1 270) 745 00 99
Fax: (+1 270) 745 00 95
e-mail: sales@microsensorsystems.com
Web: http://www.microsensorsystems.com

MoD Institute of Military Technology
HM HTI, Szilágyi Erzsébet fasor 20, Pf. 26, H-1525,
   Budapest, Hungary
Tel: (+36 1) 394 17 33
Fax: (+36 1) 394 30 14

OI Analytical/CMS Field Products
Building 400, 2148 Pelham Parkway, Pelham,
   Alabama 35124-1192, USA
Tel: (+1 205) 733 69 00
Fax: (+1 205) 733 69 19
e-mail: cmssales@oico.com
Web: http://www.oico.com

OMNIPOL as
Nekázanka 11, 112 21 Praha 1, CZ-Czech Republic
Tel: (+42 02) 401 11 76
Fax: (+42 02) 401 21 99
e-mail: omni22@omnipol.cz
Web: http://www.omnipol.cz

Orbital Sciences Corporation
Space and Electronics Systems Group, 2771 North
   Garey Avenue, Pomona, California 91767, USA
Tel: (+1 909) 593 35 81
Fax: (+1 909) 593 28 43
e-mail: cmyer@orbital.com
Web: http://www.orbital.com

ORITEST Limited
Na Belidle 21, CZ-150 00 Prague 5, Czech Republic
Tel: (+420 2) 57 31 16 39
Fax: (+420 2) 57 31 38 20
e-mail: oritest@mbox.vol.cz
Web: http://www.oritest-group.cz

Powertronic Systems Inc
13700 Chef Menteur Highway, New Orleans,
   Louisiana 70129-1907, USA
Tel: (+1 504) 254 03 83
Fax: (+1 504) 254 03 93
e-mail: psieng@bellsouth.net

PROENGIN SA
1 rue de l'Industrie, F-78210 Saint Cyr l'Ecole, France
Tel: (+33 1) 30 58 47 34
Fax: (+33 1) 30 58 93 51
e-mail: PROENGIN@wanadoo.fr

Protechnik Laboratories
PO Box 8854, Pretoria 001, Republic of South Africa
Tel: (+27 12) 665 02 31
Fax: (+27 12) 665 02 40
e-mail: philipc@protechnik.co.za
Web: http://www.protechnik.co.za

Rae Systems
218 Brandywine Avenue, Downingtown, Pennsylvania
   19335, USA
Tel: (+1 610) 873 48 91
Fax: (+1 610) 873 48 92
e-mail: raesales@raesystems.com
Web: http://www.raesystems.com

Research Institute for Chemical Defence
West Building, PO Box 925, PC 100083, Beijing,
   People's Republic of China
Tel: (+86 10) 62 03 37 05; 69 76 02 57
Fax: (+86 10) 62 03 37 05
e-mail: ricdcc@public3.bta.net.cn

Rheinmetall Landsysteme GmbH
Falckensteiner Straße 2, D-24159 Kiel, Germany
Tel: (+49 431) 39 99 22 92
Fax: (+49 431) 39 99 32 78
e-mail: rls-info@rheinmetall-ls.com
Web: http://www.rheinmetall-ls.de
Note: The takeover of Henschel Wehrtechnik GmbH
   by Rheinmetall Landsysteme GmbH was
   announced in October 2000.

Richmond Packaging (UK) Limited
New Road, Winsford, Cheshire CW7 2NY, UK
Tel: (+44 1606) 55 74 22
Fax: (+44 1606) 86 10 63
This company is authorised by the UK MoD to
   manufacture for overseas sales.

ROMTEHNICA
5C Timisoara Blvd, R-77311 Bucharest, Romania
Tel: (+40 1 413) 01 55
Fax: (+40 1 410) 14 67; (+40 1 413) 03 17
e-mail: rth.marketing@mb.roknet.ro

Rosoboronexport
21 Gogolevsky Boulevard, 119865 Moscow, Russian Federation
Tel: (+7 095) 202 66 03
Fax: (+7 095) 202 45 94
Web: http://www.rusarm.ru

Science & Technology Corporation
Suite #205, 500 Edgewood Road, Edgewood, Maryland 21040, USA
Tel: (+1 757) 865 04 67
Fax: (+1 757) 865 19 93
e-mail: gilligan@stcnet.com
Web: http://www.stcnet.com

Scientific Instrumentation Limited
2233 Hanselman Avenue, Saskatoon, Saskatchewan, Canada S7L 6A7
Tel: (+1 306) 244 08 81
Fax: (+1 306) 665 62 63
e-mail: s.i.l@sil.sk.ca
Web: http://www3.sk.sympatico.ca/scinltd

STR Inc
Science & Technology Research Incorporated, Suite 200, 112 Juliad Court, Fredericksburg, Virginina 22406, USA
Tel: (+1 540) 752 80 80
Fax: (+1 540) 752 80 85
e-mail: strdd@crosslink.com
Web: http://www.str-inc.com

Thales
Computer Products Division, Parc d'Activités Kléber, 160 boulevard de Valmy, PO Box 82, F-92704 Colombes Cedex, France
Tel: (+33 1) 41 30 30 00
Fax: (+33 1) 41 30 33 57
Web: http://www.tcc.thomson-csf.com

Tradeways Limited
184 Duke of Gloucester Street, Annapolis, Maryland 21401, USA
Tel: (+1 410) 295 08 13
Fax: (+1 410) 295 08 21
e-mail: JHtradeway@msn.com
Web: http://www.tradewaysltd.com

Triquint Semiconductors
Sawtek Division, 1818 S Highway 441, Apopka, Florida 32703, USA
Tel: (+1 407) 886 88 60
e-mail: info-sawtek@tqs.com
Web: http://www.triquint.com/company/divisions/sawtek/

Veridian Corporation
PO Box 400, 4455 Genesee Street, Buffalo, New York 14225, USA
Tel: (+1 716) 631 69 05
Fax: (+1 716) 631 68 15
e-mail: mcmahon@calspan.com
Web: http://www.calspan.com/chem-def-demil-menu.html
Note: Formerly known as The Calspan SRL Corporation. Incorporated into Veridian in 1997.

Yugoimport-SDPR pc
PO Box 89, Bulevar umetnosti 2, YU-11070 Belgrade, Yugoslavia
Tel: (+381 11) 311 27 43
Fax: (+381 11) 324 87 91
e-mail: jugoimport@sezampro.yu
Web: http://www.yugoimport-co.yu

DETECTION - C³I

AristaTek Inc
365 North 9th Street, Laramie, Wyoming 82072, USA
Tel: (+1 877) 912 22 00
Fax: (+1 307) 721 23 45
e-mail: info@aristatek.com
Web: http://www.aristatek.com

BRUHN NewTech A/S
Vasekaer 12, DK-2730 Herlev, Denmark.
Tel: (+45 44) 88 22 55
Fax: (+45 44) 53 17 55
e-mail: info@newtech.dk
Web: http://www.newtech.dk
Note: BRUHN NewTech A/S is part of the BRUHN NewTech Group.

BRUHN NewTech Inc
10420 Little Patuxent Pkwy, Suite 301, Columbia, Maryland 21044-3636, USA
Tel: (+1 410) 884 17 00
Fax: (+1 410) 844 61 71
e-mail: info@bruhn-newtech.com
Web: http://www.bruhn-newtech.com
Note: BRUHN NewTech Inc is part of the BRUHN NewTech Group.

BRUHN NewTech Limited
1 Allenby Road, Winterbourne Gunner, Salisbury, Wiltshire SP2 6HZ, UK
Tel: (+44 1980) 61 17 76
Fax: (+44 1980) 61 13 30
e-mail: info@bruhn-newtech.co.uk
Web: http://www.bruhn-newtech.co.uk
Note: BRUHN NewTech Limited is part of the BRUHN NewTech Group.

Industrieanlagen-Betriebsgesellschaft GmbH
Einsteinstrasse 20, D-85521, Ottobrunn, Germany
Tel: (+49 89) 608 80
Fax: (+49 89) 60 88 24 17
e-mail: iabg@iabg.de
Web: http://www.iabg.de

MGP Instruments
route d'Eyguires, BP1,
F-13113 Lamanon, France
Tel: (+33 0) 490 59 59 59
Fax: (+33 0) 490 59 55 18
e-mail: vely@mgpi.com
Web: http://www.mgpi.com

Odenwald - Werke Rittersbach GmbH
Oberscheflenzer Strasse, D-74834 Elztal-Rittersbach, Germany
Tel: (+49 62) 937 31
Fax: (+49 62) 937 32 19
e-mail: OWR.GMBH@t-online.de
Web: http://www.owr.de

PLG Inc
Suite 200, 300 Commerce Drive, Irvine, California 92602, USA
Tel: (+1 714) 734 42 42
Fax: (+1 714) 734 42 52
e-mail: rridley@plg.com or midas-at@absconsulting.com
Web: http://www.plg.com or http://www.absconsulting.com/midas/index.html#
Note: PLG Inc is owned by ABS Group (ABS Consulting)

Rheinmetall Landsysteme GmbH
Falckensteiner Straße 2, D-24159 Kiel, Germany
Tel: (+49 431) 39 99 22 92
Fax: (+49 431) 39 99 32 78
e-mail: rls-info@rheinmetall-ls.com
Web: http://www.rheinmetall-ls.de
Note: The takeover of Henschel Wehrtechnik GmbH by Rheinmetall Landsysteme GmbH was announced in October 2000.

SAFER Systems, LLC
Suite B, 5141 Verdugo Way, Camarillo, California 93012, USA
Tel: (+1 805) 383 97 11
Fax: (+1 805) 383 63 44
e-mail: Safer@safersystem.com
Web: http://www.safersystem.com

DETECTION - MOBILE SYSTEMS

Alvis Vehicles Limited
PO Box 106, Hadley Castle Works, Telford, Shropshire, TF1 4QW, UK
Tel: (+44 19 52) 22 45 00
Fax: (+44 19 52) 24 39 10
email: info@alvis.plc.uk
Web: http://www.alvis.plc.uk

Daewoo Heavy Industries Limited
Land Systems Division, 22nd floor, Daewoo Centre, 54 5ga Namdacmum-RO, Jung-Gu, PO Box 7955 Seoul, South Korea
Tel: (+82 32) 726 32 60
Fax: (+82 32) 726 32 69
e-mail: aero@solar.dhiltd.co.kr
Web: http://www.dhiltd.co.kr

General Dynamics Armament and Technical Products
2000 Brunswick Lane, Deland, Florida 32724, USA
Tel: (+1 904) 736 17 00
Fax: (+1 904) 736 22 50
e-mail: gmilbouer@intellitec.com
Web: http://www.gdatp.com or http://www.intellitec.com/product/chemicaldefense/
Note: General Dynamics Armament and Technical Products formed as a new company on 14 June 2002 and includes Advanced Technical Products Inc, Intellitec Division.

General Dynamics Corporation
Land Systems Division, PO Box 2074, Warren, Michigan, USA
Tel: (+1 810) 825 70 11
Fax: (+1 810) 825 70 28
e-mail: petty@gdls.cpm
Web: http://www.gdls.com/programs

Kia
15-21 Yoido-Dong, Youngdeungpo-gu, 150-706 Seoul, South Korea
Tel: (+82 2) 788 84 56/9
Fax: (+82 2) 788 84 71/06 51
e-mail: msales@kia.co.kr
Web: http://www.kia.co.kr

Kintex
PO Box 209, 66 James Boucher Street, BG-1407 Sofia, Bulgaria
Tel: (+359 2) 66 23 11
Fax: (+359 2) 963 11 23

Mil Design Bureau
MIL Moscow Helicopter Plant, 2 Sokolnichesky Val 107113, Moscow, Russia
Tel: (+7 095) 264 90 83
Fax: (+7 095) 264 55 71
e-mail: mvzmil@mvzmsk.ru
Web: http://www.mvzmsk.ru

Mitsubishi Heavy Industries Limited
5-1 Marunouchi 2-chome, Chiyoda-ku, PO Box 645, Tokyo 100, Japan
Tel: (+81 3) 32 12 31 11
Fax: (+81 3) 34 53 64 34
e-mail: mpctokyo@ppp.mindnet.or.jp
Web: http://www.mindnet.or.jp/mpc

Odenwald-Werke Rittersbach GmbH
Oberscheflenzer Strasse, D-74834 Elztal-Rittersbach, Germany
Tel: (+49 62) 937 31
Fax: (+49 62) 937 32 19
e-mail: owr.gmbh@t-online.de
Web: http://www.owr.de

RH-Alan d.o.o.
Bosanska 26, HR-10000 Zagreb, Croatia
Tel: (+385 1) 378 08 00
Fax: (+385 1) 378 08 38
e-mail: antun.persin@rh-alan.tel.hr

Rheinmetall Landsysteme GmbH
Falckensteiner Straße 2, D-24159 Kiel, Germany
Tel: (+49 431) 39 99 22 92
Fax: (+49 431) 39 99 32 78
e-mail: rls-info@rheinmetall-ls.com
Web: http://www.rheinmetall-ls.de
Note: The takeover of Henschel Wehrtechnik GmbH
by Rheinmetall Landsysteme GmbH was
announced in October 2000.

Rosoboronexport
21 Gogolevsky Boulevard, 119865 Moscow, Russian
Federation
Tel: (+7 095) 202 66 03
Fax: (+7 095) 202 45 94
Web: http://www.rusarm.ru

Sanders
65 Spit Brook Road, Nashua, New Hampshire 03061,
USA
Tel: (+1 603) 885 28 17
Fax: (+1 603) 885 28 13
email: john.h.measell@LMCO.com
Web: http://www.sanders.com

Thales
Computer Products Division, Parc d'Activités Kléber,
160 boulevard de Valmy, PO Box 82, F-92704
Colombes Cedex, France
Tel: (+33 1) 41 30 30 00
Fax: (+33 1) 41 30 33 57
Web: http://www.tcc.thomson-csf.com

Tradeways Limited
184 Duke of Gloucester Street, Annapolis, Maryland
21401, USA
Tel: (+1 410) 295 08 13
Fax: (+1 410) 295 08 21
e-mail: jhtradeway@msn.com
Web: http://www.tradewaysltd.com

TRW Inc
One Federal Systems Park Drive, Fairfax, Virginia
22033, USA
Tel: (+1 703) 968 12 22
Fax: (+1 703) 803 51 44
e-mail: sharon.whittaker@trw.com
Web: http://www.trw.com

TRW Tactical Systems
TRW Systems & Information Technology Group
One Space Park Redondo Beach, California 90278
Tel: (+1 310) 814 03 21
email: sharon.whittaker@trw.com
Web: http://www.trw.com

Tula Plant JSC
28F Smirnov Street, Tula 300041, Russian Federation
Tel: (+7 872) 20 13 32

Volsk Metalworking Plant
Volsk 412680, Saratov Region,
Russian Federation
Tel: (+7 84593) 325 15

DETECTION - MARKING SYSTEMS

INSYS Limited
Reddings Wood, Ampthill, Bedford MK45 2HD, UK
Tel: (+44 1525) 84 10 00
Fax: (+44 1525) 40 58 61
e-mail: insys-ltd.marketing@dial.pipex.com
Web: http://www.insys-ltd.co.uk

Maadi Company for Engineering Industries (F.54)
Cairo, Egypt
Tel: (+20 2) 350 17 22
Fax: (+20 2) 350 18 55
Telex: 92167 NOMPUN.ATT SMARMS

Pearson Engineering Ltd
Wincomblee Road, Walker,
Newcastle-upon-Tyne NE6 3QS, UK
Tel: (+44 191) 234 00 01
Fax: (+44 191) 262 04 02
e-mail: pearson@pearson-eng.com
Web: http://www.pearson-eng.com

Theodor Rapp GmbH & Co KG
PO Box 1364, Alte Hausacher Straße 2-4, D-77712
Haslach iK, Germany
Tel: (+49 7832) 70 30
Fax: (+49 7832) 703 22
e-mail: info@rapp-praezisionsmechanik.de
Web: http://www.rapp-praezisionsmechanik.de

Yugoimport-SDPR pc
PO Box 89, Bulevar umetnosti 2, YU-11070 Belgrade,
Yugoslavia
Tel: (+381 11) 311 27 43
Fax: (+381 11) 324 87 91
e-mail: jugoimport@sezampro.yu
Web: http://www.yugoimport-co.yu

MASKS - GENERAL

Aero Sekur SpA
Via Delle Valli, snc, I-04011 Aprilia (Latina), Italy
Tel: (+39 06) 92 01 61
Fax: (+39 06) 92 72 71 65
e-mail: commerciale@aerosekur.it
Web: http://www.aerosekur.it

Air Master Technology Limited
Amtech House, 120 Faraday Park, Swindon, Wiltshire
8N3 6JF, UK
Tel: (+44 1793) 71 65 01
Fax: (+44 1793) 71 65 02
e-mail: dalemccollum@amtech.uk.com
Web: http://www. amtech.uk.com

Alfred Kärcher GmbH & Co
Department VPS, Alfred Kärcher Strasse 23-4J,
D-71364 Winnenden, Germany
Tel: (+49 7195) 14 22 62
Fax: (+49 7195) 14 27 80
e-mail: vps@de.kaercher.com
Web: http://www.karcher-vps.com

Avon Technical Products
Protection Group, Hampton Park West, Melksham,
Wiltshire SN12 8NB, UK
Tel: (+44 1225) 89 64 57
Fax: (+44 1225) 89 63 01
e-mail: protection@avonrubber.co.uk
Web: http://www.avon-rubber.com

BIANA SA
Production of NBC Masks and Filters, 10 Apolloniou
Street, PO Box 5, GR-19400 Koropi, Greece
Tel: (+30 1) 662 39 10
Fax: (+30 1) 662 47 24
e-mail: anargyrou@biana.gr
Web: http://www.biana.gr

BIOMARINE Inc
456 Creamery Way, Exton, Philadelphia 19341-2532,
USA
Tel: (+1 610) 524 88 00
Fax: (+1 610) 524 88 07
e-mail: info@neutronicsinc.com
Web: http://www.neutronicsinc.com

BIOMARINE Inc is a subsidiary company of
Neutronics Inc.

Carleton Life Support Technologies, Limited
1-1200 Aerowood Drive, Mississauga, Ontario,
Canada L4W 2S7
Tel: (+1 905) 629 32 45
Fax: (+1 905) 629 83 06
e-mail: info@carltech.com
Web: http://www.carltech.com

Charcoal Cloth (International) Limited
Division of Chemviron Carbon Limited, 1 South Link,
Oldham, Lancashire OL4 1DE
Tel: (+44 161) 628 50 00
Fax: (+44 161) 628 51 11
e-mail: sross@calgcarb.com
Web: http://www.calgcarb.com
Note: This company is authorised by the UK MoD to
manufacture for overseas sales.

China North Industries Corporation
7A, Yue Tan Nan Jie, PO Box 2137 Beijing, People's
Republic of China
Tel: (-86 10) 68 51 22 54
Fax: (+86 10) 68 53 32 36

Civil Defence Supply
Hunt House, The Green, Welbourn, Lincolnshire
LN5 0JF, UK
Tel: (+44 1400) 27 38 50/51/52/53
Fax: (-44 1522) 81 13 53
e-mail: info@civil-defence.org
Web: http://www.civil-defence.org

CQC Ltd
Riverside Road, Barnstaple, Devon EX31 1NB, UK
Tel: (+44 1271) 34 56 78
Fax: (+44 1271) 34 50 90
e-mail: pjg@cqc.co.uk
Web: http://www.cqc.co.uk

Drägerwerk AG Lübeck
PO Box 1339, Moislinger Allee 53/55, D-2400 Lübeck
1, Germany
Tel: (+49 451) 88 20
Fax: (+49 451) 882 20 80

Engicom Systems NV
Binnensteenweg 172, B-2530 Boechout, Belgium
Tel: (+32 3) 460 04 90
Fax: (+32 3) 460 03 90

Forsheda Polymer Engineering
SE-330 12 Forsheda, Sweden
Tel: (+46 370) 890 94; 890 00
Fax: (+46 370) 817 71
e-mail: dahn.emanuelsson@forsheda.se
Web: http://www.forsheda.se

Giat Industries
13 route de la Minière, F-78034 Versailles Cedex,
France
Tel: (+33 1) 30 97 37 37
Fax: (+33 1) 30 97 39 00
e-mail: sales@giat-industries.fr
Web: http://www.giat-industries.fr

Gumárny Zubří
CZ-756 54 Zubří, Czech Republic
Tel: (+420 651) 66 21 11
Fax: (+420 651) 65 87 44
e-mail: marketing@guzu.cz
Web: http://www.guzu.cz

Helsatech GmbH
Bayreuther Strasse 11, D-95482 Gefrees, Germany
Tel: (+49 9254) 800
Fax: (+49 9254) 804 02
e-mail: helsatech@de.helsa.com
Web: http://www.helsa.com/com/static/helsatech

Hsing Hua Company Limited
PO Box 87-46, Nan-Gang, Taiwan

Huber + Suhner AG
Components and Systems Division, CH-8330
Pfäffikor/ZH, Switzerland
Tel: (+41 1) 952 26 43
Fax: (+41 1) 952 25 52
e-mail: pmoser@hubersuhner.com
Web: http://www.hubersuhner.com

ILC Dover Inc
One Moonwalker Road, Frederica, Delaware 19946-
2080, USA
Tel: (+1 302) 335 39 11
Fax: (+1 302) 335 07 62
e-mail: lawrer@ilcdover.com
Web: http://www.ilcdover.com

Interspiro Inc
31 Business Park Drive, Branford, Connecticut, 06405
USA
Tel: (+1 203) 481 38 99
Fax: (+1 203) 483 18 79
e-mail: info@interspiro.com
Web: http://www.interspiro.com

Irvin Aerospace Canada Limited
PO Box 280, 479 Central Avenue, Fort Erie, Ontario,
Canada L2A 5M9
Tel: (+1 905) 871 65 10
Fax: (+1 905) 871 65 34
e-mail: marketing@irvincanada.com

Kintex
PO Box 209, 66 James Boucher Street, BG-1407
Sofia, Bulgaria
Tel: (+359) 266 23 11
Fax: (+359) 29 63 11 23

Military Technical Institute of Protection
PO Box 547, CZ-602-00 Brno, Czech Republic
Tel: (+420) 541 18 30 01
Fax: (+420) 541 18 31 52
e-mail: vtuo_chem@telecom.cz

MKE ELSA AS
Abdülhak Hamit Cad, TR-06470 Mamak-Ankara,
Turkey
Tel: (+90 312) 368 76 55; 78 70
Fax: (+90 312) 368 16 58

MoD Institute of Military Technology
HM HTI, Szilágyi Erzsébet fasor 20, Pf. 26, H-1525,
Budapest, Hungary
Tel: (+36 1) 394 17 33
Fax: (+36 1) 394 30 14

MSA Defense Products
Safety Products Division, Mine Safety Appliances
Company, PO Box 428, Pittsburgh, Pennsylvania
15230-0428, USA
Tel: (+1 412) 733 92 74
Fax: (+1 412) 733 85 73
e-mail: evan.erickson@msanet.com
Web: http://www.msanet.com

National Organisation for Military Production
3rd Floor, 23 Talaat Harb Street, PO Box 9, Heliopolis,
Cairo, Egypt
Tel: (+20 2) 281 08 38
Fax: (+20 2) 281 08 28

North Safety Products Limited
The Court Yard, Green Lane, Heywood, Lancashire
OL10 2EX, UK
Tel: (+44 170) 669 38 00
Tel: (+44 170) 669 38 01
e-mail: sales@northsafety.co.uk
Web: http://www.northsafety.co.uk

NPO 'NEOGANIKA'
Karl Marx Street 4, 144000 Electrostal, Russian
Federation
Telex: 346323 ASTRA

OMNIPOL as
Nekázanka 11, 112 21 Praha 1, Czech Republic
Tel: (+42 02) 401 11 76
Fax: (+42 02) 401 21 99
e-mail: omni22@omnipol.cz
Web: http://www.omnipol.cz

Pao-Chang Company
PO Box 90584, Chiao-Hsi, I-Lan 262, Taiwan
Tel: (+886) 39 88 53 29
Fax: (+886) 39 88 11 16
e-mail: joeyliaw@ms41.url.com.tw

Plant KINAP JSC
23 Lesnaya Street, Samara 443071, Russian
Federation
Tel: (+7 8462) 3 40 82

Protector Technologies Group
Pimbo Road, West Pimbo, Skelmersdale, Lancashire
WN8 9RA, UK
Tel: (+44 1695) 71 17 11
Fax: (+44 1695) 71 17 64
e-mail: info@protectornet.com
Web: http://www.protectornet.com/eurafind/
protect.htm

Remploy Limited
Remploy House, 415 Edgware Road, Cricklewood,
London NW2 6LR, UK
Tel: (+44 208) 235 05 32
Fax: (+44 208) 235 05 01
Note: This company is authorised by the UK MoD to
manufacture for overseas sales.

Respirátor Rt
H-1097, Budapest, Illatos út. 9, Budapest, Hungary
Tel: (+36 1) 280 57 93
Fax: (+36 1) 280 64 28
e-mail: info@respirator.hu
Web: http://www.respirator.hu

ROMTEHNICA
5C Timisoara Blvd, R-77311, Bucharest, Romania
Tel: (+40 1 413) 01 55
Fax: (+40 1 410) 14 67/(+40 1 413) 03 17
e-mail: rth.marketing@mb.roknet.ro

Samgong Industrial Company Limited
Samgong Building, 956 Dogok-Dong, Kangnam-Ku,
Seoul, KPO Box 76, South Korea
Tel: (+82 2) 34 62 30 92
Fax: (+82 2) 34 62 38 87
e-mail: webmaster@samgong.com
Web: http://www.samgong.com

Scott Health & Safety Oy
PO Box 501, FIN-65101 Vaasa, Finland
Tel: (+358) 63 24 45 41
Fax: (+358) 63 24 45 91
e-mail: anja.jarvinen@tycoint.com
Web: http://www.scottsafety.com/contacts.htm

Service Industries Limited
Servis House, 2-Main Gulberg, Lahore, Pakistan
Tel: (+92 42) 571 19 90; 2; 3
Fax: (+92 42) 571 18 27

Seyntex nv/sa
Seyntexlaan 1, Industriepark Zuid, B-8700 Tielt,
Belgium
Tel: (+32 51) 42 37 33
Fax: (+32 51) 42 37 99
e-mail: sales@seyntex.com
Web: http://www.seyntex.com

SHALON Chemical Industries Limited
25 Nachmani Street, IL-65794 Tel Aviv, Israel
Tel: (+972 3) 629 12 25; 6/7
Fax: (+972 3) 629 16 15
e-mail: shalon@shalon.co.il
Web: http://www.doryanet.co.il/shalon/1.htm

Shanxi Xinhua Chemical Factory
PC 030008, Taiyuan, Shanxi, People's Republic of
China
Tel: (+86 351) 305 94 95
Fax: (+86 351) 305 93 20

SP Défense
ZI Paris Nord II, 13 rue de la Perdix, F-95943 Roissy
Charles de Gaulle Cedex, France
Tel: (+33 1) 49 90 70 80
Fax: (+33 1) 49 90 70 90
e-mail: sp.defense@bacou.com
Web: http://www.bacou.comSP Défense is part of
Groupe Bacou

Supergum Ltd
PO Box 54, Barkan Industrial Area, PN Ephraim
44820, Israel
Tel: (+972 3) 936 56 92
Fax: (+972 3) 936 77 42
e-mail: blaun@netvision.net.il
Web: http://www.supergum.com

Trayal Corporation
Miloša Obilića St BB, Kruševac, Serbia
Tel: (+381 37) 264 88
Fax: (+381 37) 235 17
e-mail: trayal@ptt.yu
Web: http://www.trayal.co.yu

Yugoimport-SDPR pc
PO Box 89, Bulevar umetnosti 2, YU-11070 Belgrade,
Yugoslavia
Tel: (+381 11) 311 27 43
Fax: (+381 11) 324 87 91
e-mail: jugoimport@sezampro.yu
Web: http://www.yugoimport-co.yu

MASKS - TEST EQUIPMENT

Air Techniques International
Division of Hamilton Associates Inc, 1144003
Cronridge Drive, Owings Mills, Maryland 21117-
2247, USA
Tel: (+1 410) 363 96 96
Fax: (+1 410) 363 96 95

Avon Technical Products
Protection Group, Hampton Park West, Melksham,
Wiltshire SN12 6NB, UK
Tel: (+44 1225) 89 64 57
Fax: (+44 1225) 89 63 01
e-mail: protection@avonrubber.co.uk
Web: http://www.avon-rubber.com

SFP Services Limited
Sea Vixen Industrial Estate, Wilverley Road,
Christchurch, Dorset BH23 3RU, UK
Tel: (+44 1202) 49 63 13
Fax: (+44 1202) 49 63 63
e-mail: jim@sfpservices.com
Web: http://www.sfpservices.com

SP Défense
ZI Paris Nord II, 13 rue de la Perdix, F-95943 Roissy
Charles de Gaulle Cedex, France
Tel: (+33 1) 49 90 70 80
Fax: (+33 1) 49 90 70 90
e-mail: sp.defense@bacou.com
Web: http://www.bacou.comSP Défense is part of
Groupe Bacou

Technology Marketing
PO Box 987, Millersville, Maryland 21108, USA
Tel: (+1 410) 987 91 11
Fax: (+1 410) 987 63 92
e-mail: technologymarketing@juno.com

TSI Incorporated
PO Box 64394, St Paul, Minnesota 55164, USA
Tel: (+1 651) 483 09 00
Fax: (+1 651) 483 27 48
e-mail: tsiinfo@tsi.com
Web: http://www.tsi.com

MASKS - COMMUNICATIONS

AudioPack Technologies Inc
10011 Walford Avenue, Cleveland, Ohio 44102, USA
Tel: (+1 216) 651 00 66
Fax: (+1 216) 961 12 57
e-mail: audiopack@audiopack.com
Web: http://www.audiopack.com

MSA Defense Products
Safety Products Division, Mine Safety Appliances
Company, PO Box 428, Pittsburgh, Pennsylvania
15230-0428, USA
Tel: (+1 412) 733 92 74
Fax: (+1 412) 733 85 73
e-mail: evan.erickson@msanet.com
Web: http://www.msanet.com

Tradeways Limited
184 Duke of Gloucester Street, Annapolis, Maryland
21401, USA
Tel: (+1 410) 295 08 13
Fax: (+1 410) 295 08 21
e-mail: jhtradeway@msn.com
Web: http://www.tradewaysltd.com

MASKS - AIRCREW

3M Canada Company
1360 California Avenue, Brockville, Ontario, Canada
K6V 5V8
Tel: (+1 613) 345 01 11
Fax: (+1 613) 345 26 39

Cam Lock (UK) Limited
Unit 10, Springlakes Industrial Estate, Deadbrook
 Lane, Aldershot, Hants GU12 4UH, UK
Tel: (+44 1252) 33 00 17
Fax: (+44 1252) 33 02 18
e-mail: info@camlockuk.com
Web: http://www.camlockuk.com

Carleton Life Support Technologies, Limited
1-1200 Aerowood Drive, Mississauga, Ontario,
 Canada L4W 2S7
Tel: (+1 905) 629 32 45
Fax: (+1 905) 629 83 06
e-mail: info@carltech.com
Web: http://www.carltech.com

Edgewood Chemical Biological Center (ECBC)
Aberdeen Proving Ground, Maryland, USA
Tel: (+1 410) 436 43 37
Fax: (+1 410) 436 65 29
e-mail: cet@sbccom.apgea.army.mil
Web: http://www.sbccom.apgea.army.mil/RDA
 /ecbc

Giat Industries
13 route de la Minière, F-78034 Versailles Cedex,
 France
Tel: (+33 1) 30 97 37 37
Fax: (+33 1) 30 97 39 00
e-mail: sales@giat-industries.fr
Web: http://www.giat-industries.fr

ILC Dover Inc
One Moonwalker Road, Frederica, Delaware 19946-
 2080, USA
Tel: (+1 302) 335 39 11
Fax: (+1 302) 335 07 62
e-mail: lawrer@ilcdover.com
Web: http://www.ilcdover.com

NBC Aerotech A/S
Vansjøveien 4, N-1640 Råde, Norway
Tel: (+47 692) 856 44
Fax: (+47 692) 856 45
e-mail: ik@nbc/aerotech.no

SP Défense
ZI Paris Nord II, 13 rue de la Perdix, F-95943 Roissy
 Charles de Gaulle Cedex, France
Tel: (+33 1) 49 90 70 80
Fax: (+33 1) 49 90 70 90
e-mail: sp.defense@bacou.com
Web: http://www.bacou.comSP Défense is part of
 Groupe Bacou

Supergum Ltd
PO Box 54, Barkan Industrial Area, PN Ephraim
 44820, Israel
Tel: (+972 3) 936 56 92
Fax: (+972 3) 936 77 42
e-mail: blaun@netvision.net.il
Web: http://www.supergum.com

MASKS - FILTERS

Aero Sekur SpA
Via Delle Valli, snc, I-04011 Aprilia (Latina), Italy
Tel: (+39 06) 92 01 61
Fax: (+39 06) 92 72 71 65
e-mail: commerciale@aerosekur.it
Web: http://www.aerosekur.it

Arbin
Huiskensstraat 50, NL-5916 PN Venlo, Netherlands
Tel: (+3177) 320 37 00
Fax: (+3177) 320 37 05
e-mail: info@arbin-pp.com
Web: http://www.arbin-pp.com

BIANA SA
Production of NBC Masks and Filters, 10 Apolloniou
 Street, PO Box 5, GR-19400 Koropi, Greece
Tel: (+30 1) 662 39 10
Fax: (+30 1) 662 47 24
e-mail: anargyrou@biana.gr
Web: http://www.biana.gr

J Blaschke Wehrtechnik GmbH
Wienerbergstrasse 42-44, A-1120 Vienna, Austria
Tel: (+43 1) 810 09 09
Fax: (+43 1) 810 09 09 33
e-mail: office@blaschke.com

Paul Boyé
53 quai de Bosc, F-34200 Sète Cedex, France
Tel: (+33 4) 67 46 87 87
Telex: 480087 POLBOY F
Fax: (+33 4) 67 74 22 75
e-mail: paulboye@mnet.fr

Giat Industries
13 route de la Minière, F-78034 Versailles Cedex,
 France
Tel: (+33 1) 30 97 37 37
Fax: (+33 1) 30 97 39 00
e-mail: sales@giat-industries.fr
Web: http://www.giat-industries.fr

Hazmat Protective Systems (Pty) Limited
PO Box 2177, Silvertown, Pretoria 0127, South
 Africa
Tel: (+27 12) 665 07 88
Fax: (+27 12) 665 07 89
e-mail: sales@hazmat.co.za
Web: http://www.hazmat.co.za

MICRONEL AG
Zürcherstrasse 51, CH-8317 Tagelswangen/Zürich,
 Switzerland
Tel: (+41 52) 355 16 16
Fax: (+41 52) 355 16 20
e-mail: info@micronel.ch
Web: http://www.micronel.ch

Protector Technologies Group
Pimbo Road, West Pimbo, Skelmersdale, Lancashire
 WN8 9RA, UK
Tel: (+44 1695) 71 17 11
Fax: (+44 1695) 71 17 64
e-mail: info@protector-tech.com
Web: http://www.protector-tech.com

Research Institute for Chemical Defence
West Building, PO Box 925, PC 100083, Beijing,
 People's Republic of China
Tel: (+86 10) 62 03 37 05; 69 76 02 57
Fax: (+86 10) 62 03 37 05
e-mail: ricdcc@public3.bta.net.cn

Respirátor Rt
H-1097, Budapest, Illatos út. 9, Budapest, Hungary
Tel: (+36 1) 280 57 93
Fax: (+36 1) 280 64 28
e-mail: info@respirator.hu
Web: http://www.respirator.hu

Scott Health & Safety Oy
PO Box 501, FIN-65101 Vaasa, Finland
Tel: (+358) 63 24 45 41
Fax: (+358) 63 24 45 91
e-mail: anja.jarvinen@tycoint.com
Web: http://www.scottsafety.com/contacts.htm

SHALON Chemical Industries Limited
25 Nachmani Street, IL-65794 Tel Aviv, Israel
Tel: (+972 3) 629 12 25; 6/7
Fax: (+972 3) 629 16 15
e-mail: shalon@shalon.co.il
Web: http://www.doryanet.co.il/shalon/.htm

SP Défense
ZI Paris Nord II, 13 rue de la Perdix, F-95943 Roissy
 Charles de Gaulle Cedex, France
Tel: (+33 1) 49 90 70 80
Fax: (+33 1) 49 90 70 90
e-mail: sp.defense@bacou.com
Web: http://www.bacou.comSP Défense is part of
 Groupe Bacou

ST Safety Technologies Limited
PO Box 188, IL-80600 Mitzpe Ramon, Israel
Tel: (+972 57) 58 81 57
Fax: (+972 57) 58 81 61

Sundström Safety AB
Box 10056, SE-181 10 Lidingö, Sweden
Tel: (+46 8) 767 90 85
Fax: (+46 8) 767 98 12
e-mail: ewa.sundstrom@srsafety.se
Web: http://www.srsafety.se

TNO-PML
Prinz Mauritz Laboratory, PO Box 6050, NL-2600 JA,
 Delft, Netherlands
Tel: (+31 15) 269 69 69
Fax: (+31 15) 261 24 03
e-mail: infodesk@tno.nl
Web: http://www.tno.nl

Trayal Corporation
Miloša Obilića St BB, Kruševac, Serbia
Tel: (+381 37) 264 88
Fax: (+381 37) 235 17
e-mail: trayal@ptt.yu
Web: http://www.trayal.co.yu

Yugoimport-SDPR pc
PO Box 89, Bulevar umetnosti 2, YU-11070
 Belgrade, Yugoslavia
Tel: (+381 11) 311 27 43
Fax: (+381 11) 324 87 91
e-mail: jugoimport@sezampro.yu
Web: http://www.yugoimpor-co.yu

NBC CLOTHING

Aero Sekur SpA
Via Delle Valli, snc, I-04011 Aprilia (Latina), Italy
Tel: (+39 06) 92 01 61
Fax: (+39 06) 92 72 71 65
e-mail: commerciale@aerosekur.it
Web: http://www.aerosekur.it

Alfred Kärcher GmbH & Co
Department VPS, Alfred Kärcher Strasse 28-40,
 D-71364 Winnenden, Germany
Tel: (+49 7195) 14 22 62
Fax: (+49 7195) 14 27 80
e-mail: vps@de.kaercher.com
Web: http://www.karcher-vps.com

J Blaschke Wehrtechnik GmbH
Wienerbergstrasse 42-44, A-1120 Vienna, Austria
Tel: (+43 1) 810 09 09
Fax: (+43 1) 810 09 09 33
e-mail: office@blaschke.com

B.O.I.S. Engineering Praha sro
Holeckova 103/31, CZ-150 95 Praha 5, Czech
 Republic
Tel: (+420 2) 57 32 44 02/04
Fax: (+420 2) 57 32 44 05
e-mail: info@bois-praha.cz
Web: http://www.bois-praha.cz

Paul Boyé
53 quai de Bosc, F-34200 Sète Cedex, France
Tel: (+33 4) 67 46 87 87
Telex: 480087 POLBOY F
Fax: (+33 4) 67 74 22 75
e-mail: paulboye@mnet.fr

Blücher GmbH
Protective Clothing Division, Parkstrasse 10, D-40699
 Erkrath, Germany
Tel: (+49 211) 924 40
Fax: (+49 211) 924 42 11
e-mail: vertrieb@bluecher.com
Web: http://www.bluecher.com

Charcoal Cloth (International) Limited
Division of Chemviron Carbon Limited, 1 South Link,
 Oldham, Lancashire OL4 1DE
Tel: (+44 161) 628 50 00
Fax: (+44 161) 628 51 11
e-mail: sross@calgcarb.com
Web: http://www.calgcarb.com
This company is authorised by the UK MoD to
 manufacture for overseas sales.

Chemfab MENH
701 Daniel Webster Highway, Merrimack,
New Hampshire 03054, USA
Tel: (+1 603) 424 90 00
Fax: (+1 603) 424 90 12
e-mail: marksinofsky@chemfab.com
Web: http://www.chemfab.com
Chemfab MENH is part of Saint-Gobain Performance
    Plastics

Chemoplast Industries Limited
PO Box 2110, IL-18100 Afula, Israel
Tel: (+972 4) 652 30 44/45
Fax: (+972 4) 659 74 22
e-mail: onil@netvision.net.il
Web: http://www.chemoplast.co.il

Civil Defence Supply
Hunt House, The Green, Welbourn, Lincolnshire
    LN5 0JF, UK
Tel: (+44 1400) 27 38 50/51/52/53
Fax: (+44 1522) 81 13 53
e-mail: info@civil-defence.org
Web: http://www.civil-defence.org

Compton Webb
(Member of the Vermilion plc Group / Vermilion
    Corporatewear) Gosforth Road, Derby DE24 8HU,
    UK
Tel: (+44 1332) 34 26 16
Fax: (+44 1332) 38 14 92
e-mail: enquiries@comptonwebb.co.uk
Web: http://www.comptonwebb.co.uk/frm1.htm

Consumer Fuels Inc
7250 Governors Drive West, Huntsville, Alabama
    35806, USA
Tel: (+1 205) 837 56 60

CQC Ltd
Riverside Road, Barnstaple, Devon EX31 1NB, UK
Tel: (+44 1271) 34 56 78
Fax: (+44 1271) 34 50 90
e-mail: pjg@cqc.co.uk
Web: http://www.cqc.co.uk

DuPont Engineering Products SARL
L-2984, Luxembourg
Web: http://www.dupont.com

Eurodéfhi
Européenne de Développement et d'Études des
    Facteurs Humains et des Interfaces (Eurodéfhi), 12
    Avenue de l'Europe, BP 20, 78142 Velizy, Cedex,
    France
Tel: (+33 1) 39 35 47 12
Fax: (+33 4) 39 35 47 13
e-mail: robert.schegerin@eurodefhi.com
Web: http://www.eurodefhi.com

J & S Franklin Limited
Franklin House, 151 Strand, London WC2R 1HL, UK
Tel: +44 (0207) 836 57 46
Fax: +44 (0207) 836 27 84
e-mail: defence@franklin.co.uk
Web: http://www.franklin.co.uk

Freudenberg
Adsorptive Products Group, PO Box 3, Greetland,
    Halifax, West Yorkshire HX4 8NJ, UK
Tel: (+44 1422) 31 32 02
Fax: (+44 1422) 31 32 32

Gentex Corporation
PO Box 315, Carbondale, Philadelphia 18407, USA
Tel: (+1 570) 282 85 11
Fax: (+1 570) 282 85 55
e-mail: infom@gentexcorp.com
Web: http://www.gentexcorp.com

GEOMET Technologies Inc
20251 Century Boulevard, Germantown, Maryland
    20874, USA
Tel: (+1 301) 428 98 98
Fax: (+1 301) 428 94 82
E-mail: jharris@geomet.com
Web: http://www.nbcprotect.com

Goetzloff GmbH
Schirmerstrasse 28, A-4060 Leonding-Linz, Austria
Tel: (+43 732) 38 40 27
Fax: (+43 732) 384 02 75
e-mail: office@goetzloff.at
Web: http://www.goetzloff.at

Goetzloff (USA)
2121 Jamieson Avenue #1505, Alexandria, Virginia
    22314, USA
Tel: (+1 703) 504 02 60
Fax: (+1 703) 504 06 12
e-mail: office@goetzloff.at
Web: http://www.goetzloff.com

Guilin Rubber Products Factory
Guilin, Guangxi, People's Republic of China

Helsatech GmbH
Bayreuther Strasse 11, D-95482 Gefrees, Germany
Tel: (+49 9254) 800
Fax: (+49 9254) 804 02
e-mail: helsatech@de.helsa.com
Web: http://www.helsa.com/com/static/helsatech

Herbelein Textiles
at SARATOGA (Wattwil) AG, Ebnaterstrasse, CH-9630
    Wattwil, Switzerland
Tel: (+41 746) 11 11
Fax: (+41 746) 15 11; 12
(SARATOGA (Wattwil) AG is a joint-venture company
    comprising Herbelein Textiles AG and Blücher
    GmbH)

Hi-Tech Polymer Proofings Limited
Unit 9, Millingford Industrial Estate, Bridge Street,
    Golborne, Lancashire WA3 3QE, UK
Tel: (+44 1942) 27 17 77
Fax: (+44 1942) 27 17 78e-mail: info@hi-
    techpolymer.co.uk
Web: http://www.hi-techpolymer.co.uk

Huajing Machinery Plant
PO Box 508, Yichang, Hubei, People's Republic of
    China

Indian Springs Specialty Products Inc
PO Box 469, 2095 W Genesee Road, Baldwinsville,
    New York 13027-0469, USA
Tel: (+1 315) 635 62 43
Fax: (+1 315) 635 74 73

Industrias y Confecciones SA (INDUYCO)
Tomás Bréton 62, E-28045, Madrid, Spain
Tel: (+34 91) 774 82 00
Fax: (+34 91) 774 83 00
e-mail: info@inuyco.es
Web: http://www.induyco.es/web_induyco.htm

Irvin Aerospace Canada Limited
PO Box 280, 479 Central Avenue, Fort Erie, Ontario,
    Canada L2A 5M9
Tel: (+1 905) 871 65 10
Fax: (+1 905) 871 65 34
e-mail: marketing

ISCO
Beim Bahnhof, CH-8172 Niederglatt, Switzerland
Tel: (+41 1) 851 50 50
Fax: (+41 1) 851 50 51
e-mail: info@isco.ch
Web: http://www.isco.ch/nbc_laminates.htm

Kappler Protective Apparel & Fabrics
PO Box 218 Guntersville, Alabama 35976, USA
Tel: (+1 800) 633 24 10
Fax: (+1 205) 582 27 06
e-mail: info@kappler.com

Kintex
PO Box 209, 66 James Boucher Street, BG-1407
    Sofia, Bulgaria
Tel: (+359) 266 23 11
Fax: (+359) 29 63 11 23

Lantor (UK) Limited
St Helen's Road, Bolton, Lancs BL3 3PR, UK
Tel: (+44 1204) 85 50 00
Fax: (+44 1204) 617 22
e-mail: sales@lantor.co.uk
Web: http://www.lantor.co.uk

LANX Fabric Systems
DuPont Fabrics and Separations, PO Box 6101, 200
    GBC Building, Newark, Delaware 10714-3231, USA
Tel: (+1 302) 451 32 31
Fax: (+1 302) 451 02 08
e-mail: robert.weaver@xymid.com

MoD Institute of Military Technology
HM HTI, Szilágyi Erzsébet fasor 20, Pf. 26, H-1525,
    Budapest, Hungary
Tel: (+36 1) 394 17 33
Fax: (+36 1) 394 30 14

Natick Soldier Centre
Kansas Street, Natick, Massachusetts 01760-5017,
    USA
Tel: (+1 508) 233 55 71
e-mail: william.haskell@natick.army.mil
Web: http://www.natick.army.mil
(Also known as the Soldier Systems Centre and
    formerly, as the US Army Natick Research,
    Development and Engineering Center).

National Organisation for Military Production
3rd Floor, 23 Talaat Harb Street, PO Box 9, Heliopolis,
    Cairo, Egypt
Tel: (+20 2) 281 08 38
Fax: (+20 2) 281 08 28

New Pac Safety AB
PO Box 174, SE-566 23 Habo, Sweden
Tel: (+46 36) 411 39
Fax: (+46 36) 410 31
e-mail: info@newpac.se

North Safety Products Limited
The Court Yard, Green Lane, Heywood, Lancashire
    OL10 2EX, UK
Tel: (+44 170) 669 38 00
Tel: (+44 170) 669 38 01
e-mail: sales@northsafety.co.uk
Web: http://www.northsafety.co.uk

OMNIPOL as
Nekázanka 11, 112 21 Praha 1, Czech Republic
Tel: (+42 02) 401 11 76
Fax: (+42 02) 401 21 99
e-mail: omni22@omnipol.cz
Web: http://www.omnipol.cz

PAVAG AG
Bahnhofstrasse 33, CH-6244 Nebikon, Switzerland
Tel: (+41 62) 748 93 00
Fax: (+41 62) 748 93 84
Web: http://www.igpe.ch/firmenf.html
Note: The web address is that of AIPE, of which
    PAVAG AG is a member. AIPE (L'Association
    d'intérêts de l'Industrie suisse du PolyéthylènE) - is
    the association for the Swiss polyethylene industry.

Protechnik Laboratories
PO Box 8854, Pretoria 001, Republic of South Africa
Tel: (+27 12) 665 02 31
Fax: (+27 12) 665 02 40
e-mail: philipc@protechnik.co.za
Web: http://www.protechnik.co.za

The Protective Clothing Company Limited
28 Harcourt Street, Dublin 2, Ireland
Tel: (+353 1) 289 72 35
Fax: (+353 1) 289 75 14

Remploy Limited
Remploy House, 415 Edgware Road, Cricklewood,
    London NW2 6LR, UK
Tel: (+44 208) 235 05 32
Fax: (+44 208) 235 05 01
This company is authorised by the UK MoD to
    manufacture for overseas sales.

Research Institute for Chemical Defence
West Building, PO Box 925, PC 100083, Beijing,
 People's Republic of China
Tel: (+86 10) 62 03 37 05; 69 76 02 57
Fax: (+86 10) 62 03 37 05
e-mail: ricdcc@public3.bta.net.cn

Research Institute of Rubber and Plastics Technology
 State Corporation
CZ-764 22 Zlin-Louky, Czech Republic

Respirex International Limited
Unit F, Kingsfield Business Centre, Philanthropic
 Road, Redhill, Surrey RH1 4DP, UK
Tel: (+44 1737) 77 86 00
Fax: (+44 1737) 77 94 41
e-mail: info@respirex.co.uk
Web: http://www.respirex.co.uk

Samgong Industrial Company Limited
Samgong Building, 956 Dogok-Dong, Kangnam-Ku,
 Seoul, KPO Box 76, South Korea
Tel: (+82 2) 34 62 38 88
Fax: (+82 2) 34 62 38 87
e-mail: lee4601@chollian.net
Web: http://www.samgong.com

SARATOGA (Wattwil) AG
Ebnaterstrasse, CH-9630 Wattwil, Switzerland
Tel: (+41 746) 11 11
Fax: (+41 746) 15 11; 12
(Joint-venture company comprising Herbelein
 Textiles AG and Blücher GmbH)

Seyntex nv/sa
Seyntexlaan 1, Industriepark Zuid, B-8700 Tielt,
 Belgium
Tel: (+32 51) 42 37 33
Fax: (+32 51) 42 37 99

Slavyanskaya Clothes Factory
1 Pobedy Street, Slavyansk-Na Kubani 353840,
 Krasnodar, Russian Federation
Tel: (+7 522) 297 13

Statens Konfektion
100 Lyngbyvej, DK-2100 Copenhagen, Denmark
Tel: (+45 31) 29 21 00
Fax: (+45 31) 20 51 33
Telex: 27 204 STAKON

K Stormark Konfeksjonsfabrikk A/S
Jevnakerveien 20, N-2670, Dokka, Norway
Tel: (+47 61) 11 06 11
Fax: (+47 61) 11 18 40

ST Safety Technologies Limited
PO Box 188, IL-80600 Mitzpe Ramon, Israel
Tel: (+972 57) 58 81 57
Fax: (+972 57) 58 81 61

Supergum Ltd
PO Box 54, Barkan Industrial Area, PN Ephraim
 IL-44820, Israel
Tel: (+972 3) 936 56 92
Fax: (+972 3) 936 77 42
e-mail: blaun@netvision.net.il
Web: http://www.supergum.com

Swedish Emergency Disaster Equipment
SWEDE, Henriksdalsvägen 2, SE-386 92, Färjestaden,
 Sweden
Tel: (+46 485) 484 80
Fax: (+46 485) 484 90
e-mail: info@frenatus.com
Web: http://www.frenatus.com

Tex-Shield Inc
Suite 800, 2300 M Street NW, Washington DC 20039,
 USA
Tel: (+1 202) 973 28 58
Fax: (+1 202) 973 28 50
e-mail: info@texshield.com
Web: http://www.texshield.com

Textil Skyddsteknik AB
Prästgatan 12, SE-511 54, Kinna, Sweden
Tel: (+46 320) 166 00
Fax: (+46 320) 166 10
e-mail: info@tst-sweden.se

Tradeways Limited
184 Duke of Gloucester Street, Annapolis, Maryland
 21401, USA
Tel: (+1 410) 295 08 13
Fax: (+1 410) 295 08 21
e-mail: jhtradeway@msn.com
Web: http://www.tradewaysltd.com

Trayal Corporation
Miloša Obilica St BB, Kruševac, Serbia Tel: (+381 37)
 264 88
Fax: (+381 37) 235 17
e-mail: trayal@ptt.yu
Web: http://www.trayal.co.yu

Trelleborg Industri AB
Protective Products Division, PO Box 1520, SE-271 00
 Ystad, Sweden
Tel: (+46 411) 679 40
Fax: (+46 411) 152 85
e-mail: protective@trelleborg.com
Web: http://www.trelleborg.com

UK address:
Trelleborg Beadle, Unit 30, Bergen Way, Sutton Fields
 Industrial Estate, Hull HU7 0YQ, UK
Tel: (+44 1482) 83 91 19
Fax: (+44 1482) 87 94 18

Typhoon International Limited
Limerick Road, Dormanstown Industrial Estate,
 Redcar, Cleveland TS10 5JU, UK
Tel: (+44 16 42) 48 61 04
Fax: (+44 16 42) 48 72 04
e-mail: sales@typhoon-int.co.uk
Web: http://www.typhoon-int.co.uk

Yugoimport-SDPR pc
PO Box 89, Bulevar umetnosti 2, YU-11070 Belgrade,
 Yugoslavia
Tel: (+381 11) 311 27 43
Fax: (+381 11) 324 87 91
e-mail: jugoimport@sezampro.yu
Web: http://www.yugoimport-co.yu

NBC GLOVES AND FOOTWEAR

Acton International Inc
881 Landry Street, Acton Vale, Québec, Canada
 J0H 1A0
Tel: (+1 450) 546 27 76
Fax: (+1 450) 546 02 13
e-mail: earl.laurie@airboss-acton.com
Web: http://www.airboss-acton.com
Acton International Inc is a subsidiary of AirBoss-
 Defense.

Armavir Rubber Articles Factory
2/4 Novorossiyskaya Street, Armavir 352931,
 Krasnodar, Russian Federation

Avatech BV
Ettensweg 6, 4706 PB Roosendaal, Roosendaal,
 Netherlands
Tel: (+31 165) 59 51 50
Fax: (+31 165) 53 78 05

Paul Boyé
53 quai de Bosc, F-34200 Sète Cedex, France
Tel: (+33 4) 67 46 87 87
Telex: 480087 POLBOY F
Fax: (+33 4) 67 74 22 75
e-mail: paulboye@mnet.fr

Butyl Products Limited
11 Radford Crescent, Billericay, Essex CM12 0DW,
 UK
Tel: (+44 1277) 65 32 81
Fax: (+44 1277) 65 79 21
e-mail: rod@butylproducts.co.uk
Web: http://www.butylproducts.co.uk

Civil Defence Supply
Hunt House, The Green, Welbourn, Lincolnshire
 LN5 0JF, UK
Tel: (+44 1400) 27 38 50/51/52/53
Fax: (+44 1522) 81 13 53
e-mail: info@civil-defence.org
Web: http://www.civil-defence.org

Dätwyler AG
Rubber Works, CH-6460 Altdorf, Switzerland
Tel: (+41 41) 418 75 11 22
Fax: (+41 41) 418 75 15 46

DRDC/RDDC Suffield
Defence R&D Canada/R&D pour la Défense Canada,
 PO Box 4000, Station Main, Medicine Hat, Alberta,
 Canada T1A 8K6
Tel: (-1 403) 544 46 02
Fax: (+1 403) 544 33 88
e-mail: garfield.purdon@drdc-rddc.gc.ca
Web: http://www.suffield.drdc-rddc.gc.ca

Guardian Manufacturing Company
302 Conwell Avenue, Willard, Ohio 44890-9525, USA
Tel: (+1 419) 933 27 11
Fax: (-1 419) 935 89 61
e-mail: susanl@willard-oh.com
Web: http://www.guardian-mfg.com

Impermaplast AG
CH-2805 Soyhières, Switzerland
Tel: (+41 66) 22 02 12
Fax: (+41 66) 22 02 50

LONSTROFF AG
Industriestrasse 31, CH-5001 Arau, Switzerland
Tel: (+41 62) 823 32 32
Fax: (+41 62) 823 60 10

National Organisation for Military Production
3rd Floor, 23 Talaat Harb Street, PO Box 9, Heliopolis,
 Cairo, Egypt
Tel: (+20 2) 281 08 38
Fax: (+20 2) 281 08 28

North Safety Products Limited
The Court Yard, Green Lane, Heywood, Lancashire
 OL10 2EX, UK
Tel: (+44 170) 669 38 00
Tel: (+44 170) 669 38 01
e-mail: sales@northsafety.co.uk
Web: http://www.northsafety.co.uk

Piercan
11-11 bis, rue Charbonnel, F-75013 Paris, France
Tel: (+33 1) 45 88 66 27
Fax: (+33 1) 45 80 98 30
e-mail: Piercan@Piercan.fr

Seyntex nv/sa
Seyntexlaan 1, Industriepark Zuid, B-8700 Tielt,
 Belgium
Tel: (+32 51) 42 37 33
Fax: (+32 51) 42 37 99

Silvertown UK Ltd
Horninglow Road, Burton on Trent, Staffordshire
 DE13 0SN, UK
Tel: (+44 1283) 51 05 10
Fax: (+44 1283) 51 00 52
e-mail: paul.stray@silvertown.co.uk
Web: http://www.silvertown.co.uk

Supergum Ltd
PO Box 54, Barkan Industrial Area, PN Ephraim
 44820, Israel
Tel: (+972 3) 936 56 92
Fax: (+972 3) 936 77 42
e-mail: blaun@netvision.net.il
Web: http://www.supergum.com

Yugoimport-SDPR pc
PO Box 89, Bulevar umetnosti 2, YU-11070 Belgrade,
 Yugoslavia
Tel: (+381 11) 311 27 43
Fax: (+381 11) 324 87 91
e-mail: jugoimport@sezampro.yu
Web: http://www.yugoimport-co.yu

MEDICAL - CASUALTY HOODS/ BAGS

J Blaschke Wehrtechnik GmbH
Wienerbergstrasse 42-44, A-1120 Vienna, Austria
Tel: (+43 1) 810 09 09
Fax: (+43 1) 810 09 09 33
e-mail: office@blaschke.com

Carleton Life Support Technologies Limited
1-1200 Aerowood Drive, Mississauga, Ontario,
 Canada L4W 2S7
Tel: (+1 905) 629 32 45
Fax: (+1 905) 629 83 06
e-mail: info@carltech.com
Web: http://www.carltech.com

Compton Webb
(Member of the Vermilion plc Group / Vermilion
 Corporatewear) Gosforth Road, Derby DE24 8HU,
 UK
Tel: (+44 1332) 34 26 16
Fax: (+44 1332) 38 14 92
e-mail: enquiries@comptonwebb.co.uk
Web: http://www.comptonwebb.co.uk/frm1.htm

Freudenberg
Adsorptive Products Group, PO Box 3, Greetland,
 Halifax, West Yorkshire HX4 8NJ, UK
Tel: (+44 1422) 31 32 02
Fax: (+44 1422) 31 32 32

Gentex Corporation
PO Box 315, Carbondale, Philadelphia 18407, USA
Tel: (+1 570) 282 85 11
Fax: (+1 570) 282 85 55
e-mail: info@gentexcorp.com
Web: http://www.gentexcorp.com

Goetzloff GmbH
Schirmerstrasse 28, A-4060 Leonding-Linz, Austria
Tel: (+43 732) 38 40 27
Fax: (+43 732) 38 40 27 5
e-mail: office@goetzloff.at
Web: http://www.goetzloff.at

Goetzloff (USA)
2121 Jamieson Avenue #1505, Alexandria, Virginia
 22314, USA
Tel: (+1 703) 504 02 60
Fax: (+1 703) 504 06 12
e-mail: office@goetzloff.at
Web: http://www.goetzloff.com

Helsatech GmbH
Bayreuther Strasse 11, D-95482 Gefrees, Germany
Tel: (+49 9254) 800
Fax: (+49 9254) 804 02
e-mail: helsatech@de.helsa.com
Web: http://www.helsa.com/com/static/helsatech

North Safety Products Limited
The Court Yard, Green Lane, Heywood, Lancashire
 OL10 2EX, UK
Tel: (+44 170) 669 38 00
Tel: (+44 170) 669 38 01
e-mail: sales@northsafety.co.uk
Web: http://www.northsafety.co.uk

Pneupac Limited
Bramingham Business Park, Enterprise Way, Luton,
 Bedfordshire LU3 4BU, UK
Tel: (+44 1582) 43 00 00
Fax: (+44 1582) 43 00 01
e-mail: sales@pneupac.co.uk
Web: http://www.pneupac.co.uk
Note: Pneupac Limited (formerly SIMS PneuPAC) was
 acquired by Smiths Medical in 1997 and is part of
 Smiths Group plc

Remploy Limited
Remploy House, 415 Edgware Road, Cricklewood,
 London NW2 6LR, UK
Tel: (+44 208) 235 05 32
Fax: (+44 208) 235 05 01
This company is authorised by the UK MoD to
 manufacture for overseas sales.

Techniques Michel Brochier SA
ZI rue des Chartinières, PO Box 79, F-01120 Dagneux
 Cedex, France
Tel: (+33 4) 78 06 32 22
Fax: (+33 4) 72 25 98 25
e-mail: michelbrochier@walter.fr
Web: http://www.tm-brochier.com

MEDICAL - SURVIVAL/RESCUE

Charcoal Cloth (International) Limited
Division of Chemviron Carbon Limited, 1 South Link,
 Oldham, Lancashire OL4 1DE
Tel: (+44 161) 628 50 00
Fax: (+44 161) 628 51 11
e-mail: sross@calgcarb.com
Web: http://www.calgcarb.com
This company is authorised by the UK MoD to
 manufacture for overseas sales.

ELKOM TSN
Kriska 7a, Beograd, Yugoslavia, Federal Republic
Tel: (+381 11) 42 44 98
Telex: 12124

Giat Industries
13 route de la Minière, F-78034 Versailles Cedex,
 France
Tel: (+33 1) 30 97 37 37
Fax: (+33 1) 30 97 39 00
e-mail: sales@giat-industries.fr
Web: http://www.giat-industries.fr

Odenwald-Werke Rittersbach GmbH (OWR)
D-74834 Elztal-Rittersbach, Germany
Tel: (+49 62) 937 31
Fax: (+49 62) 937 32 19
e-mail: owr.gmbh@t-online.de
Web: http://www.owr.de

MEDICAL - THERAPY

AEROCHEM
Herbert Lettko GmbH & Co KG,
Am Becherweg 2, D-55270 Ober-Olm, Germany
Tel: (+49 61) 369 91 80
Fax: (+49 61) 36 99 71 18

Alfred Kärcher GmbH & Co
Department VPS, Alfred Kärcher Strasse 28-40,
D-71364 Winnenden, Germany
Tel: (+49 7195) 14 22 62
Fax: (+49 7195) 14 27 80
e-mail: vps@de.kaercher.com
Web: http://www.karcher-vps.com

Astra Tech AB
PO Box 14, SE-431 21 Mölndal, Sweden
Tel: (+46 31) 776 30 00
Fax: (+46 31) 776 30 10
e-mail: info@astratech.com
Web: http://www.astratech.com

Defence Medical Supplies Limited
Trewillis Farm, Coverack, Helston, Cornwall TR12
 6SF, UK
Tel: (+44 1326) 28 07 76
Fax: (+44 1326) 28 01 75
e-mail: mjmuk@compuserve.com
Web: http://ourworld.compuserve.com/homepages/
 mjmuk

Meridian Medical Technologies, Inc
STI Military Systems, 10240 Old Columbia Road,
 Columbia, Maryland 21046-2371, USA
Tel: (+1 410) 309 68 30
Fax: (+1 410) 309 14 75
e-mail: tmasiuk@meridianmt.com
Web: http://www.meridianmeds.com/home.html

Roche
F Hoffmann-La Roche Ltd, Group Headquarters,
 Grenzacherstrasse 124, CH-4070 Basel,
 Switzerland
Tel: (+41) 616 88 11 11
Fax: (+41) 616 91 93 91
e-mail: (via form on website)
Web: http://www.roche.com/home/company/com_
 contact.htm

Seba Srl
Via Tolosano 60, I-48018, Faenza, Italy
Tel: (+39 0546) 466 80
Fax: (+39 0546) 465 35
e-mail: seba@sorind.it

SHALON Chemical Industries Limited
25 Nachmani Street, IL-65794 Tel Aviv, Israel
Tel: (+972 3) 629 12 25; 6; 7
Fax: (+972 3) 629 16 15
e-mail: shalon@shalon.co.il
Web: http://www.doryanet.co.il/shalon/1.htm

Solvay Duphar BV
Injectables Division, Veerweg 12, NL-8121 AA Olst,
 Netherlands
Fax: (+31 294) 43 09 55

COLPRO - SHELTERS

AEA Technology plc
551 Harwell, Didcot, Abingdon, Oxfordshire OX11
 0QJ, UK
Tel: (+44 1235) 43 41 83
Fax: (+44 1235) 43 45 57
e-mail: andrew.turner@accentus.co.uk
Web: http://www.aeat-prodsys.com/prodsys/
 divisions/OCD.html

Aero Sekur SpA
Via Delle Valli, snc, I-04011 Aprilia (Latina), Italy
Tel: (+39 06) 92 01 61
Fax: (+39 06) 92 72 71 65
e-mail: commerciale@aerosekur.it
Web: http://www.aerosekur.it

Alfred Kärcher GmbH & Co
Department VPS, Alfred Kärcher Strasse 28-40,
D-71364 Winnenden, Germany
Tel: (+49 7195) 14 22 62
Fax: (+49 7195) 14 27 80
e-mail: vps@de.kaercher.com
Web: http://www.karcher-vps.com

Andair AG
Schaubenstrasse 4, CH-8450 Andelfingen,
 Switzerland
Tel: (+41 52) 304 24 24
Fax: (+41 52) 304 24 25
e-mail: andair-ag@bluewin.ch

BCB International Limited
Clydesmuir Road, Cardiff, CF24 2QS, UK
Tel: (+44 292) 04 43 37 00
Fax: (+44 292) 04 43 37 01
e-mail: info@bcbin.com
Web: http://www.bcbin.com or http:/
 /www.ngoaid.com

BERICO AG
Südstrasse 22, CH-8172 Niederglatt, Switzerland
Tel: (+41 1) 851 52 52
Fax: (+41 1) 851 52 53
e-mail: info@berico.ch
Web: http://www.@berico.ch

J Blaschke Wehrtechnik GmbH
Wienerbergstrasse 42-44, A-1120 Vienna, Austria
Tel: (+43 1) 810 09 09
Fax: (+43 1) 810 09 09 33
e-mail: office@blaschke.com

Bonna Sabla
Tour Ariane, 5 Place de la Pyramide, F-92088 Paris,
 La Défense, France
Tel: (+33 1) 46 53 24 00
Fax: (+33 1) 46 53 24 11
e-mail: s.ducos@bonna.com
Web: http://www.bonna.com

Chemfab MENH
701 Daniel Webster Highway, Merrimack,
 New Hampshire 03054, USA
Tel: (+1 603) 424 90 00
Fax: (+1 603) 424 90 12
e-mail: marksinofsky@chemfab.com
Web: http://www.chemfab.com
Chemfab MENH is part of Saint-Gobain Performance
 Plastics

Engineered Air Systems Inc
1270 North Price Road, St Louis, Missouri 63132,
    USA
Tel: (+1 314) 993 58 80
Fax: (+1 314) 567 40 52
e-mail: easi@engineeredsupport.com
Web: http://www.engineeredsupport.com/easi.htm

J & S Franklin Limited
Franklin House, 151 Strand, London WC2R 1HL, UK
Tel: (+44 207) 836 57 46
Fax: (+44 207) 836 27 84
e-mail: defence@franklin.co.uk
Web: http://www.franklin.co.uk

INSYS Limited
Reddings Wood, Ampthill, Bedford MK45 2HD, UK
Tel: (+44 1525) 84 10 00
Fax: (+44 1525) 40 58 61
e-mail: marketing@insys-ltd.co.uk
Web: http://www.insys-ltd.co.uk

ILC Dover Inc
1 Moonwalker Road, Frederica, Delaware 19946-
    2080, USA
Tel: (+1 302) 335 39 11
Fax: (+1 302) 335 07 62
e-mail: lawrer@ilcdover.com
Web: http://www.ilcdover.com

Irvin Aerospace Canada Ltd
PO Box 280, 479 Central Avenue, Fort Eire, Ontario,
    Canada L2A 5M9
Tel: (+1 905) 871 65 10
Fax: (+1 905) 871 65 34
e-mail: marketing@irvin.co.uk
Web: http://www.irvin.co.uk

Monarflex Limited
Lyon Way, St Albans, Hertfordshire AL4 OLB, UK
Tel: (+44 1727) 83 01 16
Fax: (+44 1727) 86 80 45
e-mail: eng@monarflex.co.uk
Web: http://www.monarflex.co.uk

Natick Soldier Centre
Kansas Street, Natick, Massachusetts 01760-5017,
    USA
Tel: (+1 508) 233 55 71
e-mail: william.haskell@natick.army.mil
Web: http://www.natick.army.mil
(Also known as the Soldier Systems Centre and
formerly, as the US Army Natick Research,
Development and Engineering Center).

Nippon Tokuso Co Ltd
2-12-2 Fujimori Building, Hirakawa-Cho, Chiyodaku,
    102-0093 Tokyo, Japan
Tel: (+81 3) 32 88 33 39
Fax: (+81 3) 32 88 33 59
e-mail: info@n-tokuso.co.jp
Web: http://www.n-tokuso.co.jp

Nor Environmental Ltd
138 Shallot Crescent, North Bay, Ontario, Canada
    P1A 3XS
Tel: (+1 705) 497 03 57
Fax: (+1 705) 497 85 78
e-mail: nbcd@nbcdefence.net
Web: http://www.nbcdefence.net

SHALON Chemical Industries Limited
25 Nachmani Street, IL-65794 Tel Aviv, Israel
Tel: (+972 3) 629 12 25; 6; 7
Fax: (+972 3) 629 16 15
e-mail: shalon@shalon.co.il
Web: http://www.doryanet.co.it/shalon/1.htm

Techniques Michel Brochier
ZI rue des Chartinières, PO Box 79, F-01120 Dagneux
    Cedex, France
Tel: (+33 4) 78 06 32 22
Fax: (+33 4) 72 25 98 25
e-mail: michelbrochier@walter.fr
Web: http://www.tm-brochier.com

TEMET Oy
Asentajankatu 3, FIN-00810 Helsinki, Finland
Tel: (+358 9) 75 90 01
Fax: (+358 9) 78 59 67
e-mail: info@temet.com
Web: http://www.temet.com

Tradeways Limited
184 Duke of Gloucester Street,
Annapolis, Maryland 21401, USA
Tel: (+1 410) 295 0813
Fax: (+1 410) 295 0821
e-mail: JHtradeway@msn.com

Trelleborg Industri AB
Protective Products Division, PO Box 1520, SE-271 00
    Ystad, Sweden
Tel: (+46 411) 679 40
Fax: (+46 411) 152 85
e-mail: protective@trelleborg.com
Web: http://www.trelleborg.com

UK address:
Trelleborg Beadle, Unit 30, Bergen Way, Sutton Fields
    Industrial Estate, Hull HU7 0YQ, UK
Tel: (+44 1482) 83 91 19
Fax: (+44 1482) 87 94 18

COLPRO - FILTER SYSTEMS

AAF-International BV
Egelenburg 2, PO Box 7928, NL-1008 AC,
    Amsterdam, Netherlands
Tel: (+31 20) 549 44 11
Fax: (+31 20) 644 43 98

Aircontrol Technologies Limited
Hawthorne Road, Staines, Middlesex TW18 3AY, UK
Tel: (+44 1784) 46 61 66
Fax: (+44 1784) 46 58 94
e-mail: sales@aircontroltechnologies.co.uk
Web: http://www.airtechnologygroup.co.uk

Carleton Life Support Technologies, Limited
1-1200 Aerowood Drive, Mississauga, Ontario,
    Canada L4W 2S7
Tel: (+1 905) 629 32 45
Fax: (+1 905) 629 83 06
e-mail: info@carltech.com
Web: http://www.carltech.com

Daloc Sheltec AB
Upplagsvgen 10, SE-117 43 Stockholm, Sweden
Tel: (+46 8) 645 81 25
Fax: (+46 8) 645 33 48
e-mail: bengt.jakstad@daloc.se
Web: http://www.daloc.se

Engicom Systems NV
Binnensteenweg 172, B-2530 Boechout, Belgium
Tel: (+32 3) 460 04 90
Fax: (+32 3) 460 03 90

Engineered Air Systems Inc
1270 North Price Road, St Louis, Missouri 63132,
    USA
Tel: (+1 314) 993 58 80
Fax: (+1 314) 567 40 52
e-mail: easi@engineeredsupport.com
Web: http://www.engineeredsupport.com/easi.htm

FILTRATOR AB
Bruttovägen 14, SE-175 43 Järfälla, Sweden
Tel: (+46 8) 580 204 10
Fax: (+46 8) 580 204 05
e-mail: filtrator@telia.com

Genano Limited
Sinikellonpolku 3, FIN-01300 Vantaa, Finland
Tel: (+358 9) 774 38 70
Fax: (+358 9) 753 31 30
Web: http://www.genano.fi

General Dynamics Armament and Technical Products
2000 Brunswick Lane, Delard, Florida 32724, USA
Tel: (+1 904) 736 17 00
Fax: (+1 904) 736 22 50
e-mail: gmilbouer@intellitec.com
Web: http://www.gdatp.com   or   http://
    www.intellitec.com/product/chemicaldefense/
Note: General Dynamics Armament and Technical
    Products formed as a new company on 14 June
    2002 and includes Advanced Technical Products
    Inc, Intellitec Division.

Giat Industries
13 route de la Minière, F-78034 Versailles Cedex,
    France
Tel: (+33 1) 30 97 37 37
Fax: (-33 1) 30 97 39 00
e-mail sales@giat-industries.fr
Web: http://www.giat-industries.fr

Guild Associates Inc
5750 Shier-Rings Road, Dublin, Ohio 43016-2013,
    USA
Tel: (+1 614) 798 82 15
Fax: (+1 614) 798 19 72
e-mail: guild-associates.com
Web: http://www.guild-associates.com

Honeywell
ECS Enterprise, 2525 West 190th Street, PO Box
    2960, Torrance, California 90504, USA
Tel: (+1 310) 323 95 00
Tel: (+1 310) 512 22 21
NOTE: Formerly AlliedSignal Aerospace

Hunter Protective Systems
616 Marsat Court, Chula Vista, California 91911-4646,
    USA
Tel: (+1 619) 429 58 00
Fax: (+1 619) 429 59 00
e-mail: wpoage@hunterprotect.com
Web: http://www.hunterprotect.com

Irvin Aerospace Canada Ltd
PO Box 280, 479 Central Avenue, Fort Eire, Ontario,
    Canada L2A 5M9
Tel: (+1 905) 871 65 10
Fax: (+1 905) 871 65 34
e-mail: marketing@irvincanada.com

KINETICS Limited
11 Hamlacha Street, PO Box 10, Or-Yehuda, IL-60251,
    Israel
Tel: (+972 3) 533 50 30
Fax: (+972 3) 533 50 28
e-mail: market@kinetics.co.il
Web: http://www.kinetics.co.il

Kintex
PO Box 209, 66 James Boucher Street, BG-1407
    Sofia, Bulgaria
Tel: (+359) 266 23 11
Fax: (+359) 29 63 11 23

Lockheed Martin
Lockheed Martin Naval Electronics & Surveillance
    Systems - Undersea Systems (Lockheed Martin NE
    & SS-Manassas), 9500 Godwin Drive, Manassas,
    Virginia 20110, USA
Tel: (+1 703) 367 19 64
Fax: (+1 703) 367 54 41
e-mail: richard.read@lmco.com
Web: http://www.lockheedmartin.com/manassas

MDH Defence
34a Walworth Road, Andover, Hampshire SP10 5AA,
    UK
Tel: (+44 1264) 83 58 14
Fax: (+44 1264) 83 58 89
e-mail: sales@bioquelldefence.com
Web: http://www.bioquelldefence.com
Note: MDH Defence is part of Bioquell Group Plc

Microgenix Limited
Water House, Thames Road, Crayford, DA1 4TF, UK
Tel: (+1 322) 40 20 40
Fax: (+1 322) 41 14 00
e-mail: customer@microgenix.net
Web: http://www.microgenix.net

Military Technical Institute of Protection
PO Box 547, CZ-602-00 Brno, Czech Republic
Tel: (+420) 541 18 30 01
Fax: (+420) 541 18 31 52
e-mail: fr.oplustil@telecom.cz

Nor Environmental Ltd
138 Shallot Crescent, North Bay, Ontario, Canada
P1A 3XS
Tel: (+1 705) 497 03 57
Fax: (+1 705) 497 85 78
e-mail: nbcd@nbcdefence.net
Web: http://www.nbcdefence.net

Parmatic Filter Corporation
88 Ford Road, Denville, New Jersey 07834-1634, USA
Tel. (+1 973) 586 92 00
Fax (+1 973) 586 92 91
e-mail: parmatic@worldnet.att.net

Research Institute for Chemical Defence
West Building, PO Box 925, PC 100083, Beijing,
People's Republic of China
Tel: (+86 10) 62 03 37 05; 69 76 02 57
Fax: (+86 10) 62 03 37 05
e-mail: ricdcc@public3.bta.net.cn

Rosoboronexport
21 Gogolevsky Boulevard, 119865 Moscow, Russian
Federation
Tel: (+7 095) 202 66 03
Fax: (+7 095) 202 45 94
Web: http://www.rusarm.ru

Samgong Industrial Company Limited
Samgong Building, 956 Dogok-Dong, Kangnam-Ku,
Seoul, KPO Box 76, South Korea
Tel: (+82 2) 34 62 38 88
Fax: (+82 2) 34 62 38 87
e-mail: lee4601@chollian.net
Web: http://www.samgong.com

Shanxi Xinhau Chemical Factory
PC 030008, Taiyuan, Shanxi, People's Republic of
China
Tel: (+86 351) 351 305 94 95
Fax: (+86 351) 351 305 93 20

SP Défense
ZI Paris Nord II, 13 rue de la Perdix, F-95943 Roissy
Charles de Gaulle Cedex, France
Tel: (+33 1) 49 90 70 80
Fax: (+33 1) 49 90 70 90
e-mail: sp.defense@bacou.com
Web: http://www.bacou.comSP Défense is part of
Groupe Bacou

Stork Bronswerk BV
PO Box 494, NL-3800 AL Amersfoort, Netherlands
Tel: (+31 33) 467 83 00
Fax: (+31 33) 463 71 61
e-mail: info.bronswerk@stork.com
Web: http://www.stork.com/bronswerk

VTÚPV
V. Nejedlého 691, 682 03 Vyškov, Czech Republic
Tel: (+420) 507 30 31 00
Fax: (+420) 507 30 31 05
e-mail: vtupv@vtupv.cz
Web: http://www.vtupv.cz

DECONTAMINATION

Aboukir Engineering Industries Co
Aboukir, 21999 Alexandria, Egypt
Tel: (+20 3) 562 14 10
Fax: (+20 3) 562 33 92
e-mail: fac010@iscc.gov.eg
Web: http://isccnet.iscc.gov.eg

ACD Salvage Techniek BV
Postbus 239, NL-6880 AE Velp, Netherlands
Tel: (+31 26) 362 96 26
Fax: (+31 26) 362 96 36
e-mail: info@acd.nl
Web: http://www.airshelter.com

ACMAT (Ateliers de Construction Mécanique de
l'Atlantique)
Le Point du Jour, F-44600 Saint-Nazaire, France
Tel: (+33 2) 40 22 33 71
Fax: (+33 2) 40 66 30 96

AEROSTAR SA
Str Condorilor nr 9, Bacau, Cod 5500, Romania
Tel: (+40 34) 17 50 70
Fax: (+40 34) 17 22 59
e-mail: aerostar@aerostar.ro
Web: http://www.aerostar.ro

Airshelta Limited
Woodlands, Dale Street, Longwood, Huddersfield,
West Yorkshire HD3 4TG, UK
Tel: (+44 1484) 64 65 59
Fax: (+44 1484) 64 44 50
e-mail: sales@airshelta.co.uk
Web: http://www.airshelta.com

Alfred Kärcher GmbH & Co
Department VPS, Alfred Kärcher Strasse 28-40,
D-71364 Winnenden, Germany
Tel: (+49 7195) 14 22 62
Fax: (+49 7195) 14 27 80
e-mail: vps@de.kaercher.com
Web: http://www.karcher-vps.com

Anachemia Canada Inc
255 Rue Norman, Lachine, Montreal, Quebec,
Canada H8R 1A3
Tel: (+1 514) 489 57 11
Fax: (+1 514) 363 52 81
e-mail: msds@anachemia.com
Web: http://www.anachemia.com

J Blaschke Wehrtechnik GmbH
Wienerbergstrasse 42-44, A-1120 Vienna, Austria
Tel: (+43 1) 810 09 09
Fax: (+43 1) 810 09 09 33
e-mail: office@blaschke.com

Budapesti Vegypari Gépgyár Rt
PO Box H-1475 Budapest Pf 69, Gyömrõi út 76-80,
H-1103, Budapest X, Hungary
Tel: (+36 126) 144 44
Fax: (+36 126) 193 86
e-mail: bvgrt@mail.datanet.hu

The Centech Group Inc
4600 N. Fairfax Drive, Fourth Floor, Arlington, Virginia
22203, USA
Tel: (+1 703) 525 44 44
Fax: (+1 703) 525 23 49
e-mail: fgalaviz@centechgroup.com
Web: http://www.centechgroup.com

Cenzin Limited
ul Frascati, PL-00-489 Warsaw, Poland
Tel: (+48 22) 629 63 96
Fax: (+48 22) 628 63 56

CKD Blansko, Plc
Letrostroj Letovice sro, Praská 333, CZ-679 61
Letovice, Czech Republic
Tel: (+420 501) 93 57 94
Fax: (+420 501) 93 51 77

Cristanini SpA
I-37010 Rivoli, Verona, Italy
Tel: (+39 045) 626 94 00
Fax: (+39 045) 626 94 11
e-mail: cristanini@cristanini.it
Web: http://www.cristanini.it

Dornier GmbH
Systems and Defence Electronics, D-88039,
Friedrichshafen, Germany
Tel: (+49 7545) 899 06
Fax: (+49 7545) 894 90
e-mail: Haag.Oliver@Dornier.Dasa.de

DRDC/RDDC Suffield
Defence R&D Canada / R&D pour la Défense Canada,
PO Box 4000, Station Main, Medicine Hat, Alberta,
Canada T1A 8K6
Tel: (+1 403) 544 46 02
Fax: (+1 403) 544 33 88
e-mail: garfield.purdon@drdc-rddc.gc.ca or
cascad@drdc-rddc.gc.ca
Web: http://www.suffield.drdc-rddc.gc.ca

DEW Engineering and Development Limited
3429 Hawthorne Road, Ottawa, Ontario, Canada K1G
4G2
Tel: (+1 613) 736 51 00
Fax: (+1 613) 736 13 48
e-mail: tdear@dewengineering.com
Web: http://www.dewengineering.com

Edgewood Chemical Biological Center (ECBC)
Aberdeen Proving Ground, Maryland, USA
Tel: (+1 410) 436 43 37
Fax: (+1 410) 436 65 29
e-mail: cet@sbccom.apgea.army.mil
Web: http://www.sbccom.apgea.army.mil/RDA/
ecbc

ENGESA Engenheiros Espacializados SA
Avenida Tucunaré 125/211, PO Box 152/154, 06400
Barueri, São Paulo, Brazil
Tel: (+55 11) 421 47 11
Fax: (+55 11) 421 44 45
e-mail: engesa@engesa.com
Web: http://www.engesa.com.br

Engineered Air Systems Inc
1270 North Price Road, St Louis, Missouri 63132,
USA
Tel: (+1 314) 993 58 80
Fax: (+1 314) 567 40 52
e-mail: easiinq@aol.com

EST + a.s.
Podolí 1237, CZ-584 01 Ledeč nad Sázavou,
Czech Republic
Tel: (+42 452) 62 08 71
Fax: (+42 452) 62 08 02
e-mail: est@ledeč-net.cz
Web: http://www.ledeč-net.cz

J & S Franklin Limited
Franklin House, 151 Strand, London WC2R 1HL, UK
Tel: (+44 0207) 836 57 46
Fax: (+44 0207) 836 27 84
e-mail: defence@franklin.co.uk
Web: http://www.franklin.co.uk

Frenatus International AB
Edvard Orms väg 2, SE-386 90 Frjestaden, Sweden
Tel: (+46 485) 355 70
Fax: (+46 485) 355 80
e-mail: infor@frenatus.com
Web: http://www.frenatus.com

Giat Industries
13 route de la Minière, F-78034 Versailles Cedex,
France
Tel: (+33 1) 30 97 37 37
Fax: (+33 1) 30 97 39 00
e-mail: sales@giat-industries.fr
Web: http://www.giat-industries.fr

HAZ/MAT DQE Inc
5732 West 71st Street, Indianapolis, Indiana 46278,
USA
Tel: (+1 800) 355 46 28
e-mail: info@hazmatdqe.com
Web: http://www.hazmatdqe.com

International Celomer SFPV
BP 168, 75 boulevard Winston Churchill, F-76052, Le
Havre Cedex, France
Tel: (+33 2) 35 53 54 00
Fax: (+33 2) 35 53 54 50
e-mail: ghislaine.vanhoeke@cpn.akzonobel.com
Web: http://www.akzonobel.fr
Note: International Celomer is owned by Akzo Nobel
(HQ: Arnhem The Netherlands)

Interpro Inc
PO Box 1823, 15 Hale Street, Haverhill, Maryland,
01830, USA
Tel: (+1 978) 373 48 13
Fax: (+1 978) 373 24 38
e-mail: interact@seacoast.com

Karl H. Høie & Co A/S
Storgt. 37, N-0182 Oslo, Norway
Tel: (+47 22) 99 76 80
Fax: (+47 22) 20 28 91

Kintex
PO Box 209, 66 James Boucher Street, BG-1407
Sofia, Bulgaria
Tel: (+359) 266 23 11
Fax: (+359) 29 63 11 23

Kraneks Machinery Company Limited.
Settlement Mineevo, 153007 Ivanovo, Russian
Federation
Tel: (+7 0932) 37 65 06
Fax: (+7 0932) 37 65 07
e-mail: kraneks@power.indi.ru

Minimax GmbH
Industriestraße 10/12, D-23840 Bad Oldesloe,
Germany
Tel: (+49 453) 180 30
Fax: (+49 453) 180 32 48
e-mail: info@minimax.de
Web: http://www.mimimax.de

Modec Inc
4725 Oakland Street, Denver, Colorado 80239, USA
Tel: (+1 800) 967 78 87
e-mail: info@deconsolutions.com
Web: http://www.deconsolutions.com

MoD Institute of Military Technology
HM HTI, Szilágyi Erzsébet fasor 20, Pf. 26, H-1525,
Budapest, Hungary
Tel: (+36 1) 394 17 33
Fax: (+36 1) 394 30 14

NBC Team Limited
479 Central Avenue, PO Box 190, Fort Erie, Ontario,
Canada L2A 5M9
Tel: (+1 905) 994 88 01
Fax: (+1 905) 994 88 04
e-mail: info@nbcteam.com
Web: http://www.nbcteam.com

Nor E First Response Inc
Suite 104, PMB 335, 1050 Larrabee Avenue,
Bellingham, Washington 98225-7367, USA
Tel: (+1 360) 647 52 77
Fax: (+1 360) 647 59 06
e-mail: clive@nor-e.com
Web: http://www.nor-e.com

Ocher Engineering Plant
1 Malyshev Street, Ocher 617140, Perm Region,
Russian Federation

O'Dell Engineering Limited
28 Hilborn Avenue, Cambridge, Ontario N1T 1M7,
Canada
Tel: (+1 519) 740 86 20
Fax: (+1 519) 740 94 83
e-mail: rsdl@odell.ca
Web: http://www.odell.ca

Odenwald-Werke Rittersbach GmbH (OWR)
Oberscheflenzer Strasse, D-74834 Elztal-Rittersbach,
Germany
Tel: (+49 62) 937 31
Fax: (+49 62) 937 32 19
e-mail: owr.gmbh@t-online.de
Web: http://www.owr.de

OMNIPOL as
Nekázanka 11, CZ-112 21 Praha 1, Czech Republic
Tel: (+42 02) 401 11 76
Fax: (+42 02) 401 21 99
e-mail: omni22@omnipol.cz
Web: http://www.omnipol.cz

Pao-Chang Company
PO Box 90584, Chiao-Hsi, I-Lan 262, Taiwan
Tel: (+886) 39 88 53 29
Fax: (+886) 39 88 11 16
e-mail: joeyliaw@ms41.url.com.tw

Plysu Protection Systems Limited
Woburn Sands, Milton Keynes, Buckinghamshire,
MK17 8SE, UK
Tel: (+44 1908) 58 23 11
Fax: (+44 1908) 58 37 41
e-mail: ppssales@plysu.com
Web: http://www.plysu.com

Research Institute for Chemical Defence
West Building, PO Box 925, PC 100083, Beijing,
People's Republic of China
Tel: (+86 10) 62 03 37 05; 69 76 02 57
Fax: (+86 10) 62 03 37 05
e-mail: ricdcc@public3.bta.net.cn

Richmond Packaging (UK) Limited
New Road, Winsford, Cheshire CW7 2NY, UK
Tel: (+44 1606) 55 74 22
Fax: (+44 1606) 86 10 63
This company is authorised by the UK MoD to
manufacture for overseas sales.

Rosoboronexport
21 Gogolevsky Boulevard, 119865 Moscow, Russian
Federation
Tel: (+7 095) 202 66 03
Fax: (+7 095) 202 45 94
Web: http://www.rusarm.ru

SEDAB, Safety Equipment Development AB

Head Office:
Bromsvägen 3, SE-891 60 Örnsköldsvik, Sweden
Tel: (+46 660) 129 38
Fax: (+46 660) 824 10
e-mail: sedab@sedab.nu
Web: http://www.sedab.nu

Sales Office:
Vasagatan 3, SE-291 53 Kristianstad, Sweden
Tel: (+46 44) 124 110
Fax: (+46 44) 124 115
e-mail: sedab@sedab.nu
Web: http://www.sedab.nu

SHALON Chemical Industries Limited
25 Nachmani Street, IL-65794 Tel Aviv, Israel
Tel: (+972 3) 629 12 25; 6/7
Fax: (+972 3) 629 16 15
e-mail: shalon@shalon.co.il
Web: http://www.doryanet.co.il/shalon.htm

Stella-Meta Filtration Systems
Laverstoke Mill, Whitchurch, Hampshire RG28 7NR,
UK
Tel: (+44 1256) 89 59 59
Fax: (+44 1256) 89 20 74
e-mail: stellameta@pcimem.com
Web: http://www.stella-meta.com
Note: Stella-Meta Filtration Systems is part of PCI
Membranes, which in turn is owned by Thames
Water.

TATRA
Štefánikova 1163, 21 Kopřivnice CZ-742, Czech
Republic
Tel: (+420 656) 89 11 11/89 37 06
Fax: (+420 656) 89 37 06/72 11 48
e-mail: ou@tatra.cz
Web: http://www.tatra.cz

Tovama Avtomobilov Motorjev
Maribor, Slovenia

Tradeways Limited
184 Duke of Gloucester Street,
Annapolis, Maryland 21401, USA
Tel: (+1 410) 295 08 13
Fax: (+1 410) 295 08 21
e-mail: jhtradeway@msn.com

VOP 025 Nový Jičín, sp
Dukelská 102, CZ-742 42 Šenov u Nového Jičína,
Czech Republic
Tel: (−420 656) 70 17 40
Fax: (+420 656) 70 17 48
e-mail: marketing@vop025.cz
Web: http://www.vop025.cz

Wojskowe Zaklady Lotnicze NR 2
ul Szubinska 107, PL-85-915 Bydgoszcz, Poland
Tel: (+48 52) 364 41
Fax: (−48 52) 329 37

Yugoimport-SDPR pc
PO Box 89, Bulevar umetnosti 2, YU-11070 Belgrade,
Yugoslavia
Tel: (+381 11) 311 27 43
Fax: (+381 11) 324 87 91
e-mail: jugoimport@sezampro.yu
Web: http://www.yugoimport-co.yu

Zenon Environmental Systems Inc
3239 Dundas Street West, Oakville, Ontario L6M 4B2,
Canada
Tel: (+1 905) 465 30 30
Fax: (+1 905) 465 30 50
e-mail: manderson@zenonenv.com
Web: http://www.@zenonenv.com

DEMILITARISATION PROCESSES

Accentus plc
Advanced Process Systems, 551 Harwell,
Oxfordshire OX11 0QJ, UK
Tel: (+44 1235) 43 41 83
Fax: (+44 1235) 43 48 62
e-mail: andrew.turner@accentus.co.uk
Web: http://www.accentus.co.uk
Note: Accentus plc is a wholly-owned subsidiary of
AEA Technology plc.

Aircontrol Technologies Limited
Hawthorne Road, Staines, Middlesex TW18 3AY, UK
Tel: (+44 1784) 46 61 66
Fax: (+44 1784) 46 58 94
e-mail: sales@aircontroltechnologies.co.uk
Web: http://www.airtechnologygroup.co.uk

ARCTECH Inc
14100 Park Meadow Drive, Chantilly, Virginia 20151-
2217, USA
Tel: (+1 703) 222 02 80
Fax: (+1 703) 222 02 99
e-mail: actosol@arctech.com
Web: http://www.arctech.com

Bruker Daltonics®
Bruker Saxonia Analytik GmbH
Permoserstrasse 15, D-04318 Leipzig, Germany
Tel: (+49 341) 24 31 30
Fax: (+49 341) 24 34 04
e-mail: sales@bsax.de
Web: http://www.bsax.de
Note: Bruker Saxonia Analytik GmbH, together with
Bruker Daltonik (Germany) and Bruker Daltonics
Inc (USA), is part of Bruker Daltonics®.

CH2M Hll
6060 South Willow Drive, Greenwood Village,
Colorado 80111-5142, USA
Tel: (+1 303) 771 09 00
Fax: (+1 303) 846 22 31
e-mail: ldavis@ch2m.com
Web: http://www.ch2m.com

El Dorado Engineering Inc
Suite 109, 2964 West 4700 South, Salt Lake City, Utah
84118, USA
Tel: (+1 801) 966 82 88
Fax: (+1 801) 966 84 99
e-mail: info@ga.com
Web: http://www.ga.com

General Atomics
PO Box 85608, San Diego, California 921186-5608,
    USA
Tel: (+1 858) 455 21 52
Fax: (+1 858) 455 36 21
e-mail: info@ga.com
Web: http://www.ga.com

Lockheed Martin
Lockheed Martin Naval Electronics & Surveillance
    Systems - Undersea Systems (Lockheed Martin NE
    & SS-Manassas), 9500 Godwin Drive, Manassas,
    Virginia 20110, USA
Tel: (+1 703) 367 19 64
Fax: (+1 703) 367 54 41
e-mail: richard.read@lmco.com
Web: http://www.lockheedmartin.com/manassas

SAIC
SAIC Corporate Headquarters, 10260 Campus Point
    Drive, San Diego, California 92121, USA
Tel: (+1 505) 842 78 91
e-mail: james.j.karns@saic.com
Web: http://www.saic.com

TECHNIP GERMANY GmbH
Theodorstrasse D-90-40472 Düsseldorf, Germany
Tel: (+49 211) 659 31 81
Fax: (+49 211) 659 32 64
e-mail: unagel@technip.com
Web: http://www.technip.com

Teledyne Brown Engineering
Main Facilities, PO Box 070007, Huntsville, Alabama
    35807-7007, USA
Tel: (+1 256) 726 10 00
Fax: (+1 256) 726 34 34
e-mail: mark.gradkowski@tbe.com
Web: http://www.tbe.com

TRAINING AND SIMULATION

AEA Technology
551   Harwell,   Didcot,   Abingdon,   Oxfordshire
    OX11 0QJ, UK
Tel: (+44 1235) 43 41 83
Fax: (+44 1235) 43 45 57
e-mail: andrew.turner@accentus.co.uk
Web: http://www.aeat-prodsys.com/prodsys
    /divisions/OCD.html

Alfred Kärcher GmbH & Co
Department VPS, Alfred Kärcher Strasse 28-40,
    D-71364 Winnenden, Germany
Tel: (+49 7195) 14 22 62
Fax: (+49 7195) 14 27 80
e-mail: vps@de.kaercher.com
Web: http://www.karcher-vps.com

AME as
Kongeveien 79, N-3188 Horten, Norway
Tel: (+47 33) 03 03 00
Fax: (+47 33) 04 45 70
e-mail: rv@ame.no
Web: http://www.ame.no

AMS
Mountbatten House, Fitzherbert Road, Farlington,
    Portsmouth, Hampshire PO6 1RU, UK
Tel: (+44 23 92) 70 17 01
Fax: (+44 23 92) 70 18 00
e-mail: info@amsjv.com
Web: http://www.amsjv.com
Note: AMS (Alenia Marconi Systems) is a joint venture
    between BAE Systems and Finmeccanica of Italy.

Anachemia Canada Inc
255 Rue Norman, Lachine, Montreal, Quebec,
    Canada H8R 1A3
Tel: (+1 514) 489 57 11
Fax: (+1 514) 363 52 81
e-mail: msds@anachemia.com
Web: http://www.anachemia.com

Argon Electronics
Unit 16, Ribocon Way, Progress Business Park,
    Luton, Bedfordshire LU4 9UR, UK
Tel: (+44 1582) 49 16 16
Fax: (+44 1582) 49 27 80
e-mail: argonelec@dial.pipex.com
Web: http://www.argonelec.dial.pipex.com

Paul Boyé
53 quai de Bosc, F-34200 Sète Cedex, France
Tel: (+33 4) 67 46 87 87
Telex: 480087 POLBOY F
Fax: (+33 4) 67 74 22 75
e-mail: paulboye@mnet.fr

Brocks Explosives Limited
Gateside, Sanquhar, Dumfriesshire DG4 6JP, UK
Tel: (+44 1659) 505 31
Fax: (+44 1659) 505 26

Calspan SRL Corporation
PO Box 400, 4455 Genesee Street, Buffalo, New York
    14225, USA
Tel: (+1 716) 631 69 05
Fax: (+1 716) 631 67 22
e-mail: mcmahon@calspan.com
Web: http://www.calspan.com/chem-def-demil-
    menu.html

DNBCC
(Defence   NBC   Centre),   Winterbourne   Gunner,
    Salisbury, Wiltshire SP4 0E3, UK
Tel: (+44 1722) 43 62 65
Fax: (+44 1722) 43 62 79
e-mail: djwnbc2@aol.com

Étienne LACROIX Défense SA
6 Boulevard de Joffrery, PO Box 213, F-31607 Muret
    Cedex, France
Tel: (+33 5) 61 56 65 00
Fax: (+33 5) 61 51 42 77
e-mail: smv@etienne-lacroix.fr
Web: http://www.lacroix-defense.com

Hands Fireworks Inc
PO Box 128, Milton, Ontario, Canada L9T 2Y3
Tel: (+1 905) 878 28 31
Fax: (+1 905) 878 80 89
e-mail: hands@handsfireworks.com
Web: http://www.handsfireworks.com

Kintex
PO Box 209, 66 James Boucher Street, BG-1407
    Sofia, Bulgaria
Tel: (+359) 266 23 11
Fax: (+359) 29 63 11 23

MEVATEC Corporation
Suite 500, 1525 Perimeter Parkway, Huntsville,
    Alabama 35806, USA
Tel: (+1 256) 890 80 71
Fax: (+1 256) 890 00 00
e-mail: bill.moore@mevatec.com
Web: http://www.mevatec.com/pegem

Nammo LIAB AB
Box 154, SE-711 23 Lindesberg, Sweden
Tel: (+46 581) 871 00
Fax: (+46 581) 872 00
e-mail: info@liab.nammo.com
Web: http://www.nammo.com

PW Defence Limited
Unit 6, Minton Distribution Park, London Road,
    Amesbury, Salisbury, Wiltshire SP4 7EN, UK
Tel: (+44 1980) 62 46 71
Fax: (+44 1980) 62 57 30
e-mail: info@pwdefence.com
Web: http://www.pwdefence.com

Safe Training Systems Limited
Holly   House,   Maidenhead   Road,   Wokingham,
    Berkshire RG40 5RR, UK
Tel: (+44 1344) 48 35 63
Fax: (+44 1344) 48 51 75
e-mail: dwardsafetrainingsystems@compuserve.com
Web: http://www.sales@safetrainingsystems.com

Saratovkij Higher School of Military Engineering for
    Chemical Defence
Details unavailable

Tracer ES&T Inc
#100, 970 Los Vallecitos Boulevard, San Marcos,
    California 92069, USA
Tel: (+1 760) 744 96 11
Fax: (+1 760) 744 66 16
e-mail: tjrapp@tracer-est.com
Web: http://www.tracer-est.com

Tradeways Limited
184 Duke of Gloucester Street,
    Annapolis, Maryland 21401, USA
Tel: (+1 410) 295 0813
Fax: (+1 410) 295 0821
e-mail: jhtradeway@msn.com

Yugoimport-SDPR pc
PO Box 89, Bulevar umetnosti 2, YU-11070 Belgrade,
    Yugoslavia
Tel: (+381 11) 311 27 43
Fax: (+381 11) 324 87 91
e-mail: jugoimport@sezampro.yu
Web: http://www.yugoimport-co.yu

INDEXES

Manufacturers' Index

Z

Alphabetical Index

To help users of this title evaluate the published data, *Jane's Information Group* has divided entries into three categories.

N NEW ENTRY   Information on new equipment and/or systems appearing for the first time in the title.
V VERIFIED   The editor has made a detailed examination of the entry's content and checked its relevancy and accuracy for publication in the new edition to the best of his ability.
U UPDATED   During the verification process, significant changes to content have been made to reflect the latest position known to *Jane's* at the time of publication.

FREE ENTRY/CONTENT IN THIS PUBLICATION

Having your products and services represented in our titles means that they are being seen by the professionals who matter – both by those involved in procurement and by those working for the companies that are likely to affect your business. We therefore feel that it is very much in the interests of your organisation, as well as Jane's, to ensure your data is current and accurate.

■ **Don't forget** – You may be missing out on business if your entry in a Jane's book, CD-ROM or Online product is incorrect because you have not supplied the latest information to us.

■ **Ask yourself** – Can you afford not to be represented in Jane's printed and electronic products? And if you are listed, can you afford for your information to be out of date?

■ **And most importantly** – The best part of all is that your entries in Jane's products are TOTALLY FREE OF CHARGE.

Please provide (using a photocopy of this form) the information on the following categories where appropriate:

1. Organisation name: _____

2. Division name: _____

3. Location address: _____

    _____

4. Mailing address if different: _____

    _____

5. Telephone (please include switchboard and main department contact numbers, for example Public Relations, Sales, and so on):

    _____

6. Facsimile: _____

7. E-mail: _____

8. Web sites: _____

9. Contact name and job title: _____

    _____

10. A brief description of your organisation's activities, products and services: _____

    _____

    _____

11. Jane's publications in which you would like to be included: _____

    _____

Please send this information to:
Jacqui Beard, Information Collection, Jane's Information Group,
Sentinel House, 163 Brighton Road, Coulsdon, Surrey, CR5 2YH, UK
Tel: (+44 20) 87 00 38 08
Fax: (+44 20) 87 00 39 59
E-mail: yearbook@janes.co.uk

Copyright enquiries:
Contact: Keith Faulkner
Tel/Fax: (+44 1342) 30 50 32
E-mail: keith.faulkner@janes.co.uk

Please tick this box if you do not wish your organisation's staff to be included in Jane's mailing lists ☐

JNBC